Pesticide Toxicity to Non-target Organisms

Johnson Stanley • Gnanadhas Preetha

Pesticide Toxicity to Non-target Organisms

Exposure, Toxicity and Risk Assessment Methodologies

Springer

Johnson Stanley
Indian Council of Agricultural Research
Vivekananda Institute of Hill Agriculture
Almora, Uttarakhand, India

Gnanadhas Preetha
Tamil Nadu Agricultural University
Floriculture Research Station
Kanyakumari, Tamil Nadu, India

ISBN 978-94-017-7750-6 ISBN 978-94-017-7752-0 (eBook)
DOI 10.1007/978-94-017-7752-0

Library of Congress Control Number: 2016945036

Printed on acid-free paper

This Springer imprint is published by Springer Nature
The registered company is Springer Science+Business Media B.V. Dordrecht

Foreword

TAMIL NADU AGRICULTURAL UNIVERSITY
Directorate of Open and Distance Learning (ODL)
Coimbatore - 641 003, Tamil Nadu, INDIA

Dr. S. CHANDRASEKARAN, Ph.D.
Professor of Agricultural Entomology

Office : 0422-6611229
Mobile : 94430 61252
E-mail : tnausc@gmail.com
shubaanu@yahoo.co.uk

14.10.2015

Pesticide toxicology as a whole is comparatively an advanced science especially of the residue analysis but the subject of pesticide toxicity assessment still lags behind. Though there are many studies reporting non-target pesticide toxicity, things are somehow unorganized to arrive at with proper conclusions. This is because of the variations in the results of toxicity testing apparently due to the use of different methodologies by different scientists. At this juncture, this book comes as a compilation of different methods available for lethal and sublethal toxicity assessment to facilitate the researchers to have a choice of methods so as the best and most relevant ones can be used. The book is broadly divided as toxicity in plants comprising of predators, parasitoids, pollinators and silkworm; toxicity in soil with microorganisms and earthworms; and toxicity in water with fish as a representative organism, covering all the important non-target organisms in agro-ecosystem.

The introduction on the importance of the organism/group in all the seven chapters itself is quite informative. A small subchapter about the routes where these non-target organisms are exposed to pesticides is interesting because of some unconventional routes where in the pollinators (bees) exposed to pesticides through guttation drops, small puddle water in the field and dust from pneumatic sowing machines while sowing of treated seeds; volatilization and adsorption of pesticides in mulberry leaves to silkworms, dumping of obsolete pesticides in soil and leaf fall from pesticide treated plants to microbes and earthworms, etc. A detailed review on the lethal and sublethal toxic effects on the non-target organisms was made available along with different methods available to evaluate pesticide toxicity. A compiled report on risk assessment methodologies and especially the one which is detailed for assessing pesticide toxicity to some non-target organisms reveals the knowledge of the authors in this aspect. I hope this book will be a boon for all who are interested in pesticide toxicity studies both as researchers and students to enrich their knowledge in the subject of toxicity evaluation along with risk assessment methods.

(S. CHANDRASEKARAN)

Foreword

TAMIL NADU AGRICULTURAL UNIVERSITY
Department of Agricultural Entomology
Centre for Plant Protection Studies
Coimbatore – 641 003, Tamil Nadu, India

Dr. T. Manoharan, Ph.D.	Ph. Office	:	91 422 6611414/ 6611214
Professor & Head cum	Fax	:	91 422 6611414
Director, Centre for Advanced Faculty	E-mail	:	entomology@tnau.ac.in
Training in Entomology			trmano_1991@yahoo.com
	Mobile	:	09842040335

14.10.2015

Pesticides play an integral part in agriculture and most of the farming community relies on it as an easy, efficient and quick pest management strategy. The use of pesticides is accompanied by a variety of undesirable environmental effects. The environmental problems encountered in India owing to large-scale use of pesticides are the insecticide resistance, resurgence, adverse effect on non-target organisms, presence of pesticide residues in food commodities and persistence in soil, water and air. Nature favours the sustenance of many beneficial organisms which play different roles in crop production. So if anything affects the natural ecosystem equilibrium should be tested properly before use. Hazardous effects of pesticides on non-targets were realized long before and the Rachel Carson's "Silent Spring" published in 1962 has made the public get aware of it. Much emphasis was made later to test verify the toxic effects of the pesticides on non-target organisms even before its widespread usage in the agro-ecosystem and at the time of registration.

Pesticide toxicity to non-target organism is not a new issue to be discussed and plenty of literatures are available. But different researchers use different methods and thus the results are variable even with the same pesticide to a particular test organism. So a proper and uniform methodology is the need of the hour. In this juncture, it is praiseworthy to have a book comprising of different methodologies for testing pesticide toxicity to different non-target organisms of agro-ecosystem. The authors have included different kinds of non-target organisms present in crops ecosystem like predators, parasitoids, pollinators and beneficial insects like silkworms, earthworms and microorganisms of soil ecosystem and fishes as a representative organism for aquatic ecosystem.

This book highlights the importance of non-target organisms, effects of pesticides on them, mode of exposure, toxicity bioassay methods and risk assessment of pesticides on non-target organisms. This book also appears to be a utility guide for researchers, teachers and students to have knowledge on different methods available and to select a suitable, more realistic and ecologically relevant method to conduct their research experiments on different aspects of non-target toxicity. If a suitable method is used for assessing the pesticide toxicity, the pesticides are properly screened and a selective one can be used in our cropping system with less harmful effects on non-target organisms. The proposed risk assessment methodologies wherever not available can also be tried and test verified in future to have a proper risk assessment procedure.

I take this opportunity to congratulate the authors, J. Stanley and G. Preetha (my students), for their strenuous efforts in bringing out this book, which would be a guiding tool useful to students and scientists in their future research.

(T. MANOHARAN)

Preface

Agro-ecosystem consists of plants, soil and water as its main component. Pesticides sprayed on plants affect the plants and its target pest species besides affecting the bio-control agents (predators and parasitoids), pollinators and other beneficial organisms. When pesticides are applied in the soil or get into the soil as spray drift, it affects soil-dwelling organisms like microbes and macro organisms including earthworms. Likewise aquatic pesticides or pesticides accidentally or inadvertently get into the aquatic ecosystem affects many non-target organisms including fish. Thus, the seven chapters of this book deals with the above said organisms individually and in detail.

Voluminous literatures are already available on non-target effects of pesticides. While reviewing the literature, we felt that different methods are used to evaluate the non-target toxicity of pesticides and some methods have no relation between them to enable us to correlate and have a proper comparison or interpretation. Even the median lethal doses/concentrations reported are very different and difficult to compare if the data obtained are from different methods. Moreover, a compilation of different methods is necessary to give the researchers to choose the proper and suitable method of relevance to be used. So the purpose of this book is to compile and present the different methods of pesticide toxicity with its merits and demerits. There are many studies on the lethal toxicity of pesticides but sublethal toxicity studies are generally ignored especially in higher tiers viz., semi-field and field conditions because of its difficulty in evaluating, analyzing and interpreting. So a careful inclusion of sublethal toxic effects and methods to evaluate sublethal toxicity are made available. Laboratory studies on toxicity of pesticides are mostly done by estimating the median lethal concentrations or dose with some exceptional studies on field recommended/realistic dose. The missing link is the comparison of toxic doses/concentrations with the field realistic concentrations. Field recommended dose is the dose in which the pest and non-target organisms actually get exposed in the field. So, the effect on field dose or the comparison with field dose by means of risk estimates will give the actual field effect. Risk assessment is a holistic approach by which one can get the overall toxicity of the pesticide to the non-target organism which is studied upon. Many different risk assessment procedures are described in each chapter for the particular organism. It is surprising to know that the risk assessment of pesticides on microorganisms, silkworms, etc. is not been evolved and given due importance. So, risk assessment methodologies are proposed which are used to study risk in other non-target organisms and relevant besides some new.

The book comprises seven chapters exclusively dedicated for seven groups of the most important non-target organisms viz., predators, parasitoids, silkworms, earthworms, microorganisms and fishes. Each chapter starts with an introduction stating the importance of the non-target organism group which we deal in the chapter followed by a subchapter on exposure routes stating how they generally get exposed to the pesticides. The lethal and sublethal effects of pesticides and different methods to assess the toxicity are given in Subchaps. 3 and 4. A separate subchapter is given for risk assessment methods for pesticide toxicity on the non-target organisms.

The work was accomplished by the Divine Grace of God Almighty, who bestowed His blessings not only as knowledge and wisdom but also as time and health to carry out this task. The help, support and encouragement given by our parents and relatives are thankfully acknowledged. The knowledge and attitude passed on to us by our great teachers of Tamil Nadu Agricultural University, Coimbatore, and specially our research guides are greatly acknowledged. Thanks are due to the heads of the departments/stations and directors of our present institutions viz., ICAR – Vivekananda Institute of Hill Agriculture, Almora, and Agricultural Research Station, Thirupathisaram, for their support. We want to place our sincere thanks to both our beloved Guides of our doctoral programme at Tamil Nadu Agricultural University, Dr. S. Chandrasekaran and Dr. T. Manoharan, who with their sincere effort and encouragement imparted the knowledge to us in this field of pesticide toxicology. We thank them earnestly for being with us and made us to grow in our scientific careers and also for gracefully writing forewords to this book. Special thanks to our colleague and dear friend, Dr. Anuradha Bhartiya, for her corrections and constant encouragement to accomplish this task. Our sincere thanks are due to Springer Science+Business Media B.V., The Netherlands, for inviting us and kindling our inner interest to write a book in this aspect and for giving all logistic support in due course, proper typesetting and publishing it in a good format. We hope the compilation will help the researchers and students working on pesticide toxicity and also the authorities responsible for regulation, registration and use of pesticides to have a proper evaluation before widespread usage.

Almora, Uttarakhand, India Johnson Stanley
Nagercoil, Kanyakumari, India Gnanadhas Preetha

Contents

Abbreviations

a.i.	Active ingredient
ACP	Annealing control primer
AMF	Arbuscular mycorrhizal fungi
BSR	Basal soil respiration
CLPP	Community level physiological profiling
DAI	Days after introduction
DAT	Days after treatment
DBP	Dichlorobenzophenone
DDD	Dichlorodiphenyldichloroethane
DDT	Dichlorodiphenyltrichloroethane
DEG	Differentially expressed genes
DGGE	Denaturing gradient gel electrophoresis
dia	Diameter
EC	Emulsifiable concentrate
fg	Femtogram (10^{-15} g)
GUS	Groundwater ubiquity score
h	Hour
ha	Hectare
HAT	Hours after treatment
ht	Height
IGR	Insect growth regulator
JH	Juvenile hormone
LD	Lethal dose
LH-PCR	Length heterogeneity polymerase chain reaction
m	Meter
mg	Milligram (10^{-3} g)
min	Minute
ng	Nanogram (10^{-9} g)
PGPR	Plant growth promoting bacteria
ppm	Parts per million

RISA	Ribosomal intergenic spacer analysis
SCGE	Single cell gel electrophoresis
s	Second
SIR	Substrate induced respiration
SSCP	Single strand chain polymorphism
TGGE	Temperature gradient gel electrophoresis
T-RFLP	Terminal restriction fragment length polymorphism
WP	Wettable powder
μg	Microgram (10^{-6} g)

List of Figures

Introduction

Pesticide toxicity in agro-ecosystem is long been realized and studied upon. No doubt, pesticides helped much in achieving green revolution, which turned the developments in crop genetics, inexpensive pesticides and fertilizers and mechanization into greater yields (Tilman 1998). Without pesticides, crop loss would have been much more than what is being perceived. Pesticides being cheap, easy and effective means of managing pests, diseases and weeds are used extensively. But the extensive, indiscriminate, excessive and wrong use of pesticides caused heavy damage to ecosystem leading to toxicity and pollution. Indiscriminate use of insecticides leads to resistance and resurgence of insect pests besides leaving residues causing environmental pollution. Though pesticide toxicity to non-target organisms are realized earlier, it gained momentum after the publication of 'Silent Spring' by Rachel Carson in 1962, which brought environmental concern to general public. Many cases of non-target mortality and mass destructions were later related to pesticide toxicity. A few examples being, mass death of Swainson's hawks (>5000) due to monocrotophos poisoning in pampas of Argentina (Goldstein et al. 1999), death of fish due to nabam and endrin in Prince Edward Island (Saunders 1969) and due to fenamiphos in Florida (Schmidt 2006) along with many cases of colony collapse disorders in honeybees (Watanabe 2008; Van-Engelsdorp et al. 2010; Henry et al. 2012; Farooqui 2013).

At times, non-targets are exposed to pesticide more than the target organisms. Target weeds may be exposed less to weedicides applied for aquatic weed control than the fish (non-target) which lives entirely in the water in addition to the fat accumulations. Many pesticides starting from DDT are withdrawn or banned because of their non-target toxicity. Stringent measures are made to test non-target toxicity before registration of a new pesticide molecule. Literature on non-target pesticide toxicity is largely available. But the effect of toxicity are misunderstood or confused many a time. However, toxic effects of pesticides are very clear and if not, can be confirmed. In general, lethal effects are used to study and interpret pesticide toxicity ignoring the sublethal effects even by researches and regulative authorities. Sublethal effects are more important than lethal toxicity. A pesticide which cause a 50 % reduction of non target population may be much safer that the one which impairs its activities and reproduction. Risk assessment of pesticide application on non-targets is generally used for assessing and comparing toxic effects and exposure levels and thus considered as a holistic approach.

Pesticide Toxicity in Agro-ecosystem

Agro-ecosystem comprises of plants/crops, soil and water as its major component. The dynamic interaction between these components makes the ecosystem active and sustainable (Burke et al. 1998; Dunbabin et al. 2011). When the sustainability breaks by means of severe outbreak of pest or diseases, pesticides are used to bring down the individual organism to economic threshold levels or equilibrium position. In this sense, pesticides act as an agent to maintain the ecosystem sustainability but only when they are selective and act only on the target species. While affecting the target species sometimes pesticides have an effect on the non-target individuals, disrupting the ecosystem sustainability. For pesticide toxicity studies, target and non-target individuals are not fixed and universal. In the case of weedicides application, plants become the target and its biological control agent, beneficial organism, detritivores and organisms which depend on them for food and shelter become non-targets. On the other hand, when insecticidal application on plants is considered, the target become the pest insect and the non-target being its biological control agents, beneficials which feed on plants like pollinators etc. and plant become a passive individual. As mentioned above, plants, soil and water are the three important components of agro-ecosystem and when they are exposed to pesticides, they are affected along with the individuals present or depend on them at varying degrees. The susceptibility of different non-target organisms to an insecticide (diafenthiuron) tested on its field recommended dose is given here as an example. The 24 h mortality of non-target organism is taken as end point and a per cent mortality of 0–30 is considered harmless, 31–79 as slightly harmful, 80–99 as moderately harmful and >99 as harmful in laboratory tests (Hassan 1994).

Based on this classification, diafenthiuron was found harmless to earthworm (*Perionyx excavatus*), parasitoids (*Bracon hebetor* and *Trichogramma chilonis*) and predators (*Menochilus sexmaculatus* and *Chrysoperla carnea*). All the three Apis bees viz., *Apis dorsata, A. florea* and *A. cerana* along with parasitoid, *Chelonus blackburni* fall on slightly harmed category. Diafenthiuron is moderately harmful to stingless bee, *Trigona iridipennis* and silkworm, *Bombyx mori* and harmful to fish, *Cyprinus carpio*.

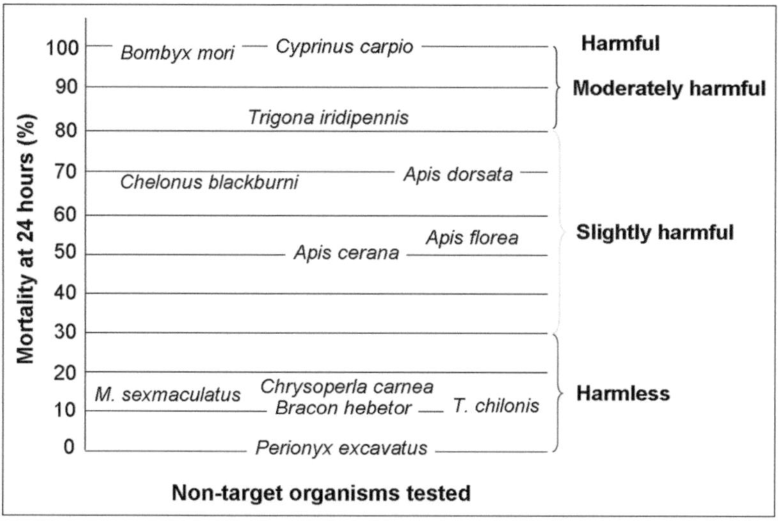

Toxicity of diafenthiuron at field recommended concentration to non-target organisms in the laboratory (Data from Preetha et al. 2009; Stanley et al. 2009, 2016). (Fish, *C. carpio* at larval stage (~1 g); Silkworm, *B. mori* as fourth instar larva; Stingless bee (*T. iridipennis*), Giant honey bee (*A. dorsata*), Dwarf honey bee (*A. florea*) and Indian bees (*A. cerana*) as foraging bees; predatory Coccinellid, *M. sexmaculatus* and Lacewing, *C. carnea* at grub stage; Parasitoids *B. hebetor and C. blackburni* at adult and *T. chilonis* as egg stages and Earthworms, *P. excavatus* as adult worms with clitellum)

Pesticide Toxicity in Plants: Affecting the Associated Organisms

Pesticides sprayed on the plants affect the plants and the organisms present in the plants and those depend on them apart from passing on residue to soil and water by means of leaf fall, stubble incorporation etc. Organisms present in the plants are pests, biocontrol agents like predators, parasitoids and pollinators which pollinate the plant and other beneficials. So non-target effects of pesticide toxicity in plants are given as follow:

Pesticide Toxicity to Predators

Arthropod predators, which include insects, mites and spiders cause pest suppression by natural predation. Biological control by introducing the predators for pest management is also an important part of pest management. But these predators either naturally present or introduced get exposed to pesticides by direct contact, indirect contact to the sprayed parts or by ingestion of toxicated preys. Local extinction or mass death of natural enemies may affect the ecosystem and cause an imbalance in the favour of pest infestation and resurgence. So pesticides are to be tested for their non-target toxicity to the predators. Though there are many methods to find the acute toxicity, determining the median lethal concentration and assessing the mortality at field recommended doses are widely used. Sublethal toxicity especially for longevity, fecundity and predation capacity is also very important for pesticide point of view. Tier II toxicity evaluation in semi-field and tier III evaluation in field conditions are generally carried out to confirm the effects obtained in the laboratory studies in a more realistic manner. Many risk assessment methodologies are used to assess the risk of pesticides on predators and among which, comparison of toxicity of test predator along with its associated pest, sequential testing scheme, total effect of pesticide (lethal and reproduction) and testing the population growth seems to be more realistic for predators. A predator specific risk estimate including the lethality, reproduction and predatory potential can be more relevant in pest management point of view and thus proposed.

Pesticide Toxicity to Parasitoids

Parasitoids are the most effective natural enemy of insect pests which includes egg, egg-larval, egg-pupal, larval, larval-pupal, pupal, adult, nymphal and nymphal-adult parasitoids attacking different stages of the host. Some of the parasitoids used against target pests are highly successful. The natural occurrence of insect parasitoids in all production systems should be encouraged to keep the target pests under control as they are simple and cost effective. But the indiscriminate use of insecticides for the target pests resulted in destruction of the effective natural parasitoids and creates imbalance in the agroecosystems. The chemical pesticide reaches the insect parasitoids through direct exposure to spray droplets, uptake of residues by contact with contaminated surfaces and uptake from contaminated food sources (Longley and Stark 1996). The predicament of the detrimental effect of insecticidal compounds on parasitoids can be determined if we could devise selective use of insecticides. Hence, the selective insecticide should be screened and identified for use in the agroecosystems thereby resulting in conservation of parasitoids for natural parasitization. Different stages of insect parasitoid can be screened for assessing the toxicity of insecticide on them. The acute, chronic, persistent, sub lethal and field toxicity studies are conducted in identifying an effective insecticide selective to parasitoids. By adopting methodologies to assess pesticide toxicity to parasitoids the effects of pesticides can be rightly determined and risk to them can be avoided.

Pesticide Toxicity to Pollinators

Pollinators play a very important role in ecosystem in performing pollination (Allsopp et al. 2008) especially in cross pollinated, self incompatible plants helping in seed set and thus sustaining the ecosystem and conserving biodiversity (Gordon et al. 1998). Of the approximately 300 commercial crops grown all over the world (Richards 1993), about 84 % are insect pollinated (Williams 1996), which reveals the importance of pollinating insects. Pesticides sprayed on plants or taken up by plants from soil can be available in all the plant parts and pollen and nectar are not exceptional. So the pollinators get exposed to pesticides through the food they take apart from direct contact to sprays, contact to sprayed surface, spray drifts etc. Effects of pesticides on pollinators especially bees is well studied and voluminous literature are available. Toxic effects of pesticides on pollinators are mostly reported as mass death or disappearance as 'colony collapse disorder'. Mortality of pollinators is studied as acute toxicity tests in the laboratory whereas disappearance is realized in field conditions. Though there are many methods developed to evaluate the acute toxicity, contact toxicity assessment as topical bioassay, spray tower, indirectly through sprayed leaves or flowers and oral toxicity as food contamination are practiced. Sublethal effects of pesticides on pollinators though are well defined, still needs more confirmations especially before making conclusions. Tests on sublethal toxicity are also complex, time consuming and difficult to interpret. However, many new methods with advanced and sophisticated instruments to study the sublethal toxicity are being developed and reported. Reports on the different methods used to assess the risk of pesticides on pollinators are available and find a place in the respective chapter.

Pesticide Toxicity to Silkworm

Silkworms are beneficial insects reared in different plants for the production of silk. They are exposed to pesticides mostly through leaf contamination. Lethality or acute mortality of silkworms due to pesticide poisoning is widely reported. Unlike other non-targets, a parameter known as safe waiting period i.e., the minimum time period needed for safe harvesting of the leaf after pesticide application is studied and reported for many pesticides to silkworm. Of all the acute toxicity tests reported, leaf contamination bioassay is found realistic to the natural exposure. Unlike other non-targets, which are exposed to pesticide sprays directly, silkworms are exposed to a quantity of pesticide per quantity of leaf, as they consume the pesticide treated leaves. So the experimental results also should be as quantity of pesticide per unit weight of leaf (mg/kg leaf) rather than quantity of pesticide per unit of pesticide solution (mg/mL of water). Silkworms are highly sensitive to insect growth regulators (IGRs), so they are studied especially on its growth and metamorphosis. Effects of pesticides on enzymes and hormones of silkworm larva are also studied and reported. But a few field studies are reported so as the studies on risk assessment. Risk of pesticide to silkworms is necessary to be studied to know the holistic effect of pesticide on this beneficial species. Keeping this in view, some methods which are relevant and used to study risk on other non-targets are proposed for use in silkworms.

Pesticide in Soil: Affecting Soil Organisms

Pesticide entry into soil may be through soil pest control programmes, spray drifts from plants, residues of plant parts etc. Owing to the physico-chemical and biological properties of soil, the pesticides in the soil undergo transformational changes and get degraded but some transformational products are highly toxic to soil dwelling organisms. Soil is considered as a living and dynamic ecosystem teeming with micro and macroscopic living organisms that perform vital functions and ecosystem services. Microbes produce enzymes and have a bigger say in the soil properties besides influencing soil formation from parent material. Earthworms are the most important soil macro-organism, also called as soil ecosystem engineers (Holdsworth et al. 2007; Eisenhauer 2010), which make the locked up nutrients in the soil available to the plant. So pesticide toxicity in soil is determined by toxicity of earthworms and microorganisms.

Pesticide Toxicity to Earthworms

Earthworms improve soil properties, enhance composting of organic matter, remediate soil toxicity and improve plant growth besides being a good source of food to many predators in the ecosystem. They get exposed to pesticide toxicity by means of contact and ingesting contaminated soil and food materials. Pesticide toxicity to earthworms is mainly expressed as median lethal concentration studied through soil contamination bioassays rather than dose. The effective concentration studied and reported in earthworms for abundance, reproduction and growth is more meaningful than acute toxicity values. The use of native and representative earthworm species is very much necessary because of vast variations in the susceptibility. There are only limited studies on chronic toxicity and toxicity in field conditions. Sublethal effects on growth and development, locomotion, respiration, reproduction and effect on enzymes are studied and reported. Pesticide effects on behaviour like avoidance, burrowing and cast production are also studied using advanced technologies. Techniques like 2D terraria or 3D soil core x-ray tomography are used to trace the burrowing pattern of pesticide treated worms. Semifield experiments involving mesocosm, terrestrial model ecosystem (TEM) and also using mesh bags and worm socks were developed and used to evaluate sublethal long term toxicity effects. Pesticide risk assessment in soil ecosystem is mainly based on earthworms

(EC 2002; Rodriguez-Castellanos and Sanchez-Hernandez 2007). Hence many methodologies are developed, studied and reported for risk of pesticides on earthworms as a representative organism of soil ecosystem.

Pesticide Toxicity on Microorganisms

Microbes perform many ecosystem services starting from soil formation to recycling of nutrients, apart from its use as biofertilizer, biopesticide and bio stress remediator in agro-ecosystem. Role of microbes in medical technology, genetic engineering and industry has revolutionized the world. Microbes are ubiquitous (Finlay and Esteban 2001) and thus exposed to pesticides in all ecosystems. But unlike other non-target organisms, all microbes are not susceptible to many of the pesticides and some even use pesticides as food source. However, pesticide toxicity on non-target and beneficial microbes cannot be ignored. Effect of pesticides on microbial biomass, population, growth, activities etc. are studied and reported. Indirect methods of toxicity assessment like basal respiration, substrate induced respiration, fatty acid profiling and other activities are being studied. Biochemical studies on enzymes and DNA based biotechnological studies are being carried out extensively and reported. However, literatures on semifield and field toxicity are scanty and risk assessment of microbes has not received enough attention. USEPA ecological risk assessment guidance does not recommend risk assessment using soil microbes (OEHHA 2009). Though microbes are included among the taxonomic groups under European environmental risk assessment (ERA) procedures for pesticides, it is not substantially covered as that of other groups (Nienstedt et al. 2012). Some risk assessment methods like risk based on effect on microbial activities especially with weighted activity approach, susceptibility comparison with pathogenic and beneficial microbes and hazard concentration 5 % are proposed based on the relevance and information available with other non-target organisms.

Pesticide in Aquatic Ecosystem

Pesticides enters the aquatic ecosystem by drift, wash off and drains from farm lands or by deliberate application for weeds and other pest management in water. Many organisms live in aquatic ecosystem and fish gain its importance not only because they are present everywhere but also due to its usage as good source of food to humans and many other predators.

Pesticide Toxicity on Fish

Fish is an important source of animal protein (FAO 2013) and valuable source of many essential and vital nutrients (Sheeshka and Murkin 2002; Thilsted 2012; Bene et al. 2015). Unlike other non-targets, fish are constantly exposed to pesticides, since they live and breathe in water. They take toxicant through breathing, through the food they take, from the medium they live and so on (Kerr and Vass 1973). So fish are highly susceptible to pesticides even at low concentrations and cause quick mortality. Acute toxicity bioassays are done by static, static-renewal and flow-through systems by contaminating the water with known levels of pesticides. Many sublethal effects are reported and studied as changes in behaviour, physiology, biochemistry, histology and on carcinogenicity, mutagenicity etc. Fish are used as bioindicator for pesticide toxicity in aquatic ecosystem and thus risk assessment methods are well developed. Fish accumulates toxicity as bioaccumulation (Mayer et al. 1977) and pass on toxicant to higher organism in food chain and cause biomagnification (Goerke et al. 2004). Human health assessments (OPP 1990; Jiang et al. 2005; Fianko et al. 2011) based on pesticide residue estimation and by means of estimating reference dose, acceptable daily intake is also included.

Thus the following chapters deal with the pesticide toxicity on non-target organisms categorized as subchapters stating, the route of pesticide exposure, effects of pesticides to those non-target organisms, methods of toxicity testing and risk assessment of pesticides on non-targets.

References

M.H. Allsopp, W.J. De-Lange, R. Veldtman, Valuing insect pollination services with cost of replacement. PLoS ONE **3**(9), e3128 (2008)

C. Bene, M. Barange, R. Subasinghe et al., Feeding 9 billion by 2050 – putting fish back on the menu. Food Secur. **7**, 261–274 (2015)

I.C. Burke, W.K. Lauenroth, M.A. Vinton et al., Plant-soil interactions in temperate grasslands, in *Plant-Induced Soil Changes: Processes and Feedbacks*, ed. by N. Van-Breemen. Dev. Biogeochem. **4**, 121–143 (1998)

V.M. Dunbabin, M. Airey, A.J. Diggle, M. Renton, Z. Rengel, R. Armstrong, Y. Chen, K.H.M. Siddique, Simulating the interaction between plant roots, soil water and nutrient flows and barriers and objects in soil using ROOTMAP, in *19th International Conference on Modelling and Simulation*, Perth, Australia, 12–16 December, 2011. Retrieved from http://mssanz.org.au/modsim2011 (2011)

EC, European Commission (Health & Consumer Protection Directorate – General). Guidance Document on Terrestrial Ecotoxicology under Council Directive 91/414/EEC. SANCO/10329. (2002)

N. Eisenhauer, The action of an animal ecosystem engineer: identification of the main mechanisms of earthworm impacts on soil microarthropods. Pedobiology **53**(6), 343–352 (2010)

FAO, The role of sustainable fisheries and aquaculture for food security and nutrition (2013). www.fao.org/fsnforum/cfs-hlpe/sites/cfs-hlpe/files/files/Fisheries%20and%20Aquaculture/PROCEEDINGS_Fisheries_Aquaculture_v0.docx

T. Farooqui, A potential link among biogenic amines-based pesticides, learning and memory and colony collapse disorder: a unique hypothesis. Neurochem. Int. **62**(1), 122–136 (2013)

J.R. Fianko, A. Donkor, S.T. Lowor, P.O. Yeboah, E.T. Glover, T. Adom, A. Faanu, Health risk associated with pesticide contamination of fish from the Densu river basin in Ghana. J. Environ. Protect. **2**, 115–123 (2011)

B.J. Finlay, G.F. Esteban, Ubiquitous microbes and ecosystem function. Limnetica **20**(l), 31–43 (2001)

H. Goerke, K. Weber, H. Bornemann, S. Ramdohr, J. Plotz, Increasing levels and biomagnification of persistent organic pollutants (POPs) in Antarctic biota. Mar. Pollut. Bull. **48**(3&4), 295–302 (2004)

M.I. Goldstein, T.E. Lacher, B. Woodbridge, Monocrotophos induced mass mortality of Swainson's hawks in Argentina, 1995–96. Ecotoxicology **8**(3), 201–214 (1999)

A. Gordon, P. Bernhardt, R. Bitner et al., The potential consequences of pollinator declines on the conservation of biodiversity and stability of food crop yields. Conserv. Biol. **12**(1), 8–17 (1998)

S.A. Hassan, Activities of the IOBC/WPRS working group 'Pesticides and beneficial organisms'. IOBC/WPRS Bull. **17**(10), 1–5 (1994)

M. Henry, M. Beguin, F. Requier et al., A common pesticide decreases foraging success and survival in honey bees. Science **336**(6079), 348–350 (2012)

A.R. Holdsworth, L.E. Frelich, P.B. Reich, Regional extent of an ecosystem engineer: earthworm invasion in northern hardwood forests. Ecol. Appl. **17**, 1666–1677 (2007)

Q.T. Jiang, T.K.M. Lee, K. Chen et al., Human health risk assessment of organochlorines associated with fish consumption in a coastal city in China. Environ. Pollut. **136**, 155–165 (2005)

S.R. Kerr, W.P. Vass, Pesticide residues in aquatic invertebrates, in *Environmental Pollution by Invertebrates*, ed. by C.A. Edwards (Plenum Press, London, 1973). 542 p

M.A. Longley, J.D. Stark, Analytical techniques for quantifying direct, residual and oral exposure of an insect parasitoid to an organophosphate insecticide. Bull. Environ. Contam. Toxicol. **57**, 683–690 (1996)

F.L. Mayer, P.M. Mehrle, H.O. Sanders, Residue dynamics and biological effects of polychlorinated biphenyls in aquatic organisms. Arch. Environ. Contam. Toxicol. **5**(1), 501–511 (1977)

K.M. Nienstedt, T.C.M. Brock, J. Van-Wensem et al., Development of a framework based on an ecosystem services approach for deriving specific protection goals for environmental risk assessment of pesticides. Sci. Total Environ. **415**(1), 31–38 (2012)

OEHHA, Office of Environmental Health Hazard Assessment. Soil toxicity and bioassessment test methods for ecological risk assessment (M. Hooper, M. Anderson, Prep.). Retrieved from http://oehha.ca.gov/ecotox/pdf/SoilTox120208.pdf (2009)

OPP, Risk assessment methodology for fish. Office of pesticide programs. Retrieved from http://nepis.epa.gov/Exe/ZyPDF.cgi/9100HM79.PDF?Dockey=9100HM79.PDF (1990)

G. Preetha, J. Stanley, T. Manoharan, S. Chandrasekaran, S. Kuttalam, Toxicity of imidacloprid and diafenthiuron to *Chrysoperla carnea* (Stephens) (Neuroptera: Chrysopidae) in the laboratory conditions. J. Plant Protect. Res. **49**(3), 290–296 (2009)

K.W. Richards, Non-Apis bees as crop pollinators. Rev. Suisse Zool. **100**, 807–822 (1993)

L. Rodriguez-Castellanos, J.C. Sanchez-Hernandez, Earthworm biomarkers of pesticide contamination: current status and perspectives. J. Pestic. Sci. **32**(4), 360–371 (2007)

J.W. Saunders, Mass mortalities and behaviour of brook trout and juvenile Atlantic salmon in a stream polluted by agricultural pesticides. J. Fish. Res. Board Can. **26**(3), 695–699 (1969)

C.W. Schmidt, Putting the earth in play: environmental awareness and sports. Environ. Health Perspect. A **114**(5), 286–295 (2006)

J. Sheeshka, E. Murkin, Nutritional aspects of fish compared with other protein sources. Comments Toxicol. **8**(4–6), 375–397 (2002)

J. Stanley, G. Preetha, S. Chandrasekaran, S. Kuttalam, Honey bees of the cardamom ecosystem and the selective toxicity of diafenthiuron to four different bee species in the laboratory. J. Apicult. Res. **48**(2), 91–98 (2009)

J. Stanley, S. Chandrasekaran, G. Preetha, S. Kuttalam, R.S. Jasmine, Selective toxicity of diafenthiuron to non-target organisms: honeybees, coccinellids, chelonus, earthworm, silkworm and fish. J. Plant Protect. Res. **56**(1) (2016)

S.H. Thilsted, The potential of nutrient-rich small fish species in aquaculture to improve human nutrition and health, in *Farming the Waters for People and Food. Proceedings of the Global Conference on Aquaculture*, eds. by R.P. Subasinghe, J.R. Arthur, D.M. Bartley et al., Phuket, Thailand, Rome: FAO and Bangkok: NACA, pp. 57–73 (2012)

D. Tilman, The greening of the green revolution. Nature **396**, 211–212 (1998)

D. Van-Engelsdorp, N. Speybroeck, J.D. Evans et al., Weighing risk factors associated with bee colony collapse disorder by classification and regression tree analysis. J. Econ. Entomol. **103**(5), 1517–1523 (2010)

M.E. Watanabe, Colony collapse disorder: many suspects, no smoking gun. Bioscience **58**(5), 384–388 (2008)

I.H. Williams, Aspects of bee diversity and crop pollination in the European Union, in *The Conservation of Bees*, ed. by A. Matheson, S.L. Buchmann, C. O'Toole, P. Westrich, I.H. Williams (Academic, New York, 1996), pp. 63–80

Chapter 1
Pesticide Toxicity to Arthropod Predators: Exposure, Toxicity and Risk Assessment Methodologies

Abstract Arthropod predators perform ecosystem service by natural pest suppression. Since they are mostly seen along with the pests, they are affected by the pesticidal sprays or by ingesting intoxicated preys. So pesticides are to be tested for their non-target toxic effects. Acute toxicity of pesticides to arthropod predators is being done by calculating the median lethal concentrations. Apart from acute toxicity, testing of chronic, persistent and sublethal toxicities are to be done because sublethal effects especially if affects the reproduction of predators are dangerous. Tier II toxicity evaluation through semi-field experiments are needed to find the toxic effects in a realistic manner than that of laboratory experiments. Finally field experiments are being done to find the real effect of pesticides on these natural enemies. Pesticide risk assessment for predators are being done by categorizing the pesticides based on the mortality in the laboratory and semi-field trials and reduction in field studies. Apart from this, hazard ratio/risk quotient, comparison of LC_{50} with field recommended concentrations are explained. Toxicity of pesticides to predators in comparison with their associated pests are being done by calculating selectivity ratio and probit substitution to find which one of them is more vulnerable. A tiered approach or sequential testing scheme starting from laboratory, proceeding with semi-field and field studies seems to be useful in risk assessment. To find the sublethal toxicity especially on the reproduction of the predators, calculation of total effect of the pesticide, coefficient of toxicity and population growth rate are found promising for assessing the risk of pesticides to predators in agro-ecosystem.

1 Importance of Arthropod Predators in Pest Management

In an ecosystem, predation is a biological interaction where a predator (hunter) feeds on its prey (Begon et al. 1996). Predators are mostly free-living and consume a large number of preys during their lifetime. In general, carnivores are termed as predators but have to prey on other organism. Predators are of different hierarchy in food chains (primary, secondary and tertiary) and many predators eat from multiple levels of the food chain. Arthropod predators on crop pests include beetles, bugs, flies, wasps, spiders and predatory mites. Some predators are so effective in managing the pest problems by themselves naturally. But in some cases, natural pest suppression alone cannot be sufficient to bring pests below economic threshold levels,

J. Stanley, G. Preetha, *Pesticide Toxicity to Non-target Organisms*,
DOI 10.1007/978-94-017-7752-0_1

particularly when the pest populations reach extreme levels. At those critical times, pest management tactics has to be done and even a pesticide may be needed to bring the pest back to equilibrium or economic threshold level. Thus, integration of biological and chemical control is the fundamental tenet on which integrated pest management (IPM) is based (Stern et al. 1959). Approaches to this integration include reducing pesticide usage, use of selective pesticides and modifying natural enemies to reduce their susceptibility to pesticides (Weinzierl 2008; Naranjo and Ellsworth 2009).

1.1 Arthropod Predators

The phylum Arthropoda includes insects, arachnids, myriapods and crustaceans, of which insects and arachnids (mites and spiders) are used for pest management in cultivated systems. Insect predators are distributed in about 167 families belonging to 14 orders of class Insecta (Sathe and Bosale 2001). With regards to significance of biological control, Coleoptera, Neuroptera, Hymenoptera, Diptera and Hemiptera are outstanding (Sahayaraj 2004). Of all the predatory arthropods in rice ecosystem, spiders and coccinellids occupied 26.9 % each, followed by Odonata (19.3 %), other Coleoptera (15.4 %), Orthoptera (7.7 %) and Hemiptera (3.8 %) (Bhattacharyya et al. 2006). In tea plantations, spiders occupied 43 % followed by Coleoptera (31 %). Other predators such as Hemiptera (8), Neuroptera (5 %), Mantodea (7 %), Odonata (4 %) and others (2 %) collectively comprised of 26 % (Das et al. 2010). Johnson et al. (2000) recorded a total of 123 species of predators in Australian farming systems with varying predatory efficiencies. Sometimes even the coexisting pests become important predators (Rosenheim et al. 1993). Though strict herbivores, lygus bugs feed opportunistically on moth eggs (Ehler 1977) and western flower thrips, *Frankliniella occidentalis* on spider mite eggs in cotton fields (Trichilo and Leigh 1986).

Characteristics of Arthropod Predators

- Predators are generally larger than their prey.
- Mostly predators are polyphagous or oligophagous and consume more than one prey organism.
- A single predator kills and eats large numbers of prey in its lifetime.
- Generally, immatures and adults of both sexes are predatory. Some adults like lacewings are exceptional.
- Predators are efficient in hunting their prey and use various strategies.
- They have high host searching ability. Mostly the larvae are active searchers with sensory and locomotory organs.
- Predators kill and consume their prey quickly and use some extra oral digestion also.
- Predators develop independently from their prey but lives in the same habitat or adjacent.

- They may be active during day and night.
- They possess the tendency to multiply faster in relation to pest densities.

1.2 Classification of Predators

Predators can be classified according to the life stage of prey they attack (e.g., egg predators, larval predators etc.), their foraging strategy (e.g., active searchers, ambush or filter feeders, web or bolas builders). Some predators dwell in the vegetation and others in the ground. Among the plant dwelling predators also, some exhibit sit and wait (damsel bugs, certain lacewings etc.) and actively roaming (lady beetle, minute pirate bug and big-eyed bug) foraging strategies (Straub and Snyder 2006; Long and Finke 2015). On the basis of prey consumption, predators can be classified as monophagous, oligophagous and polyphagous. The vedalia beetle, *Rodolia cardinalis* is almost associated only with cottony cushion scales and the green lacewing, *Chrysopa slossonae* with woolly alder aphids. The oligophagous predators (feeding to a range of related taxa), *Hippodamia convergens* and *Adalia bipunctata* feeds on aphids. Some polyphagous (general feeders) feed on a wide variety of prey and non-prey items like plant fluid and pollen also e.g. the bug, *Podisus maculiventris* and the lady beetle, *Coleomegilla maculata*. Predators are functionally categorized as,

1. **Ambushers**: These predators wait for the prey to approach within a striking distance and attack them suddenly with their raptorial legs like Phymatids and preying mantids.
2. **Attractors and trappers**: These predators make a trap and wait there (ant lion grubs and spiders) or they attract their prey towards them by some means.
3. **Searchers for inactive prey**: These predators go in search for their less mobile or sessile preys and consume them (predatory bugs, lady bird beetles, syrphids etc.)
4. **Pursuers**: They pursue for an active prey and get them (Dragon and damsel flies, some beetles etc.)
5. **Nest provisioners**: These predators include some wasps which take prey to provide food for their young ones in their nest.

1.3 Arthropod Predators in Pest Suppression

The overall pest suppression by natural enemies (predators and parasitoids) was estimated to be 33 % of cultivated systems (Hawkins et al. 1999). In nature, some predators are more effective at controlling the pests and others may appear too late and could not suppress the burgeoning pest populations. Some kinds of predators have only a minor impact as individual but contribute much in the overall pest suppression. The value of the crop losses which are prevented by native parasitoids and predators is estimated to be $ 4.49 billions annually in US (Losey and Vaughan 2006). The Natural Pest Suppression Index (Bennett and Gratton 2012) or the Biocontrol Services Index as studied by keeping exclusion cages for coccinellids on aphids revealed that the presence of predators reduced the need for insecticide treatment by 25–43 % (Gardiner et al. 2009). These not only reduce the cost of cultivation but also reduce environmental pollution to a greater extent. Generalist predators (mainly beetles, spiders and ants) were reportedly reduced the pests of cereals and other annual crops (aphids, leafhoppers and bugs) in 78 % of all the reported cases (Winder 1990; Holland et al. 1996; Landis and Van-der-Werf 1997; Snyder and Wise 2001; Symondson et al. 2002). In an exclusion experiment, aphid population was found to get increased by 28 % in ground predator exclusion treatment and 97 % increase in vegetation dwelling predator exclusions (Thies et al. 2011). The percentage actual yield increase due to the activities of ground-living natural enemies of aphids in barley was reported to be 23 % (Ostman et al. 2003). These results reveal aerial/vegetation dwellers and ground dwellers have greater impact on pest suppression as natural enemies.

In other cases, where natural enemies are not available in sufficient numbers for natural pest suppression, then there is a need to raise them and depute for pest management. If there is no effective natural enemy which could suppress the pest, then there may be a need to bring in from other ecosystem or places and introduce to solve the pest problems. About 24 species of exotic predators have been introduced into India for pest management as reported by Singh in 1994.

2 Major Arthropod Predators and Pest Management

2.1 Coccinellids

The family Coccinellidae contains approximately ~6000 species (Canepari 1990) of ladybird beetles of which, more than 90 % are beneficial as predators (Iperti 1999). Pervez (2004) catalogued prey record of 261 known predaceous coccinellids of India belonging to 57 genera. Predaceous coccinellids feed on various phytophagous insect pests, viz., aphids, scale insects, mealy bugs, mites, whiteflies, thrips, etc. and are thus important biocontrol agents (Omkar and Pervez 2002). All coccinellids, known till now are predacious except the genus *Epilachna*. As stated above, some species of lady beetles prefer only certain aphids while others attack many aphid species on a variety of crops and some others prefer mite or scales too. If aphids are scarce, coccinellids feed on thrips and other small insects, eggs of moths, beetles and mites, as well as pollen and nectar. In Coccinellidae family, most of the species of subfamily Coccinellinae are aphid predators, the Chilocorninae prey on homopteran scale insects and the Stethorinae is specialized on phytophagous mite species (Gordon 1985). A female lady beetle may lay from 20 to >1000 eggs over a period of 1–3 months. The mean total fecundity of a female *Hippodamia variegata* is reported to be 960 eggs (Kontodimas and Stathas 2005). The convergent lady beetle grubs feeds on aphids as much as its own weight every day and its adults feeds about 50 aphids per day. Some lady beetles like the seven spotted lady beetle feeds on hundreds of aphids per day.

Insect pests preyed upon by Coccinellids

Coccinellid	Pests	References
Coccinellid, *Adalia tetraspilota*	Green peach aphid, *Myzus persicae*	Joshi et al. (2012)
Coccinellid, *Chilocorus nigrita*	Guava whiteflies, *Aleurodicus dispersus*	Geetha (2000)
Coccinellid, *Chilocorus nigritus*	Coconut scale insect, *Aspidiotus destructor*	Kinawy (1991)
Coccinellid, *Coccinella* spp.	Sunflower aphid, leafhopper and head borer	Ahmed et al. (2013)
Coccinellid, *Delphastus pusillus*	Silver leaf whitefly, *Bemisia tabaci*	Heinz et al. (1994)
Coccinellid, *Serangium* spp.	Silver leaf whitefly, *B. tabaci*	Kapadia and Puri (1992) and Asiimwe et al. (2007)
Coccinellid, *Serangium parcesetosum*	Citrus whitefly, *Dialeurodes citri*	Uygun et al. (1997)
Coccinellid, *Stethorus gilvifrons*	Date palm spider mite, *Oligonychus afrasiaticus*	Latifian (2012)
Coccinellids, *Coccinella septempunctata* and *Harmonia axyridis*	Spirea aphid, *Aphis citricola* and twospotted spider mite, *Tetranychus urticae*	Lucas et al. (1997)
Coccinellids, *Coccinella transversalis* and *Cheilomenes sexmaculata*	Tobacco aphid, *Myzus nicotianae*	Jagadish et al. (2010)
Coccinellids, *Delphastus catalinae* and *Coleomegilla maculata lengi*	Greenhouse whitefly, *Trialeurodes vaporariorum*	Lucas et al. (2004)
Coccinellids, *Delphastus pusillus* and *Menochilus sexmaculatus*	Sweet potato whitefly, *B. tabaci*	Hoelmer et al. (1994)
Coccinellids, *C. septempunctata*, *H. variegata*	Apple woolly aphid, *Eriosoma lanigerum*	Mani and Krishnamoorthy (2004)
Coccinellids, *Hyperaspis maindroni*, *Nephus regularis* and *Scymnus coccivora*	Cotton mealy bug, *Phenacoccus solenopsis*	Fand et al. (2010)
Coccinellids, *Chelomenes vicina*, *Diomers flavipes*, *Diomers hottentota and C. septempunctata*	Cassava whiteflies, *B. tabaci* and *A. dispersus*	Atuncha et al. (2013)

(continued)

Coccinellid	Pests	References
Coccinellids, *Anegleis cardoni*, *A. perrotteti*, *Cryptolaemus montrouzieri*, *Axinoscymnus puttarudriahi* and *C. sexmaculata*	Spiraling whitefly, *A. dispersus*	Mani and Krishnamoorthy (1997)
Coccinellids (many species)	Wheat aphids, *Ropalosiphum padi*, *Metapolophium dirhodum* and *Sitobion avenae*	Nyaanga et al. (2014)
Convergent lady beetle, *H. convergens*	Many aphids including mustard aphid, *Lipaphis erysimi*	Lohar et al. (2012)
Eleven-spot ladybird, *Coccinella undecimpunctata*	Alfalfa aphid, *Therioaphis trifolii*	Mari et al. (2005)
Lady beetle, *M. sexmaculatus*	*Aphis gossypii*, *B. tabaci* and *Amrasca biguttula biguttula*	Bukero et al. (2014)
	Rose aphid, *Macrosiphum rosae*	Saleem et al. (2014)
	Alfalfa aphid, *T. trifolii*	Mari et al. (2005)
Lady bird beetle, *Micraspis discolor*	Red spider mites, *Oligonychus coffeae* and tea aphid, *Toxoptera aurantii*	Roy et al. (2010)
	Bean aphid, *Aphis craccivora*	Chowdhury et al. (2008)
Mealybug ladybird, *C. montrouzieri*	Mealybug, *P. solenopsis*	Kaur and Virk (2012)
	Coffee green scale, *Coccus viridis*	Kumar and Prakasam (1984)
	Green scale, *Chloropulvinaria psidii*	Mani and Krishnamoorthy (1990)
	Citrus mealybug, *Planococcus citri*	Singh (1978)
	Pink mealybug, *Maconellicoccus hirsutus*	Reddy and Narayan (1986) and Mani and Thontadarya (1988)
Multicolored Asian lady beetle, *H. axyridis*	Aphids, psyllids and scales	Koch (2003)
Pink spotted lady beetle, *C. maculata*	Cotton aphid, *A. gossypii* and two-spotted spider mite, *T. urticae*	Rondon et al. (2004)
Seven spotted lady beetle, *C. septempunctata*	Mustard aphid, *L. erysimi*	Singh and Singh (2013), Singh (2013), and Sharma and Joshi (2010)
Spider mite destroyer lady beetle, *Stethorus picipes*	Avacado brown mite, *Oligonychus punicae*	McMurtry et al. (1969) and Tanigoshi and McMurtry (1977)
Twice-stabbed lady beetle, *Chilocorus orbus*	Pine needle scale, *Chionaspis pinifoliae*	Luck and Dahlstein (1975)
Vedalia beetle, *R. cardinalis*	Cottony cushion scales, *Icerya purchasi*	Caltagirone and Doutt (1989)

2.1.1 Prey Preference and Consumption in Laboratory Tests

Preference is an important factor for the success of a polyphagous predator in biocontrol programmes. It is particularly important that the predator should prefer the target pests or the pest to be managed (Waseem et al. 2009). George (2000) reported that *Coccinella transversalis* consumed the greatest number of *A. gossypii* followed by *A. nerii* and *Pentalonia* sp. Coccinellids, *C. septempunctata* and *H. axyridis* showed a significant preference for *A. spiraecola* in the presence of *T. urticae* (Lucas et al. 1997). Coccinellids, *C. vicina*, *D. flavipes*, *D. hottentota* and *C. septempunctata* prefer nymphs of *B. tabaci* to *A. dispersus* (Atuncha et al. 2013). *Coccinella septempunctata* preferred mustard aphid, *L. erysimi* than *M. persicae* and the fourth instar grubs consumed 69.4 and 61.5 aphids, respectively (Jandial and Malik 2006).

Coccinella septempunctata, *M. sexmaculatus* and *Coccinella repanda* consumed 39.7, 31.3 and 26.9 aphids of *A. craccivora* adults/day (Das and Sagar 2001). The one, two, three and fourth instar larvae of *C. septempunctata* efficiently consumed 21.4, 46.9, 72.6 and 102.6 mustard aphids/day (Singh and Singh 2013). *Menochilus sexmaculatus* fourth instar larvae consumed 79.7, 23.4 and 21.5 nymphs of *A. gossypii*, *B. tabaci* and *A. biguttula biguttula*/day, respectively (Bukero et al. 2014). The third and fourth instar larva and adults of *A. tetraspilota* consumed about 39.9, 20.9 and 35.7 green peach aphids, *M. persicae*/day, respectively (Joshi et al. 2012). Adult beetles of *C. montrouzieri* feeds about 15.5 and 613.8 first and fourth instar nymphs of mealybug, *P. solenopsi*, respectively (Kaur and Virk 2012). Grubs of *C. montrouzieri* were reported to consume *Maconellicoccus hirsutus* at the rate of 259 nymphs or 27.5 adult females (Mani and Thontadarya 1987). *Micraspis discolor* consumed on an average 280.3 red spider mites and 188.6 tea aphids during its larval period (Roy et al. 2010).

2.1.2 Field Efficacy of Coccinellids

Chrysoperla carnea is reported to be an effective biological control agent in field crops, orchards and in green houses (Hagley and Miles 1987). *Coccinella transversalis* was reported to cause a significant high reduction in average tobacco aphid density to a tune of 58.2 % in about 1 week after release in the field (Jagdish et al. 2010). Coccinellid, *C. montrouzieri* was found very effective in managing the cotton mealybug, *P. solenopsis* in the field conditions (Rashid et al. 2012). Release of *S. picipes* @ 400–500 adult beetles/tree has significantly reduced the number of *O. punicae* along with bronzing of leaves in avocado orchards in southern California (McMurtry et al. 1969).

The introduction of vedalia beetle, *R. cardinalis* for the control of cottony cushion scales in California (Caltagirone and Doutt 1989), South Africa (Lounsbury 1940), Chile (Gonzalez and Rojas 1966), Sri Lanka (Hutson 1920), West Indies (Bennett 1971) and then in Egypt, Cyprus, the Soviet Union, Portugal, Puerto Rico, Venezuela, Peru, Hawaii, Philippines, Uruguay, Argentina etc. (Caltagirone and Doutt 1989) has given great success. The coccinellid beetle, *C. nigritus* was successfully introduced from India during 1985, as a biocontrol agent for the scale insect, *A. destructor* which infests coconut palms in south Oman. After 24 months of release, 65–100 % reduction in scale insects was reported (Kinawy 1991).

2.2 Lacewings

Neuroptera comprised of nearly 6000 species and those species belong to 3 families viz., Chrysopidae, Coniopterygidae, Hemerobiidae are important predators (Hoy 2011). The genus *Chrysoperla* of Chrysopidae includes several important species of predatory insects. The green lacewing, *Chrysoperla carnea* is a potential biocontrol agent that is used in augmentation programme for sustainable pest suppression. The females lay about 500–700 stalked eggs and the larvae voraciously feed on aphids and thus called as 'aphid lions'. The adults' feeds on pollen, nectar and aphid honey dew. Chrysopid larvae mostly feed on aphids but also on mealybugs, whiteflies, thrips and mites to a limited extent. The brown lacewings, *Hemerobius stigma* and *Sympherobius fallax navas* are reported to feed on pine bast scale, *Matsucoccus* spp. and long-tailed mealy bug, *Pseudococcus longispinus*, respectively (Branco et al. 2001; Waseem et al. 2009).

Insect pests preyed upon by Chrysopids

Pests	References
Canola aphids, *Brevicoryne brassicae*, *L. erysimi*, *M. persicae*	Khan et al. (2013)
Lettuce aphids, *Nasonovia ribisnigri*	Shrestha and Enkegaard (2013)
Russian wheat aphid, *D. noxia*	Messina and Sorenson (2001)
Tobacco aphid, *M. nicotianae*	Jagadish et al. (2010)
Strawberry aphids, *Rhodobium porosum* and cotton aphids, *A. gossypii*	Turquet et al. (2009)
Cotton aphid *A. gossypii*, safflower aphid *Uroleucon compositae*, mustard aphid *L. erysimi*, bean aphid *A. craccivora*, oleander aphid *Aphis nerii* and cabbage aphid, *B. brassicae*	Chakraborty and Korat (2010)
Cotton and sunflower sucking pests (leafhopper, thrips, aphids and whiteflies) and *Helicoverpa armigera*	Hanumantharaya et al. (2008)
Sugarcane leafhopper, *Pyrilla perpusilla*	Zia-ul-Hussnain et al. (2007)
Cotton mealybug, *P. solenopsis*	Sattar et al. (2007), Rashid et al. (2012), and Hameed et al. (2013)
Cotton whitefly, *B. tabaci*	Zia et al. (2008)
Western flower thrips, *Frankliniella occidentalis*	Shrestha and Enkegaard (2013)
Citrus red mite, *Panonychus citri*, Strawberry mites, *Tetranychus kanazwai* and *T. urticae*	Cheng et al. (2010)
Red spider mite, *Tetranychus ludeni*	Reddy (2002)
Tea red spider mite, *Oligonychus coffeae*	Vasanthakumar et al. (2012)

2.3 Predatory Bugs

Among the many predatory bugs, the mirids, reduviids (assassin bugs), nabids, lygaeids (big-eyed bugs), anthocorids (pirate bugs), pentatomids (stink bugs) are some of the important predators of insect pests. The green mirid bug, *Cyrtorhinus lividipennis* is widely distributed in rice fields and feeds on leaf and planthoppers (Sigsgaard 2007; Preetha et al. 2010). It is considered as a good predator of hoppers since a single bug can consume 66 hoppers in its lifetime of 24 days (Reyes and Gabriel 1975). *Chrysoperla carnea* was found very effective against cotton mealy bugs, *P. solenopsis* in the field conditions (Rashid et al. 2012). The feeding rate of one, two and third instar lacewings, *Mallada* sp. were 1.5, 5.1 and 5.3 nymphs of tea mosquito bug, *Helopeltis theivora*, respectively in 24 h (Borah et al. 2012). The reduviid bug, *Reduviolus roseipennis* are found feeding on *Heliothis zea* in cotton. The one, three and fifth instar nymphs consumed 1.9, 7.9 and 38.2 larvae/day, respectively (Donahoe and Pitre 1977). The nabids, *Tropiconabis capsiformis* and *Hoplistoscelis deceptivus*, the lygaeids *Geocoris punctipes* and *G. uliginosus* were found to feed on the small sized larva of soybean looper, *Pseudoplusia includens*, whereas, the pentatomid, *Stiretrus anchorago* and the reduviid, *Arilus cristatus* feeds on the medium sized larva (Richman et al. 1980). Adult *Geocoris ochropterus* can feed on 13 tea aphids per day for a period of 22 days (Mukhopadhyay and Sarker 2007).

Insect pests preyed upon by predatory bugs

Predatory bugs	Pests	References
Mirid bugs		
Green mirid bug, *C. lividipennis*	Rice brown planthopper, *Nilaparvata lugens*	Reissig et al. (1982), Chua and Mikil (1989), and Preetha et al. (2010)
Green mirid, *C. lividipennis* and brown mirid, *Tytthus parviceps*	Rice brown planthopper, *N. lugens*	Saritha et al. (2008)
Mirids, *Deraeocoris* sp. and *Campylomma nicolasi*	Whiteflies, *B. tabaci*	Kapadia and Puri (1991)
Mirid bug, *Deraeocoris brevis*	Pear psylla, *Cacopsylla pyricola*	Arthurs et al. (2007)
Mirid, *Spanogonicus albofasciatus*	Velvetbean caterpillar, *Anticarsia gemmatalis*	Godfrey et al. (1989)
Mirids, *Nesidiocoris tenuis*, *Macrolophus pygmaeus* and *Dicyphus maroccanus*	Sweet pepper aphid, *Myzus persicae*	Perez-Hedo and Urbaneja (2015)
Assassin bugs (Reduviids)		
Assassin bug, *Coranus spiniscutis*	Rice meal moth, *Corcyra cephalonica* and leaf armyworm, *Spodoptera litura*	Kumar et al. (2011)

(continued)

Predatory bugs	Pests	References
Assassin bug/Reduviid, *Acanthaspis pedestris*	Cotton pests, *Helicoverpa armigera*, *Pectinophora gossypiella*, *Spodoptera litura*, *Earias insulana*	Sahayaraj and Ambrose (1994)
Assassin bugs, *Rhynocoris marginatus*	Teak skeletonizer, *Eutectona machaeralis*	Ambrose et al. (2013)
	Aphis gossypii, *P. solenopsis* and *Dysdercus cingulatus*	Sahayaraj et al. (2012)
Harpactorine assassin bug, *Rhynocoris longifrons*	Termite, *Odontotermes obesus*, caterpillars, *H. armigera* and *Exelastis atomosa*. Bugs, *Clavigralla gibbosa*, *Nezara viridula* and *Riptortus pedestris*	Kumar et al. (2009)
Lygaeid bugs (Big-eyed bugs)		
Big-eyed bugs, *Geocoris* spp.	Peach aphid, *M. persicae* and the Colorado potato beetle, *Leptinotarsa decemlineata*	Koss et al. (2004)
Big-eyed bug, *G. punctipes*	Cotton aphid, *A. gossypii* and two-spotted spider mite, *T. urticae*	Rondon et al. (2004)
	Velvetbean caterpillar, *A. gemmatalis*	Godfrey et al. (1989)
Big-eyed bug, *G. ochropterus*	Tea pests, *Oligonychus coffeae*, *Scirtothrips dorsalis* and *Euproctis latisfascia*	Sannigrahi and Mukhopadhyay (1992)
Big-eyed bug, *G. ochropterus*	Tea aphids, *T. aurantii*	Mukhopadhyay and Sannigrahi (1993)
Anthocorid bugs (Pirate bugs)		
Pirate bug, *Orius* spp.	Aphids and thrips, *Thrips palmi*	Rutledge et al. (2004) and Ohno and Takemoto (1997)
Minute pirate bug, *Orius insidiosus*	Cotton aphid, *A. gossypii* and two-spotted spider mite, *T. urticae*	Rondon et al. (2004)
Pirate bug, *Orius indicus*	Bean blossom thrips, *Taeniothrips nigricornis*	Rajasekhara and Chatterji (1970)
Anthocorids, *Orius maxidentes* and *Carayanocoris indicus*	Thrips, *Scirtothrips dorsalis*	Muraleedharan and Ananthakrishnan (1978)
Anthocorid bug, *Anthocoris* spp.	Apple psyllid, *Psylla mali* and black bean aphid, *Aphis fabae*	Anderson (1962)
Anthocorid bug, *Anthocoris nemoralis* and *A. nemorum*	Pear psyllid, *Cacopsylla pyri*	Sigsgaard et al. (2006)
Other bugs		
Damsel bugs, *Nabis* spp.	Peach aphid, *M. persicae* and the Colorado potato beetle, *Leptinotarsa decemlineata*	Koss et al. (2004)
Spined soldier bug, *Podisus maculiventris*	Caterpillars, *Anticarsia gemmatalis*, *Heliothis virescens*, *Spodoptera frugiperda* and *Heliothis zea*	Yu (1987)

(continued)

Predatory bugs	Pests	References
Predatory stink bug, *Eocanthecona furcellata*	Cutworm, *Spodoptera litura*	Yasuda (1997)
Nabid bugs, *Nabis roseipennis*, *Tropiconabis capsiformis* and *Hoplistoscelis deceptivus*	Soybean looper, *Pseudoplusia includens*	Richman et al. (1980)
Pentatomid, *Stiretrus anchorago* and reduviid, *Arilus cristatus*	Soybean looper, *P. includens*	Richman et al. (1980)

2.4 Syrphids

The predatory Dipterans of the family Syrphidae is one of the important group of aphidophagous predators which take part in natural predation of aphid population. Majority of aphidophagous syrphids belong to the subfamily Syrphinae. There are >4700 species reported worldwide with 312 species under 71 genera known from the Indian subcontinent. Syrphids are voracious feeders of aphid and it was reported that *Syrphus confrater*, *S. balteatus* and *Ischiodon scutellaris* feeds on 34.8, 32.2, 27.7 aphids/day (Singh and Singh 2013). Though they are found effective in managing the pest, relatively less importance has been given to their field evaluation and utilization in pest management (Joshi and Ballal 2013).

Pests controlled by syrphids

Syrphids	Pests	References
Betasyrphus cerarius	Rose aphid, *Macrosiphum rosae*; Spirea aphid, *Aphis spiraecola*	Agarwala et al. (1984)
Episyrphus balteatus	Mustard aphid, *L. erysimi*	Devi et al. (2011) and Singh (2013)
Episyrphus griseocincta	Maize aphid, *Rhopalosiphum maidis*	Agarwala et al. (1984)
Episyrphus alternans	Sugarcane aphid, *Melanaphis sacchari*	Agarwala et al. (1984)
Eristalis obscuritarsus and *E. tenax*	Mustard aphid, *L. erysimi*	Atwal et al. (1971)
Eupeodes confrater	Apple woolly aphid, *E. lanigerum*	Mani and Krishnamoorthy (2004)
Ischiodon scutellaris	Mustard aphid, *L. erysimi*	Boopathi and Pathak (2011) and Singh and Singh (2013)
	Pea aphid, *Acyrthosiphum pisum*, bean aphid, *A. craccivora*, cotton aphid, *A. gossypii*, crucifer aphid, *B. brassicae* and peach aphid, *M. persicae*	Agarwala et al. (1984)
Syrphus sp.	Tea aphids, *T. aurantii*	Das et al. (2010)
Syrphus confrater and *Syrphus balteatus*	Mustard aphid, *L. erysimi*	Singh and Singh (2013)

2.5 Predatory Wasps

Many wasps are among the main predators of phytophagous insects in the agro-ecosystem (Richter 2000; Prezoto et al. 2006; Pereira et al. 2007). Because of their behaviour and have large number of individuals in colonies, some species can cause great impact on insect-pest (Richter 2000) and regulate the population dynamics of insect pests (Pereira et al. 2007). *Ancistrocerus gazella* wasp is reported as a predator of lepidopterous larvae, especially the pest tortricids and leaf folders and found effective against case bearer, *Coleophora* spp. to a limited extent (Wearing and Harris 1999). *Polistes jadwigae* and *Polistes chinensis* foragers hunt late instar *Spodoptera litura* larva (Nakasuji et al. 1976). The solitary wasps, *Symmorphus allobrogus* are a good predator of many chrysomelid larvae (Budriene and Budrys 2004). Subsocial Eumeninae wasps provision its nest with stung and paralyzed caterpillars or curculionid larvae (Spradbery 1973). The predatory wasps, *Brachygastra lecheguana*, *Protonectarina sylveirae*, *Polybia scutellaris* and *Polybia fastidiosuscula* have great potential for reducing populations of *Plutella xylostella* in the field (Bacci et al. 2009).

2.6 Predatory Beetles

The beetles belong to Carabidae, Staphylinidae and Coccinellidae are predacious on different insect pests in agro-ecosystem. Carabid beetles are generalist predators which reduce the abundance of many herbivore pests (Best and Beegle 1977; Clark et al. 1994; Menalled et al. 1999). Predaceous ground beetles, *Calleida decora* feeds on soil dwelling insect larvae, pupae, snails and slugs. The ability of the carabid, *Abax parallelepipedus* was used against slugs damaging the lettuce crop and found effective (Symondson 1993). The small arboreal ground beetle, *C. decora* and *Lebia analis* were found to feed on soybean pests, *Pseudoplusia includens* and *Heliothis zea* with a consumption rate of 25.4 and 23.0 first instar larvae in 24 h period (Brown and Goyer 1984). *Calledia decora* is also found to prey extensively on velvet bean caterpillar, *Anticarsia gemmatalis* (Godfrey et al. 1989). The ground-inhabiting species *Calosoma alternans* consumed about 16 and 10 larvae of fifth and sixth instar *P. includens* in 24 h time (Brown and Goyer 1984). Another beetle, the staphylinid or rove beetles was found to feed on many rice pests and thus reported as a good predator in rice ecosystem (Ghahari et al. 2009). Staphylinid beetle, *Oligota pygmaea* feeds on red spider mites in tea gardens (UPASI 2005).

2.7 Other Insect Predators

Many generalist predatory insects also play a major role in pest management in specific ecosystems. Ancient Chinese used ants to control pests of citrus and constructed bamboo walkways (Huang and Yang 1987; Symondson et al. 2002), while the date growers used similar practices in Yemen (Doutt 1964). Seven species of ants were reported to reduce many hemipteran and lepidopteran pests of annual and orchard crops (Peng et al. 1995; Van-Mele and Cuc 2001). The robber fly, *Promachus yesonicus* (Diptera: Asilidae) was reported to reduce significantly the density of white grubs, *Anomala* spp. by 21–99 % compared with control plots and reduced damage to wheat significantly by 68–96 % (Wei et al. 1995). Some of the important general insect predators are given in the list.

Pests preyed upon by other insect predators

Predator	Family: order	Predacious stage	Prey	References
Praying mantis, *Mantis religiosa*	Mantidae: Mantodea	Adults and immatures	Wide variety of pests	–
			Red-legged grasshopper, *Melanoplus femurrubrum*	Mooka and Davies (1966)
Dragon flies	Odonata	Adults	Stem borers and leafhoppers	Corbet (1999)
Earwig, *Labidura riparia*	Labiduridae: Dermaptera	Adults	Velvet bean caterpillar, *A. gemmatalis*	Godfrey et al. (1989)
Veliid bugs, *Microvelia atrolineata*	Veliidae: Hemiptera	Nymphs and adults	Rice brown planthopper, *N. lugens*	Reissig et al. (1982)
Predaceous aphid midge, *Aphidoletes aphidimyza*	Cecidomyiidae: Diptera	Larvae	Aphids, mites, and other small soft-bodied insects	–
			Macrosiphum euphorbiae, *M. rosae*, *M. persicae* and *A. gossypii*	Markkula et al. (1979)
Bee flies, *Systoechus* sp.	Bombyliidae: Diptera	Grub	Desert locust, *Schistocerca gregaria*	Greathead (1958)
Robber fly, *P. yesonicus*	Asilidae: Diptera	Adults	*Anomala* spp.	Wei et al. (1995)
Six spotted thrips, *Scolothrips sexmaculatus*	Thripidae: Thysanoptera	Larvae and adults	Phytophagous mites, *T. pacificus*	Gilstrap and Oatman (1976)
			Two spotted spider mites, *T. urticae*	Coville and Allen (1977)
African weaver ant, *Oecophylla longinoda*	Formicidae: Hymenoptera	Adults	Fruit flies, *Ceratitis* spp., *Bactrocera invadens*	Mele et al. (2007)
Ant, *Pheidole morrisi*	Formicidae: Hymenoptera	Adults	Velvet bean caterpillar, *A. gemmatalis*	Godfrey et al. (1989)

2.8 Predatory Mites

Predatory mites may be of (1) specialized predators of *Tetranychus* sp. represented by the *Phytoseiulus* sp.; (2) selective predators of tetranychid mites (frequently associated with those produce webbing) represented by *Galendromus*, some *Neoseiulus* and a few *Typhlodromus* sp.; (3) generalist predators represented by some *Neoseiulus* sp. and most *Typhlodromus* and *Amblyseius* sp. and (4) specialized pollen feeders/generalist predators represented by *Euseius* sp. (McMurtry and Croft 1997). In general, the predatory mites are voracious feeders and have high fecundity than the phytophagous mites. The phytoseiid mite, *Typhlodromus floridanus* preys on brown mite, *O. punicae*. At an initial ratio of 10 preys to 1 predator, the predatory mite would annihilate the pest within 13 days (Tanigoshi and McMurtry 1977).

Pests preyed upon by predatory mites

Predatory mites	Pest preys	References
Predatory mite, *Galendromus helveolus*	*Eotetranychus sexmaculatus*, *T. urticae*, *Oligonychus perseae*, and *O. punicae*	Tanigoshi and McMurtry (1977), McMurtry (1982), and Caceres and Childers (1991)
Western predatory mite, *Galendromus occidentalis*	Eriophyids such as *Aculus schlechtendali*, *Tydeids* and *Tarsonemids*	McMurtry and Rodriquez (1987)
Predatory mites, *Amblyseius andersoni* and *Typhlodromus pyri*	Rust mites, tetranychids and *P. ulmi*	Dicke et al. (1988, 1990) and Schausberger (1992)
Spider mite predator, *Phytoseiulus persimilis*	Spider mites, *Tetranychus* sp.	McMurtry (1991), Steinberg and Cohen (1992), and Van-de-Vrie and Price (1994)
Spider mite predator, *Neoseiulus fallacis* and *N. californicus*	Tetranychid mites and *P. ulmi* in apple, strawberry, caneberry, corn, sorghum, soybean, mint and hops	Croft and McGroarty (1977), Croft (1990), Nyrop et al. (1994), and Morris et al. (1996)
Predatory mite, *Neoseiulus barkeri* and *N. cucumeris*	*Frankliniella occidentalis* and *Thrips tabaci*	Gillespie (1989), Hansen (1989), and Ramakers et al. (1989)
Phytoseiid mite, *Amblyseius andersoni*	*Panonychus* sp.	Duso (1992), Ibrahim and Palacio (1994), and Croft and Zhang (1996)
Predatory mite, *Kampimodromus aberrans*	Many eriophyid mites and *Eotetranychus carpini*	Hadam et al. (1986) and Duso et al. (1991)
Phytoseiid mite, *Euseius tularensis*	Citrus red mite, *Panonychus citri*	Jones and Morse (1995) and McMurtry and Croft (1997)
	Citrus thrips, *Scirtothrips citri*	Grafton-Cardwell and Ouyan (1995)
Phytoseiid mite, *Euseius addoensis*	Thrips, *Scirtothrips aurantii*	Grout and Richards (1992)

2.9 Predatory Spiders

Spiders are carnivorous arthropods with approximately 40,000 species found all over the world in almost every kind of habitat (Turnbull 1973; Tanaka 1989; Riechert 1981). *Lycosa pseudoannulata*, *Atypena formosana*, *Tetragnatha javanas*, *Callitrichia formosana* and *Clubiona japonicola*, *Argiope catenulata* and *Plexippus* sp. are the predominant spiders in rice ecosystem (Sahu et al. 1996; Jayakumar and Sankari 2010). The spiders, *Oxyopes javanus*, *O. rufisternum*, *Peucetia viridanus*, *Salticus* sp., *Phidippus* sp., *Thomisus* sp., *Araneus* sp., *Argiope* sp. and *Clubiona* sp. were recorded in cotton. Spiders, *Oxyopes rufisternum*, *P. viridanus*, *Gasteracantha* sp., *Clubiona* sp., *Thomisus* sp. and *Phidippus* sp. were reported in okra (Preetha et al. 2009a). Spiders such as *Argiope luzona*, *Cyrtophora cicatrosa*, *Chrysso argyrodiformis*, *Hipossa pantherina*, *Oxyopes lineatipes*, *O. javanus*, *Peucetia viridana* and *L. pseudoannulata* are reported in brinjal and guard ecosystem (Sankari and Thiyagesan 2010). Altogether, 18 families of spiders in banana, 11 families in cotton, 9 families in castor and 13 families in paddy and pigeon pea were reported by Kumar (2007). Spiders of 10 families, 22 genera and 37 species were reported in peanut ecosystem (Trivedi 2009). Spiders representing 14 families, 29 genera and 40 species with predominant species like *Oxyopus* sp., *Plexippus* sp., *Phidippus* sp., *Marpissa* sp. were reported in tea ecosystem (Das et al. 2010). About 17 species of spiders were found to predate on teak defoliator, *Hyblaea puera* (Loganathan and David 1999).

Spiders consume large number of pests and considered as an important predators which help in regulating the population densities of insect pests (Pickett et al. 1946; Kajak et al. 1968; Fox and Dondale 1972; Tanaka 1989). They have higher host finding ability and capacity to consume greater number of prey than other predators. These qualities make spiders a good predator in agro-ecosystem for the management of insect pests. List of spiders, useful in the management of different insect pests is given below.

Pests preyed upon by Spiders

Spiders	Pests	References
Pardosa pseudoannulata	Leafhoppers, planthoppers, whorl maggot flies, leaf folders, case worms and stem borers in rice	Kenmore et al. (1984), Ooi and Shepard (1994), Sahu et al. (1996), Preap et al. (2001), and Jayakumar and Sankari (2010)
P. pseudoannulata, *Tetragnatha* sp. and *Araneus* sp.	Rice brown planthopper, *N. lugens*	Reissig et al. (1982)
Atypena formosana	Planthoppers, leafhoppers, small dipterans in rice	Barrion and Litsinger (1984), Shepard et al. (1987), and Sigsgaard and Villareal (1999)
Thomisus pugilis, *Clubiona filicate*, *Cheiracanthium melanostomum*, *Pardosa birmanica* and *Oxyopes shweta*	Larvae of butterflies and moths in cotton	Solanki and Kumar (2014)

(continued)

Spiders	Pests	References
Phidippus punjabensis, *Thomisus* sp., *Xysticus* sp., *Pardosa sumatrana* and *Argiope anasuja*	Aphids, whiteflies and leaf hoppers in cotton	Solanki and Kumar (2014)
Lycosa tista and *L. kempi*	Leaf hopper, white flies and thrips of cotton	Khuhro et al. (2012)
Hippasa agelenoides, *Cheiracanthium danieli*, *Argyrodes argentatus* and *Drassodes* sp.	Okra jassid *A. bigutella bigutella*, whitefly *B. tabaci*, thrips, *Thrips tabaci* and mites, *Tetranychus cinnabarinus*	Sahito et al. (2013)
Many spiders of Araneida, Salticidae, Thomisidae	Okra aphid *A. gossypii* and whitefly, *B. tabaci*	Bilal and Satti (2012)
Many spiders	Brinjal and snakegourd pests	Sankari and Thiyagesan (2010)
Pardosa altitudis, *Leucage celebesiana*, *Neoscona rumpfi* and *Theridion manjithar*	Cabbage aphid, *B. brassicae*	Khan (2013)
Many spiders	Tea pests	Das et al. (2010)

2.9.1 Arthropod Predators in Ecosystem Service

The presence, abundance and diversity of predatory arthropods have significant impacts on the functioning of ecosystems (Snyder et al. 2006; Schmitz 2007; Bruno and Cardinale 2008; Letourneau et al. 2009). Natural pest suppression and stabilizing an ecosystem balance itself is an ecosystem service. Though some varying results are reported especially with that of coccinellids (Smyrnioudis et al. 2001; Belliure et al. 2011), it is generally believed and reported that the predators indirectly reduce the prevalence of insect-vectored plant diseases in crop plants (Long and Finke 2015). Predator-mediated alteration in plant community structure and diversity indirectly by affecting the herbivory are reported (Schmitz 2009). No doubt, farming system influences the abundance and species composition of predators but the crop and the crop related factors are the main structuring factor (Booij and Noorlander 1992). Interactions between predators and detritivores also affect decomposition dynamics as explained as the effect of spiders on rate of litter disappearance (Lawrence and Wise 2004). Some aquatic arthropod predators are reported to consume many unwanted pest species also.

3 Exposure Routes of Pesticides to Predators

When a pesticide is released into the environment, some portion of it reaches the target plants/soil and the remaining is broken down in the air, deposited on non-target area or drift to far off places. Basically, there are three different routes of pesticide exposure viz., direct, residual and oral. Insect predators may get exposed to pesticides by direct contact to spray applications or by contact with fresh or dry residues in sprayed surfaces or by ingestion of contaminated preys (De-Clercq et al. 1995a, b) or by consuming contaminated water (De-Cock et al. 1996).

Though the arthropod predators happen to be exposed in many ways, their susceptibility to toxic effects of pesticide depends on the nature of the pesticide, its concentration, application method, susceptibility and behaviour of organism, apart from some environmental factors. Diafenthiuron was reported as non toxic to predatory bug, *P. maculiventris* by topical application (LC_{50}= >10,000 mg a.i./L) but highly toxic through drinking water or by residual contact whereas imidacloprid as toxic in all routes (De-Cock et al. 1996). Thiamethoxam is about 200 times more toxic to *P. nigrispinus* by ingestion through water than by residual contact (Torres and Ruberson 2004). Mullie and Everts (1991) studied the uptake of deltamethrin by studying ^{14}C isotopes in erigonid spider, *Oedothorax apicatus* and found 56 % as residual uptake, 32 % by direct uptake and 12 % by oral uptake. This may explain the fact that the residual contamination with lambda cyhalothrin causing higher mortality of spiders, *O. apicatus* and *Erigone atra* than other routes and the insecticide sprayed onto adults in webs had stronger effects than sprayed onto sitting or walking spiders on the soil surface (Dinter and Poehling 1995). In general, spiders that make webs on the plants and stay there to get their food viz., Tetragnatha, Argiope etc. may not be highly affected by contact to sprayed surface, if their webs do not hold spray particles but may be by direct contact and by ingestion of intoxicated prey. However, some other predators like coccinellids feeding on aphids in crop plants, carabaids in turf grasses are affected through all the three routes viz., topical, residual and dietary exposures.

3.1 Contact While Application

Most of the pesticide applications are directed to the pest infestation in the plants. Predators are usually found in the vicinity of the pests and thus affected by direct sprays along with the pest insects. Direct overspray of pesticides on non-target organisms occurs at-least for a portion of predator populations in the field conditions (Bernard et al. 2004). Exposure of coccinellid predators to foliar applications of clothianidin, thiamethoxam and imidacloprid caused high mortality (Mizell and Sconyers 1992; James 2003; Cloyd and Dickinson 2006). Among the different methods of pesticide application, foliar sprays make the highest exposure to the predators especially that are found in plant foliage and near the pests. The hiding behaviour of some of the actively moving predators may make them to go behind the foliage while spraying and allow them to get hid and get less contact to direct sprays also. Soil applied insecticides do not affect the predators in the foliage by contact. Most of the soil insecticides are systemic, thus absorbed by plant tissue and are therefore assumed to have few negative effects at-least as contact poisons on insect predators (Pfluger and Schuck 1991; Mizell and Sconyers 1992; Ishaaya and

Horowitz 1998). Seed dressing chemicals also do not have contact exposure to predators but affects them when they take pollen or nectar because of their systemic properties and presence in the entire plant system or through intoxicated preys.

3.2 Contact to Treated Material/Plant Parts

Insect predators which are found in the plant foliage may get exposed to pesticides indirectly through contact to treated surfaces. This exposure of insecticides to predators may be higher than the pest species in many cases. Most of the insect pest of crops remains in one place or sessile with limited mobility and found feeding on the plant. In contrary, most of the predators are active searches and move around the plant in search of prey, during which they get more exposure to the sprayed particles in the plant as contact. If they had an irritation effect by contact to the sprayed surfaces they tend to move more and if they do not sense the pesticide through the tarsal contact, the case may be much worst.

Ground dwelling predators like carabids and staphylinids may also get contact to treated surface since some portion of pesticide directed towards plant may drift and fall on soil surface. Soil applied pesticides may cause still further exposure to ground dwelling predators. The toxicity of the pesticide is also influenced by environmental parameters. Even the moisture content in the soil was reported to play a major role in pesticide toxicity to ground predators and spiders (Everts et al. 1991b). Irrigation after pesticide application is found to reduce the exposure and toxicity of pesticides to carabid predators in turfgrass (Kunkel et al. 2001). In paddy fields, residual uptake through contact may be the major exposure source especially for spiders. Some spiders that walk on the water surface such as Lycosid spiders may absorb pesticide from contaminated water (Tanaka et al. 2000). However, the spiders which web on the plants and stays on the web may get less contact to the sprayed surfaces, if the webs do not carry pesticide residues. Among all pesticide application methods, seed treatment may have the least effect to predators by contact means.

3.3 Feeding of Intoxicated Insects

Insecticides are sprayed to control the pest infestation in plants. The insect pests that survive due to less exposure or low susceptibility levels may cause threat to the predators. Predators which feed on the intoxicated preys get exposed to pesticides through ingestion. Predatory arthropods which are of active searchers/hunters/pursuers go in search for their prey and the pesticide intoxicated preys may be an easy target for them. Some important carabid predators like *E. alternans*, *P. lucublandus* and *H. pensylvanicus* prefer dead or dying invertebrates over live ones. Applications of pesticides, which kill large numbers of insects, may indirectly poison these potential natural enemies because of the willingness of predators to feed on dead or dying insects (Best and Beegle 1977). The predators which make traps/webs and ambushers also tend to get a higher number of preys which are intoxicated than active uncontaminated ones. Predators with voracious feeding habit or those which feed on preys found in colonies are more exposed, since they feed more on the pesticide exposed preys. If some refugia are set apart without pesticide sprays may help the predators to get a small relief if at all, they feed on the uncontaminated preys intermittently. Bioaccumulation, biomagnification and biotransfer of pesticides by arthropod predators though studied or reported less than that of the metals and other contaminants, but cannot be ignored.

Mortality and sublethal toxic effects on many arthropod predators due to the ingestion of pesticide intoxicated preys are already been reported (Kiritani and Kakiya 1975). The lady beetle, *R. cardinalis* and *Hippodamia undecimnotata* showed reduced survival, longevity and egg production following predation on pesticide toxicated cottony cushion scales and aphids (Grafton-Cardwell and Gu 2003; Papachristos and Milonas 2008). Spiders getting toxicity through leafhoppers fed on treated rice plants were reported by Kiritani and Kawahara (1973). Pesticide toxicity to predatory bugs through different stages of intoxicated preys are well demonstrated (Elzen 2001; Torres et al. 2002; Kim et al. 2006). Even the soil applied insecticide, imidacloprid resulted in >50 % mortality of *H. undecimnotata* larvae due to toxification of their aphid prey; sublethal effects (reduced fecundity, reduced adult longevity) were also observed (Papachristos and Milonas 2008).

3.4 Feeding of Nectar/Pollen of Treated Plants

Many predators and parasitoids feed on nectar and a some predators feeds on pollen also (Lundgren and Seagraves 2009). Adult stages of the predatory grubs like chrysopids and syrphids generally feed on nectar or honeydew and do not feed on insects (Hoy 2011). However, some others like coccinellids feed on alternative plant food i.e., nectar or pollen (Nalepa et al. 1992), when food (aphids) is lacking (Hodek et al. 2012). The aphid feeding coccinellid, *C. maculata* is capable of completing its life cycle on pollen alone (Hodek and Honek 1996).

Pesticides which are sprayed on the plants at flowering time may contaminate the pollen and nectar by direct contact/contamination and the systemic insecticides which are sprayed earlier may also cause contamination by being in the plant system including pollen and nectar. Even the insecticides which are applied as soil treatment get translocated in flower nectar as confirmed by residue analysis. Imidacloprid was found in the levels of 15 and 29 ppbs in buckwheat flowers from recommended dose and double dose applied fields, respectively (Krischik et al. 2007). Soil application of imidacloprid caused 38 % mortality in *C. maculata* in sunflowers (Smith and Krischik 1999). The exposure of pesticides through nectar and pollen is still worse in free living adults that feed only on nectar and pollen. Green lacewings, *C. carnea* adults which feed on nectar from buckwheat, *Fagopyrum esculentum* and Mexican milkweed, *Asclepias curassavica* treated with imidacloprid were found to be affected and the resulting mortalities were 86% and 94% in the label rate and double dose, respectively (Rogers et al. 2007). The only pesticide treatment which is reported to be safe for this route of toxicity exposure can be of seed treatment. The pesticide concentration in the seed gets distributed throughout the plant as it grows and gets greatly reduced by the time the plant enters its flowering phase. Thus, the seed treated plants when it comes to flowering will have a minimum or no toxicity and usually safe to non-target nectar and pollen feeders (Moser and Obrycki 2009).

3.5 Feeding on Treated Plants

Ingestion of leaf tissue by insect predators especially of hemipterans and a few larval coccinellids at times of less prey availability are also been reported (Hodek and Honek 1996; Eubanks and Denno 1999; Coll and Guershon 2002; Moser and Obrycki 2009). Predatory mirids and minute pirate bugs, *Orius tristicolor* are found feeding on plant material in the absence of preys (Askari and Stern 1972; Salas-Aguilar and Ehler 1977). The spined soldier bug, *P. maculiventris* require moisture for optimal development which they may acquire by feeding on plant juices (De-Cock et al. 1996). *Geocoris* sp. and *Nabis* sp. also supplement their arthropod diet with direct feeding on plant material (Ehler 1977). Many carabid predators like *Harpalus pennsylvanicus*, *Evarthrus alternans* feeds on various seeds also (Kirk 1973; Best and Beegle 1977).

These predators may get exposed to pesticides while they feed on the plant parts especially treated with pesticides. Larvae of *H. axyridis* that fed on seedlings raised from clothianidin or thiamethoxam treated seeds exhibited trembling, paralysis, or loss of coordination and had significantly higher mortality (Moser and Obrycki 2009). High mortalities were reported in predaceous stink bug nymphs that consumed plant sap from thiamethoxam-treated cotton plants (Torres et al. 2003). Increased mortality of *O. tristicolor* was reported when it pierced leaves of *Tagetus erecta* treated with imidacloprid (Sclar et al. 1998). Omnivorous carabids (18 different species) which fed on corn seedlings grown from seeds treated with neonicotinoid were reported to have nearly 100% mortality (Mullin et al. 2005).

3.6 Contact to Soil and Plant Debris

Predators may get exposed to pesticides through contact of soil and plant debris in the earth's surface. Pesticides which are sprayed on plants may miss the target and fall on the ground either close or far off from the target plants as drift. They normally settle on the upper crust of earth and assumed to be in the upper 0.5 cm layer (De-Schampheleire et al. 2007). These pesticides which are in soil may cause toxicity to soil dwelling predators especially of carabids and some spiders. Soil application of pesticides for the management of soil insect pests, weeds and soil borne pathogens may cause severe damage because of more exposure rates. Many predators which hide on the plant debris in the ground for light and bright sun may also get contacted to the pesticide particles settled on them.

3.7 *Exposure Through Drifts in Off-Crop Habitats*

Spray application in arable fields endangers not only non-target arthropods within treated areas but also the off-crop habitats adjacent to the treated crop, mainly through drift (Langhof et al. 2003). Relatively larger droplets may settle near the target area but smaller particles (<100 μ) are carried away by wind and settle far off (Salyani and Cromwell 1992). Off-crop habitats with diverse vegetation offers alternative prey or hosts and can be a good refuge after harvest or tillage and can be over-wintering sites (Landis et al. 2000) especially for generalist predators (Langhof et al. 2003). These off-crop boundaries harbour source populations and contribute to the recolonization of predators in pesticide treated fields (Longley et al. 1997). But if those refugia sites are also contaminated through drifts then it affects the predators drastically and thus recolonization takes a longer time.

4 Effect of Pesticides on Predators in Agro-ecosystem

Pesticide exposure to predators in agro-ecosystem may cause acute toxicity by killing them instantly or by causing indirect sublethal toxic effects. Acute toxicity cause immediate death or toxic effects either by direct contact to sprays, indirectly to the sprayed parts or by feeding on intoxicated preys. Non-target individuals may get exposed to sublethal doses of pesticides also in field conditions (Desneux et al. 2007) mainly by contacting to the sprayed parts or by feeding on intoxicated pests. Sublethal effects of pesticides may cause detrimental effects on the development and longevity of the predators affecting the fecundity and also their physiology. Many pesticides are reported to affect the potential of predators to feed on the pest species, which is the most important criteria for pest management. If a toxic pesticide is used, it kills the predator or forces it to move away from that place allowing the residual population of pest to resurge and colonize quickly when the toxic effects get decline. Thus a low dose (50 %) of pesticide is reported to render benefit the pests more than their predators (Schumacher and Freier 2008). So a selective pesticide, which causes maximum mortality to pest and minimum harm to predator, is very much necessary to combine with predators for pest control in integrated pest management.

4.1 Acute Toxicity

4.1.1 Median Lethal Values

Acute mortality of insect predators by pesticides are normally studied and reported as median lethal doses or concentrations. A low median lethal value implies a highly toxic pesticide and vice versa. So a comparison can also be made between pesticides on its toxicity when the bioassay is made on the same test organism and stage using similar test method. Median lethal values are generally assessed at 24 and or 48 h after treatment unless stated otherwise.

Median lethal dose of different pesticides to insect predators

Pesticide	Test organism	Bioassay	Median lethal dose (LD_{50})	References
Coccinellids				
Alphamethrin	*Harmonia axyridis* (adults)	Topical	100.0 ppm	Cho et al. (1997)
Deltamethrin	*H. axyridis* (adults)	Topical	89.3 ppm	Cho et al. (1997)
Esfenvalerate	*H. axyridis* (adults)	Topical	8.09 ppm	Cho et al. (1997)
Fenpropathrin	*H. axyridis* (adults)	Topical	263.4 ppm	Cho et al. (1997)
Imidacloprid	*Serangium japonicum* (adults)	Contact	11.54 mg a.i./L	He et al. (2012)
Methomyl	*H. axyridis* (adults)	Topical	34.9 ppm	Cho et al. (1997)
Monocrotophos	*H. axyridis* (adults)	Topical	366.7 ppm	Cho et al. (1997)
Phosphamidon	*H. axyridis* (adults)	Topical	44.0 ppm	Cho et al. (1997)
Predatory bugs				
Abamectin	*Orius laevigatus* (5th instar)	Residual	30.9 mg a.i./L	Van-de-Veire et al. (2002a)
Diafenthiuron	*P. maculiventris* (5th instar)	Topical	>1,00,000 mg a.i./L	De-Cock et al. (1996)
Diafenthiuron	*P. maculiventris* (♀ adults)	Topical	48,400 mg a.i./L	De-Cock et al. (1996)
Imidacloprid	*P. maculiventris* (5th instar)	Topical	0.07 mg a.i./L	De-Cock et al. (1996)
Imidacloprid	*P. maculiventris* (♀ adults)	Topical	0.79 mg a.i./L	De-Cock et al. (1996)
Spinosad	*O. laevigatus* (5th instar)	Residual	317 mg a.i./L	Van-de-Veire et al. (2002a)
Predatory mites				
Abamectin	*Amblyseius largoensis* (adults)	Residual	0.09 mg a.i./L	De-Assis et al. (2013)
Fenpyroximate	*A. largoensis* (adults)	Residual	1.47 mg a.i./L	
Milbemectin	*A. largoensis* (adults)	Residual	0.17 mg a.i./L	
Propargite	*A. largoensis* (adults)	Residual	896 mg a.i./L	
Spirodiclofen	*A. largoensis* (adults)	Residual	12.6 mg a.i./L	

Median lethal concentrations of different insecticides to insect predators

Pesticide	Test organism	Bioassay	Median lethal concentration (LC_{50})	References
Coccinellids				
Acetamiprid	*Coccinella undecimpunctata* (4th instar)	Topical	263.4 mg a.i./L	Amirzade et al. (2014)
Acetamiprid	*Adalia bipunctata* (4th instar)	Topical	222.6 mg a.i./L	Amirzade et al. (2014)
Azinphos-methyl	*Micromus tasmaniae* (2nd instar)	Residual	0.004 %	Rumpf et al. (1997a)
Cypermethrin	*M. tasmaniae* (2nd instar)	Residual	0.018 %	Rumpf et al. (1997a)
Imidacloprid	*C. undecimpunctata* (4th instar)	Topical	447.8 mg a.i./L	Amirzade et al. (2014)
Imidacloprid	*A. bipunctata* (4th instar)	Topical	218.8 mg a.i./L	Amirzade et al. (2014)
Thiamethoxam	*A. bipunctata* (4th instar)	Topical	232.3 mg a.i./L	Amirzade et al. (2014)
Chrysopids				
Endosulfan	*C. carnea* (1st instar)	Residual	251 mg a.i./L	Golmohammadi et al. (2009)
Imidacloprid	*C. carnea* (1st instar)	Residual	24.6 mg a.i./L	Golmohammadi et al. (2009)
Indoxacarb	*C. carnea* (1st instar)	Residual	133 mg a.i./L	Golmohammadi et al. (2009)
Predatory bugs				
Acephate	*C. lividipennis* (nymphs)	Residual	32.1 mg a.i./L	Preetha et al. (2010)
Cartap	*C. lividipennis* (nymphs)	Dipping	427 mg a.i./L	Tanaka et al. (2000)
Chlorpyrifos	*C. lividipennis* (nymphs)	Residual	36.6 mg a.i./L	Preetha et al. (2010)
Chlorantraniliprole	*C. lividipennis* (nymphs)	Residual	5.9 mg a.i./L	Preetha et al. (2010)
Clothianidin	*C. lividipennis* (nymphs)	Residual	0.08 mg a.i./L	Preetha et al. (2010)
Deltamethrin	*P. maculiventris* (4th instar)	Ingestion	158.8 mg a.i./L	Mohaghegh et al. (2000)
Deltamethrin	*P. maculiventris* (♀ adults)	Ingestion	43.4 mg a.i./L	Mohaghegh et al. (2000)
Deltamethrin	*C. lividipennis* (nymphs)	Residual	4.22 mg a.i./L	Preetha et al. (2010)
Deltamethrin	*C. lividipennis* (nymphs)	Dipping	0.38 mg a.i./L	Tanaka et al. (2000)
Diafenthiuron	*P. maculiventris* (5th instar)	Residual	179.3 mg a.i./L	De-Cock et al. (1996)

(continued)

Pesticide	Test organism	Bioassay	Median lethal concentration (LC_{50})	References
Diafenthiuron	*P. maculiventris* (5th instar)	Ingestion	169.1 mg a.i./L	De-Cock et al. (1996)
Endosulfan	*C. lividipennis* (nymphs)	Residual	212.5 mg a.i./L	Preetha et al. (2010)
Imidacloprid	*P. maculiventris* (5th instar)	Ingestion	4.15 mg a.i./L	De-Cock et al. (1996)
Imidacloprid	*P. maculiventris* (♀ adults)	Ingestion	0.46 mg a.i./L	De-Cock et al. (1996)
Imidacloprid	*P. maculiventris* (5th instar)	Residual	20.3 mg a.i./L	De-Cock et al. (1996)
Imidacloprid	*P. maculiventris* (♀ adults)	Residual	2.5 mg a.i./L	De-Cock et al. (1996)
Imidacloprid	*C. lividipennis* (nymphs)	Residual	1.39 mg a.i./L	Preetha et al. (2010)
Imidacloprid	*P. nigrispinus* (5th instar)	Residual	147.6 mg a.i./L	Torres and Ruberson (2004)
Imidacloprid	*P. nigrispinus* (5th instar)	Ingestion	0.44 mg a.i./L	Torres and Ruberson (2004)
Methomyl	*P. maculiventris* (4th instar)	Ingestion	5.4 mg a.i./L	Mohaghegh et al. (2000)
Methomyl	*P. maculiventris* (♀ adults)	Ingestion	10.6 mg a.i./L	Mohaghegh et al. (2000)
Methyl-parathion	*M. tasmaniae* (2nd instar)	Residual	0.007 %	Rumpf et al. (1997a)
Pymetrozine	*C. lividipennis* (nymphs)	Residual	3.29 mg a.i./L	Preetha et al. (2010)
Teflubenzuron	*P. maculiventris* (4th instar)	Ingestion	14.7 mg a.i./L	Mohaghegh et al. (2000)
Thiamethoxam	*P. nigrispinus* (5th instar)	Residual	98.84 mg a.i./L	Torres and Ruberson (2004)
Thiamethoxam	*P. nigrispinus* (5th instar)	Ingestion	0.06 mg a.i./L	Torres and Ruberson (2004)
Predatory wasps				
Carbaryl	*B. lecheguana* (adults)	Residual	0.37 mg a.i./L	Bacci et al. (2009)
Cartap	*B. lecheguana* (adults)	Residual	0.72 mg a.i./L	
Deltamethrin	*P. sylveirae* (adults)	Residual	0.003 mg a.i./L	
Methamidophos	*P. sylveirae* (adults)	Residual	0.24 mg a.i./L	
Methyl parathion	*P. scutellaris* (adults)	Residual	0.001 mg a.i./L	
Permethrin	*P. scutellaris* (adults)	Residual	0.004 mg a.i./L	
Trichlorfon	*P. fastidiosuscula* (adults)	Residual	0.10 mg a.i./L	

(continued)

Pesticide	Test organism	Bioassay	Median lethal concentration (LC_{50})	References
Predatory spiders				
Carbaryl	*P. pseudoannulata* (nymphs)	Dipping	109 mg a.i./L	Tanaka et al. (2000)
Carbaryl	*Tetragnatha maxillosa* (nymphs)	Dipping	1275 mg a.i./L	
Carbaryl	*Ummeliata insecticeps* (nymphs)	Dipping	449 mg a.i./L	
Cartap	*P. pseudoannulata* (nymphs)	Dipping	7549 mg a.i./L	
Cartap	*T. maxillosa* (nymphs)	Dipping	1660 mg a.i./L	
Diazinon	*P. pseudoannulata* (nymphs)	Dipping	592 mg a.i./L	
Diazinon	*T. maxillosa* (nymphs)	Dipping	1.9 mg a.i./L	
Deltamethrin	*P. pseudoannulata* (nymphs)	Dipping	0.04 mg a.i./L	
Deltamethrin	*T. maxillosa* (nymphs)	Dipping	0.03 mg a.i./L	
Deltamethrin	*U. insecticeps* (nymphs)	Dipping	1.1 mg a.i./L	
Deltamethrin	*Gnathonarium exsiccatum* (nymphs)	Dipping	0.83 mg a.i./L	
Imidacloprid	*P. pseudoannulata* (nymphs)	Dipping	440 mg a.i./L	
Imidacloprid	*T. maxillosa* (nymphs)	Dipping	136 mg a.i./L	
Imidacloprid	*U. insecticeps* (nymphs)	Dipping	995 mg a.i./L	
Imidacloprid	*G. exsiccatum* (nymphs)	Dipping	801 mg a.i./L	

4.1.2 Mortality in Laboratory Assays at Field Recommended Dose

Studies on mortality of beneficial insect predators when exposed with field recommended doses of pesticides is a realistic method of assessing the toxic effects, since that is the dose which they normally get in the field. Cypermethrin and deltamethrin at their field concentrations were highly toxic to mirid bug, *C. lividipennis*, the ripple bug, *Microvelia atrolineata* and the spider, *P. pseudoannulata* by contact toxicity (Fabellar and Heinrichs 1984). Azinphos methyl, phosmet, methomyl, esfenvalerate, cypermethrin and acetamiprid at their field doses was highly toxic to *O. insidiosus* up to 14 days, tested as dry residues. All the above said insecticides along with indoxacarb were found to be toxic to ladybird beetle, *H. convergens* and lacewing, *Chrysoperla rufilabris* (Roubos et al. 2014). Indoxacarb was reported to be safer to many predators like *Anthocoris nemoralis* (Pasqualini et al. 1999), *Cycloneda sanguinea*, *O. insidiosus*, *C. rufilabris*, *H. axyridis* (Michaud and Grant 2003) and Phytoseiid predatory mites, *Amblyseius andersoni* (Mattedi et al. 1998) and *Kampimodromus aberrans* (Mori et al. 1999).

Coccinellids: Pyrethroids were reported to be highly toxic to coccinellid, *Rhizobius lophanthae* followed by carbamates and organophosphates (Bellows and Morse 1993). All life stages of a lady beetle, *H. axyridis* were susceptible to topical treatment of label rates of acetamiprid, thiamethoxam and imidacloprid (Youn et al. 2003). Field recommended dose of imidacloprid, acetamiprid, cypermethrin, deltamethrin and profenofos tested for residual toxicity to coccinellid adults revealed profenofos as the most and imidacloprid as the least toxic insecticide (Ahmad et al. 2011). Among the ten insecticides tested, dichlorvos was highly toxic to eggs of *C. sexmaculata* causing a mortality of 97 % and acetamiprid and endosulfan being the least toxic registering 10 % mortality. The 48 h larval mortality was also highest (>70 %) in dichlorvos, fenvalerate and cypermethrin. Dichlorvos gave 100 % mortality to adults by 48 h followed by cypermethrin, fenvalerate and phosphamidon with >80 % mortality (Tank et al. 2007). However, acetamiprid was reportedly toxic to lady beetle, *Stethorus japonicus* (Mori and Gotoh 2001). Chlorantraniliprole and novaluron are highly toxic to *H. axyridis* and *C. maculata* causing >98 % mortality by residual contact. Diazinon and imidacloprid were moderate toxic (33–66 %) to the eggs of *C. montrouzieri* by dipping method (Aghabaglou et al. 2013). Novaluron caused a mortality of 96.7 % in *H. axyridis* in 6 days when fed by intoxicated prey whereas *C. maculata* was not found to be affected (Cabrera et al. 2014). Diafenthiuron at the field recommended dose caused 13.33 % mortality of *M. sexmaculatus* grubs by contact whereas monocrotophos registered a mortality of 80.0 % (Stanley et al. 2016). Endosulfan was reported safer to the adults of *C. septempunctata* (Sharma and Adlakha 1986; Meena et al. 2002), *C. transversalis* (Chaudhary and Ghosh 1982) and *M. sexmaculatus* (Babu 1988; Patil and Lingappa 2000; Tank et al. 2007). Copper oxychloride did not exert immediate toxic effects on the mite predator, *Stethorus nigripes* (Edwards and Hodgson 1973). Copper oxychloride and Bordeaux mixture were reportedly safe to *Scymnus coccivora* (Mani and Thorntakarya 1988).

Chrysopids: Chlorantraniliprole was reported as highly toxic to lacewing adults by direct contact (Amarasekare and Shearer 2013) but least toxic in indirect dry residual contact (Roubos *et al.* 2014). Imidacloprid and acetamiprid showed a low toxicity in the dipping test and high toxicity in the residual contact test to *C. carnea* larva (Toda and Kashio 1997). Imidacloprid at 0.2 mL/L caused mortality of *C. carnea* grubs up to 48.7 %, with reduced pupation and adult emergence (Mathirajan and Regupathy 2002). When *C. carnea* larvae were exposed to 10 ppm imidacloprid through its diet, a mortality of 60 % was observed (Kumar and Santharam 1999). Imidacloprid at the recommended dose 0.28 mL/L caused 15.3 % egg mortality, 26.7 and 33.3 % larval mortality by ingestion and contact and 50.0 % adult mortality (Preetha et al. 2009b). Cole and Horne (2006) found that imidacloprid as highly toxic to the larva of *M. tasmaniae* with >85 % mortality. In a spray chamber bioassay, 83.3 % of adult *C. carnea* died in 24 h when exposed to imidacloprid (Elzen et al. 1998). Indoxacarb @ 250 ppm caused 100 % mortality in *C. carnea* (Nasreen et al. 2003). Methomyl, cyfluthrin and fenpropathrin caused about 95 % and 60–72 % mortality to first and third instar grubs of green lacewings, respectively (Nasreen et al. 2007). Primicarb, pymetrozine and imidacloprid treated lettuce aphid, *Nasonovia ribisnigri* caused 20, 40 and 96 % mortality to *M. tasmaniae* larvae three DAT (Walker et al. 2007).

Predatory bugs: Imidacloprid was very toxic to *C. carnea* grubs and found to inhibit adult emergence also (Huerta et al. 2003). Methoxyfenozide, tebufenozide and spinosad did not cause mortality to both sex of adult big-eyed bug, *G. punctipes*, whereas chlorfenapyr caused some mortality at their field recommended rates (Elzen and Elzen 1999). The anthocorid, *O. laevigatus* was found to suffer 72 % mortality of nymphs when exposed by residual contact to field dose of teflubenzuron (Van-de-Veire et al. 1996). Fipronil was toxic for *O. laevigatus* and very toxic for mirid, *Macrolophus caliginosus* tested as residues in sweet pepper and tomato plants, respectively (Sterk et al. 2002). Imidacloprid and indoxacarb affected the survival of third instars and adult *O. insidiosus* (Studebaker and Kring 2001). Significant mortalities of minute pirate bugs, *O. tristicolor* were reported when they are kept in marigold leaves treated with imidacloprid at soil application (Sclar et al. 1998). Mizell and Sconyers (1992) found significant mortality to the hemipteran predators, *G. punctipes* and *Deraeocoris nebulosus* from topically applied imidacloprid. Spinosad, methoxyfenozide, abamectin and acetamiprid at their field recommended doses caused 4.0, 6.2, 62.7 and 72.0 % mortality of second instar nymphs of *D. brevis* at 48 h after treatment (Kim et al. 2006). Profenofos and pyriproxyfen were reported to be highly toxic to *P. maculiventris* by tarsal contact to dry residues (Wilkinson et al. 1979; De-Clercq et al. 1995a). Permethrin, emamectin benzoate and methyl parathion were equally toxic to the nymphs of *P. maculiventris* by residual contact whereas the adults are slightly tolerant to permethrin and emamectin benzoate. Consumption of chlorfenapyr treated soybean looper, *Chrysodeixis includens* also caused significantly greater mortality than imidacloprid, permethrin, spinosad, and thiodicarb to adults of *G. punctipes* and *Nabis roseipennis* (Boyd and Boethel 1998). No much mortality was reported in two spotted stink bug, *Perillus bioculatus* after 24 h of contact with potato foliage sprayed with imidacloprid (Hough-Goldstein and Whalen 1993).

Other predators: Acetamiprid is toxic to a predatory thrip, *Scolothrips takahashi* (Mori and Gotoh 2001). Many of the soil applied insecticides are highly toxic to ground-dwelling Staphylinid predator, *Aleochara bilineata* and carabid, *Bembidion lampros* in vegetable gardens (Jansen et al. 2008). Chlorfenapyr, diafenthiuron and fenpyroximate were not very toxic to predatory mites, *Neoseiulus womersleyi* (Kim and Seo 2001). Abamectin was reported to be very toxic to phytoseiid predators, *Neoseiulus cucumeris* and *Phytoseius plumifer* (Kim et al. 2005; Noii et al. 2008). Chlorantraniliprole was tested in many different formulations and found to be harmless to predatory mite, *T. pyri* by glass residue bioassay (Dinter et al. 2008). Goven and Guven (2008) reported sulphur (Thiovit®) as moderately toxic but carbaryl, methyl-parathion and quinalphos as harmless to predatory mite,

Typhlodromus perbibus by glass slide residue contact. Based on mortality and impact on fecundity, the pesticides carbaryl, cypermethrin, acetamiprid, methomyl and deltamethrin were considered as harmful and diflubenzuron as slightly harmful to predatory mite, *Euseius finlandicus* (Broufas et al. 2008). Endosulfan was found to be highly toxic than azinphos-methyl and cyhexatin both by topically and by residual to the northern yellow sac spider, *Cheiracanthium mildei* (Mansour et al. 1981).

4.2 Chronic Toxicity

When LC_{10} concentration of diafenthiuron provided through drinking water for 15 days to *P. maculiventris* adults, 75 % of the females and 80 % of the males died (De-Cock et al. 1996). The average number of eggs produced per female of *P. maculiventris* per day was 13.1, 11.9 and 13.6 for control, imidacloprid, diafenthiuron and control, respectively. About 10 % reduction in fecundity was observed in *P. maculiventris* chronically treated with diafenthiuron at LC_{10} concentration (De-Cock et al. 1996).

4.3 Persistent Toxicity

Neonicotinoid insecticides viz., imidacloprid, thiamethoxam, acetamiprid and thiacloprid were highly persistent causing 100 % mortality to second instar *O. laevigatus* tested as leaf residues even after 30 days of spraying. Spinosad and indoxacarb resulted in 60 % mortality (Van-de-Veire and Tirry 2003). Thiamethoxam was persistently toxic up to 52 days post application in potted cotton plants to predatory bug, *P. nigrispinus* and this effect was seen for only 9 days in field conditions. Thiamethoxam @ 4 mg a.i./plant caused a persistent toxicity for 40 days in field conditions (Torres and Ruberson 2004). Abamectin, bifenthrin, imidacloprid and lufenuron were found to be persistent up to 30 days in sweet pepper plants and caused >75 % mortality to *O. laevigatus* (Van-de-Veire et al. 2002b). Abamectin spray deposits at the recommended rate in sweet pepper were reportedly toxic to the predatory bug, *O. laevigatus* for 2 weeks and 1 month, during summer and spring seasons, respectively. The persistent toxicity of spinosad at its recommended dose to the predator is only for 5 days (Van-de-Veire et al. 2002b).

Chlorpyrifos, formetanate and methamidophos are persistently toxic to predatory mirids, *Dicyphus tamaninii* and *Macrolophus caliginosus* for 30 days. The persistent toxicity of endosulfan was found to be moderate to *M. caliginosus* with effects for <21 days of post treatment and short lived to *D. tamaninii* (<3 days) (Figuls et al. 1999). Castane et al. (1996) reported 79.2 % mortality of *D. tamaninii* nymphs after 7 days of exposure to imidacloprid fresh residue. Acetamiprid, imidacloprid and thiamethoxam caused 43.3, 60.0 and 73.3 % mortality to *M. caliginosus* even after 30 days of spraying in sweet pepper plants (Van-de-Veire and Tirry 2003). Short persistence (<7 days) was reported for methomyl, dinotefuran, indoxacarb and carbendazim + mancozeb as tested against predatory mites, *Neoseiulus longispinosus* (Kongchuensin and Takafuji 2006). However, abamectin, malathion and phosalone were reported to be persistently toxic to the predatory mite, *P. plumifer* by causing 100 % morality even up to 15 days (Noii et al. 2008). Fenpyroximate was reportedly persistent even up to 37 days after treatment to *Phytoseiulus persimilis* and *Galendromus occidentalis* (Irigaray et al. 2007) but found short lived to show toxic effect on *Amblyseius californicus* (Van-de-Veire et al. 2001). The persistent toxicity of carbaryl, cypermethrin, acetamiprid and methomyl was found to be >2 weeks to predatory mite, *Euseius finlandicus* (Broufas et al. 2008). Cartap caused paralysis of *L. pseudoannulata* for about 15 days after treatment (Takahanshi and Kiritani 1973).

4.4 Sublethal Toxicity

Sublethal doses of insecticides may not cause mortality to predators but affects them indirectly by altering their physiological and/or behavioural traits (Desneux et al. 2007). A pesticide which kills 50 % of insect predators can be more acceptable than the one which reduces its fecundity and make the surviving individuals malformed and reduce its predatory potential. A reduction in predatory potential of these beneficial organisms when exposed to insecticides is well documented (Lo et al. 1992; Desneux et al. 2007; He et al. 2012; Gholamzadehchitgar et al. 2014). It was reported that imidacloprid though did not cause much mortality to *Harpalus pennsylvanicus* by direct contact, residual and ingestion, it made the beetles lethargic and susceptible to ant predation (Kunkel et al. 2001). Sublethal toxic effects edpecially affecting the reproduction and predation ability are very much important in pest management point of view. Some of the sublethal effects of pesticides on predators of agro-ecosystem are given below.

4.4.1 Alteration in Developmental Time and Longevity

A significant reduction in adult lifespan of *C. carnea*, *C. septempunctata* and *Syrphus* spp. after treatment with diafenthiuron was reported (EI-Sayed and El-Ghar 1992). Larvae of *Olla v-nigrum* treated with field rates of copper sulfate as contact experienced a significant increase in developmental time. The resulting adults also get increased pre-reproductive period and laid fewer eggs compared to control. Copper sulfate was not found to affect *Curinus coeruleus* and *H. axyridis* either in terms of larval development or adult longevity (Michaud and Grant 2003). Azoxystrobin at double the field dose was found to accelerate the larval development of ladybird beetles, *Cycloneda sanguinea* and *H. axyridis*, whereas ferbam slowed down the development of *H. axyridis* (Michaud 2001). Indoxacarb and spinosad extended the developmental time of *H. axyridis* to become adults when exposed as first instar grubs (Galvan et al. 2005). Pirimicarb and pymetrozine had no adverse effects on the developmental time of *C. undecimpunctata* (Cabral et al. 2008, 2011). Grubs of *H. axyridis* exposed to leaf residues of fenpropathrin were found to have shorter larval duration (Michaud 2002).

The IGRs, fenoxycarb and pyriproxyfen prolonged the developmental time of chrysopid, *C. rufilabris* (Liu and Chen 2001; Chen and Liu 2002). Methomyl, cyfluthrin and fenpropathrin significantly affected the longevity and fecundity of lacewing adults (Nasreen et al. 2007). Adult longevity of *C. carnea* is reported to be unaffected by spinosad (Medina et al. 2001) whereas, imidacloprid was reported to have an adverse affect. *Chrysoperla carnea* adults had 60 % shorter longevity after feeding on sugar water containing 10 ppm imidacloprid (Kumar and Santharam 1999). Delayed developmental rate from third instar larvae to pupae in *M. tasmaniae* fed with imidacloprid treated lettuce aphid, *Nasonovia ribisnigri* was reported (Walker et al. 2007). A delay in 5 days in the prepupal development of *M. tasmaniae* due to the exposure of fenoxycarb 0.05 % was also reported (Rumpf et al. 1997a).

Spined soldier bug, *P. maculiventris* treated with deltamethrin (12.5 mg a.i./L) via ingestion was found to have prolonged preovipositional period (Mohaghegh et al. 2000). Permethrin exposure to pentatomid bug, *Supputius cincticeps* was found to have a decrease in the developmental time for females and vice versa in males (Zanuncio et al. 2003). Thus some of the fungicides and insecticides are reported to alter (either prolong or shorten) the developmental time of different insect predators. Shortening the life stages seems to be advantageous but these type of pesticide induced accelerations may well have negative effects on the resulting adult (Michaud and Grant 2003) which are need to be studied in detail.

4.4.2 Fecundity

Systemic insecticides were found to have an adverse impact on the reproduction potential of vedalia beetle, *R. cardinalis* (Grafton-Cardwell and Gu 2003). A reduction in the reproduction potential of coccinellid, *Eriopis connexa* developed from fourth instar larvae topically dosed with acetamiprid is reported with 40 % reduction in fecundity and only 55 % fertile eggs (Fogel et al. 2013). Indoxacarb and spinosad reduced the fecundity and fertility of female *H. axyridis* (Galvan et al. 2005). Feeding with aphids contaminated by fenpropathrin, clofentezine, hexythiazox, brompropylate and vinclozolin decreased the fecundity of *A. bipunctata* (Olszak 1999). However, primicarb and pymetrozine had no adverse effects on the fecundity, fertility, egg hatching of *C. undecimpunctata* (Cabral et al. 2008, 2011). Pyriproxyfen was not found to affect the fecundity of whitefly lady beetle, *Delphastus catalinae* when fed with insecticide treated *Bemisia tabaci* eggs (Liu and Stansly 2004). No significant variation in egg laying was found in coccinellids, *Curinus coeruleus, H. axyridis and O. v-nigrum* treated with copper sulfate fungicides as residual contact at the larval stage (Michaud and Grant 2003). Exposure of mealybug predator, *C. montrouzieri* to copper oxychloride had no negative effects on its longevity or reproduction (Mani et al. 1997).

Chrysoperla externa grubs exposed to tebufenozide affected the viability and fertility of eggs (Carvalho et al. 2003). Tebufenozide and chlorfenapyr at recommended doses significantly reduced the fecundity of big-eyed bug, *G. punctipes* (Elzen and Elzen 1999). Spinosad was not found to affect the fecundity of *C. carnea* (Medina et al. 2001). Abamectin and spinosad were reportedly affecting the fecundity of *D. brevis* (Kim et al. 2006) whereas *G. punctipes* was not affected and surprisingly enhanced the fecundity of *O. insidiosus* (Elzen 2001). The reproductive rate of female mites, *P. persimilis* was reduced by about 50 % when feed with *T. urticae* intoxicated with abamectin (Zhang and Sanderson 1990). Abamectin was not found to affect the reproduction, longevity and sex ratio of *Neoseiulus longispinosus* (Ibrahim and Yee 2000).

4.4.3 Egg Hatchability

Systemic insecticide, acetamiprid was found to affect the hatchability of coccinellids, *E. connexa* to a greater extent (Fogel et al. 2013). No significant adverse effect was observed due to imidacloprid on egg hatchability of *C. carnea* (Kumar 1998; Suganthy 2003). A marked decline in egg hatching to a tune of 63 and 54 % were observed as *P. maculiventris* when treated with teflubenzuron (200 mg a.i./L) and deltamethrin (12.5 mg a.i./L) via ingestion, respectively (Mohaghegh et al. 2000). Methoxyfenozide, spinosad, abamectin and acetamiprid at their field recommended doses did not affect the hatching of *D. brevis* eggs when topically treated (Kim et al. 2006). Spinosad did not significantly affect the egg hatchability of predatory mites, *P. persimilis* (Ahn et al. 2004; Duso et al. 2008) and *N. cucumeris* (Kim et al. 2005). Though thiamethoxam was found to reduce the oviposition rate of *P. persimilis*, it was not found to affect the egg hatching (Duso et al. 2008).

4.4.4 Malformation

Systemic insecticides were reported to inhibit the development of *R. cardinalis* from larvae to adult (Grafton-Cardwell and Gu 2003). Acetamiprid was found to disrupt the embryogenesis of coccinellid predator, *Eriopis connexa* at early embryo stage (Fogel et al. 2013). Malformations of nymphs of soldier bug, *P. maculiventris* were reported when treated with diflubenzuron and teflubenzuron. Severely malformed nymphs died within a few days, whereas slightly deformed nymphs died in the next moult (De-Clercq et al. 1995a; Mohaghegh et al. 2000). Adults of *P. maculiventris* emerging from diafenthiuron treated nymphs at concentrations >50 mg a.i./L (administered in drinking water) had malformed wings and legs (De-Cock et al. 1996). Adults of reduviid predator, *Rhynocoris kumarii* were reported to develop severe abnormalities in the alimentary canal, testis and ovary when exposed to monocrotophos, dimethoate, methyl parathion and quinalphos at sublethal doses (George and Ambrose 2004).

4.4.5 Behaviour Alteration

The predatory and reproductive behaviour of insect predators were found to get affected when exposed to pesticides (Samu and Vollrath 1992). Deltamethrin was reported to increase the grooming behaviour of *C. septempunctata* (Wiles and Jepson 1994) mainly because of irritation caused by the pesticide (Desneux et al. 2007). Smith and Krischik (1999) reported reduced survival and mobility in pink spotted coccinellid, *C. maculata* exposed to imidacloprid. The movements of *C. sanguinea* adults were reportedly reduced by azoxystrobin, ferbam and mefenoxam. Benomyl and the copper and petroleum oil combination was found to affect the movements of *H. axyridis* adults (Michaud 2001). Cessation of feeding and slow movements was observed in *P. maculiventris* nymphs treated with imidacloprid (De-Cock et al. 1996). Acetamiprid caused immobility in 88.5 % of *D. brevis* adults after topical treatment whereas, methoxyfenozide or spinosad did not affect their mobility (Kim et al. 2006). An increase in preening behaviour was found in mirid bugs, *Macrolophus pygmaeus* exposed to thiacloprid (Martinou et al. 2014). Grubs of predatory ground beetle, *Pterostichus cupreus* reared on copper-contaminated food and soil caused reduced adult locomotion (Bayley et al. 1995). Imidacloprid intoxicated ground beetles, *H. pennsylvanicus* were vulnerable to ant predation in fields mostly because of slow running (Kunkel et al. 2001). Deltamethrin treatment to erigonid spiders, *Oedothorax apicatus* reduced considerably the walking speed and made them vulnerable to predation by carabid beetles (Everts et al. 1991b).

Diazinon, fenitrothion and chlorpyrifos caused leg and proboscis tremor on predatory bugs, *Andrallus spinidens* (Gholamzadehchitgar et al. 2014). *Chrysoperla carnea* adults fed on nectar from flowering plants treated with imidacloprid as soil application exhibited trembling, inability to fly and inability to right themselves when placed upside down (Rogers et al. 2007). Spined soldier bug, *P. maculiventris* exposed to imidacloprid through drinking water also exhibited similar kind of symptoms (De-Cock et al. 1996). Lack of coordination and wing fluttering were reported in *P. maculiventris* treated with diafenthiuron mixed in drinking water (De-Cock et al. 1996). *Chrysoperla carnea* larva exhibited lack of coordination and cessation of feeding and inability to cast the old head capsule when treated with indoxacarb (Golmohammadi et al. 2009). Exposure of fenoxycarb, a juvenile hormone analog was reportedly disrupted *C. carnea* larva to produce silk for cocooning (Bortolotti et al. 2005). However, the IGR diflubenzuron allowed the *C. carnea* larva to spin cocoon but the insect died inside itself with no adult emergence at all (Medina et al. 2003). In spiders, alphacypermethrin suppressed web-building frequency and severely affected the web size and building accuracy of cross spider, *Araneus diadematus* (Samu and Vollrath 1992). Topical application of deltamethrin and fenvalerate resulted in up to a week's delay of web-building in spiders, *Oedothorax apicatus* and *Erigone atra* (Dinter and Poehling 1995).

4.4.6 Effect on Predation

A reduced predation of coccinellids, *Orcus chalybeus* on soft scale insects, when foraging on surfaces treated with a copper sulfate fungicide is reported (Lo et al. 1992). Increased handling time and a reduced consumption of *B. tabaci* eggs was reported when *Serangium japonicum* was exposed to sublethal doses of imidacloprid (5 ppm) as dry residues (He et al. 2012). Thiacloprid significantly reduced the predation rate of mirid bug, *M. pygmaeus* when exposed as residues or as treated food whereas chlorantraniliprole did not cause any effect on predation (Martinou et al. 2014). Antifeedancy and regurgitation were observed in the predatory carabid, *Nebria brevicollis* fed on deltamethrin treated aphids (Wiles and Jepson 1993). Diafenthiuron was found to reduce the predatory potential of *Podisus nigrispinus* on cotton leaf worm, *Alabama argillaceae* (Torres et al. 2002). Predatory bugs, *A. spinidens* exposed to diazinon, fenitrothion and chlorpyrifos, decreased the attack rate and increased the handling time of prey (Gholamzadehchitgar et al. 2014). Field recommended dose of cypermethrin reduced the attack rate of assassin bug, *Acanthaspis pedestris* up to 6.4 times and had decreased ability to paralyze the prey (Claver et al. 2003). Imidacloprid was found to reduce the predation rate of *Neoseiulus californicus* and *Phytoseiulus macropilis* on *T. urticae* to about 55 % and 87 %, respectively (Poletti et al. 2007).

4.4.7 Life Table Parameters at Sublethal Doses

Life-table experiments can assess the toxic effects of pesticides more accurately than any other estimates (Forbes and Calow 1999) and used for many natural enemies (Stark and Banks 2001; Acheampong and Stark 2004). Sublethal concentrations (LC_{10} and LC_{30}) of thiamethoxam were found to alter the life expectancy (e_x), declined age-specific fecundity (m_x) and decreased pre-adult development time of black-spotted lady beetles, *H. variegata*. However, the fecundity and fertility are found unaffected (Rahmani et al. 2013). Imidacloprid caused a significant difference in pupal period, post oviposition, longevity and gross fecundity rate (GFR), mean eggs per day, net reproductive rate (R_0) and doubling time (DT) of *C. montrouzieri* (Aghabaglou et al. 2013). Indoxacarb caused a simultaneous reduction in adult fertility and also the survivorship of the 1st instar larva of *H. axyridis* in laboratory (Galavan et al. 2005). The gross reproductive rate (GRR), intrinsic rate of increase (r_m), doubling time (DT), mean generation time (T) and finite rate of increase (λ) along with the sex ratio, adult longevity and adult fertility were not affected in *C. carnea* treated with imidacloprid, indoxacarb and endosulfan at their LC_{25} concentrations. Only the net reproductive rate (R_0) was significantly altered (Golmohammadi et al. 2009).

4.5 Field Effects

4.5.1 Field Toxicity

In field conditions, diafenthiuron was reported highly toxic to insect predators on aphids in cotton (El-Zahi and Abd-Elhady 2013). Diafenthiuron and imidacloprid are highly harmful to coccinellids, *Brumus suturalis* in okra (Solangi and Lohar 2007). Azinphos methyl was found to be highly toxic to coccinellids like *A. bipunctata*, *C. septempunctata* and *Oenopia conglobata* in apple, peach and pear orchards whereas, cartap is selective (Pasqualini and Civolani 2003). Smith and Krischik (1999) opined that the use of imidacloprid may not be compatible with coccinellid predator, *C. maculata* for the management of pests of interior land scapes. Patil and Lingappa (2001) reported the toxic effect of imidacloprid at 40 g a.i./ha on *C. carnea* grubs also. Toxic effect of some insecticides on the predacious lady bird beetles in country bean, *Lablab purpureus* was studied by Mollah et al. (2012). Emamectin benzoate @ 1 g, esfenvalerate @ 1 mL, cypermethrin @ 1 mL and deltamethrin @ 1 mL/L were found to cause mortality of 19.0 %, 33.8 %, 31.7 % and 28.4 %, respectively 1 day after spraying. Biswas et al. (2006) studied the toxicity of cypermethrin and carbofuran on predators of brown planthopper, *Micraspis discolor* and *P. pseudoannulata* and concluded cypermethrin 10 EC 0.05 kg a.i./ha was highly toxic than carbofuran 5G at 0.5 kg a.i./ha. Imidacloprid was reported as toxic to predatory spiders and mirid bugs in rice ecosystem (Tanaka et al. 2000; Manjunatha and Shivanna 2001). Nagata et al. (1997) stated that diazinon, fenobucarb, carbaryl, dichlorvos and imidacloprid were toxic to spiders in cotton. Copper fungicides were not compatible with aphid predatory midges, *Aphidoletes aphidimyza* and thus cannot be used in IPM programmes (Havelka and Bartova 1991).

Imidacloprid and thiamethoxam cause low toxicity to insect predators on aphids in cotton (El-Zahi and Abd-Elhady 2013). Likewise, imidacloprid was found to be least toxic to predators (spiders, ants and beetles) and effective on okra pests when compared to acetamiprid, methamidophos and diafenthiuron (Solangi and Lohar 2007). Recommended doses of acetamiprid, thiamethoxam, imidacloprid and abamectin were found safe to the predatory ladybird beetles (Acharya et al. 2002). Spinosad did not affect the number of *G. puncticeps*, *H. convergens* and *C. maculata* in cotton fields whereas lambda cyhalothrin did (Tillman and Mulrooney 2000). Imidacloprid up to 950 g a.i./kg of lettuce seed did not affect the aphid predatory syrphids (Ester and Brantjes 1999). Imidacloprid treated as seed pelleting in sugarbeet did not affect spiders, ground beetles and rove beetles (Epperlein and Schmidt 2001). Staphylinid beetle, *Oligota pygmaea*, a red spider mite predator was not found to get affected by the application of diafenthiuron in tea gardens (UPASI 2005). Abamectin was found to be less toxic to predatory mites in rose (Jasmine et al. 2008).

4.5.2 Reduction in Population After Pesticide Application in Fields

Acetamiprid, clothianidin and dinotefuran were reported to be highly toxic to *C. montrouzieri* (Cloyd and Dickenson 2006). Soybean plots treated with lambda cyhalothrin was found to harbour significantly less number of *H. axyridis* than chlorpyrifos treated plots or control plots. Densities of *H. axyridis* larvae in sweet corn plots treated with chlorpyrifos, carbaryl, bifenthrin, or lambda cyhalothrin did not differ significantly from each other but lower than in spinosad, indoxacarb, or untreated plots (Galvan et al. 2005). A temporary reduction of populations of juvenile stages of heteropterous predators (Anthocoridae and Miridae) in cotton field trials with diafenthiuron with a recovery within 10 days was reported. However, the adults of Heteroptera, Coccinella and Chrysoperla were not affected (Streibert et al. 1988). Imidacloprid and ethofenprox reduced the mirid bug, *C. lividipennis* abundance in rice fields to a point of elimination (Tanaka et al. 2000). However, imidacloprid was reported to cause only a minor decline in the abundance of predatory larvae of carabids and staphylinids in lawn based on pit fall trap catches (Kunkel et al. 1999).

Diafenthiuron did not negatively affect the mortality, fecundity or fertility of anthocorid, *Orius niger* as studied in vegetables (Van-de-Veire and Degheele 1993). After the application of dimethoate to winter wheat, Araneae were reportedly reduced by 90 % in 7 days and predatory Carabidae by 76 % even by 6 weeks after treatment along with many dead adult and larval Coccinellidae and Syrphidae (Vickerman and Sunderland 1977). Dimethoate and deltamethrin registered 60–80 % mortality to carab beetle, *Bembidion lampros* when they are confined in flag leaf after spraying in winter wheat fields (Unal and Jepson 1991). Dimethoate and vamidothion sprays in apple trees caused a significant reduction of phytoseiid mites (*Typhlodromus phialatus* and *T. pyri*) with no and 0.1 motile stages per leaf, respectively after 4 days of spraying and did not recover even after 35 days (Cavaco et al. 2003). In phenthoate and deltamethrin sprayed rice plots lycosid spider populations decreased soon after application and remained less throughout the experimentation period. Deltamethrin eliminated Tetragnatha spider, *T. maxillosa* and reduced the linyphiid spiders, *G. exsiccatum* in rice. Deltamethin was also reported to reduce the linyphiid spiders in wheat fields (Everts et al. 1989). Azinphos methyl and methidathion when sprayed in apple trees has significantly reduced the spider populations (Mansour et al. 1981). Acetamiprid, imidacloprid and thiacloprid eliminated *D. brevis* in treated pear trees for at least 5 weeks (Brooks et al. 2004).

5 Methods to Assess Pesticide Toxicity to Arthropod Predators

Many methods are developed or being developed to evaluate the toxicity of pesticides to different kinds of predators. Studies on acute toxicities of pesticides are usually carried out through laboratory tests to arrive at with median lethal dose/concentrations (Desneux et al. 2007; Biondi et al. 2013). Calculation of median lethal concentration (LC_{50}) and or dose (LD_{50}) are the widely used technique to assess the acute toxicity of a pesticide either to target pests or non-target beneficials. A lower the value of LC_{50} is desirable for target pests and a higher for beneficial organisms to make the pesticide selective toward the target pest with less harm to the non-targets. Another term LR_{50} (median lethal residue) is also used to express the residue concentration, which kills 50 % of the test organism and expressed as μg/cm^2 (Suckling et al. 1986). Acute indirect contact toxicity is assessed using dry film residues in different objects including leaves and exposed to the predators. In residue bioassays using an inert material, the exposure is limited only to the dry residues picked up via tarsal contact. Toxicity due to direct overspray of animals and their food are also needed to be analyzed (Bernard et al. 2004), since the predators get exposed to direct sprays in the fields. Laboratory acute direct contact toxicity assays is mainly done by dipping the test organism or spraying over them or as localized topical applications. Acute ingestion bioassays consist of dosing/contaminating the prey or food and then fed to the predators. Treating the preys of predatory insects like eggs, aphids, mites, hoppers, fruit flies (Preetha et al. 2009b; Walker et al. 2007; Kim and Yoo 2002; Kiritani and Kawahara 1973; Everts et al. 1991a) and treating their food like honey, water etc. are also reported (Preetha et al. 2009b; Mohaghegh et al. 2000).

Apart from acute toxicity, predators may get sublethal chronic toxicities especially after spray applications, when they move over the sprayed parts and ingest preys exposed to pesticides. This may cause a kind of low but accumulated toxicity. Chronic toxicity assessments are being done with exposing the predators to low doses but continuously for many days. Yet another important assessment is the persistence of toxic effects. It means how long it is toxic (persistent toxicity) to the predator, which is normally assessed by exposing them in treated leaves or Petri dishes at different time intervals (days) after spraying.

An accurate assessment of sublethal effects is also required to evaluate the selectivity of pesticides towards non-target organisms (Biondi et al. 2012a). In addition to acute toxicity, sublethal effects on arthropod physiology and behaviour are to be considered (Desneux et al. 2007; Biondi et al. 2013). Sublethal effects on predatory ability, reproduction capacity, locomotion, longevity and susceptibility to predation are being studied. Apart from laboratory experiments, semi-field and glass house experiments of tier II toxicity evaluation and a more realistic tier III toxicity evaluation of field effects are also being done for predators. Some of the methodologies for lethal and sublethal effects under laboratory, semi-field and field conditions are given below.

5.1 Tier I Toxicity Evaluation: Laboratory Experiments

5.1.1 Acute Toxicity Assays

Acute toxicity assays are done for testing the immediate effect of pesticides on the organism. This is a high level dose causing immediate kill or related effect to the predators in a short span of time after exposure. Acute toxicity is generally detected by arriving at a median lethal concentration/dose. Predators may get the acute pesticide toxicity through direct contact to the pesticide sprays, or by indirectly contacting to the sprayed plants or soil or by ingesting the pesticide contaminated food. Before doing the actual acute toxicity bioassays, a preliminary range finding tests were carried out by exposing the test organisms to different concentrations starting from field application rate and then tenfold dilutions to get a concentration which gives approximately 50 % mortality. This is kept as the central concentration with three concentrations above that and three below to get a set of six concentrations which would give 10–90 % mortality. The concentrations which give 10–90 % mortality are included in the actual acute toxicity experiments and are used for analysis to get median lethal values.

5.1.1.1 Direct Contact Toxicity Assays

5.1.1.1.1 Dipping the Test Organism

Coccinellid Eggs (Aghabaglou et al. 2013)

In an experiment to find the contact toxicity of pesticides on predator eggs, newly laid eggs of coccinellids were dipped on insecticide solution and tested for hatching. Insecticide solutions were prepared at specific concentrations using distilled water. Newly laid eggs of *C. montrouzieri* collected from the laboratory colony were dipped for 10 s. in each concentration. The eggs are then kept in controlled laboratory conditions for hatching. The mortality of eggs was observed after 72 h and corrected for control mortality if any, using Abbot's formula (Abbott 1925).

Predatory Mite Eggs (Castagnoli et al. 2005; Duso et al. 2008)

Gravid *P. persimilis* females were allowed in bean leaves to lay eggs for 24 h and the leaves with eggs were dipped in the test solution for 30 s. and dried. The leaves were then placed on a wet cotton pad and observed for hatching. The observations on hatching are done for 5 days after the first egg hatched in the control. A total of 100 eggs were used per treatment with 20 eggs per replication and in five replicates.

Spiders and Mirid Bugs (Tanaka et al. 2000)

In this experiment, the arthropods were exposed by dipping them directly in pesticide solutions and then kept in untreated containers for observing the mortality. To achieve dipping, a glass tube (3.5 mm dia. and height) with bottom covered with nylon gauze is used. About ten pre-anesthetized test spiders were put into the glass tube and the bottom end dipped in the insecticide solution in a Petri dish for 20 s., the test arthropods get the insecticide through the nylon net. After specific time of exposure, the excess insecticide solution was removed using filter papers. The treated spiders were kept individually in glass vials whereas mirids in plastic containers with rice seedling. In control treatment, the test individuals were dipped in distilled water. Observations on mortality were taken at 24 and 48 h after treatment and median lethal concentrations worked out.

Mites Pasted in Slides (Magalhaes and Bakker 2002; Nomikou et al. 2003)

Slide dip bioassays are used to find the toxicity of predatory mites to different pesticides (Busvine 1971). Mites were glued by their backs to a double-sided sticky tape on a microscope slide. The slides with five to ten female predatory mites were then dipped for 15 s in pesticide solutions and take out to dry. The slides were then kept in control climatic conditions and mortality observed at 24 h. The mites not moving their legs when dabbed with a brush were assessed as dead.

The slide dip assay had been the principal method used to test the acute effects of pesticides on predaceous mites. Unlike the methods in which a pesticide is topically applied or contacted as a residue on a treated surface, the slide dip is unsuitable for rearing the test mites further for sublethal toxicity and resistant studies, because the survivors were glued and cannot be taken back without damage (Amano and Haseeb 2001).

5.1.1.1.2 Micro-immersion for Predatory Mites (Duso et al. 2008)

This method is based on the one reported by Dennehy et al. (1993) and Castagnoli et al. (2005). Predatory mites were treated as micro-immersion and then transferred to holding cells containing bean leaves having *T. urticae*. This micro-immersion is achieved by drawing the predatory mites into a small pipette tip and then drawing the pesticide solution, making the mites get immersed in pesticide solution for 30 s. The mites were then ejected from the pipette and dried on a filter paper. The dried mites were then transferred singly into holding cells made of PVC plates as described by Dennehy et al. (1993) using a fine brush. This procedure can be used to evaluate the topical, residual and oral toxicities, if the predatory mites are immersed (topical), put in treated leaves (residual) and fed with the preys which are fed on treated leaves (oral).

5.1.1.1.3 Direct Sprays Using Potter's Precision Tower

Coccinellid and Chrysopid Larvae (Michaud and Grant 2003)

Predatory larvae of Coccinellidae and Chrysopidae were placed in groups in a plastic Petri dishes (5.5 cm dia.) for topical spray treatments. A total of 20 grubs were sprayed with the pesticide solution (1 mL). After the direct spray exposure the predators were kept in another uncontaminated dish and provisioned with *Ephestia* eggs and water beads. For coccinellid larvae in addition to *Ephestia* eggs, water beads, diet beads and bee pollen are also provided (Michaud 2001).

Predatory Mites (Kim and Yoo 2002)

In an experiment to test the acute toxicity of acaricides on adult predatory mites, leaf discs with mites were sprayed and mortality counted. About ten adult predatory mites, *P. persimilis* were transferred from the source colony to the leaf discs in Petri dish with the aid of a fine brush. Some two spotted spider mites were also kept in the leaf to make the predatory mites to stay in the leaf. Acaricide solutions prepared in distilled water are sprayed on the leaves with mites and allowed for dry. After the treatment, surplus numbers of two spotted spider mites were added as food. Observations on the survival of predators were evaluated at 1, 3, 5 and 7 days after treatment.

Experiments on the ovicidal action of acaricides, predatory mites were allowed to lay eggs first on the leaf discs for 24 h and the eggs were sprayed with the acaricide solution as stated above. The initial numbers of eggs kept while application and the immature predators hatched were counted to find the survival of predatory mites.

Predatory Bugs (Kim et al. 2006)

In this experiment, second instar nymphs of predatory bugs, *D. brevis* or adults (10 nos) were placed on Petri dishes with filter paper lining at the bottom and sprayed with pesticides. After drying, test insects were placed in clean glass Petri dishes and supplied with unsprayed *E. kuehniella* eggs and green beans. Mortality observations were made 24, 48 and 96 h after spraying. For egg bioassay, green beans were placed on the rearing chamber for the adults to lay eggs on the beans. Beans with eggs (<3 day old) were transferred to a glass Petri dish and then sprayed in the spray tower. Just before hatching, *E. kuehniella* eggs were added as food and observations on egg hatch and nymphal survival checked daily.

Inference: Potter's spray tower is a precise method to know the accurate amount of spray solution to get deposited in the plate. A different spray volume than described above i.e., a volume of 2 mL of pesticide per application which result in a spray deposit of 2.1 mg/cm^2, similar to the IOBC recommendation is also used in many experiments (Candolfi et al. 2000; Kim et al. 2006).

5.1.1.1.4 Direct Spray Using Motorized Spray Chamber (Galvan et al. 2005)

In this experiment, motorized spray chambers with fan nozzles which delivers 234 L/ha spray fluid at 35 psi was used to treat the test organism. About, 25 larvae were placed into Petri dish bottom which were then placed in the spray chamber and sprayed over it. After drying the test organisms were transferred to untreated Petri dishes.

5.1.1.1.5 Direct Spray Using Atomizer (Krishnamoorthy 1985)

This method is reportedly used to assess the effect of insecticides on the eggs of *C. carnea*. The brown paper strips containing stalked eggs were uniformly sprayed with insecticides using an atomizer. Approximately 20 eggs were used per replication and with three replications per treatment. The treated eggs and control eggs were kept for hatching and the per cent hatchability estimated.

5.1.1.1.6 Topical Application Using Hamilton Syringe

Coccinellid (Cho et al. 1997)

In this experiment, pesticides were tested for their contact toxic effects on adults and grubs of *H. axyridis*. Test coccinellids were collected from laboratory cultures and treated topically with 0.5 μL of insecticide solution prepared in acetone on the thoracic dorsum. Control insects were treated only with acetone. After treatment, the insects were maintained in plastic Petri dishes and provided with cotton aphids as a food and mortality determined 48 h after treatment. Moribund insects which failed to move its legs when stimulated with a fine brush were considered dead. Results were analyzed by probit analysis (Finney 1971).

Spiders

Test solutions were prepared using analytical grade acetone and used for bioassays. The range of concentrations for bioassay was obtained by the preliminary range finding test. Field collected adult spiders of uniform size were selected and placed in the refrigerator for 2 min. to calm their movement. Then they were dosed by placing 1 μL of the insecticide prepared in acetone on the cephalothorax of each spider using 1 μL repeating dispensor fitted with a 50 μL syringe and Rheodyne needle (Fig. 1.1). At least six doses excluding control with three replications and with a minimum of 30 spiders per treatment were used. Control spiders were treated with acetone alone. After the treatment, the spiders were allowed individually in boiling tubes of 3.8×20 mm dia. with brown planthoppers as feed. Mortality was recorded at 24 and 48 h after treatment and the moribund test insects considered as dead.

The same method was used to treat *C. carnea* adults also (Medina et al. 2003). Newly emerged adults (≤24 h old) were topically treated with 0.5 μL of an acetonic solution of each insecticide in the pronotum. Mortality was observed at regular intervals after treatment.

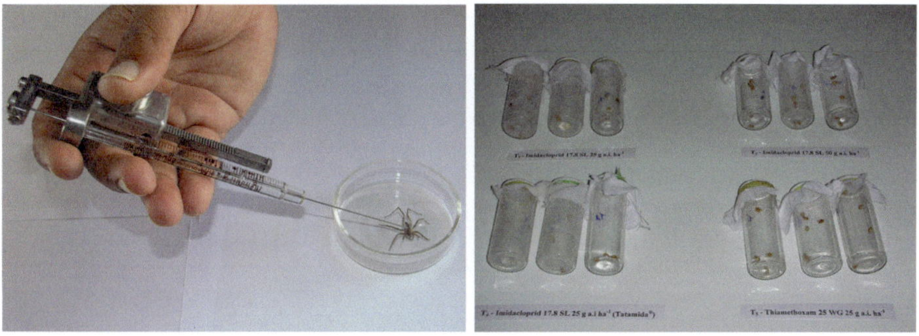

Direct contact- Topical bioassay *Dry residue -scintillation vial*

Fig. 1.1 Contact toxicity bioassays for predators

5.1.1.1.7 Topical Application Using Electric Microsyringe (Everts et al. 1991b)

This experiment is primarily used to find the lethal and sublethal effects of deltamethrin on spiders with relation to temperature and humidity. Adult female spiders were selected and dosed with 50 nL of aqueous solution of 50 mg a.i./L deltamethrin solution (2.5 ng) individually. The spiders were dosed topically on the dorsal abdomen by means of a Burkard electric microsyringe applicator. The dose used is equivalent to the field dose of 5 g a.i./ha considering the spiders surface area as 5 mm^2. Mortality was observed at specific intervals. Apart from mortality, paralysis and coordination in walking were also noted.

5.1.1.1.8 Topical Application Using Automatic Micro-syringe Pump (Amirzade et al. 2014)

In this experiment, the fourth instar grubs of coccinellid predators were tested for their susceptibility to different insecticides by topical application. The topical application was performed using an automatic micro-syringe pump. Syringe pumps are ideal for delivering accurate and precise amounts of fluids and thus used in the experiment. Distilled water was used as a control. About 30 insects were assayed per concentration (ten larvae per replication and with three replications). Treated larvae were provided with feed and maintained in a growth chamber and survival checked after 24 h.

5.1.1.2 Indirect Contact Toxicity

Apart from direct spray exposures, predators get indirect exposures by contacting to the sprayed parts. This residual toxicity is assessed mainly by exposing the predators to the sprayed materials. Inert sprayed materials like glass Petri dishes or slides or plates are used commonly. However using pesticide treated leaves or plant material may be more realistic than inert material that cannot absorb or react with the spray compound. Moreover, the toxic effect of formulations where the active constituent has low initial toxicity and has to undergo plant reactions to make it potent may go undetected in residue assays using inert material (Bernard et al. 2004).

5.1.1.2.1 Dry Film Residue: Scintillation Vial Bioassay (McCutchen and Plapp 1988)

Mirid Bugs (Preetha et al. 2010)

A dry film residue method was followed to assess the indirect contact toxicity of different insecticides to mirids, *C. lividipennis*. Test solutions were prepared using acetone and water in the ratio of 80:20 and used for bioassays. Glass scintillation vials of 15 mL capacity with an internal surface area of 50 cm^2 were cleaned by soaking overnight in soap water, rinsed with acetone and air-dried for at least 4 h before use. The vials were coated evenly with 0.5 mL of different insecticide solution and dried thoroughly by rotating the vials between the palms. Mirid bug nymphs of uniform size (10 nos) were taken from the culture and introduced into the treated vials and the mouth closed with a piece of muslin cloth fastened with a rubber band. After 1 h exposure, the mirid bugs were transferred into test tubes containing brown planthopper along with rice tillers. Observation on the mortality of mirid bugs were taken at 24 and 48 h, corrected for control mortality and analyzed to arrive at median lethal concentration.

Chrysoperla Grubs (Preetha et al. 2009b)

Scintillation vials as described above were treated and second instar larva of chrysoperla released at the rate of 10 per vial and covered with muslin cloth. After 1 h exposure, the grubs were transferred to test tubes and fed with eggs of *Corcyra cephalonica*. Observation on the mortality at 24 and 48 h along with percent pupation and adult emergence were noted.

Coccinellids

Here, third instar grubs or adults of coccinellids were released at the rate of 5 per vial and covered with muslin cloth. After an hour of exposure, the test insects were taken, kept in uncontaminated container and aphids were given as feed. Mortality observations were taken at 12, 24 and 48 h after treatment (Fig. 1.1).

Mites (Kwon et al. 2010)

Though this method is used for detecting resistance in mite pest, it can be comfortably used for toxicity testing on predatory mites. About 0.1 mL of pesticide solution in acetone was pipetted out into a 5 mL transparent glass vial (scintillation vial) and horizontally rotated using a rolling wave rotator. After complete evaporation of acetone, 15 female mites were introduced and closed with cap. Mortality of mites was examined under a portable convex lens (×10) at various time intervals. Mites showing immobility or irregular trembling were considered as dead.

Though this method is found as an easy technique with fewer escapees, the test vial should be ventilated. It is also reported by the same authors that no significant differences in mortality with the tightly closed and loosely closed cap. But it may not be true with all the pesticide chemicals especially with the ones with some fumigation toxicity. So, cloth fastened with rubber band to close the mouth of the vial as reported for other predators can be used.

5.1.1.2.2 Petri Dishes Sprayed Using Potter's Tower (Michaud and Grant 2003)

In this study, Petri dishes (5.5 cm dia.) sprayed using a Potter's tower was used to evaluate the contact toxicity of pesticides to the grubs of predatory coccinellids. The grubs were treated at first instar (24 h old) and again at third instar (1 or 2 days after second moult). An emulsion of 0.12 % copper sulfate and 1.0 % petroleum oil were prepared in distilled water and applied directly to the bottom of Petri dishes using a Potter's tower. The Petri dishes were sprayed with 1 mL of the solution and air dried for 30 min. A single coccinellid grub was transferred to the Petri dish along with a small measure of Ephestia eggs for food. The grubs were exposed for 24 h and then removed to a clean dish and provisioned with fresh food and water.

5.1.1.2.3 Glass Plates Sprayed Using Comelis Spray Chamber (De-Cock et al. 1996)

Plexiglas framed cages of 9×3.5 cm with two removable glass plates (9×9 cm) were used in the experiment. The insecticide solutions prepared in distilled water was applied on one side of the two glass plates @ 0.5 mL/plate using a Comelis spray chamber (Van-Laecke and Degheele 1993). The treated plates were left to dry and then fastened to the frame. As the cages were assembled, a moisture source, consisting of a soaked paper plug in a 2 cm dia. cup and *Galleria mellonella* larva were placed at the base plate. Test insects (predatory bugs) were released in the cages and mortality checked.

5.1.1.2.4 Drum Cell Cages for Predatory Bugs (Van-de-Veire 1992)

The experimental apparatus 'drum cell test cage' consists of a Plexiglas cylinder of size 9 cm dia. and 3.5 cm height with two round glass plates of 9 cm dia. One of the two glass plates was sprayed with water and prey eggs and pollen sprinkled to make it stuck with the plate. Both the glass plates were then treated with pesticide diluted in water using a Potter's tower. After drying, the drum cells were assembled and predatory bug nymphs put in and closed. Water sprayed glass plates were used as control. The cells were kept in incubator and prey eggs added whenever needed and mortality counted at specific time intervals.

5.1.1.2.5 Glass Plates for Predatory Mites

Acute toxicity to predatory mites is generally done through glass plate residue technique. The mite escapes while bioassay is really needed to be taken into consideration and the technique/method which is realistic and with least escapes can be used. Further the glass residue assay is not appropriate for granular formulations, seed dressings or insect growth regulators due to technical reasons (Forster and Kula 2003). The cover slip method is good, since it provides water to the mites inside the test arena, but mites were reportedly escaping. The cover slips with washer and covered above can be a better technique to prevent migration, if a plastic cover with holes for ventilation is provided.

Open Glass Plate Method (Overmeer 1988; Miles and Dutton 2003)

Open glass plate method is used to find the residual toxicity of pesticides to predatory mites. Each test unit consisted of a glass plate with a barrier of damp filter paper and a sticky non-toxic gel. Insecticides were sprayed on the glass plate and the spray deposits were allowed to get dried. After proper drying, 20 proto-nymphs were introduced and fed with pollen. Each treatment was replicated five times so that 100 mites were exposed per treatment. Observations on mortality were made 1 and 7 days after treatment.

Though the open glass plate bioassay method is easier than other assay methods, it has some disadvantages. Some of the test mites were reported to be dead on the plate, some as drowned and some even escape and found missing, but all these except the ones found alive are to be counted as dead.

Floating Island Method (Joisten 2000)

In this method, the mites kept in treated glass plates are placed floating on water. The water forms a natural barrier ensuring the mites presence in the plate and prevents escapes from the treated area. It is reported as more advantageous than open glass plate method since mite escapes before taking readings were found less than the open glass plate method.

Microscopic Cover Slips Method (Louis and Ufer 1995; Blumel et al. 2000; DEFRA 2002)

The test arena was made by joining two glass cover slips (each measuring 2.2×4.0 cm) side by side by means of glue at the top and the bottom portion. The resultant plate was therefore 4.4×4.0 cm. This plate is treated with pesticides to assess the indirect contact toxicity to the predators. When the plates were kept in a filter paper saturated with water, the water from paper was drawn into the channel by means of capillary action to provide water to the test mites. A square arena of 2.5×2.5 cm delineated in the slip by means of a gel was the actual test arena for the mites. As a source of food, untreated pollen was placed on both sides of the groove. Test mites were introduced in the area to expose them to the dry residues and mortality counted at specific intervals.

Microscopic Cover Slips, Rubber Washer and a Closed Arena (DEFRA 2002)

The cover slip method is modified to prevent escape of mites during the bioassay using rubber 'O' rings or flat fibre/rubber washers. The dia. of the hole in the washers were approximately 1 cm and glued onto the cover slip base. The cover slips with washers were then sprayed and allowed to dry. Mites were introduced into the cover slip inside the washer. Another unsprayed cover slip was placed above to prevent the escape of mites. But it was reported that water get condensed inside which promoted fungal growth on pollen and thus necessitates proper ventilation.

Tight Fit Petri Plate Method (Miles and Dutton 2003)

Tight fit Petri plates of 5 cm dia. were sprayed with the test solution and predatory mites were introduced along with natural substrates. Bran was used as substrate for *Amblyseius cucumeris* and peat/vermiculite for *Hypoaspis aculeifer* and *H. miles*. Number of live mites per replicate was counted 3 days after application using a binocular microscope.

In this method, the mites may not get the pesticide residues if they remain over the substrate. This might be the reason for very less mortality obtained in substrate method in comparison with the glass plates as reported by Miles and Dutton (2003).

5.1.1.2.6 Sprayed Soil and Filter Paper for Spiders (Mansour et al. 1992)

Pure dry quartz sand (70 g) was put in a plastic Petri dish of 8.6 cm dia. to a depth of 6 mm. The sand was moistened to 70 % of maximum water holding capacity by adding 12.9 mL water. Pesticide formulations were diluted in water to the desired concentrations and applied using a gas-chromatographic sprayer at 0.465 mL per Petri dish (8.6 cm dia.) on the sand. Controls were sprayed with water only. After 1 h of pesticide application, each sprayed Petri dish was covered with a plastic lid with hole covered with gauze to allow aeration at the centre. Adult lycosid (1 no) or five micryphantid spider were allowed in the sand to get exposed and mortality observed at specified time intervals. The residual effects of the pesticides are strongly related to the water content of the soil and soil with moisture content at field capacity had the highest toxicity (Everts et al. 1991b).

In case of filter paper bioassay, the filter papers of 9 cm dia. were kept in Petri dishes and sprayed with 0.509 mL of pesticide solution as described above. The test insects were released and observations taken as mentioned above.

5.1.1.2.7 Treated Nylon Bags for Spiders (Mansour et al. 1981)

Porous nylon bags (7 × 11 cm; 21 pores/cm^2) were dipped in aqueous solution of insecticides at field recommended concentrations. The bags were dipped properly to ensure that the inner surfaces of the bags were treated. The bags were dried at room temperature and 10 adult female spiders which had been fed 1 day before were enclosed in them @ 1 per bag. After 48 h of forced contact with the pesticide residues, the spiders were taken and kept in rearing chambers, fed and mortality noted at specific intervals.

5.1.1.2.8 Leaf Residue Bioassay Method

Sprayed Using Potter's Tower (Michaud 2001)

Leaf discs of 3 cm dia. were made from freshly picked and cleaned leaves. The leaf discs were sprayed with 1 mL aqueous solution of the test material in a Potter's precision spray tower. Leaf disc of control series were sprayed with 1 mL distilled water. After proper drying, the leaf discs were placed in plastic Petri dishes (3.5 cm dia.) and a single first instar coccinellid larva introduced. A small measure of *Ephestia* eggs was given for feed. The larva was exposed for 24 h, transferred to clean dishes and provisioned with *Ephestia* eggs, bee pollen, diet and water beads. The same method was also used to find the toxicity of pesticides to *Orius insidiosus* (Michaud and Grant 2003). Similarly, tomato leaves sprayed using Potter's tower is used for leaf residue assay with mirids, *Macrolophus pygmaeus* nymphs (Martinou et al. 2014).

In an experiment to find the residual toxicity on predatory mites, Bernard et al. (2004) used a leaf residue assay as stated below. The test unit was constructed by two open Petri dishes placed one inside the other. A glass dish of 10 cm dia. and a plastic dish of 9 cm dia. with base drilled centrally to allow the insertion of a cotton dental wick, which connect the plastic dish giving additional water supply from the glass dish. Both dishes were lined with cotton wool and wetted. An upturned bean leaf was embedded in the cotton wool in the plastic dish and the prey and predatory mites introduced. The leaf with cotton in the plastic dish along with the test organism and the prey were sprayed with pesticides in a Potter's tower. Because both cotton wool and leaf were sprayed, the live mites that strayed into the cotton wool returned to the leaf surface rather than escaping. Mortality was scored 48 h and then after 4 and 7 days. Cumulative mortality was worked out by adding all dead mites (on leaf, in cotton wool, in sticky barrier) and unaccounted escapees and dividing it with the total present during spraying (Bluemel et al. 2000). This measure may over-estimate the mortality since it includes the escapees also as dead mites. Thus, a second mortality estimate was obtained by summing all the dead mites and dividing it by the total number of live and dead mites at the time of counting thus excluding the escapees (Bluemel et al. 1993).

A modified excised leaf disc method is also given for mite bioassays (Bostanian et al. 2009) basically for pest mites but can be used for predatory mites also. The bioassay set up is a tight fit Petri dish of 5 cm with a hole in the bottom and filled with cotton wool. A treated leaf is kept over the moist cotton wool and the petiole protruding out through the hole. It is reported that this method reduces the escapees and allows observations to be made as long as 9 days after treatment.

Leaf Dipped in Pesticide Solution

In this experiment, fresh cotton leaves were taken and dipped in insecticide solution for 6 s and left to dry for 1 h at room conditions. After proper drying, the leaves were kept in Petri dishes and predatory bugs introduced (Torres and Ruberson 2004). This method implies the effect of dry residues of pesticide sprays in the field to predators but do not take into account of the toxicity of direct contact while

Intoxicated aphids for coccinellids *Topically dosed larva for bugs*

Fig. 1.2 Ingestion toxicity bioassays for predators

spraying. The method is suitable for active flying and fast-moving insects which are unsuitable for treatment by direct topical spray application in the tower (Michaud and Grant 2003).

5.1.1.3 Ingestion Toxicity

5.1.1.3.1 Prey Contamination Bioassay (Fig. 1.2)

Treated Eggs for Chrysopids (Preetha et al. 2009b)

Eggs of *C. cephalonica* were exposed to UV radiation to kill the embryo. The UV killed eggs were taken in muslin cloth and dipped in pesticide solutions, shade dried and about 1 cm^3 eggs were transferred to each test tube. The untreated check was maintained by dipping the eggs in distilled water. Altogether, ten *C. carnea* grubs of second instar stage were transferred into the test tubes containing treated *Corcyra* eggs. When the larva finished feeding the treated eggs, untreated eggs were provided until pupation. The treatments were replicated thrice. Observations on larval mortality, pupation and adult emergence were recorded. In another experiment, insecticides sprayed *Ephestia kuehniella* eggs through Potters tower were given as feed to the nymphs of mirids, *M. pygmaeus* to find the oral toxicity (Martinou et al. 2014).

Treated Aphids to Lacewings (Walker et al. 2007)

In this experiment, aphids were exposed to insecticide treated leaves and the toxicated aphid given to lacewings to find the effect of insecticides on the natural enemy.

Treating aphids: Leaf dip bioassay is used to treat the aphids. Crucifer leaf discs of 3.5 cm dia. were made and immersed in test solution for 10 s, or in distilled water for control. Leaf discs were air dried and kept in Petri dish with agar. Aphids were introduced in the leaf discs and allowed to feed for 24 h. Pesticide treatment for aphids is achieved in another way also. In this method, the lettuce plants were treated with systemic insecticide, imidacloprid as soil drenching. After 48 h of insecticide treatment, aphids were introduced in the treated plants and allowed to feed for 24 h before used as feed to predator in the actual experiment.

Treating lacewings: Lacewing grubs (5 nos) are kept in Petri dishes of 5 cm dia. and aphids given as feed. The aphids given as feed for lacewing grubs are the ones which were fed on insecticide treated lettuce for 24 h. An average of 20, 30 and 40 aphids were added daily to each dish when the lacewing larvae were at first, second and third instars. Since the predatory larva feeds on live and dead aphids, both intoxicated live and dead but not dried aphids can be given as feed. Mortality of larva was observed in stipulated time.

Treated Mites to Predatory Mites (Kim and Yoo 2002)

The experiment starts with the treatment of two spotted spider mite eggs with acaricides. Adult female of *T. urticae* were placed in leaf discs for 24 h to ensure egg laying and then removed. The leaf discs were then treated with acaricide solutions made using distilled water by spraying over the leaf discs. The leaf discs thus treated were air dried and the eggs transferred to fresh uncontaminated leaf discs. Now, predatory mites, *P. persimilis* were added to these treated eggs in untreated leaf discs to allow them to feed on the eggs. Mortality of predatory mites was observed in specified time intervals. This method evaluates only the ingestion toxicity because the treated eggs are transferred to untreated leaves to exclude residual contact.

Treated Hoppers to Spiders (Kiritani and Kawahara 1973)

Rice plants were grown in pots and pesticides treated as soil application according to the field recommended dosage. The rice plants were covered with cylindrical nylon cages and about 30 green leafhoppers introduced per pot. After allowed to feed on treated rice plants for 48 h, the hoppers were given as feed to female spiders, *P. pseudoannulata* kept in an ice cream cup with rice seedlings and mortality assessed.

5.1.1.3.2 Food Contamination Bioassay

Treated Honey to Chrysopid Adults (Preetha et al. 2009b)

In this experiment, ten freshly emerged adults of *C. carnea* were kept in clean containers and fed with pesticide contaminated food. The food consists of one part of honey and one part of Protinex® in water. The adults were fed with uncontaminated food in control. Likewise, three replications were made with a total of 30 insects per treatment. Mortality of the adults was recorded 12, 24 and 48 h after treatment and the percent mortality calculated.

Treated Drinking Water to Predatory Bugs (Mohaghegh et al. 2000)

In this experiment, insecticides were tested against spined soldier bug, *P. maculiventris* at fourth instar nymph and female adults. The test organisms were kept in Petri dishes (9 cm dia.) lined with absorbent paper and exposed to insecticides for

48 h through drinking water. A paper plug saturated with 2 mL insecticide solution prepared in distilled water was kept in a small dish (2.5 cm dia.). Control groups were supplied with distilled water alone. Predators were fed ad libitum with *G. mellonella* larva. About, 20 insects were tested per concentration and at a minimum of 8 concentrations. Survival was monitored after 48 h of exposure. For growth inhibitors like teflubenzuron mortality counts were taken after 2 days of adult emergence.

5.1.2 Chronic Toxicity Assays

5.1.2.1 Through Drinking Water to Bugs (De-Cock et al. 1996)

This experiment is to find the chronic toxicity of pesticides on adult *P. maculiventris* exposed through drinking water. A pair of *P. maculiventris* adults were kept in cages made of glass and Plexiglas frames (9 × 3.5 cm) and insecticide (LC_{10}) mixed water is given in a cup placed on the base of the cage. Fresh insecticide solution was offered at 3 days intervals. The mortality, fecundity and egg viability were monitored daily for a period of 15 days.

5.1.2.2 Through Treated Hoppers to Spiders (Kiritani and Kawahara 1973)

Rice plants were grown in pots and pesticides treated as soil application according to the field recommended dosage. The rice plants were covered with cylindrical nylon cages and green leafhoppers introduced. After allowed to feed on treated rice plants for 48 h, the hoppers were given as feed to female spiders, *P. pseudoannulata* kept in an ice cream cup with rice seedlings. The spiders were fed over a period of 10 days with five leafhoppers/day reared successively for 2 days on the treated rice plants. The feeding potential and the condition of the spiders, i.e., dead, paralyzed and normal were daily examined and recorded.

5.1.3 Persistent Toxicity Assays

5.1.3.1 Persistent Residual Toxicity in Sprayed Plants (Van-de-Veire and Tirry 2003)

Sweet pepper plants were used to find the persistent toxicity of pesticides on *O. laevigatus* and *M. caliginosus*. The lower leaf surface was sprayed with the insecticides when the plant heights were approximately 40 cm. Sweet pepper leaves were collected at 1 day after treatment and tested for toxicity by keeping in a test cage and releasing ten nymphs of *O. laevigatus* and *M. caliginosus* (second instar). The same testing was repeated at 5, 15 and 30 days after treatment to find the persistent toxicity. Mortality were scored after 4 days and corrected for control mortality using Abbott's correction (Abbott 1925).

In another experiment, tomato plants were sprayed with insecticides on both sides of the leaves and kept in glass house. Leaves of the sprayed plants were tested at different intervals (1, 3, 8, 21 and 30 days after treatment) and tested for persistent toxicity to predators (Figuls et al. 1999). In an experiment, 3 year old plum trees were sprayed till run-off with a hand sprayer and leaves detached at regular intervals of 3, 7, 10, 15, 20 and 25 days to test the persistent toxicity of pesticides to predatory mite, *E. finlandicus* (Broufas et al. 2008).

5.1.3.2 Persistent Residues in Sprayed Petri Dishes (Roubos et al. 2014)

In an experiment to find the effect of aged residues on the predators, Petri dishes treated with insecticides using Potter's tower were aged in polyhouses for 0, 3, 7 and 14 days after treatment (DAT). A total of ten replicates per treatment for each residue age were made. Dishes were kept in a greenhouse at inverted position and exposed to sunlight filtered through the glass of the greenhouse. Insect predators (*O. insidiosus*, *C. rufilabris* and *H. convergens*) were released according to the treatment at the appropriate day and food and water were provided on the lid as honey smear and cotton wick. Observations on mortality were taken at 24, 48 and 72 h of treatment.

Persistence classes developed by the IOBC (Sterk et al. 1999) are as follows:

Category	Persistence	Toxic effects
Class A	Short-lived	<5 days
Class B	Slightly persistent	5–15 days
Class C	Moderately persistent	15–30 days
Class D	Persistent	>30 days

5.1.4 Sublethal Toxicity Assays

Sublethal pesticide toxicity to predators is generally being done by measuring the predatory potential of treated predator or the impact on their longevity or reproduction capacity besides behavioural changes. Assessing the predatory potential as sublethal toxicity is an important criteria but should be done systematically and interpreted carefully. Measuring the predatory potential of pesticide treated predators in the laboratory may exclude the time spend by them in finding the pest in complex field conditions apart from interaction with other organisms and environmental impacts on them.

5.1.4.1 Life Table Studies (Golmohammadi et al. 2009)

Life table studies are usually carried out to find the effect of insecticides on the life history traits of natural enemies. In this experiment, a fertility life table was constructed using a *C. carnea* cohort (60 eggs) and the fate of the cohort was pursued until the last female died. The experiment get started by treating 1st instar larvae with LC_{25} concentration of insecticides. After 24 h of treatment, surviving larvae were kept in clean Petri dishes (6 cm dia.) maintained till adult emergence. The adults were kept in pairs of male and female and eggs produced were counted daily and hatchability estimated.

The population growth rate can be calculated using the Lotka equation (Andrewartha and Birch 1954).

$$\sum L_x m_x e^{-rmx} = 1$$

Where,

x is the age of cohort
L_x is the proportion of individuals survives to age x
m_x is the number of females produced per female of age x
r_m is the intrinsic rate of increase for the population

The other fertility life table parameters including the gross reproduction rate (GRR), net reproduction rate (R_0), generation time (T), doubling time (DT) and finite rate of increase (λ) were computed as per Cary (1993), Maia et al. (2000), Stark and Banks (2003), and Sundaram et al. (2006).

5.1.4.2 Predation, Reproduction and Longevity

5.1.4.2.1 Predatory Bugs (Mohaghegh et al. 2000)

This experiment is to find the effect of insecticides on the reproduction and longevity of adults of predatory bugs, *P. maculiventris*. About seven or nine pairs of adults were collected from stock colony and placed in separate Petri dishes (14 cm dia.) furnished with absorbent paper. The test insects were exposed to insecticides at field recommended dose for 48 h via drinking water for 2 days. After a period of 2 days exposure, uncontaminated water was given along with *G. mellonella* larva as food. Oviposition and survival were monitored daily and ovipositional rate was calculated as number of eggs per female per day. Daily hatching were also taken into consideration to find the hatching percentage.

5.1.4.2.2 Coccinellids (Michaud and Grant 2003)

Pesticide treated grubs were reared until adult emergence. The emerged adults were kept together to allow them for mating. Coccinellids which could be separated by sex by colouration such as *C. coeruleus* were sexed based on colour. Coccinellids which cannot be discriminated based on colour such as *H. axyridis*, *O. v-nigrum* were sexed by external features while copulation. Females after mating were kept in Petri dishes and egg laying observed daily. Eggs were labeled, placed in the incubator and hatching recorded.

5.1.4.2.3 *Predatory Mites (Kim and Yoo 2002)*

In this experiment, gravid two spotted spider mites were kept in leaf discs to allow them to lay eggs for 24 h. The eggs were then sprayed with acaricides and air dried. After complete air drying, the treated eggs were transferred to untreated leaf discs by counting. Then the predatory mites *P. persimilis* females were added and allowed to feed on the treated eggs. Observations on prey consumption and reproduction of *P. persimilis* were taken daily up to 8 days.

5.1.4.3 Predation by a Combination of Predators (Provost et al. 2005)

In this experiment, the impact of sublethal doses of insecticide on the predation efficacy of three predators was evaluated separately and in combination. Apple leaves were taken and kept in plastic Petri dishes with agar and about 20 preys (adults of *T. urticae*) and prestarved predators introduced. Predation efficacy was evaluated in five treatments:

(T1) one IIIrd instar nymph of mirid, *Hyaliodes axyridis*
(T2) one IIIrd instar grub of coccinellid, *H. axyridis*
(T3) one adult of predatory mite, *A. fallacies*
(T4) one nymph of mirid along with one grub of coccinellid
(T5) one nymph of mirid along with one adult predatory mite

The insecticide, lambda-cyhalothrin was sprayed on the leaves after the release of predators using a thin layer chromatography sprayer. After 4 h of exposure, predator(s) was (were) removed and *T. urticae* counted. Theoretical values were calculated by summing the results of monospecific treatments (mirid + mite + coccinellid) in order to assess the impact of intraguild predation on predation efficacy. Two-way ANOVA was used to evaluate the impact of intraguild predation and insecticide application on predation efficacy.

5.1.4.4 Functional Response of Predators (Ambrose et al. 2010)

This experiment was carried out to assess the functional response of intoxicated predators such as, the predation, searching time, attack ratio and handling time. The reduviids, *R. marginatus* were reared in plastic containers with corcyra larva. Plastic containers of 16 × 11.5 × 4.0 cm were used as bioassay containers. The predator was exposed to the sublethal dose (1/10th of the 48 h LC_{50}) of the insecticide under test by treating absorbent papers and placing in the bioassay containers. Water treated absorbent paper was used as a control. Laboratory reared fourth instar bugs were released in the bioassay container individually. The food (pest larva) was introduced into the container in different densities i.e., 1, 2, 4, 8 and 16 per container before predator introduction. After every 24 h, the prey consumed were counted and replaced with fresh prey to maintain the prey densities. Analysis was carried out to determine the relationship between the prey density and the functional response of the predators (Daniel 1987).

Various parameters according to Holling (1959) can also be analyzed to describe the functional response. Regression analysis can also be made to determine the relationship between the prey density and the prey consumed, searching time and attack ratio as per the following equation (Ambrose et al. 2013).

$$y = a(Tt - by)x$$

Where,

x – prey density
y – total number of prey killed in given period of time (Tt)
y/x – the attack ratio
Tt – total time in days when prey was exposed to the predator
b – time spent for handling each prey by the predator (Tt/k)
a – rate of discovery per unit of searching time [(y/x)/Ts]

The modified functional response model given by Holling (1959) describes the variation in prey consumption by the predator (Na) as a function of density of the prey (N). This is used to find the toxicity of neonicotinoids on the functional response of predatory mites. The number of preys consumed by the predator was estimated as follow:

$$Na = \frac{a \times T \times N}{1 + a(Th + C \times N)N}$$

Where,

Na – Number of preys consumed by the predator
a – Attack coefficient *
T – Experimentation time (days)
N – Prey density
C – Parameter related to the function shape **
Th – Time taken by the predator to identify, capture, attack, consume and digest the prey

(*a indicates how fast the functional-response curve reaches plateau
**if c = 0 means that handling time is density-independent and c >0 implies that handling time increases with N)

5.1.4.5 Repellency (Michaud 2001)

Half treated circular filter papers were used to find the repellency of insecticides to predators. Circular filter papers (9.0 cm dia.) were divided in half and the treatment half of each filter paper were sprayed with 1 mL of aqueous solution of pesticide in a Potter's precision spray tower. The other half was covered while spraying to avoid pesticide contact to the control half. Filter papers were then placed individually in plastic Petri dishes (9.0 cm dia.) and the walls coated with a fluon® solution to prevent climbing of the insects. Only one adult coccinellid beetle was introduced in the Petri dish and observed periodically at intervals of 15 min. for a total of 12 observations. A total of 20 such dishes and insects are used for a pesticide under test. Beetles resting on either the treated or untreated side of the filter paper were observed and noted. Beetles those are moving or straddled on the boundary are excluded from observation.

5.1.4.6 Locomotion (Everts et al. 1991b)

A round arena of 25 cm dia. with moist cardboard bottom was selected to test the walking speed of intoxicated spiders. Before the treatment, the spiders were acclimatized with the arena. Spiders were topically treated with insecticides in different doses and allowed individually in the arena and observed for 3 min. Walking behaviour was recorded using a video camera and pattern redrawn on transparent paper. The drawing was analyzed with a specific programme in computer.

5.1.4.7 Susceptibility to Predation

5.1.4.7.1 Intoxicated Spiders to Carabids in Lab (Everts et al. 1991b)

In this experiment, field collected spiders were dosed and tested for vulnerability to predation by carabids. Field collected spiders were used in this experiment because they are generally quicker than laboratory reared ones and selected for predation in the field. For testing the capacity of spiders to escape from the predators, the spiders were confronted in boxes with carabids. Soil was put in the floor of the boxes to facilitate running and simulate field conditions. The test organisms were dosed topically with insecticides at less than field recommended dose and the other treatment being double the field recommended dose. This is to find the effect of low and high level exposures to the test organisms. Four different treatments tested were given as follows.

1. Beetles untreated and spiders treated
2. Beetles treated and spiders untreated
3. Both beetles and spiders treated
4. Both untreated

5.1.4.7.2 Intoxicated Carabids to Ants in the Lawn (Kunkel et al. 2001)

This experiment was carried out to find the susceptibility of insecticide exposed carabids to ant predation in the field. The dog food (size of 5 mm dia.) 30 numbers were mixed with pesticide and placed in plastic container with 15 beetles. Control beetles were also kept in similar conditions but fed with pellets treated with distilled water. After 4 h of treatment, pairs consisting of an intoxicated and normal beetle were released in the lawn within 7.5 cm distance from an ant mound of *Formica schaufussi*. The fate of each beetle released was recorded. The time taken for capture or escape of beetles from ants was also noticed by using a stopwatch.

5.1.4.8 Enzyme Assays as a Biomarker for Sublethal Toxicity (Rumpf et al. 1997b)

This experiment examines two enzyme systems i.e., acetylcholinesterase (AChE), a target enzyme for organophosphate and carbamate poisoning and glutathione-S-transferase (GST), a detoxifying enzyme as possible biomarkers for sublethal toxicity of different pesticides on predators. Sublethal toxicity can be well correlated with the responses in the activity of these enzymes and thus used as biomarkers (Day and Scott 1990; Lagadic et al. 1994). Pesticides with specific mode of action bind with acetyl choline esterase resulting in the reduction of its original activity. An elevation in the levels of GST may be attributed to tolerance in the predator because of its degradation ability.

Lacewing larvae, *C. carnea* and *M. tasmaniae* were treated by exposing them to insecticide treated Petri dishes as dry film. The larvae were then supplied with pea aphids and were left in Petri dishes for 2 or 10 h for AChE and GST enzyme activity determinations, respectively. After the specific period, the larvae were anaesthetised and enzyme measured and protein determined (Bradford 1976). The method by Ellman et al. (1961) was modified for lacewing larvae and used for AChE analysis. A spectrophotometric method described by Habig et al. (1974) was adapted to determine the GST and compared with control samples.

5.2 *Tier II Evaluation: Semi-Field Experiments*

Semi-field toxicity evaluation on predators mostly involves the use of cages to confine the treated predator and prey and assessing the potential of predator on pest suppression or the mortality of the predator itself. The cages used for toxicity testing of predators can be very small plastic containers to big of 2 m^2 size. Cages are reported to alter the microclimate (Nelson and Rieske 2014), behaviour of the predator (Hand and Keaster 1967) and rate of predation (Bahar et al. 2012). Predation was reportedly decreased from small cups to open trays and finally to whole plants. Interestingly the predation was very less in Petri dishes as the predator move around the brim most of the times (Bahar et al. 2012). So care should be taken to minimize the influence of cages on the end points which are to be observed.

5.2.1 Pest Infested Plants Sprayed and Predator Introduced (Torres and Ruberson 2004)

Cotton plants were grown in pots @ one plant per pot and allowed it to get infested with whiteflies. The plants were sprayed with insecticides at the recommended concentrations and control plants with water. The impact of insecticides on *P. nigrispinus* nymphs was evaluated by confining the second instar nymphs on treated and untreated plants. About, 18 nymphs per treatment were confined in sleeve cage bags

(30 × 25 cm organdy bags) in groups of three on the cotton foliage. The caging of predators was done in different intervals and after caging mortalities scored daily. Likewise, the whiteflies were also sampled to find the effect of insecticides at different intervals by counting eggs, nymphs and pupae in an area of 1 cm^2 under a microscope. Adults were counted by gently turning the leaves at early morning hours when they generally rest in groups.

5.2.2 Soil Applied Insecticide with Coccinellids on Flowers (Smith and Krischik 1999)

Flowering plants, sunflower and chrysanthemum were grown in pots and applied with imidacloprid in the soil at labeled rate and double the rate. Coccinellid beetles, *C. maculata* were exposed to insecticide by confining them in the flowers by means of a muslin cloth and tied with rubber band at the peduncle of the inflorescence. After 7 days of confinement, the mortality if any, behaviour alterations, time taken for oviposition and fecundity were studied. The beetles were placed in a paper and distance covered per unit of time determined and noted as walking rate in cm/s. Another parameter, flip time was tested by placing the beetle dorsally on a paper and measuring the time taken to right itself. After this bioassay, the test beetles were reared by providing artificial diet. The days to first oviposition and daily egg production thereof were determined for 30 days.

5.2.3 Field Sprayed Food for Laboratory Tests (Kunkel et al. 2001)

This experiment was conducted to find the dietary toxicity of pesticides to predatory carabid beetles. Food pellets were made by adding 20 mL distilled water in 20 g powdered dog food and pressed through a series of 3 mm holes drilled in a 2 mm thick sheet of Plexiglas. About 30 food pellets were kept in plastic Petri dishes and kept open on the surface of the experimental plots. Turf grass fields were sprayed with pesticides at prescribed concentrations and allow the feed pellets in the Petri dish to get the spray droplets as that of the field. Altogether, 30 treated pellets were placed in plastic boxes containing a moistened filter paper and dental wick and ten pre-starved beetles introduced. Observations on the mortality were recorded in specific intervals.

5.2.4 Insects Exposed in Field and Observed in Lab (Kunkel et al. 2001)

In this experiment, separate field plots of turf were made at a size of 2 × 2 m as per the treatments and with three replications. Predatory carabid beetles, *H. pennsylvanicus* were kept in eight gauge wire mesh cages of size 12 × 6 × 6 cm and positioned in the turf in each plot. The plots were sprayed with pesticides at prescribed concentrations. After spraying, the cages were brought into the laboratory and emptied into plastic containers and the test insects fed with dog food and water. Observations on dead, moribund, intoxicated and normal beetles were noted.

5.2.5 Soil Applied Pesticides on Nectar Feeding Adults (Rogers et al. 2007)

Buckwheat, *Fagopyrum esculentum* and Mexican milkweed, *Asclepias curassavica* were grown in green house. At the time of flowering, imidacloprid granules were applied to the soil as three treatments viz., label rate, twice label rate and a control, 3 weeks prior to feeding experiments. Adults of *C. carnea* were kept in mesh cages of 30 × 30 × 30 cm and commercial *C. carnea* diet was provided in Petri dishes along with water and 20 % honey solution in tubes. Two tubes with untreated flowers of *F. esculentum* and *A. curassavica* were also given for acclimatization. At the start of the experiment, all the food provided are removed and flowers were given as per the treatment. Observations on dead and trembling *C. carnea* were taken each day.

Cold anthrone test: Cold anthrone test was performed to determine the presence of fructose sugars from nectar within the guts of *C. carnea* (Van-Handel 1967). *Chrysoperla carnea* from all treatments and one extra treatment with starved adults crushed separately and 5 mL anthrone reagent pipetted on to them. Dark green colour if developed indicates a positive reaction confirming the presence of nectar sugar, fructose in the gut of chrysoperla.

This is a good methodology for testing the effect of soil applied pesticides on the free living adults of predators. The confirmation of food taken by the test organism adds value to the methodology since it removes a doubt whether the insect dies by starvation due to repulsion or killed by the pesticide? If the residue of the particular insecticide were analyzed in flowers and test animals, it becomes more meaningful to confirm the toxicity.

5.3 Tier III Evaluation: Field Experiments

Tier III evaluation estimates the effect of pesticides on predators in field realistic conditions which include a simple observation of the presence of predators in the treated field. Many other experiments like trapping and assessing the predatory efficiency by using sentinel preys are also been carried out. The important thing in the field experimentation is how best it is done realistically and interpreted.

5.3.1 Predators of Plant Hoppers in Rice (Tanaka et al. 2000)

Rice plants were grown in the field following proper agronomic practices and divided into plots for insecticide spray. The plots were divided by polythene sheets and spray was made when the planthopper populations were found increasing.

The number of planthoppers and predators were determined using sticky boards (Nagata and Masuda 1978) prior to and 1, 3, 7, 17, 28 and 40 days after insecticide application. The sticky board (25 × 18 cm) sprayed with adhesive was held at the bottom of the rice hill and the plants beaten twice to allow the hoppers and predators to fall and get stuck on the board. The samples were collected from ten hills per board and two boards per plot. Additionally web making spiders such as Tetragnatha were counted in 20 hills per plot at morning hours.

5.3.2 Predators in Maize and Soybean (Galvan et al. 2005)

Sweet corn plants were grown with proper agronomic practices and insecticides applied on the crop as per the recommendations for corn borer, *Ostrinia nubilalis* management. A total of three sprays were given and generalist predator population was observed at specific intervals. On each sample date, ten randomly selected plants from the middle six rows of each plot were sampled using visual whole-plant inspection. The adult and immature stages of the predators were identified using different guides or specimens and recorded.

In case of soybean, the crop was grown with proper agronomic practices and experiment carried out in randomized block design. The spray schedule was followed mimicking the management of soybean aphids. Recordings were done as per the procedure stated in sweet corn.

5.3.3 Predators in Soybean with Direct Observation, Sweep Nets and Sticky Cards (Ohnesorg et al. 2009)

Soybean crop was grown and the experiment laid out in randomized block design. Different insecticidal treatments were given as foliar sprays and soil applications. To determine the predator abundance in the crop, three different methods were employed i.e., direct observation, sweep nets and yellow sticky cards. Sweep net and sticky cards may collect a greater portion of the active predators whereas the direct plant observation gives a better estimate on the sessile predators. Direct observations were made in ten plants per plot, net sampling with 20 pendulum sweeps and four yellow sticky cards placed per plot.

5.3.4 Predatory Bugs Caged on Cotton Plants in Field (Studebaker and Kring 2003; Torres and Ruberson 2004)

In this field trial, cotton plants were sprayed with respective insecticides to study the effect on the pest and the predator. The experiment was carried out in a completely randomized block design in a continuous field with four replicates. Predatory bug, *O. insidiosus* were caged on the fourth leaf down from the plant's terminal point as soon as sprays had dried. Cages were made using 6 cm dia. polystyrene Petri dishes clipped together using hair clips. Each cage was constructed of two Petri dish bases so that the edges would meet forming an enclosure for which foam strips were glued to the edges. An arrangement for ventilation was also made on the plastic Petri dishes by cutting and gluing with a muslin cloth. Males, females and third instar nymphs were evaluated separately to determine the effects on gender and insect stage.

In another experiment, sleeve cage bags were used to cage predatory bug, *P. nigrispinus* in cotton plants (Torres and Ruberson 2004). Insecticides were sprayed accordingly and data on whiteflies and aphids taken using standard pest count methods at different intervals. Impact of insecticides on second instar *P. nigrispinus* was determined by caging nymphs on plants using sleeve cage bags. Caging with predatory bugs was carried out in different intervals and mortality scored daily.

5.3.5 Coccinellids in Cotton and Canola (Jasmine et al. 2009; Arif et al. 2012)

Field experiments were carried on cotton with plots of 4×5 m in a randomized block design to study the effect of insecticides on coccinellids. The insecticide treatments were imposed four times at 14 days interval commencing from 60th day after sowing of the crop using pneumatic knapsack sprayer with 1000 L of spray fluid/ha. Observations on the population of coccinellids (adults and grubs) were taken a day before spray and 3, 7, 10 and 14 days post treatment in ten randomly tagged plants per plot and compared with the averages made per treatment. A similar procedure, with observations on aphid and coccinellid predator population at 1, 2, 3 and 7 days after spraying was reported in canola (Arif et al. 2012).

5.3.6 Spiders, Coccinellids and Ants in Okra (Solangi and Lohar 2007)

Okra plants were grown in field and experiments laid in randomized block design with three replications. Insecticides were sprayed at their field recommended concentrations when the insect pests are at peak. Observations on the insect pests were done at 20 plants with three leaves (upper, middle and the bottom). Predators (spiders, beetles and ants) were taken as whole plant sampling. The observations were taken 1 day prior and 1, 2, 3, 7 and 14 days after treatment.

5.3.7 Mite Predators in Apple (Cavaco et al. 2003)

In this experiment, the test plot consists of four trees in a row and have five replicates treated with insecticides at field recommended rates using a knapsack sprayer at a volume of 1000 L/ha. Leaf samples were taken before and after treatment to assess the population densities of the motile stages of phytoseiids, *Typhlodromus pyri*, *T. phialatus* and pest, *P. ulmi*. A leaf sample consisting of 25 expanded leaves selected randomly from the two central trees of each plot. Sampling were done before and 4, 7, 14, 21 and 35 days after treatments and assessed in the laboratory with a stereoscopic microscope. The percentage mortality of predatory mites was calculated using the Henderson-Tilton formula as follows:

$$\%\,\text{reduction} = \left(1 - \frac{T_a \times C_b}{T_b \times C_a}\right)100$$

Where,

T_a – number of predatory mites in the treatment after spraying
T_b – number of predatory mites in the treatment before spraying
C_a – number of predatory mites in the control after spraying
C_b – number of predatory mites in the control before spraying

According to the principles of IOBC, four evaluation categories (% mortality or reduction in beneficial capacity) were used: 1 = harmless (<25 %), 2 = slightly harmful (25–50 %), 3 = moderately harmful (51–75 %) and 4 = harmful (>75 %) (Hassan 1994).

5.3.8 Coccinellids in Apple and Pear (Pasqualini and Civolani 2003)

This experiment was made in randomized block with four trees per plot and in four replications. Pesticides were sprayed on the trees as per the recommended concentrations. For sampling of dead coccinellids due to pesticide sprays, two trees in the centre of each plot were spread with cotton sheets (2 × 1.5 m) pegged beneath them. The coccinellids died and fallen on the cotton sheet 24 (S_1) and 48 h (S_2) after spray were collected and counted. For inventory spray sampling, application of deltamethrin @ 2.5 g a.i./ha was given 48 h after spray treatment. Specimens were removed from the collecting sheets after 7 h (S_j) of inventory spray. The degree of harmfulness (%) was evaluated using the following formula given by Staubli et al. (1985).

$$\text{Degree of harmfulness}\,(\%) = \frac{(S_1 + S_2)}{(S_1 + S_2 + S_j)} \times 100$$

Where,

S_1 – mean number of individuals killed by a treatment in 24 h
S_2 – mean number of individuals killed by a treatment in 48 h
S_j – mean number of individuals surviving after 48 h

Thus, $S_1 + S_2 + S_j$ is the mean population of the tree, given by the sum of those killed and those which survived after treatment of test chemical.

5.3.9 Predatory Arthropods in Turf Using Pit Fall Traps (Kunkel et al. 1999)

The field was divided into individual plots (10 × 10 m) arranged in a randomized block with eight total replications on two sites and 1.3 m untreated alleys between plots. Insecticides were applied on the turf mimicking grub control and irrigated after treatment. Pitfall traps (50 mL centrifuge tubes containing 10 mL preservative) were placed at the rate of nine traps arranged in 'x pattern' within each plot. Pretreatment samples were taken in all plots for 2 weeks before insecticide application. When the pesticides are applied, these traps were removed and replaced once the application was over and thereafter replaced at every 2 weeks interval. In the test plots, laboratory reared prey insect (eggs, pupae and frozen larvae of black cutworms) were placed to find the effect of spray on the predatory potential of naturally occurring predators in the lawn. The numbers of preys found missing and partially consumed were recorded.

5.3.10 Sentinel Prey Items Placed in the Field (Howe et al. 2015)

Cotton crop was grown in field and the following treatments were given. Predation levels were assessed by placing artificial caterpillars (Loiselle and Farji-Brener 2002; Howe et al. 2009) of 3.5 mm dia. and 25 mm, made of light green plasticine similar to *Helicoverpa armigera*. The dummy caterpillars were glued in the leaf at about 15–30 cm apart. They are kept at the rate of 0.33 caterpillar per plant and 30 per treatment. Predation rates were calculated as the ratio of number of preys showing signs of predation and total number exposed. Comparisons on predation were made between insecticide treatment (sprayed/nonsprayed) and cropping type (monocrop/intercrop) etc.

1. Cotton monoculture sprayed with insecticide
2. Cotton monoculture, not sprayed
3. Cotton intercropped with beans (*Phaseolus* spp.) and intercrop sprayed
4. Cotton intercropped with beans and cotton sprayed.

6 Pesticide Risk Assessment for Arthropod Predators

Risk of pesticide use in agro-ecosystem are to be assessed on the predators, since they are non-target organisms acting hand-in-hand with the pesticides to achieve effective and sustainable pest management, if at all not affected by the pesticides. Several methods have been proposed for assessing the risk to predators and some are being used regularly. But there is no standardized approach for assessing the risk of pesticides on predators. The first and easier one is by determining the median lethal concentration or dose through laboratory bioassays and categorizing the pesticides based on IOBC classifications as harmless or slightly or moderately harmful or harmful. The disadvantage of this method is by not including the field recommended dosage. Field recommended concentration is the one, which the pest and beneficial organisms get exposed in the field conditions. Risk assessments through hazard ratio or risk quotient overcome this disadvantage by including the field recommended dosage of the particular pesticide and the median lethal concentration of the test organisms (DEFRA 2002). Instead of calculating a ratio such as risk quotient, some studies reported a simple comparison of median lethal concentration with that of the field concentrations and classify the pesticides.

Pesticides are used for pest management in the field, so a comparison of susceptibility of the pest and predator (beneficial organism) to the same pesticide can be more meaningful to categorize the pesticide as selective or non-selective. Thus, selectivity ratio is more frequently used to assess the pesticide risk on beneficial organism by comparing the acute toxicity of pesticide to the predator and the target or associated pest species using a similar bioassay method. Similar to selectivity ratio, probit substitution is used to determine the relative toxicity of pesticides to beneficial species at particular levels of pest mortality, such as LC_{90} of the pest. In the laboratory, predators are tested for pesticide toxicity at the the most susceptible life stage or at stages easier to handle. But when the pesticide is applied in the field, all the life stages i.e., egg, larva/grub/nymph and adults may be available and get exposed to the pesticide sprays. So an assessment of all life stages may also add value for assessing the risk of pesticide spray to the beneficial organism. All the above mentioned risk assessment methodologies (pesticide classification based on acute toxicity, hazard ratio or risk quotient, selectivity ratio, probit substitution for comparing pest and beneficial) uses laboratory tests to find the risk of pesticide application in field conditions. EPPO protocol on risk assessment is based on the assumption that a pesticide found harmless to predators in the laboratory bioassay will also be harmless in the field (EPPO/OEPP 1990).

Testing the pesticides at field recommended concentrations at laboratory conditions do not exactly reveal how the pesticides behave in the complex field conditions. So a tiered approach/sequential testing scheme which start from initial screening in laboratory (Tier I) mostly by estimating median lethal concentration/dose and proceeds with the extended laboratory, semi-field and more realistic field tests are found promising to assess the pesticide risk. If a pesticide is found harmless in tier I, then no additional testing is required (Hassan 1998; Vogt 2000; Boller et al. 2005). If a pesticide is found slightly harmful (30–79 % mortality), it may also be suitable for IPM, but additional testing is recommended. Pesticides found moderately harmful and harmful need additional testing in semi-field conditions (tier II). In the tier II system of toxicity evaluation, pesticides found harmless or slightly harmful are readily acceptable for pest management without further tests. If a pesticide is found moderately harmful, then field trials are needed to derive a conclusion. The harmful ones in the tier II are not acceptable for field use and further test with them is not required (Stark et al. 2007). But again, all the risk estimates discussed till now, use the mortality data ignoring the sublethal effects and impacts on the life history traits. In addition to death, exposure to a toxicant can result in shortened life span, reduced number of offspring, changes in the time to first reproduction, longer generation times, weight loss and mutations in offspring (Stark and Banks 2003). So risk assessment to beneficial organisms should include mortality and also the sub-lethal effects. Some times, sublethal effects cause greater damage to the predators than mere mortality especially when the reproduction is affected.

A risk assessment approach popularly known as coefficient of toxicity or total effect of pesticide which include the mortality of juveniles and reproduction capacity of adults may be more realistic. Here also, the risk is assessed based on the response of the individual not on populations and multiple effects are not considered (Stark et al. 2007). Risk assessment should also consider whether the pesticide has disrupted the biological control at the population level or not. One way to improve risk assessment is to compare life-history variables for organisms that are most likely to be exposed to a toxicant (Stark et al. 2004a). Modeling based on demographic parameters provides a clearer picture of actual pesticide impacts (Stark et al. 2004b). Studies on the effect of pesticide sprays on natural enemies in delaying the recovery of the population of the exposed predator compared to control ones, gives a more ecologically relevant risk estimation (Wennergren and Stark 2000; Stark et al. 2004a, 2007). At the last, a very specific risk estimate for a particular predator species can be calculated by giving differential weights to different parameters like mortality, fecundity and predation capacity can well suit the biological control point of view, for which the predator is mainly meant for and thus seems to be holistic approach. The different methods to assess the risk of pesticides on arthropod predators in agro-ecosystem are described hereunder.

6.1 *Risk Assessment Methodologies*

6.1.1 Based on Mortality or Reduction in Predators

Pesticides can be classified based on the acute residual contact toxicity to predators determined in laboratory bioassays as harmless, slightly harmful, moderately harmful and harmful (Hassan et al. 1994). Classification of pesticides based on the data obtained from semi-field experiments and the reduction of predators in the field experiments are also proposed by the IOBC working group (Sterk et al. 1999) and are as follows:

		Mortality		
Class	Category	Acute toxicity experiments	Semi-field experiments	Reduction of predators in field[a]
Class 1	Harmless	<30 %	<25 %	<25 %
Class 2	Slightly harmful	30–79 %	25–50 %	25–50 %
Class 3	Moderately harmful	80–99 %	51–75 %	51–75 %
Class 4	Harmful	>99 %	>75 %	>75 %

[a]In field conditions after spray with respect to pre treatment counts

6.1.2 Hazard Ratio or Risk Quotient

Risk quotient or hazard ratio is used to assess the ecological risk of pesticides (Peterson 2006) and used to assess the safety of predators like coccinellids (Peveling and Ely 2006) and mirid bugs (Preetha et al. 2010).

$$\text{Hazard ratio} = \frac{\text{Recommended field rate}\left(\text{g a.i. / ha}\right)}{\text{LC}_{50}\text{ of predator species}\left(\text{mg a.i. / L}\right)}$$

The hazard ratio of <50 for a pesticide is considered as safer, 50–2500 as slightly to moderately toxic and >2500 as dangerous.

6.1.3 Comparison of Acute Toxicity Values with Label Concentrations

In this method, the acute toxicities of different pesticides as median lethal concentrations were first established for predators with standard laboratory methods. Then the LC_{50} values are compared with the field recommended concentrations. The logic behind is the LC_{50} reveals the concentration at which 50 % of tested predator will die and the field recommended concentration is the one which usually the predator get in the field conditions (exposure). In this comparison, if the LC_{50} value is higher or near to field recommended concentrations, the pesticides can be regarded as safe, whereas if the LC_{50} values are much lower than field concentrations means the predators are at risk and further analysis in different tiers are triggered. This method was utilized to find the risk of 6 insecticides to predatory mite, *Galendromus occidentalis* (Lefebvre et al. 2011). Preetha et al. (2010) compared the contact LC_{90} values of 11 insecticides to mirid bug. *C. lividipennis* with their field recommended concentrations to assess the pesticide risk.

6.1.4 Toxicity to Predators in Comparison with Pests (Preetha et al. 2010)

In this method of risk assessment, a comparison of the susceptibility of the predator with the associated pest is made to find which one of them is more susceptible. In an experiment by Preetha et al. (2010), the risk of pesticides on green mirid bug, *C. lividipennis* was analyzed by comparing with the associated pest, brown plant hopper (*N. lugens*). Risk is generally determined through selectivity ratio and probit substitution method to find the relative toxicity.

6.1.4.1 Selectivity Ratio

The selectivity ratio is the ratio of acute toxicity value of the predator to that of the associated pest. The higher the value for this ratio, the safer the pesticide is for the predator. The selectivity ratio is calculated based on Tanaka et al. (2000) as follows:

$$\text{Selectivity ratio} = \frac{\text{LC}_{50}\text{ of predator species}\left(\mu\text{g a.i.}/\text{L}\right)}{\text{LC}_{50}\text{ of pest species}\left(\mu\text{g a.i.}/\text{L}\right)}$$

Values of 1 and <1 indicate that the chemical is nonselective to predator.
Values of >1 indicate that the chemical is selective/harmless to predator.

6.1.4.2 Probit Substitution Method

This method assesses the relative toxicity of pesticide to beneficial species at particular levels of pest mortality, say LC_{90} of the pest. Probit substitution method can be made as per Stark et al. (1995), Kumar and Regupathy (2005), and Preetha et al. (2010) as below:

$$y = 5 + m\left[x - \left(\log \text{LC}_{50}\text{ of predator species}\right)\right]$$

Where,

'y' is the probit value
'm' is the slope of the probit line for the predator
'x' is the log of the fiducial limits for LC_{90} of the pest species

Solving for 'y' gives a probit value that is then converted to the percentage mortality using a conversion table (Finney 1971). The pesticide is considered to be selective if it kills less than 90 % of predators at the concentration that kills 90 % of the pest.

6.1.5 Pesticide Toxicity to All the Life Stages of Predator

Bioassay on eggs: In an experiment to find the pesticide toxicity to *C. carnea*, Preetha et al. (2009b) used all the life stages of the predator. Pesticide toxicity was tested at the egg stage of the predator, by allowing the predator to lay eggs on paper and the eggs sprayed with the pesticides at different concentrations and per cent hatchability worked out.

Bioassay on larva: Pesticide toxicity to larvae of *C. carnea* was assessed by diet contamination and through dry residues to find the ingestion and contact toxicity. For diet contamination bioassays, eggs of *Corcyra cephalonica* were UV treated to kill the embryo and treated with pesticides by keeping them in a muslin cloth and dipping in insecticide solutions. The eggs dipped in pesticide solution are shade dried and then given to the larva as food. Mortality of larva was observed at specific intervals.

For dry residue assays, the insecticides were prepared in acetone: water mixture and coated as thin film in scintillation vials and larva released. The vials were covered with muslin cloth and the larva allowed to get exposed to pesticides for 1 h. After specified time, the larva was removed and kept in uncontaminated vials and provided with food. Mortality observations were taken at specified intervals.

Bioassay on adults: *Chrysoperla carnea* adults feed on nectar in the fields, so honey contamination bioassay was used to find the toxicity of pesticides to the predator adults. Food consists of honey and protein were contaminated with insecticide solution and given to the adults to feed. Mortality observations on adults were noted.

Pesticides were categorized based on the mortality data of egg (unhatched eggs), larva (ingestion and contact) and adults using IOBC classification. Pesticides gave <30 % mortality were categorized as harmless, >80 % as moderately harmful and >99 % as harmful.

6.1.6 Tiered Approach: Laboratory, Semi-field and Field Studies (Miles and Dutton 2003)

In this experiment, a glass plate method was used to find the residual toxicity of insecticides to the predatory mites. The glass plates were sprayed with insecticides and mites were allowed in it after complete drying. Observations on the mortality were taken after 1 and 7 days after treatment. Another experiment with tight fit Petri dishes sprayed with insecticides and predatory mites introduced along with natural substrate like bran, peat or vermiculite is also made. A semi-field experiment with French bean plants was conducted in glass house conditions. The plants of 20 cm height were sprayed with insecticides and allowed to dry. After complete drying the plants were trimmed to one leaf and prey mites, *T. urticae* introduced in the leaves. Then the predatory mites, *P. persimilis* and *N. californicus* were introduced and live predators counted after 6 days of introduction. Another experiment as direct spray was made in the plants having pest mites and predatory mites was also performed. After 7 days of application ten leaves were sampled from each replicate and the number of live predator mites and eggs counted. In case of field trials, insecticides were applied in apple orchards and mites sampled before and at different times after application. Leaves were collected and the number of mites determined in the laboratory. The effect of insecticides was categorized according to the IOBC classification proposed for laboratory, semi-field and field studies (Hassan 1992).

6.1.7 Sequential Testing Scheme (Vand-de-Veire et al. 2002b)

The sequential testing scheme is principally similar to that of tiered test, which starts with a worst case laboratory test and proceed with extended laboratory tests, semi field experiments and field trials. The first step is to determine the acute toxicity and in this experiment by Vand-de-Veire et al. (2002b) the sequential testing scheme is applied to find the toxicity of pesticides to predatory bug, *O. laevigatus*.

Laboratory tests: Acute indirect contact toxicity for predatory nymphs was estimated by using a 'drum cell method' as described in the methodology subchapter. To determine the effect of pesticides on ovipositional ability of the bugs, the nymphs survived in the drum cell experiment were allowed to oviposit in a pepper plant; eggs laid on the plants were counted and mean number per female calculated. The total effect of the pesticide (Ex) was calculated using the formula given in 6.1.8 of this subchapter using the data obtained on the mortality of nymphs and fecundity of the adult females. The pesticides were classified based on the categories suggested by IOBC on the basis of values obtained for total effect (Ex) (Hassan 1992; Sterk et al. 1999).

Extended laboratory tests: Pesticides were sprayed on potted chrysanthemum plants at blooming stage. Test cages of size 40 cm height and 30 cm dia. made of Plexiglas with proper ventilation were made and the sprayed chrysanthemum plants kept inside it. About 15 adults of *O. laevigatus* (3–5 days old) were introduced in a cage and untreated *E. kuehniella* eggs (50 mg/cage) were provided as additional food. The mortality percent determined and pesticides categorized based on IOBC classification (Hassan 1992).

Persistence test in laboratory: Spanish pepper plants (five leaves stage) were sprayed with pesticides and the predator exposed to pesticide by means of 'drum cells' described above. Drum cells were placed on the plants in such a way the leaves were inside the drum and the stem left outside. The treated plants were introduced in the drum cell cages at 5, 10, 15 and 30 days after spraying and 20 s instar nymphs introduced. Food was provided *ad libidum* and mortality scored at specific intervals. The pesticides were categorized based on IOBC classification for persistence (Hassan 1992).

Semi-field test: A semi-field test was also conducted using potted sweet pepper plants sprayed with insecticides. Some flower pollen and prey eggs were spread on the plant leaves and five L1 and L2 predatory nymphs were introduced. Mortality of nymphs were counted at specified time and pesticide categorized as per IOBC classification.

Semi-field test for persistence: This experiment is similar to the semi-field experiment described just above, with a difference of using sprayed sweet pepper plant on the same day of spray, 7 and 14 days after spraying to find the persistent toxicity.

Field (glass house) test: Sweet pepper plants were grown in glass house and *O. laevigatus* adults were released on the plants when the flower thrips incidence was observed. There were about four introductions at 2 weeks intervals which resulted in a density of $1/m^2$ population. The total field was divided into three blocks of 6×8 m size and pretreatment count taken by collecting 30 flowers per block and checking for the presence of predatory nymphs. The insecticides were sprayed as designated to the blocks and water (reference control) in one of the blocks. The flowers (30/block) were collected at weekly intervals and checked for the number of bugs and pesticides categorized as harmful and harmless based on the data from reference standards. The difference in the number of bugs per 30 flowers in control and treatment was taken for calculating the per cent reduction.

6.1.8 Total Effect of a Pesticide/Coefficient of Toxicity

6.1.8.1 Survival and Fecundity (Kavousi and Talebi 2003)

In this method, the total effect of pesticide was determined by observing the mortality and fecundity of the treated predator. This method is followed to find the risk of pesticides on predatory mite, *P. persimilis* by assessing the pesticide induced mortality of juvenile mites and the pesticidal impact on egg production. Different bioassays were performed to find the mortality of mites due to pesticide treatment and the impact of pesticide on egg production. The data obtained from these bioassays were used to find the total effect through the following formula.

$$\mathrm{Ex} = 100 - \frac{100 - \mathrm{Mt}}{100 - \mathrm{Mc}} \times \frac{\mathrm{Rt}}{\mathrm{Rc}} \times 100$$

Where,

Mt – % mortality in the treated group
Mc – % mortality in the control group
Rt – average egg production per treated female
Rc – average egg production per female in the control group

In this estimate, risk is assessed based on IOBC recommendation by categorizing pesticides as harmful if Ex is >75 % and harmless if it is <25 %. The moderately harmful pesticides have an Ex ranging from 51 % to 75 % and slightly harmful ones with 25–50 % (Sterk et al. 1999).

6.1.8.2 Survival, Fecundity and Hatchability (Duso et al. 2008)

In this method, the survival/toxicity, fecundity and egg-hatching of pesticide treated predatory mites were analyzed by calculating coefficient of toxicity.

$$\mathrm{E} = 100 - (100 - \mathrm{M}) \times \mathrm{R}$$

Where,

E – Coefficient of toxicity

M – Percent mortality
R – Ratio between the average numbers of hatched eggs produced by treated females and the average number of hatched eggs produced by females in the control group

6.1.9 Delay in Population Growth Index (Wennergren and Stark 2000; Stark et al. 2004a)

This delay index is a measure of population recovery by comparing the control populations with that of those exposed to pesticide. It includes the measurement of mortality and also the reductions in the number of offspring. Here, the population number is taken as the end point and the time taken by the control population to reach a particular number is compared with the time taken by the exposed population to reach the same number. This method by Stark et al. (2004a), involves life table parameters and a matrix projection model to predict the time taken for a treated and untreated population to grow from 10 to 100,000 individuals. The matrix is multiplied by an initial condition vector, containing information on the age distribution of the population. Survival and fecundity parameter values were obtained from life tables. In this experiment, many pests and natural enemies including pea aphid and its predator, *C. septempunctata* were studied. Life table data were taken from previously published literature for pea aphid (Walthall and Stark 1997) and *C. septempunctata* (Stark et al. 2004b). Survival and fecundity of each species was manipulated in the model in such a way that the population get reduced to 50 % (depicting the lethal effect of the pesticide) and the number of offspring produced get reduced by 50 % (sublethal effect). Delay in population growth was calculated by calculating the difference in the number of days taken by the treated population and control population to reach 100,000 from an initial population of 10 individuals. This model seems to be good since it involves not only the lethal and sublethal effects but also studies the population as a whole and not individual effects like other risk approaches.

6.1.10 Species Specific Risk Estimation Based on Biological Pest Control Parameters

This estimate of risk assessment is proposed with an aim on testing the toxic effects of pesticides on the most important parameter of the predator i.e., biological pest control by predation. Pesticide may affect the predator by causing mortality or by causing sublethal effects including behavioural alterations. This risk assessment is thus proposed to include different parameters like mortality, fecundity and predation capacity including locomotion (most important for active searchers) host searching and prey consumption. These parameters and any other, which are appropriate for biological control point of view, can be considered. Different weightages can be given to these parameters to reflect the important parameter of that particular predator thus making the risk estimate more specific to the predator and its biological control ability. The ability of natural enemies to search for prey is critical for biological control success (Kean et al. 2003; Snyder et al. 2006). The Working Group 'Pesticides and Beneficial Organisms' of the International Organization for Biological Control and Integrated Control of Noxious Animals and Plants/West Palaearctic Regional Section (IOBC/WPRS) has developed laboratory methods in which a decline in beneficial capacity such as predatory activity is also investigated in addition to mortality (Amano and Haseeb 2001).

6.2 *Risk of Pesticides on Arthropod Predators*

Risk of pesticide application can be calculated in many ways but a realistic and ecologically relevant risk estimate is more important to take decisions on use of pesticides along with predators for effective pest management. The calculated and reported risks of different pesticides to arthropod predators using different methods are summarized here.

6.2.1 Classification of Pesticides Based on Mortality and Reduction in Field

Nymphs of mirid predator, *Macrolophus pygmaeus* were exposed to the pesticides by all the three routes viz., direct, residual and oral. According to IOBC rating scheme, chlorantraniliprole and emamectin-benzoate (<25 % mortality) were classified as harmless, indoxacarb and spinosad (30 % mortality) as slightly harmful and thiacloprid and metaflumizone (80–100 % mortality) as harmful (Martinou et al. 2014). Methomyl was found to be harmful, teflubenzuron as slightly harmful and bromopropylate, dicofol + tetradifon, fenpyroximate, pirimicarb and tau-fluvalinate were harmless to predatory mirid bug, *Dicyphus tamaninii* tested in tomato leaf residue bioassay (Castane et al. 1996). Based on IOBC classification, insecticides like imidacloprid, bifenthrin and thiamethoxam are harmful but, pyriproxyfen, chlorfenapyr and spinosad are harmless to *O. laevigatus*. Chlorfenapyr, imidacloprid and thiamethoxam, fipronil and bifenthrin were harmful to *M. caliginosus*. Chlorfenapyr, imidacloprid, bifenthrin, fipronil and abamectin are reported to be harmful to the adults of predatory mites, *P. persimilis* whereas spinosad is harmless (Sterk et al. 2002).

6.2.2 Risk Based on No Observed Effect Levels (NOEL)

The no observed effect application rates (NOER) of hexaflumuron to *C. septempunctata* was reported to be 1.52 g a.i./ha for survival and developmental duration whereas it was 3.04 g a.i./ha for hatching and pupation. These NOERs are very low compared to field application rate of hexaflumuron (135 g a.i./ha) in cotton cultivation, suggesting potential risks to beneficial arthropods (Yu et al. 2014).

6.2.3 Hazard Ratio or Risk Quotient

The hazard quotient (HQ) of imidacloprid to coccinellid, *S. japonicum* by contact is found to be 3.47 and thus found to be in risk category (He et al. 2012). Insecticide, BPMC with a hazard quotient of >300,000 was found dangerous to predatory mirid, *C. lividipennis* whereas endosulfan, acephate, chlorpyrifos, methyl parathion, deltamethrin, imidacloprid, pymetrozine and chlorantraniliprole were safe. Ethofenprox and clothianidin were found slightly to moderately toxic to *C. lividipennis* as revealed by hazard ratio calculations (Preetha et al. 2010).

6.2.4 Comparison of LC_{50} with Field Recommended Concentrations

In an experiment to find the risk of pesticides to predatory mites, *Galendromus occidentalis* by comparing the LC_{50} values with the field recommended concentrations, novaluron, clothianidin and chlorantraniliprole were found slightly toxic and recommended for higher tier studies. Insecticides, spinetoram and spirotetramat with LC_{50} values of 34.3 and 7.7-fold less than that of their respective field concentrations were considered highly toxic. Flubendiamide was harmless to all growth stages of the mite and it is recommended for inclusion in IPM programs without additional tier II field evaluations (Lefebvre et al. 2011). In a similar experiment to assess risk of insecticides on mirid bugs, *C. lividipennis*, Preetha et al. (2010) compared the LC_{90} values with the field recommended concentrations and found only chlorantraniliprole, deltamethrin, chlorpyrifos and endosulfan as less toxic insecticides with ≤2-folds lesser LC_{90} values than field recommended concentrations. BPMC was found to have 4140-folds less LC_{90} value than the field recommended concentration. Other insecticides like ethofenprox, pymetrozine, methyl parathion and acephate were highly toxic (Preetha et al. 2010).

6.2.5 Toxicity to Predators in Comparison with Pests

The effect of some insecticides on pests and their natural enemies in cotton was evaluated by Ahmed et al. (2014) and concluded that nitenpyram, thiacloprid and imidacloprid as less toxic to predators and more toxic to the sucking pests. Selectivity ratio was used to assess the risk of pesticides on coccinellid, *H. axyridis* in comparison with its aphid preys, *A. citricola* and *Myzus malisuctus* and found esfenvalerate as highly toxic to both *A. citricola* and its predator, *H. axyridis*. Among the eight insecticides tested, alphamethrin was found to be safer to *H. axyridis* as revealed by selectivity ratio with *A. citricola* whereas methomyl, monocrotophos and phosphomidon were toxic (Cho et al. 1997). Calculation of selective toxicity ratio (STR) for tobacco aphid, *M. nicotianae* and its predator, *C. sexmaculata* revealed acephate as toxic with low STR, followed by imidacloprid, while endosulfan is found safer (Patil and Lingappa 2000).

Most organophosphate insecticides and methomyl were highly toxic to the predator, *C. carnea* than tobacco budworm, *Heliothis virescens* and thus non-selective (Plapp and Bull 1978). Insecticides, BPMC, deltamethrin, imidacloprid, clothianidin and pymetrozine were found non selective to mirid bugs, *C. lividipennis* when compared to brown plant hopper, *N. lugens* through selectivity ratio whereas endosulfan, acephate, chlorpyrifos, ethofenprox and chlorantraniliprole were selective to the predator (Preetha et al. 2010). Deltamethrin was more toxic to prey insects, *Dione juno juno* and eucalyptus caterpillars than the predatory *Podisus nigrispinus* (Zanuncio et al. 1993; Picanco et al. 1996).

Imidacloprid was reported as selective to natural enemies (spiders) of BPH, *N. lugens* (Jian et al. 1996). Among the nine insecticides tested for selective toxicity to spiders, *P. pseudoannulata*, *U. insecticeps* and *G. exsiccatum* with respect to the prey *N. lugens*, only deltamethrin (selectivity ratio >0.6) was found non selective to the predators. Insecticides such as diazinon, ethofenprox, deltamethrin and cartap were non-selective to *T. maxillosa*. All the nine insecticides were non selective to the Dryinid wasp, *Haplogonatopus apicalis*. All the insecticides except for diazinon and fenobucarb were non selective to mirid bug, *C. lividipennis* (Tanaka et al. 2000). Abamectin was more toxic to two spotted spidermites than predatory mite (*P. persimilis*) and thus can be recommended for use in spidermite control (Zhang and Sanderson 1990). In an experiment with acaricides viz., bifenazate, acequinocyl, chlorfenapyr and fenbutatin oxide, 78–86 % of adult female predatory mites, *P. persimilis* were found to survive after 5 days of treatment whereas all *T. urticae* adult females died within 1–3 days of treatment. Milbemectin and fenazaquin were found to be highly toxic to both the predatory mite, *P. persimilis* and the pest, *T. urticae* (Kim and Yoo 2002). Acaricides, fenpyroximate and spirodiclofen were reportedly selective for predatory mite, *Amblyseius largoensis* than the pest, *Raoiella indica* (De-Assis et al. 2013). Bifenazate and acequinocyl were found to affect the reproduction of pest, *T. urticae* with only 4.2 and 2.4 eggs per 10 females whereas the fecundity of predatory mites, *P. persimilis* were not affected registering 235 and 216 eggs per 10 females, respectively. Fenazaquin affected the fecundity of predatory mite more than that of pest mite registering 0.6 and 21.6 eggs/10 females, respectively (Kim and Yoo 2002).

6.2.6 Pesticide Toxicity to All the Life Stages of Predator

Recommended doses of imidacloprid and diafenthiuron were reported to cause only 15.38 and 9.96 % mortality to the eggs of *C. carnea* (Preetha et al. 2009b) and thus both the pesticides falls under harmless category of IOBC classification. In ingestion bioassays, both diafenthiuron and imidacloprid registered only 23.33 and 26.67 % mortality to *C. carnea* grubs and thus harmless. In dry residue bioassay, the grubs of *C. carnea* was found to get affected by diafenthiuron and imidacloprid causing a mortality of 23.33 % and 33.33 %, respectively (Preetha et al. 2009b) revealing the insecticides as harmless and slightly harmful. Though imidacloprid and diafenthiuron caused a differential mortality of 6.7 and 26.7 % to *C. carnea* adults through honey contamination bioassay, both the insecticides falls under harmless category. Ethofenprox and acetamiprid were highly toxic to most developmental stages and also adults of *H. axyridis*. Abamectin was highly toxic to eggs, larvae, pupae and adult ladybirds whereas acaricides were safe to all stages except for the eggs (Youn et al. 2003).

6.2.7 Tiered Approach and Sequential Testing Scheme

Diafenthiuron was not selective to immature Anthocorids and Mirids and found harmful to the predatory bugs, *O. laevigatus* (Fieber) in sequential testing scheme (Van-de-Veire et al. 2002b). All the fungicides, captan, carbendazim, sulfur and tolylfluanide were found to be harmless based on the results obtained from laboratory, extended laboratory and semi-field experiments. Among the 11 insecticides tested, only pyriproxyfen and tebufenozide and one among the five acaricides, hexythiazox were found to be harmless in all the three tests. Insecticides like dichlorvos, imidacloprid and thiocyclam and two acaricides (abamectin and diafenthiuron) were found to be harmful in all the three tests i.e., laboratory, extended laboratory and semi-field experiments (Vand-de-Veire et al. 2002b).

6.2.8 Total Effect of a Pesticide/Coefficient of Toxicity

The sublethal effects on the reproductive capacity of *O. laevigatus* were studied based on toxicity coefficient (Ex) and the pesticides were classified according to the IOBC toxicity categories. Based on this, abamectin was classified as harmful and spinosad, emamectin, metaflumizone as moderately harmful insecticides (Biondi et al. 2012b). Hexythiazox was reported harmless (E <30) to mite, *P. plumifer*, while abamectin and fenpyroximate at field rates were harmful with total toxic effect, E >99 (Nadimi et al. 2009). Risk assessment by calculating coefficient of toxicity (E) showed azoxystrobin (54.64) and tolylfluanide (52.93) as slightly harmful, spinosad (77.32) and thiamethoxam (92.37) as moderately harmful, pyrethrins (100) and abamectin (100) as harmful to predatory mite, *P. persimilis* (Duso et al. 2008). Based on the calculation of total effects of pesticide, heptenophos was found to be harmless with $E=-3.7$ and in fact increased the fecundity with very low rate of mortality to *P. persimilis*. Malathion and primiphos methyl were found to be slightly and moderately harmful to the predatory mite, *P. persimilis* (Kavousi and Talebi 2003).

6.2.9 Delay in Population Growth Index

In a population of pea aphid and its predator, *C. septempunctata*, 50 % mortality causes a delay in population growth for 8 and 31 days, respectively revealing the predator as more susceptible than the pest. In case, a pesticide caused 50 %

mortality combined with 50 % reduction offspring, then the delay in population growth is 20 and 67 days, respectively for the pest and predator. A delay of one generation time interval for the coccinellid, *C. septempunctata* gives a time for its prey, *A. pisum* to complete seven generations on pea (assuming the prey is not affected by the pesticide) before the ladybird beetle get recovered (Stark et al. 2004a). In another study, 50 % mortality and 50 % reduction in offspring in *C. septempunctata* caused by a pesticide will lead to a delay in the population growth for 31 and 23 days, respectively. A mortality of 43 % population and 43 % reduction in offspring caused a delay of one generation time interval for the predator, *C. septempunctata* (Stark et al. 2007).

References

W.S. Abbott, A method of computing the effectiveness of an insecticide. J. Econ. Entomol. **18**, 265–267 (1925)

S. Acharya, H.P. Mishra, D. Dash, Efficacy of insecticides against okra jassid, *Amrasca biguttula biguttula* Ishida. Ann. Plant Protect. Sci. **10**(2), 230–232 (2002)

S. Acheampong, J.D. Stark, Effects of the agricultural adjuvant Sylgard 309 and the insecticide pymetrozine on demographic parameters of the aphid parasitoid, *Diaeretiella rapae*. Biol. Control **31**, 133–137 (2004)

B.K. Agarwala, P. Laska, D.N. Raychudhuri, Prey records of aphidophagous syrphid flies from India (Diptera: Syrphidae). Acta Entomol. Bohemoslov. **81**, 15–21 (1984)

S. Aghabaglou, S. Alvandy, S. Goldasteh, Z.R. Karahroudi, Study on ovicidal and side effects of diazinon and imidacloprid on *Cryptolaemus montrouzieri* Mulsant (Coleoptera: Coccinellidae). J. Entomol. Zool. Stud. **1**(6), 22–26 (2013)

M. Ahmad, M. Rafiq, M.I. Arif, A.H. Sayyed, Toxicity of some commonly used insecticides against *Coccinella undecimpunctata* (Coleoptera: Coccinellidae). Pak. J. Zool. **43**(6), 1161–1165 (2011)

R. Ahmed, J. Yousaf, I. Nadeem, M. Saleem, A. Ali, Response of sunflower (*Helianthus annuus* L.) hybrids to population of different insect pests and their bio-control agents. J. Agric. Res. **51**(1), 31–39 (2013)

S. Ahmed, M.S. Nisar, M.M. Shakir, M. Imran, K. Iqbal, Comparative efficacy of some neonicotinoids and traditional insecticides on sucking insect pests and their natural enemies on BT-121 cotton crop. J. Anim. Plant Sci. **24**(2), 660–663 (2014)

K.S. Ahn, S.Y. Lee, K.Y. Lee, Y.S. Lee, G.H. Kim, Selective toxicity of pesticides to the predatory mite, *Phytoseiulus persimilis* and control effects of the two-spotted spider mite. Tetranychus urticae by predatory mite and pesticide mixture on rose. Korean J. Appl. Entomol. **43**, 71–79 (2004)

H. Amano, M. Haseeb, Recently-proposed methods and concepts of testing the effects of pesticides on the beneficial mite and insect species: study limitations and implications in IPM. Appl. Entomol. Zool. **36**(1), 1–11 (2001)

K.G. Amarasekare, P.W. Shearer, Comparing effects of insecticides on two green lacewings species, *Chrysoperla johnsoni* and *Chrysoperla carnea* (Neuroptera: Chrysopidae). J. Econ. Entomol. **106**, 1126–1133 (2013)

D.P. Ambrose, S.J. Rajan, J.M. Raja, Impacts of Synergy-505 on the functional response and behaviour of the reduviid bug, *Rhynocoris marginatus*. J. Insect Sci. **10**, 187 (2010)

D.P. Ambrose, K. Nagarajan, A.G. Kumar, Interaction of reduviid predator, *Rhynocoris marginatus* (Fabricius) (Hemiptera: Reduviidae) with its prey teak skeletonizer, *Eutectona machaeralis*

Walker (Lepidoptera: Pyralidae) as revealed through functional response. J. Entomol. Res. **37**(1), 55–59 (2013)

N. Amirzade, H. Izadi, M.A. Jalali, H. Zohdi, Evaluation of three neonicotinoid insecticides against the common pistachio psylla, *Agonoscena pistaciae* and its natural enemies. J. Insect Sci. **14**, 35 (2014)

N.H. Anderson, Growth and fecundity of *Anthocoris* spp. reared on various prey (Heteroptera: Anthocoridae). Entomol. Exp. Appl. **5**(1), 40–52 (1962)

H. Andrewartha, L. Birch, *The Distribution and Abundance of Animals* (University of Chicago Press, Chicago, 1954). 827 p

M.J. Arif, Q. Abbas, M.D. Gogi, M. Ashfaq, H.A. Sayyad, M.A. Khan, H. Karar, Performance of some insecticides against canola aphids and associated coccinellid predators under field conditions. Pak. Entomol. **34**(1), 37–41 (2012)

S.P. Arthurs, L.A. Lacey, E.R. Miliczky, Evaluation of the codling moth granulovirus and spinosad for codling moth control and impact on non-target species in pear orchards. Biol. Control **41**, 99–109 (2007)

P. Asiimwe, J.S. Ecaat, M. Guershon, S. Kyamanywa, D. Gerling, J.P. Legg, Evaluation of *Serangium* sp. (Col., Coccinellidae), a predator of *Bemisia tabaci* (Hom., Aleyrodidae) on cassava. J. Appl. Entomol. **131**, 76–80 (2007)

A. Askari, V.M. Stern, Biology and feeding habits of *Orius tristicolor* (Hemiptera: Anthocoidae). Ann. Entomol. Soc. Am. **65**, 96–100 (1972)

H. Atuncha, E. Ateka, R. Amata, R. Mwirichia, M. Kasina, B. Mbevi, E. Wakoli, Evaluation of predation potential of coccinellids on cassava whiteflies. J. Entomol. Nematol. **5**(7), 84–87 (2013)

A.S. Atwal, J.P. Chaudhary, M. Ramzan, Mortality factors in the natural population of cabbage-aphid, *Lipaphis erysimi* (Kalt.) (Aphididae: Homoptera), in relation to parasites, predators and weather conditions. Indian J. Agric. Sci. **41**(5), 507–510 (1971)

P.C.S. Babu, Toxicity of insecticides to the aphid, *Aphis craccivora* Koch and its coccinellid predator, *Menochilus sexmaculatus* F. on cowpea and hyacinth bean. Madras Agric. J. **75**, 409–413 (1988)

L. Bacci, M.C. Picanco, E.C. Barros, J.F. Rosado, G.A. Silva, V.F. Silva, N.R. Silva, Physiological selectivity of insecticides to wasps (Hymenoptera: Vespidae) preying on the diamondback moth. Sociobiology **53**(1), 151–167 (2009)

M.H. Bahar, J.N. Stanley, P.C. Gregg, A.P.D. Socorro, P. Kristiansen, Comparing the predatory performance of green lacewing on cotton bollworm on conventional and Bt cotton. J. Appl. Entomol. **136**, 263–270 (2012)

A.T. Barrion, J.A. Litsinger, *Riceland Spiders of South and Southeast Asia* (C.A.B. International/International Rice Research Institute, Wallingford/Manila, 1984), p. 700

M. Bayley, E. Baatrup, U. Heimbach, P. Bjerregaard, Elevated copper levels during larval development cause altered locomotor behaviour in the adult carabid beetle, *Pterostichus cupreus* L. (Coleoptera: Carabidae). Ecotoxicol. Environ. Saf. **32**, 166–170 (1995)

M. Begon, C. Townsend, J. Harper, *Ecology: Individuals, Populations and Communities*, 3rd edn. (Blackwell Science, London, 1996). 1068 p

B. Belliure, R. Amoros-Jimenez, A. Fereres, M.A. Marcos-Garcia, Antipredator behaviour of *Myzus persicae* affects transmission efficiency of broad bean wilt virus. Virus Res. **159**(2), 206–214 (2011)

T.S. Bellows, J.G. Morse, Toxicity of insecticides used in citrus to *Aphytus melinus* DeBach (Hymenoptera: Aphelinidae) and *Rhizobius lophanthae* (Blaisd.) (Coleoptera: Coccinellidae). Can. Entomol. **125**, 987–994 (1993)

F.D. Bennett, Some recent successes in the field of biological control in the West Indies. Rev. Peru. Entomol. **14**, 369–373 (1971)

A.B. Bennett, C. Gratton, Measuring natural pest suppression at different spatial scales affects the importance of local variables. Environ. Entomol. **41**, 1077–1085 (2012)

M.B. Bernard, P.A. Horne, A.A. Hoffmann, Developing an ecotoxicological testing standard for predatory mites in Australia: acute and sublethal effects of fungicides on *Euseius victoriensis* and *Galendromus occidentalis* (Acarina: Phytoseiidae). J. Econ. Entomol. **97**(3), 891–899 (2004)

R.L. Best, C.C. Beegle, Food preferences of five species of carabids commonly found in Iowa cornfields. Environ. Entomol. **6**(1), 9–12 (1977)

B. Bhattacharyya, A. Basit, D.K. Saikia, Parasitoids and predators of rice insect pests of Jorhat districts of Assam. J. Biol. Control. **20**(1), 37–44 (2006)

N.A.A. Bilal, A.A. Satti, The major predators and their seasonal abundance in okra fields at Elgorair scheme, Northern Sudan. Experiment **4**(4), 271–276 (2012)

A. Biondi, V. Mommaerts, G. Smagghe, E. Vinuela, L. Zappala, N. Desneux, The non-target impact of spinosyns on beneficial arthropods. Pest Manag. Sci. **68**(12), 1523–1536 (2012a)

A. Biondi, N. Desneux, G. Siscaro, L. Zappala, Using organic-certified rather than synthetic pesticides may not be safer for biological control agents: selectivity and side effects of 14 pesticides on the predator, *Orius laevigatus*. Chemosphere **87**(7), 803–812 (2012b)

A. Biondi, Z. Lucia, J.D. Stark, N. Desneux, Do biopesticides affect the demographic traits of a parasitoid wasp and its biocontrol services through sublethal effects? PLoS ONE **8**(9), e76548 (2013)

T.K. Biswas, M.A. Rahman, M.M.H. Khan, M.M. Alam, M. Jahan, Toxic effects of two insecticides on brown plant hopper, *Nilaparvata lugens* and its predators, *Micraspis discolor* and *Lycosa pseudoannulata*. Int. J. Zool. Res. **2**, 192–203 (2006)

S. Bluemel, F. Bakker, A. Grove, Evaluation of different methods to assess the side-effects of pesticides on *Phytoseiulus persimilis*. Exp. Appl. Acarol. **17**, 161–169 (1993)

S. Bluemel, F.M. Bakker, B. Baier et al., Laboratory residual contact test with the predatory mite, *Typhlodromus pyri* Scheuten (Acari: Phytoseiidae) for regulatory testing of plant protection products, in *Guidelines to Evaluate Side Effects of Plant Protection Products to Non-target Arthropods* (IOBC/WPRS, Gent, 2000), pp. 121–143

E.F. Boller, H. Vogt, P. Ternes, C. Malavolta, Working document on selectivity of pesticides. Retrieved from www.IOBC.Ch/2005/Working%20Document%20Pesticides_Explanations.pdf

C.J.H. Booij, J. Noorlander, Farming systems and insect predators. Agric. Ecosyst. Environ. **40**, 125–135 (1992)

T. Boopathi, K.A. Pathak, Efficacy of bio and synthetic pesticides to *Lipaphis erysimi* Kalt. and its predator, *Ischiodon scutellaris* (Fabricius) in broccoli ecosystem. J. Biol. Control. **25**(4), 294–297 (2011)

T. Borah, A. Rahman, M. Borthakur, B.K. Barthakur, Biology and predatory potential of a newly recorded green lacewing, *Mallada* sp. (Neuroptera: Chrysopidae) on tea mosquito bug, *Helopeltis theivora* (Waterhouse). Two Bud **59**, 60–62 (2012)

L. Bortolotti, A.M. Sbrenna, G. Sbrenna, Action of fenoxycarb on metamorphosis and cocoon spinning in *Chrysoperla carnea* (Neuroptera: Chrysopidae): identification of the JHA-sensitive period. Eur. J. Entomol. **102**, 27–32 (2005)

N.J. Bostanian, S. Beudjekian, E. McGregor, G. Racette, A modified excised leaf disc method to estimate the toxicity of slow and fast acting reduced risk acaricides to mites. J. Econ. Entomol. **102**(6), 2084–2089 (2009)

M.L. Boyd, D.J. Boethel, Susceptibility of predaceous hemipteran species to selected insecticides on soybean in Louisiana. J. Econ. Entomol. **91**, 401–409 (1998)

M.M. Bradford, A rapid and sensitive method for the quantitation of microgram quantities of protein utilizing the principle of protein dye-binding. Anal. Biochem. **72**, 248–254 (1976)

M. Branco, J.C. Franco, C.J. Carvalho, Z. Mendel, Occurrence of *Hemerobius stigma* in pine bast scale (*Matsucoccus* spp.) populations: opportunistic predation or obligatory association? Bull. Zool. Bachic. **33**(3), 397–407 (2001)

D.J. Brooks, A.T. Walston, J. Farnsworth, A. Farnsworth, J. Smith, H. Riedl, Impact of foliar insecticides on pear psylla and natural enemies. Arthropod. Manag. Tests **29**, A30 (2004)

G.D. Broufas, M.L. Pappas, G. Vassiliou, D.S. Koveos, Toxicity of certain pesticides to the predatory mite, *Euseius finlandicus* (Acari: Phytoseiidae). IOBC/WPRS Bull. **35**, 85–91 (2008)

D.W. Brown, R.A. Goyer, Comparative consumption rates for selected insect predators of two defoliating caterpillar species of soybean in Louisiana. J. Georgia Entomol. Soc. **19**(3), 299–303 (1984)

J.F. Bruno, B.J. Cardinale, Cascading effects of predator richness. Front. Ecol. Environ. **6**, 539–546 (2008)
A. Budriene, E. Budrys, Applicability of the 'locomotor ganglia' hypothesis to the stinging behaviour of *Symmorphus Allobrogus*, a predatory wasp hunting chrysomelid larvae. Acta Zool. Lituan. **14**(4), 23–33 (2004)
A. Bukero, M.A. Rustamani, G.H. Abro, A.M. Lodhi, Feeding potential of *Menochilus sexmaculatus* (Fabricius) against sucking insect pests. J. Basic Appl. Sci. **10**, 554–558 (2014)
J.R. Busvine, *A Critical Review of the Techniques for Testing Insecticides* (Commonwealth Agricultural Bureaux, London, 1971). 345 p
S. Cabral, P. Garcia, A.O. Soares, Effects of pirimicarb, buprofezin and pymetrozine on survival, development and reproduction of *Coccinella undecimpunctata* (Coleoptera: Coccinellidae). Biocontrol Sci. Technol. **18**(3), 307–318 (2008)
S. Cabral, A.O. Soares, P. Garcia, Voracity of *Coccinella undecimpunctata*: effects of insecticides when foraging in a prey/plant system. J. Pest. Sci. **84**, 373–379 (2011)
P. Cabrera, D. Cormier, M. Fournier, E. Lucas, Lethal effects of two reduced risk insecticides on *Harmonia axyridis* and *Coleomegilla maculata* (Col., Coccinellidae) following two routes of exposure. IOBC-WPRS Bull. **103**, 41–45 (2014)
S. Caceres, C.C. Childers, Biology and life tables of *Gelendromus helveolus* (Acari: Phytoseiidae) on Florida citrus. Environ. Entomol. **20**, 224–229 (1991)
L.E. Caltagirone, R.L. Doutt, The history of the vedalia beetle importation to California and its impact on the development of biological control. Annu. Rev. Entomol. **34**, 1–16 (1989)
M. Candolfi, F. Bigler, P. Campbell et al., Principles for regulatory testing and interpretation of semi-field and field studies with non-target arthropods. J. Pest. Sci. **73**, 141–147 (2000)
C. Canepari, New data on some coccinellidae (Coleoptera) from the Mediterranean region. Zootaxa **2318**, 394–399 (1990)
J.R. Carey, *Applied Demography for Biologists with Special Emphasis on Insects* (Oxford University Press, New York, 1993). 206 p
G.A. Carvalho, D. Bezerra, B. Souza, C.F. Carvalho, *Efeitos de Inseticidas usados na cultura do algodoeiro sobre Chrysoperla externa* (Hagen) (Neuroptera: Chrysopidae). Neotrop. Entomol. **32**, 699–706 (2003)
M. Castagnoli, M. Liguori, S. Simoni, C. Duso, Toxicity of some insecticides to *Tetranychus urticae*, *Neoseiulus californicus* and *Tydeus californicus*. BioControl **50**, 611–622 (2005)
C. Castane, J. Arino, J. Arno, Toxicity of some insecticides and acaricides to the predatory bug, *Dicyphus tamaninii* (Het.: Miridae). Entomophaga **41**(2), 211–216 (1996)
M. Cavaco, M. Goncalves, A. Nave, J. Santos, P. Silvino, C. Veiga, R. Rodrigues, Evaluation of the side effects of five insecticides on predatory mites (Acari: Phytoseiidae) in apple orchards in two different regions of Portugal. IOBC/WPRS Bull. **26**(5), 1–8 (2003)
D. Chakraborty, D.M. Korat, Feeding efficiency of green lacewing, *Chrysoperla carnea* (Stephens) on different species of aphids. Karnataka J. Agric. Sci. **23**(5), 793–794 (2010)
S.K. Chaudhary, M.R. Ghosh, Influence of some modem insecticides on the incidence of *Coccinella transversalis* Fab., a predator of *Lipaphis erysimi* Kalt. Sci. Cult. **48**, 214–216 (1982)
T.Y. Chen, T.X. Liu, Susceptibility of immature stages of *Chrysoperla rufilabris* (Neurop., Chrysopidae) to pyriproxyfen, a juvenile hormone analog. J. Appl. Entomol. **126**(2 & 3), 125–129 (2002)
L.L. Cheng, J.R. Nechols, D.C. Margolies, J.F. Campbell, P.S. Yang, Assessment of prey preference by the mass-produced generalist predator, *Mallada basalis* (Walker) (Neuroptera: Chrysopidae) when offered two species of spider mites, *Tetranychus kanzawai* Kishida and *Panonychus citri* (McGregor) (Acari: Tetranychidae) on papaya. Biol. Control **53**, 267–272 (2010)
J. Cho, K.J. Hong, J.K. Yoo, J.R. Bang, J.O. Lee, Comparative toxicity of selected insecticides to *Aphis citricola*, *Myzus malisuctus* (Homoptera: Aphididae) and the predator *Harmonia axyridis* (Coleoptera: Coccinellidae). J. Econ. Entomol. **90**(1), 11–14 (1997)

S.P. Chowdhury, M.A. Ahad, M.R. Amin, M.R. Amin, N.A. Rasel, Bean aphid predation efficiency of lady bird beetle, *Micraspis discolor* F. (Coleoptera: Coccinellidae). J. Soil Nat. **2**(3), 40–45 (2008)

T.H. Chua, E. Mikil, Effects of prey number and stage on the biology of *Cyrtorhinus lividipennis* (Hemiptera: Miridae): a predator of *Nilaparvata lugens* (Homoptera: Delphacidae). Environ. Entomol. **18**(2), 251–255 (1989)

M.S. Clark, J.M. Luna, N.D. Stone, R.R. Youngman, Generalist predator consumption of armyworm (Lepidoptera: Noctuidae) and effect of predator removal on damage in no-till corn. Environ. Entomol. **23**(3), 617–622 (1994)

M.A. Claver, B. Ravichandran, M.M. Khan, D.P. Ambrose, Impact of cypermethrin on the functional response, predatory and mating behaviour of a non-target potential biological control agent, *Acanthaspis pedestris* (Stal) (Het., Reduviidae). J. Appl. Entomol. **127**, 18–22 (2003)

R.A. Cloyd, A. Dickinson, Effect of insecticides on mealybug destroyer (Coleoptera: Coccinellidae) and parasitoid *Leptomastix dactylopii* (Hymenoptera: Encyrtidae), natural enemies of citrus mealybug (Homoptera: Pseudococcidae). J. Econ. Entomol. **99**, 1596–1604 (2006)

P.G. Cole, P.A. Horne, The impact of aphicide drenches on *Micromus tasmaniae* (Walker) (Neuroptera: Hemerobiidae) and the implications for pest control in lettuce crops. Aust. J. Entomol. **45**, 244–248 (2006)

M. Coll, M. Guershon, Omnivory in terrestrial arthropods: mixing plant and prey diets. Annu. Rev. Entomol. **47**, 267–297 (2002)

P.S. Corbet, *Dragonflies: Behaviour and Ecology of Odonata* (Harley Books, Martins, Colchester, 1999). 830 p

P.L. Coville, W.W. Allen, Life table and feeding habits of *Scolothrips sexmaculatus* (Thysanoptera: Thripidae). Ann. Entomol. Soc. Am. **70**(1), 11–16 (1977)

B.A. Croft, *Arthropod Biological Control Agents and Pesticides* (Wiley Interscience, New York, 1990). 723 p

B.A. Croft, D.L. McGroarty, The role of *Amblyseius fallacis* in Michigan apple orchards. Res. Rept. 33, Michigan Agric. Exp. Stn. 48 (1977)

B.A. Croft, Z. Zhang, Assessing roles of individual species in a phytoseiid complex using adult female and immature traits: relevance to biological control. *Proceedings of the IX International Congress Acarology* (1996)

W.W. Daniel, *Biostatistics: A Foundation for Analysis in the Health Sciences* (Wiley, PAU, Ludhiana, 1987) 734 p

L. Das, P. Sagar, Feeding pattern of three coccinellid beetles on *Aphis craccivora* Koch-A pest of chrysanthemum. *Proceedings of the Symposium Biocontrol Based Pest Management for Quality Crop Protection in the Current Millennium* on July 18–19 (PAU, Ludhiana, 2001), 44 p

S. Das, S. Roy, A. Mukhopadhyay, Diversity of arthropod natural enemies in the tea plantations of North Bengal with emphasis on their association with tea pests. Curr. Sci. **99**(10), 1457–1463 (2010)

K.E. Day, I.M. Scott, Use of acetylcholinesterae activity to detect sublethal toxicity in stream invertebrates exposed to low concentrations of organophosphate insecticides. Aquat. Toxicol. **18**, 101–114 (1990)

C.P.O. De-Assis, E.G.F. De-Morais, M.G.C. Gondim, Toxicity of acaricides to *Raoiella indica* and their selectivity for its predator, *Amblyseius largoensis* (Acari: Tenuipalpidae: Phytoseiidae). Exp. Appl. Acarol. **60**(3), 357–365 (2013)

P. De-Clercq, E. Vinuela, G. Smagghe, D. Degheele, Transport and kinetics of diflubenzuron and pyriproxyfen in the beet armyworm *Spodoptera exigua* and its predator, *Podisus maculiventris*. Entomol. Exp. Appl. **76**, 189–194 (1995a)

P. De-Clercq, A. De-Cock, L. Tirry, E. Vinuela, D. Degheele, Toxicity of diflubenzuron and pyriproxyfen to the predatory bug, *Podisus maculiventris*. Entomol. Exp. Appl. **74**, 17–22 (1995b)

A. De-Cock, P. De-Clercq, L. Tirry, D. Degheele, Toxicity of diafenthiuron and imidacloprid to the predatory bug, *Podisus maculiventris* (Heteroptera: Pentatomidae). Environ. Entomol. **25**(2), 476–480 (1996)

DEFRA, Department for Environment, Food and Rural Affairs. Protocols for laboratory, extended laboratory and semi-field bioassays in pesticide risk assessment schemes for non-target arthropods. Final Project Report. (2002). Assessed through www.pesticides.gov.uk/Resources/CRD/.../R/Research_PN0928.pdf

T.J. Dennehy, A.W. Farnham, I. Denholm, The microimmersion bioassay: a novel method for topical application of pesticides to spidermites. Pestic. Sci. **39**, 47–54 (1993)

M. De-Schampheleire, P. Spanoghe, E. Brusselman, S. Sonck, Risk assessment of pesticide spray drift damage in Belgium. Crop Protect. **26**(4), 602–611 (2007)

N. Desneux, A. Decourtye, J. Delpuech, The sublethal effects of pesticides on beneficial arthropods. Annu. Rev. Entomol. **52**, 81–106 (2007)

Y.R. Devi, J. Kalita, T.K. Singh, Biological control potential of an aphidophagous syrphid, *Episyrphus balteatus*, De-Geer (Diptera: Syrphidae) on mustard aphid, *Lipaphis erysimi* (Kalt.) (Homoptera: Aphididae) on cabbage ecosystem in Manipur. J. Exp. Sci. **2**(12), 13–16 (2011)

M. Dicke, M.W. Sabelis, M. De-Jong, An analysis of prey preference in phytoseiid mites by using an olfactometer, predation models and electrophoresis. Exp. Appl. Acarol. **5**, 225–241 (1988)

M. Dicke, M.W. Sabelis, M. De-Jong, Do phytoseiids select the best prey species in terms of reproductive success? Exp. Appl. Acarol. **8**, 161–173 (1990)

A. Dinter, H.M. Poehling, Side-effects of insecticides on two erigonid spider species. Entomol. Exp. Appl. **74**(2), 151–163 (1995)

A. Dinter, K. Brugger, A. Bassi, N.M. Frost, M.D. Woodward, Chlorantraniliprole (DPX-E2Y45, DuPont™ Rynaxypyr®, Coragen® and Altacor® insecticide) – a novel anthranilic diamide insecticide – demonstrating low toxicity and low risk for beneficial insects and predatory mites. IOBC/WPRS Bull. **35**, 128–135 (2008)

M.C. Donahoe, H.N. Pitre, *Reduviolus roseipennis* behavior and effectiveness in reducing numbers of *Heliothis zea* on cotton. Environ. Entomol. **6**(6), 872–876 (1977)

R.L. Doutt, The historical development of biological control, in *Biological Control of Insect Pests and Weeds*, ed. by P. DeBach (Chapman and Hall, London, 1964), pp. 21–42

C. Duso, Biological control of tetranychid mites in peach orchards of northern Italy: role of *Amblyseius andersoni* (Chant) and *Amblyseius finlandicus* (Oud.) (Acari: Phytoseiidae). Acta Phytopathol. Entomol. Hung. **27**, 211–217 (1992)

C. Duso, C. Pasqualetto, P. Camporese, Role of the predatory mites *Amblyseius aberrans* (Oud.), *Typhlodromus pyri* Scheuten and *Amblyseius andersoni* (Chant) (Acari, Phytoseiidae) in vineyards. II. Minimum releases of *A. aberrans* and *T. pyri* to control spidermite populations (Acari, Tetranychidae). J. Appl. Entomol. **112**, 298–308 (1991)

C. Duso, V. Malagnini, A. Pozzebon, F.M. Buzzetti, P. Tirello, A method to assess the effects of pesticides on the predatory mite, *Phytoseiulus persimilis* (Acari Phytoseiidae) in the laboratory. Biocontrol Sci. Technol. **18**(10), 1027–1044 (2008)

B.A.B. Edwards, P.J. Hodgson, The toxicity of commonly used orchard chemicals to *Stethorus nigripes* (Coleoptera: Coccinellidae). J. Aust. Entomol. Soc. **12**, 222–224 (1973)

L.E. Ehler, Natural enemies of cabbage looper on cotton in the San Joaquan Valley. Hilgardia **45**(3), 73–106 (1977)

A.E.G.M. El-Sayed, G.E.S.A. El-Ghar, The influence of normal and low-rate application of insecticides on populations of the cotton whitefly and melon aphid and associated parasites and predators on cucumber. Anz. SchacUingskde Pflanzenschutz Umweltschutz **65**, 54–57 (1992)

G.L. Ellman, K.D. Courtney, A.J. Vulentino, R.M. Featherstone, A new rapid colorimetric determination of acetylcholinesterase activity. Biochem. Pharmacol. **7**, 88–95 (1961)

Z.S. El-Zahi, H.K. Abd-Elhady, Insect predators and control of *Aphis gossypii* comparing to certain insecticides under caged-cotton plants conditions. Pak. J. Biol. Sci. **16**(5), 233–238 (2013)

G.W. Elzen, Lethal and sublethal effects of insecticide residues on *Orius insidiosus* (Hemiptera: Anthocoridae) and *Geocoris punctipes* (Hemiptera: Lygaeidae). J. Econ. Entomol. **94**, 55–59 (2001)

G.W. Elzen, P.J. Elzen, Lethal and sub lethal effects of selected insecticides on *Geocoris punctipes*. Southwest. Entomol. **24**(3), 199–205 (1999)

G.W. Elzen, P.J. Elzen, E.G. King, Laboratory toxicity of insecticide residues to *Orius insidiosus*, *Geocoris punctipes*, *Hippodamia convergens*, and *Chrysoperla carnea*. Southwest. Entomol. **23**, 335–342 (1998)

K. Epperlein, H.W. Schmidt, Effects of pelleting sugarbeet seed with Gaucho® (imidacloprid) on associated fauna in the agricultural ecosystem. Pflanzenschutz-Nachrichten Bayer **54**(3), 369–398 (2001)

EPPO/OEPP, Guideline for evaluation of side effects of plant protection products (*Phytoseiulus persimilis*). EPPO/OEPP Bull. **20**, 531–550 (1990)

A. Ester, N.B.M. Brantjes, Controlling aphids in iceberg lettuce by pelleting the seeds with insecticides. (1999). Retrieved from agris.fao.org/agris-search/search.do?recordID=BE2001000113

M.D. Eubanks, R.F. Denno, The ecological consequences of variation in plants and prey for an omnivorous insect. Ecology **80**, 1253–1266 (1999)

J.W. Everts, B. Aukema, R. Hengeveld, J.H. Koeman, Side effects of pesticides on ground dwelling predatory arthropods in arable ecosystems. Environ. Pollut. **59**, 203–225 (1989)

J.W. Everts, B. Aukema, W. Mullie, C.A.M. Van-Gestel, Exposure of the ground dwelling spider *Oedothorax apicatus* (Blackwall) (Erigonidae) to spray and residues of deltamethrin. Arch. Environ. Contam. Toxicol. **20**(1), 13–19 (1991a)

J.W. Everts, I. Willemsen, M. Stulp, L. Simons, B. Aukema, J. Kammenga, The toxic effects of diafenthiuron on Linyphiid and Erigonid spiders in connection with ambient temperature, humidity and predation. Arch. Environ. Contam. Toxicol. **20**(1), 20–24 (1991b)

L.T. Fabellar, E.A. Heinrichs, Toxicity of insecticides to predators of rice brown planthoppers, *Nilaparvata lugens* (Stal) (Homoptera: Delphacidae). Environ. Entomol. **13**(3), 832–837 (1984)

B.B. Fand, R.D. Gautam, S.S. Sachin, Effect of developmental stage and density of *Phenacoccus solenopsis* Tinsley (Hemiptera: Pseudococcidae) on the predatory performance of four coccinellid predators. J. Biol. Control. **24**(2), 110–115 (2010)

M. Figuls, C. Castane, R. Gabarra, Residual toxicity of some insecticides on the predatory bugs *Dicyphus tamaninii* and *Macrolophus caliginosus*. BioControl **44**, 89–98 (1999)

D.J. Finney, *Probit Analysis*, 3rd edn. (Cambridge University Press, Cambridge, 1971). 333 p

M.N. Fogel, M.I. Schneider, N. Desneux, B. Gonzalez, A.E. Ronco, Impact of the neonicotinoid acetamiprid on immature stages of the predator *Eriopis connexa* (Coleoptera: Coccinellidae). Ecotoxicology **22**(6), 1063–1071 (2013)

V.E. Forbes, P. Calow, Is the per capita rate of increase a good measure of population-level effects in ecotoxicology? Environ. Toxicol. Chem. **18**, 1544–1556 (1999)

F. Forster, C. Kula, New soil invertebrate testing scheme proposed by EPPO with special reference to formulations of plant protection products like granular formulations and seed dressings. IOBC/WPRS Bull. **26**(5), 93–97 (2003)

C.J.S. Fox, C.D. Dondale, Annotated list of spiders (Araneae) from hayfields and their margins in Nova Scotia. Can. Entomol. **104**, 1911–1915 (1972)

T.L. Galvan, R.L. Koch, W.D. Hutchison, Effects of spinosad and indoxacarb on survival, development and reproduction of the multicolored Asian lady beetle (Coleoptera: Coccinellidae). Biol. Control **34**, 108–114 (2005)

M.M. Gardiner, D.A. Landis, C. Gratton et al., Landscape diversity enhances biological control of an introduced crop pest in the north-central USA. Ecol. Appl. **19**, 143–154 (2009)

B. Geetha, Biology and management of spiraling whitefly *Aleurodicus dispersus* Russell (Homoptera: Aleyrodidae). Ph.D. Thesis, Tamil Nadu Agricultural University, Coimbatore, India, 196 p (2000)

P.J.E. George, Prey preference of a ladybird beetle *Coccinella transversalis* Fab. Insect Environ. **6**(3), 124–125 (2000)

P.J.E. George, D.P. Ambrose, Toxic effects of insecticides in the histomorphology of alimentary canal, testis and ovary in a reduviid, *Rhynocoris kumarii* Ambrose and Livingstone (Hemiptera: Reduviidae). J. Adv. Zool. **25**, 46–50 (2004)

H. Ghahari, S. Anlas, H. Sakenin, H. Ostovan, M. Tabari, A contribution to the rove beetles (Coleoptera: Staphylinoidea: Staphylinidae) of Iranian rice fields and surrounding grasslands. Linzer Biol. Beitr. **41**(2), 1959–1968 (2009)

M. Gholamzadehchitgar, J. Hajizadeh, M. Ghadamyari, A. Karimi-Malati, H. Hoda, Sublethal effects of diazinon, fenitrothion and chlorpyrifos on the functional response of predatory bug,

Andrallus spinidens Fabricius (Hem.: Pentatomidae) in the laboratory conditions. J. King Saud Univ. Sci. **26**, 113–118 (2014)
D.R. Gillespie, Biological control of thrips on greenhouse cucumber by *Amblyseius cucumeris*. Entomophaga **34**, 185–192 (1989)
F.E. Gilstrap, E.R. Oatman, The bionomics of *Scolothrips sexmaculatus* (Pergande) (Thysanoptera: Thripidae), an insect predator of spider mites. Hilgardia **44**(2), 27–59 (1976)
K.E. Godfrey, W.H. Whitcomb, J.L. Stimac, Arthropod predators of velvetbean caterpillar, *Anticarsia gemmatalis* Hubner (Lepidoptera: Noctuidae), eggs and larvae. Environ. Entomol. **18**(1), 118–123 (1989)
G. Golmohammadi, M. Hejazi, S. Iranipour, S.A. Mohammadi, Lethal and sublethal effects of endosulfan, imidacloprid and indoxacarb on first instar larvae of *Chrysoperla carnea* (Neu.: Chrysopidae) under laboratory conditions. J. Entomol. Soc. Iran **28**(2), 37–47 (2009)
R.H. Gonzalez, S. Rojas, Estudio analftico del control biologico de plagas agrfcolas en Chile. Agrie. Tee. Santiago **26**, 133–147 (1966)
R.D. Gordon, The Coccinellidae (Coleoptera) of America north of Mexico. J. N. Y. Entomol. Soc. **93**, 1–12 (1985)
M.A. Goven, B. Guven, Side effects of pesticides used in vineyards in the Aegean region on the predatory mite *Typhlodromus perbibus* Wainstein and Arutunjan (Acari: Phytoseiidae) under laboratory conditions. IOBC/WPRS Bull. **35**, 92–95 (2008)
E.E. Grafton-Cardwell, P. Gu, Conserving vedalia beetle, *Rodolia cardinalis* (Mulsant) (Coleoptera: Coccinellidae), in citrus: a continuing challenge as new insecticides gain registration. J. Econ. Entomol. **96**, 1388–1398 (2003)
E.E. Grafton-Cardwell, Y. Ouyan, Augmentation of *Euseius tularensis* (Acari: Phytoseiidae) in citrus. Environ. Entomol. **24**(3), 738–747 (1995)
D.J. Greathead, Observations on two species of Systoechus (Diptera: Bombyliidae) preying on the desert locust, *Schistocerca gregaria* (Forskal) in Eastern Africa. Entomophaga **3**(1), 3–22 (1958)
T.G. Grout, G.I. Richards, *Euseius addoensis addoensis*, an effective predator of citrus thrips, *Scirtothrips aurantii*, in the eastern Cape Province of South Africa. Exp. Appl. Acarol. **15**, 1–13 (1992)
W.H. Habig, M.J. Pabst, W.B. Jakoby, Glutathione S-transferases, the first enzymatic step in mercapturic acid formation. J. Biol. Chem. **249**, 7130–7139 (1974)
J.J. Hadam, M.T. AliNiazee, B.A. Croft, Phytoseiid mites of major crops in the Willamette Valley and pesticide resistance in *Typhlodromus pyri* Scheuten. Environ. Entomol. **15**, 1255–1262 (1986)
E.A.C. Hagley, N. Miles, Release of *Chrysoperla carnea* Stephens (Neuroptera: Chrysopidae) for control of *Tetranychus urticae* Koch (Acarina: Tetranychidae) on peach grown in a protected environment structure. Can. Entomol. **119**(2), 205–206 (1987)
A. Hameed, M. Saleem, S. Ahmad, M.I. Aziz, H. Karar, Influence of prey consumption on life parameters and predatory potential of *Chrysoperla carnea* against cotton mealy bug. Pak. J. Zool. **45**(1), 177–182 (2013)
L.F. Hand, A.J. Keaster, Environment of an insect field cage. J. Econ. Entomol. **60**, 910 (1967)
L.S. Hansen, The effect of initial thrips density *Thrips tabaci* Lind. (Thysanoptera: Thripidae) on the control exerted by *Amblyseius barkeri* Hughes (Acarina: Phytoseiidae) on glasshouse cucumber. J. Appl. Entomol. **107**, 130–135 (1989)
L. Hanumantharaya, K.B. Goud, G.K. Ramegowda, Use of green lacewing, *Chrysoperla carnea* (Stephens) and neem seed kernel extract for management of insect pests on cotton. Karnataka J. Agric. Sci. **21**(1), 41–44 (2008)
S.A. Hassan, Guidelines for testing the effects of pesticides on beneficial organisms: description of test methods. IOBC/WPRS Bull. **15**(3), 1–3 (1992)
S.A. Hassan, Activities of the IOBC/WPRS working group, pesticides and beneficial organisms. IOBC/WPRS Bull. **17**(10), 1–5 (1994)
S.A. Hassan, Standard laboratory methods to tests the side effects of pesticides (initial and persistent) on *Trichogramma cacoeciae* Marchal (Hym., Trichogrammatidae), in *Ecotoxicology*

Pesticides and Beneficial Organisms, ed. by P.T. Haskell, P. McEwen (Kluwer Academic Publishers, Dordrecht, 1998), pp. 71–79

S.A. Hassan, F. Bigler, H. Bogenschutz et al., Results of the sixth joint pesticide testing programme carried out by the IOBC/WPRS working group, pesticides and beneficial organisms. Entomophaga **39**, 107–119 (1994)

J. Havelka, P. Bartova, Toxicities of greenhouse pesticides to the 3rd instar larvae of the aphidophagous gall midge, *Aphidoletes aphidimyza* (Rondani) (Diptera, Cecidomyiidae). Ochrana Rostlin **27**, 293–300 (1991)

B.A. Hawkins, N.J. Mills, M.A. Jervis, P.W. Price, Is the biological control of insects a natural phenomenon? Oikos **86**, 493–506 (1999)

Y. He, J. Zhao, Y. Zheng, N. Desneux, K. Wu, Lethal effect of imidacloprid on the coccinellid predator *Serangium japonicum* and sublethal effects on predator voracity and on functional response to the whitefly, *Bemisia tabaci*. Ecotoxicology **21**(5), 1291–1300 (2012)

K.M. Heinz, J.R. Brazzle, C.H. Pickett, E.T. Natwick, J.M. Nelson, M.P. Parrell, Predatory beetle may suppress silverleaf whitefly. California Agric. **48**(2), 35–40 (1994)

I. Hodek, A. Honek, *Ecology of Coccinellidae* (Springer Science + Business media B.V, Dordrecht, 1996). 464 p

I. Hodek, A. Honek, H.F. Van-Emden, *Ecology and Behaviour of the Ladybird Beetles (Coccinellidae)* (Wiley-Blackwell, Chichester, West Sussex, UK, 2012), 561 p

K. Hoelmer, L.S. Osborne, R.K. Yokomi, Interactions of the whitefly predator, *Delphastus pusillus* (Coleoptera: Coccinellidae) with parasitized sweetpotato whitefly (Homoptera: Aleyrodidae). Environ. Entomol. **23**(1), 136–139 (1994)

J.M. Holland, S.R. Thomas, A. Hewitt, Some effects of polyphagous predators on an outbreak of cereal aphid (*Sitobion avenae* F.) and orange wheat blossom midge (*Sitodiplosis mosselana* Gehin). Agric. Ecosyst. Environ. **59**, 181–190 (1996)

C.S. Holling, Some characteristics of simple type of predation and parasitism. Can. Entomol. **91**, 385–398 (1959)

J. Hough-Goldstein, J. Whalen, Inundative release of predatory stink bugs for control of Colorado potato beetle. Biol. Control **3**, 343–347 (1993)

A. Howe, G.L. Lovei, G. Nachman, Dummy caterpillars as a simple method to assess predation rates on invertebrates in a tropical agroecosystem. Entomol. Exp. Appl. **131**, 325–329 (2009)

A.G. Howe, G. Nachman, G.L. Lovei, Predation pressure in Ugandan cotton fields measured by a sentinel prey method. Entomol. Exp. Appl. **154**, 161–170 (2015)

M.A. Hoy, *Agricultural Acarology: Introduction to Integrated Mite Management* (CRC Press, Boca Raton, FL, 2011), 430 p

H.T. Huang, P. Yang, The ancient cultured citrus ant. BioScience **37**(9), 665–671 (1987)

A. Huerta, P. Medina, P. Castanera, E. Vinuela, Residual effects of some modern pesticides on *Chrysoperla carnea* (Stephens) adults under laboratory conditions. IOBC/WPRS Bull. **26**, 165–170 (2003)

J.C. Hutson, The fluted scale and the vedalia beetle. Trop. Agric. Mag. Ceylon Agric. Soc. **55**, 95–97 (1920)

Y.B. Ibrahim, V.B. Palacio, Life history and demography of the predatory mite, *Amblyseius longispinosus* Evans. Exp. Appl. Acarol. **18**(6), 361–369 (1994)

Y.B. Ibrahim, T.S. Yee, Influence of sublethal exposure to abamectin on the biological performance of *Neoseiulus longispinosus* (Acari: Phytoseiidae). J. Econ. Entomol. **93**, 1085–1089 (2000)

G. Iperti, Biodiversity of predaceous coccinellidae in relation to bioindication and economic importance. Elsevier Agric. Econ. Environ. **74**, 323–342 (1999)

F.J. Irigaray, F.G. Zalom, P.B. Thompson, Residual toxicity of acaricides to *Galendromus occidentalis* and *Phytoseiulus persimilis* (Acari: Phytoseiidae) reproductive potential. Biol. Control **40**, 153–159 (2007)

I. Ishaaya, A.R. Horowitz, *Insecticides with Novel Modes of Action: Mechanisms and Applications* (Springer, Berlin, 1998). 289 p

K.S. Jagadish, M. Jayaramaiah, B. Shivayogeshwara, Bioefficacy of three promising predators on *Myzus nicotianae* Blackman (Homoptera: Aphididae). J. Biopestic. **3**(1), 62–67 (2010)

D.G. James, Pesticide susceptibility of two coccinellids (*Stethorus punctum* and *Harmonia axyridis*) important in biological control of mites and aphids in Washington hops. Biocontrol Sci. Technol. **13**, 253–259 (2003)

V.K. Jandial, K. Malik, Feeding potential of *Coccinella septempunctata* Linn. (Coccinellidae: Coleoptera) on mustard aphid, *Lipaphis erysimi* Kalt. and potato peach aphid, *Myzus persicae* Sulzer. J. Entomol. Res. **30**(4), 291–293 (2006)

J.P. Jansen, L. Hautier, N. Mabon, B. Schiffers, Pesticides selectivity list to beneficial arthropods in four field vegetable crops. IOBC/WPRS Bull. **35**, 66–77 (2008)

S.R. Jasmine, S. Kuttalam, J. Stanley, Bioefficacy and safety of abamectin against two spotted spider mite and its predatory mite on rose. Ann. Plant Protect. Sci. **16**(1), 46–52 (2008)

S.R. Jasmine, S. Kuttalam, P.S. Subramaniam, J. Stanley, Effect of abamectin 1.9 EC- microbial insecticide to coccinellids in cotton ecosystem, in *Ecofriendly Insect Pest Management*, ed. by S. Ignacimuthu, B.V. David (Elite Publishing House Pvt Ltd, Chennai, 2009), pp. 74–80

S. Jayakumar, A. Sankari, Spider population and their predatory efficiency in different rice establishment techniques in Aduthurai. Tamil Nadu J. Biopestic. **3**(1), 20–27 (2010)

M.L. Johnson, S. Pearce, M. Wade, A. Davies, L. Silberbauer, P. Gregg, M.P. Zalucki, *Review of Beneficials in Cotton Farming Systems* (Cotton Research and Development Corporation, Narrabri, 2000). 82 p

M. Joisten, The 'island method' – a new laboratory test method to determine side-effects of plant protection products on the predatory mite *Typhlodromus pyri* Scheuten (Acari, Phytoseiidae). J. Appl. Entomol. **124**, 267–268 (2000)

S.A. Jones, J.G. Morse, Use of isoelectric focusing electrophoresis to evaluate citrus thrips (Thysanoptera: Thripidae) predation by *Euseius tularensis* (Acari: Phytoseiidae). Environ. Entomol. **24**(5), 1040–1051 (1995)

S. Joshi, C.R. Ballal, Syrphid predators for biological control of aphids. J. Biol. Control. **27**(3), 151–170 (2013)

P.C. Joshi, L. Khamashon, B.R. Kaushal, Predation of *Adalia tetraspilota* (Hope) (Coleoptera: Coccinellidae) on green peach aphid (*Myzus persicae*. Sulzer). Entomol. Ornithol. Herpetol. **1**(1), 101 (2012)

A. Kajak, L. Andrzejeuska, Z. Wojcik, The role of spiders in the decrease of damages caused by Acridoidea on meadows – experimental investigations. Ekol. Pol. **16**, 755–764 (1968)

M.N. Kapadia, S.N. Puri, Biology and comparative predation efficacy of three heteropteran species recorded as predators of *Bemisia tabaci* in Maharashtra. Entomophaga **36**(4), 555–559 (1991)

M.N. Kapadia, S.N. Puri, Biology of *Serangium parcesetosum* as a predator of cotton whitefly. J. Maharashtra Agric. Univ. **17**, 162–163 (1992)

H. Kaur, J.S. Virk, Feeding potential of *Cryptolaemus montrouzieri* against the mealybug, *Phenacoccus solenopsis*. Phytoparasitica **40**, 131–136 (2012)

A. Kavousi, K. Talebi, Side-effects of three pesticides on the predatory mite, *Phytoseiulus persimilis* (Acari: Phytoseiidae). Exp. Appl. Acarol. **31**, 51–58 (2003)

J. Kean, S. Wratten, J. Tylianakis, N. Barlow, The population consequences of natural enemy enhancement and implications for conservation biological control. Ecol. Lett. **6**, 604–612 (2003)

P.E. Kenmore, F. Carino, C. Perez, V. Dyck, Gutierrez, Population regulation of the rice brown planthopper (*Nilaparavata lugens* Stal.) within rice fields in the Philippines. J. Plant Protect. Trop. **1**, 1–37 (1984)

A.A. Khan, Evaluation of the biological control efficiency of four spiders using functional response experiments. Bioscan **8**(4), 1123–1128 (2013)

M.H. Khan, N. Ahmad, S.M.M.S. Rashdi, M. Ismail, I. Rauf, M. Tofique, Studies on the compatibility of neem oil with predator, *Chrysoperla carnea* for the management of aphids (Homoptera: Aphididae) in canola (*Brassica napus* L.). J. Cereals Oilseeds **4**(6), 85–88 (2013)

R. Khuhro, A. Ghafoor, A. Mahmood et al., Assessment of potential of predatory spiders in controlling the cotton jassid (*Amrasca devastans*) under laboratory conditions. J. Anim. Plant Sci. **22**(3), 635–638 (2012)

S.S. Kim, S.G. Seo, Relative toxicity of some acaricides to the predatory mite, *Amblyseius womersleyi* and the twospotted spider mite, *Tetranychus urticae* (Acari: Phytoseiidae, Tetranychidae). Appl. Entomol. Zool. **36**, 509–514 (2001)
S.S. Kim, S.S. Yoo, Comparative toxicity of some acaricides to the predatory mite, *Phytoseiulus persimilis* and the two spotted spider mite, *Tetranychus urticae*. BioControl **47**, 563–573 (2002)
S.K. Kim, S.G. Seo, J.D. Park, S.G. Kim, D.I. Kim, Effect of selected pesticides on predatory mite, *Amblyseius cucumeris* (Acari: Phytoseiidae). J. Entomol. Sci. **40**, 107–114 (2005)
D. Kim, D.J. Brooks, H. Riedl, Lethal and sublethal effects of abamectin, spinosad, methoxyfenozide and acetamiprid on the predaceous plant bug, *Deraeocoris brevis* in the laboratory. BioControl **51**(4), 465–484 (2006)
M.M. Kinawy, Biological control off the coconut scale insect (*Aspidiotus destructor* Sign, Homoptera: Diaspididae) in the southern region of Oman (Dhofar). Trop. Pest Manag. **37**(4), 387–389 (1991)
K. Kiritani, N. Kakiya, An analysis of the predator-prey system in the paddy field. Res. Popul. Ecol. **17**, 29–38 (1975)
K. Kiritani, S. Kawahara, Food-chain toxicity of granular formulations of insecticides to a predator, *Lycosa pseudoannulata* of *Nephotettix cincticeps*. Botyu-Kagaku **38**(2), 69–75 (1973)
V.M. Kirk, Biology of a ground beetle, *Harpalus pennsylvanicus*. Ann. Entomol. Soc. Am. **66**, 513–518 (1973)
R.L. Koch, The multicolored Asian lady beetle, *Harmonia axyridis*: a review of its biology, uses in biological control and non-target impacts. J. Insect Sci. **3**, 32 (2003)
M. Kongchuensin, A. Takafuji, Effects of some pesticides on the predatory mite, *Neoseiulus longispinosus* (Evans) (Gamasina: Phytoseiidae). J. Acarol. Soc. Jpn. **15**(1), 17–27 (2006)
D.C. Kontodimas, G.J. Stathas, Phenology, fecundity and life table parameters of the predator *Hippodamia variegata* reared on *Dysaphis crataegi*. BioControl **50**, 223–233 (2005)
A.M. Koss, G.C. Chang, W.E. Snyder, Predation of green peach aphids by generalist predators in the presence of alternative, Colorado potato beetle egg prey. Biol. Control **31**(2), 237–244 (2004)
V.A. Krischik, A.L. Landmark, G.E. Heimpel, Soil-applied imidacloprid is translocated to nectar and kills nectar-feeding *Anagyrus pseudococci* (Girault) (Hymenoptera: Encyrtidae). Environ. Entomol. **36**(5), 1238–1245 (2007)
A. Krishnamoorthy, Effect of several pesticides on eggs, larvae and adults of green lacewing, *Chrysopa scelestes* (Banks). Entomon **10**, 21–28 (1985)
D. Kumar, Role of spiders as bio-control agents to minimize the use of pesticides in the agricultural fields of Vadodara, Gujarat. J. Ecophysiol. Occup. Health **7**, 101–106 (2007)
M.G. Kumar, C.B. Prakasam, The recovery of *Cryptolaemus montrouzieri* Muls. on the coffee green scale, *Coccus viridis* (Green) on the Shevaroy Hills. J. Coffee Res. **14**, 34–35 (1984)
C.M.S. Kumar, A. Regupathy, Risk assessment of neonicotinoids applied to coffee ecosystem. Int. Pest. Control. **47**, 82–87 (2005)
K. Kumar, Studies on bioefficacy and determination of residues of imidacloprid applied against sucking pests of cotton. Ph.D. Thesis, Tamil Nadu Agricultural University, Coimbatore, India, 114 p (1998)
K. Kumar, G. Santharam, Laboratory evaluation of imidacloprid against *Trichogramma chilonis* Ishii and *Chrysoperla carnea* (Stephens). J. Biol. Control. **13**, 18–23 (1999)
S.P. Kumar, A.G. Kumar, D.P. Ambrose, Impact of intraspecific competition in the predation of *Rhynocoris longifrons* (Stål) (Hemiptera: Reduviidae) on camponotine ant, *Camponotus compressus* Fabricius. Hexapoda **16**(1), 1–4 (2009)
A.G. Kumar, K. Rajan, S.J. Rajan, D.P. Ambrose, Predatory behaviour of an assassin bug, *Coranus spiniscutis* (Reuter) on rice meal moth, *Corcyra cephalonica* (Stainton) and leaf armyworm, *Spodoptera litura* (Fabricius). *National Conference on Insect Pest Management, A Current Scenario*, St. Xavier's College, Palayamkottai, India, pp. 404–408 (2011)

B.A. Kunkel, D.W. Held, D.A. Potter, Impact of halofenozide, imidacloprid and bendiocarb on beneficial invertebrate and predatory activity in turf grass. J. Econ. Entomol. **92**, 922–930 (1999)

B.A. Kunkel, D.W. Held, D.A. Potter, Lethal and sublethal effects of bendiocarb, halofenozide and imidacloprid on *Harpalus pennsylvanicus* (Coleoptera: Carabidae) following different modes of exposure in turf grass. J. Econ. Entomol. **94**, 60–67 (2001)

D.H. Kwon, D.Y. Song, S. Kang et al., Residual contact vial bioassay for the on-site detection of acaricide resistance in the two-spotted spider mite. J. Asia-Pacific Entomol. **13**(4), 333–337 (2010)

L. Lagadic, T. Caquat, F. Ramade, The role of biomarkers in environmental assessment. Invertebrate populations and communities. Ecotoxicology **3**, 193–209 (1994)

D.A. Landis, W. Van-der-Werf, Early season predation impacts the establishment of aphids and spread of beet yellows virus in sugar beet. Entomophaga **42**, 499–516 (1997)

D.A. Landis, S.D. Wratten, G.M. Gurr, Habitat management to conserve natural enemies of arthropod pests in agriculture. Annu. Rev. Entomol. **45**, 175–201 (2000)

M. Langhof, A. Gathmann, H. Poehling, Field study on the effects of insecticide drift on non-target arthropods: residual toxicity, effect on populations and decolonization processes. IOBC/WPRS Bull. **26**(5), 73–82 (2003)

M. Latifian, Voracity and feeding preferences of larvae and adult stages of *Stethorus gilvifrons* Mulsant. (Coleoptera: Coccinellidae) on larvae and adult of *Oligonychus afrasiaticus* McGregor (Acarina: Tetranychidae). Int. J. Agric. Crop Sci. **4**(9), 540–546 (2012)

K.L. Lawrence, D.H. Wise, Unexpected indirect effect of spiders on the rate of litter disappearance in a deciduous forest. Pedobiologia **48**, 149–157 (2004)

M. Lefebvre, N.J. Bostanian, H.M.A. Thistlewood, Y. Mauffette, G. Racette, A laboratory assessment of the toxic attributes of six 'reduced risk insecticides' on *Galendromus occidentalis* (Acari: Phytoseiidae). Chemosphere **84**(1), 25–30 (2011)

D.K. Letourneau, J.A. Jedlicka, S.G. Bothwell, C.R. Moreno, Effects of natural enemy biodiversity on the suppression of arthropod herbivores in terrestrial ecosystems. Annu. Rev. Ecol. Evol. Syst. **40**, 573–592 (2009)

T.X. Liu, T.Y. Chen, Effects of the insect growth regulator fenoxycarb on immature *Chrysoperla rufilabris* (Neuroptera: Chrysopidae). Florida Entomol. **84**, 628–633 (2001)

T. Liu, P.A. Stansly, Lethal and sublethal effects of two insect growth regulators on adult *Delphastus catalinae* (Coleoptera: Coccinellidae), a predator of whiteflies (Homoptera: Aleyrodidae). Biol. Control **30**, 298–305 (2004)

P.L. Lo, R.H. Blank, A.J. Popay, Effect of pesticides on predation of soft wax scale by the steel-blue ladybird. *Proceedings of the 45th New Zealand Plant Protection Conference*, Wellington, New Zealand, 11–13 August, pp 99–102 (1992)

J. Loganathan, P.M.M. David, Predator complex of the teak defoliator, *Hyblaea puera* Cramer (Lepidoptera: Hyblaeidae) in an intensively managed teak plantation at Veeravanallur, Tamil Nadu. Entomon **24**(3), 259–263 (1999)

M.K. Lohar, S.N. Khuhro, M.H. Lakho, G.A. Magsi, T.A. Khuhro, Biology and feeding potential of predator, *Hippodamia convergens* guir, (Coleoptera: Coccinellidae) on mustard aphid, *Lipaphis erysimi* (kalt.) in laboratory. Pak. J. Agric. Agril Engg. Vet. Sci. **28**(2), 150–159 (2012)

B.A. Loiselle, A.G. Farji-Brener, What's up? An experimental comparison of predation levels between canopy and understory in a tropical wet forest. Biotropica **34**, 327–330 (2002)

E.Y. Long, D.L. Finke, Predators indirectly reduce the prevalence of an insect-vectored plant pathogen independent of predator diversity. Oecologia **177**, 1067–1074 (2015)

M. Longley, P.C. Jepson, J. Izquierdo, N. Sotherton, Temporal and spatial changes in aphid and parasitoid populations following applications of deltamethrin in winter wheat. Entomol. Exp. Appl. **83**, 41–52 (1997)

J.E. Losey, M. Vaughan, The economic value of ecological services provided by insects. Bioscience **56**, 311–323 (2006)

F. Louis, A. Ufer, Methodical improvements of standard laboratory tests for determining the side-effects of agrochemicals on predatory mites (Acari: Phytoseiidae). Anz. Schladlingskde Pflanzenschutz Umweltschutz **68**, 153–154 (1995)

C.P. Lounsbury, The pioneering period of economic entomology in South Africa. J. Entomol. Soc. South Afr. **3**, 9–29 (1940)

E. Lucas, D. Coderre, C. Vincent, Voracity and feeding preferences of two aphidophagous coccinellids on *Aphis citricola* and *Tetranychus urticae*. Entomol. Exp. Appl. **85**, 151–159 (1997)

E. Lucas, C. Labrecque, D. Coderre, *Delphastus catalinae* and *Coleomegilla maculata* (Coleoptera: Coccinellidae) as biological control agents of the greenhouse whitefly, *Trialeurodes vaporariorum* (Homoptera: Aleyrodidae). Pest Manag. Sci. **60**(11), 1073–1078 (2004)

R.F. Luck, D.L. Dahlstein, Natural decline of a pine needle scale (*Chionaspis pinifoliae* [Fitch]), outbreak at South Lake Tahoe, California following cessation of adult mosquito control with malathion. Ecology **56**(4), 893–904 (1975)

J. Lundgren, M. Seagraves, Physiological benefits of nectar feeding by a predatory beetle. Biol. J. Linn. Soc. **104**(3), 661–669 (2009)

S. Magalhaes, F.M. Bakker, Plant feeding by a predatory mite inhabiting cassava. Exp. Appl. Acarol. **27**, 27–37 (2002)

A.H.N. Maia, J.B.L. Alferdo, C. Campanhola, Statistical inference on associated fecundity life table parameters using jackknife technique: computational aspects. J. Econ. Entomol. **93**, 511–518 (2000)

M. Mani, A. Krishnamoorthy, Role of predators in control of horticultural crop pests of India, in *Indian Insect Predators in Biological Control*, ed. by K. Sahayaraj (Daya Books, New Delhi, 2004), pp. 297–309

M. Mani, A. Krishnamoorthy, Evaluation of the exotic predator, *Cryptolaemus montrouzieri* Muls (Coccinellidae, Coleoptera) in the suppression of green scale, *Chloropulvinaria psii* (Mask.) (Coccidae, Homoptera). Entomon **15**, 45–48 (1990)

M. Mani, A. Krishnamoorthy, Discovery of Australian ladybird beetle (*Cryptolaemus montrouzieri*) on spiraling whitefly (*Aleurodicus dispersus*) in India. Insect Environ. **3**, 5–6 (1997)

M. Mani, T.S. Thontadarya, Development and feeding potential of coccinellid predator, *Cryptolaemus montrouzieri* Muls. on the grape mealybug, *Maconellicoccus hirsutus* (Green). J. Biol. Control. **1**, 19–22 (1987)

M. Mani, T.S. Thorntakarya, Studies on the safety of different pesticides to the grape mealybug natural enemies, *Anagyrus dactylopii* (How.) and *Scymnus coccivora* Ayyar. Indian J. Plant Protect. **16**, 205–210 (1988)

M. Mani, V.J. Lakshmi, A. Krishnamoorthy, Side effects of some pesticides on the adult longevity, progeny production and prey consumption of *Cryptolaemus montrouzieri* Mulsant (Coccinellidae, Coleoptera). Indian J. Plant Protect. **25**, 48–51 (1997)

M. Manjunatha, B.K. Shivanna, Field evaluation of RIL 18 SL (Imidacloprid) against rice brown planthopper and green leafhopper. Insect Environ. **6**, 177–178 (2001)

F. Mansour, D. Rosen, H.N. Plaut, A. Shulov, Effects of commonly used insecticides on *Cheiracanthium mildei* and other spiders occurring on apple. Phytoparasitica **9**(2), 139–144 (1981)

F. Mansour, U. Heimbach, A. Wehling, Effects of pesticide residues on ground-dwelling lycosid and micryphantid spiders in laboratory tests. Phytoparasitica **20**(3), 195–202 (1992)

J.M. Mari, N.H. Rizvi, S.M. Nizamani, K.H. Qureshi, M.K. Lohar, Predatory efficiency of *Menochilus sexmaculatus* Fab. and *Coccinella undecimpunctata* Lin., (Coccinellidae: Coleoptera) on alfalfa aphid, *Therioaphis trifolii* (Monell.). Asian J. Plant Sci. **4**(4), 365–369 (2005)

M. Markkula, K. Tiitanen, M. Hamalainen, A. Forsberg, The aphid midge *Aphidoletes aphidimyza* (Diptera, Cecidomyiidae) and its use in biological control of aphids. Ann. Entomol. Fenn. **45**(4), 89–98 (1979)

A.F. Martinou, N. Seraphides, M.C. Stavrinides, Lethal and behavioral effects of pesticides on the insect predator, *Macrolophus pygmaeus*. Chemosphere **96**, 167–173 (2014)

V.G. Mathirajan, A. Regupathy, Effect of thiamethoxam 25 WG (Actara®) on *Chrysoperla carnea*. Ann. Plant Protect. Sci. **10**(2), 374–375 (2002)

L. Mattedi, F. Forno, C. Rizzi, D. Forti, G. Angeli, Indoxacarb: evaluation of the efficacy on codling moth (*Cydia pomonella*) and side-effects on *Amblyseius andersoni. Proceedings of the Giornate Fitopatologiche*. Scicli e Ragusa, Italy, 3–7 May, pp. 179–184 (1998)

B.F. Mccutchen, F.W. Plapp, Monitoring procedure for resistance to synthetic pyrethroids in tobacco bollworm larvae. *Proceedings, Beltwide Cotton Production Research Conference*, New Orelans. Memphis Teen, 3–8 January, pp. 356–358 (1988)

J.A. McMurtry, The use of phytoseiids for biological control: progress and uture prospects, in *Recent Advances in Knowledge of the Phytoseiidae*, ed. by M.A. Hoy (University of California Press, Berkeley, 1982), pp. 23–48

J.A. McMurtry, Augmentative releases to control mites in agriculture, in *Modern Acarology*, ed. by F. Dusbabek, V. Bukva (Academia/SPB Academic Publishing, Prague/The Hague, 1991), pp. 151–157

J.A. McMurtry, B.A. Croft, Life-styles of phytoseiid mites and their roles in biological control. Annu. Rev. Entomol. **42**, 291–321 (1997)

J.A. McMurtry, J. Rodriquez, Nutritional ecology of phytoseiid mites, in *Nutritional Ecology of Insects, Mites and Spiders*, ed. by F. Slansky, J. Rodriquez (Wiley, New York, 1987). 1016 p

J.A. McMurtry, H.G. Johnson, G.T. Scriven, Experiments to determine effects of mass releases of *Stethorus picipes* on the level of infestation of the avocado brown mite. J. Econ. Entomol. **62**(5), 1216–1221 (1969)

P. Medina, F. Budia, G. Smagghe, E. Vinuela, Activity of spinosad, diflubenzuron and azadirachtin on eggs and pupae of *Chrysoperla carnea* (Stephens) under laboratory conditions. Biocontrol Sci. Technol. **11**, 597–506 (2001)

P. Medina, F. Budia, P. Estal, A. Adan, E. Vinuela, Side effects of six insecticides on different developmental stages of *Chrysoperla carnea* (Neuroptera: Chrysopidae). IOBC/WPRS Bull. **26**(5), 33–40 (2003)

B.L. Meena, S.R. Dadhich, R.L. Kumawat, Efficacy of some insecticides against ladybird beetle, *Coccinella septempunctata* Linn. feeding on fenugreek aphid, *Acrythosiphum pisum* (Harris). Ann. Biol. **18**, 171–173 (2002)

P.V. Mele, J. Vayssieres, E.V. Tellingen, J. Vrolijks, Effects of an African weaver ant, *Oecophylla longinoda* in controlling mango fruit flies (Diptera: Tephritidae) in Benin. J. Econ. Entomol. **100**(3), 695–701 (2007)

F.D. Menalled, J.C. Lee, D.A. Landis, Manipulating carabid beetle abundance alters prey removal rates in corn fields. BioControl **43**(4), 441–456 (1999)

F.J. Messina, S.M. Sorenson, Effectiveness of lacewing larvae in reducing Russian wheat aphid populations on susceptible and resistant wheat. Biol. Control **21**, 19–26 (2001)

J.P. Michaud, Responses of two ladybeetles to eight fungicides used in Florida citrus: implications for biological control. J. Insect Sci. **1**, 6 (2001)

J.P. Michaud, Relative toxicity of six insecticides to *Cycloneda sanguinea* and *Harmonia axyridis* (Coleoptera: Coccinellidae). J. Entomol. Sci. **37**, 82–93 (2002)

J.P. Michaud, A.K. Grant, IPM-compatibility of foliar insecticides for citrus: indices derived from toxicity to beneficial insects from four orders. J. Insect Sci. **3**, 18 (2003)

M. Miles, R. Dutton, Testing the effects of spinosad to predatory mites in laboratory, extended laboratory, semi-field and field studies. IOBC/WPRS Bull. **26**(5), 9–20 (2003)

R.F. Mizell, M.C. Sconyers, Toxicity of imidacloprid to selected arthropod predators in the laboratory. Florida Entomol. **75**, 277–280 (1992)

J. Mohaghegh, P. De-Clercq, L. Tirry, Toxicity of selected insecticides to the spined soldier bug, *Podisus maculiventris* (Heteroptera: Pentatomidae). Biocontrol Sci. Technol. **10**(1), 33–40 (2000)

M.M.I. Mollah, M.M. Rahman, M.Z. Alam, Toxic effect of some insecticides on predatory lady bird beetles (Coleoptera: Coccinellidae) in country bean (*Lablab purpureus* L.) field. World J. Zool. **7**(4), 347–350 (2012)

L.J. Mooka, D.M. Davies, The European praying mantis (*Mantis religiosa* L.) as a predator of the red-legged grasshopper (*Melanoplus femurrubrum* (De Geer). Can. Entomol. **98**(9), 913–918 (1966)

K. Mori, T. Gotoh, Effects of pesticides on the spider mite predators, *Scolothrips takahashii* (Thysanoptera: Thripidae) and *Strethorus japonicus* (Cloeoptera: Coccinellidae). Int. J. Acarol. **27**(4), 299–302 (2001)

N. Mori, G. Posenato, G. Sancassani, L. Tosi, V. Girolami, Insecticides for the control of cicadellids in vineyards. Informatore Agrario **55**, 93–97 (1999)

M.A. Morris, B.A. Croft, R.E. Berry, Overwintering and effects of fall habitat manipulation and carbofuran on *Neoseiulus fallacis* and *Tetranychus urticae* in peppermint. Exp. Appl. Acarol. **20**, 249–257 (1996)

S.E. Moser, J.J. Obrycki, Non-target effects of neonicotinoid seed treatments: mortality of coccinellid larvae related to zoophytophagy. Biol. Control **51**, 487–492 (2009)

A. Mukhopadhyay, S. Sannigrahi, Evaluation of predatory potential of *Geocoris ochropterus* (Hemiptera: Lygaeidae) on tea aphid, *Toxoptera aurantii* (Hemiptera: Aphididae). Tea **14**(1), 44–49 (1993)

A. Mukhopadhyay, M. Sarker, *Natural Enemies of Some Tea Pests with Special Reference to Darjeeling, Terai and the Dooras* (NTRF, Tea Board, Calcutta, 2007), pp. 7–49

W.C. Mullie, J.W. Everts, Uptake and elimination of (^{14}C) deltamethrin by *Oedothorax apicatus* (Arachnida: Erigonidae) with respect to bioavailability. Pestic. Biochem. Physiol. **39**, 27–34 (1991)

C.A. Mullin, M.C. Saunders, T.W. Leslie, D.J. Biddinger, S.J. Fleischer, Toxic and behavioral effects to carabidae of seed treatments used on Cry3Bb1- and Cry1Ab/c-protected corn. Environ. Entomol. **34**, 1626–1636 (2005)

N. Muraleedharan, T.N. Ananthakrishnan, Bioecology of four species of Anthocoridae predacious on thrips. Rec. Zool. Surv. India Occ. Paper **11**, 32 (1978)

A. Nadimi, K. Kamali, M. Arbabi, F. Abdoli, Selectivity of three miticides to spider mite predator, *Phytoseius plumifer* (Acari: Phytoseiidae) under laboratory conditions. Agric. Sci. China **8**(3), 326–331 (2009)

T. Nagata, T. Masuda, Efficiency of sticky boards for population estimation of the brown planthopper, *Nilaparvata lugens* (Stal) (Hemiptera: Delphacidae) on rice hills. Appl. Entomol. Zool. **13**, 55–62 (1978)

K. Nagata, G. Shinjo, H. Okuda, M. Yoshida, Efficacy of several insecticides to red-back widow spider, *Latrodectus hasseltii*, collected in Osaka and Mie Prefectures. Jpn. J. Med. Entomol. Zool. **48**(2), 135–139 (1997)

F. Nakasuji, H. Yamanaka, K. Kiritani, Predation of larvae of the tobacco cutworm, *Spodoptera litura* (Lepidoptera, Noctuidae) by Polistes wasps. Kontyu **44**, 205–213 (1976)

C.A. Nalepa, S.B. Bambara, A.M. Burroughs, Pollen and nectar feeding by *Chilocorus kuwanae* (Silvestri) (Coleoptera: Coccinellidae). Proc. Entomol. Soc. Wash. **94**, 596–597 (1992)

S.E. Naranjo, P.C. Ellsworth, Fifty years of the integrated control concept: moving the model and implementation forward in Arizona. Pest Manage. Sci. **65**, 1267–1286 (2009)

A. Nasreen, G. Mostafa, M. Ashfaq, Selectivity of some insecticides to *Chrysoperla carnea* (Stephen) (Neuroptera: Chrysopidae) in laboratory. Pak. J. Biol. Sci. **6**, 536–538 (2003)

A. Nasreen, M. Ashfaq, G. Mustafa, R.R. Khan, Mortality rates of five commercial insecticides on *Chrysoperla carnea* (Stephens) (Chrysopidae: Neuroptera). Pak. J. Agric. Sci. **44**(2), 266–271 (2007)

L.A. Nelson, L.K. Rieske, Microclimatic variation within sleeve cages used in ecological studies. J. Insect Sci. **14**(1), 167 (2014)

S. Noii, K. Talebi, A. Saboori, H. Allahyari, Q. Sabahi, A. Ashouri, Study on the side-effects of three pesticides on the predatory mite, *Phytoseius plumifer* (Canestrini & Fanzago) (Acari: Phytoseiidae) under laboratory conditions. IOBC/WPRS Bull. **35**, 146–151 (2008)

M. Nomikou, A. Janssen, M.W. Sabelis, Phytoseiid predator of whitefly feeds on plant tissue. Exp. Appl. Acarol. **31**, 27–36 (2003)

J.G. Nyaanga, A.W. Kamau, R.S. Pathak, R.K. Tuey, Effect of insecticide application regimes on cereal aphids and their coccinellid predators on wheat. Int. J. Sci. Nat. **5**(4), 658–662 (2014)

J.P. Nyrop, J. Minns, C. Herring, Influence of ground cover on dynamics of *Amblyseius fallacis* Garman (Acarina: Phytoseiidae) in New York apple orchards. Agric. Ecosyst. Environ. **50**, 61–72 (1994)

W.J. Ohnesorg, K.D. Johnson, M.E. Oneal, Impact of reduced-risk insecticides on soybean aphid and associated natural enemies. J. Econ. Entomol. **102**(5), 1816–1826 (2009)

K. Ohno, H. Takemoto, Species composition and seasonal occurrence of *Orius* spp. (Heteroptera: Anthocoridae), predacious natural enemies of *Thrips palmi* (Thysanoptera: Thripidae), in egg-plant fields and surrounding habitats. Appl. Entomol. Zool. **32**(1), 27–35 (1997)

R.W. Olszak, Influence of some pesticides on mortality and fecundity of the aphidophagous coccinellid, *Adalia bipunctata* L. (Col., Coccinellidae). J. Appl. Entomol. **123**(1), 41–45 (1999)

F. Omkar, A. Pervez, Coccinellids in insect pest management, in *Bioagents and Biopesticides in Insect Pest Management*, ed. by R.P. Srivastav (International Book Distributing Co, Lucknow, 2002)

P.A.C. Ooi, B.M. Shepard, Predators and parasitoids of rice insect pests, in *Biology and Management of Rice Insects*, ed. by E.A. Heinrich (Wiley Eastern Limited, New Delhi, 1994), pp. 585–612

O. Ostman, B. Ekbom, J. Bengtsson, Yield increase attributable to aphid predation by ground-living polyphagous natural enemies in spring barley in Sweden. Ecol. Econ. **45**, 149–158 (2003)

W.P.J. Overmeer, Laboratory method for testing side-effects of pesticides on the predacious mites, *Typhlodromus pyri* and *Amblyseius potentillae* (Acari: Phytoseiidae). IOBC. WPRS. Bull. **21**, 65–69 (1988)

D.P. Papachristos, P.G. Milonas, Adverse effects of soil applied insecticides on the predatory coccinellid, *Hippodamia undecimnotata* (Coleoptera: Coccinellidae). Biol. Control **47**, 77–81 (2008)

E. Pasqualini, S. Civolani, Studies on side effects of some insecticides on aphid-feeding Coccinellidae in Emilia-Romagna fruit crops. IOBC/WPRS Bull. **26**(5), 51–55 (2003)

E. Pasqualini, S. Civolani, S. Vergnani, C. Cavazza, M. Ardizzoni, Selectivity of some insecticides on *Anthocoris nemoralis*. Informatore Agrario **55**, 71–74 (1999)

C.S. Patil, S. Lingappa, Persistent toxicity of imidacloprid to *Cheilomenes sexmaculata* (Fab). Insect Environ. **6**, 68–69 (2000)

C.S. Patil, S. Lingappa, Integrated management of tobacco aphid. Ann. Plant Protect. Sci. **9**, 220–224 (2001)

R.K. Peng, K. Christian, K. Gibb, The effect of the green ant, *Oecophylla smaragdina* (Hymenoptera: Formicidae), on insect pests of cashew trees in Australia. Bull. Entomol. Res. **85**, 279–284 (1995)

E.J.G. Pereira, M.C. Picanco, L. Bacci, T.M.C. Della-Lucia, E.M. Silva, F.L. Fernandes, Natural mortality factors of *Leucoptera coffeella* (Lepidoptera: Lyonetiidae) on *Coffea arabica*. Biocontrol Sci. Technol. **17**, 441–455 (2007)

M. Perez-Hedo, A. Urbaneja, Prospects for predatory mirid bugs as biocontrol agents of aphids in sweet peppers. J. Pest. Sci. **88**(1), 65–73 (2015)

A. Pervez, Predaceous coccinellids in India: predator-prey catalogue (Coleoptera: Coccinellidae). Orient. Insects **38**(1), 27–61 (2004)

R.K. Peterson, Comparing ecological risks of pesticides: the utility of a risk quotient ranking approach across refinements of exposure. Pest Manag. Sci. **62**(1), 46–56 (2006)

R. Peveling, S.O. Ely, Side-effects of botanical insecticides derived from Meliaceae on coccinellid predators of the date palm scale. Crop Protect. **25**, 1253–1258 (2006)

W. Pfluger, R. Schuck, Ecotoxicological profile of imidacloprid. PflSchutz. Nachr. Bayer **44**, 145–158 (1991)

M.C. Picanco, R.N.C. Guedes, V.C. Batalha, R.P. Campos, Toxicity of insecticides to *Dione juno juno* (Lepidoptera: Heliconidae) and selectivity to two of its predaceous bugs. Trop. Sci. **36**, 51–53 (1996)

A.D. Pickett, N.A. Patterson, H.T. Stultz, F.T. Lord, The influence of spray programs on the fauna of apple orchards in Nova Scotia: I. An appraisal of the problem and a method of approach. Sci. Agric. **26**, 590–600 (1946)

F.W. Plapp, D.L. Bull, Toxicity and selectivity of some insecticides to *Chrysopa carnea*, a predator of the tobacco budworm. Environ. Entomol. **7**(3), 431–434 (1978)

M. Poletti, A.H.N. Maia, C. Omoto, Toxicity of neonicotinoid insecticides to *Neoseiulus californicus* and *Phytoseiulus macropilis* (Acari: Phytoseiidae) and their impact on functional response to *Tetranychus urticae* (Acari: Tetranychidae). Biol. Control **40**, 30–36 (2007)

V. Preap, M.P. Zalucki, G.C. Jahn, H.J. Nesbite, Effectiveness of brown planthopper predators: population suppression by two species of spider, *Pardosa pseudoannulata* (Araneae: Lycosidae) and *Araneus inustus* (Araneae: Araneidae). J. Asia-Pacific Entomol. **4**(2), 93–97 (2001)

G. Preetha, T. Manoharan, J. Stanley, Diversity of spiders in cotton and bhendi ecosystems, in *International Congress on Global Warming on Biodiversity of Insects: Management and Conservation*, February 9–12. (Bharathiar University, Coimbatore, 2009a), p. 32

G. Preetha, J. Stanley, T. Manoharan, S. Chandrasekaran, S. Kuttalam, Toxicity of imidacloprid and diafenthiuron to *Chrysoperla carnea* (Stephens) (Neuroptera: Chrysopidae) in the laboratory conditions. J. Plant Protect. Res. **9**(3), 290–296 (2009b)

G. Preetha, J. Stanley, S. Suresh, R. Samiyappan, Risk assessment of insecticides used in rice on miridbug, *Cyrtorhinus lividipennis* Reuter, the important predator of brown planthopper, *Nilaparvata lugens* (Stal.). Chemosphere **80**(5), 498–503 (2010)

F. Prezoto, H.H. Santos-Prezoto, V.L.L. Machado, J.C. Zanuncio, Prey captured and used in *Polistes versicolor* (Olivier) (Hymenoptera: Vespidae) nourishment. Neotrop. Entomol. **35**, 707–709 (2006)

C. Provost, D. Coderre, E. Lucas, G. Chouinard, N.J. Bostanian, Impact of intraguild predation and lambda-cyhalothrin on predation efficacy of three acarophagous predators. Pest Manag. Sci. **61**, 532–538 (2005)

S. Rahmani, A.R. Bandani, Q. Sabahi, Population statistics and biological traits of *Hippodamia variegata* (Goeze) (Coleoptera: Coccinellidae) affected by LC_{30} of thiamethoxam and pirimicarb. Arch. Phytopathol. Plant Protect. **46**(15), 1839–1847 (2013)

K. Rajasekhara, S. Chatterji, Biology of *Orius indicus* (Hemiptera: Anthocoridae), a predator of *Taeniothrips nigricornis* (Thysanoptera). Ann. Entomol. Soc. Am. **63**, 364–367 (1970)

P.M.J. Ramakers, M. Dissevelt, K. Peeters, Large scale introductions of phy toseiid predators to control thrips on cucumber. Int. Symp. Crop Protect. Gent. Belg. Meded. Fac. Landbouwwet. Rijksuniv. Gent. **54**, 923–930 (1989)

M.M.U. Rashid, M.K. Khattak, K. Abdullah, M. Amir, M. Tariqand, S. Nawaz, Feeding potential of *Chrysoperla carnea* and *Cryptolaemus montrouzieri* on cotton mealybug, *Phenacoccus solenopsis*. J. Anim. Plant Sci. **22**(3), 639–643 (2012)

G.V.P. Reddy, Plant volatiles mediate orientation and plant preference by the predator *Chrysoperla carnea* Stephens (Neuroptera: Chrysopidae). Biol. Control **25**(1), 49–55 (2002)

R.A. Reddy, L.K. Narayan, Biology and control of the grape mealybug. Indian Grape J. **2**, 30–39 (1986)

W.H. Reissig, E.A. Heinrichs, S.L. Valencia, Effects of insecticides on *Nilaparvata lugens* and its predators: spiders, *Microvelia atrolineata* and *Cyrtorhinus lividipennis*. Environ. Entomol. **11**(1), 193–199 (1982)

T.M. Reyes, B.P. Gabriel, The life history and consumption habits of *Cyrtorhinus lividipennis* Reuter (Hemiptera: Miridae). Philipp. Ent. **3**(2), 79–88 (1975)

D.B. Richman, R.C. Hemenway, W.H. Whitcomb, Cage evaluation of predators of the soybean looper, *Pseudoplusia includens* (Lepidoptera: Noctuidae). Environ. Entomol. **9**(3), 315–317 (1980)

M.R. Richter, Social wasp (Hymenoptera: Vespidae) foraging behavior. Annu. Rev. Entomol. **45**, 121–150 (2000)

S.E. Riechert, The consequences of being territorial: spiders, a case study. Am. Nat. **117**(6), 871–892 (1981)

M.A. Rogers, V.A. Krischik, L.A. Martin, Effect of soil application of imidacloprid on survival of adult green lacewing, *Chrysoperla carnea* (Neuroptera: Chrysopidae), used for biological control in greenhouse. Biol. Control **42**, 172–177 (2007)

S.I. Rondon, D.J. Cantliffe, J.F. Price, The feeding behavior of the bigeyed bug, minute pirate bug and pink spotted lady beetle relative to main strawberry pests. Environ. Entomol. **33**(4), 1014–1019 (2004)

J.A. Rosenheim, L.R. Wilhoit, C.A. Armer, Influence of intraguild predation among generalist insect predators on the suppression of an herbivore population. Oecologia **96**, 439–449 (1993)

C.R. Roubos, C. Rodriguez-Saona, R. Holdcraft, K.S. Mason, R. Isaacs, Relative toxicity and residual activity of insecticides used in blueberry pest management: mortality of natural enemies. J. Econ. Entomol. **107**(1), 277–285 (2014)

S. Roy, A. Mukhopadhyay, S. Das, G. Gurusubramanian, Bioefficacy of coccinellid predators on major tea pests. J. Biopestic. **3**(1), 33–36 (2010)

S. Rumpf, C. Frampton, B. Chapman, Acute toxicity of insecticides to *Micromus tasmaniae* (Neuroptera: Hcmerobiidae) and *Chrysoperla carnea* (Neuroptera: Chrysopidae): LC_{50} and LC_{90} estimates for various test durations. J. Econ. Entomol. **90**(6), 1493–1499 (1997a)

S. Rumpf, F. Hetzel, C. Frampton, Lacewings (Neuroptera: Hemerobiidae and Chrysopidae) and integrated pest management: enzyme activity as biomarker of sublethal insecticide exposure. J. Econ. Entomol. **90**(1), 102–108 (1997b)

C.E. Rutledge, R.J. Oneil, T.B. Fox, D.A. Landis, Soybean aphid predators and their use in integrated pest management. Ann. Entomol. Soc. Am. **97**(2), 240–248 (2004)

K. Sahayaraj, Insect predators in pest control, in *Indian Insect Predators in Biological Control,* ed. by K. Sahayaraj (Daya Books, New Delhi, 2004), pp. 1–23

K. Sahayaraj, D.P. Ambrose, Stage, host preference and functional response of Reduviid predator. *Acanthaspis pedestris* Stal. to four cotton pests. J. Biol. Control. **8**(1), 23–26 (1994)

K. Sahayaraj, S. Kalidas, M. Tomson, Stage preference and functional response of *Rhynocoris longifrons* (Stal) (Hemiptera: Reduviidae) on three hemipteran cotton pests. Braz. Arch. Biol. Technol. **55**(5), 732–740 (2012)

H.A. Sahito, M.A. Talpur, M.A. Soomro, A.H. Mastoi, K.H. Dhiloo, Feeding efficacy of spiders on sucking complex of okra, *Abelmoschus esculentus* L. J. Agric. Sustain. **2**(2), 142–159 (2013)

S. Sahu, R. Shatrughnga, S. Kumar, Pawan, Host preference and feeding potential of spiders predaceous in insect pests of rice. J. Entomol. Res. **20**(2), 145–150 (1996)

J. Salas-Aguilar, L.E. Ehler, Feeding habits of *Orius tristicolor*. Ann. Entomol. Soc. Am. **70**, 60–62 (1977)

M. Saleem, D. Hussain, H. Anwar, M. Saleem, G. Ghouse, M. Abbas, Predation efficacy of *Menochilus sexmaculatus* Fabricus (Coleoptera: Coccinellidae) against *Macrosiphum rosae* under laboratory conditions. J. Entomol. Zool. Stud. **2**(3), 160–163 (2014)

M. Salyani, R.P. Cromwell, Spray drift from ground and aerial applications. Trans. Am. Soc. Agric. Eng. **35**(4), 1113–1120 (1992)

F. Samu, F. Vollrath, Spider orb web as bioassay for pesticide side effects. Entomol. Exp. Appl. **62**(2), 117–124 (1992)

A. Sankari, K. Thiyagesan, Population and predatory potency of spiders in brinjal and snakegourd. J. Biopestic. **3**(1), 28–32 (2010)

S. Sannigrahi, A. Mukhopadhyay, Laboratory evaluation of predatory efficiency of *Geocoris ochropterus* Fieber (Hemiptera: Lygaeidae) on some common tea pests. Sri Lanka J. Tea Sci. **61**(2), 39–44 (1992)

P. Saritha, K.L. Reddy, N. Krishnaiah, A.S. Prasad, V.R. Babu, Fertilization in rice and its impact on mirid bugs predating on brown planthopper nymphs. Indian J. Plant Protect. **36**(2), 208–211 (2008)

T.V. Sathe, Y.A. Bhosale, *Insect Pest Predators* (Daya Publishing House, New Delhi, 2001). 169 p

M. Sattar, M. Hamed, S. Nadeem, Predatory potential of *Chrysoperla carnea* (Stephens) (Neuroptera: Chrysopidae) against cotton mealy bug. Pak. Entomol. **29**(2), 103–106 (2007)

P. Schausberger, Comparative studies on the influence of different food on the reproduction of *Amblyseius aberrans* Oud. and *Amblyseius finlandicus* Oud. (Acarina: Phytoseiidae). J. Appl. Entomol. **113**, 476–486 (1992)

O.J. Schmitz, Predator diversity and trophic interactions. Ecology **88**, 2415–2426 (2007)

O.J. Schmitz, Effects of predator functional diversity on grassland ecosystem function. Ecology **90**, 2339–2345 (2009)

K. Schumacher, B. Freier, Who benefits from low-input pesticide use within the tritrophic system: crop – aphid – predator? IOBC/WPRS Bull. **35**, 10–17 (2008)

D.C. Sclar, D. Gerace, W.S. Cranshaw, Observations of population increases and injury by spider mites (Acari: Tetranychidae) on ornamental plants treated with imidacloprid. J. Econ. Entomol. **91**, 250–255 (1998)

H.C. Sharma, R.L. Adlakha, Toxicity of some insecticides to the adults of *Coccinella septempunctata* L. after predating upon poisoned cabbage aphid, *Brevicoryne brassicae* L. Indian J. Entomol. **48**, 204–211 (1986)

P.K. Sharma, P.C. Joshi, Feeding potential of seven spotted ladybeetle, *Coccinella septumpunctata* (Linnaeus) on mustard aphid, *Lipaphis erysimi* (Kaltenbach) and cotton aphid, *Aphis gossypii* (Glover). Nat. Sci. **8**(12), 198–202 (2010)

B.M. Shepard, A.T. Barrion, J.A. Litsinger, *Friends of the Rice Farmer. Helpful Insects, Spiders and Pathogens* (International Rice Research Institute, Manila, 1987). 137 p

G. Shrestha, A. Enkegaard, The green lacewing, *Chrysoperla carnea*: preference between lettuce aphids, *Nasonovia ribisnigri* and western flower thrips, *Frankliniella occidentalis*. J. Insect Sci. **13**(1), 94 (2013)

L. Sigsgaard, Early season natural control of the brown planthopper, *Nilaparvata lugens*: the contribution and interaction of two spider species and a predatory bug. Bull. Entomol. Res. **97**(5), 533–544 (2007)

L. Sigsgaard, S. Villareal, Predation rates of *Atypena formosana* (Araneae: Linyphiidae) on brown planthopper and green leafhopper. Int. Rice Res. Notes **24**(3), 18 (1999)

L. Sigsgaard, P. Esbjerg, H. Philipsen, Experimental releases of *Anthocoris nemoralis* F. and *Anthocoris nemorum* (L.) (Heteroptera: Anthocoridae) against the pear psyllid, *Cacopsylla pyri* L. (Homoptera: Psyllidae) in pear. Biol. Control **39**, 87–95 (2006)

K. Singh, Preying propensity of larvae/grubs of syrphid and coccinellid predators on mustard aphids, *Lipaphis erysimi*. Int. J. Agric. Food Sci. Technol. **4**(7), 687–694 (2013)

K. Singh, N.N. Singh, Preying capacity of different established predators of the aphid, *Lipaphis erysimi* (Kalt.) Infesting rapeseed-mustard crop in laboratory conditions. Plant Protect. Sci. **49**(2), 84–88 (2013)

S.P. Singh, Propagation of a coccinellid beetle for the biological control of citrus and coffee mealybugs. Scientific Conference CPA. Bangalore, Indian Institute of Horticulture Research India (1978)

S.F. Smith, V.A. Krischik, Effects of systemic Imidacloprid on *Coleomegilla maculata* (Coleoptera: Coccinellidae). Environ. Entomol. **28**, 1189–1195 (1999)

I.N. Smyrnioudis, R. Harrington, S.J. Clark, N. Katis, The effect of natural enemies on the spread of barley yellow dwarf virus (BYDV) by *Rhopalosiphum padi* (Hemiptera: Aphididae). Bull. Entomol. Res. **91**, 301–303 (2001)

W.E. Snyder, D.H. Wise, Contrasting trophic cascades generated by a community of generalist predators. Ecology **82**, 1571–1583 (2001)

W.E. Snyder, G.B. Snyder, D.L. Finke, C.S. Straub, Predatory biodiversity strengthens herbivore suppression. Ecol. Lett. **9**, 789–796 (2006)

B.K. Solangi, M.K. Lohar, Effect of some insecticides on the population of insect pests and predators on okra. Asian J. Plant Sci. **6**(6), 920–926 (2007)

R. Solanki, D. Kumar, Effect of pesticides on spider population in cotton agro-system of Vadodara, Gujarat. IIS Univ. J. Sci. Tech. **3**(1), 48–52 (2014)

J.P. Spradbery, *Wasps: An Account of the Biology and Natural History of Solitary and Social Wasps* (University of Washington Press, Seattle, 1973). 408 p

J. Stanley, S. Chandrasekaran, G. Preetha, S. Kuttalam, R.S. Jasmine, Selective toxicity of diafenthiuron to non-target organisms: honeybees, coccinellids, chelonus, earthworm, silkworm and fish. J. Plant Protect. Res. **56**(1), 1–5 (2016)

J.D. Stark, J.E. Banks, Selective pesticides: are they less hazardous to the environment? Biosciences **51**(11), 980–982 (2001)
J.D. Stark, J.E. Banks, Population-level effects of pesticides and other toxicants on arthropods. Annu. Rev. Entomol. **48**, 505–519 (2003)
J.D. Stark, P.C. Jepson, D.F. Mayer, Limitations to use of topical toxicity data for predictions of pesticide side effects in the field. J. Econ. Entomol. **88**(5), 1081–1088 (1995)
J.D. Stark, J.E. Banks, R.I. Vargas, How risky is risk assessment? The role that life history strategies play in susceptibility of species to stress. Proc. Natl. Acad. Sci. U. S. A. **101**, 732–736 (2004a)
J.D. Stark, J.E. Banks, S. Acheampong, Estimating susceptibility of biological control agents to pesticides: influence of life history strategies and population structure. Biol. Control **29**(3), 392–398 (2004b)
J.D. Stark, R. Vargas, J.E. Banks, Incorporating ecologically relevant measures of pesticide effect for estimating the compatibility of pesticides and biocontrol agents. J. Econ. Entomol. **100**(4), 1027–1032 (2007)
S. Steinberg, M. Cohen, Biological control of the two-spotted spider mite (Tetranychus urticae) in apple orchards by inundative releases of the predatory mite *Phytoseiulus persimilis* – a feasibility study. Phytoparasitica **20**, 37–44 (1992)
G. Sterk, S. Hassan, M. Baillod et al., Results of the seventh joint pesticide testing programme carded out by the IOBC/WPRS working group, pesticides and beneficial organisms. Biocontrol **44**, 99–117 (1999)
G. Sterk, F. Heuts, N. Merck, J. Bock, Sensitivity of non-target arthropods and beneficial fungal species to chemical and biological plant protection products: results of laboratory and semi-field trials. *Proceedings of the 1st International Symposium Biological Control of Arthropods*, Honolulu, Hawaii, 14–18th January, pp. 308–313 (2002). Assessed through www.bugwood.org/arthropod/day4/sterk.pdf
V.M. Stern, R.F. Smith, R. Van-den-Bosch, K.S. Hagen, The integration of chemical and biological control of the spotted alfalfa aphid. Hilgardia **29**, 81–101 (1959)
C.S. Straub, W.E. Snyder, Species identity dominates the relationship between predator biodiversity and herbivore suppression. Ecology **87**, 277–282 (2006)
H.P. Streibert, J. Drabek, A. Rindlisbacher, CGA 160630-a new type of acaricide/insecticide for the control of sucking pests complex in cotton and other crops. *Brighton Crop Protection Conference – Pests and Diseases,* Brighton, UK, 1: 25–33 (1988)
E. Studebaker, T.J. Kring, Lethal and sublethal effects of early season insecticides on insidious flower bug (*Orius insidiosus*); AAES research series 497 (2001). Retrieved from http://arkansasagnews.uark.edu/497–45.pdf
G.E. Studebaker, T.J. Kring, Effects of insecticides on *Orius insidiosus* (Hemiptera: Anthocoridae) measured by field, greenhouse and petri dish bioassays. Florida Entomol. **86**(2), 178–185 (2003)
D.M. Suckling, R.B. Chapman, D.R. Penman, Toxicity of insecticides to *Epiphyas postvittana* (Walker) and *Planotortrix excessana* (Walker) (Lepidoptera: Tortricidae). New Zeland. J. Exp. Agric. **14**, 89–95 (1986)
M. Suganthy, Bioefficacy and residues of Confidence® (Imidacloprid 17.8% SL) on cotton, vegetables and mango. Ph.D. Thesis. Tamil Nadu Agricultural University, Coimbatore, India, 197 p (2003)
A.M. Sundaram, N. Dhandapani, N. Sathiah, J. Stanley, Life table studies of *Spodoptera litura* Fabricius on cauliflower leaves treated with gibberellic acid and *Pesudomonas fluorescens* (Migula). J. Plant Protect. Environ. **3**(1), 15–20 (2006)
W.O.C. Symondson, The effects of crop development upon slug distribution and control by *Abax parallelepipedus* (Coleoptera: Carabidae). Ann. Appl. Biol. **123**, 449–457 (1993)
W.O.C. Symondson, K.D. Sunderland, M.H. Greenstone, Can generalist predators be effective biocontrol agents? Annu. Rev. Entomol. **47**, 561–594 (2002)
Y. Takahanshi, K. Kiritani, The selective toxicity of insecticides against insect pests of rice and their natural enemies. Appl. Entomol. Zool. **8**, 220–226 (1973)
K. Tanaka, Movement of the spiders in arable land. Plant Protect. **43**(1), 34–39 (1989)

K. Tanaka, S. Endo, H. Kazano, Toxicity of insecticides to predators of rice planthoppers: spiders, the mirid bug and the dryinid wasp. Appl. Entomol. Zool. **35**(1), 177–187 (2000)
L.K. Tanigoshi, J.A. McMurtry, The dynamics of predation of Stethorus picipes (Coleoptera: Coccinellidae) and *Typhlodromus floridanus* on the prey *Oligonychus punicae* (Acarina: Phytoseiidae, Tetranychidae). Hilgardia **45**, 237–261 (1977)
B.D. Tank, D.M. Korat, P.K. Borad, Relative toxicity of some insecticides against *Cheilomenes sexmaculata* (Fab) in laboratory. Karnataka J. Agric. Sci. **20**(3), 639–641 (2007)
C. Thies, S. Haenke, C. Scherber et al., The relationship between agricultural intensification and biological control: experimental tests across Europe. Ecol. Appl. **21**, 2187–2196 (2011)
P.G. Tillman, J.E. Mulrooney, Effect of certain insecticides on the natural enemies *Coleomegilla maculata* and *Hippodamia convergens* (Coleoptera: Coccinellidae), *Geocoris punctipes* (Hemiptera: Lygaeidae) and *Bracon mellitor*, *Cardiochiles nigriceps* and *Cotesia marginiventris* (Hymenoptera: Braconidae) in cotton. J. Econ. Entomol. **93**(6), 1638–1643 (2000)
S. Toda, T. Kashio, Toxic effect of pesticides on the larvae of *Chrysoperla carnea*. Proc. Assoc. Plant Protect. Kyushu **43**, 101–105 (1997)
J.B. Torres, J.R. Ruberson, Toxicity of thiamethoxam and imidacloprid to *Podisus nigrispinus* (Dallas) (Heteroptera: Pentatomidae) nymphs associated to aphid and whitefly control in cotton. Neotrop. Entomol. **33**(1), 99–106 (2004)
J.B. Torres, C.S.A. Silva-Torres, M.R. Oliveira, J. Ferreira, Compatibility of insecticides and acaricides to the predatory stinkbug, *Podisus nigrispinus* (Dallas) (Heteroptera: Pentatomidae) on cotton. Neotrop. Entomol. **31**, 311–317 (2002)
J.B. Torres, C.S.A. Silva-Torres, R. Barros, Relative effects of the insecticide thiamethoxam on the predator, *Podisus nigrispinus* and the tobacco whitefly, *Bemisia tabaci* in nectaried and nectariless cotton. Pest Manag. Sci. **59**, 315–323 (2003)
P.J. Trichilo, T.F. Leigh, Predation on spider mite eggs by the western flower thrips, *Frankliniella occidentalis* (Thysanoptera: Thripidae), an opportunist in a cotton agroecosystem. Environ. Entomol. **15**, 821–825 (1986)
V. Trivedi, Diversity of spiders in ground nut crop fields in village area of Sourashtra region. J. Bombay Nat. Hist. Soc. **106**(2), 184–189 (2009)
A.L. Turnbull, Ecology of the true spiders (Araneomorphae). Annu. Rev. Entomol. **18**, 305–348 (1973)
M. Turquet, J.J. Pommier, M. Piron, E. Lascaux, G. Lorin, Biological control of aphids with *Chrysoperla carnea* on strawberry. Acta Horticult. **842**, 641–644 (2009)
G. Unal, P.C. Jepson, The toxicity of aphicide residues to beneficial invertebrates in cereal crops. Ann. Appl. Biol. **118**(3), 493–502 (1991)
UPASI, Seventy nine annual report, eds. by N. Muraleedharan, R. Rajkumar (United Planters' Association of Southern India, 2005), Tea Research Foundation, pp. 28–85
N. Uygun, M. Ulusoy, Y. Karaca, U. Kersting, Approaches to biological control of *Dialeurodes citri* (Asmead) in Turkey. IOBC/WPRS Bull. **20**, 52–56 (1997)
M. Vand-de-Veire, G. Sterk, M. Van-der-Staaij, P.M.J. Ramkers, L. Tirry, Sequential testing scheme for the assessment of the side-effects of plant protection products on the predatory bug *Orius laevigatus*. BioControl **47**, 101–113 (2002)
M. Van-de-Veire, Laboratory methods for testing side-effects of pesticides on the predatory bug, *Orius niger* Wolff. IOBC/WPRS Bull. **15**(3), 89–95 (1992)
M. Van-de-Veire, D. Degheele, Side effects of diafenthiuron on the greenhouse whitefly parasitoid *Encarsia formosa* and the predatory bug *Orius niger* and its possible use in IPM in greenhouse vegetables. Meded. Fac. Landbouwwet. Univ. Gent. **53**, 509–514 (1993)
M. Van-de-Veire, L. Tirry, Side effects of pesticides on four species of beneficials used in IPM in glasshouse vegetable crops: 'Worst case' laboratory tests. IOBC/WPRS Bull. **26**(5), 41–50 (2003)
M. Van-de-Veire, G. Smagghe, D. Degheele, Laboratory test method to evaluate the effect of 31 pesticides on the predatory bug, *Orius laevigatus* (Het.: Anthocorid ae). Entomophaga **41**, 235–243 (1996)
M. Van-de-Veire, W. Cornelis, L. Tirry, Development of a laboratory test method to determine the duration of pesticide effects on predatory mites. IOBC/WPRS Bull. **24**, 61–66 (2001)

M. Van-de-Veire, M. Klein, L. Tirry, Residual activity of abamectin and spinosad against the predatory bug, *Orius laevigatus*. Phytoparasitica **30**(5), 525–528 (2002)

M. Van-de-Vrie, J.F. Price, Manual for biological control of twospotted spider mites on strawberry in Florida. Univ. Flor. Agric. Exp. Stn. Rep. DOV 1994-1, 10 p (1994)

E. Van-Handel, Determination of fructose and fructose-yielding carbohydrates with cold anthrone. Anal. Biochem. **19**(1), 193–194 (1967)

K. Van-Laecke, D. Degheele, Effect of insecticide synergist combinations on the survival of *Spodoptera exigua*. Pestic. Sci. **37**, 283–288 (1993)

P. Van-Mele, N.T.T. Cuc, Farmers perceptions and practices in use of *Dolichoderus thoracicus* (Smith) (Hymenoptera: Formicidae) for biological control of pests of *Sapodilla*. Biol. Control **20**, 23–29 (2001)

D. Vasanthakumar, A. Roobakkumar, V.J. Rahman, P. Kumar, C. Sundaravadivelan, A. Babu, Enhancement of the reproductive potential of *Mallada boninensis okamoto* (Neuroptera: Chrysopidae), a predator of red spider mite infesting tea: an evaluation of artificial diets. Arch. Biol. Sci. **64**(1), 281–285 (2012)

G.P. Vickerman, K.D. Sunderland, Some effects of dimethoate on arthropods in winter wheat. J. Appl. Ecol. **14**, 767–777 (1977)

H. Vogt, Sensitivity of non-target arthropod species to plant protection products according to laboratory results of the IOBC, Pesticides and beneficial organisms. IOBC/WPRS Bull. **23**, 3–15 (2000)

M.K. Walker, M.A.W. Stufkens, A.R. Wallace, Indirect non-target effects of insecticides on Tasmanian brown lacewing (*Micromus tasmaniae*) from feeding on lettuce aphid (*Nasonovia ribisnigri*). Biol. Control **43**, 31–40 (2007)

W.K. Walthall, J.D. Stark, Comparison of two population-level ecotoxicological endpoints: the intrinsic (rm) and instantaneous (ri) rates of increase. Environ. Toxicol. Chem. **16**(5), 1068–1073 (1997)

A.G. Waseem, C. Mike, R. Shazia, Studies on the feeding preference of brown lacewing (*Sympherobius fallax navas*) larvae for different stages of long-tailed mealy bug (*Pseudococcus longispinus*). Pak. Entomol. **3**, 1–5 (2009)

C.H. Wearing, A.C. Harris, Evaluation of the predatory wasp, *Ancistrocerus gazella*, for biological control of leafrollers in Otago fruit crops: 1. Prey composition, nest structure and wasp productivity from artificial nests. Biocontrol Sci. Technol. **9**(3), 315–325 (1999)

X. Wei, X. Xu, C.J. DeLoach, Biological control of white grubs (Coleoptera: Scarabaeidae) by larvae of *Promachus yesonicus* (Diptera: Asilidae) in China. Biol. Control **5**, 290–296 (1995)

R.A. Weinzierl, Integrating pesticides with biotic and biological control for arthropod pest management, in *Integrated Pest Management: Concepts, Tactics, Strategies and Case Studies*, ed. by E.B. Radcliffe, W.D. Hutchison, R.E. Cancelado (Cambridge University Press, Cambridge, 2008), pp. 179–191

U. Wennergren, J.D. Stark, Modeling long-term effects of pesticides on populations: beyond just counting dead animals. Ecol. Appl. **10**, 295–302 (2000)

J.A. Wiles, P.C. Jepson, The dietary effects of deltamethrin upon *Nebria brevicollis* (F.) (Coleoptera: Carabidae). Pestic. Sci. **38**, 329–334 (1993)

J.A. Wiles, P.C. Jepson, Sub-lethal effects of deltamethrin residues on the within crop behaviour and distribution of *Coccinella septempunctata*. Entomol. Exp. Appl. **72**, 33–45 (1994)

J.D. Wilkinson, K.D. Biever, C.M. Ignoffo, Synthetic pyrethroid and organophosphate insecticides against the parasitoid, *Apanteles marginiventris* and the predators, *Geocoris punctipes*, *Hippodamia convergens* and *Podisus maculiventris*. J. Econ. Entomol. **72**(4), 473–475 (1979)

L.H. Winder, Predation of the cereal aphid, *Sitobion avenae* by polyphagous predators on the ground. Ecol. Entomol. **15**, 105–110 (1990)

T. Yasuda, Chemical cues from *Spodoptera litura* larvae elicit prey-locating behaviour by the predatory stink bug, *Eocanthecona furcellata*. Entomol. Exp. Appl. **82**(3), 349–354 (1997)

Y.N. Youn, M.J. Seo, J.G. Shin, C. Jang, Y.M. Yu, Toxicity of greenhouse pesticides to multicolored Asian lady beetles, *Harmonia axyridis* (Coleoptera: Coccinellidae). Biol. Control **28**(2), 164–170 (2003)

S.J. Yu, Biochemical defense capacity in the spined soldier bug (*Podisus maculiventris*) and its lepidopterous prey. Pestic. Biochem. Physiol. **28**(2), 216–223 (1987)

C. Yu, M. Fu, R. Lin, Y. Zhang, L. Yongquan, H. Jiang, T.C.M. Brock, Toxic effects of hexaflumuron on the development of *Coccinella septempunctata*. Environ. Sci. Pollut. Res. **21**, 1418–1424 (2014)

J.C. Zanuncio, R.N.C. Guedes, J.F. Garcia, L.A. Rodrigues, Impact of two formulations of deltamethrin in aerial application against eucalyptus caterpillars and their predaceous bugs. Mededelingen Faculteit Landbouwwetenschappen Universiteit Gent **58**, 477–481 (1993)

T.V. Zanuncio, J.E. Serrao, J.C. Zanuncio, R.N.C. Guedes, Permethrin-induced hormesis on the predator, *Supputius cincticeps* (Stal, 1860) (Heteroptera: Pentatomidae). Crop Protect. **22**, 941–947 (2003)

Z.Q. Zhang, J.P. Sanderson, Relative toxicity of abamectin to the predatory mite, *Phytoseiulus persimilis* (Acari: Phytoseiidae) and twospotted spider mite (Acari: Tetranychidae). J. Econ. Entomol. **83**, 1783–1790 (1990)

K. Zia, F. Hafeez, R.R. Khan, M. Arshad, U. Naeem-Ullah, Effectiveness of *Chrysoperla carnea* (Stephens) (Neuroptera: Chrysopidae) on the population of *Bemisia tabaci* (Homoptera: Aleyrodidae) in different cotton genotypes. J. Agric. Soc. Sci. **4**, 112–116 (2008)

R. Zia-ul-Hussnain, A. Naheed, S. Rizwana, Biocontrol of insect pests of sugarcane (Saccharum sp.). Pak. Sugar J. **21**(5), 14–23 (2007)

Index

O

Other predatory insects

P

Predatory beetles

Predatory bugs

Chapter 2
Pesticide Toxicity to Parasitoids: Exposure, Toxicity and Risk Assessment Methodologies

Abstract Parasitoids occur naturally as well as reared and released against target pests, which is included as an important component in integrated pest management (IPM) programme. Mostly parasitoids are host specific and they are exposed to pesticides directly while spraying or through contaminated host insects or by consuming nectar of the crop plants treated with pesticides. Hence, it is essential to assess the impact of pesticides on natural enemies like parasitoids. Mostly acute toxicity bioassays are conducted using eggs, immature stages (cocoons/ mummies) and adults to determine the median lethal concentrations to assess the effect of pesticides. Ingestion toxicity assays are also being carried out since the adult parasitoids feeds on the nectar of the flowering plants. Apart from these bioassays, sublethal studies are also important to assess the chronic effects of pesticides on the fecundity, adult emergence, host foraging ability, longevity, generation time, sex ratio and reproduction of parasitoids. In field conditions also, pesticide toxicity on parasitoids are being assessed by examining their parasitization efficiency. Risk assessment of insecticides for parasitoids were studied mostly using LC_{50} values of parasitoids and the risk quotient was derived based on which the pesticide was categorized from harmless to dangerous. Thus, the insecticide effective against target pests and selective to parasitoids can be identified and included in the IPM packages.

1 Importance of Insect Parasitoids

Insect parasitoids play a vital role in suppression of insect pests in agro ecosystem. Parasitoids are nothing but insects that feed on another insect during different stages of their life cycle resulting in death of the host organism and after completing its life cycle it turns to be a free living organism, independent on host. In general, it is believed by Entomologists that around 10% insect species alone are known to science and also 800,000 species of parasitoids are in existence (http://www.entomology.wisc.edu/mbcn/fea506.html). Parasitoids occur in different insect orders; however majority of them belongs to the orders, Diptera and Hymenoptera. Hymenoptera (320,000 species) is one of the four hyper diverse insect groups, the other three are Coleoptera (350,000 species), Lepidoptera (150,000 species) and Diptera (120,000 species). The success of Hymenoptera is due to their body structure with small hindwings that are linked to the forewings by the hamuli, the

J. Stanley, G. Preetha, *Pesticide Toxicity to Non-target Organisms*,
DOI 10.1007/978-94-017-7752-0_2

haplodiploid sex determination and the parasitoid mode of life (Bonet 2009). Parasitic wasps are found in the following 12 super families of Hymenoptera: Orussoidea (70 species), Stephanoidea (100 species), Trigonalyoidea (100 species), Megalyroidea (100 species), Evanioidea (1200 species), Ceraphronoidea (2000 species), Proctotrupoidea (6000 species), Platygastroidea (10,000 species), Cynipoidea (4000 species), Chalcidoidea (100,000 species), Ichneumonoidea (100,000 species), Chrysidoidea (6348 species) and Vespoidea (11,124 species) (Aldrey and Fontal-Cazalla 1999; Pennacchio and Strand 2006).

Lenteren (1986) stated that more than 80 % of the effective entomophagous insects are parasitoids, and among the remaining percentage, 17 and 1 % were contributed by predators and pathogens, respectively. He also estimated that 5000 species are used in biological control and partial, substantial and complete control was achieved by 100, 100 and 70 species, respectively. The most effective egg parasitoids in pest suppression belong to the families Scelionidae, Mymaridae and Trichogrammatidae (Greathead 1986).

1.1 Insect Parasitoids

Insect parasitoids are about 64,000, 15,000 and 3400 described species belonging to the orders Hymenoptera, Diptera and Coleoptera, respectively. The parasitoids are also seen in the insect orders like Lepidoptera, Neuroptera and Trichoptera (Eggleton and Belshaw 1992; Godfray 1994). The insect orders comprising parasitoids are tabulated below.

Order	Family	Parasitoid	Host
Hymenoptera	Trichogrammatidae	*Trichogramma chilonis*	Lepidopterans
	Braconidae	*Bracon hebetor*	Lepidopteran pests of stored products
	Ichneumonidae	*Itoplectis naranyae*	Lepidopterans
	Aphelinidae	*Eretmocerus mundus*	Whitefly
Diptera	Tachinidae	*Sturmiopsis inferens*	Sugar cane shoot borer
Coleoptera	Carabidae	Ground beetle, *Lebia grandis*	Colorado potato beetle pupae
	Staphylinidae	Rove beetle, *Aleochara bilineata*	Cabbage maggot
	Rhipiphoridae	*Rhipidius* sp.	Cockroaches
	Rhipiceridae	*Sandalus* sp.	Cicadas
	Meloidae	Blister beetle	Egg cases of grasshopper
Strepsiptera	Halictophagidae	Twisted wing insects	Bees, wasps, leafhoppers and planthoppers
Neuroptera	Mantispidae	Mantid flies	Egg sacs of spiders

Major Characteristics of Parasitoids

- Parasitoids are specialized in their choice of host
- They usually destroy their hosts during development
- Usually the parasitoids are smaller than the host
- The female alone searches for the host
- Parasitoid adults are free living while only the immature stages are parasitic
- Eggs or larvae are usually laid in, on, or near host
- Immatures itself almost kill host usually
- Parasitoids require only one host for their development

1.2 Mode of Development of Parasitoids

(i) **With respect to the host**

(a) *Endoparasitoid*: Parasitoids that lives inside their host body and ultimately kills it. e.g. *Chelonus blackburni*.
(b) *Ectoparasitoid*: Parasitoids that lives externally on the host body and kills its host. e.g. *Bracon brevicornis*.

(ii) **Number of immatures per individual host**

(a) *Solitary parasitoid*: A single individual develops in a host. e.g. *Chelonus blackburni* on potato tuber moth.
(b) *Gregarious parasitoid*: Several progeny parasitises a single host. e.g. *Copidosoma koehleri* on potato tuber moth.

(iii) **With respect to host stage**

1. *Egg parasitoid*: The parasitoids that attack the egg stage of the host. e.g. *Trichogramma chilonis*
2. *Egg larval parasitoid*: The parasitoid that lays the eggs in the eggs of host insects and the parasitoid larvae develops in the host larva. e.g. *Chelonus blackburni*
3. *Egg pupal parasitoid*: If the parasitoids lay its egg in the host eggs and emerge our as adult from a host pupa, it is called as egg pupal parasitoid. e.g. *Fopius arisanus* of tephritid fruit flies.
4. *Larval parasitoid*: The parasitoids that attack the larval stage of the host. e.g. *Apanteles* spp.
5. *Larval pupal parasitoid*: Parasitoids lay eggs on the host larvae and come out as adult from host pupae. e.g. *Pleurotropis epilachnae*
6. *Pupal parasitoid*: The parasitoids that attack the pupal stage of the host. e.g. *Brachymeria nephantidis*
7. *Adult parasitoids*: Parasitoids of adult hosts. e.g. *Blaesoziphae kellyi*
8. *Nymphal parasitoids*: The parasitoids that attack the nymphal stage of the host e.g. *Epipyrops fuliginosa* on nymphs of *Idioscopus clypealis*.

9. *Nymphal adult parasitoids*: The parasitoids that lay its eggs on host nymphs but the parasitoid continues to develop and emerge from the dead adult host. e.g. *Epiricania melanoleuca* parasitic on *Pyrilla perpusilla*.

(iv) **With respect to effect of parasitization on host**

1. *Idiobionts*: Those parasitoids in which the development of the host is prevented upon parasitization (e.g. egg parasitoids)
2. *Koinobionts*: Those parasitoids in which the development of the host is continued after parasitization (e.g. larval-pupal parasitoids).

(v) **With respect to food web relationships**

1. *Primary parasitoid*: Parasitism on a normal host.
2. *Secondary parasitoid*: The parasitoids develop on the primary parasitoid. e.g. *Opisina arenosella* (pest) *Bracon brevicornis* (primary parasitoid) and *Pleurotropis* sp. (secondary parasitoid).
3. *Tertiary parasitoid*: The parasitoid develops on a secondary parasitoid.

(vi) **Number of hosts attacked**

(a) *Monophagous*: Those parasitoids specific to one particular host. e.g. The ichneumonid parasitoid, *Mesolicus tenthredinis* is specific for a saw fly.
(b) *Oligophagous or Stenophagous*: Parasitoids that develops only on closely related hosts e.g. *Exerterus amictoriius* parasitizes saw fly of Diprion and Neodiprion genera.
(c) *Polyphagous*: Parasitoids parasitizing multiple hosts, e.g. Tachnid fly, *Compsilura concinnata* have about 20 hosts.

(vii) **Number of host individuals essential for attack**

(a) *Heteroxenous*: Require more than one host e.g. The tachinid, *Ceromasla auricaudata*.
(b) *Monoxenous*: Those parasitoids that need only one host for its development, e.g. *Drino bohemica*, a tachinid parasitoid of European spruce sawfly.

(viii) **Competition among the immature stages**

1. Intraspecific competition: Superparasitism
2. Interspecific competition: Multiple parasitism

1.3 Pest Suppression by Parasitoids

The abundance of natural enemies is an important criterion for short term pest control than species richness (Duelli and Obrist 2003). But for a long term pest control, high diversity of natural enemy species is more important than abundance (Bengtsson et al. 2003; Tilman 1996). Among the wide assemblage of parasitoids, selection of effective parasitoid from its place of origin for introduction against a target pest is one of the important strategies for successive pest suppression (Waage and Mills 1992).

The major step in classical biological control program is the selection of right species of parasitoid for introduction. Most of those parasitoids are from the centre of origin of the target pest and coevolved along with the pest. In a biocontrol program, when the target pest is of exotic origin we prefer importation of natural enemies. Such introductions may not establish in the new places or if so they may not become pests. But still at certain cases due to the lack of natural enemies for suppressing the pest population, the introduced insect may gain the pest status. When an introduced insect become a pest, then natural enemies are also have to be imported (Caltagirone 1981). The control of alfalfa weevil, *Hypera postica* was successfully managed by importing its natural enemies (Bryan et al. 1993). The first deliberate movement of parasitoids from one location to another was conducted by C.V. Riley, who distributed parasitoids of the weevil *Conotrachelus nenuphar* around the state of Missouri in 1870 (Doutt 1964). The first parasitoid successfully moved and established from one continent to another, however, was *Cotesia* (=*Apanteles*) *glomeratus*, which was shipped from England to the United States for suppression of *Pieris rapae* by the U.S. Department of Agriculture in 1883 (Riley 1885, 1893).

Augmentation refers to mass culturing and release of natural enemies to increase its population. Inoculative release is done for *Encarsia formosa* for the management of whitefly, *Trialeurodes vaporariorum* (Hussey and Scopes 1985; Parrella 1990). Release of Trichogramma is an excellent example of an augmentative method that has been successful in many agricultural systems. Trichogramma is mass produced and field released innundatively for decades in biological control programme (Li 1994). Conservation of natural enemies involves actions that preserve and increase natural enemies by environmental manipulation.

1.4 Feeding Habits of Insect Parasitoids

Host feeding by parasitoid wasps has been regarded as a positive attribute in biological control point of view. Here, the host insects are get killed due to feeding along with parasitism (Ueno 1998). The parasitoids feed on the fluids exudating from the ovipositional wound of the host insect. In some cases, the hosts are killed by host feeding rather than parasitisation (Heimpel and Collier 1996). Host feeding resulted in the destruction of the host, or deteriorates its quality for egg laying (Jervis and Kidd 1986). The indirect benefits may be more important than the immediate cost to offspring. Host feeding is reported to increase the fecundity of the parasitoids (Heimpel and Collier 1996) may be supplying essential nutrients (Flanders 1953).

1.5 Major Insect Parasitoids

List of pests controlled by *Trichogramma* spp.

Parasitoids	Pests	References
Trichogramma dendrolimi	Pine moth, citrus swallow tail	Hirose (1986)
	Spodptera litura	Hamada (1992)
T. exiguum	*Helicoverpa zea*	Suh et al. (2000)
T. oleae, *T. cacoeciae*, *T. bourarachae*	Olive moth, *Prays oleae*	Blibech et al. (2015)
T. chilonis	Tomato fruitworm, *Helicoverpa armigera*	Usman et al. (2012), Ballal and Singh (2003), and Puneeth and Vijayan (2014)
	Sugarcane early shoot borer, *Chilo infuscatellus*; top shoot borer, *Scirpophaga excerptalis*	Sankar et al. (2014)
	Sugarcane borers, *S. excerptalis*, *C. infuscatellus*, *Emmalocera depresella*, *Acigona steniellus*	Hussnain et al. (2007)
	Sugarcane stem borer, *C. Infuscatellus*	Muhammad et al. (2012) and Ahmad et al. (2012)
	Cotton bollworms, *H. armigera* and *Earias insulana*	Masood et al. (2011)
T. pretiosum	Cotton leafworm, *Alabama argillacea*	Almeida (2001)
	H. armigera	Ballal and Singh (2003)
	Plutella xylostella	Vasquez et al. (1997)
T. minutum	Codling moth, *Cydia pomonella*; Oriental fruit moth, *Grapholita molesta*	Pinto et al. (2002)
	Chilotraea auricilia and *Sesamia inferens*	Saxena (1977)
T. evanescens	*P. xylostella*	Tabone et al. (2010)
	Rice stem borer	Sherif et al. (2005)
	H. armigera	Timus and Croitoru (2006) and El-Wakeil (2007)
	European corn borer, *Ostrinia nubilalis*	Tancik and Cagan (2004)
	H. zea	Atwa and Atwa (2014)
T. japonicum	*Scirpophaga incertulas*	Kim et al. (1986) and Varma et al. (2013)
T. galloi	Sugarcane borer, *D. saccharalis*	Consoli et al. (2001)
Trichogramma spp.	Sorghum stem borer, *H. armigera*	Romeis et al. (1999)
	Diamond backmoth, *P. xylostella*	Tabone et al. (2002)
T. bactrae	*P. xylostella*	Vasquez et al. (1997)

Crop pests controlled by other insect parasitoids

Common parasitoids	Insect prey	References
Anaphes iole	Tarnished plant bug, *Lygus lineolaris*	Williams III et al. (2003)
Anagrus nilaparvatae	Brown planthopper, *Nilaparvata lugens*	Wang et al. (2008)
Eretmocerus mundus	Whitefly, *Bemisia tabaci*	Sohrabi et al. (2014)
Eretmocerus paulistus	*Aleurothrixus floccosus*	Tello et al. (2013)
Trissolcus grandis	*Eurygaster integriceps*	Radjabi (1995) and Critchley (1998)
Encarsia formosa	*Trialeurodes vaporariorum*	Lenteren et al. (1996)
Aphidius colemani	Aphids	Prado et al. (2015)
Aphidius ervi	*Acyrthosiphon pisum*	He et al. (2005)
Diaeretiella rapae	*Brevicoryne brassicae*, *Aphis craccivora*, *A. nerii*	Saleh (2014)
Tetrastichus schoenobii	*Chilotrea infuscatellus*	Saxena (1977)
	S. incertulas	Kim et al. (1986) and Varma et al. (2013)
Tetrastichus howardi	Sugarcane early shoot borer, *C. infuscatellus*; top shoot borer, *S. excerptalis*	Sankar et al. (2014)
Telenomus dingus	*Scirpophaga nivella*	Saxena (1977)
	S. incertulas	Varma et al. (2013)
Telenomus rowani	*S. incertulas*	Kim et al. (1986)
Telenomus spp.	*C. infuscatellus*, *Proceras indicus*, *S. nivella*	Saxena (1977)

1.6 Biological Efficiency of Parasitoids in Field Conditions

The parasitoids are described as more valuable than predators since they are more host specific, well adapted and require less food per individual. Parasitoids play a major role in biological control and the food obtained from flowering plants can have a positive impact on survival, searchability and rate of parasitism. The most successful egg parasitoid, *Trichogramma* species was polyphagous attacking several lepidopterans and many others (Thomson and Stinner 1989). Diaz et al. (2012) suggested that *Trichogramma atopovirilia* and *T. pretiosum* might be potential parasitoids for the control of *Spodoptera frugiperda* and *Copitarsia decolora*, with parasitism percentage of 30–60 %, respectively. The studies of Ayvaz et al. (2008) revealed that one release point as adequate to achieve sufficient parasitisation in grape plants whereas multiple release points are needed for corn.

The egg parasitoid of *Nezara viridula*, *Trissolcus basalis* (Hymenoptera: Platygastridae) attained greater impact on the hosts, even on multiparasitoid species

(Peri et al. 2014). *Encarsia formosa* is an effective parasitoid for the control of greenhouse whiteflies and factors contributing for its success have been identified by Hoddle et al. (1998). The *E. formosa* limits the whitefly population growth when the intrinsic rate of increase is greater than the host's intrinsic rate of increase. In corn fields treated by Trichogramma egg parasitoid wasp, the range of egg parasitism of corn stem borer, *Ostrinia nubilalis* was 10–28 % (Movahedi et al. 2014).

Trichogrammatoidea sp. nr. *lutea* and *T.* sp. nr. *mwanzai* showed higher parasitism compared to several other strains in high-mid and mid-low altitudes, respectively (Kalyebi et al. 2005). Egg parasitoids of vegetable ecosystem consist of Trichogrammatids (*Trichogramma*, *Trichogrammatoidea*) parasitizing on Lepidoptera, Scelionids (*Telenomus*, *Trissolcus*) on Lepidoptera and Heteroptera and Mymarids on leafhopper and thrips. *Trichogramma chilonis* is a promising bio control agent for used against a large number of lepidopteran pests (Krishnamoorthy 2012). Parasitism of European corn borer egg masses was greater in release plots than in control plots by the egg parasitoid *Trichogramma ostriniae* (Hoffmann et al. 2006).

The releases of *T. bourarachae*, *T. cordubensis* and *T. euproctidis* have a greater improvement in the management of lepidopterous pests of olive. *Trichogramma bourarachae* was found to be capable in dispersal and foraging (Hegazi et al. 2007). Studies on the efficacy of inundative releases of *T. evanescens* for the management of *Maruca vitrata* revealed that parasitism increased by 53 and 43 % in release plots, during dry and rainy season, respectively (Ulrichs and Mewis 2004). The field release techniques and parasitisation of *T. chilonis* and *T. evanescens* on *Sitotroga cerealella*, *Corcyra cephalonica* and *Leucinodes orbonalis* were also assessed. In micro-plot technique, *T. evanescens* parasitized 75.5 and 38.83 % of host eggs by adult release and paper strip method. In open field condition, *T. chilonis* parasitized 78.6 and 40.2 % of host eggs by adult release and paper strip method (Chowdhury et al. 2016).

2 Exposure Routes of Pesticides to Parasitoids

The successful integration of biological control in pest management is based on the impacts of pesticides on natural enemies (Croft 1990; Johnson and Tabashnik 1999). The toxic effects of pesticides on natural enemies can be through direct (direct contact with poison) or indirect (via host insect) means. The direct impacts are exhibited as acute mortality or long-term sublethal effects (Johnson and Tabashnik 1999).

The threat due to pesticide exposure on natural enemies not only depends on the group of pesticides, application methods or mode of action but also with respect to the development of the parasitoid within the host and its stage. During pesticide application, the adult parasitoids come into contact directly with the pesticide droplets or they receive the toxin from the treated surfaces while searching hosts (Longley and Jepson 1996). Parasitoids can be exposed to these residues while feeding on nectar from flowers and contaminated honeydew secreated by toxicated insects (Longley and Stark 1996; Stapel et al. 1999). The young ones (immature stages) may get exposed to pesticides during spray or indirectly through contaminated hosts on their development (Suss 1983; Hsiech and Allen 1986; Longley 1999).

2.1 Exposure via Direct Exposure to Spray Droplets

Pesticides on direct sprays may reduce survival of the adult parasitoids and or kill the individuals while developing inside the hosts (Croft 1990). Spray treatment with thiacloprid in the field did not show a notable impact on the biocontrol efficiency of parasitoid fauna in the target crops (Schuld and Schmuck 2000). Fenitrothion, spinosad and milbemectin caused 100 % mortality of *Trissolcus nigripedius*, an egg parasitoid of pentatomid bug, *Dolycoris baccarum* within 24 h both by direct contact and by indirect contact to residues. Thiamethoxam was found to be less susceptible insecticide to the parasitoid, *T. nigripedius* via topical application or residual exposure when compared to ingestion toxicity (Lim and Mahmoud 2008).

2.2 Exposure via Uptake of Residues by Contact with Contaminated Surfaces

Chlorpyrifos followed by imidacloprid had the highest toxicity to the wasp, *A. nilaparvatae*, while insect growth regulators (IGR) had the lowest toxicity through acute contact toxicity tests (Wang et al. 2008). Reduction in the life time of parasitoid species viz., *Hyposoter didymator* and *Chelonus inanitus* was observed in the insecticide treated individuals, irrespective of the route of uptake (residual, topical and ingestion bioassays on adults), with the exception of *C. inanitus* adults treated with imidacloprid (Medina et al. 2008). Two bioassay methods viz., contact of parasitoids with fresh and dried residues of the insecticides that were significantly toxic in the bioassay I were conducted on *Opius scabriventris*, parasitoid of *Liriomyza* sp. The results of bioassay I, among eight insecticides tested, cartap hydrochloride and abamectin + mineral oil were harmful (Class 4) and deltamethrin was slightly harmful (class 2). Only abamectin + mineral oil were harmful (Class 4) to *O. scabriventris* in the second bioassay method (Araujo et al. 2015). Although certain armoured-scale parasitoids may be secluded from residues of neonicotinoids (Rosenheim and Hoy 1988b), these parasitoids may be exposed to toxic residues while trying to emerge out of scale (Rill et al. 2008).

2.3 Exposure via Oral Uptake from Contaminated Food Sources

The hosts or prey species feed on their host plants and get the toxic products within them and when the natural enemies feeding on the prey hosts they were exposed to residues (Cardwell and Gu 2003). Natural enemies may get affected if the active ingredients get distributed in the flower parts such as petals and sepals also (Hagen 1986). Natural enemies foraging on plant surfaces may be exposed to concentrations of pesticides present in the guttation drops. They get contaminated even via soil or growing medium (Girolami et al. 2009). Dichlorvos was the most toxic insecticide, which generated 100 % mortality of *Anagrus nilaparvatae* 2 h after treatment through oral toxicity tests. Isoprocarb, imidacloprid and thiamethoxam were the most toxic insecticides and killed all wasps within 4 h (Wang et al. 2008). Stapel et al. (2000) reported that the wasp, *Microplitis croceipes* when fed on extra floral nectar of cotton treated with imidacloprid reduced foraging ability and longevity of the parasitoid. Fenitrothion was highly toxic to *T. nigripedius* when ingested (Lim and Mahmoud 2008).

3 Effects of Pesticides on Parasitoids in Agroecosystem

The use of non-selective insecticides resulted in the decrease of natural enemies which bring serious consequences in the pest population dynamics. The negative effects of pesticides on organisms can be classified as acute and chronic. In the acute intoxication, results will be concluded after an exposure with a single dose of insecticidal treatment. The symptoms appear very fast for the products extremely or highly toxic, some hours after the excessive exposure for a shorter period. It may be mild, moderate or severe, based on the quantum of chemical absorbed (Walker et al. 1978). The sub acute intoxication occurs by moderate or small exposure to products highly or moderately toxic. The chronic intoxication appears after months or years and which may be due to moderate exposure to single or multiple toxic products (Fernandes et al. 2010).

3.1 Acute Toxicity

3.1.1 Median Lethal Values

Acute toxicity tests are based on mortality and observed within shorter time (Walthall and Stark 1997). The preliminary step in the assessment of toxicity of a pesticide is determination of median lethal values (the dose that causes 50 % mortality of the test individuals) which can be expressed as LD_{50} (lethal dose 50) or LC_{50}

Median lethal concentrations of different insecticides to insect parasitoids

Pesticide	Test organism	Bioassay	Median lethal concentration (LC_{50})	References
***Trichogramma* spp.**				
Imidacloprid	*Trichogramma chilonis* (adults)	Contact	0.0027 mg a.i./L	Preetha et al. (2009)
	T. ostriniae (adults)	Contact	502.13 mg a.i./L	Wang et al. (2012)
	T. confusum (adults)	Contact	752.62 mg a.i./L	Wang et al. (2012)
	T. cacoeciae (adults)	Contact	6.25 ppm	Saber (2011)
Thiamethoxam	*T. chilonis* (adults)	Contact	0.0014 mg a.i./L	Preetha et al. (2009)
	T. confusum (adults)	Contact	0.24 mg a.i./L	Wang et al. 2012
	T. japonicum (adults)	Contact	0.40 mg a.i./L	Wang et al. (2012)
Chlorantraniliprole	*T. chilonis* (adults)	Contact	1.95 mg a.i./L	Preetha et al. (2009)
Clothianidin	*T. chilonis* (adults)	Contact	0.0113 mg a.i./L	Preetha et al. (2009)
Pymetrozine	*T. chilonis* (adults)	Contact	0.96 mg a.i./L	Preetha et al. (2009)
Ethofenprox	*T. chilonis* (adults)	Contact	0.0045 mg a.i./L	Preetha et al. (2009)
BPMC	*T. chilonis* (adults)	Contact	0.03 mg a.i./L	Preetha et al. (2009)
Endosulfan	*T. chilonis* (adults)	Contact	1.85 mg a.i./L	Preetha et al. (2009)
Acephate	*T. chilonis* (adults)	Contact	4.47 mg a.i./L	Preetha et al. (2009)
Abamectin	*T. chilonis* (adults)	Contact	1.72 ppm	Madhusudan (2015)
	T. japonicum (adults)	Contact	0.49 mg a.i./L	Wang et al. (2012)
Chlorpyrifos	*T. chilonis* (adults)	Contact	11.34 ppm	Madhusudan (2015)
Cypermethrin	*T. chilonis* (adults)	Contact	1.30 ppm	Madhusudan (2015)
Indoxacarb	*T. chilonis* (adults)	Contact	176.09 ppm	Madhusudan (2015)
Malathion	*T. chilonis* (adults)	Contact	1.05 ppm	Madhusudan (2015)
Quinalphos	*T. chilonis* (adults)	Contact	271.47 ppm	Madhusudan (2015)
Spinosad	*T. chilonis* (adults)	Contact	2.86 ppm	Madhusudan (2015)
Triazophos	*T. chilonis* (adults)	Contact	1.29 ppm	Madhusudan (2015)
Emamectin benzoate	*T. confusum* (adults)	Contact	21.76 mg a.i./L	Wang et al. (2012)
Fenpyroximate	*T. cacoeciae* (adults)	Contact	1949 ppm	Saber (2011)
Nitenpyram	*T. confusum* (adults)	Contact	0.83 mg a.i./L	Wang et al. (2012)
	T. japonicum (adults)	Contact	0.72 mg a.i./L	Wang et al. (2012)
Anagrus nilaparvatae				
Chlorpyrifos	*A. nilaparvatae (adults)*	Contact	0.002 mg a.i./L	Wang et al. (2008)
Imidacloprid	*A. nilaparvatae (adults)*	Contact	0.021 mg a.i./L	
Fipronil	*A. nilaparvatae (adults)*	Contact	0.180 mg a.i./L	
Methamidophos	*A. nilaparvatae (adults)*	Contact	0.191 mg a.i./L	
Thiamethoxam	*A. nilaparvatae (adults)*	Contact	0.520 mg a.i./L	
Isoprocarb	*A. nilaparvatae (adults)*	Contact	1.071 mg a.i./L	
Triazophos	*A. nilaparvatae (adults)*	Contact	1.253 mg a.i./L	
Abamectin	*A. nilaparvatae (adults)*	Contact	8.499 mg a.i./L	
Silafluofen	*A. nilaparvatae (adults)*	Contact	14.22 mg a.i./L	
Dichlorvos	*A. nilaparvatae (adults)*	Contact	15.95 mg a.i./L	

(continued)

Pesticide	Test organism	Bioassay	Median lethal concentration (LC_{50})	References
Cotesia plutellae				
Rynaxypyr	*C. plutellae* (adults)	Contact	0.0004 %	Halappa et al. (2012)
Indoxacarb	*C. plutellae* (adults)	Contact	0.0005 %	
Flubendiamide	*C. plutellae* (adults)	Contact	0.0006 %	
Emamectin benzoate	*C. plutellae* (adults)	Contact	0.0468 %	
Spinosad	*C. plutellae* (adults)	Contact	0.0475 %	
Novaluron	*C. plutellae* (adults)	Contact	0.0621 %	
Neochrysocharis okazakii				
Imidacloprid	*N. okazakii* (adults)	Contact	0.0035 mg a.i./L	Tran and Ueno (2012)
Pymetrozine	*N. okazakii* (adults)	Contact	8.7790 mg a.i./L	
Lufenuron	*N. okazakii* (adults)	Contact	0.0508 mg a.i./L	
Ethofenprox	*N. okazakii* (adults)	Contact	0.0085 mg a.i./L	
Clothianidin	*N. okazakii* (adults)	Contact	0.0231 mg a.i./L	
Habrobracon hebetor				
Chlorpyrifos	*H. hebetor* (larvae)	Contact	3.69 ppm	Mahdavi et al. (2015)
	H. hebetor (adults)	Contact	1.75 ppm	
Spinosad	*H. hebetor* (larvae)	Contact	151.37 ppm	
	H. hebetor (adults)	Contact	117.37 ppm	
	H. hebetor (adults – ♀)	Residual	15.64 mg a.i./L	Dastjerdi et al. (2008)
	H. hebetor (adults – ♂)	Residual	11.73 mg a.i./L	
Profenofos	*H. hebetor* (adults – ♀)	Residual	12.44 mg a.i./L	
	H. hebetor (adults – ♂)	Residual	6.91 mg a.i./L	
Thiodicarb	*H. hebetor* (adults – ♀)	Residual	81.04 mg a.i./L	
	H. hebetor (adults – ♂)	Residual	40.39 mg a.i./L	

(lethal concentration 50). These median lethal values are useful in comparing the toxicities of various chemical products. When the lethal values of a chemical were found to be lower; it indicates greater toxicity.

3.1.2 Mortality in Laboratory Assays at Field Recommended Dose

3.1.2.1 *Trichogramma* spp.

Flubendiamide is found to be safe, indoxacarb and lufenuron are mildly toxic, while spinosad and emamectin are highly toxic to the egg parasitoid, *Trichogramma chilonis* (Sattar et al. 2011). Fenoxycarb was non-toxic or low toxic to *T. evanescens* by not affecting the immature stages and not preventing adult emergence. On the other hand, a significant reduction in adult emergence and parasitism was caused by fenvalerate and thus highly toxic. Thiacloprid was slightly toxic (class 2) to

T. evanescens on direct exposure to chemical residues, while, fenvalerate was moderately toxic and the mortality was found to be more than fenoxycarb and thiacloprid tested as direct sprays on the host eggs or by indirect contact to residues in a glass surface (Abdulhay and Rathi 2014). The parasitisation of *T. chilonis* was found to get affected by spinosad (Saljoqi et al. 2012). The adult emergence and parasitization of *T. chilonis* is found to be 90.67 and 85.32 %, respectively when exposed to the field recommended dose of imidacloprid (25 g a.i./ha) (Preetha et al. 2010).

Imidacloprid, carbosulfan, methamidophos and thiodicarb were toxic to *T. chilonis* at all concentrations by leaf dip bioassay method. The recommended and higher doses of acetamiprid and thiamethoxam were found to be moderately harmful and harmful, respectively whereas, the lower doses were found to be slightly harmful to *T. chilonis*. Buprofezin was found to be harmless at all doses (Nasreen et al. 2004). Imidacloprid was regarded as highly toxic to *T. platneri* (Brunner et al. 2001). The least emergence of *T. chilonis* was observed on exposure to spinosad at all parasitism situations. At 3 HAT (hour after treatment), maximum survival was recorded by chlorantraniliprole (42 %) followed by lufenuron (36 %) and minimum survival was observed in emamectin benzoate (18 %) and then by imidacloprid (22 %) (Hussain et al. 2012).

Emamectin benzoate 1.9 EC affected the adult emergence and survival of *T. chilonis* (Hussain et al. 2010). Abamectin was designated as harmful to *T. pretiosum* (Carvalho et al. 2003) and it was also reported by Alexandre et al. (2006) as harmful, slightly harmful and moderately harmful to *T. pretiosum* adults, larvae and pupae, respectively. Deltamethrin (Decis® 100 mL/ha) and spinosad (Tracer® 20 mL/ha) were found to be moderately harmful and harmless to moderately harmful, respectively to all *Trichogramma* species. All species of Trichogramma showed differences in the adult emergence time and parasitism viability. Deltamethrin and spinosad residues affected parasitism viability 31 and 24 days after the treatment for almost all the species (Blibech et al. 2015).

Esfenvalerate and spinosad at 7.5 and 24 g/ha were categorized as class 4 (harmful) and chlorfluazuron (10 g/ha), methoxyfenozide (19.2 g/ha), lactofen (165 g/ha), fomesafen (250 g/ha), fluazifop (125 g/ha), glyphosate (960 g/ha), azoxistrobin + ciproconazol (60+24 g/ha), azoxystrobin (50 g/ha) and myclobutanil (125 g/ha) were grouped as harmless to all immature stages of *T. pretiosum* (Bueno et al. 2008). Imidacloprid and fenpyroximate at recommended doses caused 100 and 32 % mortality of *T. cacoeciae* adults (Saber 2011). Endosulfan and ethofenprox were reported to be extremely toxic (class 4) to *T. pretiosum* and *T. exiguum*. Triflumuron was identified as selective insecticide to the parasitoids in the eggs of *Ephestia kuehniella*, *Plutella xylostella* and *Spodoptera frugiperda* (Goulart et al. 2012).

The LC_{50} values ranged from 1.29 to 2.57 for avermectins, 2.26–14.03 for pyrethroids and 1.12–239.1 mg a.i./L for neonicotinoids and were found to be less toxic to *T. evanescens* and insect growth regulators (IGRs) exhibited lowest toxicity with the LC_{50} values of 3383–5650 mg a.i./L (Wang et al. 2014). Naled and chlorfenapyr recorded 100 and 95 % adult mortality of *T.* nr. *brassicae* when directly applied and exposed to residues within 24 h of treatment (Kapuge et al. 2003). Imidacloprid was found to be most toxic to *T. chilonis*, at 30 min, 1, 2 and 6 h after treatment with their lowest LC_{50} value of 0.07, 0.03, 0.075 and 0.00004 %, respectively followed by cypermethrin. Chlorpyrifos and monocrotophos were least toxic to *T. chilonis* with its higher LC_{50} value (Khulbe and Kumar 2015).

3.1.2.2 *Anagyrus* sp.

Fipronil and α-cypermethrin caused significant acute toxicity of vine mealybug parasitoids, *Anagyrus* species near *pseudococci* and *Coccidoxenoides perininutus* (Hymenoptera: Encyrtidae). Buprofezin and mancozeb were non toxic to parasitoids. The mortality of parasitoids emerging from mummies was found to be low which is evident that mummy case serves as a barrier to pesticides (Mgocheki and Addison 2009). *Anagyrus pseudococci* (a nectar-feeding wasp) when fed on imidacloprid treated buckwheat flowers, after 1 DAT (days after treatment), the survival was found to be 38 % when compared to untreated flowers (98 %) and which decreased to 0 and 57 % on treated and untreated flowers, respectively at 7 DAT (Krischik et al. 2007).

3.1.2.3 *Aphidius* sp.

Cypermethrin showed a high survival percentage of aphid parasitoid, *Aphidius colemani* when exposed 72 h after application whereas, chlorpyrifos was extremely toxic to the parasitoid pupae and adults at all post-spraying introducing periods (Irshaid and Hasan 2011). Lufenuron was harmless to *Aphidius gifuensis* through contact and ingestion toxicity bioassays (Kobori and Amano 2004). The residues of thiacloprid severely affected *Aphidius rhopalosiphi* adults whereas, pre-imaginal stages remained unaffected (Schuld and Schmuck 2000).

3.1.2.4 *Chelonus blackburni*

Diafenthiuron is moderately harmful to the adults of *C. blackburni* tested by insecticide coated scintillation vial bioassay causing 86.67 % mortality at the recommended dose of 1.6 g/L. The 48 h LC_{50} of diafenthiuron was 0.89 g/L (Stanley et al. 2016). The recommended dose of imidacloprid (25 g a.i./ha) caused 56 % mortality and thus moderately toxic to *C. blackburni* adults (Preetha et al. 2010). A minimum adverse effect was found to cause by abamectin on *C. blackburni* adult emergence (Sheebajasmine et al. 2008).

3.1.2.5 *Bracon* spp.

Imidacloprid was found to be toxic to the parasitoid *Bracon hebetor*, causing 70 % mortality at 48 h after treatment (HAT) (Preetha et al. 2010). Imidacloprid and thiacloprid had least impact on *Habrobracon hebetor* and could be used along with the parasitoid in integrated pest management programs (Mahdavi 2013).

3.1.2.6 *Diadegma insulare*

Tebufenozide and spinosad were non toxic to *Diadegma insulare* at 24 h and 30 min after treatment, respectively. However, 100 % mortality was observed at 8 h after treatment with spinosad (Hill and Foster 2000).

3.1.2.7 *Diaeretiella rapae*

Phosphomidon, dichlorvos and monocrotophos were harmful to *D. rapae* causing 100 % mortality followed by methamidophos which was moderately harmful causing 97 % mortality after 24 h of application in glass plate bioassay method. The adult parasitoid emergence was reduced by about 9 and 7 %, when directly sprayed on *D. rapae* pupae with Dimecron and Nogos, respectively. Monofos and Tamaron reduced adult parasitoid emergence to 3 % compared to control (78 %) within 1 week of treatment (Kakakhel et al. 1998). The residues of deltamethrin had lower toxic effect on *D. rapae* which could limit populations of *Myzus persicae* (Desneux et al. 2005).

3.1.2.8 *Cotesia plutellae*

Indoxacarb (53 mg a.i./L), λ-cyhalothrin (28 mg a.i./L) and spinosad (53 mg a.i./L) caused 100, 88.5 and 50 % mortalities of *C. plutellae* adults, respectively. Ten day old residues of indoxacarb and λ-cyhalothrin caused 39 and 44 % mortality, respectively and spinosad residues of 7 and 10 days old resulted in 24 and 0 % mortality of *C. plutellae* adults, respectively (Haseeb et al. 2004).

3.1.2.9 Other Parasitoids

Chlorpyrifos was the most toxic insecticide to four parasitoid species viz., *Aphytis melinus*, *Eretmocerus eremicus*, *Encarsia formosa* and *Gonatocerus ashmeadi* based on LC_{50} values and among the four test parasitoids, *A. melinus* was the most susceptible (Prabhaker et al. 2007). Organophosphate and carbamate insecticides were highly toxic to apple leaf roller parasitoids, *Colpoclypeus florus* and *Trichogramma platneri* (Hymenoptera: Eulophidae) in topical applications but foliar residues of some products were non toxic after 7 days. Among the newer molecules, imidacloprid and abamectin were classified as highly toxic to both parasitoids on topical application but were non toxic as 1 day old residues (Brunner et al. 2001). Azadirachtin residues were non toxic to stink bug parasitoid, *Trichopoda pennipes* whereas, spinosad was as highly toxic to this parasitoid (Tillman 2008). When *A. nilaparvatae* fed with honey contaminated with imidacloprid, the parasitism rates were significantly lower than control (9.58 %) (Liu et al. 2010). Spinetoram, imidacloprid+cyfluthrin, abamectin and tolfenpyrad caused 100 % mortality in 72 h of a hymenopteran parasitoid of potato, *Tamarixia triozae* or tomato psyllid, *Bactericera cockerelli* (Hemiptera: Trizoidae) in glass surface residue bioassay whereas, spinetoram was extremely toxic to *T. triozae* adults as 15 days old residues caused 100 % mortality in leaf residue bioassay. Chlorantraniliprole, fenpyroximate, pymetrozine, spiromesifen and spirotetramat had less adverse effect on *T. triozae* adults, tested through glass surface or leaf-residue bioassays. Spinetoram, abamectin and imidacloprid + cyfluthrin caused 100 % mortality in adults of *T. triozae* when taken by ingestion. Tolfenpyrad caused 75.0 % mortality in 12 h, whereas pymetrozine showed moderate effects on the survival of adults (Liu et al. 2012). Among various treatments, novaluron followed by spinetoram exhibited lowest parasitoid diversity index (Hemandez et al. 2011).

Aphytis melinus, was not reportedly affected by spirotetramat applied to the host (California red scale, *Aonidiella aurantii*) when the parasitoid was in the egg or larval stage, while the adults on exposure to leaves with residues showed 2 weeks of moderate reductions in survival (Garcera et al. 2013). Endosulfan, monocrotophos, profenofos and dimethoate caused 100 % mortality within 1 h in mealybug parasitoids, *Aenasius bambawalei* and *A. advena* and this was found to be more than untreated check. Imidacloprid resulted in 100 % mortality of both the sexes of the parasitoids after 3 h (Nalini and Manickavasagam 2011). Acetamiprid and emamectin benzoate showed less reduction percentages to the parasitoid, *Aphytis lepidosaphes* (Hymenoptera: Aphelinidae) on citrus (Dewer et al. 2012). Organophosphates proved to be more toxic than pyrethroids or carbamates (Longley 1999). The maximum recommended field concentration and half the dose of amitraz were found to be harmful and one-fourth dose was moderately harmful to *Encarsia formosa* (Chitgar and Ghadamyari 2012). Abamectin was reported to reduce the survivability of larvae and pupae of *Hemiptarsenus varicornis* and *Diglyphus isaea* (leafminer parasitoids) (Bjorksten and Robinson 2005).

3.2 Chronic Toxicity

Apart from lethal toxicity, pesticides may also cause sub lethal toxicity on test individuals that survived on exposure to the toxic residues. These effects are not much considered while conducting acute toxicity bioassays (Laskowski 2001). Hence, we also evaluate chronic exposure studies to assess their toxicity. The mortality of treated pupae, longevity of emerged adults, fecundity, egg hatchability, per cent pupation, per cent adult emergence and sex ratio of offspring of *T. chilonis* were adversely affected due to exposure to chlorantraniliprole. Duration of life cycle of the offsprings was not adversely affected by insecticidal treatments (Mallick and Mandal 2014).

3.3 Persistent Toxicity

Abamectin and pymetrozine are classified as short lived (Class A) as their persistence is <5 days and imidacloprid as slightly persistent (Class B) with persistence between 5 and 15 days against adult of a thelytokous parasitoid, *Lysiphlebus fabarum* (Hymenoptera: Aphidiidae) of *Aphis fabae* (Sabahi et al. 2009). Cartap hydrochloride 75 % SG was found to be moderately persistent (16–30 days), chlorfenapyr 10 % F and emamectin benzoate 1 % EC as slightly persistent (5–10 days) against female wasps of *C. plutellae* using leaf disc bioassay. The synthetic pyrethrod, permethrin was found to be short lived with toxicity for only less than 5 days (Haseeb and Amano 2002). Emamectin benzoate and spinosad were moderately persistent (16–30 days; Class C), indoxacarb was slightly persistent (5–15 days; Class B) and lufenuron and flubendiamide were short lived (<5 days; Class A) against *T. chilonis* adults (Sattar et al. 2011). The pupae of *T. bourarachae* on exposure to residues on olive leaves caused reduction in the parasitism viability of 35.5 % and 25.5 % for deltamethrin and spinosad treatment, respectively after 31 days (Blibech et al. 2015).

Fresh residues of flonicamid, flubendiamide and spirotetramat were harmless to *Aphidius ervi* and *Chelonus inanitus*. Residual toxicity of metaflumizone was moderately persistent and persistent to *C. inanitus* and *A. ervi*, respectively. However, deltamethrin was slightly persistent and persistent, respectively for *A. ervi* and *C. inanitus* (Medina et al. 2012). The persistent toxicity of endosulfan, monocrotophos, cypermethrin and dimethoate on per cent parasitization and emergence of *T. brasiliensis* lasted for 9, 9, 5 and 3 days after treatment, respectively (Amandeep et al. 2012). Dimethoate, phosalone and dichlorvos at 0.05, 0.07 and 0.10 %, respectively were found to be less toxic to adults of *Anicetus ceylonensis*. Dichlorvos (0.10 %), methyl demeton (0.05 %) and endosulfan (0.07 %) showed least persistence whereas fenvalerate at 0.01 % and monocrotophos at 0.05 % had higher persistent toxicity to the adult parasitoids (Mani and Krishnamoorthy 1997). The 1-day residues of lambda-cyhalothrin caused a significant population reduction of *Eretmocerus mundus* and found to be harmful from 7 days after introduction (DAI) until the last evaluation at 38 DAI which was graded as moderately harmful (Bengochea et al. 2012).

3.4 Sublethal Toxicity

Sublethal toxicity is an important strategy effective against the continuous use of conventional insecticides which may have lethal and sublethal effects on natural enemies. Effects of insecticides on natural enemies are mostly determined based on the results of the acute toxicity bioassays (Desneux et al. 2007). In the field conditions, parasitoids are exposed to a high concentration of pesticide at the time of application and then the concentration get decreased as the pesticides get degraded. So, the rates equivalent to the LC_{50} are usually tested against parasitoids. Insecticides affect the efficacy of potential natural enemies by interfering with their biological traits viz., fertility, fecundity, emergence, developmental rates, longevity and also by modifying behavioural traits. Several authors reported the reduction in the efficacy of parasitoids (Cross et al. 1999; Sterk et al. 1999; Brunner et al. 2001; Kapuge et al. 2003).

Sublethal effects of insecticides cause the parasitoids to be less effective in parasitizing the hosts in field conditions. So, sublethal effects should also be done as these can be expressed several days after treatment in addition to acute mortalities (Tipping and Burbutis 1983). The long term sublethal effects normally happen after 24 h of pesticide exposure. Insecticides at sublethal doses can have positive or negative impact on beneficials (Croft 1977; Elzen 1989; Messing and Croft 1990; Wright and Verkerk 1995). The positive effects include increased fecundity (Fleschner and Scriven 1957; Attallah and Newsom 1966), enhanced parasitoid efficiency (Irving and Wyatt 1973), increase in movement or searching (Dempster 1968; Critchley 1972) and reduction in developmental periods (Adams 1960; Lawrence et al. 1973). The sublethal effects of various insecticides on different parasitoids are detailed below.

3.4.1 Fecundity

Bastos et al. (2006) showed that the fecundity of *T. pretiosum* is highly reduced by pyrethroids irrespective of the host species, *Sitotroga cerealella* and *Ephestia kuehniella*. Fecundity of *T. pretiosum* was found to get reduced by the exposure of betacyfluthrin and esfenvalerate. The fecundity of parasitoids was found to get affected by these pyrethroids even for two subsequent generations. The higher concentrations of deltamethrin resulted in >50 % reduction in potential fecundity (Meilin et al. 2012).

3.4.2 Adult Emergence

Fenitrothion and deltamethrin had a significantly reduced the adult emergence of egg parasitoids, *Trissolcus grandis*. They also reduced the emergence rates of egg parasitoid by 18.0 and 34.4 %, respectively on comparison with untreated check

(Saber et al. 2005). Lambda cyhalothrin, thiodicarb, profenofos, cypermethrin, spinosad adversely affected the emergence of *Trichogramma* from *Helicoverpa zea* eggs when exposed as larval, prepupal, or pupal stages. Spinosad and profenofos were the most toxic products to *T. exiguum* adult females and then by lambda cyhalothrin, cypermethrin and thiodicarb (Suh et al. 2000). Dimethoate reduced the emergence of *D. rapae* adults from treated mummies. The Mean number of offspring per female ranged between 49.3 and 93.3 for primicarb and pymetrozine, respectively (Kheradmand et al. 2012). Dimethoate and chlorpyrifos were reported to cause a reduction in the emergence of parasitoids, *Tamarixia radiata* (Beloti et al. 2015). *Encarsia formosa* exposed to pyriproxyfen at various concentrations exhibited different sublethal effects (Liu and Stansly 1997). The *Eretmocerus tejanus* pupae treated with buprofezin showed shorter life span of emerging adults than those treated with water, endosulfan and thiodicarb (Jones et al. 1998).

3.4.3 Host Foraging Ability and Longevity

Reduced foraging ability and longevity was reported in *Microplitis croceipes* when fed upon extrafloral nectar from insecticide treated cotton plants (Stapel et al. 2000). It was also found that the foraging ability of parasitoids was severely affected for 2 days in imidacloprid and 18 days in aldicarb treatment. Elzen (1989) reported that cotton plants sprayed with fenvalerate-chlordimeform mixture resulted in decrease in flight activity (indicator of foraging efficiency) of the parasitoid females, *M. croceipes*. The odours from some insecticide treated plants may cause avoidance behaviour in parasitoids and thus affects the effectiveness as biological control agents.

The survivors of the parasitoid, *Aphytis melinus* on exposure to carbaryl, exhibited no significant sublethal effects whereas, to organophosphates resulted in reduced longevity by 73–85 %. A temporary reduction in progeny production was also found due to OP exposure (Rosenheim and Hoy 1988a). Spinosad caused 100 % mortality of *T. chilonis* adults within 15 min. of exposure (Saljoqi et al. 2012). The adult longevity of *Telenomus busseolae* was reported to be affected by deltamethrin (Bayram et al. 2010), *Aphidius ervi* (Desneux et al. 2006b) and *Habrobracom hebetor* (Sarmadi et al. 2010). The endoparasitoid, *Hyposoter didymator* when exposed to diflubenzuron decreased female longevity and reduced the parasitization rate (Schneider et al. 2004). Longevity of *Eretmocerus tejanus* that emerged after applications of buprofezin and azinphos-methyl are slightly shorter (3.32 and 3.09 days) when compared to control (4.25 days) (Jones et al. 1998).

3.4.4 Mobility and Orientation

A significant sublethal effects (arrestment response and walking behavior) of deltamethrin was found on *Trissolcus basalis* to a contact kairomone from *Nezara viridula* (Salerno et al. 2002). The orientation behaviour of *A. ervi* and *D. rapae* toward plants infested with aphids is not altered on exposure of parasitoid to deltamethrin (Desneux et al. 2006a, b). The walking behaviour of *T. chilonis* was observed for 3 min, first in absence and then in presence of spinosad (0.005 % and 0.001 %) and observed that lower distance (16.22 cm) was covered by the parasitoid when treated with 0.005 % concentration, while it travelled 25.4 cm when treated with 0.001 % showing that the higher concentrations affect parasitoid behaviour and search ability (Saljoqi et al. 2012).

3.4.5 Communication

Deltamethrin at sublethal dose exposed to only males induced an increase of the arrestment of *T. brassicae* males by the sex pheromone and deltamethrin enhanced the response of non insecticide-treated males when the treated females emits the pheromone (Delpuech et al. 1999).

3.4.6 Behaviour

Tebufenozide and gamma-cyhalothrin affected the parasitism of the F_0 generation of *Tamarixia radiata*, but the effect was not found in next generations (Beloti et al. 2015). When *Anagyrus pseudococci* fed on buckwheat flowers treated with imidacloprid (soil application) altered the behaviour and reduction in the survivorship of the parasitoid (Krischik et al. 2007).

3.4.7 Generation Time and Sex Ratio of Parasitoids

Hexaflumuron, profenofos and spinosad was found to reduce the generation time of *H. hebetor* and also affected its sex ratio (Dastjerdi et al. 2009). Development of both males and females of parasitoid, *Cotesia marginiventris* on cotton was delayed (<1 day) by λ-cyhalothrin, dicrotophos and emamectin benzoate. The lowest survival was observed in the emamectin benzoate and λ-cyhalothrin treatments (Ruberson and Roberts 2005). Chlorpyrifos shifted the sex ratio of the parasitoid, *Aphytis melinus* offspring toward more males (Rosenheim and Hoy 1988a).

Abamectin irrespective of developmental stage and the exposed parasitoid generation affected emergence and sex ratio of *Trichogramma pretiosum*. Abamectin, lufenuron and primicarb exposed at egg-larval stage had an influence on the adult female parasitoids by affecting their longevity (Carvalho et al. 2003). The developmental time of *Apanteles galleriae* reared on cypermethrin treated larvae of *Achroia grisella* was increased by more than 50 % when compared to *Apanteles galleriae* reared on untreated hosts (Ergin et al. 2007). Fenoxycarb increased larval duration of *Pseudoperichaeta nigrolineata* irrespective of doses (Grenier and Plantevin 1990).

3.4.8 Life Table Parameters

Pymetrozine reduced parasitoid's life table parameters in comparison to the control and caused 27 % mortality in adult stage (Kheradmand et al. 2012). Saber (2011) stated that fenpyroximate was compatible with parasitoid, *T. cacoeciae*; whereas imidacloprid exhibited toxic effects on parasitoid adults.

3.4.9 Offspring Production

An insect growth regulator, diflubenzuron completely blocks the production of *Colpoclypeus florus* offspring thereby causing severe sublethal effects (Brunner et al. 2001). Methoxyfenozide had small effect on the number of progeny of *Arrhenophagus chionaspidis* whereas, clothianidin, acetamiprid and emamectin benzoate treated progenies never reach the pupal or adult stage (Yoshioka and Takeda 2006). Chlorpyrifos, carbaryl and thiamethoxam lowered the number of *Tiphia vernalis* reaching to cocoon stage and carbaryl and chlorpyrifos also reduced the number of *Popillia japonica* larvae parasitized (Oliver et al. 2005).

3.5 *Field Toxicity*

In field trials, methoxyfenozide showed no impacts on Trichogramma adult emergence from treated parasitized eggs, whereas indoxacarb exhibited <8 % emergence (Kapuge et al. 2003). Diafenthiuron is relatively less toxic to *Trichogramma* sp. (www.bioresources.com/pretisoumchemicals). Acetamiprid and emamectin benzoate showed less reduction percentages to the parasitoid, *Aphytis lepidosaphes* (Hymenoptera: Aphelinidae) on citrus (Dewer et al. 2012). Parasitism of aphids by *Aphidius* sp. and of pod borers by *Argyrophylax* sp. (tachinid flies), *Phanerotoma leucobasis* and *Braunsia kriegeri* (braconid) and *Pristomerus* sp. (ichneumonid) were significantly reduced by insecticides viz., cypermethrin and dimethoate (Munyuli et al. 2009). Spinosad at labelled rate was moderately to highly toxic to *Aphelinus mali* whereas, other compounds and spinosad at lower dose had no detrimental effects. Carbaryl had the greatest residual toxicity to *A. mali* causing 85 % mortality at 21 DAP (days after application) and declining to 40 % by 28 days (Rogers et al. 2011).

The residues of insecticides such as fipronil, spinosad, cyfluthrin etc. were most deleterious to *Anaphes iole* (95 %) (Williams III et al. 2003). The effect of abamectin on *C. plutellae* revealed that abamectin at 9, 11 and 13 g/ha was relatively safer to the parasitoid populations (Sheebajasmine et al. 2008). Thiacloprid was classified as a slightly dangerous insecticide to *Lysiphlebus fabarum* in immature stages (Purhematy et al. 2013). Under field conditions, fenvalerate 5 EC and chlorpyrifos 48 EC were categorized as moderately harmful (80–99 %), while, flufenoxuron 10 EC was categorized as slightly harmful (30–79 %) according to the IOBC ranking to *Trichogramma evanescens* (Mohamed 2015).

4 Methods to Assess Pesticide Toxicity to Parasitoids

Bioassays will be used based on the mode in which the test insect was harmed by the chemicals. The pesticide selectivity to natural enemies can be estimated through toxicity bioassays and the methods for assessing the toxicity of insecticides to parasitoids are detailed hereunder.

4.1 Acute Contact Toxicity Bioassays

4.1.1 Bioassays with Parasitized Eggs

4.1.1.1 Egg Card Bioassay: Dip Method (Hussain et al. 2010)

In this method, five egg cards containing 40 parasitized host (*Sitotroga cerealella*) eggs were dipped in the respective insecticidal treatment for 1 s on 1, 2, 4, 5 and 7 day after oviposition during egg, larval, pre-pupal, early pupal and pupal stage after ovi-position. The treated egg cards were dried and transferred to small plastic Petri dish and maintained in the laboratory until parasitoid emergence.

In another experiment by Bueno et al. (2008), the host (*Anagasta kuehniella*) eggs (approximately 250 nos) in 1 cm^2 card board squares were placed on vials and newly emerged parasitoids (*T. pretiosum*) were introduced into the egg card containing vials for 24 h. Later these parasitized egg cards were removed and transferred to another vials and treated with insecticides at 72 h (eggs), 144 h (larvae) and 192 h (pupae) by dip method and dried for an hour to eliminate excess moisture. Then the card were transferred to transparent polythene bags and observed for the emergence of parasitoids from the corresponding treatments.

4.1.1.2 Egg Card Bioassay: Spray Method (Preetha et al. 2010)

The parasitized egg cards (16 × 32 cm) were sprayed using atomizer with insecticides of respective treatments and water was used in the control (Fig. 2.1). After shade drying for 10 min, three 7 × 2 cm cards, representing three replications were cut and kept in test tubes. The parasitoids emerged/cm2 area at 24 and 48 h after treatment (HAT) were recorded. Per cent emergence was calculated as per the formula given below

$$\text{Percent emergence} = \frac{\text{No. of wasps emerged}}{\text{Total no. of eggs in 1 cm}^2} \times 100$$

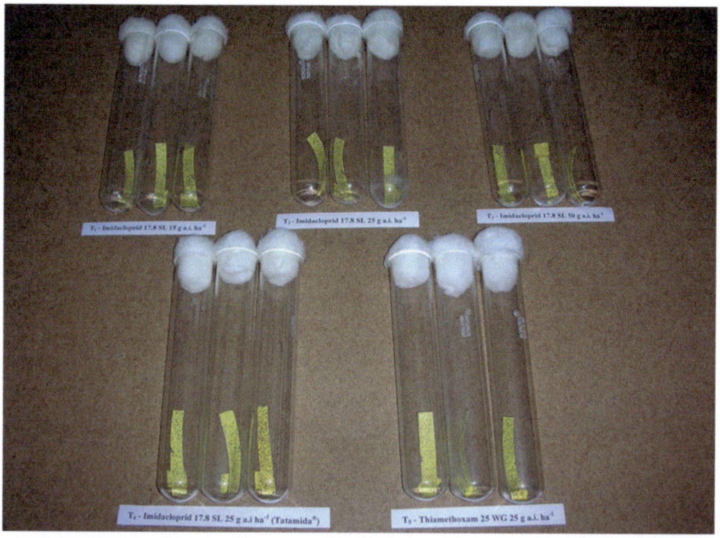

Fig. 2.1 Toxicity on emergence and parasitization -egg card bioassay

Fresh eggs were provided to these parasitoids at 6:1 ratio. After 24 and 48 h of provision of eggs, the parasitized eggs were counted and percent parasitisation recorded. The eggs appearing black and plumpy were denoted as parasitized eggs.

$$\text{Parasitization}\ (\%) = \frac{\text{No. of parasitized eggs}}{\text{Total no. of } \textit{Corcyra} \text{ eggs}} \times 100$$

4.1.1.3 Oviposition Preference Test (Saljoqi et al. 2012)

In this bioassay method, the desired insecticidal concentrations were sprayed on the host eggs (*Sitotroga cereallela*) and glued on cards each of around 50–100 eggs and placed on vials containing freshly emerged virgin females of parasitoids (*T. chilonis*) with individual virgin male and allowed for mating for a day and the females were left as such till death. Observations were recorded on the number of eggs being parasitized, number of adults emerged from the parasitized eggs and adult longevity.

4.1.1.4 Direct Spray on Host Eggs (Abdulhay and Rathi 2014)

In this method, parasitized *E. cautella* eggs (30 nos) were placed in a Petri dish containing filter paper (5.5 cm dia.) and sprayed with the respective field recommended insecticidal dose. Filter papers and eggs were shade dried for 1 h and then 30 eggs from each insecticidal treatment and control (water alone) were transferred individually into plastic vials. Honey solution (50 %) was streaked on the vials and observations recorded daily on the emergence or death of the adult parasitoids (*T. evanescens*). Three parameters viz., parasitisation, parasitoid emergence and the reduction in capacity of parasitism were evaluated. After 14 days of initial parasitism, a final evaluation was made on the emergence and the reduction in parasitism capacity was calculated using the formula:

$$PR = \left[1 - (P / p) \times 100\right]$$

Where,

PR – Per cent reduction in parasitism
P – average value for each parasitism for insecticide
p – average parasitism rate for the control treatment

4.1.2 Bioassays with Parasitoid Pupae

4.1.2.1 Topical Bioassays (Mgocheki and Addison 2009)

Twenty mummies with parasitoid were selected and placed on sticky tape which was placed on glass plates. The recommended field concentration of insecticides were sprayed on the mummies using Potter's spray tower and the sticky tapes were allowed to dry for an hour and then sprinkled with fine soil to prevent emerging parasitoids to come in contact with the residues and also from getting stuck on the adhesive. The tape was kept in ventilated Petri dish and observed for emergence of the parasitoids daily and the emerged parasitoids were transferred to ventilated vials and provided with 50 % honey: water solution.

4.1.2.2 Pesticides Sprayed on Parasitoid Cocoon (Ruberson and Roberts 2005)

In this method, fresh parasitoid cocoons (*Cotesia marginiventris* cocoons of <12 h old) were sprayed with the respective insecticidal concentration, air dried and placed individually in tissue culture tubes and observed daily twice for adult emergence. The emerged adults were paired within the insecticidal treatments and each pair was placed in a plastic Petri dish streaked with honey as feed. Fifty 24–48 h old host caterpillar (beet armyworm) were introduced into each Petri dish every day for the female parasitoid to oviposit them and the previous day released caterpillars were removed and transferred to diet and kept until parasitoid emergence or pupation of caterpillar. Assessment was made on wasp fecundity and the ability of the parasitoids to locate and attack hosts.

4.1.3 Bioassays with Parasitoid Adults

4.1.3.1 Potter's Spray Tower (Brunner et al. 2001)

In this method, 1–2 (*Colpoclypeus florus*) and 2–4 day old (*Trichogramma platneri*) adult females of parasitoids were used for the tests. Five test insects selected at random were anaesthetized with CO_2 and placed on filter paper. Test pesticide was sprayed topically using Potter's spray tower to the test parasitoids and were transferred immediately to small Petri dish. Nutrient and moisture sources were given and mortality was recorded at 48 h after treatment.

4.1.3.2 Preval Sprayer (Tillman 2008)

In this method, the adult insects (*T. pennipes*) were cooled down in the refrigerator for 1–2 min. The anesthetized adults were kept in Petri dish and applied topically with insecticidal solution using Preval sprayer. Then the treated insects were transferred to clean Petri dish and sugar solution was given as feed. Mortality was recorded at 1, 24 and 48 h after treatment.

4.1.3.3 Using Exposure Chamber (Mgocheki and Addison 2009)

Exposure chamber is made of two glass plates treated with pesticides (10 × 10 cm) fitted to a Munger cell (10 × 10 × 2 cm) with six holes (0.8 cm dia.) through the side of the walls for ventilation. The pesticidal doses were applied to glass plates using Potter's spray tower and were air dried. The Munger cells were assembled with treated glass surfaces facing inwards and the holes covered with fine gauze. The parasitoids 20 one day old (*Coccidoxenoides perminutus*) and 1–2 days old (*Anagyrus* sp. near *pseudococci*) were introduced into each cell through the hole left uncovered and was plugged with cotton wool soaked in 50 % honey-water solution, as food source. Observations were recorded at 6, 12, 18 and 24 h after introduction.

4.1.3.4 Using Modified Exposure Chamber (Longley and Jepson 1997a)

Exposure chambers (modified from Mead-Briggs 1992) consisted of two insecticide treated glass plates fitted to a section of plastic drain pipe (10.5 cm dia. and 2.5 cm ht), with five holes (1 cm dia.) for ventilation and covered with fine gauze leaving one hole as uncovered and through which the parasitoids were released and plugged with 1:1 honey: water solution soaked cotton wool.

In this bioassay, the glass plates were not sprayed but covered with leaves from untreated host plants i.e., winter wheat (Tottman and Broad 1987). Leaves with their adaxial surfaces uppermost were fixed to the glass using adhesive tapes. Different concentrations of test pesticide solutions were prepared. Treatments were imposed using Potter's spray tower and were shade dried for 1 h. The chambers were then assembled and ten parasitoids (five male and five female) were carefully introduced to each chamber. A 2 cm wide strip of nontransparent tape was sticked all over the edges of the glass plate to keep the parasitoids in the illuminated area. Parasitoid responses were assessed at different intervals upto 24 h after exposure.

In another bioassay, a glass plate (12 × 12 cm) and a glass dish (6 cm dia.) were sprayed with the field concentration of insecticide. Spray on both glass plate and dish is to give maximum exposure of parasitoids to pesticide residues. After drying, the two components were kept together and were sealed by which humidified air were introduced using hypodermic needles. Individual female *Aphidius rhopalosiphi* were placed inside the exposure chambers for 2, 3, 5, 10, 15, 30 and 45 min and control chamber for min only. Fifteen test insects were used for each treatment. After the treatment time, parasitoids were transferred to Petri dish and honey solution soaked cotton was given as feed. Observations were recorded on the parasitoids response immediately, 1 h and 24 h after exposure over the following 4 days.

4.1.3.5 Using Exposure Cages (Mahdavi 2013)

The experimental setup has a frame and two glass plates as ceiling and floor. Both the sides of frame was drilled with five ventilation holes (5 mm in dia.), covered with black netting. The water tube and food for the wasps are placed on the two openings in the fourth side of the frame. The glass plates were sprayed with 3 mL of the field recommended concentrations of test pesticides using a Potter's spray tower. The plates were allowed to dry completely for 1 h and ten CO_2 anesthetized young female adults (<24 h old) kept in each exposure cages and the ceiling was fixed. Honey was placed on a small strip of paper and provided as food for wasps. The number of dead and alive wasps in each cage was counted at 12, 24, 36 and 48 h after the treatment.

4.1.3.6 Using Experimental Cages (Sohrabi et al. 2014)

Young adult parasitoids (<24 h old) were exposed to the field-recommended concentration of insecticide for assessing the parasitoid functional response. The host (cucumber) leaves were dipped for 10 s into insecticidal concentrations and into distilled water for control. After drying for an hour, the treated leaves were placed in the experimental cage. The experimental cages consisted of round plastic containers (5.2 cm ht, 4 cm dia.) with four lateral screened holes (1 cm dia.) for ventilation. Leaf discs (4 cm dia.) from treated leaves were placed with their adaxial side on a layer of 1 % agar. About, 40–50 pairs of young adult parasitoids (*Eretmocerus mundus*) were exposed to treated leaves. After 48 h, from the alive females, six were randomly selected and placed individually to clip cages (4 cm dia.) on fresh, undipped leaves with different prey densities of 2, 4, 8, 16, 32 and 64 *B. tabaci* (second instar nymphs). Wasps were removed after 24 h. Observations were recorded on the number of parasitized whitefly nymphs with the aid of a stereoscopic microscope.

4.1.3.7 Cage Bioassay (Carmo et al. 2010)

The egg masses (150 eggs) of *Spodoptera frugiperda* parasitized by *Telenomus remus* were transferred to glass vials which contains honey droplets on its inner wall. Then the glass vials were sealed with plastic film and at parasitoid emergence, glass plates (13 × 13 cm) were sprayed with the respective test chemical and with distilled water in case of control, ensuring a deposit of 1.25 mg/cm^2. After drying, the plates were set as per Hassan (1992). The parasitoids were exposed to the glass plates by fitting the vials with parasitoids to the cages. After 24 h, cards with *S. frugiperda* egg (400 eggs) were placed inside along with honey feed. A second card was kept inside the cages 24 h after the first card and then by the third card. Percent parasitism and parasitoid emergence were assessed.

4.1.3.8 Using Bioassay Chamber (Williams III et al. 2003)

A bioassay chamber was made using a transparent flexible tube of about 2.5 cm dia. and 3.5 cm long which have holes covered with organdy cloths for ventilation. Two scintillation vial caps (2.3 cm dia.), containing agar and a leaf disc treated with insecticides were kept inside the tubing. The abaxial side of the leaf disc formed the chambers ceiling, while the adaxial side of another leaf disc formed the floor. After sealing the chambers with dialysis membrane, 25 parasitoids (mixed gender) were released, inserted the feeding tube and approximately 25 μL of honey solution added. After a 48 h of exposure, wasps in each chamber were assessed for mortality using a microscope.

4.1.3.9 Nonchoice and Residual Tests Bioassay (Tello et al. 2013)

In this method, the internal glass surfaces of vials were swabbed with cotton containing insecticide. The vials were allowed to dry for specific periods such as 1, 48 and 96 h and 10–30 adults of parasitoids (*Eretmocerus paulistus*) were introduced. The insects were exposed for 12 h and mortality was recorded.

4.1.3.10 Toxicity Bioassay with Fresh Residues (Gonzalez-Zamora et al. 2013)

In this method, pesticides were applied to tops and bottoms of Petri dishes with a Potter's spray tower which was calibrated to leave 1.5 mg of solution/cm^2. Pesticides were left to dry for an hour, and then six to eight adult females of parasitoid (*Aphytis melinus*) that were 24–48 h old were introduced into each Petri dish and honey was provided as feed. Parasitoid mortality on each Petri dish was evaluated after 24 h.

4.1.3.11 Toxicity Bioassay with 7-day Old Residues (Gonzalez-Zamora et al. 2013)

This method is in continuation with the previous bioassay with fresh residues. When the mortality was 100 % then the insecticide was classified as harmful and when the mortality was <100 %, they are less harmful. After the insecticides were sprayed on Petri dishes were kept upside open and covered using filter paper. The experimental set up was kept at room temperature for 7 days to allow residues to age and evaporate. After this period, six to eight parasitoids (*A. melinus*) females (24–48 h old) were introduced in each Petri dish. Honey was given as food. Adult mortality was evaluated after 24 h.

4.1.3.12 Dry Film Residue Bioassay (Preetha et al. 2009)

Different insecticidal concentrations were prepared in acetone: water (80:20) and 0.5 ml of the respective concentration was coated on glass scintillation vials of 15 mL capacity and dried. The control vials were treated with acetone: water. Twenty adult wasps just emerging and moving from the parasitized egg cards (*T. chilonis*) were allowed in the vial and covered. A muslin cloth was used to cover the vial and secured with rubber bands. The wasps were removed after 4 h and transferred to clean test tubes. Observations on mortality was recorded at 24 and 48 h after treatment.

4.1.3.13 Filter Paper Bioassay (Rogers et al. 2011)

This bioassay procedure consists of a filter paper disc placed into the lid of a Petri dish with 0.5 mL of pesticide solution applied on the paper and allowed to dry for 20–25 min and water was used for the control. Approximately 30 adult parasitoids (*Aphelinus mali*) were collected and used for a replication and then quickly covered with bottom of the Petri dish and sealed. Mortality was recorded after 16 h.

4.1.3.14 Dipped Surface Residue Bioassay (Hussain et al. 2010)

This bioassay method consists of ventilated glass bioassay chambers (15 cm long and 4 cm dia.). Filter paper (15 × 10 cm) was dipped in the respective insecticidal treatments, air dried and put into the glass bioassays chamber. Approximately, 20 newly emerged parasitoid adults (*T. chilonis*) were released in each bioassay chamber lined with treated filter paper and mortality was recorded after 4 and 24 h.

4.1.3.15 Leaf Disc Bioassay Using Glass Jars (Hill and Foster 2000)

In this bioassay method, leaf discs (8.75 cm in dia.) were taken from uninfested cabbage plants and were immersed individually in the respective insecticidal concentration for 3 s and air dried for 2 h. Control leaves were treated with water. Treated leaf discs were lined inside the glass jar and ten randomly selected, newly emerged adult parasitoids (*Diadegma insulare*) were released into each jar and covered with nylon netting secured by an elastic band. A 50 % honey solution was streaked on the net as food source for the parasitoids. Mortality of parasitoids was recorded at 30 min, 8 and 24 h. In this method leaves are used as substrate in which pesticides are applied and exposed to parasitoids rather than inert glass surface. Substrate plays a vital role in determining the contact toxicity of pesticides (Croft 1990; Wright and Verkerk 1995).

4.1.3.16 Leaf Disc Bioassay in Petri Dish with Agar Beds (Prabhaker et al. 2007)

In this Petri dish bioassay technique, agar medium was prepared and spread on a Petri dish to maintain moisture for test leaves (citrus for *Gonatocerus ashmeadi* and *A. melinus* or cotton for *Eretmocerus eremicus* and *Encarsia formosa*) for up to 7 days. Freshly excised leaf discs of test plants were dipped in respective insecticidal concentration for 30 s. Treated leaf discs were shade dried for an hour and placed on the agar beds in the Petri dish. Then 15 respective test insects were released in each Petri dish. A thin strip of honey was given as feed for the test insects by smearing on the inner side of the Petri dish lid. Mortality of insects was recorded at 24 and 48 h for general insecticides and 96 h for the insect growth regulators like buprofezin and pyriproxyfen since there may be delay in mortality impacts.

4.1.3.17 Leaf Disc Bioassay Using Disposable Petri Dish (Aydogdu and Guner 2012)

In this method, the leaves (almond/apple/plum/cherry) were treated with insecticides in 1 mL of application dose, dried and then placed in a disposable Petri plate (60 mm dia.). Then four alive *Itoplectis maculator* adults obtained from *Archips rosana* larvae were released into the Petri plate and 50 % honey solution soaked cotton was used as food source. Mortality rates were recorded at 1, 2, 4, 8, 12, 16, 20 and 24 h after treatment.

4.1.3.18 Residual Toxicity Bioassay Using Petri Dish (Tillman 2008)

In residual toxicity tests, the insecticidal concentrations were sprayed on the top and bottom of a Petri dish (100 × 15 mm) using a sprayer and dried for 1 h. After drying, the test insects (*Trichopoda pennipes*) were released singly in the Petri dishes. The test insects were transferred to uncontaminated dishes after the exposure period and fed with sugar solution. Mortality was recorded at 1, 24 and 48 h after treatment.

In another method given by Abdulhay and Rathi (2014), the lid and bottom of glass Petri dish (1 cm in depth and 10 cm in dia.) were sprayed with 1 mL of field recommended concentrations of respective insecticides and water in case of control. The treated Petri dishes were dried for 2 h and then 20–30 unsexed parasitoid adults (24–48 h old) were introduced into each Petri dish for 24 h. The mortality was recorded at 2, 4, 6 and 24 h after treatment.

4.2 Acute Ingestion Toxicity Bioassays

4.2.1 Insecticide Honey Mixture in Glass Cover Slip (Wang et al. 2008)

In this method, plastic Petri dish (3 cm ht and 4.8 cm dia.) drilled with three holes of 1 cm dia., one at the base and two on the sides and covered with mesh (0.1 mm) for ventilation was used. Each insecticidal formulation was made to the required concentration with 10 % honey solution. The honey water mixture alone was used as control. Droplets of insecticide honey mixture solution was kept in microscopic glass cover slips and kept in Petri dish as food for the parasitoids. Four droplets (1 mL each) of were kept on the coverslip as food. Female wasps (*Anagrus nilaparvatae*) of about ten numbers were introduced to each Petri dish. Mortality was recorded after 1, 2, 4 and 8 h.

4.2.2 Insecticide Sugar Mixture in Feeding Wells (Tillman 2008)

The respective insecticidal concentration was mixed with sugar solution and kept in feeding wells in Petri dishes and the adult test insects (*T. pennipes*) that had been starved for 24 h prior to the assay were placed singly into these feeding wells and were allowed to feed once. After the insects have taken the treated food, they were transferred individually into Petri dishes. The time taken for an insect to die after feeding was recorded if applicable. Mortality was recorded at 24 and 48 h after treatment.

4.3 *Persistent Toxicity*

4.3.1 Field-Aged Residues Tested in Laboratory (Rogers et al. 2011)

In this method, the trees (apple) were sprayed with the respective insecticide using a Knapsack sprayer. The leaves of a similar size from the treatments were collected on day '0' immediately after the spray deposits had dried and on additional days after spraying. The leaves from an untreated tree were used for control. The leaves were collected from the treated trees and the experiment was continued until there was little toxicity. Thirty test insects (*A. mali*) adults were exposed to the treated leaf with residues within a Petri dish for 24 h, after which the insects were assessed for mortality.

4.3.2 Residual Toxicity Bioassay (Wang et al. 2008)

In this bioassay method, three pots per treatment containing rice plants were sprayed with insecticides. Treated plants were allowed to dry and leaves were removed from the respective treatment and a leaf blade of approximate size of 7.0 × 1.0 cm cut out. Test tubes were taken and 1 % agar medium was poured and allow to get solidify. About two leaf blades were kept on the agar bed to keep the leaf blade turgid. Ten adult parasitoids (*A. nilaparvatae*) were introduced into each tube. Mortality was recorded after 8 h. The residual toxicity of the insecticides to the adult parasitoids was periodically assessed on 0, 1, 3 and 7 days after treatment.

4.3.3 Bioassay Using Aged Residue in Petri Dish (Roubos et al. 2014)

In this method, insecticidal treatments were applied to the inner surface of Petri dish (4.7 cm dia.) bottoms by using a Potter's precision spray tower. Petri dishes were allowed to get dry and test insects were introduced. For aged residue experiments, the Petri dishes were treated and aged in a greenhouse for 0, 3, 7 and 14 days after treatment. After residues aged for the appropriate period of time, ten adult insects (mixture of females and males of *Aphidius colemani*) were introduced into each Petri dish. Honey solution was smeared on the inner surface of lid along with a cotton wig for water. Mortality of test insect were observed at 24, 48 and 72 h exposure time.

4.3.4 Clip Cage Bioassay (Longley and Jepson 1997b)

In this bioassay method, the live female parasitoids belonging to Aphidius genus were kept individually in ventilated clip cages (2 × 1.75 cm) for 24 h before conducting experiments and are fed with 50 % honey solution. After application of deltamethrin on wheat leaves and dried for 1 h, ten clip cages, each containing a single live Aphidius female were placed on randomly selected flag leaves and similarly placed for the control treatment also. Mortality of parasitoids was recorded after 24 h and the same procedure was repeated on third and fifth day after application.

4.4 Sublethal Bioassay

4.4.1 Behavioural Test (Desneux et al. 2004)

In this method, the female parasitoids of *D. rapae* and *Aphidius matricariae* were exposed to insecticides for 24 h via glass residual bioassay method. The survivors of the parasitoids after 24 h were collected and placed individually in Petri dishes and used for behavioural studies. The parasitoid females were placed individually on an aphid patch (*M. persicae*) and observations were recorded using a microscope. Various observations such as antennal contact, antennal examination, stinging attempt and stinging were recorded. The frequency and duration of walking on and out of aphid patch, grooming and time spent immobile are also noted. The frequencies of these behaviours were recorded using the Observer software. The observations were recorded on 3–4 females/species/dose till the parasitoid flew away or left the patch for 60 s.

4.4.2 Assessment of Sublethal Effects via Ingestion Toxicity (Wang et al. 2008)

The wasps (*A. nilaparvatae*) surviving from the ingestion toxicity bioassay method were removed and transferred to uncontaminated tubes and fed with honey solution. Equal males for mating were introduced into the tube containing females for an hour and five females were selected for assessing the sublethal effects like longevity, fecundity and emergence. A single female wasp was allowed to get into a tube containing a rice stem with approx. 40–50 brown planthopper (BPH) eggs. The longevity of the wasp was recorded every 12 h. The host eggs and feed were replaced daily. Observations were recorded on the total number of parasitized eggs and the rate of emergence.

4.4.3 Standard Leslie Matrix Model for Population Growth (Banks et al. 2011)

This model is used to find out the possibility of predicting the impact of pesticides on a guild of parasitoids by assessing one representative species. A standard Leslie matrix model (Leslie 1945) was used to represent the population growth through time. It includes four life stages for four different parasitoid species.

The population growth may be described by the matrix equation 1:

$$X(t+1)=\begin{pmatrix} X1(t+1) \\ X2(t+1) \\ X3(t+1) \\ X4(t+1) \end{pmatrix} = \begin{pmatrix} 0 & 0 & 0 & f4 \\ a1 & 0 & 0 & 0 \\ 0 & a2 & 0 & 0 \\ 0 & 0 & a3 & a4 \end{pmatrix} = \begin{pmatrix} x1(t) \\ X2(t) \\ X3(t) \\ X4(t) \end{pmatrix} = AX(t)$$

xi – the number of individuals in each of the four life stages; i = (1, 2, 3, 4), The population expressed as a vector X = [x1, x2, x3, x4] T.
ai – the rate survivability to the next stage ($0<ai<$, $i=1, 2, 3$ and $0 \le a4 \le 1$)
f4 – Rate of reproduction of adult.

If the dominant eigenvalue (λ) of the transition matrix $(A)>1$ = population will grow (Caswell 2001; Cushing 1998).

The expression derived relating the dominant eigenvalue to the net reproductive rate of the population, R_0 is the number of progenies produced by an insect in its lifetime (Banks et al. 2010).

This rate is given by:

$$R_0 = \frac{f_4 a_1 a_2 a_3}{1 - a4} = f_4 a_1 a_2 a_3 \left(1 + a_4 + a_4^{\ 2} + a_4^{\ 3} + \ldots\right)$$

$R_0 > 1$ = growing population.
$R_0 < 1$ = population going to extinction (Dyer et al. 2008).

4.4.4 Simulation Model to Estimate the Impact of Pesticide Exposure on Parasitoids (Stark and Bamfo 2002)

In this method, the crucifer leaves each containing batches of 50 mummified *Brevicoryne brassicae* were kept on a moistened paper towel in Petri dishes. The field dose and three different concentrations of field dose (one-half, one-fourth and one-eighth) were applied using Potter's spray tower in addition to water for control. The data was recorded for 2 weeks on the emergence of parasitoid, *D. rapae*. The experiment was replicated thrice on different day with different generations of parasitoid. To estimate the effects of insecticide exposure on the parasitoid populations, a simulation model was developed (Leslie 1945; Caswell 1989; Carey 1993).

The primary matrix of the model – the life history characteristics of a parasitoid, a starting vector, n(t) – information on age distribution of the population multiplied against the primary matrix resulting in a secondary vector, n(t+1). Again multiplied against the matrix, thus could project the growth of population by the time step of the matrix. The simulation was run twice, once with a starting vector consisting adults and again with the stable age distribution.

4.4.5 Functional Response Bioassay (Rezaei et al. 2014)

The parasitoids were exposed to sub lethal insecticidal concentration (LC_{25}) employing glass vial residue bioassay method. Seven parasitoids treated with insecticides and distilled water was introduced individually to transparent cylindrical containers (7.5 cm dia.; 8 cm length) containing 4–5 leaf stage canola seedlings infested by third instar nymphs of *Lipaphis erysimi* at different densities (2, 4, 6, 8, 16, 32, and 64) having ten replicates each. The parasitoids were removed after 24 h and the aphids were maintained for about 10 days in order to count the mummies.

To determine the type of functional response,

$$\frac{N_a}{N_0} = \frac{\exp\left(P_0 + P_0N_0 + P_0N_0^{\ 2} + P_0N_0^{\ 3}\right)}{1 + \exp\left(P_0 + P_0N_0 + P_2N0_2 + P_3N0_3\right)}$$

(Juliano 2001).

Where,

Na – No. of parasitized hosts
N_0 – Initial host density
P_0,P_1,P_2,P_3 – Intercept, linear, quadratic and cubic

Type of functional response was determined by examining the sign of P1 and P2. A negative and positive linear parameter (P_1) indicates type II and III, respectively (Juliano 2001).

To compare the search rates (a) and the handling time (Th) of the parasitoid on host, the Rogers' type II random parasitism equation (Rogers 1972) can be fitted.

$$N_a = N_0\left\{1 - \exp\left[-a\left(T - ThN_0\right)\right]\right\}$$

Where,

T – Total time (24 h)
a – Attack rate (h – 1)
Th – Time of handling

Pairwise comparisons were made using indicator variable method (Juliano 2001)

$$N_a = N_0\left\{1 - \exp\left[-\left(a + Da_{(j)}\right)\left(T - \left(Th + DTh_{(j)}\right)\right)Na\right]\right\}$$

Where,

j – Indicator variable
Da and DTh – the difference between 'a' and 'Th' parameters in different treatments (Juliano 2001; Allahyari et al. 2004)

4.4.6 Toxicity Assessment on Immature Parasitoids (Mummies) (Longley and Jepson 1997a)

Newly emerged female *A. rhopalosiphi* (mated) were released into the cages containing second instar *Sitobion avenae* for parasitization for a period of 24 h. The parasitized aphids were allowed on the barley seedlings for 10 days for mummification. Then the leaf sections containing mummies were cut and glued to glass plates. The mummies used for the experiment was 80 and replicated eight times and treated with recommended dose and half of recommended dose of insecticides and water for control. Then the plates were allowed to dry for 1 h and then covered with a lid of fine mesh gauze and maintained at controlled conditions. Emerging parasitoids were fed with honey.

The parasitoids emerging were noted and transferred to an untreated chamber and fed with food and assessed for longevity. For assessing the fecundity, ten surviving females from control and treatment were transferred to potted barley seedlings. About 40 *S. avenae* were introduced on the plants and kept inside the cylinders of clear acetate sheeting. The plant chambers were maintained at controlled environment with constant illumination for 24 h and then the parasitoids were removed. Observations were recorded on the mummified aphids (Nos) after 12 days. Levene test was used for testing the homogeneity of variance (Day and Quinn 1989). The fecundity data was analysed using a two-sample t test.

4.4.7 Egg Card Dip Bioassay (Carvalho et al. 2003)

The newly emerged *T. pretiosum* females (15 adults) were individualized in glass tubes and provided with blue paper cards containing UV killed *Anagasta kuehniella* eggs for 24 h. The paper cards with eggs were maintained at controlled environment for the development of *T. pretiosum*. Eggs of *A. kuehniella* containing *T. pretiosum* at egg larval, pre pupal and pupal with 24, 96 and 192 h after parasitism respectively were treated by dipping them in the respective insecticidal solution for 5 s and after elimination of excess moisture, paper cards were individualized in glass tubes. Then 15 newly emerged females were selected at random from treated F1 generation eggs from each treatment, individualized in glass tubes and provided with non-treated and inviabilized eggs of *A. kuehniella* for 48 h for assessing the pesticide impacts in the next generation. The following parameters were evaluated.

1. The number of eggs parasitized by F1 generation insects
2. Emergence of parasitoid in F1 and F2 generations
3. The parasitoid sex ratio in the egg-larval, prepupal and pupal stages
4. The F1 female longevity.

A paper card containing about 125 parasitized host eggs constitute the experimental unit. The comparisons between treatments were analysed by Scott and Knott's groupment analysis test at 5 % probability (Scott and Knott 1974).

4.4.8 Assessment of Longevity, Progeny Production, Size and Sex Ratio (Rosenheim and Hoy 1988a)

In this method, young female parasitoids (*Aphytis melinus*) were exposed to pesticides for 24 h that would cause about 50 % mortality. Afterwards the live *A. melinus* parasitoids were removed and provided with host insects (*Aspidiotus nerii*), to assess their longevity, progeny production and sex ratio of their offspring. On day 1, adult parasitoids were allowed to emerge from parasitized *Aspidiotus nerii* by keeping them in an emergence cage. The emerged parasitoids were given with honey as food. On the day after parasitoid emergence, the host materials were removed and the parasitoids allowed for mating. On day 3, plastic cups (30 mL) capped with gauze which was dipped for 5 s into the insecticidal solution and air dried, were used as exposure surface. Approximately ten female of *A. melinus*, were added to each plastic cups with treated guaze. The vials were held for 24 h under constant light and observations recorded. A total of ten surviving females from treatment and control were transferred to glass jar containing mature oleander scales for parasitoid oviposition. Mortality was assessed on every successive day and the dead parasitoids were removed daily and honey solution given as food. This procedure was continued till the death of all parasitoids. Potatoes removed from the glass jars were maintained for parasitoid development and emergence.

The development rate of the parasitoid, *A. melinus* was studied by keeping 20 female parasitoids with excess *A. nerii* for oviposition for 24 h. The parasitoids were removed and the host insects monitored daily for parasitoid emergence until 21 days. All the parasitoids emerged were collected, counted and sex determined. Parasitoids were mounted on slides according to the procedure given by Rosen and DeBach (1979). Hind tibial length was also measured under micrometer as an index of parasitoid size.

4.4.9 Sublethal Effect Bioassay (Jones et al. 1998)

In this experiment, about 50 *Bemisia argentifolii* adults were confined on the abaxial side of a sweet potato leaf with the help of a 23 mm dia. clip cage for 4–5 h for egg laying. After 3 days, two female parasitoids (already mated) were confined in this experimental leaves for 8 h to allow parasitisation. After 5 days of parasitisation (parasitoids in first larval stage), the leaves were sprayed with test insecticides. The sprayed leaves were excised and kept in the experimental unit for parasitoid emergence. The experimental unit consists of 60/150 × 25 mm tissue culture dish covered on top with organdy for ventilation. Another spray was given to other plants at 12 days after parasitisation (parasitoid at pupal stage).

On the 15th day of parasitisation, healthy parasitoid pupae were collected individually and kept in small capsules containing a droplet of pure honey for assessing the parasitoid emergence. Mating and reproduction of the survivors of the pupal spray treatment was assessed. Parasitoids were paired according to the treatment and allowed for mating. Each pair was then allowed in culture dish containing 200–300 IInd instar *B. argentifolii* in a leaf. The progeny production was assessed, counted and sexed.

4.4.10 Bioassay to Measure Host Foraging Ability Using Wind Tunnel (Stapel et al. 2000)

In this method, the parasitoid, *Microplites croceipes* females starved for 2 days were allowed individually to feed on 2 μL extrafloral nectar droplet collected after insecticidal treatment until satiation. The wasps were transferred individually to glass vials with honey/water solution. The next day, they were exposed to hosts and frass individually and made them to sting once before evaluating their flight response. To stimulate a flight response in the wind tunnel, a cotton leaf fed overnight by *Helicoverpa zea* was placed upwind and wasps were released 1 m downwind individually. The response of wasps was recorded if it initiates the flight within 5 min. and those landing on leaf. Actually for completion of flight three chances were given. After completion of wind tunnel experiments, longevity was assessed by exposing the wasps individually into vials with honey/water solution.

4.4.11 Walking Behaviour in Response to Chemical Stimuli (Saljoqi et al. 2012)

Walking behavior of the parasitoid, *Trichogramma chilonis* is determined with the help of computer based software. Different sizes of grids were made on an arena and the parasitoid wasps were set free to walk over these grids. The accuracy of the result depends upon the size of the grids as smaller the grids size, result will be accurate and the grids were numbered. A transparent cover slips of length 5 cm and width 6 cm having very thin boundaries was used for avoiding the escape of wasps. The test insecticide was sprayed evenly on the adult parasitoids in different insecticidal concentrations and also control with water. The same procedure was carried out without insecticide also. The software was started as soon as the insect start moving on the grid and numbers of those grids were entered in the computer from which the wasp passed away and the observation was recorded for 3 min.

5 Pesticide Risk Assessment for Parasitoids

Parasitoids are considered as the most effective natural enemy for biological control of insect pests (Murdoch 1994). However, many of the parasitoid populations are be susceptible to chemical pesticides (Theiling and Croft 1998) and thus can't be integrated. The toxicity effects can be studied through laboratory toxicity studies, field safety of insecticides to parasitoids and also risk assessment methods. Risk quotient is a good indicator for evaluating the safety of insecticides to natural enemies in the field (Stark et al. 1995). Risk quotient is used for assessing the risk of pesticides to non-targets (Campbell et al. 2000) and can be used as an indicator of ecological risk (Peterson 2006). Risk quotient is used to assess the safety of pesticides to parasitoids such as *Bracon hebetor* (Danfa et al. 1998), *Trichogramma cacoeciae* (Hassan et al. 1998) and *T. chilonis* (Preetha et al. 2009).

Pesticide risk assessment to biocontrol agents and natural enemies should be assessed scientifically and the safer insecticide molecules can be identified and its compatibility with IPM can be determined (Stark et al. 1995; Jepson and Croft 1998). Low or no toxic effect levels are to be estimated to study the risk of pesticides to beneficial organisms (Leeuwen and Hermens 1995; Koojman et al. 1996; Bruijn and Hof 1997; Hoeven 1997). Some of the pesticide risk assessment methods available for parasitoids are given below

5.1 Risk Assessment Methodologies

5.1.1 Risk Assessment Based on Mortality (Miyata et al. 2001)

Dry film residue bioassay method was adopted for assessing the mortality of parasitoid adults. The insecticides are categorized based on mortality percentage as follows:

Mortality	Category
<50 %	Harmless
50–79 %	Slightly harmful
80–99 %	Moderately harmful
>99 %	Harmful

In topical application method using Potter's spray tower conducted by Brunner et al. (2001), the insecticides were graded based on the corrected mortality of the parasitoid as follows:

<20 % – low toxicity
>20 % and <70 % – moderate toxicity
>70 % – highly toxic.

5.1.2 Risk Assessment Based on the Reduction in Parasitism (Abdulhay and Rathi 2014)

In the residual acute toxicity bioassay method, according to IOBC criteria (Grutzmacher et al. 2004), classified the insecticide as follows:

Class	Reduction in parasitisation capacity	Category
I	<30 %	Non-toxic
II	30–79 %	Slightly toxic
III	80–99 %	Moderately toxic
IV	>99 %	Toxic

5.1.3 Risk Quotient (Preetha et al. 2009)

Risk quotients can be estimated from the LC_{50} values based on the formula given below. The LC_{50} values at 48 h after treatment are usually used for calculating the risk quotient.

$$\text{Risk quotient} = \frac{\text{Recommended field rate g a.i./ha}}{LC_{50} \text{ of beneficial insect mg a.i./L}}$$

Risk quotient	Category
<50	Harmless
50–2500	Slightly to moderately toxic
>2500	Dangerous

5.1.4 Environmental Impact Quotient (EIQ) (Biddinger et al. 2014)

The environmental impacts of reduced-risk and conventional insecticides were compared by EIQ analysis which is calculated as follows:

$$\text{EIQ Field Rating} = \sum \left(\text{EIQi} * \text{RTi} * \text{APi} \right)$$

Where,

EIQi = EIQ value of pesticide i
RTi = rate of pesticide i
APi = number of applications of pesticide i

5.1.5 Tiered Approach for Risk Assessment (Miles and Alix 2012; http://www.pesticides.gov.uk)

In Tier I or laboratory tests, the parasitoids are usually exposed to pesticide residues through glass plates. The Tier I, Hazard Quotient (HQ) was calculated using median lethal rate (LR_{50}) values for the parasitoids. Tier II tests for parasitoids are done by exposing them to pesticide residues on leaf surface and mortality assessed. In the Tier II risk assessment, a 50 % effect trigger is applied that is equal to an HQ trigger of 1 (depending on LR_{50} values for parasitoids derived from extended laboratory tests). If the HQ value calculated is well below the trigger value for the parasitoids, it was concluded that the use was of low risk. For the parasitoids, HQ trigger = 2.

5.2 *Risk of Pesticides on Parasitoids*

5.2.1 Risk Assessment Based on Acute Toxicity

Organophosphates and carbamates showed high toxicity to *Trichogramma confusum* with LC_{50} values ranging from 0.037 to 0.29 and from 0.17 to 1.61 mg a.i./L, respectively. Among the seven different classes of insecticides, the least toxicity was exhibited by insect growth regulators with LC_{50} values ranging from 3907 to 10,154 mg a.i./L. Based on risk quotient analysis, the neonicotinoids tested except thiamethoxam, pyrethroids and phenylpyrazoles are reported safe to *T. confusum*. Avermectins and IGRs were also found safer to the parasitoids. Organophosphates and carbamates tested are either slightly to moderately toxic or highly toxic (Wang et al. 2013).

The median lethal concentration of thiamethoxam was 0.0014 mg a.i./L to *T. chilonis* and reported as highly toxic followed by imidacloprid (0.0027 mg a.i./L) tested on insecticide coated vial (scintillation) residue bioassay. The values of acephate and endosulfan showed lower toxicity with an LC_{50} of 4.4703 and 1.8501 mg a.i./L, respectively. Based on risk quotient, among the tested chemicals chlorantraniliprole was harmless and the remaining insecticides viz., thiamethoxam, imidacloprid, ethofenprox and BPMC were dangerous to *T. chilonis* (Preetha et al. 2009).

Organophosphates viz., chlorpyrifos, fenitrothion, phoxim, profenofos and triazophos and carbamates viz., carbaryl, carbsulfan, isoprocarb, metolcarb, and primicarb are found to exhibit high toxicity to *T. japonicum*, with an LC_{50} of 0.035–0.49 mg a.i./L, followed by antibiotics (abamectin, ivermectin and emamectin), phenylpyrazoles (ethiprole, and fipronil), pyrethroids (cypermethrin, fenpropathrin and lambda-cyhaothrin), and neonicotinoids (imidaclothiz, nitenpyram, thiacloprid, acetamiprid, imidacloprid and thiamethoxam). The insect growth regulators like hexaflumuron, tebufenozide, chlorfluazuron and fufenozide are found to be least toxic to *Trichogramma japonicum* with an LC_{50} of 3383–30,206 mg a.i./L by dry film residue bioassay method. Based on risk quotient analysis, phenylpyrazoles, pyrethroids, insect growth regulators, neonicotinoids except thiamethoxam and antibiotics except abamectin are classified as safe to *T. japonicum* whereas most of the OPs and carbamates tested were moderately or highly toxic (Zhao et al. 2012).

Organophosphates and carbamates were highly toxic to *T. evanescens*. Among the phenylpyrazoles, ethiprole was less toxic. Based on LC_{50} values, avermectins (1.29–2.57 mg a.i./L), pyrethroids (2.26–14.03 mg a.i./L) and neonicotinoids (1.12–239.1 mg a.i./L) were found to be less toxic, whereas insect growth regulators exhibited a low toxicity to the parasitoid, with very high LC_{50} values (3383–5650 mg a.i./L). Based on risk quotient, avermectins, neonicotinoids, pyrethroids and IGRs are safe or low toxic to the parasitoid, *T. evanescens*. Phenylpyrazoles except ethiprole, most of the OPs and carbamates falls on slightly to moderate toxic category or highly toxic (Wang et al. 2014).

For *T. chilonis*, lambda cyhalothrin, carbosulfan and indoxacarb were reported to be dangerous based on risk quotient and bifenthrin, thiamethoxam, imidacloprid, acetamiprid, pymetrozine and buprofezin were slightly to moderately toxic. Thiamethoxam, pymetrozine and buprofezin were slightly to moderately toxic to *T. brasiliensis* and the remaining insecticides viz., lambda cyhalothrin, carbosulfan, indoxacarb, imidacloprid and acetamiprid were toxic to this parasitoid (Shankarganesh et al. 2013). Based on risk quotient analysis, chlorantraniliprole was reported to be harmless to *T. chilonis* while thiamethoxam was dangerous whereas, buprofezin, flubendiamide and spinosad were slightly to moderately toxic to *T. chilonis* (Uma 2013). The acute toxicity of abamectin, emamectin benzoate, indoxacarb and spinosad against *T. chilonis* revealed that they were less toxic to the parasitoid and can be recommended for IPM programme. Among the different methods evaluated, hazard ratio is found to be the best as it accounted for the field dose as criteria for determining the toxicity of the insecticides tested (Sampathkumar and Krishnamoorthy 2013).

Tran and Ueno (2012) stated that the LC_{50} values were 0.0035, 0.0085, 0.0231, 0.0508 and 8.779 mg a.i./L for imidacloprid, ethofenprox, clothianidin, lufenuron, and pymetrozine, respectively to *Neochrysocharis okazakii*, leafminer parasitoid on vegetables. Based on risk quotient, imidacloprid and ethofenprox are highly toxic, lufenuron and clothianidin were slightly to moderately toxic. Pymetrozine was reported to be safe to *N. okazakii*. Among the four parasitoids viz., *Diaeretiella rapae*, *Psyttalia fletcheri*, *Diachasmimorpha longicaudata* and *Fopius arisanus*, *D. rapae* is far more robust to pesticide exposure (Banks et al. 2011). Fipronil, chlorfenapyr and diafenthiuron were reported harmful, whereas, abamectin and cypermethrin as moderately harmful and slightly harmful to *Cotesia plutellae* adults by dry film residue bioassay (Miyata et al. 2001).

5.2.2 Tiered Evaluation of Pesticide Toxicity

The median lethal rates (LR_{50}) was reported to be much higher at Tier II than at Tier I but the ranking of LR_{50}s for the three parasitoids, *Aphidius colemani*, *A. ervi* and *A. rhopalosiphi* for insecticides viz., dimethoate, λ-cyhalothrin and imidacloprid were the same at both the tiers. At Tier I, a hazard quotient (HQ) of more than 2, revealed all the three pesticides as harmful to *A. rhopalosiphi*. At tier II, dimethoate exceeded this trigger value of 2 and imidacloprid with HQ 1.9, reveals the need for further testing. Since the median lethal values vary very much between 24 and 48 h data, 48 h data can be used to find the risk assessments. The probit and logit models were found to be nicely correlated with each other and thus both can be used for finding out the median lethal values, provided goodness of fit to be reported. For finding the median lethal values, a geometric series of doses based on the field recommended dose and a factor of two can be used (http://www.pesticides.gov.uk). The residues of deltamethrin dissipated over 5 days after treatment. Though deltamethrin was reported to control aphid at very low dose than the recommended field rate, some aphids are found to survive inside some crop refuges (Longley and Jepson 1997b). So, persistence of chemical should be also taken into consideration while assessing the risk.

References

H.S. Abdulhay, M.H. Rathi, Effect of some insecticides on the egg parasitoid, *Trichogramma evanescens* Westwood (Hymenoptera: Trichogrammatidae). J. Al-Nahrain Univ. **17**(3), 116–123 (2014)

J.B. Adams, Effects of spraying 2,4-D amine on coccinellid larvae. Can. J. Zool. **38**, 285–288 (1960)

S. Ahmad, M. Ashfaq, M. Hassan, S.T. Sahi, Potential of parasitoid *Trichogramma chilonis* (Ishii) (Hymenoptera: Trichogrammatidae) against the sugarcane stem borer, *Chilo infuscatellus* (Lepidoptera: Pyralidae) under field conditions. Intl. J. Biodiv. Conserv. **4**(1), 36–38 (2012)

J.L.N. Aldrey, F.M. Fontal-Cazalla, Filogenia y evolucion del orden Hymenoptera. Bol. S.E.A. **26**, 459–474 (1999)

P.M. Alexandre, A.C. Geraldo, E.P. Adriano, D.R.L. Carlos, Selectivity evaluation of insecticides used to control tomato pests to *Trichogramma pretiosum*. J. Biol. Cont. **51**, 769–778 (2006)

H. Allahyari, P.A. Fard, J. Nozari, Effects of host on functional response of offspring in two populations of *Trissolcus grandis* on the sunn pest. J. Appl. Entomol. **128**, 39–43 (2004)

R.P. Almeida, Cotton insect pest control on a small farm: an approach of successful biological control using *Trichogramma*. Proc. Exp. Appl. Entomol. **12**, 81–84 (2001)

K. Amandeep, N.N. Singh, K. Mukesh, Persistent toxicity of insecticides on biotic potential of *Trichogramma brasiliensis* Ashmead (Hymenoptera: Trichogrammatidae), an egg parasitoid of tomato fruit borer. Indian J. Entomol. **74**(4), 359–365 (2012)

E.L. Araujo, C.H.F. Nogueira, C.E.S. Bezerra, E.M. Costa, Toxicity of insecticides used in melon crops to *Opius scabriventris* (Hymenoptera: Braconidae). Biosci. J. **31**(5), 1370–1377 (2015)

Y.H. Attallah, L.D. Newsom, Ecological and nutritional studies on *Coleomegilla maculata* De Geer (Coleoptera: Coccinellidae). III. The effect of DDT, toxaphene and endrin on the reproductive and survival potentials. J. Econ. Entomol. **59**, 1181–1187 (1966)

Atwa, A. Atwa, Can *Trichogramma* and entomopathogenic nematodes serve as effective biocontrol options for *Helicoverpa zea* in tomato? Acta Adv. Agric. Sci. **02**(10), 09–21 (2014)

M. Aydogdu, U. Guner, Effects of 5 different insecticides on mortality of the Leafroller parazitoid *Itoplectis maculator* (Fabricius, 1775) (Ichneumonidae: Hymenoptera). Turk. Entomol. Bult. **2**(4), 243–249 (2012)

A. Ayvaz, E. Karasu, S. Karaborklu, S. Yilmaz, Dispersal ability and parasitization performance of egg parasitoid *Trichogramma evanescens* Westwood (Hymenoptera: Trichogrammatidae) in field and storage conditions. Turk. J. Biol. **32**, 127–133 (2008)

C.R. Ballal, S.P. Singh, The effectiveness of *Trichogramma chilonis*, *Trichogramma pretiosum* and *Trichogramma brasiliense* (Hymenoptera: Trichogrammatidae) as parasitoids of *Helicoverpa armigera* (Lepidoptera: Noctuidae) on sunflower (*Helianthus annuus*) and redgram (*Cajanus cajan*). Biocontrol Sci. Technol. **13**(2), 231–240 (2003)

J.E. Banks, A.S. Ackleh, J.D. Stark, The use of surrogate species in risk assessment: using life history data to safeguard against false negatives. Risk Anal. **30**, 175–182 (2010)

J.E. Banks, J.D. Stark, R.I. Vargas, A.S. Ackleh, Parasitoids and ecological risk assessment: can toxicity data developed for one species be used to protect an entire guild? Biol. Control **59**(3), 336–339 (2011)

C.S. Bastos, R.P. Almeida, F.A. Suinaga, Selectivity of pesticides used on cotton (*Gossypium hirsutum*) to *Trichogramma pretiosum* reared on two laboratory- reared hosts. Pest Manag. Sci. **62**, 91–98 (2006)

A. Bayram, G. Salerno, A. Onofri, E. Conti, Sub-lethal effects of two pyrethroids on biological parameters and behavioral responses to host cues in the egg parasitoid *Telenomus busseolae*. Biol. Control **53**, 153–160 (2010)

V.H. Beloti, G.R. Alves, D.F. Araujo, M.M. Picoli, R.A. Moral, C.G. Demetrio, P.T. Yamamoto, Lethal and sublethal effects of insecticides used on citrus, on the ectoparasitoid *Tamarixia radiata*. PLoS One **10**(7), e0132128 (2015)

P. Bengochea, P. Medina, F. Amor et al., The effect of emamectin benzoate on two parasitoids, *Aphidius colemani* Viereck (Hymenoptera: Braconidae) and *Eretmocerus mundus* Mercet

(Hymenoptera: Aphelinidae), used in pepper greenhouses. Span. J. Agric. Res. **10**(3), 806–814 (2012)

J. Bengtsson, P. Angelstam, T. Elmqvist, U. Emanuelsson, C. Folke, M. Ihse, F. Moberg, M. Nystrom, Reserves, resilience and dynamic landscapes. Ambio **32**, 389–396 (2003)

D.J. Biddinger, T.W. Leslie, N.K. Joshi, Reduced-risk pest management programs for eastern U.S. peach orchards: effects on arthropod predators, parasitoids and select pests. J. Econ. Entomol. **107**(3), 1084–1091 (2014)

T.A. Bjorksten, M. Robinson, Juvenile and sublethal effects of selected pesticides on the leafminer parasitoids *Hemiptarsenus varicornis* and *Diglyphus isaea* (Hymenoptera: Eulophidae) from Australia. J. Econ. Entomol. **98**(6), 1831–1838 (2005)

I. Blibech, M. Ksantini, T. Jardak, M. Bouaziz, Effect of insecticides on *Trichogramma* parasitoids used in biological control against *Prays oleae* insect pest. Adv. Chem. Eng. Sci. **5**, 362–372 (2015)

A. Bonet, Parasitoid wasps, natural enemies of insects, in *Tropical Biology and Conservation Management*, ed. by K. Del-Claro, P.S. Oliveira, V. Rico-Gray, vol. VII (Eolss Publishers, Paris, 2009)

J.H.M.D. Bruijn, M. Hof, How to measure no effect. Part IV: how acceptable is the ECx from an environmental policy point of view? Environmetrics **8**, 263–267 (1997)

J.F. Brunner, J.E. Dunley, M.D. Doerr, E.H. Beers, Effect of pesticides on *Colpoclypeus florus* (Hymenoptera: Eulophidae) and *Trichogramma platneri* (Hymenoptera: Trichogrammatidae), parasitoids of leafrollers in Washington. J. Econ. Entomol. **94**(5), 1075–1084 (2001)

M.D. Bryan, R.J. Dysart, T.L. Burger, Releases of introduced parasites of the alfalfa weevil in the United States, 1957–88. USDA APHIS Misc. Publ. No. 1504. 203 p (1993)

A.F. Bueno, R.C.O.F. Bueno, J.R.P. Parra, S.S. Vieira, Effects of pesticides used in soybean crops to the egg parasitoid, *Trichogramma pretiosum*. Cienc. Rural. Santa Maria **38**(6), 1495–1503 (2008)

L.E. Caltagirone, Landmark examples in classical biological control. Annu. Rev. Entomol. **26**, 213–232 (1981)

P.J. Campbell, K.C. Brown, E.G. Harrison, F. Bakker, K.L. Barrett, M. Candolfi, A hazard quotient approach for assessing the risk to non-target arthropods from plant protection products under 91/414/ EEC: hazard quotient trigger value proposal and validation. Anz. Schadlingskunde **73**, 117–124 (2000)

E.E.G. Cardwell, P. Gu, Conserving vedalia beetle, *Rodolia cardinalis* (Mulsant) (Coleoptera: Coccinellidae), in citrus: a continuing challenge as new insecticides gain registration. J. Econ. Entomol. **96**, 1388–1398 (2003)

J.R. Carey, *Applied Demography for Biologists* (Oxford University Press, New York, 1993)

E.L. Carmo, A.F. Bueno, R.C.O.F. Bueno, Pesticide selectivity for the insect egg parasitoid *Telenomus remus* Carmo. BioControl **55**, 455–464 (2010)

G.A. Carvalho, P.R. Reis, L.C.D. Rocha, J.C. Moraes, L.C. Fuini, C.C. Ecole, Side-effects of insecticides used in tomato fields on *Trichogramma pretiosum* (Hymenoptera, Trichogrammatidae). Acta Sci. **25**(2), 275–279 (2003)

H. Caswell, *Matrix Population Models* (Sinauer Associates, Sunderland, 1989)

H. Caswell, *Matrix Population Models: Construction, Analysis and Interpretation*, 2nd edn. (Sinauer Associates Press, Sunderland, 2001)

M.G. Chitgar, M. Ghadamyari, Effects of amitraz on the parasitoid *Encarsia formosa* (Gahan) (Hymenoptera:Aphelinidae) for control of *Trialeurodes vaporariorum* Westwood (Homoptera: Aleyrodidae): IOBC methods. J. Entomol. Res. Soc. **14**(2), 61–69 (2012)

Z.J. Chowdhury, S.N. Alam, C.K. Dash, M.A. Maleque, A. Akhter, Determination of parasitism efficacy and development of effective field release technique for *Trichogramma* spp. (Trichogrammatidae: Hymenoptera). Am. J. Exp. Agric. **10**(1), 1–7 (2016)

F.L. Consoli, P.S.M. Botelho, J.R.P. Parra, Selectivity of insecticides to the egg parasitoid *Trichogramma galloi* Zucchi, (Hym., Trichogrammatidae). J. Appl. Ent. **125**, 37–43 (2001)

B.R. Critchley, A laboratory study of the effects of some soil applied organophosphorus pesticides on Carabidae (Coleoptera). Bull. Entomol. Res. **62**, 229–242 (1972)

B.R. Critchley, Literature review of sunn pest, *Eurygaster integriceps* (Hemiptera, Scutelleridae). Crop Prot. **17**, 271–287 (1998)

B.A. Croft, Susceptibility surveillance to pesticides among arthropod natural enemies: modes of uptake and basic responses. Z. Pflanzenkrankh. Pflanzenschutz **84**, 140–157 (1977)

B.A. Croft, *Arthropod Biological Control Agents and Pesticides* (Wiley, New York, 1990), 723 p

J.V. Cross, M.G. Solomon, D. Babandreier et al., Biocontrol of pests of apples and pears in Northern and Central Europe: 2. Parasitoids. Biocontrol Sci. Technol. **9**(3), 277–314 (1999)

J.M. Cushing, *An Introduction to Structured Population Dynamics, Conference Series in Applied Mathematics*, vol. 71 (SIAM, Philadelphia, 1998)

A. Danfa, B. Fall, H.V. Valk, Acute toxicity tests with *Bracon hebetor* Say (Hymenoptera: Braconidae) using different locust control insecticides in the Sahel, in *Environmental Side-Effects of Locust and Grasshopper Control*, ed. by J.W. Everts, D. Mbaye, O. Barry, W. Mullie (FAO, Dakar, 1998), pp. 117–136

H.R. Dastjerdi, M.J. Hejazi, G.N. Ganbalani, M. Saber, Toxicity of some biorational and conventional insecticides to cotton bollworm, *Helicoverpa armigera* (Lepidoptera: Noctuidae) and its ectoparasitoid, *Habrobracon hebetor* (Hymenoptera: Braconidae). J. Entomol. Soc. Iran **28**(1), 27–37 (2008)

H.R. Dastjerdi, M.J. Hejazi, G.N. Ganbalani, M. Saber, Sublethal effects of some conventional and biorational insecticides on ectoparasitoid, *Habrobracon hebetor* Say (Hymenoptera: Braconidae). J. Entomol. **6**(2), 82–89 (2009)

R.W. Day, G.P. Quinn, Comparisons of treatments after an analysis of variance in ecology. Ecol. Monogr. **59**, 433–463 (1989)

J.M. Delpuech, B. Legallet, O. Terrier, P. Fouillet, Modifications of the sex pheromonal communication of *Trichogramma brassicae* by a sublethal dose of deltamethrin. Chemosphere **38**(4), 729–739 (1999)

J.P. Dempster, The sublethal effect of DDT on the rate of feeding by the ground beetle *Harpalus rufipes*. Entomol. Exp. Applic. **11**, 51–54 (1968)

N. Desneux, E. Wajnberg, X. Fauvergue, S. Privet, L. Kaiser, Oviposition behaviour and patch-time allocation in two aphid parasitoids exposed to deltamethrin residues. Entomol. Exp. Applic. **112**, 227–235 (2004)

N. Desneux, X. Fauvergue, F.O. Dechaume-Moncharmont, L. Kerhoas, Y. Ballanger, L. Kaiser, *Diaeretiella rapae* limits *Myzus persicae* populations after applications of deltamethrin in oilseed rape. J. Econ. Entomol. **98**(1), 9–17 (2005)

N. Desneux, R. Ramirez-Romero, L. Kaiser, Multistep bioassay to predict recolonization potential of emerging parasitoids after a pesticide treatment. Environ. Toxicol. Chem. **25**(10), 2675–2682 (2006a)

N. Desneux, R. Denoyelle, L. Kaiser, A multi-step bioassay to assess the effect of the deltamethrin on the parasitic wasp *Aphidius ervi*. Chemosphere **65**, 1697–1706 (2006b)

N. Desneux, A. Decourtye, J.M. Delpuech, The sublethal effects of pesticides on beneficial arthropods. Ann. Rev. Entomol. **52**, 81–106 (2007)

Y. Dewer, S. Abdel-Razak, A. Barakat, Comparative efficacy of some insecticides against purple scale insect, *Lepidosaphes beckii* (Hemiptera: Coccoidea) and its parasitoid in citrus orchard in Egypt. Egypt. Acad. J. Biol. Sci. **5**(3), 121–127 (2012)

M.F. Díaz, A. Ramirez, K. Poveda, Efficiency of different egg parasitoids and increased floral diversity for the biological control of noctuid pests. Biol. Control **60**, 182–191 (2012)

R.L. Doutt, The historical development of biological control, in *Biological Control of Insect Pests and Weeds*, ed. by P. DeBach (Chapman and Hall, London, 1964), pp. 21–42

P. Duelli, M.K. Obrist, Biodiversity indicators: the choice of values and measures. Agric. Ecosyst. Environ. **98**, 87–98 (2003)

S.D. Dyer, D.J. Versteeg, S.E. Belanger, J.G. Chaney, S. Raimondo, M.G. Barron, Comparison of species sensitivity distributions derived from interspecies correlation models to distributions used to derive water quality criteria. Environ. Sci. Technol. **42**, 3076–3083 (2008)

P. Eggleton, R. Belshaw, Insect parasitoids: an evolutionary overview. Philos. Trans. R. Soc. London B. **337**(1279), 1–20 (1992)

N.E. El-Wakeil, Evaluation of efficiency of *Trichogramma evanescens* reared on different factitious hosts to control *Helicoverpa armigera*. J. Pest. Sci. **80**, 29–34 (2007)

G.W. Elzen, Sublethal effects of pesticides on beneficial arthropods, in *Pesticides and Non-target Invertebrates*, ed. by P.C. Jepson (Intercept, Wimborne, 1989), pp. 129–150

E. Ergin, A. Er, F. Uckan, D.B. Rivers, Effect of cypermethrin exposed hosts on egg-adult development time, number of offspring, sex ratio, longevity and size of *Apanteles galleriae* Wilkinson (Hymenoptera: Braconidae). Belgian J. Zool. **137**, 27–31 (2007)

F.L. Fernandes, L. Bacci, M.S. Fernandes, Impact and selectivity of insecticides to predators and parasitoids. EntomoBrasilis **3**(1), 1–10 (2010)

S.E. Flanders, Predatism by the adult hymenopterous parasite and its role in biological control. J. Econ. Entomol. **46**, 541–544 (1953)

C.A. Fleschner, G.T. Scriven, Effect of soil-type and DDT on ovipositional responses of *Chrysopa californica* (Coq.) on lemon. J. Econ. Entomol. **50**, 221–222 (1957)

C. Garcera, Y. Ouyang, S.J. Scott, E. Molto, G.E.E. Cardwell, Effects of spirotetramat on *Aonidiella aurantii* (Homoptera: Diaspididae) and its parasitoid, *Aphytis melinus* (Hymenoptera: Aphelinidae). J. Econ. Entomol. **106**(5), 2126–2134 (2013)

V. Girolami, L. Mazzon, A. Squartini, N. Mori, M. Marzaro, A.D. Bernardo, Translocation of neonicotinoid insecticides from coated seeds to seedling guttation drops: a novel way of intoxication for bees. J. Econ. Entomol. **102**, 1808–1815 (2009)

H.C.J. Godfray, *Parasitoids: Behavioural and Evolutionary Ecology* (Princeton University, Princeton, 1994), 465 p

J.E. Gonzalez-Zamora, M.L. Castillo, C. Avilla, Side effects of different pesticides used in citrus on the adult stage of the parasitoid, *Aphytis melinus* DeBach (Hymenoptera: Aphelinidae) and its progeny. Span. J. Agric. Res. **11**(2), 494–504 (2013)

R.M. Goulart, H.X.L. Volpe, A.M. Vacari, R.T. Thuler, S.A.D. Bortoli, Insecticide selectivity to two species of *Trichogramma* in three different hosts, as determined by IOBC/WPRS methodology. Pest Manag. Sci. **68**, 240–244 (2012)

D.J. Greathead, Parasitoids in classical biological control, in *Insect Parasitoids*, ed. by J. Waage, D. Greathead (Academic, London, 1986), pp. 289–318

S. Grenier, G. Plantevin, Development modifications of the parasitoid *Pseudoperichaeta nigrolineata* (Dipt., Tachinidae) by fenoxycarb, an insect growth regulator applied onto its host *Ostrinia nubilalis* (Lep., Pyralidae). J. Appl. Entomol. **110**, 462–470 (1990)

A.D. Grutzmacher, O. Zimmermann, A. Yousef, S.A. Hassan, The side-effects of pesticides used in integrated production of peaches in Brazil on the egg parasitoid *Trichogramma cacoeciae* Marchal (Hym., Trichogrammatidae). J. Appl. Entomol. **128**, 377–383 (2004)

K.S. Hagen, Ecosystem analysis: plant cultivars (HPR), entomophagous species and food supplements, in *Interactions of Plant Resistance and Parasitoids and Predators of Insects*, ed. by D.J. Boethel, R.D. Eikenbary (Wiley, New York, 1986), pp. 151–197

B. Halappa, J.S. Awaknavar, D. Archana, Toxicity of insecticides to larval endoparasitoid, *Cotesia plutellae* (Kurdj.) of diamondback moth, *Plutella xylostella* (L.) under laboratory condition. Pestic. Res. J. **24**(2), 231–234 (2012)

R. Hamada, Egg parasitoids of common cutworm, *Spodoptera litura* (Fabricius) (Lepidoptera: Noctuidae). Jpn. J. Appl. Entomol. Zool. **36**(4), 258–259 (1992)

M. Haseeb, H. Amano, Effects of contact, oral and persistent toxicity of selected pesticides on *Cotesia plutellae* (Hym., Braconidae), a potential parasitoid of *Plutella xylostella* (Lep., Plutellidae). J. Appl. Entomol. **126**(1), 8–13 (2002)

M. Haseeb, T.X. Liu, W.A. Jones, Effects of selected insecticides on *Cotesia plutellae*, endoparasitoid of *Plutella xylostella*. Biocontrol **49**(1), 33–46 (2004)

S.A. Hassan, Guideline for the evaluation of side-effects of plant protection product on *Trichogramma cacoeciae*. Guidelines for testing the effects of pesticides on beneficial organisms: description of test methods. IOBC/WPRS Bull. **15**, 18–39 (1992)

S.A. Hassan, B. Hafes, P.E. Degrande, K. Herai, The side-effects of pesticides on the egg parasitoid *Trichogramma cacoeciae* Marchal (Hym., Trichogrammatidae), acute dose-response and persistence tests. J. Appl. Entomol. **122**, 569–573 (1998)

X.Z. He, Q. Wang, D.A.J. Teulon, The effect of parasitism by *Aphidius ervi* on development and reproduction of the pea aphid, *Acyrthosiphon pisum*. N. Z. Plant Prot. **58**, 202–207 (2005)

E. Hegazi, A. Herz, S.A. Hassan, W.E. Khafagi, E. Agamy, A. Zaitun, G.A. El-Aziz, S. Showeil, S. El-Said, N. Khamis, Field efficiency of indigenous egg parasitoids (Hymenoptera, Trichogrammatidae) to control the olive moth (*Prays oleae*, Lepidoptera, Yponomeutidae) and the jasmine moth (*Palpita unionalis*, Lepidoptera, Pyralidae) in an olive plantation in Egypt. Biol. Control **43**(2), 171–187 (2007)

G.E. Heimpel, T.R. Collier, The evolution of host-feeding behaviour in insect parasitoids. Biol. Rev. **71**, 373–400 (1996)

R. Hemandez, M. Harris, T.X. Liu, Impact of insecticides on parasitoids of the leafminer, *Liriomyza trifolii*, in pepper in South Texas. J. Insect Sci. **11**, 61 (2011)

T.A. Hill, R.E. Foster, Effect of insecticides on the diamondback moth (Lepidoptera: Plutellidae) and its parasitoid, *Diadegma insulare* (Hymenoptera: Ichneumonidae). J. Econ. Entomol. **93**(3), 763–768 (2000)

Y. Hirose, Biological and ecological comparison of *Trichogramma* and *Telenomus* as control agents of lepidopterous pests. J. Appl. Entomol. **101**, 39–47 (1986)

M.S. Hoddle, R.G.V. Driesche, J.P. Sanderson, Biology and use of the whitefly parasitoid, *Encarsia formosa*. Annu. Rev. Entomol. **43**, 645–669 (1998)

N.V. Hoeven, How to measure no effect? Part III: statistical aspects of NOEC, ECx AND NEC estimates. Environmetrics **8**, 255–261 (1997)

M.P. Hoffmann, S.A. Pitcher, S.A. Cheever, J. Gardner, J.E. Losey, T.P. Kuhar, C.A. Laub, R.R. Youngman, Efficacy of inoculative releases of *Trichogramma ostriniae* (Hymenoptera: Trichogrammatidae) against eEuropean corn borer *Ostrinia nubilalis* (Lepidoptera: Crambidae) in field corn. Biol. Control **36**(3), 345–349 (2006)

C.Y. Hsiech, W.W. Allen, Effects of insecticides on emergence, survival, longevity and fecundity of the parasitoid *Diaeretiella rapae* (Hymenoptera: Aphidiidae) from mummified *Myzus persicae* (Homoptera: Aphididae). J. Econ. Entomol. **79**, 1599–1602 (1986)

D. Hussain, M. Akram, Z. Iqbal, A. Ali, M. Saleem, Effect of some insecticides on *Trichogramma chilonis* Ishii (Trichogrammatidae: Hymenoptera) immature and adult survival. J. Agric. Res. **48**(4), 531–537 (2010)

D. Hussain, A. Ali, M. Mushtaq-ul-Hassan, S. Ali, M. Saleem, S. Nadeem, Evaluation of toxicity of some new insecticides against egg parasitoid *Trichogramma chilonis* (Ishii) (Hymenoptera: Trichogrammitidae). Pak. J. Zool. **44**(4), 1123–1127 (2012)

N.W. Hussey, N. Scopes, *Biological Pest Control: The Glasshouse Experience* (Cornell Univ Press, Ithaca, 1985), 240 p

Z. Hussnain, A. Naheed, S. Rizwana, Biocontrol of insect pests of sugarcane (*Saccharum* sp.). Pak. Sugar J. **22**(5), 14–23 (2007)

L.A. Irshaid, H.S. Hasan, Bioresidual effect of two insecticides on melon aphid, *Aphis gossypii* Glover (Homoptera: Aphididae) and its parasitoid, *Aphidius colemani* Verick (Hymenoptera: Brachonidae). Am.-Eur. J. Agric. Environ. Sci. **11**(2), 228–236 (2011)

S.N. Irving, I.J. Wyatt, Effects of sublethal doses of pesticides on the oviposition behavior of *Encarsia formosa*. Ann. Appl. Biol. **75**, 57–62 (1973)

P.C. Jepson, B. Croft, The issues: introduction, in *Ecotoxicology: Pesticides and Beneficial Organisms*, ed. by P.T. Haskelland, P. McEwen (Kluwer Academic Publishers, Dordrecht, 1998), pp. 191–201

M.A. Jervis, N.A.C. Kidd, Host-feeding strategies in hymenopteran parasitoids. Biol. Rev. **61**, 395–434 (1986)

M.W. Johnson, B.E. Tabashnik, Enhanced biological control through pesticide selectivity, in *Handbook of Biological Control*, ed. by T.S. Bellows, T.W. Fisher, L.E. Caltagirone, D.L. Dahlsten, C. Huffaker, G. Gordh (Academic, San Diego, 1999), pp. 297–317

W.A. Jones, M.A. Ciomperlik, D.A. Wolfenbarger, Lethal and sublethal effects of insecticides on two parasitoids attacking *Bemisia argentifolii* (Homoptera: Aleyrodidae). Biol. Control **11**, 70–76 (1998)

S.A. Juliano, Nonlinear curve-fitting: predation and functional response curves, in *Design and Analysis of Ecological Experiments*, ed. by S.M. Scheiner, J. Gurevitch, 2nd edn. (Oxford University Press, New York, 2001), pp. 178–196

S.A. Kakakhel, K. Ahad, M. Amjad, S.A. Hassan, The side-effects of pesticides on *Diaeretiella rapae*, a parasitoid of the turnip aphid (*Lipaphis erysimi*). *J. Pest Sci.* **71**(3), 61–63 (1998)

A. Kalyebi, W.A. Overholt, F. Schulthess, J.M. Mueke, S.A. Hassan, S. Sithanantham, Functional response of six indigenous trichogrammatid egg parasitoids (Hymenoptera: Trichogrammatidae) in Kenya: influence of temperature and relative humidity. Biol. Control **32**, 164–171 (2005)

S.H. Kapuge, S. McDougall, A.A. Hoffmann, Effects of methoxyfenozide, indoxacarb and other insecticides on the beneficial egg parasitoid *Trichogramma* nr. *brassicae* (Hymenoptera: Trichogrammatidae) under laboratory and field conditions. J. Econ. Entomol. **96**(4), 1083–1090 (2003)

K. Kheradmand, M. Khosravian, S. Shahrokhi, Side effect of four insecticides on demographic statistics of aphid parasitoid, *Diaeretiella rapae* (McIntosh) (Hym., Braconidae). Ann. Biol. Res **3**(7), 3340–3345 (2012)

P. Khulbe, A. Kumar, Toxicity determination of some insecticides against *Trichogramma chilonis* Ishii. Indian For. **141**(3), 333–337 (2015)

H.S. Kim, E.A. Heinrichs, P. Mylvaganam, Egg parasitism of *Scirpophaga incertulas* Walker (Lepidoptera: Pyralidae) by Hymenopterous parasitoids in IRRI rice fields. Korean J. Plant Prot. **25**(1), 37–40 (1986)

Y. Kobori, H. Amano, Effects of agrochemicals on life-history parameters of *Aphidius gifuensis* Ashmead (Hymenoptera: Braconidae). Appl. Entomol. Zool. **39**, 255–261 (2004)

S.A.L.M. Koojman, J.J.M. Bedaux, W. Slob, No-effect concentration as a basis for ecological risk assessment. Risk Anal. **16**, 445–447 (1996)

V.A. Krischik, A.L. Landmark, G.E. Heimpel, Soil applied imidacloprid is translocated to nectar and kills nectar feeding *Anagyrus pseudococci* (Girault) (Hymenoptera: Encyrtidae). Environ. Entomol. **36**, 1238–1245 (2007)

A. Krishnamoorthy, Exploitation of egg parasitoids for control of potential pests in vegetable ecosystems in India. Comunicata Sci. **3**(1), 1–15 (2012)

R. Laskowski, Why short-term bioassays are not meaningful? Effects of a pesticide (imidacloprid) and a metal (cadmium) on pea aphids (*Acyrthosiphon pisum* Harris). Ecotoxicology **10**, 177–183 (2001)

P.O. Lawrence, S.H. Kerr, W.H. Whitcomb, *Chrysopa rufilabris*: effect of selected pesticides on duration of third larval stadium, pupal stage and adult survival. Environ. Entomol. **2**, 477–480 (1973)

C.J.V. Leeuwen, J.L.M. Hermens, *Risk Assessment of Chemicals: An Introduction* (Kluwer Academic Publishers, Dordrecht, 1995), 374 p

J.C.V. Lenteren, Parasitoids in the greenhouse: successes with seasonal inoculative release system, in *Insect Parasitoids*, ed. by J. Waage, D. Greathead (Academic, London, 1986), pp. 341–374

J.C.V. Lenteren, H.J.W.V. Roermund, S. Sutterlin, Biological control of greenhouse whitefly (*Trialeurodes vaporariorum*) with the parasitoid *Encarsia formosa*: how does it work? Biol. Control **6**, 1–10 (1996)

P.H. Leslie, On the use of matrices in certain population dynamics. Biometrika **33**, 184–212 (1945)

L.Y. Li, Worldwide use of *Trichogramma* for biological control on different crops: a survey, in *Biological Control with Egg Parasitoids*, ed. by E. Wajnberg, S.A. Hassan (CAB Intl, Wallingford, 1994), pp. 37–54

U.T. Lim, A.M.A. Mahmoud, Ecotoxicological effect of fenitrothion on *Trissolcus nigripedius* (Hymenoptera: Scelionidae) an egg parasitoid of *Dolycoris baccarum* (Hemiptera: Pentatomidae). J. Asia-Pac. Entomol. **11**(4), 207–210 (2008)

T.X. Liu, P.A. Stansly, Effects of pyriproxyfen on three species of *Encarsia* (Hymenoptera: Aphelinidae), endoparasitoids of *Bemisia argentifolii* (Homoptera: Aleyrodidae). J. Econ. Entomol. **90**(2), 404–411 (1997)

F. Liu, S.W. Bao, Y. Song, H.Y. Lu, J.X. Xu, Effects of imidacloprid on the orientation behavior and parasitizing capacity of *Anagrus nilaparvatae*, an egg parasitoid of *Nilaparvata lugens*. BioControl **5**, 473–483 (2010)

T.X. Liu, Y.M. Zhang, L.N. Peng, P. Rojas, J.T. Trumble, Risk assessment of selected insecticides on *Tamarixia triozae* (Hymenoptera: Eulophidae), a parasitoid of *Bactericera cockerelli* (Hemiptera: Trizoidae). J. Econ. Entomol. **105**(2), 490–496 (2012)

M. Longley, A review of pesticide effects upon immature aphid parasitoids within mummified hosts. Intl. J. Pest Manag. **45**(2), 139–145 (1999)

M.A. Longley, P.C. Jepson, Effects of honeydew and insecticide residues on the distribution of foraging aphid parasitoids under glasshouse and field conditions. Entomol. Exp. Applic. **81**, 189–198 (1996)

M. Longley, P. Jepson, Effects of life stage, substrate and crop protection on the exposure and susceptibility of *Aphidius rhopalosiphi* Destefani-Perez (Hymenoptera: Braconidae) to deltamethrin. Environ. Toxicol. Chem. **16**(5), 1034–1041 (1997a)

M. Longley, P.C. Jepson, Cereal aphid and parasitoid survival in a logarithmically diluted deltamethrin spray transect in winter wheat: field-based risk assessment. Environ. Toxicol. Chem. **16**(8), 1761–1767 (1997b)

M.A. Longley, J.D. Stark, Analytical techniques for quantifying direct, residual and oral exposure of an insect parasitoid to an organophosphate insecticide. Bull. Environ. Contam. Toxicol. **57**, 683–690 (1996)

S. Madhusudan, Selective evaluation of insecticides to control tomato pests to *Trichogramma chilonis* (Trichogrammatidae: Hymenopthera) adult survival. J. Plant Agric. Res. **1**(1), 1–9 (2015)

V. Mahdavi, Residual toxicity of some pesticides on the larval ectoparasitoid, *Habrobracon hebetor* Say (Hymenoptera: Braconidae). J. Plant Prot. Res. **53**(1), 27–31 (2013)

V. Mahdavi, M. Saber, H.R. Dastjerdi, S.G. Kamita, Lethal and demographic impact of chlorpyrifos and spinosad on the ectoparasitoid *Habrobracon hebetor* (Say) (Hymenoptera: Braconidae). Neotrop. Entomol. (2015). doi:10.1007/S13744-015-0319-0

S. Mallick, S. Mandal, Chronic toxicity of rynaxypyr 20% EC, an anthranilic diamide insecticide, on the egg parasitoid, *Trichogramma chilonis* Ishii. *Conference Paper submitted to Entomological Society of America*, Annual Meeting 2014 (2014).

M. Mani, A. Krishnamoorthy, Effects of different pesticides upon the wax scale parasitoid, *Anicetus ceylonensis* (Hym.: Encyrtidae). Int. J. Pest Manag. **43**(2), 123–126 (1997)

A. Masood, M.J. Arif, M. Hamed, M.A. Talpur, Field performance of *Trichogramma chilonis* against cotton bollworms infestation in different cotton varieties as a sustainable IPM approach. Pak. J. Agri. Agril. Eng. Vet. Sci. **27**(2), 176–184 (2011)

M. Mead-Briggs, A laboratory method for evaluating the side-effects of pesticides on the cereal aphid parasitoid *Aphidius rhopalosiphi* (DeStefani-Perez). Asp. Appl. Biol. **31**, 179–189 (1992)

P. Medina, J.J. Morales, M. Gonzalez-Nunez, E. Vinuela, Is the use of some selected insecticides compatible with two noctuid endoparasitoids: *Hyposoter didymator* and *Chelonus inanitus*? Pesticides and beneficial organisms. IOBC/WPRS Bull. **35**, 51–59 (2008)

P. Medina, G. Manzanares, R. Izarra, A. Adan, G. Smagghe, E. Vinuela, Influence of insecticide persistence on the survival of the two braconid parasitoids *Chelonus inanitus* and *Aphidius ervi. Proc. meeting at Catania, Sicily* (*Italy*). IOBC/WPRS Bull. **80**, 271–272 (2012), 9–12 Oct

A. Meilin, Y.A. Trisyono, E. Martono, D. Buchori, The effects of deltamethrin applied at sublethal concentrations on the adults of *Anagrus nilaparvatae* (Hymenoptera: Mymaridae). ARPN J. Agric. Biol. Sci. **7**(12), 1032–1037 (2012)

R. Messing, B.A. Croft, Sublethal influences, in *Arthropod Biological Control Agents and Pesticides*, ed. by B.A. Croft (Wiley, New York, 1990), pp. 157–183

N. Mgocheki, P. Addison, Effect of contact pesticides on vine mealybug parasitoids, *Anagyrus* sp. near *pseudococci* (Girault) and *Coccidoxenoides perminutus* (Timberlake) (Hymenoptera: Encyrtidae). S. Afr. J. Enol. Vitic. **30**(2), 110–116 (2009)

M.J. Miles, A. Alix, Assessing the comparative risk of plant protection products to honey bees, non-target arthropods and non-Apis bees. Julius-Kuhn-Archiv. **437**, 30–38 (2012)

T. Miyata, L. Insung, O. Sarnthoy, P. Keinmeessuke, Effects of several insecticides on the larval parasitoid, *Cotesia plutellae* Kurdjumov, of diamondback moth, *Plutella xylostella* (L.). *Proceedings of the 4th international workshop,* Melbourne, pp. 223–226 (2001)

E.M. Mohamed, Side effects of certain common insecticides used in cotton fields on the egg parasitoid, *Trichogramma evanescens* west. (Trichogrammatidae: hymenoptera). Egypt. Acad. J. Biol. Sci **8**(1), 79–87 (2015)

M. Movahedi, G.N. Ghanbalani, H.R. Aghdam, S. Imani, Evaluating the efficiency of *Trichogramma pintoii* and *Bacillus thuringiensis* var. *kurstaki* on corn stem borer, *Ostrinia nubilalis* (Lepidoptera: Pyralidae) in corn fields of Moghan Plain. Intl. J. Adv. Biol. Biom. Res. **2**(11), 2825–2829 (2014)

R. Muhammad, M.A. Rustamani, N. Suleman, N. Ahmad, Q. Ahmad, Impact of release intervals and densities of *Trichogramma chilonis* (Ishii) (Hymenoptera: Trichogrammatidae) against the sugarcane stem borer, *Chilo infuscatellus* (Lepidoptera; Pyralidae) under field conditions. J. Basic Appl. Sci. **8**, 472–477 (2012)

T. Munyuli, S. Kyamanywa, G. Luther, Effects of cropping system and insecticide application on the incidence of arthropod parasitoids of cowpea insect pests in Uganda and Democratic Republic of the Congo. Tunis. J. Plant Prot. **4**, 77–90 (2009)

W.W. Murdoch, Population regulation in theory and practice. Ecology **75**, 271–287 (1994)

T. Nalini, S. Manickavasagam, Toxicity of selected insecticides to mealybug parasitoids, *Aenasius bambawalei* Hayat and *Aenasius advena* Compere (Hymenoptera: Encyrtidae). J. Biol. Control. **25**(1), 14–17 (2011)

A. Nasreen, G.M. Cheema, M. Ashfaq, M.A. Saleem, Survival of *Trichogramma chilonis* Ishii (Hymenoptera: Trichogrammitidae) after exposure to different insecticides: laboratory studies. Pak. J. Zool. **36**(1), 79–82 (2004)

J.B. Oliver, C.M. Mannion, M.G. Klein, J.J. Moyseenko, B. Bishop, Effect of insecticides on *Tiphia vernalis* (Hymenoptera: Tiphiidae) oviposition and survival of progeny to cocoon stage when parasitizing *Popillia japonica* (Coleoptera: Scarabaeidae) larvae. J. Econ. Entomol. **98**(3), 694–703 (2005)

M.L. Parrella, Biological pest control in ornamentals: status and perspectives. SROP/WPRS Bull. **13**(5), 161–168 (1990)

F. Pennacchio, M.R. Strand, Evolution of developmental strategies in parasitic Hymenoptera. Annu. Rev. Entomol. **51**, 233–258 (2006)

E. Peri, A. Cusumano, V. Amodeo, E. Wajnberg, S. Colazza, Intraguild interactions between two egg parasitoids of a true bug in semi-field and field conditions. PLoS One **9**(6), 1–8 (2014)

R.K.D. Peterson, Comparing ecological risks of pesticides: the utility of a risk quotient ranking approach across refinements of exposure. Pest Manag. Sci. **62**, 46–56 (2006)

J.D. Pinto, A.B. Koopmanschap, G.R. Platner, R. Stouthamer, The North American *Trichogramma* (Hymenoptera: Trichogrammatidae) parasitizing certain Tortricidae (Lepidoptera) on apple and pear with ITS2 DNA characterizations and description of a new species. Biol. Control **23**, 134–142 (2002)

N. Prabhaker, J.G. Morse, S.J. Castle, S.E. Naranjo, T.J. Henneberry, N.C. Toscano, Toxicity of seven foliar insecticides to four insect parasitoids attacking citrus and cotton pests. J. Econ. Entomol. **100**(4), 1053–1061 (2007)

S.G. Prado, S.E. Jandricic, S.D. Fran, Ecological interactions affecting the efficacy of *Aphidius colemani* in greenhouse crops. Insects **6**, 538–575 (2015)

G. Preetha, J. Stanley, S. Suresh, S. Kuttalam, R. Samiyappan, Toxicity of selected insecticides to *Trichogramma chilonis*: assessing their safety in the rice ecosystem. Phytoparasitica **37**, 209–215 (2009)

G. Preetha, T. Manoharan, J. Stanley, S. Kuttalam, Impact of chloronicotinyl insecticide, imidacloprid on egg, egg-larval and larval parasitoids under laboratory conditions. J. Plant Prot. Res. **50**(4), 535–540 (2010)

P. Puneeth, V.A. Vijayan, Parasitization capacity of *Trichogramma chilonis* Ishii (Hymenoptera: Trichogrammatidae) on the eggs of *Helicoverpa armigera* (Lepidoptera: Noctuidae) under laboratory conditions. I.J.S.N. **5**(3), 462–466 (2014)

A. Purhematy, K. Ahmadia, M. Moshrefi, Toxicity of thiacloprid and fenvalerate on the black bean aphid, *Aphis fabae* and biosafety against its parasitoid, *Lysiphlebus fabarum*. J. Biopestic. **6**(2), 207–210 (2013)

G. Radjabi, Investigations on the various aspects of Hymenopterous egg parasitoids in alleviating the outbreak occurrence of *Eurygaster integriceps* in Iran. Appl. Entomol. Phytopathol. **62**, 13–14 (1995)

N. Rezaei, F. Kocheyli, M.S. Mossadegh, K.T. Jahromi, A. Kavousi, Effect of sublethal doses of thiamethoxam and pirimicarb on functional response of *Diaeretiella rapae* (Hymenoptera: Braconidae), parasitoid of *Lipaphis erysimi* (Hemiptera: Aphididae). J. Crop Prot. **3**(4), 467–477 (2014)

C.V. Riley, Fourth report of the U.S. entomological commission, in *Butterflies of Eastern United States and Canada*, ed. by S.H. Scudder (Cambridge University Press, Cambridge, 1885), 323 p

C.V. Riley, Parasitic and predaceous insects in applied entomology. Insect Life **6**, 130–141 (1893)

S.M. Rill, E.E.G. Cardwell, J.G. Morse, Effects of two insect growth regulators and a neonicotinoid on various life stages of *Aphytis melinus* (Hymenoptera: Aphelinidae). BioControl **53**, 579–587 (2008)

D. Rogers, Random search and insect population models. J. Anim. Ecol. **41**(2), 369–383 (1972)

D.J. Rogers, N. Sharma, D.C. Stretton, J.T.S. Walker, Toxicity of pesticides to *Aphelinus mali*, the parasitoid of woolly apple aphid. N. Z. Plant Prot. **64**, 235–240 (2011)

J. Romeis, T.G. Shanower, C.P. Zebitz, *Trichogramma* egg parasitism of *Helicoverpa armigera* on pigeonpea and sorghum in Southern India. Entomol. Exp. Appl. **90**(1), 69–81 (1999)

D. Rosen, P. DeBach, Species of Aphytis of the world (Hymenoptera: Aphelinidae). Ser. Entomol. **17**, 556 (1979)

J.A. Rosenheim, M.A. Hoy, Sublethal effects of pesticides on the parasitoid *Aphytis melinus* (Hymenoptera: Aphelinidae). J. Econ. Entomol. **81**(2), 476–483 (1988a)

J.A. Rosenheim, M.A. Hoy, Genetic improvement of a parasitoid biological control agent: artificial selection for insecticide resistance in *Aphytis melinus* (Hymenoptera: Aphelinidae). J. Econ. Entomol. **81**, 1539–1550 (1988b)

C.R. Roubos, C. Rodriguez-Saona, R. Holdcraft, K.S. Mason, R. Isaacs, Relative toxicity and residual activity of insecticides used in blueberry pest management: mortality of natural enemies. J. Econ. Entomol. **107**(1), 277–285 (2014)

J.R. Ruberson, P.M. Roberts, Pesticide effects on natural enemies of cotton pests. Georgia Cotton Research-Extension Report 08/2005, pp. 124–132 (2005)

Q. Sabahi, A. Rasekh, A.H. Sangaki, A.S. Garjan, The persistence toxicity of three insecticides against adult of a thelytokous parasitoid, *Lysiphlebus fabarum* (Hymenoptera: Aphidiidae). Commun. Agric. Appl. Biol. Sci. **74**(1), 159–164 (2009)

M. Saber, Acute and population level toxicity of imidacloprid and fenpyroximate on an important egg parasitoid, *Trichogramma cacoeciae* (Hymenoptera: Trichogrammatidae). Ecotoxicology **20**(6), 1476–1484 (2011)

M. Saber, J.M. Hejazi, K. Kamali, S. Moharramipour, Lethal and sublethal effects of fenitrothion and deltamethrin residues on the egg parasitoid *Trissolcus grandis* (Hymenoptera: Scelionidae). J. Econ. Entomol. **98**(1), 35–40 (2005)

A.A.A. Saleh, Efficacy of the aphid parasitoid *Diaeretiella rapae* (Mintosh) to control *Brevicoryne brassicae* L., *Aphis craccivora* (Koch) and *Aphis nerii* Boyer at Sharkia Governorate, Egypt. Egypt. J. Agric. Res. **92**(1), 21–31 (2014)

G. Salerno, S. Colazza, E. Conti, Sub-lethal effects of deltamethrin on walking behaviour and response to host kairomone of the egg parasitoid *Trissolcus basalis*. Pest Manag. Sci. **58**, 663–668 (2002)

A.U.R. Saljoqi, M. Nawaz, A. Farid, I.A. Khan, Compatibility of spinosad with *Trichogramma chilonis* (Ishii) (Hymenoptera: Trichogrammatidae) in integrated pest management of *Sitotroga cereallela*. Pak. J. Zool. **44**(1), 133–139 (2012)

M. Sampathkumar, S.V. Krishnamoorthy, Risk assessment of *Trichogramma chilonis* (Ishii) to new molecules evaluated against spotted bollworm, *Earias vittella* Fab. in cotton. J. Biol. Control. **27**(4), 272–277 (2013)

M. Sankar, S. Jaigeetha, M.S. Rao, Exploitation of biocontrol agents, *Trichogramma chilonis* and *Tetrastichus howardi* on yield improvement in sugarcane at EID Parry (India) Ltd., sugar mill command areas. Intl. J. Innov. Res. Dev. **3**(8), 314–318 (2014)

S. Sarmadi, G.N. Ganbalani, H.R. Dastjerdi, M. Hassanpour, R.F. Pourabad, The effects of imidacloprid, indoxacarb and deltamethrin on some biological and demographic parameters of *Habrobracon hebetor* Say (Hymenoptera: Braconidae) in adult stage treatment. Mun. Entomol. Zool. **5**, 646–651 (2010)

S. Sattar, Farmanullah, A.R. Saljoqi, M. Arif, H. Sattar, J.I. Qazi, Toxicity of some new insecticides against *Trichogramma chilonis* (Hymenoptera: Trichogrammatidae) under laboratory and extended laboratory conditions. Pak. J. Zool. **43**(6), 1117–1125 (2011)

A.P. Saxena, Role of egg-parasites in the control of sugarcane borers (Lepidoptera). J. Appl. Entomol. **84**(14), 299–305 (1977)

M.I. Schneider, G. Smagghe, S. Pineda, E. Vinuela, Action of insect growth regulator insecticides and spinosad on life history parameters and absorption in third-instar larvae of the endoparasitoid, Hyposoter didymator. Biol. Control **31**, 189–198 (2004)

M. Schuld, R. Schmuck, Effects of thiacloprid, a new chloronicotinyl insecticide, on the egg parasitoid *Trichogramma cacaoeciae*. Ecotoxicology **9**(3), 197–205 (2000)

A.J. Scott, M.A. Knott, A cluster analyses method for grouping means in the analyses of variance. Biometrics **30**(3), 507–512 (1974)

K. Shankarganesh, B. Paul, R.D. Gautam, Studies on ecological safety of insecticides to egg parasitoids, *Trichogramma chilonis* Ishii and *Trichogramma brasiliensis* (Ashmead). Natl. Acad. Sci. Lett. **36**(6), 581–585 (2013)

R. Sheebajasmine, S. Kuttalam, J. Stanley, Effect of abamectin sprays on *Cotesia plutellae* (Kurdjumov) population in cabbage fields. J. Biol. Control. **22**, 119–124 (2008)

M.R. Sherif, A.S. Hendawy, M.M. El-Habashy, Management of rice insect pests. Egypt. J. Agric. Res. **83**(5), 111–130 (2005)

F. Sohrabi, P. Shishehbor, M. Saber, M.S. Mosaddegh, Effects of buprofezin and imidacloprid on the functional response of *Eretmocerus mundus* Mercet. Plant Prot. Sci. **50**(3), 145–150 (2014)

J. Stanley, S. Chandrasekaran, G. Preetha, S. Kuttalam, R.S. Jasmine, Selective toxicity of diafenthiuron to non-target organisms: honeybees, coccinellids, chelonus, earthworm, silkworm and fish. J. Plant Prot. Res. **56**(1), 1–5 (2016)

J.O. Stapel, A.M. Cortesero, W.J. Lewis, Disruptive sublethal effects of insecticides on biological control: altered foraging ability and life span of a parasitoid after feeding on extrafloral nectar of cotton treated with systemic insecticides. Biol. Control **17**, 243–249 (1999)

J.O. Stapel, A.M. Cortesero, W.J. Lewis, Disruptive sublethal effects of insecticides on biological control: altered foraging ability and life span of a parasitoid after feeding on extrafloral nectar of cotton treated with systemic insecticides. Biol. Control **17**, 243–249 (2000)

J.D. Stark, P.C. Jepson, D.F. Mayer, Limitations to use of topical toxicity data for predictions of pesticide side effects in the field. J. Econ. Entomol. **88**, 1081–1088 (1995)

J.D. Stark, S. Bamfo, Population-level outcomes of differential susceptibility among life stages of the aphid parasitoid *Diaeretiella rapae* to pesticides. *1st international symposium on biological control of Arthropods*. USA, January 14–18, pp. 314–317 (2002)

G. Sterk, S.A. Hassan, M. Baillod et al., Results of the seventh joint esticide testing programme carried out by the IOBC/WPRS working group, pesticides and beneficial organisms. BioControl **44**, 99–117 (1999)

C.P. Suh, D.B. Orr, J.W.V. Duyn, Effect of insecticides on *Trichogramma exiguum* preimaginal development and adult survival. J. Econ. Entomol. **93**, 577–583 (2000)

L. Suss, Survival of pupal stage of *Aphidius ervi* Hal. in mummified *Sitobion avenae* F. to pesticide treatment, in *Aphid Antagonists*, ed. by R. Cavallaro (Balkema, Rotterdam, 1983), pp. 129–134

E. Tabone, C. Bardon, N. Lezcano, B. Pintureau, L. Lapchin, Biological control of *Plutella xylostella* (L.): selection of the host *Trichogramma* strains based on biological characteristics. *Proceedings of the international symposium,* Montpelier, 21–24 October, pp. 247–248 (2002)

E. Tabone, C. Bardon, N. Desneux, E. Wajnberg, Parasitism of different *Trichogramma* species and strains on *Plutella xylostella* L. on greenhouse cauliflower. J. Pest. Sci. **83**, 251–256 (2010)

J. Tancik, L. Cagan, Efficiency of *Trichogramma evanescens* Westwood in the control of European corn borer (*Ostrinia nubilalis* HBN) on corn. *Acta fytotechnica et zootechnica*, 7 Sp. No., *Proc. XVI Slovak and Czech Plant Protec. Conf.* organized at Slovak Agricultural University in Nitra, Slovakia. (2004)

V. Tello, L. Diaz, M. Sanchez, Side effects of the natural pesticide spinosad (GF-120 formulation) on *Eretmocerus paulistus* (Hymenoptera: Aphelinidae), a parasitoid of the whitefly *Aleurothrixus floccosus* (Hemiptera: Aleyrodidae), under laboratory conditions. Cien. Inv. Agric. **40**(2), 407–417 (2013)

K.M. Theiling, B.A. Croft, Pesticide side effects on arthropod natural enemies: a database summary. Agric. Ecosyst. Environ. **21**, 191–218 (1998)

M.S. Thomson, R.E. Stinner, *Trichogramma* spp. (Hymenoptera: Trichogrammatidae): field hosts and multiple parasitism in North Carolina. J. Entomol. Sci. **24**, 232–240 (1989)

G. Tillman, Laboratory effects of two organically-certified insecticides on *Trichopoda pennipes* (Diptera: Tachinidae). J. Entomol. Sci. **43**(4), 408–417 (2008)

D. Tilman, Biodiversity: population versus ecosystem stability. Ecology **77**, 350–363 (1996)

A. Timus, N. Croitoru, Biological method of struggle against the basic wreckers of the sweet corn in Moldova. Bull. Usamv-Cn. **62**, 21–24 (2006)

P.W. Tipping, P.P. Burbutis, Some effects of pesticide residues on *Trichgramma nubinale* (Hymenoptera: Trichogrammatidae). J. Econ. Entomol. **76**, 892–896 (1983)

D.R. Tottman, H. Broad, The decimal code for the growth stages of cereals, with illustrations. Ann. Appl. Biol. **110**, 441–454 (1987)

D.H. Tran, T. Ueno, Toxicity of insecticides to *Neochrysocharis okazakii*, a parasitoid *Liriomyza* leafminers on vegetables. J. Fac. Agric. Kyushu Univ. **57**(1), 127–131 (2012)

T. Ueno, Selective host-feeding on parasitized hosts by the parasitoid *Itoplectis naranyae* (Hymenoptera: Ichneumonidae) and its implication for biological control. Bull. Entomol. Res. **88**(04), 461–466 (1998)

C. Ulrichs, I. Mewis, Evaluation of the efficacy of *Trichogramma evanescens* Westwood (Hym., Trichogrammatidae) inundative releases for the control of *Maruca vitrata* F. (Lep., Pyralidae). J. Appl. Entomol. **128**(6), 426–431 (2004)

Uma, S. *Impact of novel insecticides on Trichogramma chilonis Ishii (Hymenoptera: Trichogrammatidae)*. M.Sc. thesis, (Vellanikkara, Department of Agricultural Entomology, College of Horticulture, KAU, 2013)

M. Usman, M. Inayatullah, A.U.K. Sohail, S.F. Shah, Effect of egg parasitoid, *Trichogramma chilonis*, in combination with *chrysoperla carnea* and neem seed extract against tomato fruitworm, *Helicoverpa armigera*. Sarhad J. Agric. **28**(2), 253–257 (2012)

N.R.G. Varma, R. Jagadeeshwar, S. Chitra, Relative composition of egg parasitoids of rice yellow stem borer, *Scirpophaga incertulas* (Walker). J. Rice Res. **6**(2), 53–58 (2013)

L.A. Vasquez, A.M. Shelton, M.P. Hoffmann, R.T. Roush, Laboratory evaluation of commercial Trichogrammatid products for potential use against *Plutella xylostella* (L.) (Lepidoptera: Plutellidae). Biol. Control **9**, 143–148 (1997)

J.K. Waage, N.J. Mills, Biological control, in *Natural Enemies: The Population Biology of Predators, Parasites and Diseases*, ed. by M.J. Crawley (Blackwell, Oxford, 1992), pp. 412–430

C.H. Walker, S.P. Hopkin, R.M. Sibly, D.B, Peakall, *Principles of Ecotoxicology* (London, 1978), 321 p

W.K. Walthall, J.D. Stark, A comparison of acute mortality and population growth rate as endpoints of toxicological effect. Ecotoxicol. Environ. Saf. **37**, 45–52 (1997)

H.Y. Wang, Y. Yang, J.Y. Su, J.L. Shena, C.F. Gao, Y.C. Zhu, Assessment of the impact of insecticides on *Anagrus nilaparvatae* (Pang et Wang) (Hymenoptera: Mymanidae), an egg parasitoid of the rice planthopper, *Nilaparvata lugens* (Hemiptera: Delphacidae). Crop Prot. **27**, 514–522 (2008)

D.S. Wang, Y.R. He, X.L. Guo, Y.L. Luo, Acute toxicities and sublethal effects of some conventional insecticides on *Trichogramma chilonis* (Hymenoptera: Trichogrammatidae). J. Econ. Entomol. **105**(4), 1157–1163 (2012)

Y. Wang, L. Chen, X. An, J. Jiang, Q. Wang, L. Cai, X. Zhao, Susceptibility to selected insecticides and risk assessment in the insect egg parasitoid *Trichogramma confusum* (Hymenoptera: Trichogrammatidae). J. Econ. Entomol. **106**(1), 142–149 (2013)

Y. Wang, C. Wu, T. Cang, L. Yang, W. Yu, X. Zhao, Q. Wang, L. Cai, Toxicity risk of insecticides to the insect egg parasitoid *Trichogramma evanescens* Westwood (Hymenoptera: Trichogrammatidae). Pest Manag. Sci. **70**(3), 398–404 (2014)

L. Williams III, L.D. Price, V. Manriqueb, Toxicity of field-weathered insecticide residues to *Anaphes iole* (Hymenoptera: Mymaridae), an egg parasitoid of *Lygus lineolaris* (Heteroptera: Miridae) and implications for inundative biological control in cotton. Biol. Control **26**, 217–223 (2003)

D.J. Wright, R.H.J. Verkerk, Integration of chemical and biological control systems for arthropods: evaluation in a multitrophic context. Pestic. Sci. **44**, 207–218 (1995)

T. Yoshioka, M. Takeda, Effects of pesticides on *Thomsonisca typica* Mercet and *Arrhenophagus chionaspidis* Girault, parasitoids of *Pseudaulacaspis pentagona* (Targioni). Bull. Fukuoka Agric. Res. Cent. **25**, 145–149 (2006)

X. Zhao, C. Wu, Y. Wang, T. Cang, L. Chen, P. Yu, Q. Wang, Assessment of toxicity risk of insecticides used in rice ecosystem on *Trichogramma japonicum* an egg parasitoid of rice lepidopterans. J. Econ. Entomol. **105**(1), 92–101 (2012)

Index

A

B

C

D

Chapter 3
Pesticide Toxicity to Pollinators: Exposure, Toxicity and Risk Assessment Methodologies

Abstract Insect pollinators are the most affected and perhaps the most studied non-target organisms for pesticide poisoning. Pesticides sprayed on plants can be toxic to the foraging pollinators when they come in contact directly, contact with the sprayed parts or when they collect and consume or store pollen or nectar. Pesticides cause acute toxicity and mortality to pollinators which are usually assessed by studying median lethal dose and or concentration. Pesticides also cause sublethal effects on pollinators affecting their growth and development, behaviour and activities, etc. Sublethal effects on homing and foraging, larval development and adult emergence, visual and olfactory learning are being reported at very low concentrations of pesticide exposures also. Unconventional exposures through spray drift, guttation, hive treatment are also gaining importance because of significant effects even at low concentrations. Semi-field toxicity tests in cages or tunnels, poly houses and hive studies for pesticides and IGRs are commonly used as second tier of toxicity evaluation. A more realistic field toxicity study on foraging, repellency and colony health are being done not only on honey bees but also on other pollinators. Estimating the acute toxicity and no effect level concentrations are the basic steps for assessing the pesticide risk to pollinators. In a deterministic risk assessment approach, characterization of risk is done by calculating the risk estimates like risk quotient, toxicity exposure ratio, hazard quotient etc. But estimation of lethal dose might be a partial means of risk assessment because of non-inclusion of sublethal effects. Risk assessments based on sublethal effects are also discussed in detail.

1 Importance of Insect Pollinators

Of all the estimated 2,40,000 species of flowering plants for which one or more pollen vectors have been recorded, 2,19,850 are pollinated by animals (Nabhan and Buchnann 2012). Among the animal pollinators, insects which have coevolved specifically for plant pollination form a very important group. A few plant families have adapted quite successfully to enable insect pollination (Crepet 1983). Insects play an important role in pollinating a plant which is known as entomophily – a process where insects transfer the pollen from the anther to the stigma. Insects are found to be responsible for around 84 % of the pollination (Williams 1996) that occurs in the approximately 300 commercial crops that are being grown (Richards

J. Stanley, G. Preetha, *Pesticide Toxicity to Non-target Organisms*,
DOI 10.1007/978-94-017-7752-0_3

1993). Insects also play a role in pollinating 80–85 % of all commercial hectares (Williams 1996), with fruits, vegetables and oilseeds (Richards 2001). These insect pollinated crops represents approximately one-third of the world's total food production (Allen-Wardell et al. 1998; Klein et al. 2007). One should keep in mind that a variety of crops, are self-pollinating or pollinated by wind, especially the staple grains that most human diets are based on like rice, wheat, maize, sorghum, millets, rye, and barley. It can be said that the pollinator's fundamental contribution to the world's caloric need is not much. But on the other hand, fruits and vegetables, which help in diversifying the average human's diet and also contributes in providing essential nutrients, tend to have a heavy dependence on pollinators (Prescott-Allen and Prescott-Allen 1990; Roubik 1995). There are quite a lot of oilseed crops that depend on pollinators, and pollination by insects is essential for the seed production of forage and hay crops, such as alfalfa and clover, that are used as animal feed. These animals inturn provide us milk and meat. Though the plants that are pollinated by animals are not necessarily the major providers of a humanbeing's diet, animal-pollinated plants still contribute mainly to improving the quality of a person's diet (Eilers et al. 2011). Plants and trees that are used for timber, wood, beverages, fibers; nutraceuticals and medicines derived from plants also depend on pollinators (Bawa 1990; Roubik 1995; Buchmann and Nabhan 1996; Valiente-Banuet et al. 1996).

1.1 Insect Pollinators

More than 1,00,000 invertebrate species are categorized as insect pollinators, and they include bees, wasps, moths, butterflies, beetles and flies. In the following four orders mentioned below, there are many important pollinating insect species:

Hymenoptera (bees, wasps and ants)
Lepidoptera (butterflies and moths)
Diptera (flies)
Coleoptera (beetles)

Hymenopterans are considered pollinators with a very high degree of efficiency. Bees (Apidae) are generally considered pollinators with the highest degree of importance, particularly, *Apis mellifera* to which an estimated 80 % of the world's agricultural pollination services are attributed to (Carreck and Williams 1998). On the other hand, feral colonies of Indian bees, *Apis cerana indica* and giant rock bees, *Apis dorsata* are important pollinators (Stanley et al. 2009) in plantations that are placed near natural ecosystems in the tropics. In Poland, honey bees are found to be the single most important species of pollinators that are responsible for 90–95 % of pollination carried out by insects (Majewski 2014). The domestication of honey bees, and their management based on need, has been quite successful, and as a result they help to serve as an important pollinator.

In more recent times, quite a lot of other non Apis bee like bumblebee (*Bombus terrestris*), have also been commercially domesticated that have been used for agricultural pollination on a large scale (Delaplane and Mayer 2000). These bumblebees have also established their importance as natural pollinators in the colder Artic zones (Kevan 1972). With respect to certain crops, it has been found that bumble bees are even more efficient pollinators than honey bees. Since they buzz-pollinate, bumble bees (*Bombus* spp.), digger bees (*Anthophora* spp.) and some other native bees are able to pollinate blueberries and tomatoes much more efficiently than honey bees (Buchmann 1983; Cane and Payne 1990; MacKenzie et al. 1996; MacKenzie 1997; Javorek et al. 2002; Greenleaf and Kremen 2006). A lot of non-Apis bees are able to pollinate hybrid sunflowers and could do it in a much more efficient manner when compared to honey bees (Parker 1981; DeGrandi-Hoffman and Watkins 2000; Greenleaf and Kremen 2006). In India, non-Apis bees such as *Trigona iridipennis* and *Amegilla* spp. were found to be pollinating cardamom plants (Stanley et al. 2009). Other bees, namely the alfalfa leafcutting bee (*Megachile rotundata*), the alkali bee (*Nomia melanderi*) and several species of mason bees (*Osmia* spp.) and bumblebees (*Bombus* spp.) have also been observed to be good pollinators (Cane 1997). The *Megachile* spp. has been found to exhibit great promise as a pollinator of alfalfa (Hobbs et al. 1961), sunflowers (Neff and Simpson 1991) and cranberries (Cane et al. 1996). The solitary bee that has been the most intensively managed with respect to pollination is the alfalfa leafcutting bee, *M. rotundata* (Pitts-Singer and Cane 2011). Different mason bees have been reported to pollinate different crops, such as the red mason bee, *Osmia bicornis* with respect to apples (Gruber et al. 2011), *O. cornuta* with regard to Japanese plums (Calzoni and Speranza 1998), and *O. rufa* with respect to oilseed rape (Steffan-dewenter 2003).

It is true that a good many species of plants are pollinated by bees quite efficiently, but, there are other common flower visitors that include flies, butterflies and beetles that also pollinate plants (Muhammad et al. 1973; Free 1993; Cutler et al. 2012; King et al. 2013; Tyler and Davis 2013). Excluding bees, some of the other major taxonomic groups of pollinators are flies which pollinate 18.8 % of crop species, bats (6.5 %), wasps (5.2 %), beetles (5.1 %), birds (4.1 %), moths (2.9 %), butterflies (1.5 %) and thrips (1.3 %) (Nabhan and Buchnann 2012). With respect to onion that are grown for seed production, wild bees, flies and wasps are found to be crucial pollinators (Long and Morandin 2011). In the high Arctic zones, there are some plants that are mainly pollinated by flies (Kevan 1972). Despite the fact that a large number of hoverflies, *Eristalis tenax* and *Episyrphus balteatus* are needed to obtain a wholesome pollination potential, and are also found to be less efficient than bees, are still regarded as a pollinator with respect to the oilseed rape (Jauker et al. 2012). Midges (Ceratopogonidae) have been identified as an important pollinator of the coco with a high degree of efficiency (Young 1985; Free 1993).

Flowers exhibiting different flower traits might have different pollinator taxa that have adapted to these flowers (Junker et al. 2013). When it comes to reproduction, there are not many plant species that rely on a single pollinator species or a group. There are about 45 species of insects of 5 orders visits flowers of *Geranium thunbergii*, out of which 11 species belonging to 3 orders are important for pollination (Kandori 2002). Wild pollination services could account for an even substantially higher proportion of pollination services than was previously thought, even in modern and intense farming systems (Winfree et al. 2008). All over the world, there are found to be at least 17,000 species of native/wild bees (Michener 2007) which pollinate crops (Delaplane and Mayer 2000; Klein et al. 2007). These native/wild bees contribute substantially in pollinating crops such as coffee (Klein et al. 2003), watermelon (Kremen et al. 2002; Winfree et al. 2007), tomato (Greenleaf and Kremen 2006), blueberry (Cane 1997; Isaacs and Kirk 2010), sunflower (Greenleaf and Kremen 2006), canola (Morandin and Winston 2005) etc. These native bees offer their services in the realm of pollination which is directly found to benefit crop production. Moreover, in umpteen ways, they complement the twofold services provided by honey bees, namely, the biological, by improving, in certain instances, the efficacy of honey bee pollination (Greenleaf and Kremen 2006), and the economical, by insuring pollination shortages if any (Winfree et al. 2011).

Insect pollinators other than bees have their own regions of influence in particular locations apart from where particular crops exist. Altitudinal variation/elevation also found to affect the relative abundance of principal taxa of pollinating insects (Hymenoptera, Diptera, Coleoptera and Lepidoptera). With respect to a particular species, the principal pollinators are found to vary both spatially as well as temporally. For example, the alpine sky pilot, *Polemonium viscosum*, is mainly pollinated by bumble bees at higher elevations and by flies at lower elevations of hills (Galen et al. 1987). A reduction in pollination by hymenopterans along a gradient of increasing elevation in the Chilean Andes was taken over by lepidopterans and more particularly, the dipterans (Arroyo et al. 1982). A similar phenomenon was identified by Muller (1880) and Warren et al. (1988), who also reported a decline in pollination by insects belonging to order Hymenoptera with higher elevations in the European Alps and Utah, respectively. These higher elevation crops were found to be pollinated by lepidopterans and dipterans. Thus a vast diversity of pollinators would be essential to pollinate different crops in diverse ecosystems.

1.2 Impact of Insect Pollination on Crop Yield

Though pollination is earlier thought as an ecosystem function, it is increasingly realized as a crucial input for crop cultivation (Allsopp et al. 2008). Crops relying on insect pollination include apple, melon, citrus, tomato, apricot, grape, peach, cherry, mango, berry, olive, carrot, bean, cucumber, potato, onion, pumpkin, sunflower, various nuts, a range of herbs, alfalfa, cotton, lavender etc. (Abrol 2012). Increase in seed yield in different crops due to bee pollination is reported worldwide. In cage trials, 15 % more soybean yield was obtained in cages with bees when compared to cages without them (Erickson et al. 1978). The seed yield of canola, *Brassica napus* was reported to get increased by 46 % in the presence of three honey bee hives per hectare, compared to fields with no hive (Sabbahi et al. 2005). An increase in seed yield of 12.0 %, 21.7 % and 29.2 % was reported in Chinese cabbage, khol-khol and broccoli, respectively in planned honey bee pollination (Sushil et al. 2013). Indian bee, *A. cerana* is reported to increase the seed yield of crops like cabbage, cauliflower, radish, mustard etc. (Verma and Partap 1994; Partap and Verma 1994). Impact of honey bee pollination (*A. cerana*) on pod and seed set and seed yield is given in the following table.

Impact of planned honey bee pollination in different crops

Crops	Percent increase		
	Pod set	Seed set	Seed yield
Radish	19.33	11.93	20.41
Broccoli	12.51	16.61	28.67
Chinese cabbage	13.49	11.73	12.01
Knol-khol	14.93	10.95	21.77
Mustard	18.60	17.75	12.23
Coriander	–	27.84	16.81
Fenugreek	16.54	18.30	18.51
Onion	–	25.19	24.40

1.3 Economic Value of Pollination

Some of the ecosystem services such as pollination cannot be valued in terms of money. However, the following estimates show the importance of pollination service provided by insect pollinators. The economic value of global pollination service is amounted to € 153 billion, among which a major share is taken by vegetable and fruit with a value of € 50 billion each. Others like oil crops, stimulants, nuts and spices share the remaining part (Gallai et al. 2009). Insect pollinated crops accounted for 20 % of cropland and 19 % of total farmgate crop value in UK during 2007 (Breeze et al. 2011). The annual social gains due to honey bee (*A. mellifera*) agricultural pollination were estimated to range between $ 1.6 and $ 5.7 billion in US (Southwick and Southwick 1992). The value of honey bee pollination is estimated as $ 443 million in Canada (Scott-Dupree et al. 1995). The loss resultant from too low number of pollinators in plantations of Poland in 2012 is estimated at € 728.5 million (Majewski 2014). However, it is difficult to measure the pollination service in terms of economic gains. Some crops depend wholly on pollinators for their seed production which cannot be measured in economic terms. Some estimates like that of Levin (1983) included 10 % of the value of cattle and dairy production because honey bees pollinate and help in seed production of fodder crops also. However, if market value is added to these indirect effects of pollination, it will exaggerate the economic value of pollination services.

1.4 Indirect Impacts of Insect Pollination

The indirect benefits of insect pollination in terms of food and forage crop production will have a direct benefit to human. Even the quality and shelf life of fruits are also correlated with insect pollination (Klatt et al. 2014). Insect pollination services are also essential in propagating numerous wild plant species (Ollerton et al. 2011) and thus maintaining the plant diversity and ecosystem sustainability (Willis and

Garrod 1993). Many insect pollinated wild plants provide forage and shelter for farmland birds and other wildlife apart from adding aesthetic value to the landscape (Jacobs et al. 2009). These plants, in turn do many ecosystem services, such as organic matter addition in soil, carbon sequestration, flood and erosion control, etc. (Daily et al. 1997).

Insect pollination not only conserves energy for the plant (less pollen/ovule produced), but also play a part in angiosperm diversity (Crepet 1979). The fecundity of plants is highly dependent on pollinators (Waser 1978; Campbell and Motten 1985). If pollination service is not there, there will be reduction in seed production leading to a smaller plant population which could enter an extinction vortex (Bond 1994; Kearns et al. 1998). Plants are primary food producers and on which, all the higher taxa depends for their food and survival. Thus pollinators play an important role in biodiversity conservation. Pollination itself can be part of a trophic cascade and pollinators are in turn a food resource for insectivorous predators. The consequences of major shifts in pollinator populations could have economic repercussions either directly in their appeal to nature lovers and collectors or indirectly in their effects on the plants they pollinate (Knight et al. 2005). Some pollinators are highly appreciated aesthetically such as bees, butterflies, hummingbirds etc.

In agriculture, use of pollinators as vectors to deliver biocontrol agents in pathosystems involving flower infection is also demonstrated. Honey bees and bumblebees have been used as vectors for beneficial microorganisms and biocontrol agents from the hive to flowers (Kevan et al. 2003). Honey bees (*A. mellifera*), bumble bees (*B. terrestris* and *B. impatiens*), megachilid bees (*O. cornuta*) are used for the dispersal of *Pseudomonas fluorescens* against *Erwinia amylovora* in apple and pear (Johnson et al. 1993); *Gliocladium roseum* for *Botrytis cinerea* in strawberry (Peng et al. 1992; Yu and Sutton 1997); *Clonostachys rosea* for grey mould in raspberries (Yu and Sutton 1997); *P. fluorescens* for fire blight in apple (Thomson et al. 1992); *Trichoderma harzianum* for grey mould of strawberry (Maccagnani et al. 1999); *Bacillus subtilis for* mummy berry disease in blue berries (Dedej et al. 2004) and fire blight in pear (Maccagnani et al. 2006). Successful dissemination was also achieved in the case of pathogenic fungi against pollen beetles (*Meligethes aeneus*) in oilseed rape (Butt et al. 1998), Bacillus thuringiensis against moth (*Cochylis hospes*) in sunflower (Jyoti and Brewer 1999) and viruses for the control of helicoverpa larva in clover (Gross et al. 1994). Bumble bees are used to spread *B. bassiana* for the management of greenhouse whitefly, western flower thrips and green peach aphid (Al-mazraawi et al. 2006; Shipp et al. 2006). Multiple biocontrol agents were made to deliver through bumble bees for the management of insect pest and disease (Shipp et al. 2006). Further, honey bees can be used to deliver gameticides or herbicides to interfere with seed set in weeds (Kevan et al. 2008). Thus, apart from pollination service, these insects act as important link in plant biodiversity conservation and even in pest management activities.

2 Routes of Pesticide Exposure to Pollinators

Bees and other pollinators can be exposed to toxic pesticides through many routes. The routes of exposure to pollinators especially of honey bees in comparison to wild bees are given in detail by Alix and Miles (2011). Pesticides sprayed on plants can be toxic to foraging honey bees when they get contact to contaminated surface (Koch and Weiber 1997) or directly, when they fly or by adsorbing pesticide dust (Prier et al. 2001) or by consuming contaminated pollen and nectar. Foraging honey bees are therefore directly get exposed, but also intoxicate the whole colony by bringing contaminated pollen and nectar back to the hive (Bos and Masson 1983; Villa et al. 2000). The colony dwellers are indirectly exposed as returning foragers store or exchange contaminated material with hive conspecifics (Rortais et al. 2005; Krupke et al. 2012).

Inhalation of fumigant toxicants near the treated areas is also but a minor route of exposure. Residues of insecticides especially of systemic nature may also get transfer into honeydew. Normally the concentrations which do not affect aphids are unlikely to affect bees but there is a risk of overspray on these plants (Alix and Miles 2011). Exposure through miticide treatments in hives and through contaminated environment (air, water etc.) cannot be over viewed. The honey bee queen is exposed to pesticides when the poisoned nurse bees' offer contaminated glandular secretions to the queens (Davis and Shuel 1988a, b). In eusocial pollinators like honey bees, different life stages perform different duties (nursing, cleaning, cell building, guarding, foraging etc.) and are exposed to different concentrations and with variation in duration of the exposure.

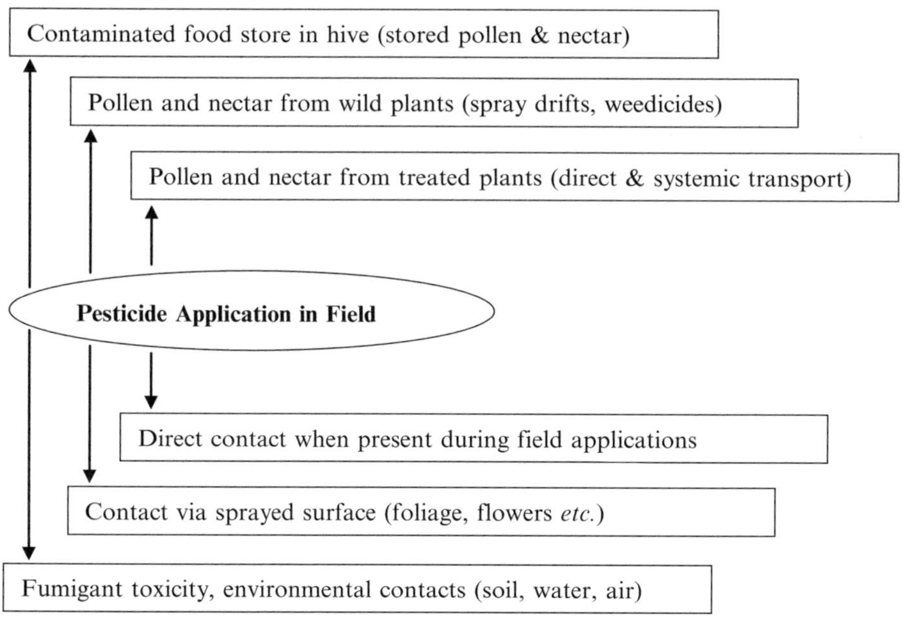

Fig. 3.1 Exposure routes of pesticides to different stages of honey bees

2.1 *Pesticide Application in Field*

Foragers are poisoned in the field itself. Hive bees are usually poisoned by contaminated nectar and pollen which are brought back and stored in the hive by foragers (Fig. 3.1). When the hive bees die, the brood will show signs of neglect and immature bees still in the cells may die. The larva may also die since they are highly sensitive to pesticides. If a colony loses its foraging bees, nectar and pollen collection will be affected but there is a chance that the colony can recover. If both the foragers and hive bees are lost, the colony may never recover (Sanford 1993).

The route of pesticide exposure may vary with the pollinating organisms/insects also. Pollinating insects that do not store food (pollen and nectar) is exposed only in the field as they do not carry food to hive. For bees also, many a times, insecticides are being sprayed during early morning or evening hours to avoid direct toxicity to honey bees. But the bumblebee which commences its activities early in the morning and stay upto late evening (Fussell and Corbet 1991; Corbet et al. 1993), may get more exposure than honey bees. Generalist pollinators may get less exposure from the pesticide treated crops, as they get a part of food only from sprayed and other part from unsprayed plants.

The pesticide formulations also play an important role in exposure and toxicity to pollinators. Dust formulations are reportedly risky than other formulations. The pesticide dusts may be carried back to the hives by foraging bees because they get stuck to their body hairs. Among the liquid pesticide formulations, ultra-low volume (ULV) formulations are generally more dangerous to bees (Sanford 1993). Microencapsulated insecticides are much more toxic to honey bees than the other formulations. Sometimes, bees collect these and take to their colony (Devillers 2002). Seed treatments once thought safer, is known to be serious because of the toxic effects of sublethal doses which remain in the nectar and pollen apart from exposure through contaminated air due to pneumatic sowing machines.

Studying the route of intoxication/exposure is an important factor for assessing the risk, as it influence on the reach of pesticide to its target site of action (Soares et al. 2015). These exposure pathways are of important concern (Henry et al. 2012) and only after knowing the routes of exposure, one can study the ways and means to reduce intentional and unintentional intoxications to the pollinators.

2.2 *Direct Contact via Crop Spraying*

Spraying of pesticides on crops at blooming stage and especially when the pollinators are foraging in that crop or close by, pose a threat to pollinators. They can be directly covered in the spray while spraying on the plant foliage or exposure to drift droplets. This exposure of bees may cause instant mortality or seriously affect foraging and homing. In honey bees, this has been observed as total loss of flying/foraging bees from a colony. During peak seasons, colonies of honey bees may be able to recover from this loss of foraging bees to some extent by the emerging brood. However, the far smaller sizes of other pollinator colonies like that of bumblebees' results in heavy loss with a potentially far greater impact (Thompson 2001). The acute contact toxicity varies with pollinators and also well correlated with the size of the pollinators. Unlike honey bees, other pollinators like bumble bees vary in weight and size even within the species and their susceptibility to pesticides also varies (Thompson 2001) with heavier bees have higher LD_{50} values compared to smaller individuals (Van-der-Steen 1994).

The loss of solitary bees through direct contact exposure is of much greater concern because the death of one bee may ruin the whole colony. Many pollinators such as bumble bees over-winter and the queen will emerge and establish the colony. When the queen come for foraging and if killed by contact the whole colony formation is ended and have more deleterious effect (Thompson and Hunt 1999). As stated earlier, honey bees can replace the loss of forager by producing new forager from the developing brood. However, honey bees have another mode of contact exposure, as acaricides like flumethrin, caumaphos are generally treated topically in maintained honey bee colonies. Though they are found safer for the bees by contact the hive products may accumulate these pesticides. Honey bees may indirectly ingest these acaricides through hygienic behaviour during the application period and receive low doses through comb wax remodeling after the application period (Oruc et al. 2012). However, liposoluble products with high wax affinity can get accumulated in wax.

2.3 *Contact via Sprayed Surface*

Pollinators may pick up traces of pesticides when get in contact with recently treated foliage or other surfaces as they crawl-over sprayed surfaces of plants or other media. If the pollinator gets a lethal dose it may die before returning to the hive. But the case of this indirect contact toxicity will become worst if the worker bee get contacted in the field, come to hive and touches the pollen or honey stores or other members or even the comb, passing on the contamination to the entire hive. This contamination may be stored in pollen or honey and cause mortality months later also; bees may bring sufficient of it into the hive on their bodies to have a contact or fumigant effect (Palmer-Jones 1958). Pollinators get exposed to pesticides indirectly from the usage of pesticides to control mosquito and ticks in the environment also (Oruc et al. 2012). Honey bees may also get contact to acaricides which are used for the control of varroatosis in bee hives. For the control of varroatosis, insecticides like flumethrin are applied to hives formulated as a plastic strip for several weeks. Acaricide contamination in beeswax may also cause serious threat to honey bees (Johnson et al. 2010).

2.4 Through Pollen or Nectar from Treated Crops

Systemic pesticides are taken up by the growing plants and transported in the sap to roots, stems, leaves and flowers. These insecticides provide a long term control over sap feeding or even root feeding insects. If the insecticide is found on the entire plant it will be there in pollen and nectar also. Direct contamination of nectar by spray applications maynot occur in flowers having nectaries protected by corolla and petals (Clinch 1967). However, indirect contamination can occur especially of the systemic pesticides which get diffused and translocated to the whole plant system. These pesticide compounds (pesticides or metabolites) exist in the entire plant system and get into the nectar also (Maurizro and Schenker 1957; Jaycox 1964). Since the systemic pesticides occur in the entire plant system, they are most likely to occur in nectar and pollen. The concentration of pesticides in nectar depends on both the amount applied, method of application and nature of nectar secretion. An apparent selective transport of the insecticides into the nectar was evident by the results of the investigations with dimethoate and carbofuran in *Ajuga reptans*, *Brassica napus* and *Vicia faba*. The concentration of these insecticides was higher in nectar than in the pesticide solution where the flowers were kept, which shows the transport as more than just passive movement with water (Davis and Shuel 1988a, b). Several reports state that not only the truly systemic pesticides but others also get into the nectar and contaminate in amounts causing toxicity (Jaycox 1964). Trunk injection of insecticides in trees especially during flowering and pre flowering periods can also lead to contamination of pollen and nectar (Alix and Miles 2011).

The contaminated food is taken by both the pollen and nectar collecting worker bees and thus possessing threat to them. In general, neonicotinoids (systemic insecticides) reported in pollen and nectar is in safe limits. No or negligible residues of imidacloprid was reported in nectar and pollen when Gaucho 70 WS was used to treat sunflower seeds (Schmuck et al. 2001). However, high concentrations of pesticides in pollen and nectar are also reported. For instance, the average concentration of imidacloprid is about 2.1 μg/kg in maize pollen (Bonmatin et al. 2005), 3 μg/kg in sunflower pollen (Bonmatin 2002; Bonmatin et al. 2003), 1.9 μg/kg in sunflower nectar (Schmuck et al. 2001), 4.4–7.6 μg/kg in rape pollen and 0.6–0.8 μg/kg in rape nectar (Scott-Dupree and Spivak 2001). A study for pesticide contamination in pollen loads collected by honey bees revealed that out of 101 samples analyzed only 9 samples were found containing no pesticide residues (Chauzat et al. 2006). Even if the contaminated pollen and nectar may not cause instant mortality to the bees, the concern is for bee larvae and for sublethal effects on the behaviour and communication of adult bees.

As stated in contact toxicity the oral toxicity also varies with size of bumble bees, with higher LD_{50}s for heavier bees. Size may also influence oral exposure of bumblebees to pesticides. An increase in body weight is correlated with increased consumption and thus an increase in the uptake of toxicants. A 30–40 % increase in uptake was found with doubling in body weight (Prys-Jones and Corbet 1991).

One of the greatest potential risks to pollinators is through the use of insecticides applied to control crop pests which infest during blooming stage of crops like clover case-bearer moths, *Coleophora* spp. in white clover (Clinch 1967), aphids and flower thrips in various crops. Application of toxic pesticides to non-flowering crops may not pose a threat to pollinators but there may be some flowering weeds in and around those fields. Some times, weed flowers are also highly attractive to pollinators (Thompson 2001). Thus the highest risk for pollinators is by ingestion of contaminated pollen and nectar, but during the foraging on flowers, bees are often covered with pollen and may cause contact toxicity too (Bonmatin et al. 2005).

2.5 Through Pollen and Nectar of Wild Plants

The flowering weeds, shrubs and bushes in and around agricultural field may get pesticide exposure by means of spray drift or by intentional sprays to avoid shift of pests from those plants to cultivated crop. Weeds and bushes in and near agricultural fields are often managed using weedicides also. At times flowering weeds and wild plants are more attractive to pollinators like bumblebees and probably an important route of pesticide exposure (Thompson 2001). Recent studies described levels of neonicotinoids up to 9 mg/kg in wildflowers such as dandelions growing near treated crops, so exposure is not limited to bees feeding on the crop (Krupke et al. 2012).

2.6 Through Contaminated Pollen, Nectar and Wax in Bee Hives

Pollinators get expose to residues present in pollen and nectar from the contaminated plants and honey and wax in the comb (Stoner and Eitzer 2012; Goulson 2013). The partially affected/unaffected foragers carry the contaminated food from the field back to the hive or nest, contaminating the colony's food resources.

Foraging worker bees feed mainly on nectar but contamination of pollen affect the newly emerging bees. Bee larvae are fed with pollen and nectar and thus contamination in pollen or nectar may pose a threat to them too. This pesticide contaminated pollen posses risk even at low concentration levels because pesticides are biologically active even at low doses. Further, many pesticides are reported as highly toxic by ingestion than contact. Apparently foraging bees feed on nectar directly from flowers than from the honey stores (Sanchez-Bayo and Goka 2014). But they carry and process far more nectar than they consume and during this process some active ingredient of toxicants may get absorbed into its body. If so, the foragers are said to be highly exposed (Sanchez-Bayo and Goka 2014).

About 98 pesticides and metabolites were detected in mixtures up to 214 ppm in bee pollen. Almost 60 % of the 259 wax and 350 pollen samples from hive were reportedly contained a minimum of one pesticide compound. About 47 % of wax and pollen samples contained chlorothalonil, a widely-used fungicide and the in-hive acaricides, fluvalinate and coumaphos. Many insecticides, the fungicide chlorothalonil and the herbicide pendimethalin were reported to present in bee pollen (Mullin et al. 2010).

Application of acaricides for treating bee hives cause contamination in wax (brood combs and honey combs) and honey (Bogdanov 2006). Residues in wax are not much to be worried upon since their availability to the bees is considered to be negligible compared to the direct exposure by contact or dietary intake of pollen and honey (USEPA 2012a). However, the pesticide residues in wax also have a greater impact on bees (Zhu et al. 2014) and especially the acaricides, which are found often in wax samples (Serra-Bonvehi and Orantes-Bermejo 2010). Fluvalinate upto 204 ppm and coumaphos upto 94 ppm were reported in 98 % of the comb as well as the foundation wax samples (Mullin et al. 2010). This acaricide contamination in beeswax is mainly due to the in-hive treatments for mites (Chauzat and Faucon 2007). The level of contamination in bee wax is correlated with the number of treatments done for mite management (Bogdanov et al. 1998b). Haarmann et al. (2002) demonstrated that the beeswax in the queen cell tended to accumulate more pesticide. Queen bees can be exposed to miticides in two ways: one by the circulation of nurse bees attending the queen cells the other by the contamination of beeswax while the queen cells are being built. Propolis was also found to contain these acaricides (Bogdanov et al. 1998a) and the fluvalinate residues found in propolis were higher than that in wax (Bogdanov 2006).

2.7 Through Inhalation of Volatile Pesticides

Inhalation of volatile pesticides during or after application to the crops is also considered as a route of pollinator exposure to pesticide (Sanchez-Bayo and Goka 2014). However, this inhalation from the treated crops is considered a minor route of exposure for most pesticides to pollinators (Geoghegan et al. 2013).

2.8 Through Unconventional Routes of Exposure

Soil: Systemic granular or seed coating insecticides can move from the soil or seed-coatings into the plant system and reach nectar and pollen (Tasei et al. 2001). Persistent pesticides can cause contamination to crops grown in soils treated at the previous or even two seasons before (Krupke et al. 2012). Even some herbicides and fungicides which are hydrophilic and are known to get translocate in plants are found in honey samples (Hsieh et al. 1998; Vieira and Sumner 1999; Sanchez-Bayo and Goka 2014).

Water: Some pesticides readily dissolve in water, representing another route of environmental contamination. Bees also drink water (Kovac et al. 2010) and were reported to collect water from contaminated paddy fields (Sanchez-Bayo and Goka 2014). There are reports that during a dry spell, weedicides paraquat treatment to a field formed small puddles and bees collect this also (Fletcher and Barnett 2003).

Air: Bee poisoning can also occur due to drift of insecticidal dusts of treated seeds during sowing on adjacent areas with flowering bee forage plants (Georgiadis et al. 2012). It contaminates the air and can get deposited on soil or plants too (Greatti et al. 2003, 2006). The tiny solid fragments of the pesticide seed coating especially of that of neonicotinoids due to drift or wind flow, fall on the pollinators or surrounding environment and cause ill effects on them. This hypothesis was proven by many scientists in different countries. Air contamination with pesticides during the sowing of maize coated with Gaucho® (imidacloprid) was demonstrated. The escape/dust-off of the pesticide active ingredients from the seed drill were monitored on its fan drain (Girolami et al. 2012). The air in the fan drain was found to get contaminated with the pesticides used for seed coating and the same pesticide was found on the nearby grasses and flowers (Greatti et al. 2003). Concentrations even more than 250 μg imidacloprid/g paper were found in filter papers kept for 240 s in sowing machines. In some places, residues of the active ingredient were found at least for 4 days after sowing (Greatti et al. 2006). Bee poisoning incidents in Germany during spring 2008 caused by abrasion of active substance from clothianidin treated seeds during sowing of maize was reported (Pistorius et al. 2009). However some reports showed very low abrasion rate of less than 4 % of the Gaucho® FS 350 seed dressing formulation which possess no threat to bees. In cage trials also it was found that the amount of residues come from seed drills maynot cause detrimental effects on honeybees (Schnier et al. 2003).

Guttation: The guttation drops also form another route of pesticide exposure to pollinators. Water drops of guttation and dew are collected by honey bees to use in dilution of honey/pollen, maintaining hive temperature etc. Neonicotinoid residues were found in guttation drops in plants treated with those pesticides as seed treatment. High levels of the neonicotinoid insecticides such as imidacloprid (up to 346 mg/L), clothianidin (102 mg/L) and thiamethoxam (146 mg/L) were obtained from guttation drops from corn plants grown from seeds coated with respective insecticides. Bees may collect guttation drops for water and these levels of residue may pose a risk to them (Tapparo et al. 2011). Corn plants grown from neonicotinoid-coated seeds have higher amount of residues in guttation drops even upto 200 mg/L for imidacloprid (Girolami et al. 2009). Thus guttation remains a potential exposure route and can be included in the risk assessment process also.

3 Effects of Pesticides on Pollinators

Among the many toxic substances in the environment, agrochemicals especially pesticides are mostly found to affect the pollinators. Pesticides especially the insecticides which are designed and developed to kill or affect insects possess greater threat to insect pollinators. Though there are insecticides with unique modes of action which affect particularly one or a group of insects, many broad spectrum insecticides affect all the insects but in various degrees. Insecticides may cause mortality to the pollinators if sprayed over the foraging pollinators while treating the plants or when they get intoxicated food. Some pesticides may cause sublethal effects such as reducing their efficiency to do routine works, their learning behaviour, reproduction etc. The colony collapse disorders and bee disappearance syndrome are often correlated with the pesticide usage. There is no conclusive data and strong evidence on the impact of these chemicals on the overall population of bees making many suspects and no smoking gun. But this drastic decline in bee population is observed in the parts of world where agrochemicals are used extensively for a long time. This correlation suggests us, that pesticides are the most important factor for declining in bee population (Sanchez-Bayo and Goka 2014).

3.1 Mortality of Pollinators

3.1.1 Acute Mortality

3.1.1.1 Acute Contact Toxicity

Acute toxicity of pesticides to different pollinators is studied by arriving at a median lethal dose. Safer pesticides have high LD_{50} values and vice versa. Contact insecticides may cause high contact toxicity whereas others may act as stomach poisons. It was reported as technical grade clothianidin as highly toxic by direct contact to *A. mellifera* followed by carbofuran, imidacloprid = spinosad and lambda cyhalothrin (Bailey et al. 2005). The topical LD_{50}s of diflubenzuron to third and fourth instar larva of *A. mellifera* were 2.42 and 6.01 mg/larva, respectively, whereas it was 1.49 and 3.65 mg/larva, respectively for *A. cerana indica* (Chandel and Gupta 1992). Contact LD_{50} values were approximately 24 ng/bee for *A. mellifera mellifera* and 14 ng/bee for *A. mellifera caucasica* for imidacloprid (Suchail et al. 2000). The topical bioassays with diafenthiuron on *A. cerana indica* showed an LD_{50} value of 6.70 μg/bee, which corresponds to 98.56 μg/g (Stanley et al. 2009). The contact

LD_{50} values of 18, 30, 22, 75 and 138 ng/bee was reported for imidacloprid, thiamethoxam, clothianidin, dinotefuran and nitenpyram (Iwasa et al. 2004).

The contact LD_{50}s of captan and dimethoate to *O. lignaria* are reported as 269.68 and 1.02 μg a.i./bee at 72 h after treatment, revealing the fungicide as safer (Ladurner et al. 2005a). The 72 h topical contact LD_{50} of chlorpyrifos-methyl, dimethoate, heptenophos and lambda cyhalothrin to *B. terrestris* was reported to be 0.09, 0.94, 2.19 and 0.11 μg/bee whereas the indirect contact toxicity (LC_{50}) was 16.90, 27.14, 400.84 and 3.05, respectively (Marletto et al. 2003). The LD_{50} of deltamethrin to male and female alfalfa leafcutting bees, *M. rotundata* were reported as 0.005 and 0.012 μg/bee, respectively (Tasei et al. 1988). Topical treatments with imidacloprid on stingless bee, *Scaptotrigona postica* resulted in LD_{50} values of 25.2 and 24.5 ng a.i./bee at 24 and 48 h, respectively (Soares et al. 2015).

3.1.1.1.1 Acute Contact Toxicity at Field Recommended Dose

Field recommended doses of benfurocarb (Oncol®) @ 300 mg a.i./L, carbosulfan (Marshal®) @ 313 mg a.i./L and furathiocarb (Deltanete®) @ 400 mg a.i./L caused 70.4, 70.2 and 66.8 % mortality to *A. mellifera* in 8 h of treatment by contact through filter paper assay. Methiocarb (Mesural®) @ 500 mg a.i./L caused 39.4 and 92.2 % mortality at 8 and 16 h of treatment (Akca et al. 2009). All bees died by contact when exposed to carbofuran treated tassels collected from field sites after 1 day of treatment whereas no mortality was recorded for lambda cyhalothrin and spinosad (Bailey et al. 2005). The field recommended dose of diafenthiuron caused 71.4, 51.7, 55.5 and 85.2 % mortality to *A. dorsata*, *A. cerana indica*, *A. florea* and *T. iridipennis*, respectively at 24 HAT by indirect contact (Stanley et al. 2009). Spirotetramat at the field dose, 60 g a.i./ha was found to cause 6.6, 13.3, 16.6 and 20.0 % mortality to *T. iridipennis*, *A. florae*, *A. cerana* and *A. mellifera* at 24 h after treatment (HAT) and thus safe to bees by contact (Vinothkumar et al. 2010). Clothianidin at 37.5 ppm caused 100 % mortality to *A. mellifera* bees within 24 h and the concentration of 15 ppm within a period of 48 h. Similarly, thiamethoxam at 100 ppm caused total mortality within 6 h (Laurino et al. 2011).

Emamectin benzoate, spinosad, indoxacarb, thiodicarb, fipronil, fenvalerate, acetamiprid, thiacloprid, imidacloprid, clothianidin, acephate and cartap hydrochloride at their field recommended doses caused 100 % mortality to *A. mellifera* by dry film bioassay (Nadaf et al. 2013). All the *A. cerana* bees were found to die at 48 h when treated topically at the field doses of chlorpyrifos, dichlorvos, malathion, profenofos, monocrotophos and methyl demeton. Besides this, deltamethrin, spinosad and thiamethoxam also caused 100 % mortality to both the bees at 48 HAT. Flubendiamide showed discrepancies in toxicity to both the bee species with 26.7 % mortality to Indian bees and 66.7 % mortality to European bees in topical bioassays at 48 HAT. Endosulfan and acetamiprid caused a mortality maximum upto 36.7 % to *A. cerana* and *A. mellifera* (Stanley et al. 2015). High mortality by parathion, high mortality with long term effect by oxydemeton methyl, high mortality with repellent effects by deltamethrin, lambda cyhalothrin and moderate mortality by dimethoate at their double field recommended doses to *B. terrestris* has been reported (Gretenkord and Drescher 1993).

3.1.1.2 Acute Oral Toxicity

Pollinators take nectar and or pollen as their food and the pesticide contaminated nectar or pollen affects the pollinators. Much of the experiments on oral toxicity are carried out by deliberately contaminating the sugar solution by a known concentration of pesticides and fed to the pollinators. Here also, median lethal dose/concentration are taken as the criteria to express the toxicity of pesticides on pollinators. Some of the insecticides like dimethoate and cypermethrin have oral LD_{50}s to *A. mellifera* as 152 and 160 ng/bee (Jaycox 1964; Bendahou et al. 1997). The oral acute LD_{50} values of imidacloprid were 57 and 37 ng/bee at 48 and 72 h (Suchail et al. 2001). The 48 h median lethal concentration (oral) of flumethrin was found to be 0.178 μg/bee (Oruc et al. 2012). The oral LD_{50} of dimethoate for 4 day old larvae of *A. mellifera* was 1.9 μg/larva at 48 h after exposure (Aupinel et al. 2007). LC_{50} of fipronil by oral bioassays at 48 HAT to *Melipona scutellaris* bees was reported as 0.011 ng a.i./L (Lourenc et al. 2012), showing that melipona bees as more sensitive than *A. mellifera* (1.27 ng/L: Roat et al. 2010) and *S. postica* (0.24 ng/L: Jacob 2012).

The oral LD_{50}s of fungicide, propiconozole to *A. mellifera* was found to be 61.67, 59.48 and 57.25 μg a.i./bee at 24, 48 and 72 h after treatment (Ladurner et al. 2005a). A 1 % concentration of N-methyl-2- pyrrolidone, solvent used in insecticide formulations was reported acutely toxic with 100 % mortality in larval honey bees within 24 h after treatment (Zhu et al. 2014). The median lethal dose (oral) of methomyl to *A. mellifera*, *B. terrestis* and *B. lapidarius* is found to be 0.08, 3.20 and 2.78 μg/g of bees at 24 h, revealing honey bee as the most susceptible (Drescher and Geusen-Pfister 1991). The oral LD_{50} of oxydemeton methyl and deltamethrin to *B. terrestris* is reported as 0.75 and 0.6 μg/g of bee at 24 h (Gretenkord and Drescher 1993). The LD_{50} values of captan, propiconozole and dimethoate to *O. lignaria* is found to be 100.45, 40.04 and 0.26 μg a.i/bee at 48 h after treatment (Ladurner et al. 2005a). The acute oral LD_{50} of acephate, carbaryl, cartap hydrochloride, chlorpyrifos-methyl, dimethoate, imidacloprid and lambda cyhalothrin to *B. terrestris* was reported as 8.36, 3.92, 2.44, 0.38, 0.44, 0.04 and 0.21 μg/bee at 24 h (Marletto et al. 2003). Dietary LC_{50} values of imidacloprid on stingless bee, *S. postica* were 42.5 and 14.3 ng a.i./μL of diet after 24 and 48 h of exposure, respectively (Soares et al. 2015).

3.1.1.2.1 Acute Oral Toxicity at Field Recommended Concentration

No significant mortality was reported when newly emerged honey bees were fed with pollen collected for sweet corn exposed to or grown from fields treated with spinosad, carbofuran and lambda cyhalothrin as foliar sprays, after 1 day of application or seed treatments (imidacloprid and clothianidin) (Bailey et al. 2005). Thiametoxam at 0.5 ppm (>100 times less than field dose) caused 100 % mortality in bees within 6 h (Laurino et al. 2011).

3.1.2 Chronic Mortality

The chronic LD_{50} of acephate, methamidophos or dimethoate to bees was 100-fold lower than the respective acute 24 h oral LD_{50} (Fiedler 1987). Imidacloprid was chronically toxic to *A. mellifera* even at doses 60–6000 times lower than those cause acute toxicity (Suchail et al. 2001). Chronic effect of pesticides in larval bees is more important because all the metabolic wastes get accumulated on their body until it moult as pupa after which they defecate as meconium (Winston 1987). So, if the brood is exposed to pesticides there is no way to excrete them off and thus may get continuous pesticide stress (Wu et al. 2011). This condition is different from that of the adult bees or other insects which excrete and remove the toxic wastes regularly (Zhu et al. 2014). Dietary chlorothalonil at 34 mg/L was reported to kill more than 50 % of larvae in 6 days. This dose of pesticide is not found to cause any toxic effects on adult bees. The fungicide chlorothalonil and the miticides, fluvalinate or coumaphos were found synergistically toxic to 4-day-old bee larva (Zhu et al. 2014).

Diafenthiuron at 30 μg/L fed to the bees in a chronic semifield studies revealed about 40 % of bees fed the pesticide were not found to forage on the third day of exposure. The median lethal time (LT_{50}) of diafenthiuron 30 μg/L to *A. cerana* was found to be 85.7 h (Stanley et al. 2010). Some studies also implicate formulation additives or adjuvants as key risk factors (Ciarlo et al. 2012). The ingredients used in the formulation of captan are more toxic to bee brood than the active ingredient of the fungicide (Everich et al. 2009). N-methyl-2- pyrrolidone used in pesticide formulations as solvent was found to be highly toxic to larval honey bees with a median lethal survival time of only 4 days when the solvent was mixed at 0.02 % in diet (Zhu et al. 2014).

3.1.3 Mortality/Effects Under Semi-field Conditions

In semi field experiments, a mortality of >75 % of *A. cerana* and *A. mellifera* bees was found to occur at field doses of chlorpyrifos, dichlorvos, profenofos, monocrotophos, malathion, deltamethrin, methyl demeton, spinosad and thiamethoxam. A mortality of <25 % was reported only for endosulfan and acetamiprid for both the bees. Fungicides such as carbendazim, mancozeb, chlorothalonil and propiconazole were also found to be less toxic to both the honeybees in semi-field studies (Stanley et al. 2015).

Out of nine insecticides tested on bumble bees, *B. terrestris* in potted plants of cucumber, dimethoate, chlorpyrifos-methyl and quinalphos (≥85 % mortality in 72 h) were found toxic, whereas a lower hazard was recorded for heptenophos and ethiofencarb (Incerti et al. 2003). In tomatoes treated with chlorantraniliprole in polyhouse, the visitation of bumble bees were not affected as compared to control while in dimethoate affected the visitation. Further no significant differences were found between pre and post application mortality data of chlorantraniliprole (Coragen®) treatment group and the control group (Dinter et al. 2009).

Fig. 3.2 Mortality of honey bees due to insecticide sprays in field

3.1.4 Effects Under Field Conditions

Malathion was reported as highly toxic to honey bees in field conditions resulting in minimum bee visits (Deshmukh 1991). Very high mortality of *A. mellifera* bees was reported in hives kept near fields sprayed with malathion (Fig. 3.2). However, no mortality was reported in hives adjacent to endosulfan applied fields (Deshmukh 1991). Chlorpyrifos is reported as highly toxic to *A. cerana* foragers (Thomos and Phadke 1994). More number of *A. mellifera* bees (3.63 times than that of control) were found dead when alfalfa treated with carbaryl 1 kg a.i./ha in the field whereas no significant effect was observed in spinosad treatment (Miles 2003). No acute mortality of bees was found in apple orchards treated with indoxacarb at field recommended doses (Van-der-Steen and Dinter 2007). Experiments with *A. cerana indica* in mustard revealed acephate as most toxic followed by imidacloprid, ethofenprox and cypermethrin resulting in minimum visits and low population of the bees. Endosulfan (0.035 %) and cartap hydrochloride (0.08 %) were found not to affect the bee visits in the field and found safer to the bees (Gour and Pareek 2004). Needham and Stevenson (1973) showed that endosulfan as safer than azinphos-methyl to honeybees. Many studies revealed endosulfan as safe to honeybees (Kapil and Lamba 1974; Misra and Verma 1982; Hasan et al. 1986; Reddy 1997; Singh et al. 1997). Johansen (1977) predicted a very minimal hazard due to the usage of endosulfan on bees.

Direct and indirect contact and ingestion studies showed acetamiprid as safe to *A. mellifera* (Takahashi et al. 1992; Suchail et al. 2000; Laurino et al. 2011) and also for bumble bees (Takahashi et al. 1992). Fields sprayed with endosulfan and acetamiprid at recommended doses had a total number of 2.25, 3.25 and 2.13, 3.00 *A. cerana* and *A. mellifera* foragers per m^2/30 s. just after the spray whereas it was 2.13 and 2.88 in untreated plots, revealing no much difference between sprayed and unsprayed fields (Stanley et al. 2015).

In an experiment to find the exposure toxicity to bees while sowing corn seeds treated with thiamethoxam, the mortality of bees were found to more than double in the sowing day and declined immediately after that. A mean reduction of pollen foragers entering the hive 1 day after sowing treatment was about 62.8 % in treatment as compared to control. This effect was noticed even after 15 days after sowing (Tremolada et al. 2010). No difference between the colonies kept in control and seed treated (clothianidin) canola fields in terms of bee mortality, brood development and honey yield was observed for a period of four and a half months (Cutler and Scott-Dupree 2007).

Bombus impatiens colonies exposed to clothianidin treated weedy turf with white clover showed reduced foraging with fewer honey pots and reduced colony weight within 5 days whereas chlorantraniliprole treatment showed no such effect. Clothianidin also caused mortality of bumble bee workers (Larson et al. 2013).

3.2 *Sublethal Effects*

A long term and low level exposure of pesticides or even acute exposures causing some effects but not mortality are regarded as sublethal effects. Changes in division of labour, inability in foraging, distorted communication, problem in homing, difficulty in odour discrimination etc. can be included in sublethal effects. Oral flumethrin even at low doses are reported to cause disruption of motor coordination in honey bees resulting in convulsions in wings, legs and antenna (Oruc et al. 2012). The sublethal behavioural effect of pesticides in bees and the potential use of this in risk assessment are reviewed by Thompson (2003). Low levels of pesticides may also act as stressors that make bees more prone to biological infections (Mommaerts and Smagghe 2011; Vidau et al. 2011; Pettis et al. 2012). Though it is very difficult to assess the sublethal toxicity, advancement of science will enable us in finding out the sublethal effects of pesticides on pollinators at molecular levels *i.e.* the response take place at molecular levels such as gene/transcriptome responses as molecular ecotoxicological markers (Blacquiere et al. 2012).

3.2.1 Effect of Pesticide on Survivability and Size

Sublethal doses of diazinon were reported to affect the longevity of honey bees with a reduction in lifespan of up to 20 % in exposed bees (MacKenzie and Winston 1989). Long term feeding of honey bee colonies with sucrose syrup containing 0.01 and 0.1 ppm of carbofuran did not cause any impediment on the survival of adult bees and the matured brood. Reduction in the survival of adult bees and sealed brood was found at 1.0 ppm carbofuran and the colonies did not survive in the succeeding winter (Stoner et al. 1982).

Imidacloprid at low doses of 10 and 6 mg a.i./kg in syrup and pollen, respectively significantly reduced the survivability of bumblebees, *B. terrestris* as revealed in a laboratory feeding test (Tasei et al. 2000). Deltamethin even at very low doses were found to have effect on the longevity of alfalfa leaf cutter bees. About 35 % of bees were found to get killed after 12 days of treatment of deltamethrin at 3.2×10^{-5} μg/

bee (Tasei et al. 1988). Bumble bees were found to be 8 and 12 % smaller than that of the control workers in *B. terrestris* colonies treated with imidaclopird through pollen and sugar solution (Whitehorn et al. 2012).

3.2.2 Effect of Pesticides on Colony Formation

Dimethoate at 1 ppm caused a significant adverse effect in comb formation and egg laying especially when no alternative forages are available. The queen ceased egg laying in colonies treated with 10 ppm dimethoate (Waller et al. 1979). However, these effects were not observed when sufficient alternative forages are available (Stoner et al. 1983). Colonies fed with carbofuran at 1 ppm survived that summer but produced significantly fewer adult bees and died during following winter (Stoner et al. 1982). Fenthion, dimethoate and acephate exposures resulted in failure of the colonies to requeen themselves (Stoner et al. 1982, 1983, 1985). The queen supersedure rate in colonies treated with very low doses of cypermethrin (80 %) was significantly greater than in controls (30 %) (Bendahou et al. 1999).

Treatment of female leafcutter bees, *Megachile rotundata* with deltamethrin at 2×10^{-3} µg/bee (six times less dose of LD_{50}) resulted in 20 % less egg laying throughout a 6 week period after dosing (Tasei et al. 1988). A laboratory feeding study with micro and queenless bumble bee colonies showed that imidacloprid had no effect on larval development time but reduced the brood production with ejection of larvae by the workers (Tasei et al. 2000).

3.2.3 Effect of Pesticides on Immature Stages

Larvae dosed with low levels of dimethoate (0.313 µg/g royal jelly) showed stimulation of growth and maturation compared to controls although some of the larvae lost their typical C shape and were stretched out dorsally (Davis et al. 1988). Deformed adult bees were obtained when the larvae are exposed to malathion, dimethoate, carbaryl and captan. Those adults are of very small in size with stunted bodies, crippled legs and wings. Wing malformations are commonly seen and in some cases wingless adults were also reported (Jay 1964; Atkins and Kellum 1986). Honey bee larvae were found to grow slowly when exposed to either carbofuran or dimethoate at 5 µg/g royal jelly. Exposure of bee larva to carbofuran and dimethoate even at low doses resulted in deformed pupa (Davis et al. 1988).

Alfalfa leaf cutter bee, *M. rotundata* larva when fed on pollen provisions contaminated at the rate of 1 mg/kg of deltamethrin affected larval development. About 40 % of the larvae exposed to deltamethrin didnot reach the last instar (Tasei et al. 1988). Larval mortality was reported when *B. terrestris* was treated with 150 ppm teflubenzuron (DeWael et al. 1995).

3.2.4 Effect of Pesticides on Queen and Fecundity

Fluvalinate was found to cause reduction in oviposition and queen loss (Sokol 1996) and sexual incompetence in drones (Rinderer et al. 1999). The mean queen weight of the fluvalinate and coumaphos treated colonies were significantly lower than that of control. However, fluvalinate was not found to affect the ovary or the number of sperm cells in the spermatheca of queen whereas coumaphos did affect both (Haarmann et al. 2002). A bad brood pattern with unhatched eggs were found in colonies treated with herbicides, 2,4 D and 2,4,5 T at 100 mg/kg (Morton and Moffett 1972). A total rejection of grafted larvae of *A. mellifera* in queen cups made of beeswax contaminated with 300, 600 or 1000 mg/kg of coumaphos was reported. The queens produced if any from queen cells made of contaminated beeswax were of 51.1 % lighter than control queens. In another experiment, it was found that 50 % of larva in queen cups made of beeswax contaminated with 100 mg/kg of coumaphos were rejected (Pettis et al. 2004). Fenoxycarb treated colonies with queens were not able to successfully mate and lay eggs. Further, no drone was produced in those colonies (Thompson et al. 2005). Honey bee queens fed with sublethal doses of bifenthrin and deltamethrin was found to lay only 30 % and 50 % of their eggs, respectively (Dai et al. 2010).

A substantial reduction in the production of *B. terrestris* queen was reported when colonies treated with imidacloprid in pollen and sugar solution. The mean reduction of queens produced by colonies was about 85.4–89.8 in treatment with respect to control (Whitehorn et al. 2012). Dietary intake of imidacloprid 1 μg/L could cause a reduction in brood production by one third in bumble bees, *B. terrestris* (Laycock et al. 2012).

3.2.5 Effect of Pesticides on Division on Labour

Worker honey bees usually progress as hive bees, guard bees and foraging bees (Kolmes and Winston 1988). These age dependent activities are regulated mostly by juvenile hormone. Sudden demographic changes such as loss of foragers in the colony may also slightly influence the division of labour (Winston and Punnett 1982). Greater effect on the division of labour occur when newly emerged bees are exposed (Smirle 1993). House cleaning was found to reduce significantly in colonies treated with pesticides (Nation et al. 1986). An early shift in the activities of worker honey bees with precocious foraging was reported when treated with juvenile hormone analogue, methoprene (Robinson 1985; Jaycox et al. 1974; Tasei 2001).

3.2.6 Effect of Pesticides on Foraging

Foraging bees usually associate smell with the resources to locate them efficiently. When the scout bees find a potential resource it communicates to the foragers by means of dancing representing the direction and distance of the resource. The process involves memory, learning, navigation and ability for proper communication and many others (Menzel 1993; Menzel et al. 1998). Bumble bees also use these clues along ability to integrate local landmarks to locate their nest site (Goulson and Stout 2001). Pyrethroids appear to affect the ability of the foraging honey bee to return to the hive at realistic exposure levels. Permethrin treatment resulted in 43 % of foragers returning once to the colony and only 4 % returned twice and none of the treated bees were retrieved in the following morning (Cox and Wilson 1984).

Treatment of foragers with a neonicotinoid insecticide, imidacloprid, was observed to have a slight effect on the preciseness of the communicated direction and distance. Bees have been reported to show transitory disruptions of foraging activity when fed with imidacloprid at doses above 50 ppb which was observed to be time dependent and persisted overnight at doses of 100 ppb (Schmuck 1999). Imidacloprid at 50 ppb in sucrose reduced the flight activity of bees (Decourtye et al. 1999). Adult worker bees of *B. impatiens* exposed to spinosad at 0.8 mg/kg during larval development were found to be slower foragers (Morandin et al. 2005).

3.2.7 Effect of Pesticides on Homing

Deltamethrin altered the homing (flight) ability of bees at levels below those that affect co-ordination of flight muscles. Deltamethrin treatment caused 81 % of foragers not to return to the colony within 30 s after release whereas control bees returned in 10 s (Vandame et al. 1995). A delay in the homing behaviour of *A. mellifera* compared to control was observed when fed with 100 ppb imidacloprid. Honey bees exposed to 500 ppb of imidacloprid was not found either in the hive or in the feeding site after 24 h of exposure (Bortolotti et al. 2003). The results of the experiment made by Henry et al. (2012) predicted a failure of 10–31 % in homing in honey bees exposed to thiamethoxam as spray in crops.

3.2.8 Effect of Pesticides on Movement and Communication

Sublethal oral exposure of honey bees to parathion prevented the bees from communicating the direction of an artificial food source to other bees. The deviations in dance language of bees affect the ability of the bee to orient relative to gravity however there was no adverse effect on the flight direction (Schricker and Stephen 1970). Tau fluvalinate at 0.3 μg/bee and imidaclorpird at 50 ppb were found to cause a reduction in movement and interaction time of *A. mellifera* bees (Teeters et al. 2012). Sublethal doses of parathion caused disturbance in dancing by altering the angle and rhythm, and thus disturbing its communication (Schricker and Stephen 1970; Schricker 1974).

3.2.9 Effect of Pesticides on Learning and Habituation

Odour perception and responses to pheromones are important for the survival of honey bee colonies. Nicotinic cholinergic systems are integral in attention, learning and memory in animals and have been reported to be responsible for the proboscis extension reflex pathway and habituation in bees (Guez et al. 2001). A number of studies have investigated the effects of insecticides on learning (Decourtye et al. 1999) and on the habituation (Guez et al. 2001).

Pyrethroid treated bees learned odour-mediated responses at a slower rate and give less positive responses after several training periods compared to control bees. Among the six pyrethroids tested, the training responses of bees are affected the most by flucythrinate and cyfluthrin with permethrin, fenvalerate and cypermethrin in between and the least by fluvalinate (Taylor et al. 1987). The authors also suggested that the decreased foraging on pyrethroid treated crops reported elsewhere may therefore not due to repellency but to sublethal toxic dysfunction. Mamood and Waller (1990) suggested that pyrethroids affect learning rather than recall of stored information in bees. Neonicotinoid, imidacloprid at 4–40 ppb (3–33 % LD_{50}) was also reported to reduce the olfactory learning performance in individual bees and affected flight activity (Decourtye et al. 1999).

3.2.10 Effect of Pesticides on Other Behavioural Activities

Permethrin treated bees showed serious disturbance of behaviour spending more time in self-cleaning, trembling dance, abdomen tucking, rotating and abdomen cleaning as compared to control (Cox and Wilson 1984). An uncoordinated running and walking of forager bees of *A. mellifera* compared to control was reported in imidacloprid treatment. The communicative capacity of the treated bees seemed to be impaired and this could cause a decline in the social behaviour (Medrzycki et al. 2003). Acute intoxication of imidacloprid in honey bees leads to neurotoxicity symptoms, such as hyperresponsiveness, hyperactivity and trembling which lead to hyporesponsiveness and hypoactivity (Suchail et al. 2001).

3.2.11 Effect of Pesticides on Irritation and Repellency

Repellency is normally been tested in semi-field conditions by studying the level of foraging of bees to an artifical feeder with pesticide contaminated sugar solution (Solomon and Hooker 1989). Sucrose solution with aldicarb even at very low doses caused a significant reduction in foraging of bees (Nigg et al. 1991). The fungicide captan when dissolved in sucrose was also found to exert repellency in honey bees (Solomon and Hooker 1989).

The neonicotinoid insecticide, fipronil significantly reduced the visitation rate of honey bees to sucrose feeders but not when applied to flowering oilseed rape at field rates (Mayer and Lunden 1999). Cypermethrin (15 g a.i./ha) sprayed on oilseed rape during periods of peak foraging activity resulted in a slight decline in the level of foraging and the levels of pollen collection. In fields sprayed at a higher rate (20 g a.i./ha) foraging was affected by 85 % immediately after spray but recovered by the following day (Shires et al. 1984a).

4 Methods to Assess Pesticide Toxicity to Pollinators

Several tests are developed to determine the impact of pesticides on pollinators and used to assess the toxicity on these beneficial insects. These tests can be classified as laboratory, semi field and field tests to determine the acute toxicity, chronic and also the sublethal effects.

4.1 Tier I Toxicity Evaluation: Laboratory Tests

Laboratory tests offer the most convenient way for rapidly estimating the toxicity of pesticides to pollinators but they may not reflect the reality observed in the fields. Many methodologies are reported to estimate the acute toxicity of pesticides to honey bees under more realistic environmental conditions (Devillers 2002).

4.1.1 Acute Toxicity Tests

Acute toxicity tests are normally carried out in laboratory conditions and expressed as a lethal dose/concentrations. The median lethal dose (LD_{50}) is the dose at which 50 % of the test insects die within a defined period. Similarly lethal concentration is the concentration which causes mortality to 50 % of test organism. To initiate with these median lethal test procedures, a set of preliminary range finding tests are being carried out to determine the range of pesticide concentrations that cause different per cent mortality to bees (Desneux et al. 2006). The pesticide concentration which causes approximately 50 % mortality to the test organism is determined. Keeping this concentration as median, three concentrations higher than this and three lower were tested for actual bioassay. The concentrations which cause a mortality of 10–90 % can be used for determining of median lethal concentrations. At times, the median lethal doses or concentrations are used to track the pesticide susceptibility of pollinators. Insecticides are classified as highly toxic (acute LD_{50}, <2 μg per bee), moderately toxic (acute LD_{50}, 2–10.99 μg per bee), slightly toxic (acute LD_{50}, 11–100 μg per bee) and essentially nontoxic (acute LD_{50}, >100 μg per bee) to adult bees (WSDA 2010; Oruc et al. 2012).

Apart from these median lethal tests, acute toxicity tests are also conducted to estimate the risk of pesticides usage in field conditions at their recommended concentrations. In this, the test organisms are exposed to the field recommended concentrations to find the actual effect in worst case exposures. Acute toxicity tests are carried out using different procedures. Some organizations like OECD, EPPO and many others recognize methodologies for testing chemicals, and many are published. Some of the important methodologies are described here under.

4.1.1.1 Contact Toxicity

As stated earlier, contact toxicity of pesticides to pollinators can be of direct contact i.e. exposed directly while spray application and indirect contact by getting contact with already sprayed surface.

4.1.1.1.1 Direct Contact Toxicity

Tests on direct contact toxicity gives an indication of bee mortality while spraying on a crop, but do not necessarily reflect the residual toxicity. The following methodologies given by several authors are used to find the direct contact toxicity to pollinators. The first three methods on direct contact toxicity given below, though provide information on the toxicity but fail to assess the actual amount of pesticides administered on the bees. Topical application tests should include a reference standard and the LD_{50} of that toxic standard need to be within specified ranges, which help in validating the methodology (Medrzycki et al. 2013).

Direct Spray Method (Palmer-Jones 1958)

In this method, pesticide is sprayed directly on the anesthetized bees in a Petri dish. Batches of 50 bees, without preliminary feeding are anaesthetized with CO_2 and spread over a filter paper in a Petri plate. While the bees are in unconscious state, they are sprayed with 1 mL of the particular dilution of pesticide under test. This 1 mL is sufficient to wet the bees thoroughly. The head of an atomizer fitted on a graduated cylinder is used from which a known volume of pesticide could be delivered as spray. Nowadays graduated atomizers or sprayers are also available, which can be used for this purpose. Four batches, each of 50 bees were sprayed for a single treatment in the test. After spraying, the bees were caged with feeders and mortality noted after 24 h. This test symbolizes the bees getting contact to the spray particles directly when it is sprayed on the plants. This method reveals the risk of spraying the crop while it is in peak blooming and pollinators are at work.

Bell Jar Duster Method (Johansen 1978; Medrzycki et al. 2013)

In this method, the contact toxicity of pesticide is assessed by placing bees in a bell-jar duster. The duster was used to administer pesticides to bees through vacuum and subsequent imploding of incoming air to disperse the pesticide homogeneously over the bees. The duster consists of a bell jar and an electric mixer with a variable speed transmission. The hopper was attached rigidly to the frame of the mixer and the shaft was mounted in a chuck on the vertical drive shaft of the motor. The dust was driven out from the hopper by centrifugal force through the slight clearance space between the revolving disc and the lower edge of the funnel. The dust thus arrived in the jar get finely dispersed and swirled by the action of a small fan.

Potter Spray Tower Method (Bailey et al. 2005)

In this method, technical grade of pesticides are generally used to test their direct contact toxicity to pollinators. Test solutions (w/v) were prepared by dissolving each pesticide in 19:1 acetone: olive oil. Some technical compounds which do not dissolve in acetone such as spinosad, reverse-osmosis (RO) water can be used. Each insecticide at four or five concentrations and controls with the solvent mixture are to be tested. Cohorts of 20 forager (>20 day old) bees are collected from the colonies, anaesthetized using CO_2 for 3 s, kept in Petri dishes and treated in Potter spray tower. The Petri dishes with bees were sprayed with 5 mL aliquots of each treatment. Treated bees were kept in paper cups and covered. They are provided with cotton wicks soaked in 1:1 (v/v) honey: water solution. Observations on the number of dead bees per cups can be counted at 24 h after exposure or any other defined interval. The advantage of this laboratory spraying apparatus is it's capability of giving good replications and an even distribution of spray over an area of 9 cm (Potter 1952). So, this method is advantageous than the direct spray method given by Palmer-Jones (1958).

Topical Application on Thoracic Dorsum (Stanley et al. 2010)

In this experiment, insecticide solutions were prepared using technical-grade pesticide diluted in analytical-grade acetone. The immobilized bees are topically dosed with a 1 μL drop of insecticide solution on the thoracic dorsum with a Hamilton repeating dispensor. At least five doses of insecticides with four replications with a minimum of 30 bees per treatment are used. Mortality was recorded at different time intervals and corrected using Abbott's formula (Abbott 1925) for mortalities in control. The data obtained was subjected to probit analysis, as per Finney (1971) and a log dose probit mortality (ldpm) line obtained. This is a standard method for testing contact toxicity of pesticides to different insects (Fig. 3.3).

Topical Application on Ventral Thorax (Van-Der-Steen 2001)

This method is used to assess the contact toxicity of insecticides on bumble bees. Similar to the above mentioned topical application, 1 μL test solution was pipetted on ventral part of thorax between second and third pairs of legs. The treated bees were housed together by dose and fed sucrose solution *ad libitum*. Mortality was recorded at 24, 48, 72 h and LD_{50} expressed in mg a.i. or formulation per bumblebee. The ventral thorax is preferred for topical dosing than dorsal thorax to avoid removal of pesticide kept on thorax using legs by the recovering individuals. It is to be noted that bumble bees have a tuft of hairs all over the body and also on the dorsal thorax.

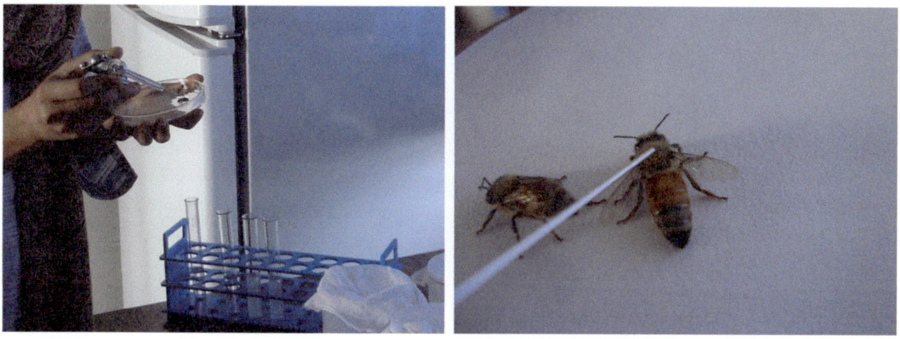

Fig. 3.3 Topical bioassay on thoracic dorsum using Hamilton syringe

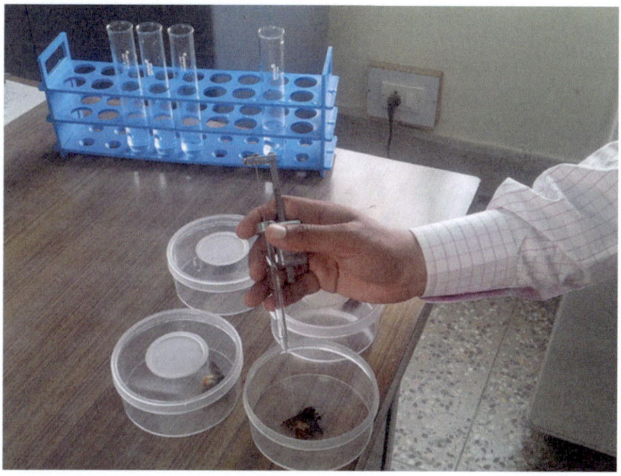

Fig. 3.4 Topical application in coxa to bumble bee

Topical Application on Coxa (Marletto et al. 2003)

In this experiment, bees were introduced in plastic containers of 20 cm height and anesthetized using dry ice or CO_2. A droplet of 10 μL of the dispersion to test or water for the control was laid between the coxae of the anesthetized bees by means of an automatic pipette. The bees were then checked for mortality at specific time. The solution applied on this method seems to be high since 10 μL is used instead of 1 μL in other experiments (Fig. 3.4).

4.1.1.1.2 Indirect Contact Toxicity

Several workers have determined the residual contact toxicity of pesticides to pollinators by applying in glass (Way and Synge 1948), grease-proof paper (Glynne and Connell 1954), cellophane (Beran and Neururer 1955), tin foil (Palmer-Jones 1958), filter paper (Stanley et al. 2009), leaves (Laurino et al. 2011) etc. as a contact medium. These materials, unlikely give an accurate indication of the residual toxicity on plants. These materials are inert and absorption in particular, cannot take

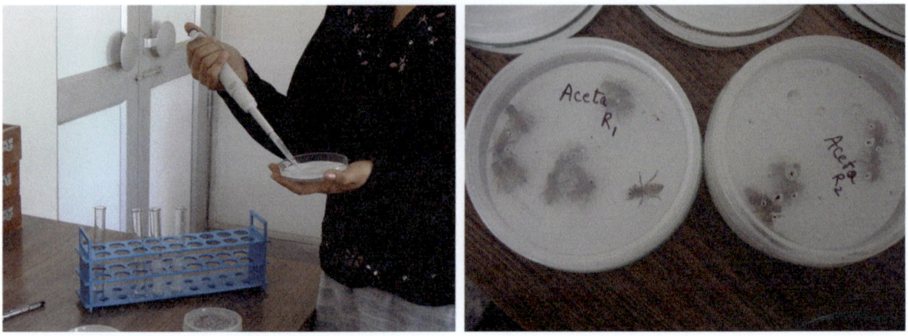

Fig. 3.5 Filter paper disc bioassay

place as that of plants (Clinch 1967) except the leaves. The nature of the surface on which the pesticide is sprayed also has an influence on its toxicity (Flynn and Schoof 1966). So indirect contact toxicity tests using plant materials especially flowers can be realistic. Way and Synge (1948) used flowers of apple, cineraria and daisy for testing indirect contact toxicity to bees. Johansen, (1965) used white Dutch clover plants sprayed with pesticides in the field and taken foliage samples in cages where bees were exposed. These tests using plant foliage or flowers may be realistic to field conditions. Some of the indirect contact tests are described as under.

Indirect Contact Tests Using Filter Paper (Stanley et al. 2009)

This method is based on the one described by Busvine (1980) with modifications. One to 3 days old worker bees of different bee species *viz*., Indian bee, *A. cerana indica*; giant honey bee, *A. dorsata*; little bee, *A. florea* and stingless bee, *T. iridipennis* were used in the study. Plastic containers with perforations were used for the experiment to allow adequate aeration for the bees. Filter paper discs were made wet with 1 mL of different concentrations of insecticides dissolved in distilled water and allowed to dry by hanging for about 10 min. These shade dried filter papers were placed in the bioassay container and honey bees were released at the rate of 10 per container. Honey bees were kept in refrigerator prior to release so as to calm them for an easy transfer. After exposure for 1 h, the bees were allowed in a big polyethylene bag perforated using pins and 40 % sucrose solution soaked in cotton wool was provided for food. The bee mortality was observed at specific intervals. Abbott's correction (Abbott 1925) was made if natural mortality was found in the control bees. This is one of the easiest methods to find the indirect contact toxicity of insecticides on pollinators (Fig. 3.5).

Indirect Contact Toxicity Using Tin Sheets in Cages (Palmer-Jones 1958)

Test pesticides were evenly sprayed over one side of a sheet of tin foil of 600 cm^2 area. These sheets are reported superior to grease paper for this purpose as it does not absorb moisture from the air. The foil was used to line completely the inside of

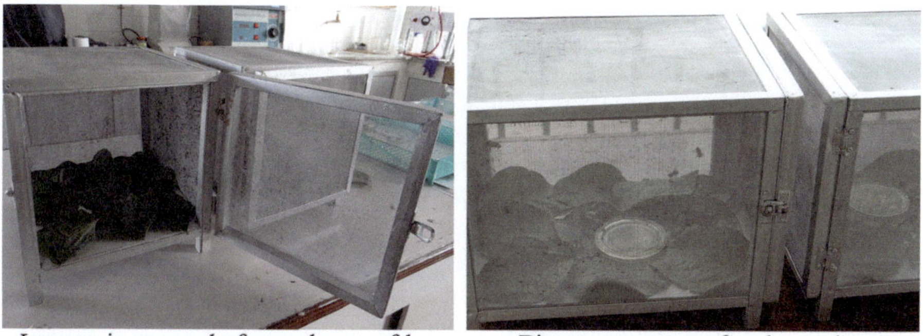

Leaves in cages before release of bees *Bioassay setup after bee release*

Fig. 3.6 Indirect contact toxicity bioassay using leaves

a square wire cage with press-on lid. Fifty bees that had first been fed for approximately 1 h are then caged and kept in contact with the dried pesticide for 1 h without syrup. During this period air was sucked into and out of the cage by a tube attached to a pump. This prevents the build up of vapour from the pesticide which could act as a fumigant. After treatment, the bees are removed, kept for 24 h in a clean cage supplied with syrup and mortality noted.

Indirect Contact Using Cages with Removable Bottom (Marletto et al. 2003)

In this experiment, cages with removable bottom were used. The bottom of the cages were removed and sprayed with the test products mixed with water. After the complete evaporation of the water, the bottoms of the containers were mounted with the upper parts. Then the bees were introduced and fed with sugar solution. After 3 h of exposure, treated bottoms were removed and replaced with uncontaminated ones. Observations on bee mortality were made at regular intervals after treatment. In this method, only the bottom was treated and bees clinging to the sides and lid can elude getting contact with the pesticide. During 3 h of exposure, the bees touching the contaminated bottom may contaminate slightly the sides also.

Indirect Contact Tests Using Leaves (Laurino et al. 2011)

The leaves of Spanish chestnut were sprayed to drip with a high-volume pneumatic hand sprayer and allowed to dry in the shade for at least 3 h. Pure water was used for untreated controls. The leaves are then introduced into cages and arranged in a manner to completely cover the floor. Honey bees were introduced into the cages (ten bees/cage) and allowed to walk freely on the bottom covered with leaves and kept for 3 h after which the leaves were removed. During the trial, the honey bees are fed with sugar candy from a feeder. Test mortality was checked at 3 and 6 h on the first day of the trial and at different time interval on the following days. As stated in the above method with removable bottom, here also only the bottom portion is treated, however leaves are used as a medium which give a realistic exposure (Fig. 3.6).

Indirect Residual Tests Using Foliage (USEPA 1996)

The test substance at its label rate is to be applied on alfalfa plots at least of 1 m^2 area using a hand sprayer. Three control plots were treated with uncontaminated water. After the residues have aged for the appropriate time period say 3, 8 and 24 h after application, alfalfa foliage was harvested from the test plots and returned to the laboratory, chopped, mixed and divided into 15 g portions. Each portion is placed into a single test chamber and 25 numbers of 1–7 days old worker bees are introduced into the test chamber. Control and test bees are to be kept under the same environmental conditions. The test insects were fed *ad libitum* with 50 % sugar-water solution. Observations on bee mortality are noted after 24 h of bee introduction. Throughout the test period, all signs of intoxication and mortality can be recorded and reported by dosage level and by time of occurrence.

Indirect Contact Tests Using Flowers (Clinch 1967)

This experiment uses white clover flowers to find the indirect contact toxicity of pesticide to pollinators. The flowers were sprayed with insecticides using a potter tower. The basic design of the spray tower used in this study is as given by Potter (1952), but the atomizing nozzle used is an Aerograph type MP spray gun. This apparatus is capable of applying an even deposit over a circular area. Thus this tower is made to enable the spray to be applied as accurately as possible. It was originally intended that the sprays would be applied to the flowers of white clover plants grown in pots. But because of insufficient plants with flowers, detached flowers picked from plants growing in the field was used. The stem of detached flowers was inserted into the perforated holes made in the lid of a container with water. Eight holes were made in equal spacing in a 9 cm dia. lid. Metal enclosures were made with top covered with nylon to keep the flowers and bees. To confine the insecticide deposits to the flowers and not on the container lids, filter paper was kept over the lid before inserting the flowers and removed after the spray was over. The spray of insecticides at desirable concentration was administered and allowed to dry before release of bees. Bees at groups of 10 were introduced in one set after slight anesthetization and allowed to feed on sprayed flowers for 1 h period. Then they are transferred to post treatment cages and fed with sugar. Observation on mortality was taken at 24 h after treatment. A set of artificial flowers are also made and used to compare the residual toxicity on a non-absorbent material. Though it is a very old method, it seems to be good and realistic.

Residual Contact Test Using Corn Tassels (Bailey et al. 2005)

Pollen shedding maize tassel are used for this bioassay. The tassels are collected from non-transgenic sweet corn treated with insecticides in the field at field recommended doses. The treatments include foliar sprays of insecticides and seed treatment. Tassels for foliar applied insecticides were collected from the field before insecticide application and 12, 36 and 60 h after treatment, to test the toxicity.

Bioassay setup before bee release *Bioassay setup after bee release*

Fig. 3.7 Contact toxicity bioassay using corn tassels

Tassels from seed treated plots were collected at the start of pollen shed and for three consecutive days. The tassels were kept in individual jars and 25 forager bees collected from the colony entrance into glass jars were released. The jars were provisioned with sugar solution and fed *ad libitum* via gravity feeders and a Bee Boost® strip and kept in laboratory conditions. The number of dead bees per cage was counted after 24 h of exposure (Fig. 3.7).

4.1.1.2 Oral or Ingestion Toxicity

4.1.1.2.1 Acute Oral Toxicity Test with Capillary Feeders (Murray 1985)

In this test, appropriate concentration of technical grade test chemicals are to be prepared in acetone then diluted with 50 % w/v sucrose solution. For test with formulated pesticides, the required concentrations are to be prepared by diluting with sucrose solution directly. These solutions are placed in glass capillary feeding tubes mounted on the lids of the test chamber. Test chambers were made using 2 mm aperture stainless steel mesh. The nominal dose taken by each bee during the test is about 20 μL/bee assuming equal distribution of feed. When all the test solutions are consumed, the capillary tubes are replaced with feeders having uncontaminated sucrose solution.

4.1.1.2.2 Oral Toxicity Tests: Bees Singly Fed (Marletto et al. 2003)

In this experiment, bumblebees were not allowed to feed *ad libitum*, but individually. The worker bees are placed singly in cylindrical containers (dia. 30 mm and height 50 mm) of black high density polyethylene in which a hole of 2 mm in dia. was made in the side wall near the bottom. The bees were starved for about 3 h and a droplet of 10 μL of the dispersion to be tested was kept near the hole to allow the bee to reach the solution with its tongue. The test insects which consumed the entire solution of 10 μL within 15 min. were assessed for their mortality.

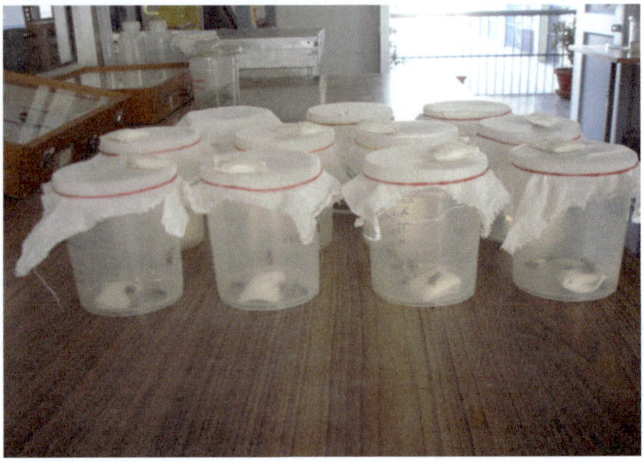

Fig. 3.8 Bioassay on ingestion toxicity of pesticides to bees

4.1.1.2.3 Ingestion Toxicity Tests Using Cotton Swabs (Stanley et al. 2009)

For this bioassay, honey bees were fed with sugar solution contaminated with appropriate amount of pesticides using a piece of cotton. The bees were kept in plastic containers with netted lid at ten bees per container. A piece of cotton soaked in pesticide contaminated sugar solution at different concentrations was given as feed for 6 h. At least 30 bees were maintained for a treatment. After 6 h of feeding, the bees were given with fresh uncontaminated sugar solution. Observations on mortality was taken at 6, 12, 24 and 48 h after treatment and corrected for mortality in control if any using Abbott's correction (Fig. 3.8).

4.1.1.2.4 Ingestion Tests After Starvation (Laurino et al. 2011)

In this method, honey bees were kept for 2 h after capture under dark and cool environment (11–13 °C) and made to starve to empty the crop content. After this starvation period, the normal procedure for the ingestion test can be followed. The amount of sucrose syrup ingested by each honey bee during oral toxicity tests can also be determined by weighing the feeder at the beginning and at the end of the allowed 1 h feeding period as well as taking into account the syrup density. In this study, on an average 35 μL was calculated per bee which is used to find the ingestion lethal dose (LD_{50}) from the relative LC_{50}.

4.1.1.2.5 Ingestion Toxicity Test Using a Flower Method (Ladurner et al. 2003)

This method uses original flower with some manipulations using ampoules. A tiny plastic ampoule (inside dia. 2 mm, outside dia. 3 mm, height 5 mm) was introduced into the calyx after removing the inner parts of the flower. The test solution is pipetted into the ampoule. To facilitate flower manipulation, large, actinomorphic flowers with open corollas are used including cherry, apple, morning glory and periwinkle. The plastic ampoule used in the experiment was built by heat sealing one side of a 5 mm section polyethylene tube. A flower and a bee were kept in a paper cup of 8 cm dia. and 5 cm height and covered. Test solution of about 10 μL was offered to the bee for 1 h and a sample size of 20 individuals was used per treatment. As it was tested by the authors on *A. mellifera*, *O. lignaria* and *M. rotundata* in different light regimes, we assume this method is good to conduct the ingestion toxicity of a pesticide with a specific dose administered to a bee.

4.1.1.2.6 *Ingestion Toxicity Test Using an Artificial Flower Method (Ladurner et al. 2005b)*

Similar to the flower method described above, this method uses an artificial flower instead of a real one. As honey bees and bumble bees seem to have an innate preference for blue and yellow colours and spontaneously fly towards symmetrical patterns; artificial blue petals (colour tapes) were arranged symmetrically with yellow nectar guides with a central hole (dia. 0.5 cm). A piece of filter paper (dia. 0.5 cm) soaked in 4 μL of diluted lavender oil was also made to give odour stimuli. The artificial flower was fitted with a 5 mL sample vial with a lid. *Osmia lignaria* and *M. rotundata* females were starved for 24 h and *A. mellifera* workers for 4 h, prior to being tested. Simultaneously natural flowers were also used to test the feeding of the bees. Feeding success on the natural flower was near 90 % in all three bee species whereas it was very low in artificial flowers especially *M. rotundata* and *O. lignaria* bees rendering the method ineffective.

4.1.1.2.7 *Oral Toxicity Test Using Pollen (Bailey et al. 2005)*

Bee frames containing sealed brood were taken from hives and incubated in the laboratory. Cohorts of 20 newly emerged bees (1 day old) were collected from hive and kept in cages of mesh bottoms and glass fronts. Pollen samples were collected from non-transgenic sweet corn plants treated with insecticides at recommended field rates in field. The samples were collected in tassel before treatment and 12, 36 and 60 h after the treatment of insecticides. Test pollen was sieved through 20,120 and 240 mesh plastic sieves to remove contaminants if any. Pollen from seed treated plots was collected at the start of pollen shed and for three consecutive days for testing. The control treatment consists of pollen collected from untreated sweet corn. These pollen samples were given as feed to the honey bees in the cage along with sugar solution and Bee Boost®. Observations on the number of dead bees per cage were counted 24 h after exposure.

4.1.1.2.8 *Oral Toxicity to Larva (Zhu et al. 2014)*

Frames of newly-hatched larvae were taken, transferred into sterile, 48 well culture plates (24 larva per plate) and kept in laboratory conditions. Diet containing 50 % royal jelly and other ingredients were prepared and the pesticide test solution was mixed thoroughly into the diet at specific concentrations. Untreated diet and solvent treated (methanol and acetone) were also used because of use of solvents in the preparation of pesticide solution from the technical grade. Sterile, push-in queen cups with 10 μL of diet treatment were placed in empty wells in the culture plates. A 00 camel hair paint brush was used to transfer each larva from the cell on the frame to the cup for feeding. The whole setup was kept under 95 % RH and 34 °C and in the dark. Diet for each larval bee was replaced daily, larval mortality assessed and dead larvae removed. If the food intake or the concentrations of pesticides consumed by each larva are not measured, this method will not provide the exact quantification of toxicity. Still it is a good technique to find the oral toxicity for a bee larva.

4.1.1.3 Fumigant Toxicity

4.1.1.3.1 Using Wire Gazed Circular Holder (Palmer-Jones 1958)

In this method, fumigant effect of pesticide was studied by using circular holders, 10 cm dia. and 6.25 cm depth with tops and bottoms made of wire gauze. Each holder was made with a flange that could be slid over the top of a tin which held a Petri plate in which fitted a 9 cm filter paper. The filter papers were soaked with 1 mL of the particular dilution of the pesticide under test and the holder, containing 50 bees, fitted above. The whole setup was then placed in a 30 °C incubator and the bees exposed to the pesticide vapour for 1 h. Afterwards they were transferred to cages with feeders and kept under observation for 24 h and mortality noted. Bees were fed before exposure to the vapour, but no syrup was given during exposure. As bees that had fed lightly before exposure to the vapours were found to be less affected than those that had fed heavily, two set of bees i.e., fed for 25 min. and another for a minimum of 1.5 h, representing light and heavy feeding were used in the study.

4.1.2 Tests for Insect Repellency

4.1.2.1 Repellency Test for Pesticides (Thompson and Wilkins 2003)

A clear plastic box (12×8 cm) was taken in which a pre-weighed glass sucrose feeder was placed in the centre of a piece of filter paper (4×7 cm). The filter paper was earlier treated with pesticide or untreated. The filter paper was treated by applying the field recommended rate of pesticides dissolved in water and allowing it to dry prior to placing in the test cage. The bees were starved for 1.5–2 h prior to the test, anesthetized with CO_2 and ten adult workers transferred to the test cages. The bees were allowed access to the feeder until the control bees consume all the test feed or for a maximum of 4 h. The feeders were weighed at the end to determine the amount of sucrose consumed in treatment and control.

4.1.3 Tests for Insect Growth Regulators

4.1.3.1 Larval Morphogenic Test (Atkins and Kellum 1986)

This test allows the assessment of pesticide effects on bee brood in the hive. A small cage was made of queen excluder material and the queen confined over about 500 worker cells of an empty comb for 24 h. Then the treatment solution (1 μL/cell) was applied to the food in the bottom of each cell with a microsyringe. A sample of 100 larvae living in several horizontal adjacent rows was treated for each dilution of pesticide and control. The number of surviving larva was assessed after cell capping. After emergence, survival and morphogenic effects were assessed.

4.1.3.2 Larval Test for Bumble Bees (Gretenkord and Drescher 1996)

The test groups were prepared by removing egg cells from colonies until hatching and equalizing the number of larvae in all the groups. They obtained standard cells with ten young larvae which were kept at 29 °C in rearing boxes ($12.5 \times 7 \times 5$ cm) with three nurse workers. These were fed with syrup and pollen dough until pupation. Trials, comprising three replicates of each treatment and control, should start with the recommended concentration for field use. If negative effects are observed, the trials can be continued with lower concentrations.

4.2 Tier II Toxicity Evaluation: Semi-field Experiments

Semi field tests denote second tier of toxicity evaluation after laboratory tests. If some effects are observed in tier I, the risk can further be exemplified in tier II in a more realistic exposure condition. Semi-field experiments using cages give a realistic exposure condition and are used to study the pesticide toxicity to honey bees (EPPO 2010a). In this second tier evaluation we include the studies made under tunnels or cages and also those made in the colonies/hives of pollinators.

4.2.1 Cage or Tunnel Studies

4.2.1.1 Phacelia Crop in Cages (Miles 2003; Dinter et al. 2009)

This semi-field (cage) study is based on the EPPO Guideline No. 170. These tests were carried out in flowering plants of *Phacelia tanacetifolia* in cages. Each cage was provided with a colony of 3000–5000 workers and a queen. A toxic reference treatment (negative control) and a water treatment are also included along with actual treatments. Two different studies were conducted as follows. The first test comprised of one time application of insecticide early morning before the onset of bee flight at two different rates as per the field recommended doses suggested for different crops/stages and target pests. The second test was made to assess repeated application of insecticide on the crops. Applications of insecticide at field recommended rate with different spray intervals e.g. 0, 7, 10 and 20 days (can be changed according to the recommendation) can be made. Another experiment with field recommended dose of insecticide sprayed at midday when bees were active in the crop was also made. In all the semi-field experiments, bee mortality, hive activity and foraging behaviour were assessed each day. At the end of the experiment, condition of brood and food reserves were also assessed.

Many other observations apart from mortality during pre and post treatments were made by Dinter et al. (2009). Some of the observations like foraging activity (numbers of foraging honey bees/m^2), behavioural effects (both the foraging bees on the crop and the bees around the hive) and brood effects (assessments of the status of the bee colony regarding visibility of queen and availability of eggs, larvae, pupae and adult bees inside the hive) were studied in the tunnel test.

4.2.1.2 Large Tunnels with Four Plots (Medrzycki et al. 2013)

Semi-field studies used for bee toxicity experiments are largely based on CEB protocol 230 (CEB 2011). Tunnels should be at least 120 m^2 and covered with a net that duplicate natural climatic conditions. Inside the tunnel, four plots of the same size (2×8 m) were separated by covering with net or film and hive placed at the centre. After the settlement of hive under the tunnels, the bees were allowed to forage on crop plots and strength parameters assessed (Delaplane et al. 2013) for 2–5 days. Then insecticides are sprayed at their field recommended doses and dead bees were collected every day in the morning in order to record the mortality of the previous day accurately.

4.2.1.3 Potted Plants of Mustard in Net (Stanley et al. 2015) (Fig. 3.9)

Potted mustard plants were used for pesticide toxicity studies in semi-field conditions. At full blooming stage of the crop (~70 days from sowing) pots were grouped and sprayed with the test insecticide at their field doses using a 2 L hand sprayer. Treated plants were allowed to dry and kept near the bee introduced open fields of mustard. Pots were placed in four replications with four pots per replication. The bees, foraging on the sprayed plants were collected using test tubes and allowed in small cages. A total of ten bees were collected and released in first quadrat with mosquito nets having four pots with treated plants inside and sealed properly at sides to prevent the escape of bees. A total of 40 bees were collected and released in all the four quadrats and kept in shade for 1 h to allow bees to forage. After 1 h, bees were collected back from each quadrat and allowed to be in containers and provided

Pots kept inside nets & ready for bee transfer *Bees feeding on flowers inside net*

Fig. 3.9 Semi-field experiments using potted plants in net

with uncontaminated sugar solution. Mortality of bees were recorded at 1, 24 and 48 h after treatment and percent mortality calculated.

4.2.1.4 Potted Plants of Phacelia in Cages (Murray 1985)

Large stainless steel mesh (2 mm aperture) cages with a glass front and base were used in this experiment. Formulated test pesticides were sprayed on potted flowering plants of *Phacelia campanularia* using hand sprayer delivering a volume of 340 L/ha. Plants were shade dried for 30 min. and four pots kept in a cage and 100 bees released for a treatment. After 2 h of foraging on sprayed plants, sugar solution was given to the bees. The absence of artificial feed on the early part of the test encourages the bees to forage on the sprayed flowers. Observations on bee mortality were carried out at 1, 2, 4, 24 and 48 h after bee introduction. On each occasion recordings on the number of flowers visited by the bees for 30 s was also carried out. In addition, number of bees foraging on the flowers was also counted at 1, 2 and 4 h after treatment. Though it is reported originally as the test to find out residual toxicity, it allows the bee to take nectar from treated plants thus may be considered for both residual and ingestion toxicity.

4.2.1.5 Test Using Wheat by Simulating Honey Dew (Shires et al. 1984b)

This experiment is based on CEB draft Guideline No. 230 to test the pesticide toxicity by simulating honey dew on wheat in enclosures. Plots of winter wheat were treated and covered with nets. Sucrose solution was sprayed on the wheat to simulate honey dew. A small bee colony was introduced in the plot and kept inside the net. Observations on the bee behaviour and mortality in front of the hives were recorded. The strength and condition of colony (adults and brood) before and after the exposure period are also studied.

4.2.1.6 Potted Plants of Cucumber for Bumble Bees (Incerti et al. 2003)

Though this experiment is stated as extended laboratory test by the authors it fits well with our semi field procedure. Adult bumble bees were taken and kept in laboratory conditions, from which medium size workers were collected operating under red light. Cages were made measuring 0.5×0.5 m bottom (plywood) and 1 m height (plastic walls) offering sufficient flying space for bees and with openings to allow the introduction of insects and plants. Cucumber plants (*Cucumis sativus*) was selected as potted plants and used in the test because they have large leaf surface, grow fast and attract bumblebees for both pollen and nectar. When the plants were of 40 days old and blooming (~12 flowers/plant), they are sprayed with test chemical at its field recommended concentration. Two plants were kept in each cage and 30 bumble bees released. About 30 g of commercial sucrose solution was also provided per cage. Two different treatments were given with release of bees after 1 and 48 h after pesticide spray. Observations on the abnormal behaviour and mortality were recorded at 24, 48 and 72 h after the introduction of bees.

4.2.1.7 Bumble Bee in Poly House Tomato (Dinter et al. 2009)

This experiment is based on SETAC/ESCORT and EPPO 170-3 recommendations (EPPO 2001; Barrett et al. 1994). Poly house grown flowering tomato plants were used to test the toxicity of pesticide on bumble bees. At least 420 m^2 was designated for each treatment with one bumble bee hive/plot (25 workers/colony). The treatments were replicated four times and plots separated by nets of mesh 5 mm. Altogether four treatments of field recommended dose of insecticide, sprayed while bumble bees were foraging or at 24, 48 and 72 h before opening of hives. The bees were exposed for 21 days and observations were made on the larval and adult mortality, brood development and condition of colonies. Foraging activity of bumble bees was assessed by estimating the number of flowers visited by counting bite marks on flowers.

4.2.1.8 IGR Test on Bumble Bees (Thompson and Barrett 1999)

In this experiment 5×3 m compartments containing tomato plants were used as a plot and a single queen-right bumble bee colony with 100–200 workers introduced. IGRs were sprayed on tomato at 10 days intervals. Colonies were fed with additional pollen which was also treated at the same rate as plants. Each IGR treatment and the control spray with water were repeated in two compartments. The bee colony was monitored for number of dead adults, dead larvae, foraging bees and the general appearance of the colonies.

4.2.2 Colony or Hive studies

Most of these colony/hive studies are of forced in-hive nutritional studies which investigates the distribution of a xenobiotic within the colony and within the hive (beeswax, pollen, honey) to estimate the effect of pesticides on bee colonies and their development (Medrzycki et al. 2013). For all the studies on the effect of pesticides on brood, observations on brood development by calculating brood index, compensation index, brood termination rate, etc. can be done (Medrzycki et al. 2013).

4.2.2.1 Colonies Fed with Contaminated Syrup (Faucon et al. 2005)

In this experiment, eight colonies were designated for one treatment and were fed with saccharose syrup with insecticide during summer till the end of winter. Each colony was provided with 1 L of saccharose syrup in 13 occasions during the test period. The treated colony was compared with the untreated or control colonies unfed or fed with only saccharose syrup. The number of bees entering the hive, number of bees with pollen entering the hive, adult population, capped brood area, frequency of diseases, mortality, number of frames with brood and a score of colony after winter were noted. The observations were taken during feeding and also post feeding.

4.2.2.2 Colonies Fed with Contaminated Syrup and Pollen (Tasei et al. 2000)

Micro colonies of three workers of bumble bees, *B. terrestris* were reared under controlled environment in the laboratory in plywood boxes (11.3×4.5×4.3 cm) with a screened bottom and a transparent cover. Commercial syrup solution and bee pollen were given as food. Candidate insecticide in two different doses were administered in pollen and syrup and fed to the bees. Control food was non-contaminated syrup and pollen. During the whole test period, observations were made every day on the mortality of workers and the number of larvae ejected from brood cells. On the 85th day, the surviving workers were frozen and taken for residue analysis.

4.2.2.3 IGRs in Sugar Solution (Oomen et al. 1992)

This is considered as the official method of the EPPO for honey bee feeding tests with pesticides especially of insect growth regulators. The test substance at the field recommended concentration was mixed in 1 L of sugar solution and each colony was fed with it. Three replicates per product and per concentration are to be carried out. The trial compared a test substance with an IGR reference and a pure sugar solution. Altogether 300 cells per colony i.e., 100 with eggs, 100 with young larvae and 100 with old larvae were marked before treatment. Observations on the brood development were made for the following 3 weeks period at weekly intervals. A dead bee trap fitted in the front of the hives is used to count dead bees. To describe and note the brood development, each cell was marked in a transparency sheet kept over the brood using a colour code for each stage of the bee. New transparencies are used at every count and compared with the previous ones.

4.2.2.4 Larval Ejection Assessment in Bumble Bees (Thompson and Barrett 2001)

This study was designed to find the effect of pesticides especially insect growth regulators on bumble bees. Tomatoes were grown in glasshouses. At full bloom they were sprayed with insecticides and bumble bee colonies (~200 workers) were introduced to assess the effect of insecticide on the viability of the bumble bee colonies. Observation on the number of dead larvae ejected from the colony were made and compared with the results from control. This method explains the pesticide toxicity of brood when fed with contaminated food collected by foragers.

4.2.2.5 Pesticides on Queen Through Beeswax Queen Cups (Pettis et al. 2004)

This experiment was designed to test the effect of pesticides on the acceptance of bee larva grafted in queen cells. The queen rearing methodology employed here is that described by Laidlaw and Page (1997). Bee wax queen cups were made by using naturally produced uncontaminated bees wax. Different concentrations of pesticide were mixed with this uncontaminated beewax. Queen cups were made using individual wooden dowels (9 mm dia.) by dipping them in heated and pesticide treated wax. Queen cups made from wax with no pesticide serve as control. Young larvae were grafted into the queen cups. Altogether, 24 queen cups were used per treatment and the individual treatment cups were randomly placed. Observations on the acceptance as per treatment and weight of all mature queens were made by carefully opening the cells.

4.2.2.6 Colonies Treated and Observed in Field (Whitehorn et al. 2012)

In this experiment, *B. terrestris* colonies were used find the oral toxicity of pesticides. There were three treatments viz., control (fed with uncontaminated food), low (fed with pollen and sugar water containing low but representing levels found in seed treated rape) and high (fed with food containing double but still close to field realistic range). Colonies were randomly allocated to one of the three treatments and fed *ad libitum* pollen and sugar-water over a period of 14 days in the laboratory according to the treatment. After 2 weeks, all colonies were allowed in the field and their performance monitored for 6 weeks. The colonies were observed for weight loss/gain, number of males, workers, pupa and queens produced.

4.2.3 Sublethal Effects

4.2.3.1 Effect on Brood Development: Digital Image Processing Test (Medrzycki et al. 2013)

The acetate method described just above to assess the larval development, though found effective but time consuming and only restricted number of cells can be assessed and cause a long 'off-hive-time'. To overcome all these disadvantages, a digital image processing method was developed (Wang and ClaBen 2011; Jeker et al. 2012). In principle, the use of digital image processing allows to evaluate the development of an unlimited number of brood cells. In this method, photographing of the brood combs are to be done at the field site and evaluating the same at the laboratory. All the photographs are to be taken under same parameters such as distance and focal length. In the laboratory, the image processing system will compare the previous photos and calculate all relevant parameters such as brood termination rate, brood index and compensation index.

4.2.3.2 Larval Development and Adult Emergence: Non Apis Bees (Abbott et al. 2008)

This experiment is used to study the effect of pesticides on larval development and adult emergence of wild bees. Over wintering adults of *O. lignaria* were taken, kept in mesh cages and incubated at 24 °C for adult emergence. Bees were kept in nest blocks which were set up adjacent to blueberry fields. Acetate sheets were placed between the top and bottom boards of nest blocks for easy inspection. Three doses of pesticides, a low dose which is likely to be encountered by bees in the field; an intermediate dose and a high dose (unlikely to be encountered) were tested.

For laboratory tests, eggs were collected with their pollen provisions from the nest blocks and placed in 24 well culture plates. Two different methodologies were performed, one with injecting 10 μL of pesticide solution in the own pollen and another with replacing pollen with pre-blend pollen mixture with pesticide. In the field also, 10 μL of pesticide was injected into the pollen provision adjacent to eggs, using a syringe through the transparent acetate cover. The eggs were left to develop in their nest in the field conditions. The larvae under study (both in the field and laboratory) were monitored for larval development, days under each larval instar and cocoon formation. After almost all of the larvae had developed into adults, the cocoons were kept in environmental chamber at 15 °C for 6 days and then into a refrigerator set at 4 °C to allow them to over winter. Next year, the cocoons were incubated at 22 °C to allow emergence. Upon emergence, observations on mortality, sex, weight of the bees etc. were recorded.

4.2.3.3 Effect on Learning Ability: Proboscis Extension Response

In the course of foraging, a learning process occurs in bees during which floral parameters such as location, shape, colour and smell of flowers are associated with a reward (Menzel and Muller 1996). These floral cues are memorized by the forager and used for flower recognition during the following trips. Under laboratory conditions, learning and memory can be analyzed using a bioassay based on the olfactory conditioning of the proboscis extension (CPE) response on restrained individuals. Effect of pesticides on the olfactory learning of honey bees was generally assessed by conditioned proboscis extension response (PER) (Abramson et al. 1999; Decourtye et al. 2003, 2004; Desneux et al. 2007). At conditioning phase, the proboscis extension is obtained by touching the antenna with sugar solution and immediately an odour was also released. Here, the sugar solution act as unconditioned stimulus and the odour, conditioned stimulus. Thus the test bees were made to get associated with the sugar solution and odour and thus they elicit a response only for odour also (Han et al. 2010). The odour of the flower was associated with the reward, nectar and pollen (Menzel and Muller 1996).

The classical odour conditioning of the proboscis extension reflex, as described for example, by Bitterman et al. (1983) and Sandoz et al. (1995), is based on the temporal paired association of a conditioned and an unconditioned stimulus. During conditioning, the proboscis extension reflex is obtained by sucrose solution (US) and simultaneously by an odour (CS) and the extension of proboscis is immediately rewarded (Reward: R). Bees can develop the proboscis extension response as a Conditioned Response (CR) to the odour alone after even a single pairing of the odour with a sucrose reward (Decourtye and Pham-Delegue 2002).

The bees can be exposed to the toxicant at various stages of the procedure to test the effect on learning ability: (1) prior to the bioassay, which may be analogous to the case of bees feeding on contaminated food stored in the hive, before the bees become foragers, eg. Nurse bees; (2) during conditioning phase, the product being administered in the reward which is comparable with the exposure of foraging bees when they visit treated plants; (3) between the conditioning and the extinction phases, to evaluate the effect of the product on the time course of memory (Pham-Delegue et al. 2002).

4.2.3.4 Test for Conditioned Response After Contact Toxicity Treatment (Taylor et al. 1987)

This assay also uses the proboscis extension response after treating the bees by indirect contact to pesticides. Filter papers (9 cm dia) were treated with 1.5 mL of acetone in which the know concentration of insecticide dissolved. The paper was dried for 1 h and put in a glass Petri dish. A flight cone was used to funnel approx. 20 forager bees exiting from the colony in to the Petri dish with filter paper. The Petri dish was covered with a ventilated lid and sugar solution given as feed. After the 24 h exposure, 15 treated survivors and 15 untreated bees were used for the actual bioassay i.e., to compare their conditioning to odour. Individual bees were restrained in a bee holder to facilitate proboscis extension conditioning. Each bee was positioned in front of the exhaust duct and thyme scent was blown over its antennae (conditioned stimulus). During the second half of a 6 s scent release, a small pipette dipped in 30 % sucrose solution (unconditioned stimulus) was touched on the antenna. Proboscis extension response was tested and the bees were categorized as positive to odour training and negative (untrained).

4.2.3.5 Test to Assess Visual and Olfactory Learning (Han et al. 2010)

As proboscis extension response (PER) bioassay is a well known method to study the olfactory learning of bees especially after sublethal doses of toxicants. An innovative T Maze was coupled with PER bioassay to study the visual and olfactory learning simultaneously. Orientation of bees is based on the bee's ability to associate the visual sign with the reward (Zhang et al. 1996) in a similar way as in the PER test. In this test, bees were collected and provided with normal food to adopt them to the experimental conditions for 2 days. Subsequently, the bees were exposed to the dietary treatments for 7 days and then were used for T-tube maze and PER test.

T- maze assay: T-tube of arm length of 20 cm (internal dia.: 1.6 cm) was used in the experiment (condition). The arms were completely covered with yellow or blue papers to give colour stimulus (conditioned stimulus). The honey bees were starved for 24 h before conditioning. During conditioning phase, the bees were allowed in the entry place individually and sugar solution was kept in the end of the blue arm as a reward. Preference of bees during the conditioning sessions was recorded to understand the associative learning experience. The tube was washed with ethanol and dried before the entry of next bee. After conditioning, two such tubes were assembled to make a T maze for evaluation of visual learning. In this, the bees have to choose two times blue colour by turning once left and then right to get a reward.

Conditioned proboscis extension reflex assay: The pipette tips were cut and the bees were trapped in it. The trapped bees can move only their antennae and mouthparts freely. Bees were starved for 3 h prior to conditioning and then placed in the main airflow (50 mL/s) to be familiarized with the experimental context. Linalool was delivered as conditional stimuli through a secondary airflow (2.5 mL/s for 6 s). During odour delivery, the PER was elicited after 3 s by touching the antennae with sugar solution as an unconditional stimuli (US). After this US, a reward was given and odour delivery stopped. Three conditioning trials were carried out with 20 min. intervals followed by five test trials. In test trials, only the conditioned stimulus (odour: linalool) was delivered for 6 s. The conditioned PER was recorded as a 'yes' or 'no' response during the test sessions.

4.2.3.6 Video Tracking to Test Sublethal Effects (Teeters et al. 2012)

In this study, the sublethal effects of pesticides on bees are assessed using a video-tracking system. The effects of pesticides on locomotion, interactions and time spent near a food source were observed over a 24 h period. Bees were either treated topically with different doses of pesticide or exposed to different concentrations (ppb) in a sugar agar cube and observed. This technology is considered as good to study the actual effects without any disturbance to the test organism.

4.2.3.7 Effect of Pesticides on Homing and Foraging (Stanley et al. 2010)

This method is developed to study the effect of pesticides on the homing and foraging activity of bees and based on the one suggested by Bortolotti et al. (2003) with modifications. Foraging bees of the test hive were trained to take sugar solution kept in a feeder, which was gradually moved from the hive up to 500 m distance over a period of 2 days. The flying path from the hive to the feeder was linear and without obstacles or barriers. In this study, three treatments (untreated sugar solution and sugar solution with pesticide at low dose and a high dose) were used. On the trial day, 30 honey bees feeding on the sugar solution were captured and labeled with green coloured fingernail polish (Abramson et al. 2007) and designated as control. The marked bees were taken to the feeder site and released there. Observations on the number of marked bees thus released returning to the hive were noted. Observations at the hive were made by two persons for a period of 1 h and the percentage of bees returning to the hive was calculated.

The sugar solution in the feeder site was contaminated with low dose of test pesticide and 30 bees were captured while feeding on the test solution. The bees were labeled with pink coloured fingernail polish and transferred into the flying cage, released after 1 h and counted at the hive entrance. The same procedure was repeated for high dose of pesticide and the bees were labeled with red coloured nail polish. The two pesticide concentrations tested were equivalent to LC_1 in the honey contamination bioassay and ten times of this concentration. The behaviour of the bees leaving the flying cage (uncoordinated flights, if any) was recorded. Another observation on the labeled bees coming for foraging on the next day (24 h after treatment) at the feeder site for 2 h was taken. These bees can be taken to the laboratory to find the median lethal time (LT_{50}). The bees can be allowed in flying cages, fed with uncontaminated sugar solution and mortality noted at 12 h interval.

4.2.3.7.1 Automatic Monitoring of Bees: Radiofrequency Identification (Schneider et al. 2012)

Radiofrequency identification (RFID) labeling is used to obtain detailed information on foraging behaviour of bees (Streit et al. 2003). For this, a direction sensitive reading device was kept in the hive entrance and another at the feeder site (Decourtye et al. 2011). These devices are used to observe the number of foraging trips from the hive to the feeder, the duration of foraging trips and the interval between each foraging. Each test bee was pasted on its thorax with a RFID transponder having a unique ID number. The test bees were treated orally with the test pesticide and released at the feeder site. The foraging behaviour of these bees was recorded at 3 h periods immediately after and between 24 and 48 h after treatment. The advantage of this method is that detailed and precise information on behaviour can be obtained with little effort and at reasonable cost.

4.2.3.7.2 Homing Behaviour Using RFID and Modeling on Population Dynamics (Henry et al. 2012)

In this study, the effect of sublethal exposure on the homing behaviour of foraging bees was studied. The mortality caused by homing failure in exposed foragers was assessed by monitoring free ranging honey bees using RFID tagging technology. The effect of this on colony dynamics was assessed by using a model on honey bee population dynamics given by Khoury et al. (2011). Altogether 653 individual free ranging foragers from 3 different colonies were tested in the course of 4 separate treatments versus control homing experiments. The test foragers were pasted with microchips, fed with field realistic sublethal dose of insecticide in sucrose solution and released almost 1 km away from the colony. RFID readers placed at the hive can detect the bees with microchip entering the colony. Mortality due to homing failure was derived from the proportion of non returning foragers. Equal numbers of control foragers were also released to determine the other causes of homing failure.

Another experiment was made to study the effect of this homing failure in familiar and unfamiliar foragers. Foragers familiar with the pathway back to the colony may be less prone to homing failure than unfamiliar foragers. This experiment with familiar and unfamiliar bees can give the fiducial limits of the effect studied in the first experiment. A simulation model was made using all the tested scenarios to find the predicted effect of sublethal toxicity of insecticide on the function of a bee colony.

Methods to Assess Toxicity Through Unconventional Routes of Pesticide Exposure

1. **Exposure through Air**
   This exposure to pollinator normally happens when the pesticides coated in the seed get aberrated/drifted from seed drills and cause pollution in air. Apart form this, volatile pesticides also contaminate air by their fumigant action but they are less likely to cause problem to pollinators.

   Filter paper test (Greatti et al. 2003)
   To test the possible aberration of active ingredient of seed coating chemical from the centrifugal fan drain of the pneumatic seed drill, paper filters (25×25 cm, surface 625 cm^2) were placed on it for different time intervals (30, 60 and 120 s) during the sowing operations. The exposed papers were

frozen and taken to lab for analysis using gas chromatography to quantify the residue levels.

Grass and flower test (Greatti et al. 2006)
Samples of grass and flowers of 1 m^2 area were collected from the borders of pesticide coated seed sown fields. The samples were collected 1 day before sowing operation, the day of sowing and 1, 2, 4, 6 and 8 days after sowing. Grass samples and flowers in bloom at that time in the sampling area were collected. The samples are frozen and taken for extraction and analysis of pesticides in the laboratory.

Acute toxicity with dusts from pesticide coated seeds (Sgolastra et al. 2012)
In this experiment the dust was extracted from seeds treated with seed dressing pesticides using Heubach cylinder (Heimbach 2008). In the Heubach method, treated seeds were mechanically stressed inside a rotating drum. A vacuum pump was used to produce an air flow in the rotating drum. This air flow makes the abraded dust particles to get transported from the rotating drum through the glass cylinder and subsequently through the filter unit. Fine dust particles get deposited in the filter while coarse particles are collected in the glass cylinder (Medrzycki et al. 2013). The dust thus obtained were sieved and the fraction <45 μm was used for bioassay. This is field realistic because the majority of the dispersed particles during sowing were reportedly smaller than 45 μm (ApeNet 2011). This dust was mixed with talc to get the concentration equivalent to that observed in the field and apple leaves were treated with it. Forager bees (ten bees/cage) were exposed to the pesticide for 3 h on treated apple leaves kept in a hoarding cage. Mortality or other effects on bees were assessed at regular intervals.

Field evaluation of drift (Georgiadis et al. 2012)
Drift experiments were carried out in wind exposed areas by Georgiadis et al. (2012) to find the bee poisoning due to drift of insecticidal dusts from treated seeds during sowing. In the preceding season, winter oilseed rape (*Brassica napus*) was grown to serve as a bee attractive crop next to the experimental field. Colonies each were placed at distances of 0, 50 and 500 m in the windward border of the oilseed rape field to represent treatment, control and remote exposures. In the experimental field, seed treated maize was sown using seed drill. The colony and brood size were evaluated 5 days before the start of experiment using the Liebefelder evaluation method. The same was done after 15 and 30 days of initial evaluation.

2. **Exposure through guttation water**

Test for pesticide toxicity through guttation water (Shawki et al. 2006)
This method is used to test the exposure of pesticides to pollinators through guttation water and dew. Field recommended doses of pesticide were applied to a 15 m^2 plot of winter rape using a hand sprayer. Samples of guttation water

and dew (20 mL each) were collected daily using micropipettes until 10 days of treatment. These samples were taken to the laboratory and tested on the same day. To induce guttation, plants were irrigated and covered with polythene sheets at night and samples collected early in the morning.

Oral toxicity: Worker bees of *A. mellifera* were slightly anaesthetized with CO_2 and placed in plastic pots of 200 mL at ten bees/pot and starved for 1 h. Then the bees were fed with 50 % sugar solution mixed with guttation water or dew (1:1 v/v) for 24 h, using the glass feeding device. The control workers were fed with sugar syrup made of uncontaminated water. Observations on mortality were taken after 24 h of exposure.

Contact toxicity: The contact toxicity was done using Petri dishes (ten bees/dish) covered with nylon mesh. The guttation water or dew (2 mL) obtained from treated plants was made to wet the filter paper and kept on the dish bottom. The bees were kept at laboratory temperature and mortality determined after 24 h.

Chronic toxicity: To test chronic toxicity, the worker bees were transferred to laboratory cages, 14 × 14 × 7 cm. They were provided with 50 % sugar syrup mixed with guttation water or dew for 10 days. At the end of experiment (10 days), bee mortality was determined.

4.3 Tier III Toxicity Evaluation: Field Studies

Field tests are designed as the higher tier for the bee risk assessment of plant protection product (EPPO 2010b). In fact, field tests provide the highly reliable data. However, field studies are not often repeated because of environmental variabilities, complexicites and the high cost incurred. Moreover, methodological limitations make it difficult to carry out (Medrzycki et al. 2013).

4.3.1 Field Assessment in Seed Treated Canola (Cutler and Scott-Dupree 2007)

This field investigation was carried out to ascertain effects of seed treated canola, *B. napus* on honey bee, *A. mellifera* colonies during and after exposure to flowering. Colonies were placed in the middle of 1 ha seed treated or control canola fields for 3 weeks during bloom and thereafter they were moved to an apiary. Altogether 4 fields were designated as treated and 4 as control fields and 4 colonies per field were used giving 32 colonies in total. Bee mortality and brood development in each colony were assessed for 130 days from initial exposure to canola. Honey, pollen and wax samples were obtained and analyzed for residues by using HPLC.

4.3.2 Test to Find Exposure While Sowing Seed Treated Maize (Tremolada et al. 2010)

This experiment assesses the pesticide drift while sowing operation and exposure to bees while flying or even in the nearby hives. Corn seeds treated with recommended dose of insecticides and fungicides were sown using pneumatic seed drills. Altogether two exposure and four control hives were used in the experiment and kept more than 200 m apart from each other. Observations on direct mortality (lethal effect) in the hive area and intensity of foraging (sublethal effect) were assessed. Under basket traps to collect dead bees kept on the ground outside the entrance of the hive are used to find the bee mortality at the hive area. Bee mortality measurements were conducted for 6 days, commencing 2 days prior to sowing daily by 4 PM. The foraging activity of bees was monitored each day for 3 days after sowing and again on 10th and 15th day. The foraging bees with pollen entering the hive were counted for assessing the foraging activity.

4.3.3 Honey Bees in Apple Orchard (Van-der-Steen and Dinter 2007)

This field assessment of acute toxicity of insecticide (indoxacarb) was evaluated in apple orchards. Before the commencement of blooming, honey bee colonies were introduced in each orchard. Munster dead bee traps were kept in all the hives to collect dead honey bees and were counted every 3 days for about 2 weeks before and after the insecticide treatment. Some orchards are designated as control without insecticide spray and others were sprayed at field recommended doses. Mortality in colonies in treated orchards are compared with control ones.

4.3.4 Field Assessment of Residual Toxicity (Miles 2003)

In this experiment, honey bees were intended to be exposed to dry residues of pesticides in various crops like alfalfa, citrus and almonds. In alfalfa, hives were kept in the field and covered before the spray applications and uncovered after 3 h of spraying to expose the bees to the treated crop. Applications of pesticides were made in almonds and citrus, the previous day night. Observations on bee mortality, foraging behaviour and brood area were taken to assess the effect of pesticide.

4.3.5 Test for Bee Foraging in Treated Mustard (Gour and Pareek 2004)

The experiment was conducted using mustard plants grown under proper agronomic practices in randomized block design. The insecticides were sprayed twice at 15 days interval on the crop at blooming stage with foot sprayer at the rate of 830 L spray fluid per hectare. Population of foraging honey bees was recorded during the peak foraging time (between 12 Noon and 1 PM) for 5 min on ten randomly selected

Fig. 3.10 Repellency studies in field

and tagged plants per plot 1 day before and 1, 3, 7, 10 and 15 days after each spraying. In this experiment, the authors correlate the number of bees present with the toxicity of pesticide to bees i.e. low incidence represents high toxicity.

4.3.6 Test for Repellency in Mustard (Stanley et al. 2015)

It is a sequential testing trial, in which the pesticides that found safer in the lower tier studies were carried forward to field. It is not to test the toxicity but to test the repellency, if any, to honeybees. Pesticides which are non toxic but highly repellent to bees are also harmful in view of pollination. Three fields of mustard per treatment were sprayed with insecticides or water at the peak time of flowering. Observations were taken on *A. cerana* and *A. mellifera* visiting the flowers before and after the treatment using 1 m^2 quadrats. Observations were made for 30 s in each treatment (Fig. 3.10).

4.3.7 Pesticide Toxicity in Lawn/Turf to Bumble Bees (Larson et al. 2013)

This experiment was carried out to evaluate the scenario of resident bees foraging on flowering weeds in a treated lawn for 6 days. Plots were made of size of 3.35 × 3.35 m (ten replicates) with similar clover density and at least 2 m apart. Screen enclosures were erected over the lawn after pesticide application. Colonies of *B. impatiens* with 20 workers and a queen were kept 1 per enclosure 2 days after pesticide application. After 6 days, the colonies were inspected for weight and transferred to uncontaminated area and allowed to forage for 6 weeks. After that, they are assessed for number of living and dead adults, queens, honey pots and living and dead larvae, pupae and weight of live adults and queens.

4.3.8 Field Toxicity of Pesticides in Non-Apis Bees (Torchio 1983)

In this field test, pesticides were tested for their effect on *M. rotundata* kept in shelters in alfalfa fields. Two shelters (86 × 50 × 39 cm) were placed in two distant parts of 1 km long and 45 m wide field which was used to test two different pesticides. Pesticides were sprayed in designated areas at proper concentrations. Shelters were supplied with boards drilled with hundreds of holes each accommodating paper soda straws 0.5 cm in diameter and 6.5 cm long where leafcutter bees were nesting. Fifty nests were marked and monitored before and after the treatment. Straws were extracted and examined at night. When the nesting period was completed, boards were returned to the laboratory and kept at 26 °C for 1 month. Straws containing the marked nests were hibernated for 4 months after that it was incubated at 26 °C for 20 days. When incubation was completed, marked nests were dissected to record larval mortality.

4.3.9 In Situ Observation on Pollinators in Sprayed Field (Robinson and Johansen 1978)

This experiment was conducted to find the short term effects of pesticides on forest pollinators. The pollinators were observed and counted on marked sight units of small blooming plots of 0.8 m^2 area in big forest covers/fields applied with pesticides. About 25 units were sampled for each plant species to be studied. Each sighting was carried out during the warm hours of the day lasted for about 10 s. Observations were done at weekly intervals before pesticide application and followed closely after the applications to compare the pre and post treatment presence of pollinators.

5 Pesticide Risk Assessment for Pollinators

Risk is an event which causes an adverse effect. Risk of pesticide application to non targets especially pollinators is to be assessed so as to reduce the adverse effect caused, if any. There are many ways and methods to assess the risk of pesticide usage on pollinating insects. Analysis of potential exposure and its adverse effect is the basis of risk assessment. A real risk assessment involves the estimation of both hazard and exposure. Since the hazard (level of threat) is the inherent quality of the pesticide, it can be estimated easily by following procedures such as bioassays. The levels of exposure have to be estimated to get proper assessment of risks. Generally, the risk assessments for acute effects are being done by median lethal dose (LD_{50}) or concentration (LC_{50}) (Poquet et al. 2014). The no observed effect level (NOEL) or the lowest observed effect level (LOEL) are also used for assessing the pesticide toxicity to pollinators but are available only for a few pesticides (Sanchez-Bayo and Goka 2014) and are irrelevant for risk assessments (Landis and Chapman 2011) and inaccuracy (Fox et al. 2012). Thus the median lethal values are mostly used for risk assessment calculations.

In a deterministic risk assessment approach, the primary outcome of the risk characterization is the calculations of the risk estimate, for example, risk quotient (RQ) or toxicity exposure ratio (TER). Initial risk assessments can be based on empirically derived relationships (HQ for sprays) and also on TERs for systemic exposure through pollen and nectar (Alix and Miles 2011). Through these estimates, when the exposure level is higher than the toxicological value, the situation is at risk which results in HQ or TER values higher or lower than 1 (Poquet et al. 2014). The risk estimate is interpreted through its comparison with a level of concern. A level of concern or a trigger value, is a number intended to demarcate a point, above or below which (depending on the estimate in usage), risks are to be considered (Fischer and Moriarty 2011).

According to the guidelines of The International Commission for Plant-Bee Relationships (ICPBR) – ‘Bee Protection Group’, pesticide risk assessment for honey bees should be based on a sequential testing scheme and hazard ratio (Lewis et al. 2004). However, toxicity determination relied mostly on measuring LD_{50} or LC_{50} (Desneux et al. 2007). The estimated lethal dose or concentration cannot give full account of the deleterious effects of pesticides (Haseeb et al. 2004), because serious effects are reported in sublethal doses also (Colin et al. 2004). More meaningful risk assessments other than taking median values can be done using one tenth of these median lethal values (Sanchez-Bayo and Goka 2014). NOAEL, NOEL, NOAEC, NOEC, as defined as the highest exposure level for which no (adverse) effects were observed can also be considered for risk assessment especially with that of long term studies (Poquet et al. 2014). Thus acute toxicity studies, sublethal assays and field experiments are required to perform a proper risk assessment and the field risk assessment studies should cover all potential routes of exposure (Blacquiere et al. 2012). Further, risk for different pollinators should be studied separately and also with the different individuals in the colony such as larva, queen, nurse and forager bees. Some of the methods to assess risk to pollinators are explained here under:

5.1 Risk Assessment Methodologies

5.1.1 Classification of Pesticides Based on Acute Toxicity Values

Pesticides can be classified based on their acute toxicity to honey bees in three categories. Group I pesticides are highly toxic with LD_{50} <2 µg/bee. In this case, severe losses may be expected, if exposed. Group II includes moderately toxic pesticides with LD_{50} 2–11 µg/bee. They can be used if, dosage, timing and method of application are correct, but cannot be used in fields having bee foragers and also in hives. Group III deals with relatively nontoxic pesticides (LD_{50} >11 µg/bee) which cause minimum injury to bees. Most of the herbicides and fungicides belong to this last class (Atkins et al. 1981).

Based on the contact LD_{50} value, the pesticide is classified as practically non toxic ($LD_{50} \geq 11$ μg/bee), moderately toxic (LD_{50} 10.9–2 μg/bee) and highly toxic ($LD_{50} < 2$ μg/bee). All the pesticides other than those falls under the category of non toxic are need to be tested for contact toxicity on foliage to honey bees (USEPA 2012a).

5.1.2 Risk Estimates

5.1.2.1 Hazard Quotient (HQ)

$$\text{Hazard Quotient (HQ)} = \frac{\text{Application rate (g a.i./ha)}}{\text{Acute LD}_{50}\ (\mu\text{g a.i./bee})}$$

(EPPO 2010b)

The hazard quotient is defined as the ratio between environmental exposure and toxicity. A hazard quotient of <50 is used to define a pesticide as harmless to honeybees, 50–2500 as slight to moderately toxic and >2500 as dangerous to bees (EPPO 1993).

In hazard quotient (HQ), application rate was used, which is only a rough indicator of exposure and do not explain the environmental fate of the chemical (Villa et al. 2000). Hazard quotient is suitable for the pesticides which are sprayed on plants but not to those applied to soil or seeds. Further, the pesticides sprayed on the plants exhibit a residual action of a few hours or a few days and the systemic insecticides penetrate into the plant system (Elbert et al. 1991). The application dose on soil or plant is not relevant as that of the presence of pesticide in the nectar and pollen for risk assessment on honey bees. Therefore, in the case of systemic insecticides, this hazard quotient should not be used (Rortais et al. 2005).

Further, this approach of risk assessment does not take into account the size/weight of the bee or the route of exposure. For example, an application rate of 15 g a.i./ha for alpha cypermethrin gives a hazard ratio of 88 for bumblebees and 500 for honey bees. Furthermore, the foraging behaviour of pollinators and thus their exposure to pesticide is not considered in hazard quotient analysis (Thompson 2001).

5.1.2.2 Pollen Hazard Quotients (PHQ) (Stoner and Eitzer 2013)

$$\text{Pollen Hazard Quotients (PHQ)} = \frac{\text{Concentration in ppb}}{\text{Acute LD}_{50}\ (\mu\text{g a.i./bee})}$$

Pollen hazard quotient (PHQ) is similar to hazard quotient but uses the concentration of pesticides in pollen instead of the field application rate. PHQ evaluates the hazard from pesticide residues in pollen in relation to acute toxicity to honey bees. The pollen hazard quotient is then correlated with the amount of pesticide consumed by the bees along with pollen. A normal nurse bee consumes about 9.5 mg pollen/bee per day (Crailsheim et al. 1992). This value is used to calculate the daily exposure in relation to contact and oral LD_{50}. Calculating all these values, it is found that a bee consuming a pollen having PHQ 10 would be eating about 0.01 % of its median lethal dose per day. Thus a PHQ of 500 would correspond to 0.5 % of LD_{50} per day (Stoner and Eitzer 2013).

The concept of Hazard Quotients (HQ) can be extended to different matrices. In case of nectar hazard quotient (NHQ), assuming a nectar foraging bee would consume 229 mg of nectar/day; if it consumes nectar with 35 % sugar content and Nectar Hazard Quotient of 50 would tend to consume 1.1 % of the LD_{50} per day (Stoner and Eitzer 2013).

5.1.2.3 Revised Hazard Quotient (Poquet et al. 2014)

The revised HQ is calculated from the field recommended dose and the exposure surface area of a bee and expressed as ng a.i./bee. For each pesticide, two scenarios were also presented: the most and the least protective that combined the highest LD_{50} and lowest field rate, for the most protective scenario, and vice-versa for the least protective scenario. The hazard quotient for acetamiprid is 0.6 and in revised HQ it is 0.006 and these values for chlorpyrifos-ethyl is 1645 and 17.27 (revised). By determining the exposed surface area of the bee the revised HQ is equal to $1.05 \times 10^{-2} \times \mathrm{HQ}$.

5.1.2.4 Hazard Quotient Based on Residues in Pollen and Honey (Naggar et al. 2015)

$$\text{Hazard Quotient based on residues} = \frac{\mathrm{TDI}(\text{honey}) + \mathrm{TDI}(\text{pollen})}{\text{Acute oral } \mathrm{LD}_{50}}$$

In this method, hazard was assessed based on the residues of pesticides in pollen and honey including the consumption rates of the same by bees. For assessing hazard quotient first we need to calculate the total dietary intake (TDI). The total food consumption rate (TFR) is used to find the total daily intake (TDI) of pesticides by bees through food. An adult worker bee is reported to take 292 mg of nectar per day (USEPA 2012b) and 9.5 mg of pollen per day (Crailsheim et al. 1992). Organophosphorus (OP) pesticides were tested in this experiment accordingly,

$$\mathrm{TDI} = \mathrm{OPH} \text{ and } \mathrm{OPP} \times \mathrm{TFR}$$

Where,

OPH is the concentration of OPs detected in honey
OPP is the concentration of OPs detected in pollen
TFR is the rate of consumption of nectar and pollen

Hazard quotient of individual pesticide is calculated by TDI of pesticide in honey and pollen divided by the LD_{50} for each pesticide. The total HQ is the sum of the HQs for individual pesticide. If the total HQs, found exceeding the levels of concern (LOC) of 0.4 set by USEPA for acute lethality (USEPA 2012b) or 0.1 over a 1-day consumption as given by European Food Safety Authority (EFSA 2013) then higher-tier assessments are needed. The margin of exposure (MOE) is also calculated which is inverse of HQ.

5.1.2.5 Risk Quotient (RQ)

$$\text{Risk Quotient}\ (\text{RQ}) = \frac{\text{Application rate}\ \left(\text{g a.i.}/\text{cm}^2\right)}{\text{Acute}\ \text{LD}_{50}\ (\mu\text{g a.i.}/\text{bee})}$$

(EPHC 2009)

Exposure to bees is determined for spray applications based on the maximum application rate. This rate is converted to a rate of chemical/cm^2 on the assumption that a honey bee is approximately 1 cm^2 in surface area (Davis and Williams 1990) and precisely 1.05 cm^2 (Poquet et al. 2014). For example, a chemical with an application rate of 150 g a.i./ha results in a rate of $1.5\ \mu g/cm^2 \left(150\ g = 1.5\times10^8\ \mu g;\ 1\ ha = 1\times10^8\ cm^2\right)$. This results in an assumed exposure of 1.5 μg/bee (EPHC 2009).

5.1.2.6 Toxicity Exposure Ratio

Toxicity Exposure Ratio (TER) tries to correlate the toxicity of pesticide (say LC_{50}) with that of the exposure (say mg of pesticide residues present in per g of pollen or nectar) (Alix et al. 2009). It is also said to be the ratio of the measure of the effects (say, median lethal values or NOEC) to the estimated exposure. TER is a good indicator of risk of pesticide toxicity for bees especially for a particular exposure route (Villa et al. 2000). However, this approach does not take into consideration the amount of food consumed by the bees (pesticide intake) but only the amount of pesticide present in their food. The differences in foraging behaviour between species are not considered in calculating TER (Thompson 2001).

5.1.2.7 PEC and PNEC Ratio

This is the ratio between the predicted exposure concentration with that of predicted no effect concentration. If the ratio is more than 1, further risk assessment is triggered. In risk assessments, the predicted effect concentration (PEC) should be less than the predicted no effect concentration (PNEC) to reveal no risk.

5.1.3 Three Tier Evaluation of Pesticide Toxicity (Stanley et al. 2015)

A three tier toxicity assessment with laboratory studies, a semi-field and field study seems to be realistic (Halm et al. 2006; Stanley et al. 2010). The laboratory experiments are carriedout to determine the contact and ingestion toxicity of the pesticides to bees in a worst case condition (Romeis et al. 2011). If some toxic effects are observed in laboratory tests, then tier II system of semi-field experiments especially the cage tests are to be conducted (EPPO 2010b) and at last a confirmation field test. In this particular study, the Tier I evaluation involved the acute contact (both filter paper and topical bioassays) and oral toxicity tests followed by tier II system with semifield experiments with potted plants kept inside net and tier III field tests to find the repellent effect if any with the safer insecticide as revealed by tier I and II studies.

5.1.4 Risk Assessment for Dietary Residual Exposure (Sanchez-Bayo and Goka 2014)

Some researches linked the pesticide residue levels of pollen and honey to the colony collapse disorder (VanEngelsdorp et al. 2010). But risk assessments based on the prevalence of residues in the food of pollinators and their toxicity are rare (Sanchez-Bayo and Goka 2014). A constant exposure to pesticides through diet may not only cause sublethal impairments but also have a chronic toxicity and a delayed mortality (Tennekes and Sanchez-Bayo 2012).

In this assessment by Sanchez-Bayo and Goka (2014), known and published pesticide toxicity data were taken from different sources like Pesticide Manual (Tomlin 2009), the ECOTOX database of the U.S. Environment Protection Agency (http://cfpub.epa.gov/ecotox/) and the Agri-Tox Database of the Agence Nationale de Se'curite' Sanitaire de l'Alimentation, de l'Environnement et du Travail in France (http://www.agritox.anses.fr/index.php). If more than one LD_{50} value was reported for a particular pesticide then a geometric mean was made for further calculations.

5.1.4.1 Standard Risk Approach

Risk quotients which are usually used to find the pesticide toxicity on bees is calculated as the ratio of predicted environmental concentration with its corresponding LD_{50}. Prevalence of pesticide residue in pollen and nectar is one of the important parameter which determines the exposure level. So the standard risk approach was modified including the probability of exposure and given as follow by Sanchez-Bayo and Goka (2014).

$$\text{Risk} = \frac{\text{Frequency}\,(\%) \times \text{residue dose}\ (\mu\text{g})}{LD_{50}\,(\mu\text{g}/\text{bee})}$$

Risks > 5 % can be considered high
1 to 5 % the risk is moderate
<1 % can be regarded as low.

They also calculated the risk at the lowest effect level, approximate estimates of LD_{10}, which was done by multiplying 0.1 with LD_{50}. This approximation by multiplying by 0.1 is based on many field studies to estimate the lowest effect levels of many pesticides (Wijngaarden et al. 2005).

Risk of synergistic mixtures: LD_{50} of the mixture *ie*. an insecticide with other compounds was estimated by Sanchez-Bayo and Goka (2014) as,

$$LD_{50} \text{ of mixture} = \frac{LD_{50} \text{ of insecticide}}{\text{Synergistic factor}}$$

After calculating the LD_{50} of mixture, the risk approach given above is calculated. Since the probability of finding residues of both the insecticide and the synergist in the same pollen or honey cannot be estimated, they considered the lowest frequency of either compound only.

5.1.4.2 A New Approach

Since the above said assessments is well suited for contact exposures and may not be appropriate to assess risks by chronic dietary exposures, Sanchez-Bayo and Goka (2014) has given a new approach of risk assessment especially applicable for the pollinators which constantly consume pollen, nectar and honey. Here, residue load, prevalence and toxicity are included in estimating the risk. Dietary risk of pesticide residues can be ideally assessed by calculating the days to reach the LD_{50}s if test bees continue to take the available amount of residues in their food. The days to reach the LD_{50}s are then compared with the life span of the test organism. In situations where the days to reach LD_{50}s is lesser than the actual life span, then there would be a serious risk.

(i) ***Fixed dose approach***: Days to reach the dietary LD_{50} (henceforth T_{50}) in a test organism which continues to consume the present level of pesticide contamination in their food is calculated as follows:

$$T_{50}\ (\text{Days}) = \frac{LD_{50}\ (\mu g / \text{bee})}{\text{Daily dose}\ (\mu g)}$$

T_{50}s of 2 days or less: High risk
2–7 days: Moderate risk
7–60 days: Low risk

(ii) ***Time-cumulative effects***: This approach holds well in cases where the dietary LD_{50}s decrease with exposure time. The rate of change of LD_{50} with respect to time is calculated as,

$$\mathrm{Ln}\,T_{50}\ (\text{days}) = a + b * \ \mathrm{Ln}\ LD_{50}\ (\mu g / \text{bee})$$

where a (intercept) and b (slope) are parameters particular for a specific chemical and the species tested (Sanchez-Bayo 2009). This holds good to chemicals like neonicotinoids especially imidacloprid and thiamethoxam which irreversibly bind to specific receptors and have decreased LD_{50} with time. But this approach does not consider the degradation and metabolism of pesticide in the body of the test organism and elimination by excretion.

5.1.5 Assessing Risk to Pollinators in Comparison with Pests (Stanley et al. 2010)

In this method of risk assessment of pesticide application, the susceptibility of pest and pollinator to the particular pesticide is taken into account to know which one of them is affected the most. This is generally calculated using selectivity ratio and probit substitution method.

(a) **Selectivity ratio**

The selectivity ratio is the ratio of acute toxicity value of the pollinator to that of the pest. This shows which of the both (either the pollinator or pest) is more susceptible to the pesticide. The higher the value for this ratio, the safer an insecticide is for the beneficial insect. The selectivity ratio is calculated as follows:

$$\text{Selectivity ratio} = \frac{LD_{50}\ \text{of beneficial species}\ (\mu g\ \text{a.i.} / \text{bee})}{LD_{50}\ \text{of pest species}\ (\mu g\,\text{a.i.} / \text{insect})}$$

Values of 1 and <1 indicate that the chemical is nonselective to bees.
Values of >1 indicate that the chemical is selective/harmless to bees

(Ruffinengo et al. 2005)

(b) **Probit substitution method**

The relative toxicity of the pesticide to the target pest and the non-target beneficial species can be assessed through this method. It predicts the mortality of bee at a concentration of pesticide which gives a particular level of pest mortality, say LD_{90} of the pest:

$$y = 5 + m\left[x - (\log\ LD_{50}\ \text{of beneficial species})\right]$$

Where,

'y' is the probit value,
'm' is the slope of the probit line for the beneficial species
'x' is the log of the fiducial limits for LD_{90} of the pest species

Solving for y gives a probit value that is then converted to the percentage mortality using a conversion table (Finney 1971). The pesticide is considered to be selective if it kills less than 90 % of bees at the dose that kills 90 % of the pest.

5.1.6 Risk Description on Sublethal Effects (USEPA 2014)

A range of behavioural effects are caused by pesticides to pollinators, from the effects of odour discrimination in the individual bee to loss of foraging bees due to disruption in their homing behaviour. These effects may have potential impact on the development and survival of pollinators. So Risk assessments should include significant behavioural effects that influence the development and survival of colony (Thompson 2003). The assessment endpoints of sublethal effects and use in risk assessment still remain qualitative. Normally laboratory test and semi-field studies and even field tests are used to assess the mortality of bees, but emphasis has to be given to detect and interpret the consequences of sublethal behavioural effects caused by pesticides. Observation of sublethal effects are carried out in Tier I laboratory toxicity studies using individuals. The ability of these studies to test social interactions between bees such as tropholaxis, foraging behaviour etc. are to be studied. Since behavioural impairment may have a strong consequence on pollination efficiency these effects are to be included in risk assessments (Pham-Delegue et al. 2002).

5.1.7 Risk Assessment for Larva (Alix et al. 2009)

Some pesticides like insect growth regulators are highly toxic to brood than the adult bees. The toxicity data obtained from adult bees especially for the pesticides which are having specific larvicidal activity donot hold good for larval effects. In those cases a stepwise approach which include laboratory studies on the intrinsic properties of the product to immature stages are to be made. Micro colonies exposed with spiked feeding solution can be assessed for the effects on the brood development. Field realistic test doses are fixed by measuring the contamination levels in the nectar or pollen.

5.1.8 Risk Assessment of Pesticides to Queen

The strength and activity of a bee colony is largely depends on the health and reproductive ability of its queen. Most of the pesticide risk assessment techniques donot take into account of the effects on queen bees (Sanchez-Bayo and Goka 2014) with a few exceptions such as Henry et al. (2012) and Whitehorn et al. (2012), who studied the effects on the queens of honey bees and bumble bees. The pesticide exposure to queen bee is similar to larval exposure because both consume pollen and royal jelly. Queens consume more quantity and thus more exposed (Sanchez-Bayo and Goka 2014). Bumble bee queens fed with imidacloprid (Gill et al. 2012) and thiamethoxam (Whitehorn et al. 2012) contaminated pollens caused serious effects on its egg laying. So, risk assessments must include the effects of pesticides on queens also.

5.2 *Risks of Pesticides on Pollinators*

In general, insecticides are highly toxic to bees followed by acaricides, fungicides and the least being weedicides (Sanchez-Bayo and Goka 2014). Imidacloprid, clothianidin and chlorpyrifos impose risk of serious effect on bumble bees also. These compounds pose high risk to both honey and bumble bees with very low topical LD_{50}s and their frequent high presence in pollen surveys worldwide. The high risk of phosmet is mainly due to the high residue levels in surveys in spite of its moderate toxicity to honey bees (Sanchez-Bayo and Goka 2014). Thiamethoxam and lindane found in honey pose risk to the larvae. The risk of clothianidin and imidacloprid to nectar foragers are found to be high whereas it was moderated in the case of larvae. A moderate risks in consumption of pollen to both nurse bees and larvae were reported for thiamethoxam, clothianidin, imidacloprid and phosmet. The dietary risk of imidacloprid to bumble bees is very high for nectar foragers and nurse bees and a moderate risk to larvae. Residues of bifenthrin, carbaryl, lambda-cyhalothrin, and dimethoate in honey pose a low risk to forager bumble bees (Sanchez-Bayo and Goka 2014).

Based on IOBC/WPRS working group classification for insecticidal effects on nontargets (Hassan 1989) diafenthiuron is reported as slightly harmful to *A. dorsata*, *A. cerana*, *A. florea* and moderately harmful to *T. iridipennis* based on acute median lethal values (Stanley et al. 2009). Endosulfan and acetamiprid causing less mortality to *A. cerana* and *A. mellifera* are reported as safe insecticides (Stanley et al. 2015). Based on LC_{50} values, methyl demeton, endrin and dieldrin are reported as moderately toxic to *A. cerana* and malathion, parathion, lindane, phorate, phosphamidon and mevinphos as toxic (Kapil and Lamba 1974). Organophosphates including chlorpyrifos, dichlorvos, profenofos, monocrotophos, malathion and methyl demeton causing >80 % mortality to both *A. cerana* and *A. mellifera* are reported as highly toxic. Deltamethrin, indoxacarb, flubendiamide, spinosad, cartap hydrochloride, thiamethoxam and imidacloprid falls under slightly to moderately toxic category with 40–90 % mortalities to both the bees (Stanley et al. 2015).

Risk assessment of diafenthiuron sprays on honey bees with respect to the pest species (*Conogethes punctiferalis*) shows the pesticide is non selective to bees. By probit substitution, it was predicted that the dose at which 90 % of the pest dies will cause 100 % mortality to bees (Stanley et al. 2010). The oral and contact NOEC (no observable effect concentration) determined for chlorantraniliprole (Coragen®) was reported as 63 and 12.5 μg/bee, respectively. The hazard quotients (HQ) for both formulated products of chlorantraniliprole (Coragen® and Altacor®) assuming the label rate of 60 g/ha were much less than 1, revealing safety (Dinter et al. 2009). Among the five insecticides tested for *M. rotundata*, only flubendiamide with less than 0.1 hazard quotient is found safe whereas spinosad with 9.9 hazard quotient is found unsafe to the bees followed by spinetoram (4.0) (Gradish et al. 2012).

In pollen hazard quotient (PHQ), it was found that even some pesticides present at very low concentrations in pollen poses greater risk than low toxic pesticides present in higher concentrations. For example, the presence of fipronil at 3.5 ppb in pollen is hazardous than the presence of carbary at 227 ppb, because LD_{50} value of fipronil is 180 times lesser than carbaryl to bees. Phosmet was found to be in very high concentrations of 16,556 ppb, and thus had high PHQ values (Stoner and Eitzer 2013). After soil treatment of imidacloprid, the squash nectar was found to contain about 10 ppb of imidacloprid (Stoner and Eitzer 2012), which corresponds to a nectar hazard quotient (NHQ) value of 2564. It is estimated that, if a nectar forager that consume 229 mg of nectar (normal feeding rate), would have taken 59 % of the oral LD_{50} for imidacloprid per day (Stoner and Eitzer 2013).

Thus there are many risk assessment methods and it is being envolved time to time including more relevance to the field or actual risk to the pollinators. As of now, many of the risk assessments are based on work done in honey bees especially *A. mellifera* but the sensitivity and exposure levels vary with species. For e.g. cyhalothrin and fipronil toxicity is greatest to the alfalfa leafcutter bee (*M. rotundata*) less to the honeybee and least to the alkali bee, *N. melanderi* (Mayer et al. 1998; Mayer and Lunden 1999). Among the insect growth regulators, diflubenzuron at 400 g a.i./ha is generally safe for honey bees but harmful to bumble bees at 300 g a.i./ha. Fenoxycarb was safe for bumble bees at 1200 g a.i./ha but dangerous to honey bees brood at 140 g a.i./ha (Tasei 2001). Differences in sensitivity to acute effects among adult bees have also been reported. Nurse bees were reported to be more susceptible to insecticides than the foragers (Davis 1989). Risk assessment of pesticides used for different applications like seed treatment, foliar sprays or dusting are to be carried out separately in relevance to the risk posed. The attractiveness of plants to pollinators should also be considered as major criteria along with others such as rate of systemic transfer to pollen and nectar (Alix et al. 2009). The residues in soil and plant can be assessed with endpoints DT_{50} and DT_{90} (disappearance time) and the succeeding crop is also to be planned accordingly (Alix and Vergnet 2007).

References

W.S. Abbott, A method of computing the effectiveness of an insecticide. J. Econ. Entomol. **18**, 265–267 (1925)

V.A. Abbott, J.L. Nadeau, H.A. Higo, M.L. Winston, Lethal and sublethal effects of imidacloprid on *Osmia lignaria* and clothianidin on *Megachile rotundata* (Hymenoptera: Megachilidae). J. Econ. Entomol. **101**(3), 784–796 (2008)

C.I. Abramson, I.S. Aquino, F.S. Ramalho, J.M. Price, The effect of insecticides on learning in the Africanized honey bee (*Apis mellifera* L.). Arch. Environ. Contam. Toxicol. **37**, 529–535 (1999)

C.I. Abramson, P.A. Wanderley, M.J.A. Wanderley, J.C.R. Silva, L.M. Michaluk, The effect of essential oils of sweet fennel and pignut on mortality and learning in africanized honeybees (*Apis mellifera* L.) (Hymenoptera: Apidae). Neotrop. Entomol. **36**, 828–835 (2007)

D.P. Abrol, *Pollination Biology: Biodiversity Conservation and Agricultural Production* (Springer, Dordrecht, 2012), p. 707

I. Akca, C. Tuncer, A. Guler, I. Saruhan, Residual toxicity of 8 different insecticides on honey bee (*Apis mellifera* Hymenoptera: Apidae). J. Anim. Vet. Adv. **8**(3), 436–440 (2009)

A. Alix, M.J. Miles, Exposure of honey bees and other pollinating species to pesticides. Julius-Kuhn. Archiv. **437**, 19–29 (2011)

A. Alix, C. Vergnet, Risk assessment to honey bees: a scheme developed in France for non-sprayed systemic compounds. Pest Manag. Sci. **63**(11), 1069–1080 (2007)

A. Alix, M.P. Chauzat, S. Duchard, G. Lewis, C. Maus, M.J. Miles, E. Pilling, H.M. Thompson, K. Wallner, Guidance for the assessment of risks to bees from the use of plant protection products applied as seed coating and soil applications – conclusions of the ICPBR dedicated working group. Julius-Kuhn. Archiv. **423**, 15–27 (2009)

G. Allen-Wardell, P. Bernhardt, R. Bitner et al., The potential consequences of pollinator declines on the conservation of biodiversity and stability of food crop yields. Conserv. Biol. **12**, 8–17 (1998)

M.H. Allsopp, W.J. De-Lange, R. Veldtman, Valuing insect pollination services with cost of replacement. PLoS One **3**(9), e3128 (2008)

M.S. Al-mazraawi, L. Shipp, B. Broadbent, P.G. Kevan, Biological control of *Lygus lineolaris* (Hemiptera: Miridae) and *Frankliniella occidentalis* (Thysanoptera: Thripidae) by *Bombus impatiens* (Hymenoptera: Apidae) vectored *Beauveria bassiana* in greenhouse sweet pepper. Biol. Control **37**, 89–97 (2006)

APENET, Effects of coated maize seed on honey bees. Report based on results obtained from the third year of activity of the ApeNet project (CRA-API, Bologna. 2011). Retrieved from http://www.reterurale.it/apenet

M.T.K. Arroyo, R. Primack, J. Armesto, Community studies in pollination ecology in the high temperate Andes of Central Chile 1. Pollination mechanisms and altitudinal variation. Am. J. Bot. **69**(1), 82–97 (1982)

E.L. Atkins, D. Kellum, Comparative morphogenic and toxicity studies on the effect of pesticides on honeybe brood. J. Apicult. Res. **25**, 242–255 (1986)

E.L. Atkins, D. Kellum, K.W. Atkins, Reducing pesticide hazards to honey bees: mortality prediction techniques and integrated management strategies. Division of Agricultural Sciences, University of California, Leaflet 2883, 22 p. (1981)

P. Aupinel, D. Fortini, B. Michaud, F. Marolleau, J.N. Tasei, J.F. Odoux, Toxicity of dimethoate and fenoxycarb to honey bee brood (*Apis mellifera*), using a new *in vitro* standardized feeding method. Pest Manag. Sci. **63**, 1090–1094 (2007)

J. Bailey, C. Scott-Dupree, R. Harris, J. Tolman, B. Harris, Contact and oral toxicity to honey bees (*Apis mellifera*) of agents registered for use for sweet corn insect control in Ontario, Canada. Apidologie **36**, 623–633 (2005)

K.L. Barrett, N. Grandy, E.G. Harrison, S. Hassan, P. Oomen, *Guidance Document on Regulatory Testing Procedures for Pesticides with Non-target Arthropods*. From the ESCORT workshop (European Standard Characteristics of Beneficial Regulatory Testing); Wageningen, 28–30 March 1994, pp. 51

K.S. Bawa, Plant pollinator interactions in tropical rain forests. Annu. Rev. Ecol. Syst. **21**, 399–422 (1990)

N. Bendahou, M. Bounias, C. Fleche, Acute toxicity of cypermethrin and fenitrothion on honeybees (*Apis mellifera mellifera*) according to age, formulations and (chronic paralysis virus)/insecticide interaction. J. Environ. Biol. **18**, 55–65 (1997)

N. Bendahou, C. Fleche, M. Bounias, Biological and biochemical effects of chronic exposure to very low levels of dietary cypermethrin (Cymbush®) on honeybee colonies (Hymenoptera: Apidae). Ecotoxicol. Environ. Saf. **44**, 7–153 (1999)

F. Beran, J. Neururer, Zur kenntnis der wirkung von pflanzen schutzmitteln auf die honigbiene (Apis mellifica). Mitteilung; Bienengiftigkeit von pflanzenscnutzmitteln. Pflanzenschutzberichte **15**, 97–160 (1955)
M.E. Bitterman, R. Menzel, A. Fietz, S. Schafer, Classical conditioning of proboscis extension in honeybees (*Apis mellifera*). J. Comp. Psychol. **97**, 107–119 (1983)
T. Blacquiere, G. Smagghe, C.A.M. Gestel, V. Mommaerts, Neonicotinoids in bees: a review on concentrations, side-effects and risk assessment. Ecotoxicology **21**, 973–992 (2012)
S. Bogdanov, Contaminants of bee products. Apidologie **37**(1), 1–18 (2006)
S. Bogdanov, V. Kilchenmann, A. Imdorf, Acaricide residues in some bee products. J. Apicult. Res. **37**, 57–67 (1998a)
S. Bogdanov, V. Kilchenmann, A. Imdorf, P. Fluri, Residues in honey after application of thymol against varroa using the Frakno Thymol Frame. Am. Bee J. **138**, 610–611 (1998b)
W.J. Bond, Do mutualisms matter? Assessing the impact of pollinator and disperser disruption on plant extinction. Philos. Trans. R. Soc. B **344**, 83–90 (1994)
J.M. Bonmatin, Insecticide et pollinisateurs: une de'rive de la chimie? Sciences **2**, 42–46 (2002)
J.M. Bonmatin, R. Charvet, M.E. Colin, Presence of systemic pesticide imidacloprid in crops and risk to nontargeted insect species, in *Proceedings of the 3rd Mediterranean Group of Pesticide Research,* ed. by P. Cabras, M. Montury, J. Cooper. (MGPR, Aixen Provence, France, 2003), p. 9
J.M. Bonmatin, P.A. Marchand, R. Charvet, I. Moineau, E.R. Bengsch, M.E. Colin, Quantification of imidacloprid uptake in maize crops. J. Agric. Food Chem. **53**, 5336–5341 (2005)
L. Bortolotti, R. Montanari, J. Marcelino, P. Medrzycki, S. Maini, C. Porrini, Effects of sub-lethal imidacloprid doses on the homing rate and foraging activity of honey bees. Bull. Insectol. **56**(1), 63–67 (2003)
C. Bos, C. Masson, Repellent effects of deltamethrin on honey bees. Agronomie **3**, 545–553 (1983)
T.D. Breeze, A.P. Bailey, K.G. Balcombe, S.G. Potts, Pollination services in the UK: how important are honeybees? Agric. Ecosyst. Environ. **142**, 137–143 (2011)
S.L. Buchmann, Buzz pollination in angiosperms, in *Handbook of Experimental Pollination Biology*, ed. by C.E. Jones, R.J. Little (Van Nostrand Reinhold, New York, 1983), pp. 73–113
S.L. Buchmann, G.P. Nabhan, *The Forgotten Pollinators* (Island Press, Covelho, 1996). 292 p
J.R. Busvine, Recommended methods for measurement of pest resistance to pesticides. FAO Plant Protection paper No. 21 (Rome, 1980)
T.M. Butt, N.L. Carreck, L. Ibrahim, I.H. Williams, Honeybee-mediated infection of pollen beetle (*Meligethes aeneus* Fab.) by the insect-pathogenic fungus, *Metarhizium anisopliae*. Biocontrol Sci. Tech. **8**, 533–538 (1998)
G.L. Calzoni, A. Speranza, Insect controlled pollination in Japanese plum (*Prunus salicina* Lindl.). Sci. Horticult. **72**(3&4), 227–237 (1998)
D.R. Campbell, A.F. Motten, The mechanism of competition for pollination between two forest herbs. Ecology **66**, 554–563 (1985)
J.H. Cane, Ground-nesting bees: the neglected pollinator resource for agriculture. Acta Horticult. **437**, 309–324 (1997)
J.H. Cane, J.A. Payne, Native bee pollinates rabbiteye blueberry. Alabama Agric. Exp. Sta. **37**, 4 (1990)
J.H. Cane, D. Schiffhauer, L.J. Kervin, Pollination, foraging and nesting ecology of the leafcutting bee *Megachile* (*Delomegachile*) *addenda* (Hymenoptera: Megachilidae) on cranberry beds. Ann. Entomol. Soc. Am. **89**(3), 361–367 (1996)
N. Carreck, I. Williams, The economic value of bees in the UK. Bee World **79**(3), 115–123 (1998)
CEB, *Methode n°230*: method for the evaluation of side-effects of plant protection products on honey bees (*Apis mellifera* L.). *Association Française de Protection des Plantes*, pp. 43 (2011)
R.S. Chandel, P.R. Gupta, Toxicity of diflubenzuron and penfluron to immature stages of *Apis cerana indica* F. and *Apis mellifera* L. Apidologie **23**, 465–473 (1992)
M.P. Chauzat, J.P. Faucon, Pesticide residues in beeswax samples collected from honey bee colonies (*Apis mellifera* L.) in France. Pest Manag. Sci. **63**(11), 1100–1106 (2007)

M.P. Chauzat, J.P. Faucon, A.C. Martel, J. Lachaize, N. Cougoule, M. Aubert, A survey of pesticide residues in pollen loads collected by honey bees in France. J. Econ. Entomol. **99**(2), 253–262 (2006)

T.J. Ciarlo, C.A. Mullin, J.L. Frazier, D.R. Schmehl, Learning impairment in honey bees caused by agricultural spray adjuvants. PLoS One **7**(7), e40848 (2012)

P.G. Clinch, The residual contact toxicity to honey bees of insecticides sprayed on to white clover (*Trifolium repens*) in the laboratory. N. Z. J. Agric. Res. **10**(2), 289–300 (1967)

M.E. Colin, J.M. Bonmatin, I. Moineau, C. Gaimon, S. Brun, J.P. Vermandere, A method to quantify and analyse the foraging activity of honey bees: relevance to the sublethal effects induced by systemicinsecticides. Arch. Environ. Contam. Toxicol. **47**, 387–395 (2004)

S.A. Corbet, M. Fussell, R. Ake, A. Fraser, C. Gunson, A. Savage, K. Smith, Temperature and pollinating activity of social bees. Ecol. Entomol. **18**, 17–30 (1993)

R.L. Cox, W.T. Wilson, Effects of permethrin on the behavior of individually tagged honey bees, *Apis mellifera* L. (Hymenoptera: Apidae). Environ. Entomol. **13**, 375–378 (1984)

K. Crailsheim, L.H.W. Schneider, N. Hrassnigg et al., Pollen consumption and utilization in worker honeybees (*Apis mellifera carnica*): dependence on individual age and function. J. Insect Physiol. **38**, 409–419 (1992)

W.L. Crepet, Insect pollination: a paleontological perspective. BioScience **29**(2), 102–108 (1979)

W.L. Crepet, The role of insect pollination in the evolution of the angiosperms, in *Pollination Biology*, ed. by L.A. Real (Academic, London, 1983), pp. 29–50

G.C. Cutler, C.D. Scott-Dupree, Exposure to clothianidin seed-treated canola has no long-term impact on honey bees. J. Econ. Entomol. **100**(3), 765–772 (2007)

G.C. Cutler, K.W. Reeh, J.M. Sproule, K. Ramanaidu, Berry unexpected: nocturnal pollination of lowbush blueberry. Can. J. Plant Sci. **92**, 707–711 (2012)

P.L. Dai, Q. Wang, J.H. Sun, F. Liu, X. Wang, Y.Y. Wu, T. Zhou, Effects of sublethal concentrations of bifenthrin and deltamethrin on fecundity, growth and development of the honey bee, *Apis mellifera ligustica*. Environ. Toxicol. Chem. **29**, 644–649 (2010)

G.C. Daily, S. Alexander, P.R. Ehrlich et al., Ecosystem services: benefits supplied to human societies by natural ecosystems. Issues Ecol. **2**, 1–16 (1997)

A.R. Davis, The study of insecticide poisoning of honeybee brood. Bee World **70**, 163–174 (1989)

A. Davis, R. Shuel, Distribution of 14^{C} labelled coarbofuran and dimethoate in royal jelly, queen larvae and nurse honey bees. Apidologie **19**, 37–50 (1988a)

A. Davis, R.W. Shuel, Distribution of carbofuran and dimethoate in flowers and their secretion in nectar as related to nectary vascular supply. Can. J. Bot. **66**, 1248–1255 (1988b)

B.N.K. Davis, C.T. Williams, Buffer zone widths for honeybees from ground and aerial spraying of insecticides. Environ. Pollut. **63**, 247–259 (1990)

A.R. Davis, K.R. Solomon, R.W. Shuel, Laboratory studies of honey bee larval growth and development as affected by systemic insecticides at adult sub-lethal levels. J. Apicult. Res. **27**, 146–161 (1988)

A. Decourtye, M.H. Pham-Delegue, The proboscis extension response: assessing the sublethal effects of pesticides on the honey bee, in *Honey Bees: Estimating the Environmental Impact of Chemicals*, ed. by J. Devillers, M. Pham-Delegue (Taylor and Francis, London, 2002), pp. 67–84

A. Decourtye, M. Le-Metayer, H. Pottiau, M. Tissuer, J.F. Odoux, M.H. Pham-Delegue, Impairment of olfactory learning performances in the honeybee after long term ingestion of imidacloprid, in *Hazards of Pesticides to Bees*, ed. by L.P. Belzunces, C. Pelissier, G.B. Lewis (INRA, Paris, 1999), pp. 113–117

A. Decourtye, E. Lacassie, M.H. Pham-Delegue, Learning performances of honeybees (*Apis mellifera* L.) are differentially affected by imidacloprid according to the season. Pest Manag. Sci. **59**, 269–278 (2003)

A. Decourtye, J. Devillers, S. Cluzeau, M. Charreton, M.H. Pham-Delegue, Effects of imidacloprid and deltamethrin on associative learning in honeybee under semi-field and laboratory conditions. Ecotoxicol. Environ. Saf. **57**, 410–419 (2004)

A. Decourtye, J. Devillers, P. Aupinel et al., Honeybee tracking with microchips: a new methodology to measure the effects of pesticides. Ecotoxicology **20**, 429–437 (2011)
S. Dedej, K.S. Delaplane, H. Scherm, Effectiveness of honey bees in delivering the biocontrol agent *Bacillus subtilis* to blueberry flowers to suppress mummy berry disease. Biol. Control **31**, 422–427 (2004)
G. DeGrandi-Hoffman, J.C. Watkins, The foraging activity of honey bees *Apis mellifera* and non-Apis bees on hybrid sunflower (*Helianthus annuus*) and its influence on cross-pollination and seed set. J. Apicult. Res. **39**, 37–45 (2000)
K.S. Delaplane, D.E. Mayer, *Crop Pollination by Bees* (CABI Publishing, Wallingford, 2000). 344 p
K.S. Delaplane, V.J. Steen, E. Guzman, Standard methods for estimating strength parameters of *Apis mellifera* colonies, in *The Coloss Bee Book*, eds. by V. Dietemann, J.D. Ellis, P. Neumann, volume I: standard methods for *Apis mellifera* research. J. Apicult. Res. **52**(1), 1–12 (2013)
P.B. Deshmukh, Toxicity of some insecticides to Indian honeybee, *Apis cerana* F. Pestology **15**, 31–33 (1991)
N. Desneux, R. Ramirez-Romero, L. Kaiser, Multistep bioassay to predict recolonization potential of emerging parasitoids after a pesticide treatment. Environ. Toxicol. Chem. **25**(10), 2675–2682 (2006)
N. Desneux, A. Decourtye, J.M. Delpuech, The sublethal effects of pesticides on beneficial arthropods. Annu. Rev. Entomol. **52**, 81–106 (2007)
J. Devillers, Acute toxicity of pesticides to honey bees, in *Honey Bees: Estimating the Environmental Impact of Chemicals*, ed. by J. Devillers, M. Pham-Delegue (Taylor and Francis, London, 2002), pp. 56–66
L. DeWael, M. DeGreef, O. VanLaere, Toxicity of pyriproxifen and fenoxycarb to bumble bee brood using a new method for testing insect growth regulators. J. Apicult. Res. **34**, 3–8 (1995)
A. Dinter, K.E. Brugger, N. Frost, M.D. Woodward, Chlorantraniliprole (Rynaxypyr®): a novel DuPont™ insecticide with low toxicity and low risk for honey bees (*Apis mellifera*) and bumble bees (*Bombus terrestris*) providing excellent tools for uses in integrated pest management. *Hazards of Pesticides to Bees – 10th International Symposium ICP-Bee Protection Group*. Julius-Kuhn. Archiv. **423**, 84–96 (2009)
W. Drescher, H. Geusen-Pfister, Comparative testing of the oral toxicity of acephate, dimethoate and methomyl to honeybees, bumblebees and syrphidae. Acta Horticult. **288**, 133–136 (1991)
EFSA, European Food Safety Authority. Guidance document on the risk assessment of plant protection products on bees (*Apis mellifera*, *Bombus* spp and solitary bees). EFSAJ **11**(7), 268 (2013)
E.J. Eilers, C. Kremen, S.S. Greenleaf, A.K. Garber, A.M. Klein, Contribution of pollinator-mediated crops to nutrients in the human food supply. PLoS One **6**(6), e21363 (2011)
A. Elbert, B. Beckert, J. Hartwig, C. Erdelen, Imidacloprid – a new systemic insecticide. Pflanzenschutz-Nachr. Bayer **44**, 113–136 (1991)
EPHC, *Environmental Risk Assessment Guidance Manual for Agricultural and Veterinary Chemicals* (Environmental Protection and Heritage Council, Canberra, 2009)
EPPO, European and Mediterranean Plant Protection Organization. Decision making scheme for the environmental risk assessments for the plant protection products – honey bees. EPPO Bull. **23**, 151–165 (1993)
EPPO, European and Mediterranean Plant Protection Organization. Guidelines for the efficacy evaluation of plant protection products, in *Hazards of Pesticides to Bees*, ed. by L.P. Belzunces, C. Pelissier, G.B. Lewis (INRA, Paris, 2001), pp. 279–286. Colloq. No. 98
EPPO, Side-effects on honey bees. EPPO Bull. **40**(3), 313–319 (2010a)
EPPO, Environmental risk assessment scheme for plant protection products. Chapter 10: honey bees. EPPO Bull. **40**(3), 323–331 (2010b)
E.H. Erickson, G.A. Berger, J.G. Shannon, J.M. Robins, Honey bee pollination increases soybean yields in the Mississippi Delta region of Arkansas and Missouri. J. Econ. Entomol. **71**(4), 601–603 (1978)

R. Everich, C. Schiller, J. Whitehead, M. Beavers, K. Barrett, Effects of captan on *Apis mellifera* brood development under field conditions in California almond orchards. J. Econ. Entomol. **102**(1), 20–29 (2009)

J. Faucon, C. Aurieres, P. Drajnudel, L. Mathieu, M. Ribiere, A. Martel, S. Zeggane, M. Chauzat, M.F.A. Aubert, Experimental study on the toxicity of imidacloprid given in syrup to honeybee (*Apis mellifera*) colonies. Pest Manag. Sci. **61**, 111–125 (2005)

L. Fiedler, Assessment of chronic toxicity of selected insecticides to honeybees. J. Apicult. Res. **26**(2), 115–122 (1987)

D.J. Finney, *Probit Analysis*, 3rd edn. (Cambridge University Press, Cambridge, 1971). 333 p

D. Fischer, T. Moriarty, Pesticide risk assessment for pollinators: summary of a SETAC Pellston workshop on pesticide risk assessment for pollinators 15–21 January 2011, Pensacola. Retrieved from http://c.ymcdn.com/sites/www.setac.org/resource/resmgr/publications_and_resources/executivesummarypollinators_.pdf?hhSearchTerms=SETAC+and+Pellston+and+Workshop

M. Fletcher, L. Barnett, Bee pesticide poisoning incidents in the United Kingdom. Bull. Insectol. **56**(1), 141–145 (2003)

A.D. Flynn, H.F. Schoof, Effect of surface on residual activity of selected compounds. J. Econ. Entomol. **59**, 678–681 (1966)

D.R. Fox, E. Billoir, S. Charles, M.L. Delignette-Muller, C. Lopes, What to do with NOECS/NOELS–prohibition or innovation? Integr. Environ. Assess. Manag. **8**, 764–766 (2012)

J.B. Free, *Insect Pollination of Crops*, 2nd edn. (Academic, London, 1993)

M. Fussell, S.A. Corbet, Forage for bumblebees and honeybees in farmland, a case study. J. Apicult. Res. **30**, 87–97 (1991)

C. Galen, K.A. Zimmer, M.E. Newport, Pollination in floral scent morphs of *Polemonium viscosum*: a mechanism for disruptive selection on flower size. Evolution **41**, 599–606 (1987)

N. Gallai, J. Salles, J. Settele, B.E. Vaissiere, Economic valuation of the vulnerability of world agriculture confronted with pollinator decline. Ecol. Econ. **68**(3), 810–821 (2009)

T.S. Geoghegan, J.H. Kimberly, M. Scheringer, Predicting honeybee exposure to pesticides from vapour drift using a combined pesticide emission and atmospheric transport model. SETAC Australasia- Multidisciplinary approaches to managing environmental pollution. Melbourne 1–3 October pp. 174 (2013)

P. Georgiadis, J. Pistorius, U. Heimbach, M. Stahler, K. Schwabe, Dust drift during sowing of maize – effects on honey bees. Julius-Kuhn. Archiv. **437**, 134–139 (2012)

R.J. Gill, O. Ramos-Rodriguez, N.E. Raine, Combined pesticide exposure severely affects individual- and colony-level traits in bees. Nature **491**, 105–108 (2012)

V. Girolami, L. Mazzon, A. Squartini et al., Translocation of neonicotinoid insecticides from coated seeds to seedling guttation drops: a novel way of intoxication for bees. J. Econ. Entomol. **102**(5), 1808–1815 (2009)

V. Girolami, M. Marzaro, L. Vivan, L. Mazzon, M. Greatti, C. Giorio, D. Marton, A. Tapparo, Fatal powdering of bees in flight with particulates of neonicotinoids seed coating and humidity implication. J. Appl. Entomol. **136**(1–2), 17–26 (2012)

J.G. Glynne, J.U. Connell, Studies of the toxicity to worker honey-bees (*Apis mellifera* L.) of certain chemicals used in plant protection. Ann. Appl. Biol. **41**, 271–279 (1954)

D. Goulson, An overview of the environmental risks posed by neonicotinoid insecticides. J. Appl. Ecol. **50**, 977–987 (2013)

D. Goulson, J.C. Stout, Homing ability of the bumble bee, *Bombus terrestris* (Hymenoptera: Apidae). Apidologie **32**, 105–111 (2001)

I.S. Gour, B.L. Pareek, Effect of insecticides on the incidence of honey bee, *Apis cerana indica* Fab. on mustard bloom under semi-arid region of Rajasthan. Madras Agric. J. **91**(7–12), 430–432 (2004)

A.E. Gradish, C.D. Scott-Dupree, G.C. Cutler, Susceptibility of *Megachile rotundata* to insecticides used in wild blueberry production in Atlantic Canada. J. Pest. Sci. **85**, 133–140 (2012)

M. Greatti, A.G. Sabatini, R. Barbattini, S. Rossi, A. Stravisi, Risk of environmental contamination by the active ingredient imidacloprid used for corn seed dressing. Preliminary results. Bull. Insectol. **56**, 69–72 (2003)

M. Greatti, R. Barbattini, A. Stravisi, A.G. Sabatini, S. Rossi, Presence of the a.i. imidacloprid on vegetation near corn fields sown with Gaucho® dressed seeds. Bull. Insectol. **59**, 99–103 (2006)

S.S. Greenleaf, C. Kremen, Wild bees enhance honey bees' pollination of hybrid sunflower. Proc. Natl. Acad. Sci. U. S. A. **103**, 13890–13895 (2006)

C. Gretenkord, W. Drescher, Development of a cage test method for the evaluation of pesticide hazards to the bumble bee *Bombus terrestris* L., in *Proceedings of the 5th International Symposium on the Hazards of Pesticides to Bees*, 26–28 October 1993, Wageningen, Appendix 20

C. Gretenkord, W. Drescher, Laboratory and cage test methods for the evaluation of the effects of growth regulators (Insegar®, Dimilin®) on the brood of *Bombus terrestris* L., in *Proceedings of the 6th International Symposium.* Hazards of pesticides to bees, 17–19 September 1996, Braunschweig, Appendix 29

H.R. Gross, J.J. Hamm, J.E. Carpenter, Design and application of a hive mounted device that uses honey bees (Hymenoptera: Apidae) to disseminate Heliothis nuclear polyhedrosis virus. Environ. Entomol. **23**, 492–501 (1994)

B. Gruber, K. Eckel, J. Everaars, C.F. Dormann, On managing the red mason bee (*Osmia bicornis*) in apple orchards. Apidologie **42**(5), 564–576 (2011)

D. Guez, S. Suchail, M. Gauthier, R. Maleszka, L.P. Belzunces, Contrasting effects of imidacloprid on habituation in 7 and 8 day-old honey bees (*Apis mellifera*). Neurobiol. Learn. Mem. **76**, 183–191 (2001)

T. Haarmann, M. Spivak, D. Weaver, B. Weaver, T. Glenn, Effects of fluvalinate and coumaphos on queen honey bees (Hymenoptera: Apidae) in two commercial queen rearing operations. J. Econ. Entomol. **95**, 28–35 (2002)

M.P. Halm, A. Rortais, G. Arnold, J.N. Tasei, S. Rault, A new risk assessment approach for systemic insecticides: the case of honeybees and imidacloprid (Gaucho®). Environ. Sci. Technol. **40**, 2448–2454 (2006)

P. Han, C. Niu, C. Lei, J. Cui, N. Desneux, Use of an innovative T-tube maze assay and the proboscis extension response assay to assess sublethal effects of GM products and pesticides on learning capacity of the honey bee *Apis mellifera* L. Ecotoxicology **19**, 1612–1619 (2010)

S.B. Hasan, P.G. Deo, S.K. Majunder, Relative toxicity of insecticides to household and beneficial insects. Indian Bee J. **48**, 42–44 (1986)

M. Haseeb, T.X. Liu, W.A. Jones, Effects of selected insecticides on *Cotesia plutellae*, endoparasitoid of *Plutella xylostella*. BioControl **49**, 33–46 (2004)

S.A. Hassan, Testing methodology and concepts of IOBC–WPRS working group, in *Pesticides and Non-target Invertebrates*, ed. by P.C. Jepson (Intercept, Wimborm, 1989), pp. 1–18

U. Heimbach, *Heubach Method to Determine the Particulate Matter of Maize Seeds Treated with Insecticides* (JK Institute for Plant Protection in Agriculture and Grassland, Braunschweig, 2008). http://www.jki.bund.de/fileadmin/dam_uploads/_A/pdf/Heubach%20Method%20english.pdf

M. Henry, M. Beguin, F. Requier et al., A common pesticide decreases foraging success and survival in honey bees. Science **336**, 348–350 (2012)

G.A. Hobbs, W.O. Nummi, J.F. Virostek, Food gathering behaviour of honey, bumble and leafcutter bees (Hymenoptera: Apoidea) in Alberta. Can. Entomol. **93**(6), 409–419 (1961)

Y.N. Hsieh, L.F. Liu, Y.S. Wang, Uptake, translocation and metabolism of the herbicide molinate in tobacco and rice. Pestic. Sci. **53**, 149–154 (1998)

F. Incerti, L. Bortolotti, C. Porrini, A.M. Sbrenna, G. Sbrenna, An extended laboratory test to evaluate the effects of pesticides on bumblebees- preliminary results. Bull. Insectol. **56**(1), 159–164 (2003)

R. Isaacs, A. Kirk, Pollination services provided to small and large highbush blueberry fields by wild and managed bees. J. Appl. Ecol. **47**(4), 841–849 (2010)

T. Iwasa, N. Motoyama, J.T. Ambrose, R.M. Roe, Mechanism for the differential toxicity of neonicotinoid insecticides in the honey bee, *Apis mellifera*. Crop. Prot. **23**, 371–378 (2004)

C.R.O. Jacob, *Efeitos do inseticida fipronil sobre os corpos pedunculados de opera'rias Scaptotrigona postica (Latreille, 1807) (Hymenoptera, Apidae, Meliponini).* Masters dissertation, Universidade Estadual Paulista, Rio Claro, 2012

J.H. Jacobs, S.J. Clark, I. Denholm, D. Goulson, C. Stoate, J.L. Osborne, Pollination biology of fruit bearing hedgerow plants and the role of flower visiting insects in fruit set. Ann. Bot. **104**(7), 1397–1404 (2009)

F. Jauker, B. Bondarenko, H.C. Becker, I. Steffan-Dewenter, Pollination efficiency of wild bees and hoverflies provided to oilseed rape. Agric. For. Entomol. **14**(1), 81–87 (2012)

S.K. Javorek, K.E. Mackenzie, S.P.V. Kloet, Comparative pollination effectiveness among bees (Hymenoptera: Apoidea) on lowbush blueberry (Ericaceae: *Vaccinium angustifolium*). Ann. Entomol. Soc. Am. **95**, 345–351 (2002)

S.C. Jay, Starvation studies of larval honey bees. Can. J. Zool. **42**, 455–462 (1964)

E.R. Jaycox, Effect on honeybees of nectar from systemic insecticide-treated plants. J. Econ. Entomol. **57**, 31–35 (1964)

E.R. Jaycox, W. Skowronek, G. Guynn, Behavioral changes in worker honey bees (*Apis mellifera*) induced by injections of juvenile hormone mimic. Ann. Entomol. Soc. Am. **67**, 529–534 (1974)

L. Jeker, L. Schmid, T. Meschberger, M. Candolfi, S. Pudenz, J.P. Magyar, Computer-assisted digital image analysis and evaluation of brood development in honey bee combs. J. Apicult. Res. **51**(1), 63–73 (2012)

C. Johansen, Bee poisoning versus clover seed weevil control in white Dutch clover grown for seed. J. Econ. Entomol. **58**, 857–863 (1965)

C.A. Johansen, Pesticides and pollinators. Annu. Rev. Entomol. **22**, 177–192 (1977)

C. Johansen, Bee poisoning test protocols for the United States. *Proceedings of the EPA Conference*, 8–9 November 1978

K.B. Johnson, V.O. Stockwell, M.J. McLaughlin, D. Sugar, J.E. Loper, R.G. Roberts, Effect of bacterial antagonists on establishment of honey bee-dispersed *Erwinia amylovora* in pear blossoms and on fire blight control. Phytopathology **83**, 995–1002 (1993)

R.M. Johnson, M.D. Ellis, C.A. Mullin, M. Frazier, Pesticides and honey bee toxicity-USA. Apidologie **41**, 312–331 (2010)

R.R. Junker, N. Bluthgen, T. Brehm et al., Specialization on traits as basis for the niche-breadth of flower visitors and as structuring mechanism of ecological networks. Funct. Ecol. **27**, 329–341 (2013)

J.L. Jyoti, G.J. Brewer, Resistance in sunflower and interaction with *Bacillus thuringiensis* for control of banded sunflower moth (Lepidoptera: Tortricidae). J. Econ. Entomol. **92**(5), 1230–1233 (1999)

I. Kandori, Diverse visitors with various pollinator importance and temporal change in the important pollinators of *Geranium thunbergii* (Geraniaceae). Ecol. Res. **17**(3), 283–294 (2002)

R.P. Kapil, D.P.S. Lamba, Toxicity of some important insecticides to *Apis cerana* Fab. Indian J. Entomol. **36**, 6–10 (1974)

C. Kearns, D. Inouye, N. Waser, Endangered mutualisms: the conservation of plant-pollinator interactions. Annu. Rev. Ecol. Evol. Syst. **29**, 83–112 (1998)

P.G. Kevan, Insect pollination of high Arctic flowers. J. Ecol. **60**(3), 831–847 (1972)

P. Kevan, M. Al-mazraawi, J. Sutton et al., Using pollinators to deliver biological control agents against crop pests, in *Pesticide Formulations and Delivery Systems: Meeting the Challenges of the Current Crop Protection Industry*, ed. by R.A. Downer, J.C. Mueninghoff, G.C. Volgas (American society for testing, West Conshohocken, 2003), pp. 148–152

P.G. Kevan, J.P. Kopongo, Al-Mazraawi, L. Shipp, Honey bees, bumble bees and biocontrol, new alliances between old friends, in *Bee Pollination in Agricultural Ecosystems*, ed. by R.R. James, T. PittsSinger (Oxford University Press, Oxford, 2008), pp. 65–79

D.S. Khoury, M.R. Myerscough, A.B. Barron, A quantitative model of honey bee colony population dynamics. PLoS One **6**, e18491 (2011)

C. King, G. Ballantyne, P.G. Willmer, Why flower visitation is a poor proxy for pollination: measuring single-visit pollen deposition, with implications for pollination networks and conservation. Methods Ecol. Evol. **4**, 811–818 (2013)

B.K. Klatt, F. Klaus, C. Westphal, T. Tscharntke, Enhancing crop shelf life with pollination. Agric. Food Secur. **3**, 14 (2014)

A.M. Klein, I. Steffan-Dewenter, T. Tscharntke, Fruit set of highland coffee increases with the diversity of pollinating bees. Proc. R. Soc. Lond. B **270**, 955–961 (2003)

A.M. Klein, B.E. Vaissiere, J.H. Cane et al., Importance of pollinators in changing landscapes for world crops. Proc. R. Soc. Lond. B **274**, 303–313 (2007)

T.M. Knight, M.W. McCoy, J.M. Chase, K.A. McCoy, R.D. Holt, Trophic cascades across ecosystems. Nature **437**, 880–883 (2005)

H. Koch, P. Weiber, Exposure of honeybees during pesticide application under field conditions. Apidologie **28**, 439–447 (1997)

S.A. Kolmes, M.L. Winston, Division of labour among worker honey bees in demographically manipulated colonies. Insect. Soc. **35**(3), 262–270 (1988)

H. Kovac, A. Stabentheiner, S. Schmaranzer, Thermoregulation of water foraging honeybees – balancing of endothermic activity with radiative heat gain and functional requirements. J. Insect Physiol. **56**, 1834–1845 (2010)

C. Kremen, N.M. Williams, R.W. Thorp, Crop pollination from native bees at risk from agricultural intensification. Proc. Natl. Acad. Sci. U. S. A. **99**, 16812–16816 (2002)

C.H. Krupke, G.J. Hunt, B.D. Eitzer, G. Andino, K. Given, Multiple routes of pesticide exposure for honey bees living near agricultural fields. PLoS One **7**(1), e29268 (2012)

E. Ladurner, J. Bosch, S. Maini, W.P. Kemp, A method to feed individual bees (Hymenoptera: Apiformes) known amounts of pesticides. Apidologie **34**, 597–602 (2003)

E. Ladurner, J. Bosch, W.P. Kemp, S. Maini, Assessing delayed and acute toxicity of five formulated fungicides *to Osmia lignaria* Say and *Apis mellifera*. Apidologie **36**, 449–460 (2005a)

E. Ladurner, J. Bosch, W.P. Kemp, S. Maini, Evaluation of a standard artificial flower design to feed individual bees known amounts of pesticides. Apidologie **36**, 379–387 (2005b)

H.H. Laidlaw, R.E. Page, *Queen Rearing and Bee Breeding* (Wicwas Press, Cheshire, 1997). 216 p

W.G. Landis, P.M. Chapman, Well past time to stop using NOELs and LOELs. Integr. Environ. Assess. Manag. **7**, 6–8 (2011)

J.L. Larson, C.T. Redmond, D.A. Potter, Assessing insecticide hazard to bumble bees foraging on flowering weeds in treated lawns. PLoS One **8**(6), e66375 (2013)

D. Laurino, M. Porporato, A. Patetta, A. Manino, Toxicity of neonicotinoids insecticides to honeybees: laboratory tests. Bull. Insectol. **64**(1), 107–113 (2011)

I. Laycock, K.M. Lenthal, A.T. Barratt, J.E. Cresswell, Effects of imidacloprid, a neonicotinoid pesticide, on reproduction in worker bumble bees (*Bombus terrestris*). Ecotoxicology **21**(7), 1937–1945 (2012)

M.D. Levin, Value of crops pollinated by honeybees. Bull. Entomol. Soc. Am. **29**(11), 50–51 (1983)

G.B. Lewis, P.A. Oomen, D. Brasse, J.H. Stevenson, The assessment of pesticide risk to bees: the work of the ICPBR Bee Protection Group. *First European Conference on Apidology – Bees and Pesticides*, Udine, 2004

R.F. Long, L. Morandin, Low hybrid onion seed yields relate to honey bee visits and insecticide use. Calif. Agric. **65**(3), 155–158 (2011)

C.T. Lourenc, S.M. Carvalho, O. Malaspina, R.C.F. Nocelli, Oral toxicity of fipronil insecticide against the stingless bee, *Melipona scutellaris* (Latreille, 1811). Bull. Environ. Contam. Toxicol. **89**, 921–924 (2012)

B. Maccagnani, M. Mocioni, M.L. Gullino, E. Ladurner, Application of *Trichoderma harzianum* by using *Apis mellifera* as a vector for the control of grey mould of strawberry, first results. IOBC Bull. **22**(1), 161–164 (1999)

B. Maccagnani, C. Bazzi, E. Biondi, D. Tesoriero, S. Maini, Potential of *Osmia cornuta* as a carrier of antagonistic bacteria in biological control of fire blight: a comparison with *Apis mellifera*. Acta Horticult. **704**, 379–386 (2006)

K.E. MacKenzie, Pollination requirements of three highbush blueberry (*Vaccinium corymbosum* L.) cultivars. J. Am. Soc. Horticult. Sci. **122**(6), 891–896 (1997)

K.E. MacKenzie, M.L. Winston, Effects of sublethal exposure to diazinon on longevity and temporal division of labour in the honey bee (Hymenoptera: Apidae). J. Econ. Entomol. **82**, 75–82 (1989)

K.E. MacKenzie, E. Eickwort, C. George, Diversity and abundance of bees (Hymenoptera: Apoidea) foraging on highbush blueberry (*Vaccinium corymbosum* L.) in Central New York. J. Kansas Entomol. Soc. **69**(4), 185–194 (1996)

J. Majewski, Economic value of pollination of major crops in Poland in 2012, in Economic science for rural development. *International Scientific Conference*, Jelgava, 24–25 April 2014, pp. 14–21

A.N. Mamood, G.D. Waller, Recovery of learning responses by honeybees following sublethal exposure to permethrin. Physiol. Entomol. **15**, 55–60 (1990)

F. Marletto, A. Patetta, A. Manino, Laboratory assessment of pesticide toxicity to bumblebees. Bull. Insectol. **56**(1), 155–158 (2003)

A. Maurizro, P. Schenker, Experiments on the secretion in nectar of plant protection substances poisonous to bees. Milt. Schweiz. Ent. Ges. **30**, 140–150 (1957)

D.F. Mayer, J.D. Lunden, Field and laboratory tests of the effects of fipronil on adult female bees of *Apis mellifera*, *Megachile rotundata* and *Nomia melanderi*. J. Apicult. Res. **38**, 191–197 (1999)

D.F. Mayer, G. Kovacs, J.D. Lunden, Field and laboratory tests on the effects of cyhalothrin on adults of *Apis mellifera*, *Megachile rotundata* and *Nomia melanderi*. J. Apicult. Res. **37**, 33–37 (1998)

P. Medrzycki, R. Montanari, L. Bortolotti, A.G. Sabatini, S. Maini, C. Porrini, Effects of imidacloprid administered in sub-lethal doses on honey bee behaviour: laboratory tests. Bull. Insectol. **56**(1), 59–62 (2003)

P. Medrzycki, H. Giffard, P. Aupinel et al., Standard methods for toxicology research in *Apis mellifera*. J. Apicult. Res. **52**(4), 1–60 (2013)

R. Menzel, Associative learning in honey bees. Apidologie **24**, 157–168 (1993)

R. Menzel, U. Muller, Learning and memory in honeybees: from behavior to neural substrates. Annu. Rev. Neurosci. **19**, 379–404 (1996)

R. Menzel, K. Geiger, J. Joerges, U. Muller, L. Chittka, Bees travel homeward routes by integrating separately acquired vector memories. Anim. Behav. **55**, 139–152 (1998)

C.D. Michener, *The Bees of the World*, 2nd edn. (Johns Hopkins University Press, Baltimore, 2007). 992 p

M. Miles, The effects of spinosad, a naturally derived insect control agent to the honeybee. Bull. Insectol. **56**(1), 119–124 (2003)

R.C. Misra, A.K. Verma, Relative toxicity of some insecticides to *Apis cerana* F. Indian Bee J. **44**, 69–71 (1982)

V. Mommaerts, G. Smagghe, Side-effects of pesticides on the pollinator Bombus: an overview, in *Pesticides in the Modern World – Pests Control and Pesticides Exposure and Toxicity Assessment*, ed. by M. Stoytcheva (InTech, Rijeka, 2011), pp. 508–552

L.A. Morandin, M.L. Winston, Wild bee abundance and seed production in conventional, organic and genetically modified canola. Ecol. Appl. **15**, 871–881 (2005)

L.A. Morandin, M.L. Winston, M.T. Franklin, V.A. Abbott, Lethal and sub-lethal effects of spinosad on bumble bees (*Bombus impatiens* Cresson). Pest Manag. Sci. **61**(7), 619–626 (2005)

H. Morton, J. Moffett, Ovicidal and larvicidal effects of certain herbicides on honey bees. Environ. Entomol. **1**, 611–614 (1972)

S. Muhammad, A. Gondal, Manzoor-ul-Haq, Studies on the role of *Apis indica* F. in the pollination of cauliflower (*Brassica oleracea* var. *botrytis* Linn.) and radish (*Raphanus sativus* Linn). S. U. Res. J. **7**, 87–93 (1973)

H. Mullar, The fertilizers of alpine flowers. Nature **21**, 275 (1880)
C.A. Mullin, M. Frazier, J.L. Frazier et al., High levels of miticides and agrochemicals in North American apiaries: implications for honey bee health. PLoS One **5**(3), e9754 (2010)
A. Murray, Acute and residual toxicity of a new pyrethroid insecticide, WL 85871 to honey bees. Bull. Environ. Contam. Toxicol. **34**, 560–564 (1985)
G.P. Nabhan, S.L. Buchnann, Services provided by pollinators, in *Nature's Services: Societal Dependence on Natural Ecosystems*, ed. by G. Daily (Island Press, Washington, DC, 2012), p. 412
H. Nadaf, G.S. Yadav, H.D. Kaushik, S.K. Sharma, Toxicity of new molecules of insecticides against honeybee, *Apis mellifera* L. Trends Biosci. **6**(4), 445–447 (2013)
Y.A. Naggar, G. Codling, A. Vogt, E. Naiem, M. Mona, A. Seif, J.P. Giesy, Organophosphorus insecticides in honey, pollen and bees (*Apis mellifera* L.) and their potential hazard to bee colonies in Egypt. Ecotoxicol. Environ. Saf. **114**, 1–8 (2015)
J.L. Nation, F.A. Robinson, S.J. Yu, A.B. Bolten, Influence upon honey bees of chronic exposure to very low levels of selected insecticides in their diet. J. Apicult. Res. **25**, 170–177 (1986)
P.H. Needham, J.H. Stevenson, The toxicity to foraging honeybees, *Apis mellifera*, of endosulfan, malathion and azinphos-methyl applied to flowering oil seed rape, *Brassica napus*. Ann. Appl. Biol. **75**, 235–240 (1973)
J.L. Neff, B.B. Simpson, Nest biology and mating behavior of *Megachile fortis* in Central Texas (Hymenoptera: Megachilidae). J. Kansas Entomol. Soc. **64**, 324–336 (1991)
H.N. Nigg, R.V. Russ, W.D. Mahon, J.H. Stamper, J.L. Knapp, Contamination of sucrose solution with aldicarb sulfoxide inhibits foraging by honeybees (Hymenoptera: Apidae). J. Econ. Entomol. **84**, 810–813 (1991)
J. Ollerton, R. Winfree, S. Tarrant, How many flowering plants are pollinated by animals? Oikos **120**(3), 321–326 (2011)
P.A. Oomen, A. Ruijter, J.V. Steen, Method for honey bee brood feeding tests with insect growth-regulating insecticides. EPPO Bull. **22**, 613–616 (1992)
H.H. Oruc, J.M. Hranitz, A. Sorucu, M. Duell, I. Cakmak, L. Aydinand, A. Orman, Determination of acute oral toxicity of flumethrin in honey bees. J. Econ. Entomol. **105**(6), 1890–1894 (2012)
T. Palmer-Jones, Laboratory methods for measuring the toxicity of pesticides to honey bees. N. Z. J. Agric. Res. **1**, 290–300 (1958)
F.D. Parker, Sunflower pollination: abundance, diversity and seasonality of bees and their effects on seed yield. J. Apicult. Res. **21**(1), 49–61 (1981)
U. Partap, L.R. Verma, Pollination of radish by *Apis cerena*. J. Apicult. Res. **33**, 237–241 (1994)
G. Peng, J.C. Sutton, P.G. Kevan, Effectiveness of honey-bees for applying the biocontrol agent *Gliocladium roseum* to strawberry flowers to suppress *Botrytis cinerea*. Can. J. Plant Pathol. **14**, 117–129 (1992)
J.S. Pettis, A.M. Collins, R. Wilbanks, M.F. Feldlaufer, Effects of coumaphos on queen rearing in the honey bee, *Apis mellifera*. Apidologie **35**, 605–610 (2004)
J. Pettis, D. Van-Engelsdorp, J. Johnson, G. Dively, Pesticide exposure in honey bees results in increased levels of the gut pathogen Nosema. Naturwissenschaften **99**, 153–158 (2012)
M. Pham-Delegue, A. Decourtye, L. Kaiser, J. Devillers, Behavioural methods to assess the effects of pesticides on honey bees. Apidologie **33**, 425–432 (2002)
J. Pistorius, G. Bischoff, U. Heimbach, M. Stahler, Bee poisoning incidents in Germany in spring 2008 caused by abrasion of active substance from treated seeds during sowing of maize. Julius-Kuhn. Archiv. **423**, 118–126 (2009)
T.L. Pitts-Singer, J.H. Cane, The alfalfa leafcutting bee, *Megachile rotundata*: the world's most intensively managed solitary bee. Annu. Rev. Entomol. **56**, 221–237 (2011)
Y. Poquet, L. Bodin, M. Tchamitchian et al., A pragmatic approach to assess the exposure of the honey bee (*Apis mellifera*) when subjected to pesticide spray. PLoS One **9**(11), e113728 (2014)
C. Potter, An improved laboratory apparatus for applying direct sprays and surface films, with data on the electrostatic charge on atomized spray fluids. Ann. Appl. Biol. **39**, 1–28 (1952)

R. Prescott-Allen, C. Prescott-Allen, How many plants feed the World? Conserv. Biol. **4**(4), 365–374 (1990)

K.R.S. Prier, B. Lighthart, J.J. Bromenshenk, Adsorption model of aerosolized bacterial spores (*Bacillus subtilis* variety niger) onto free-flying honey bees (Hymenoptera: Apidae) and its validation. Environ. Entomol. **30**, 1188–1194 (2001)

O.E. Prys-Jones, S.A. Corbet, *Bumblebees*, 2nd edn. (Richmond Publishing, Slough, 1991)

M.L.K. Reddy, Relative toxicity of some insecticides to honeybees (*Apis cerana indica* Fab.). J. Res. ANGRAU **25**, 26–28 (1997)

K.W. Richards, Non-apis bees as crop pollinators. Rev. Suisse Zool. **100**, 807–822 (1993)

A.J. Richards, Does low biodiversity resulting from modern agricultural practice affect crop pollination and yield? Ann. Bot. **88**, 165–172 (2001)

T.E. Rinderer, L. DeGuzman, V.A. Lancaster, G.T. Delatte, J.A. Stelzer, Varroa in the mating yard: the effects of *Varroa jacobsoni* and Apistan on drone honey bees. Am. Bee J. **139**, 134–139 (1999)

T.C. Roat, S.M. Carvalho, R.C.F. Nocelli, E.C.M. Silva-Zacarin, O. Malaspina, Acute toxicity of fipronil to newly-emerged honeybee, *Apis mellifera* L., 1758 (Hymenoptera, Apidae) africanized. Encontro sobre Abelhas RibeiraoPreto (2010)

G.E. Robinson, Effects of a juvenile hormone analogue on honey bee foraging behaviour and alarm pheromone production. J. Insect Physiol. **31**, 277–282 (1985)

W.S. Robinson, C.A. Johansen, Effects of control chemicals for Douglas-fir tussock moth *Orgia pseudotsugata* (McDonnough) on forest pollination (Lepidoptera: Lymantriidae). Melanderia **30**, 9–56 (1978)

J. Romeis, R.L. Hellmich, M.P. Candolfi et al., Recommendations for the design of laboratory studies on non-target arthropods for risk assessment of genetically engineered plants. Transgenic Res. **20**, 1–22 (2011)

A. Rortais, G. Arnold, M.P. Halm, F. Touffet-Briens, Modes of honeybees exposure to systemic insecticides: estimated amounts of contaminated pollen and nectar consumed by different categories of bees. Apidologie **36**, 71–83 (2005)

D.W. Roubik, *Pollination of Cultivated Plants in the Tropics*. Food and Agricultural Organization service bulletin, vol. 118 (Food and Agriculture Organization, Rome, 1995)

S. Ruffinengo, M. Eguaras, I. Floris, C. Faverin, P. Bailac, M. Ponzi, LD_{50} and repellent effects of essential oils form Argentinian wild plant species on *Vorroa destructor*. J. Econ. Entomol. **98**, 651–655 (2005)

R. Sabbahi, D. De-Oliveira, J. Marceau, Influence of honey bee (Hymenoptera: Apidae) density on the production of canola (Crucifera: Brassicacae). J. Econ. Entomol. **98**(2), 367–372 (2005)

F. Sanchez-Bayo, From simple toxicological models to prediction of toxic effects in time. Ecotoxicology **18**, 343–354 (2009)

F. Sanchez-Bayo, K. Goka, Pesticide residues and bees – a risk assessment. PLoS One **9**(4), e94482 (2014)

J. Sandoz, B. Roger, M.H. Pham-Delegue, Olfactory learning and memory in the honeybee: comparison of different classical conditioning procedures of the proboscis extension response. C. R. Acad. Sci. Paris Sci. Vie. **318**, 749–755 (1995)

M.T. Sanford, Protecting honey bees from pesticides. Circular 534, Florida Cooperative Extension Service, Institute of Food and Agricultural Sciences, University of Florida. April 25, 1993. http://edis.ifas.ufl.edu/scripts/htmlgen.exe?DOCUMENT_AA145

R. Schmuck, No causal relationship between Gaucho seed dressing in sunflowers and the French bee syndrome. Pflanzenschutz-Nachr. Bayer **52**(99), 257–299 (1999)

R. Schmuck, R. Schoning, A. Stork, O. Schramel, Risk posed to honeybees (*Apis mellifera* L, Hymenoptera) by an imidacloprid seed dressing of sunflowers. Pest Manag. Sci. **57**(3), 225–238 (2001)

C.W. Schneider, J. Tautz, B. Grunewald, S. Fuchs, RFID tracking of sublethal effects of two neonicotinoid insecticides on the foraging behaviour of *Apis mellifera*. PLoS One **7**(1), e30023 (2012)

H.F. Schnier, G. Wenig, F. Laubert, V. Simon, R. Schmuck, Honey bee safety of imidacloprid corn seed treatment. Bull. Insectol. **56**(1), 73–75 (2003)
B. Schricker, Der Einfluss subletaler Dosen von Parathion (E 605) auf das Zeitgedächtnis der Honigbiene. Apidologie **5**, 385–398 (1974)
B. Schricker, W.P. Stephen, The effect of sublethal doses of parathion on honeybee behaviour. I. Oral adminstration and the communication dance. J. Apicult. Res. **9**, 141–153 (1970)
C.D. Scott-Dupree, M.S. Spivak, The impact of GAUCHO and TI-435 seed-treated canola on honeybees, *Apis mellifera* L., in *Final Report,* ed. by Canadian Honey Council, Calgary, pp. 1–22 (http://www.honeycouncil.ca/users/getdownload.asp?DownloadID)83) (2001)
C. Scott-Dupree, M. Winston, G. Hergert, S.C. Jay, D. Nelson, J. Gates, B. Termeer, G. Otis, A guide managing bees for crop pollination. Canadian association of professional apiculturists. Canadian Department of Agriculture, p. 41 (1995)
J. Serra-Bonvehi, J. Orantes-Bermejo, Acaricides and their residues in Spanish commercial beeswax. Pest Manag. Sci. **66**(11), 1230–1235 (2010)
F. Sgolastra, T. Renzi, S. Draghetti, P. Medrzycki, M. Lodesani, S. Maini, C. Porrini, Effects of neonicotinoid dust from maize seed-dressing on honey bees. Bull. Insectol. **65**(2), 273–280 (2012)
M.A.A. Shawki, D. Titera, J. Kazda, J. Kohoutkova, V. Taborsky, Toxicity to honeybees of water guttation and dew collected from winter rape treated with Nurelle D®. Plant Prot. Sci. **42**(1), 9–14 (2006)
L. Shipp, J.P. Kapongo, P. Kevan, J. Sutton, B. Broadbend, Bumble bees: an effective delivery system for microbial control agents for arthropod pests and disease management. International organization of biological and integrated control of noxious animals and plants, West Palaearc. Region. Sect. **29**, 47–51 (2006)
S.W. Shires, J. LeBlanc, A. Murray, S. Forbes, P. Debray, A field trial to assess the effects of a new pyrethroid insecticide, WL85871 on foraging honeybees in oilseed rape. J. Apicult. Res. **23**, 217–226 (1984a)
S.W. Shires, J.L. Blanc, P. Debray, S. Forbes, J. Louveaux, Field experiments on the effects of a new pyrethroid insecticide WL-85871 on bees foraging artificial aphid honeydew on winter wheat. Pestic. Sci. **15**(6), 543–552 (1984b)
P.B. Singh, D.S. Singh, M. Naim, V.S. Srivastava, Contact toxicity of some insecticides to oriental honeybee, *Apis cerana indica foragers*. Ann. Plant Prot. Sci. **5**, 207–209 (1997)
M.J. Smirle, The influence of colony population and brood rearing intensity on the activity of detoxifying enzymes in worker honey bees. Physiol. Entomol. **18**, 420–424 (1993)
H.M. Soares, C.R.O. Jacob, S.M. Carvalho, R.C.F. Nocelli, O. Malaspina, Toxicity of imidacloprid to the stingless bee, *Scaptotrigona postica* Latreille, 1807 (Hymenoptera: Apidae). Bull. Environ. Contam. Toxicol. **94**(6), 675–680 (2015)
R. Sokol, The influence of a multimonth persistence of Fluwarol in a hive of a honey bee colony. Med. Weter. **52**, 718–720 (1996)
M.G. Solomon, K.J.M. Hooker, Chemical repellents for reducing pesticide hazard to honeybees in apple orchards. J. Apicult. Res. **28**, 223–227 (1989)
E. Southwick, L. Southwick Jr., Estimating the economic value of honey bees (Hymenoptera: Apidae) as agricultural pollinators in the United States. J. Econ. Entomol. **85**(3), 621–633 (1992)
J. Stanley, G. Preetha, S. Chandrasekaran, S. Kuttalam, Honey bees of the cardamom ecosystem and the selective toxicity of diafenthiuron to four different bee species in the laboratory. J. Apicult. Res. **48**(2), 91–98 (2009)
J. Stanley, S. Chandrasekaran, G. Preetha, S. Kuttalam, Toxicity of diafenthiuron to honey bees in laboratory, semi-field and field conditions. Pest Manag. Sci. **66**, 505–510 (2010)
J. Stanley, K. Sah, S.K. Jain, J.C. Bhatt, S.N. Sushil, Evaluation of pesticide toxicity at their field recommended doses to honeybees, *Apis cerana* and *A. mellifera* through laboratory, semi-field and field studies. Chemosphere **119**, 668–674 (2015)

I. Steffan-dewenter, Seed set of male-sterile and male-fertile oilseed rape (*Brassica napus*) in relation to pollinator density. Apidologie **34**, 227–235 (2003)

K.A. Stoner, B.D. Eitzer, Movement of soil-applied imidacloprid and thiamethoxam into nectar and pollen of squash (*Cucurbita pepo*). PLoS One **7**, e39114 (2012)

K.A. Stoner, B.D. Eitzer, Using a hazard quotient to evaluate pesticide residues detected in pollen trapped from honey bees (*Apis mellifera*) in Connecticut. PLoS One **8**(10), e77550 (2013)

A. Stoner, W.T. Wilson, H.A. Rhodes, Carbofuran: effect of long-term feeding of low doses in sucrose syrup on honey bees in standard-size field colonies. Environ. Entomol. **11**, 53–59 (1982)

A. Stoner, W.T. Wilson, J. Harvey, Dimethoate (Cygon®): effect of long-term feeding of low doses on honey bees in standard size field colonies. Southwestern Entomol. **8**, 174–177 (1983)

A. Stoner, W.T. Wilson, J. Harvey, Acephate (Orthene®): effects on honey bee queen, brood and worker survival. Am. Bee J. **125**, 448–450 (1985)

S. Streit, F. Bock, C.W.W. Pirk, J. Tautz, Automatic life-long monitoring of individual insect behaviour now possible. Zoology **106**, 169–171 (2003)

S. Suchail, D. Guez, L.P. Belzunces, Characteristics of imidacloprid toxicity in two *Apis mellifera* subspecies. Environ. Toxicol. Chem. **19**(7), 1901–1905 (2000)

S. Suchail, D. Guez, L.P. Belzunces, Discrepancy between acute and chronic toxicity induced by imidacloprid and its metabolites in *Apis mellifera*. Environ. Toxicol. Chem. **20**(11), 2482–2486 (2001)

S.N. Sushil, J. Stanley, N.K. Hedau, J.C. Bhatt, Enhancing seed production of three brassica vegetables by honey bee pollination in North-western Himalayas of India. Univ. J. Agric. Res. **1**(3), 49–53 (2013)

H. Takahashi, J. Mitsui, N. Takakusa, M. Mastuda, H. Yoneda, J. Suzuki, K. Ishimitsu, T. Kishimoto, NI-25, a new type of systemic and broad spectrum insecticide. *Brighton Crop Protection Conference*, Brighton, UK. Pests Dis. **1**, 27–36 (1992)

A. Tapparo, C. Giorio, M. Marzaro, D. Marton, L. Solda, V. Girolami, Rapid analysis of neonicotinoid insecticides in guttation drops of corn seedlings obtained from coated seeds. J. Environ. Monit. **13**(6), 1564–1568 (2011)

J.N. Tasei, Effects of insect growth regulators on honey bees and non-apis bees. A review. Apidologie **32**, 527–545 (2001)

J.N. Tasei, S. Carre, B. Moscatelli, C. Grondeau, Research of the LD_{50} of deltamethrin on the alfalfa leafcutting bee: *Megachlle rotundata* and the effects of sublethal doses on adults and larvae. Apidologie **19**, 291–306 (1988)

J.N. Tasei, J. Lerin, G. Ripault, Sub-lethal effects of imidacloprid on bumblebees, *Bombus terrestris* (Hymenoptera: Apidae) during a laboratory feeding test. Pest Manag. Sci. **56**, 784–788 (2000)

J.N. Tasei, G. Ripault, E. Rivault, Hazards of imidacloprid seed coating to *Bombus terrestris* (Hymenoptera: Apidae) when applied to sunflower. J. Econ. Entomol. **94**, 623–627 (2001)

K.S. Taylor, G.D. Waller, L.A. Crowder, Impairment of a classical conditioned response of the honey bee (*Apis mellifera* L.) by sublethal doses of synthetic pyrethroid insecticides. Apidologie **18**, 243–252 (1987)

B.S. Teeters, R.M. Johnson, M.D. Ellis, B.D. Siegfried, Using video-tracking to assess sublethal effects of pesticides on honey bees (*Apis mellifera* L.). Environ. Toxicol. Chem. **31**(6), 1349–1354 (2012)

H.A. Tennekes, F. Sanchez-Bayo, Time-dependent toxicity of neonicotinoids and other toxicants: implications for a new approach to risk assessment. J. Environ. Anal. Toxicol. **S4**, 001 (2012)

J. Thomas, K.G. Phadke, Relative toxicity of oxydemeton methyl, chlorpyriphos and quinalphos to honeybee *Apis cerana indica*. Indian J. Agric. Sci. **64**, 207–209 (1994)

H.M. Thompson, Assessing the exposure and toxicity of pesticides to bumblebees (*Bombus* sp.). Apidologie **32**, 305–321 (2001)

H.M. Thompson, Behavioural effects of pesticides in bees – their potential for use in risk assessment. Ecotoxicology **12**, 317–330 (2003)

H.M. Thompson, L.V. Hunt, Extrapolating from honeybees to bumblebees in pesticide risk assessment. Ecotoxicology **8**, 147–166 (1999)
H.M. Thompson, K.A. Barrett, *Assessing the effects of glasshouse application of a novel insect growth regulator on bumble bee colonies*, eds. by L.P. Belzunces, C. Pélissier, G.B. Lewis. Hazards of pesticides to bees, Proceeding of the 7th International Symposium on Hazards of Pesticides to Bees 7–9 Sep, Avignon, France, INRA Paris (1999), pp. 227–228
H.M. Thompson, K.A. Barrett, Assessing the effects of a glasshouse application of a novel insect growth regulator on bumblebee colonies. *Proceedings of the 7th ICPBR Bee Protection Symposium*, Avignon, INRA, France, 2001
H. Thompson, S. Wilkins, Assessment of the synergy and repellency of pyrethroid/fungicide mixtures. Bull. Insectol. **56**(1), 131–134 (2003)
H. Thompson, S. Wilkins, A. Battersby, R. Waite, D. Wilkinson, The effects of four growth regulators insecticide on honey bee colony development, queen rearing and drone sperm production. Ecotoxicology **14**, 757–769 (2005)
S.V. Thomson, D.R. Hansen, K.M. Flint, J.D. Vandenberg, Dissemination of bacteria antagonistic to *Erwinia amylovora* by honey bees. Plant Dis. **76**, 1052–1056 (1992)
C.D.S. Tomlin, *The E-Pesticide Manual*, 12th edn. (British Crop Protection Council, Surrey, 2009)
P.F. Torchio, The effects of field applications of naled and trichlorfon on the alfalfa leafcutting bee, *Megachile rotundata* (Fabricius). J. Kansas Entomol. Soc. **56**, 62–68 (1983)
P. Tremolada, M. Mazzoleni, F. Saliu, M. Colombo, M. Vighi, Field trial for evaluating the effects on honeybees of corn sown using Cruiser® and Celest xl® treated seeds. Bull. Environ. Contam. Toxicol. **85**, 229–234 (2010)
T.J. Tyler, A.R. Davis, Evaluation of inflorescence visitors as pollinators of *Echinacea angustifolia* (Asteraceae): comparison of techniques. J. Econ. Entomol. **106**, 2055–2071 (2013)
USEPA, Ecological effects test guidelines. OPPTS 850.3030. Honey bee toxicity of residues on foliage. EPA 712–C–96–148, (1996)
USEPA, Ecological effects test guidelines. OCSPP 850.3030. Honey bee toxicity of residues on foliage. EPA 712C 018. January 2012a. Retrieved from www.regulations.gov/#!documentDetail;D=EPAHQ-OPPT-2009-0154-0017
USEPA, White paper in support of the proposed risk assessment process for bees, Office of chemical safety and pollution prevention. Office of pesticide programs, Environmental fate and effects division, United States Environmental Protection Agency, Washington, DC. (2012b)
USEPA, Guidance for assessing pesticide risks to bees. Retrieved from http://www2.epa.gov/sites/production/files/2014-06/documents/pollinator_risk_assessment_guidance_06_19_14.pdf (2014)
A. Valiente-Banuet, M.D.C. Arizmendi, A.R. Martinez, L.D. Canseco, Ecological relationship between columnar cacti and nectar feeding bats in Mexico. J. Trop. Ecol. **12**(1), 103–119 (1996)
R. Vandame, M. Meled, M.E. Colin, L.P. Belzunces, Alteration of the homing-flight in the honey bee *Apis mellifera* L. exposed to sublethal dose of deltamethrin. Environ. Toxicol. Chem. **14**, 855–860 (1995)
J. Van-der-Steen, Entwicklung einer Methode zur Bestimmung der Kontakt LD_{50} von Planzenschutzmitteln bei Hummeln (*Bombus terrestris* L.). Apidologie **25**, 463–465 (1994)
J. Van-der-Steen, Review of the methods to determine the hazard and toxicity of pesticides to bumblebees. Apidologie **32**(5), 399–406 (2001)
J. Van-der-Steen, A. Dinter, A monitoring study to assess the acute mortality effects of indoxacarb on honey bees (*Apis mellifera* L.) in flowering apple orchards. Pest Manag. Sci. **63**(11), 1095–1099 (2007)
D. Van-Engelsdorp, N. Speybroeck, J.D. Evans et al., Weighing risk factors associated with 'Bee Colony Collapse Disorder' by classification and regression tree analysis. J. Econ. Entomol. **103**, 1517–1523 (2010)
L.R. Verma, U. Partap, Foraging behaviour of *Apis cerana* on cauliflower and cabbage and its impact on seed production. J. Apicult. Res. **33**, 231–236 (1994)

C. Vidau, M. Diogon, J. Aufauvre et al., Exposure to sublethal doses of fipronil and thiacloprid highly increases mortality of honeybees previously infected by *Nosema ceranae*. PLoS One **6**(6), e21550 (2011)

R.F. Vieira, D.R. Sumner, Application of fungicides to foliage through overhead sprinkler irrigation – a review. Pestic. Sci. **55**, 412–422 (1999)

S. Villa, M. Vighi, A. Finizio, G.B. Serini, Risk assessment for honeybees from pesticide-exposed pollen. Ecotoxicology **9**, 287–297 (2000)

B. Vinothkumar, N. Kumaran, N. Boomathi, P.A. Saravanan, S. Kuttalam, Toxicity of spirotetramat 150 OD to honeybees. Madras Agric. J. **97**(1–3), 86–87 (2010)

G.D. Waller, R.J. Barker, J.H. Martin, Effects of dimethoate on honeybee foraging. Chemosphere **7**, 461–463 (1979)

M. Wang, C. ClaBen, Automated evaluation of honey bee brood trials using digital image processing according to OECD 75, Oomen (1992) and beyond. Poster presentation at SETAC North America, November 2011, Boston

S.D. Warren, K.T. Harper, G.M. Booth, Elevational distribution of insect pollinators. Am. Midl. Nat. **120**(2), 325–330 (1988)

N.M. Waser, Competition for humming bird pollination and sequential flowering in two Colorado wild flowers. Ecology **59**, 934–944 (1978)

M.J. Way, A.D. Synge, The effects of DDT and of benzene hexachloride on bees. Ann. Appl. Biol. **35**, 94–109 (1948)

P.R. Whitehorn, S. O'Connor, F.L. Wackers, D. Goulson, Neonicotinoid pesticide reduces bumble bee colony growth and queen production. Science **336**, 351–352 (2012)

R.P.A.V. Wijngaarden, T.C.M. Brock, P.J. Van-den-Brink, Threshold levels for effects of insecticides in freshwater ecosystems: a review. Ecotoxicology **14**, 355–380 (2005)

I.H. Williams, Aspects of bee diversity and crop pollination in the European Union, in *The Conservation of Bees*, ed. by A. Matheson, S.L. Buchmann, C. O'Toole, P. Westrich, I.H. Williams (Academic, New York, 1996), pp. 63–80

K.G. Willis, G.D. Garrod, Valuing landscapes: a contingent valuation approach. J. Environ. Manage. **37**(1), 1–22 (1993)

R. Winfree, N.M. Williams, J. Dushoff, C. Kremen, Native bees provide insurance against ongoing honey bee losses. Ecol. Lett. **10**, 1105–1113 (2007)

R. Winfree, N.M. Williams, H. Gaines, J.S. Ascher, C. Kremen, Wild bee pollinators provide the majority of crop visitation across land-use gradients in New Jersey and Pennsylvania, USA. J. Appl. Ecol. **45**, 793–802 (2008)

R. Winfree, B.J. Gross, C. Kremen, Valuing pollination services to agriculture. Ecol. Econ. **71**, 80–88 (2011)

M.L. Winston, *The Biology of the Honey Bee* (Harvard University Press, Cambridge, 1987). 281 p

M.L. Winston, E.N. Punnett, Factors affecting temporal division of labour in honey bees. Can. J. Zool. **60**, 2947–2952 (1982)

WSDA, Washington State Department of Agriculture. Pollinator protection requirements for section 18. Emergency exceptions and section 24(c), special local need registration in Washington State, AGR PUB 631-225 (2010)

J.Y. Wu, C.M. Anelli, W.S. Sheppard, Sublethal effects of pesticide residues in brood comb on worker honey bee (*Apis mellifera*) development and longevity. PLoS One **6**(2), e14720 (2011)

A.M. Young, Studies of cecidomyiid midges (Diptera: Cecidomyiidae) as cocoa pollinators (*Theobroma cacao* L.) in Central America. Proc. Entomol. Soc. Wash. **87**, 49–79 (1985)

H. Yu, J.C. Sutton, Effectiveness of bumblebees and honeybees for delivering inoculum of *Gliocladium roseum* to raspberry flowers to control *Botrytis cinerea*. Biol. Control **10**, 113–122 (1997)

S.W. Zhang, K. Bartsch, M.V. Srinivasan, Maze learning by honeybees. Neurobiol. Learn. Mem. **66**, 267–282 (1996)

W. Zhu, D.R. Schmehl, C.A. Mullin, J.L. Frazier, Four common pesticides, their mixtures and a formulation solvent in the hive environment have high oral toxicity to honey bee larvae. PLoS One **9**(1), e77547 (2014)

Index

Chapter 4
Pesticide Toxicity to Silkworms: Exposure, Toxicity and Risk Assessment Methodologies

Abstract Sericulture is an important occupation and a source of income especially for land less poor people. Silk is used not only for making excellent clothes but also in wound dressings, surgical sutures, artificial skin and tendons. Silk proteins are used in medicine, cosmetics, food and feed etc. Silkworms are affected by pesticides in many different ways. Pesticides may cause acute toxicity or sublethal effects leading to the impairment of silk production and quality. Pesticides cause retardation of growth and development of larva, spinning ability of lava, fecundity of moth etc. Many IGRs are reported to cause a 'non-spinning syndrome' and dauer larva even at very low concentrations. Silkworms are exposed to pesticides mainly through leaf contamination and thus leaf contamination bioassays are used extensively for pesticide toxicity studies. Unlike other non-target assays, expressing the pesticide amount as quantity per weight of food is more realistic in case of silkworms than quantity per pesticide solution. Bioassays with pesticides applied as quantitative spray on leaves, leaf disc, leaf sandwich, artificial diet etc can quantify the amount of pesticide actually taken by the silkworms. Bioassays on sublethal toxicity on food utilization, growth and development, spinning and cocoon quality, fecundity and hatching, effects on enzymes and hormones and genotoxicity are being conducted. Risk of pesticide exposure to silkworms can be done by classifying the pesticides based on acute toxicity, comparing the toxicity of pest (mulberry pest) and beneficial (silkworm), by calculating predicted exposure/environmental concentration (PEC) and no effect concentration (PNEC), sequential testing, strain susceptibility distribution, Hazard concentration 5 %, etc.

1 Importance of Silkworm, Silk and Sericulture

In history, the Silk Road, which stretched from China to India, Persia and Europe has linked together many civilizations and fostered economic development. Silk was the main and important fiber of pre-industrial era. A fabric with a thousand years of history, silk is used till today for making grand clothes. Silk is the most durable article of dress ever found and thus used extensively as a high quality fabric. The knitted fabrics of silk are used for the preparation of garments, parachutes, insulation coil for telephones and wireless receivers, tyres of racing cars, medical dressings and suture materials and many more. Silk fibers and proteins have vast

J. Stanley, G. Preetha, *Pesticide Toxicity to Non-target Organisms*,
DOI 10.1007/978-94-017-7752-0_4

applications in medical field, cosmetics, textile industry and also in the field of tissue engineering. Sericin has special properties like antioxidant, antibacterial, UV resistant, unique moisture absorption and release and inhibition of tyrosine kinase with many future scopes (Joseph and Raja 2012).

1.1 Silk as a Fiber and Fabric

Silk is the strongest and toughest of all natural fibers (Ahmed et al. 2012) with smooth and soft texture but not slippery like other synthetic fibers. This natural protein fiber has excellent mechanical and optical properties, biocompatibility, biodegradability and implant ability (Omenetto and Kaplan 2008). Silk fibers from the *Bombyx mori* silkworm have a triangular cross section with rounded corners and the flat surfaces of the fibrils, through which the light get reflected in many angles, giving silk a natural sheen.

Textiles as clothing material must comfortably fit the human body and senses apart from having suitable mechanical properties. Silk is one such material which is compatible the most, with the human body and also has a pleasing look. Silk is acclaimed for its smooth, lustrous texture that makes it comfortable to wear in both winter climates and warmer seasons (Boado 2013). Silk is used by the designers in fashion clothing with elegant designs to show a sense of style and charming of women (Guo 2014). Apart form high fashion clothes, saris, lining and lingerie; silk is used in the manufacture of good quality veils, ribbons, ties, laces, upholstery etc. A specially knitted silk fabric (MICROAIR DermaSilk®) is reported to be suitable for atopic dermatitis with unique properties of allowing the human skin to breathe and do not cause any irritation (Ricci et al. 2012).

1.2 Medical Uses of Silk/Silk Protein

The fine and soft nature of silk with specific tensile strength, easiness in tying and difficulty in untying and its biocompatibility with human living tissue (Mori and Tsukada 2000) approves it's utility in medical industry. Silk sericin used along with silver zinc sulfadiazine is reported safer and effective in healing burn and scratch wounds (Aramwit et al. 2013). Sericin coated textiles had more vertical wicking and moisture regain and also exhibit improved antistat, UV protection and radical scavenging activity. Thus sericin treated fabrics are suitable for use as medical textiles especially in wound dressings. Patients of atopic dermatitis, pressure ulcers and rashes having abrasive skin injuries are treated using the sericin treated fabrics (Gupta et al. 2015). Sericin can be used as an antimicrobial finish for cotton fabrics tested especially against *Escherichia coli* and *Staphylococcus aureus* and found effective (Rajendran et al. 2012). Silk fibroin membranes are used for burn wound dressings and also as artificial skin and as artificial tendons (Mori and Tsukada

2000). A silk fibroin based wound dressing developed by Tasubouchi (1999) was reported to accelerate healing and could be peeled off easily without damaging the newly formed skin. Silk thread is used in surgical sutures also and about 45 % of the silk yarns sold in Japanese market are used for surgical purposes (Mori and Tsukada 2000). Chinese silk sutures were tested for surgical site infections in breast cancer mastectomy and found as comparable with the antibacterial suture, Polyglactin (Zhang et al. 2011).

Fibers, films and 3-D structures made by polymerizing the silk protein, sericin are used as scaffolds for complex tissue reconstructions (Sehnal 2008). The other silk protein, fibroin hydrogel scaffolds are reported to promote in situ bone regeneration (Matta et al. 2004). Bone-like tissues are being engineered in vitro using porous biodegradable silk scaffolds and human bone marrow. The stable macroporous structure and the tailorable mechanical properties matching those of natural human bone make them well suitable for the purpose (Meinel et al. 2004). Silk fibroin has a possibility of being compatible with human blood also (Sakabe et al. 1989). The use of silk fibroin microtubes either implanted directly or preseeded with cells for blood vessel repair is an attractive biomaterial for microvascular grafts with several advantages over existing scaffold materials (Lovett et al. 2007).

The high oxygen permeability, optical properties and of cource the biocompatibility of silk proteins makes them a proper material for making contact lens. A sericin-fibroin blended film can be a preffered material to make artificial corneas (Murase 1994). Silk is used in biosensors for enzyme immobilization activity. The glucose oxidase immobilized in silk films can be used as a sensor for detecting glucose in blood (Asakura et al. 1992). Like wise, the silk fibroin immobilized antigens are applicable as biosensors for disease detection also (Mori and Tsukada 2000). Silk fibroin membranes impregnated with pharmaceuticals are used in drug delivery systems (Tsukada et al. 1994). Sulfated silk fibroins have anti-HIV properties and completely blocked the binding of virus to the cells at a concentration of 100 μg/mL in laboratory conditions (Gotoh et al. 2000).

1.3 Silk Proteins in Cosmetics

Sericin may be a valuable natural ingredient for cosmetic industries because of its antioxidant property suppressing lipid peroxidation and inhibiting tyrosinase activity (Kato et al. 1998). Sericin can be used in aerosol shaving gels, powders for cosmetics, dermatitis inhibitor, wound protection film, nail cosmetics and chewing gums (Gulrajani 2005). Sericin gels were reported to cause low skin impedance, increase hydroxyproline levels and hydrating epidermal cells and allow only minimal transepidermal water loss thus can be a good moisturizer. Sericin containing gels applied on the skin surface form a moisturizing, antiwrinkle and protective film. These gels applied on skin are reported to impart immediate, smooth and silky feeling which is found long lasting than many other gels (Padamwar et al. 2005). Sericin based lotions, creams and ointments showed increased skin elasticity, antiwrinkle and antiaging effects (Voegeli et al. 1993; Padamwar and Pawar 2004). Sericin is used in foundation creams, eyeliners, lotions, sweat absorbing and UV protecting cosmetics, sunscreen powders etc. It is also a constituent in nail cosmetics which prevent chapping and brittleness. Many hair conditioners and shampoos contain sericin as an ingredient (Padamwar and Pawar 2004).

Like sericin, fibroin is also nontoxic, innocuous and thus used as a cosmetic additive, which gives good colour also (Wang et al. 2006). Fibroin powder obtained from tussah silk is found to have antibacterial property and used as antibacterial powders (Fu 2006). Fibroin alone or in combination with sericin has been used in skin, hair and nail cosmetics (Padamwar and Pawar 2004). Cosmetics meant for sweat and sebum absorption containing cellulose fibers impregnated with fibroin dispersion and aqueous sericin solution are also reported (Miyashita 1999). Powder with ingredients of 5–30 % sericin and 70–95 % fibroin showed properties such as antistaticity and moisture absorbability (Kirikawa et al. 2000).

1.4 Silkworm as Food

Silkworm pupae are taken as snack food by boiling and seasoning them. It is used as a general food material and also used in traditional medicines in some countries (Pereira et al. 2003; Mishra et al. 2003). Silkworm pupae have been added in the list of novel food resources by the Ministry of Health in China (Zhu 2004). The pupa and pre-pupa of eri, muga and mulberry silkworms are considered a delicacy (Mishra et al. 2003) and reported as a good source of protein, fat and minerals (Longvah et al. 2011). Silkworm pupae contain 45–55 % protein on dry matter basis and can significantly raise hemoglobin and serum total protein levels (Shi et al. 1990). Medicines made of silkworms are intended to relieve flatulence and bodily spasms and to dissolve phlegm.

1.5 Silk/Silk Protein as Food

Silk fibroin is composed of 18 amino acids including many essential amino acids and considered as a valuable food source. It is studied and commercialized as a source of functional food or dietary supplements (Rhee et al. 1997; Baycin et al. 2007). Intake of sericin containing food accelerates the absorption of minerals (Padamwar and Pawar 2004). Food with sericin as ingredient is found effective against constipation problem and have some anti-tumour activities (Sasaki et al. 2000a). Consumption of 3 % sericin was found to enhance the bioavailability and the intestinal absorption of Zn, Fe, Mg and Ca but did not affect serum concentrations of these elements in rats (Sasaki et al. 2000b). Silk fibroin is reported to lower blood cholesterol and glucose levels and alcohol absorption. It is reported to be a neuroprotective and added in fortified milk (Choi et al. 2008). Since direct digestion of silk fibroin is difficult for humans, it can be made as hydrolysate and can be taken as functional food (Lee et al. 2011).

1.5.1 Unconventional Uses of Silk and Silkworm

- Silkworm pupae are traditionally used as fertilizer and also as animal and poultry feed (Fagoonee 1983; Zhou et al. 1996; Zhu 2004). Silkworm pupa powder meal is reported as the cheapest protein source and has potential to replace the costly and contaminated fish meal used in poultry industry (Dutta et al. 2012).
- Sericin coated films are used in refrigeration equipments because of its antifrosting property (Tanaka and Mizuno 2001).
- Sericin products can be used as a coagulant for purification of waste waters and hygroscopic moisture releasing polyurethane foams (Mondal et al. 2007). A membrane to separate water from organics can be made by copolymerizing acrylonitrile with sericin (Yamada and Fuwa 1994; Yamada et al. 1993). An insolubilized membrane made of silk fibroin could be used to separate the mixture of water and alcohol (Chisti 1998; Mizoguchi et al. 1991).
- As an insect, silkworm (*B. mori*) is being used in many scientific studies (Miyajima et al. 1987) as a model organism and the first lepidopteran for which draft genome sequences made available in 2004 (Xia et al. 2008). Silkworm is a model animal to evaluate drug toxicity and metabolism (Hamamoto et al. 2009). *Bombyx mori* possess excellent characteristics as an experimental animal to study developmental physiology, nutrition, endocrinology, mutation and toxicological studies (Tazima 1978). Baculovirus mediated transgenesis of silkworms allow specific alterations in a target sequence which leads to a wide range of new uses of silk fibroin for the production of oxygen permeable membranes and biocompatible materials (Mori and Tsukada 2000).

1.5.2 Sericulture as an Occupation and Source of Income

Sericulture plays a role in the utilization of the natural resources in a most effective manner for socio-economic upliftment, employment and income generation (Malik et al. 2008). There are more than 58 countries practicing sericulture in the world (Dewangan et al. 2012). In India, sericulture spread over 22 states, covering 172,000 ha area. Sericulture is a labour intensive enterprise and reported to provide employment to about 72.5 lakh people in India directly or indirectly (Dewangan et al. 2013). Unlike agricultural crops, sericulture provides regular income throughout the year and requires low investment but fetch high profits. In the ancient era, silk from China was the most lucrative and sought-after and highly desired luxury item traded across the Eurasian continent (Garthwaite 2005). India, the world's second largest silk producing country has the annual domestic consumption of 26,000 MT of silk and fetches around rupees 3,000 crores as foreign exchange earnings from export of silk goods (Gupta 2008). Sericulture plays a vital role in the flow of income from the urban rich sections of the society to the rural poor, as demand for silk is largely from the higher income group (Dewangan et al. 2011). Thus provide a way for economic stabilization and lessening the gap between the rich and the poor leading to less economic disparity.

2 Routes of Pesticide Exposure to Silkworm

Silkworms are seriously affected by pesticide exposure and toxicity. Though they are reared carefully and mostly in the indoor insectaries, they get exposed to pesticides in various ways. The most important route of pesticide exposure to silkworm is through food contamination. Leaf contamination of host plants of silkworms are through two ways viz., through direct pesticide sprays for pest management in host plants and the other by means of secondary contamination by spray drifts, volatilization of pesticides etc. Pesticides used in rearing facility and on the silkworms can also cause toxicity but this kind of exposure is not widespread.

2.1 Pesticides Applied for Pest Management in Host Plants

Host plants of silkworms especially of mulberry, castor, oak *etc* are infested by large number of pests. The silkworm, *Bombyx mori* is domesticated for centuries in mulberry, *Morus alba* (Rangaswamy et al. 1976). Tasar silkworm, *Antheraea mylitta* feeds on *Terminalia arjuna*, *T. tomentosa* and *Shorea robusta* (Rai et al. 2006). The oak tasar, *A. proylei* which is a cross between *A. pernyi* and *A. roylei* feeds on various species of *Quercus* (Jolly et al. 1979; Gosh and Gosh 1995). The Chinese tasar moth, *A. pernyi* feeds on many wild host plants including oak (Yu and Sun 1993) and about 70 % of the tussah silkworm resource in the world come from China (Wang and Dong 2002; Ni et al. 2003). Muga silkworm, *A. assama* specific to Assam state of India feeds mainly on *Machilus bombycina* and *Litsea polyantha* (Thangavelu et al. 1988). The eri silkworm, *Philosamia ricini* feeds on *Ricinus communis*, *Heteropanax fragrans*, *Manihot utilissima* and *Evodia fraxinifolia* (Sarkar 1988). The eri or cynthia moth, *Samia cynthia* feeds on *Ailanthus altissima* (Miller 1990; Ding et al. 2006). The cercopia silkworm, *Hyalophora cecropia* feeds mainly on *Prunus serotina* (Grabstein and Scriber 1982). Thus there are a wide range of host plants supporting different kinds of silkworms.

There are about 150 insect pests reported to attack these food plants of silkworms and out of these, the shoot fly, gall midges, hairy caterpillars and mealy bugs are important (Singh and Saratchandra 2002). More than 12 important pests including mealy bug, leaf roller, hairy caterpillar, cut worms, grasshoppers, beetles and root grubs (Singh and Saratchandra 2002) along with 15 species of mites are reported as pests of mulberry itself (Mohananasundaram and Sivagami 1983; Narayanaswamy et al. 1996). *Terminalia arjuna*, the tasar silkworm host plant is largely infested by defoliators, leaf miners and gall insects apart from stem borers (Singh and Thangavelu 1994). Sap suckers, defoliators, meristem feeders and gall forming insects attack oak trees (Singh et al. 2000). Whiteflies, thrips, leaf hoppers, hairy caterpillars, semiloopers, capsule borers cause serious damage in castor, the eri silkworm host plant (Singh et al. 2000; Singh and Saratchandra 2002). Thus a large amount of different kinds of pesticides are used for the control of these many varieties of pest of host plants, which in turn goes to silkworm by feeding on the contaminated leaves. The intoxication of silk worm larva through pesticides applied for the management of mulberry pests or diseases is well documented (Sun et al. 2012).

2.2 *Drift from Intercrops and Nearby Cultivated Fields*

Many annual crops are cultivated as intercrops in mulberry plantations such as, chilli (Savalgi and Patil 2003), soybean (Zheng et al. 2011), marigold (Govindaiah et al. 1991) etc. Mulberry is intercropped in coconut plantations (Das et al. 2010). Upland paddy and spices like aromatic ginger are grown under *Terminalia arjuna* in agri-silviculture system (Patle et al. 2003; Panwar et al. 2013; Ghosh and Pal 2002). Castor, the host plant of eri silkworm is intercropped with ground nut, green gram, black gram, cowpea, sesamum, chickpea (Reddy et al. 1965; Tarhalkar and Rao 1975; Porwal et al. 2006; Dhimmar 2009) and vegetables like cluster bean, cucumber etc. (Padmavathi and Raghavaiah 2004). These intercrops grown under the host plants of silkworm are also attacked by pests and diseases and pesticides are used for their management. In such cases, pesticides drifts from intercrops may contaminate the host plants of silkworm. Such reports on spray drifts to the mulberry garden affecting the growth, development, fecundity and survival of *B. mori* are already been reported (Chitgupekar and Basavanagoud 2014).

Contamination of the mulberry leaves are also due to pesticide drift from crop fields that are in close vicinity to mulberry plantations (Etebari et al. 2007; Zhang et al. 2008). Sericulture fields especially in Assam are adjacent to paddy fields and tea gardens, where wide spread use of pesticides are being practiced and thus indirectly contaminating silkworm host plants (Bora et al. 2012). Since the spray drifts falls on mulberry plantations from adjacent rice areas, granulated pesticides are recommended in such areas for rice pest control instead of spray formulations (Chen and Mi 1984). In south Brazil, chlorantraniliprole applied in sugarcane crop grown in the vicinity of silkworm farms are reported to cause severe damages in silkworm production (Munhoz et al. 2013).

2.3 Drifts from Aerial Sprays

Aerial application of insecticides to agricultural fields may cause damage to silkworm production (Munhoz et al. 2013). Through aerial application and water currents, insecticides or herbicides make forays into the mulberry plantations, thereby making difficulty for the survival of silkworms (Nath 1993). Aerial application of organophosphates in agricultural fields and their drifting to nearby mulberry plantations were reported to influence food consumption and utilization of silkworms (Nath et al. 1997). Aerial spray of phoxim affecting silkworm is also reported (Gu 2013). Chlorantraniliprole when applied in sugarcane crops especially through aerial sprays near to silkworm farms in Brazil appear to cause serious damage in silkworm rearing (Munhoz et al. 2013).

2.4 Volatalization of Pesticides

The characteristics of pesticides such as recommended volume, time of treatment, spray drift and volatility can influence the exposure to silkworms. Considering the volatile property and wide spread use in the sericultural area, the herbicide molinate is a possible pesticide for which the silkworm are exposed the most, through secondary poisoning of volatilization (Park et al. 2007). Herbicides are used for the management of weeds and grasses in mulberry plantations and in-turn get volatilized and adsorb or absorb by mulberry leaves causing toxicity to silkworms.

2.5 Pesticides Applied on Silkworms and in Rearing Rooms

Silkworms are also attacked by several parasites and predators (Mondal and Singh 1990; Singh and Saratchandra 2002). Even mulberry silkworms which can be cultured in indoor condition are not free from pests and diseases (Bora et al. 2012). The dermestid beetles damaging the cocoon cause severe economic damage in silk production (Singh et al. 2000). The major predators of silkworm larva are stink bug, preying mantids, wasps and ants. Among the parasitoids, the uzi fly (Exorista and Blepharipa) and Ichneumon flies (Xanthopimpla and Apanteles) are the most important causing serious damages (Singh and Thangavelu 1991; Singh et al. 1993). Deltamethrin though effective against uzi fly, *Exorista sorbillans* is extremely toxic to muga silk worm, *A. assama* larva (Khanikor 2011). Pesticides are sprayed in rearing/culture rooms (insectary) for the management of pests, these compounds can affect silkworms through cross contamination.

3 Effects of Pesticides on Silkworm

Pesticides may cause acute toxic effects on silkworm larvae (Zhang et al. 2008; Sun et al. 2012; Chi et al. 2015) or sub-lethal effects on the larva leading to the impairment on production and quality of silk (Leonardi et al. 1998; Wang et al. 1999; Vassarmidaki et al. 2000). Mortality of silkworm due to pesticide poisoning is still regarded as the major threat to silk industry (Cui et al. 2003; Wu et al. 2006). Apart from mortality, pesticides also cause retardation of growth and development of larva, spinning ability of the larvae, fecundity of moths and also affect the economic parameters by means of low cocoon weight and fiber quality (Wang et al. 1999; Vyjayanthi and Subramanyam 2002a). Pesticide poisoning accounts for about 30 % reduction in annual silk production in China, the major producer of silk (Li et al. 2010a). Silkworm larvae are extremely susceptible to pesticides especially to the insect growth regulators which induces a developmental arrest in fifth instar larvae at very low doses and also cause 'non-spinning syndrome' (Etebari et al. 2007) e.g. diflubenzuron (Kim 2002), fenoxycarb (Cappellozza 1994; Monconduit and Mauchamp 1999) etc. Some of the toxic effects of pesticides to this beneficial insect are given in detail as below.

3.1 Lethality or Mortality of Silkworm

3.1.1 Acute Toxicity

Acute toxicity implies sudden toxicity i.e., toxicity of pesticide causing an immediate mortality to the test organism. In general, acute toxicity tests are conducted in laboratory to find median lethal concentration (LC_{50}) or median lethal dose (LD_{50}). The pesticide with less median lethal values is regarded as highly toxic and vice versa. Acute toxicity is carried out in specific laboratory conditions and at specific stage of the test organisms since these are reported to significantly influence the toxicity of the pesticide.

3.1.1.1 Ingestion Toxicity: Median Lethal Concentration

Ingestion toxicity is by which the pesticide is exposed through the food the silkworm takes especially of mulberry leaves or through artificial diet in some cases. The median lethal concentration of triazophos to third instar silkworm larva, *B. mori* is 1.03 mg/kg of mulberry leaf and thus found toxic (Chen et al. 1998). The median lethal concentration of cypermethrin, imidacloprid, phoxim, dichlorvos and dimethoate to six different strains of *B. mori* were reported as 0.016–0.069, 0.13–0.29, 0.42–0.48, 5.6–6.9 and 10.2–24.7 mg/kg of fresh mulberry leaf (Sun et al. 2012). The 48 h ingestion LC_{50} of chlorpyrifos and fenpropathrin to silkworm was

4.66 and 0.30 mg/L, respectively (Wu et al. 2003). The 48 h median lethal concentration of permethrin, tetramethrin, bifenthrin, ethofenprox, dichlorvos and phoxim were 0.75, 2.83, 0.06, 0.80, 4.11, 0.45 mg/L, respectively (Zhang et al. 2008). The 24 h LC_{50} of diafenthiuron to fourth instar larva of silkworm was 0.36 g/L (Stanley et al. 2016).

The 48 h LC_{50} of chlorpyrifos to two, three, four and fifth instar silkworm larvae was 1.11, 1.73, 3.48 and 4.12 mg/L, respectively (Zhu et al. 2006a). The LC_{50} of phoxim to first to fifth instar larvae of *B. mori* is reported as 254.8, 313.9, 1048.1, 1591.4 and 2868.8 ng/mL, respectively (Li et al. 2010a). The LC_{50} of fenpropathrin, phoxim and their mixture to silkworm larva by food intake method were 0.043, 0.703 and 0.112 μg/mL, respectively (Zhu and Cui 2000). The 48 h LC_{50} of chlorpyrifos EC and CS formulation to fourth instar silkworm larva was found to be 4.35 and 8.35 mg/L, respectively (Ji et al. 2010).

3.1.1.1.1 Median Lethal Concentration of Different Pesticides to Silkworm

Pesticides	Stage of larva (instar)	Period of mortality (hours)	Median lethal concentration (LC_{50})	Reference
Avermectin	4	96	16.01 g/L	Shen et al. (2011)
Azadirachtin	3	–	2.0×10^{-4} mg/L	Chen et al. (2008a)
Chlorantraniliprole	3	–	4.0×10^{-3} mg/L	Chen et al. (2010)
Chlorpyrifos EC	2	48	0.82 mg/L	Ji et al. (2010)
Chlorpyrifos EC	3	48	1.87 mg/L	Ji et al. (2010)
Chlorpyrifos CS	2	48	2.48 mg/L	Ji et al. (2010)
Chlorpyrifos CS	3	48	4.22 mg/L	Ji et al. (2010)
Dichlorvos	–	72	1.07 mg/L	Ma et al. (2006)
Imidacloprid	4	–	2.71 mg/L	Lu and Wu (2000)
Dimethoate	2	96	815 mg/L	Chi et al. (2015)
Emamectin benzoate	2	96	0.002 mg/kg	Chi et al. (2015)
Lambda cyhalothrin	2	96	0.004 mg/kg	Chi et al. (2015)
Permethrin	–	24	0.329 mg/L	Ma et al. (2005)
Phoxim	–	24	0.703 mg/L	Zhu and Cui (2000)
Phoxim	5	24	7.86 mg/L	Peng et al. (2011)
Phoxim	5	2	2.5 mg/L	Yu et al. (2011)

Acute toxicity of chlorpyrifos was found to increase with increase in temperature with LC_{50} of 1.7 mg/L at 20 °C and 0.4 mg/L at 35 °C tested in third instar *B. mori* larva. The duration of leaf dip also influence on the toxicity of chlorpyrifos with LC_{50} of 5.55 and 2.12 mg/L for 1 s and 10 min immersion periods, respectively tested on the fourth instar larva (Zhu et al. 2006a). Apart from the active ingredient, the pesticide formulation also influences the acute mortality of silkworms. The emulsifiable concentrate (EC) of chlorpyrifos is reported to be more toxic to silkworm than capsule suspension (CS) formulation as revealed by both ingestion and contact bioassays (Ji et al. 2010).

3.1.1.2 Contact Toxicity: Median Lethal Dose

The median lethal dose (LD_{50}) of emamectin benzoate was found to be very low to silkworms (0.011 mg/kg) and thus reported as extremely toxic (Cang et al. 2007). The median lethal dose (96 h) of triazophos to the second instar larvae of silkworm was found to be 0.015 μg/larva and 1.41 g/cm², as tested in topical bioassay on pronotum and toxicant membrane assay, respectively (Zhu et al. 2006b). The contact LD_{50} of fenpropathrin, phoxim and their mixture to silkworm larva were 0.717, 27.15 and 4.384 μg/mL, respectively (Zhu and Cui 2000). Contact toxicity (48 h LC_{50}) of chlorpyrifos to third instar larva of *B. mori* was reported to be 0.20 μg/cm² as revealed by filter paper bioassay (Zhu et al. 2006a). Contact toxicity (48 h LD_{50}) of chlorpyrifos EC and CS formulation for third silkworm larva exposed for 1 h in pesticide treated filter paper was 0.38 and 0.48 μg/cm², respectively (Ji et al. 2010).

3.1.1.2.1 Median Lethal Dose of Different Pesticides to Silkworm

Pesticides	Stage of larva (instar)	Period of mortality (hours)	Median lethal dose (LD_{50})	Reference
Azadirachtin	3	–	1.25×10^{-9} mg/larva	Chen et al. (2008a)
Chlorantraniliprole	3	–	1.14×10^{-8} mg/larva	Chen et al. (2010)
Chlorpyrifos	–	24	6.32 μg/cm²	Wu et al. (2003)
Ethion	5	–	0.037 ppm	Nath (2003)
Emamectin benzoate	–	–	0.011 mg/kg	Cang et al. (2007)
Fenitrothion	5	–	0.306 ppm	Nath (2003)
Fenpropathrin	–	24	0.92 μg/cm²	Wu et al. (2003)
Imidacloprid	4	–	0.034 μg/larva	Lu and Wu (2000)

3.1.1.3 Acute Toxicity at Field Recommended Dose

Among the six insecticides tested, carbosulfan and methomyl showed high stomach toxicity and moderate contact toxicity to silkworm larva whereas cypermethrin, fenvalerate, fipronil and abamectin had high stomach and contact toxicity, but slight fumigation toxicity (Li 2001). High toxicity of imidacloprid and very high toxicity of brofluthrinate to silkworms was reported and thus pollution of the chemical in mulberry is to be prohibited (Gong et al. 1999, 2001). Diafenthiuron is found very toxic to silkworms recording 100 % mortality at the field recommended dose (Stanley 2007). Pyriproxyfen at concentrations less than 1 mL/L showed lethal toxicity to young larvae of *B. mori* and toxic symptom on the grownup larvae without any acute mortality (Sun et al. 2008). Acetamiprid was found to be less toxic to silkworm larva as compared to chlorpyrifos, quinalphos and endosulfan (Mathirajan and Raghuraman 2003). Insecticides emamectin benzoate and abamectin at their field recommended doses caused only 10 % mortality to fourth instar silkworm larva and flubendiamide, 8.6 % mortality after 48 h after treatment (HAT) (Kordy 2014). Among the eight insecticides at their field recommended doses tested against first and third instar silkworm larva, only chlorfenapyr was found safer. Indoxacarb, spinosad, emamectin benzoate, flubendiamide, chlorantraniliprole, lambda cyhalothrin and endosulfan were found to cause 100 % mortality at 48 HAT to the first and third instar larva (Chitgupekar and Basavanagoud 2014). Mulberry leaves were safe when sprayed with chlorfenapyr at 0.33 and 1 mL/L at third and fifth day of field application in summer months, whereas it caused 100 % mortality to the third instar silkworm larva at 0.33 mL/L dose also even after 7 days in spring season (She et al. 2012).

3.1.1.4 Residual Toxicity, Safe Waiting Period and NOEC

Residual toxicity of chlorfenapyr at 250, 200, 150, 100 and 50 mg/L sprayed on mulberry leaves were 5, 5, 3, 1 and 0 days, respectively to silkworms. The safe waiting period for chlorfenapyr at 100 mg/L was 5 days for silkworms (Liu et al. 2008). Based on residual toxicity, the safe waiting period for naled, dimethoate and dichlorvos for silkworms were found to be 3, 7 and 7 days after application (Li et al. 2010b). Persistent toxicity (mortality) of dimethoate and dichlorvos to silkworm larva was found to be for 3 and 5 days, respectively (Wang 2010). The safer interval of DDVP for silkworms in mulberry is 7 days and 5.5 days for 1 and 2 mL/L concentrations (Wei et al. 2008). The safe period for chlorpyrifos 2 mL/L was reported as 15–20 days and 20–25 days for spring and autumn, respectively (Xu et al. 2007). Safe period of parathion at 1 and 2 mL/L in mulberry to silkworms is arrived at 10 and 8 days, respectively (Chen et al. 2008b). The no effect concentration of triazophos to silkworm larva is about 0.04 mg/L as revealed by leaf dip bioassays (Zhu et al. 2006b). The maximum acceptable daily dose of dimehypo for proper growth and cocooning of *B. mori* is less than 1.7×10^{-6} μg/day in spring reared larvae and less than 1.7×10^{-8} μg/day in autumn reared larvae (Wang et al. 1999).

3.1.2 Chronic Toxicity

Chlorfenapyr was not found to be acutely toxic to silkworm larvae but exhibited accumulated chronic toxicity. Continuous feeding of contaminated leaves resulted in increase in mortality which is proportionate to concentrations (Liu et al. 2008). Chronic mortality of silkworms was reported in low concentrations i.e., less than 1 mL/L of chlorpyrifos (Xu et al. 2007). Silkworms from first day of second instar were fed continuously on mulberry leaves dipped in triazophos solution to find the chronic mortality. The LC_{50} at 48 HAT at the beginning of the third, fourth and fifth instar was 1.54, 0.74 and 0.53 mg/L, respectively (Zhu et al. 2006b).

3.2 Sublethal Toxicity

3.2.1 Effect on Food Intake, Growth and Development

Food Intake and Utilization Silkworm larvae exposed to juvenile hormone analogue were reported to have a reduced feeding potential (Sakurai and Imokawa 1988). The food consumption and defecation of fifth instar silkworm larva was significantly reduced when exposed to ethion and fenitrothion at their respective lethal doses and an insignificant increase in food intake was observed in sublethal (1/5th of LD_{50}) doses (Nath and Kumar 1999; Nath 2003). Less feeding and vomiting are reported in silkworms exposed to dichlorvos and phoxim whereas less feeding and 'S' or 'C' shaped shortening of body in permethrin, tetramethrin, bifenthrin and ethofenprox treatments (Zhang et al. 2008). A reduction of 50 % water intake through food by *Philosamia ricini* fed with permethrin compared to untreated larva was reported (Naik and Delvi 1984). Fenvalerate was found to reduce the rates of feeding, assimilation and conversion along with the efficiencies of conversion of ingested and digested food into body substance as tested in final second instars of bivoltine and multivoltine silkworms. The toxic effect of fenvalerate was observed to interfere with the feeding behaviour of the silkworms and also the physiology of digestion (Vyjayanthi and Subramanyam 2002a). IGRs at doses exceeding 100 pg/larva, reduced the efficiency of utilization of ingested food (Leonardi et al. 1996; Monconduit and Mauchamp 1998; Kamimura and Kiuchi 1998). Fenoxycarb was reported to reduce the efficiency in the utilization of ingested food in silkworm larva (Assal et al. 1994; Leonardi et al. 2001). A reduction in frass production in silkworm was found when topically dosed with fenoxycarb and is highly correlated with the dose (Leonardi et al. 2001). Buprofezin treated leaves fed to fifth instar *B. mori* larva led to a significant lower larval weight compared to control (Vassarmidaki et al. 2000). Avermectin at 1–10 μg/L showed a significant effect on larval weight of silkworms (Zhang et al. 2006). Abamectin at sublethal doses decreased the food intake and body mass of silkworm larva. The efficiency of conversion of ingested food and the conversion of digested food was also reduced with increase in approximate digestibility (Zhu et al. 2008). But methoprene treated fifth instar silkworm larva has increased in weight varying from 4.65 (untreated control) to 5.70 g (20 ng a.i./insect). Methoprene treated larva produced heavy cocoons, pupa and shells. The increase in weight was found to rise with increasing dose of methoprene up to 1 ng a.i./insect and afterwards decreased drastically (Miranda et al. 2002). Hexachlorocyclohexane exposure to silkworm larva resulted in reduced pupal and shell weights and also impacted survival and adult emergence (Bhagyalakshmi et al. 1995).

Life Cycle Growth and development of silkworm larva was reported to get affected by continuous feeding of pyrethroids at 10 ng/L (Sun et al. 2002). The nereis toxin insecticide, dimehypo was found to affect the growth and development of silkworm larvae lengthening its life cycle (Wang et al. 1999). Silkworm larva exposed to juvenile hormones at early days of last instar caused prolongation in the larval stage or produced dauer larvae (Akai and Kobayashi 1971). Methoprene was found to prolong duration of fifth instar hybrid silkworm larva by 30 h (Begum et al. 2011). Mulberry leaves treated with methyl parathion at 0.00035 ppm was found to extend the total larval period by 2 days i.e., the larval period in pesticide treatment was 21.3 days whereas it was 19.1 days in control. The pupal duration of pesticide treated lot was 12.98 days as compared to 10.28 days in control (Kumutha et al. 2009).

Growth and Development The developmental duration of silkworm, weight of silkworm, cocoon weight, cocoon shell weight, pupal weight and pupation rate were found not to be affected by chlorfenapyr 10 SC and thus safe to be used at 3 mL/L in mulberry trees (Chitgupekar and Basavanagoud 2014). But herbicide, rac-metolachlor treated silkworms were found to grow slowly and weigh very less than the control larva (Zhan et al. 2006). The antijuvenile compound K-22 though did not cause mortality of silkworm larva, delayed the time of ecdysis to fourth instar by 1–5 days in different doses (Asano et al. 1984). Application of the anti JH compound (40 μg/larva) to the larva at the beginning of third ecdysis induced 100 % precocious metamorphosis. The larva treated on three or fourth day after third ecdysis were not precocious, when tested at 40 μg by topical application (Asano et al. 1986). Diflubenzuron has been known to prevent metamorphosis in silkworm from larval to pupal stage by blocking juvenile hormone degradation.

3.2.2 Effect on Cocooning and Spinning

Bombyx mori larvae are extremely susceptible to IGRs, that at very low doses (50–100 pg/larva) induces a non-spinning syndrome and a developmental arrest even in fifth instar larvae (Leonardi et al. 1996; Monconduit and Mauchamp 1998; Kamimura and Kiuchi 1998), producing dauer larvae that will die of septicemia (Leonardi et al. 2001). Low sublethal doses of fenoxycarb continuously administered to fifth instar silkworm larva significantly prolonged the larval period and resulted in no cocooning (Li et al. 2009). Fenoxycarb topically dosed at 60 pg/larva caused 50 % permanent non-spinning (dauer) larvae (Leonardi et al. 1996). But the antijuvenile compound K-22 (terpenoid imidazole) added in artificial diet and given to silkworms cause precocious spinning of cocoons in a dose dependent manner (Asano et al. 1984).

The trace feeding or fumigation of pyriproxyfen to silkworm larvae leads to significant prolongation of instar duration and retards cocoon making ability (Sun et al. 2008). It was also reported that dimehypo as extremely harmful for cocooning activity of *B. mori* (Wang et al. 1999). Continuous feeding of chlorfenapyr contaminated leaves resulted in reduced cocoon percentage and eclosion rate which is proportionate to concentrations (Liu et al. 2008). A significant dose-dependent reduction in cocoon making and pupation in silkworm was found when second instar larvae exposed to herbicide, clodinafop-propargyl (Yin et al. 2008).

3.2.3 Effect on Silk Production and Fiber Quality

Chronic exposure of insecticides at lower doses may lead to sub-lethal effects which could affect silk production and silk quality (Sun et al. 2012). Contamination of mulberry leaves with nereis toxin insecticide, dimehypo was implicated with the reduction in silk production by silkworms (Wang et al. 1999). Chlorantraniliprole contaminated mulberry leaves cause the silkworm larva to stop feeding, leading to the production of a thin-shelled cocoons (Munhoz et al. 2013). Reduction in economic parameters of silkworm after topical application of juvenile hormone analogues are also reported (Miranda et al. 2002). A significant difference was reported in the weight of cocoon and shell produced by the silkworm larva treated with buprofezin, compared to the ones produced by larva fed with untreated leaves (Vassarmidaki et al. 2000). A significant increase in the cocoon weight (10.3 %), shell weight (14.5 %) and shell percentage (3.8 %) was found with methoprene treated silkworms (Begum et al. 2011). Methoprene was found to promote weight of *B. mori* silk glands up to a rate of 1 ng a.i/insect and a decline in weight was found in higher rates (Miranda et al. 2002). Diflubenzuron inhibits the growth of silk gland and alters its colour making them white opaque (Kim and Sohn 2001). A 21 % increment of silk production in silkworms was reported by the use of the SJ-42-F juvenile hormone (Chowdhary et al. 1986).

Changes in the ultrastructure of the posterior silk gland cells were reported in silkworm, *B. mori* exposed to insecticide, dimehypo. Physiological activity of the posterior silk gland cells were affected leading to changes in biosynthesis of fibroin (Wang et al. 1999). The fibroin content of cocoons was found to be low in cocoons produced by methoprene treated larva, which indicate a negative influence of methoprene on protein synthesis (Miranda et al. 2002). Hexachlorocyclohexane was found to impair the growth of silk gland in silkworms and reduce the fibroin content, deteriorating the quality and quantity of silk thread produced (Bhagyalakshmi et al. 1995). However, increase in silk protein, fibroin synthesis in silk gland cell by the injection of juvenile hormone in silkworm larva is also demonstrated (Akai and Kiguchi 1971).

3.2.4 Effect on Reproduction and Fecundity

Organophosphorus insecticides like parathion, fenthion, isoxathion, diazinon, disulfoton were found to adversely affect the reproduction of silkworm when treated at larval stage. The acaricide chlordimeform, herbicide trifluralin and chemosterilant metepa significantly affected the reproduction of silkworm (Kuribayashi 1981a). Abnormal mating behaviour was observed in silkworms treated with parathion at 0.001 % (Kuribayashi 1981a). The organochloride insecticide, hexachlorocyclohexane treatment in silkworm larvae caused a reduction in the number of eggs laid by adult moths (Bhagyalakshmi et al. 1995). Dichlorvos was found to affect the egg laying of silk moths. The number of eggs laid by moths emerged from larvae fed with dichlorvos 0.0005 % treated leaves was 485 per moth whereas the untreated control moth laid 547 eggs, showing a 11.3 % reduction in egg laying. Dichlorvos treated silkworm moths produced only 56.6 % fertilized eggs and the hatching percentage also reduced with 79.6 % eggs hatched in pesticide exposed moths compared to 93.9 % in untreated check (Kumutha et al. 2013). Silkworm larvae exposed to parathion and disulfoton caused an ovicidal action on the eggs produced by them as adults. The embryos of these eggs develop normally just before hatching but found dead before hatching and in some cases just after hatching. Some of the insecticides entered in the larva are passed through the egg causing death to the embryo or the neonate larva (Kuribayashi 1981b). Only 50 % hatching was found in eggs laid by moths emerged out of larva treated with parathion. The hatched larvae also behaved abnormally and found vomiting excessively (Kuribayashi 1981a).

3.2.5 Effect on Hemolymph

Organophosphorus insecticides, ethion and fenitrothion are reported to reduce the amount of protein in silkworm haemolymph (Nath et al. 1997). Juvenile hormone and the juvenile hormone analogue, fenoxycarb and pyriproxyfen were found to induce inhibition of larval haemolymph protein synthesis in silkworm, *B. mori* (Monconduit and Mauchamp 1998). A significant change in the carbohydrate metabolism in the hemolymph and fat body of silkworms exposed to ethion is reported (Nath 2000). Hemolymph glucose is found to get reduced in silkworms exposed to pyriproxyfen after 24 h of exposure as 13.0 mg/dL in 1 ppm treatment compared to 18.2 mg/dL in control. The urea content was not much changed whereas the uric acid was significantly less in treated larva compared to control ones tested after 24 h of exposure (Etebari et al. 2007). An increase of 48 and 34 % of trehalose and glycogen was reported in hemolymph of hybrid silkworms treated with methoprene (Begum et al. 2011). Significant decrease in pyruvate levels and increase in lactate levels in the hemolymph and fat body were observed on fifth instar silkworm, *B. mori* exposed to lethal and sublethal doses of fenitrothion and ethion (Nath 2000). The hemolymph metal ion concentration is found to reduce significantly after insecticide exposure to silkworm larva compared to untreated control (Kordy 2014).

Severe proteolysis and transamination of amino acids with enhanced activities of protease, alanine aminotransferase, aspartate aminotransferase and elevation of glutamate dehydrogenase were reported in the haemolymph and fat body of fifth instar silkworm larva exposed to fenitrothion and ethion (Nath et al. 1997). Increases in free amino acids, urea, uric acid and lactate levels, activities of protease, alanine aminotransferase and aspartate aminotransferase are reported in the hemolymph of silkworms exposed to phoxim and TiO_2 attenuate it (Li et al. 2012). Activity of alanine aminotransferase, aspartate aminotransferase and alkaline phosphatase decreased significantly after 24 h of pyriproxyfen treatment. After 120 h of pesticide exposure, reverse changes were observed in the activity of these enzymes (Etebari et al. 2007). The acid phosphatase activity of rac-metolachlor-treated hemolymph was 44–73 % lower than that of control. Hemolymph catalase activity decreased and found to be 50 % lower than control in rac-metolachlor treated silkworms (Zhan et al. 2006). Diflubenzuron treated fifth instar silkworms were found to have reduced esterase activities in haemolymph whereas no difference in esterases in the fat body (Kim et al. 2002). Lactate dehydrogenase activities were also significantly reduced resulted in an increase in lactate levels in the hemolymph and fat bodies in silkworms exposed to fenitrothion and ethion (Nath 2000).

3.2.6 Effects on Fat Bodies and Cellular Toxicity

Sublethal doses of fenitrothion and ethion cause a significant depletion in fat body glycogen reserves in silkworms followed by concomitant increase in fat body phosphorylase and trehalase activities. These changes indicate an increased glycogenolysis at tissue levels. The hypo-trehalosemia and hypo-glycaemia in fat bodies of silkworm larva indicate the mobilization of trehalose and glucose from fat body (Nath 2003). Trehalose and glycogen was found to increase in fat body of methoprene treated hybrid silkworms by 29 and 43 % compared to untreated insects (Begum et al. 2011). Ethion and fenitrothion are reported to reduce the amount of protein in fat bodies of silkworms indicating proteolysis and transamination of amino acids (Nath et al. 1997). Malondialdehyde and glutathione transferase activity were found to increase significantly in the fat bodies and midgut of fifth instar silkworms exposed to phoxim 2.5 mg/L (Yu et al. 2011). The activity of P450 in the midgut and fat body of phoxim treated silkworm was found to increase to 1.7 and 6.7-folds, compared to the control. Further, the activity of GST had no change in midgut but increased in fat body whereas the activity of carboxylesterase get decreased in the midgut and increased in fat body (Wang et al. 2013).

Histopathological examinations and transmission electron microscopic studies on phoxim exposed silkworms showed swollen mitochondria with disappearance of mitochondrial cristae, which are the important features in insect apoptotic cells (Gu et al. 2014). Topical application of fenoxycarb increased cholesterol in midgut brush border membrane of silkworm larvae (Leonardi et al. 2001). Topical application of IGRs at doses of 2.5 μg/larva reduced the amino acid uptakes in the midgut brush border membrane vesicles whereas an increased intake was noticed in extremely low doses (2.5 fg/larva) (Leonardi et al. 1996; Monconduit and Mauchamp 1998; Kamimura and Kiuchi 1998). Leucine uptake in brush border membranes of silkworm midgut was found to reduce at 2.5 μg/larva dose of fenoxycarb whereas it increased with the same dose when orally fed (Leonardi et al. 2001). The regenerative cells presented in the midgut epithelium of silkworm larva exposed to chlorantraniliprole were found to undergo hypertrophy and hyperplasia. The longitudinal and circular muscles of walls of midgut were found disorganized (Munhoz et al., 2013).

3.2.7 Effect on Enzymes and Hormones

Digestive Enzymes Pesticides are found to affect digestive enzymes of silkworms. Insecticide, fenvalerate was reported to interfere with the metabolism of silkworms by impairing the functional abilities of digestive enzymes. A reduction in the activity of amylase, sucrase and protease and enhancement in trehalase activity in midgut of fenvalerate exposed silkworm larva was reported (Vyjayanthi and Subramanyam 2002b). Amylase and sucrase activities in the midgut decreased significantly in larvae treated with abamectin (Zhu et al. 2008). Phoxim exposure to silkworm causes reduction in the activities of lactate dehydrogenase, succinate dehydrogenase and malate dehydrogenase in the insect (Li et al. 2012). In the rac-metolachlor treated fifth instar silkworm larva, hemolymph lactate dehydrogenase activity was significantly low indicating changes in the carbohydrate metabolism for the first 4 days and then gradually returned to normal. These changes cause a low production of pyruvate from lactate and that leads to shifting of aerobic metabolism to anaerobic so as to meet the energy demands (Zhan et al. 2006). Methoprene enhances protease, aspartate aminotransaminase, alanine aminotransaminase, adenosine triphosphate synthase and cytochrome-c-oxidase activity levels in silkworm indicating overall surge of oxidative metabolism (Mamatha et al. 2008). In midgut, the alkaline phosphatase activity was found to be reduced by 46 % in silkworms treated with rac-metolachlor compared to control. Alkaline phosphatase is correlated with the cocoon quality apart from digestion, absorption and transportation of nutrients (Zhan et al. 2006) and thus can be a biochemical index of health and economic value of the silkworm (Miao 2002).

Antioxidant Enzymes Superoxide dismutase and catalase activities were found to get increased in the midgut of silkworm larva fed with mulberry leaves containing imidacloprid 2 ppm. The superoxide dismutase activity in the imidacloprid treated silkworms was 20.7 per mg protein/h and in untreated control it was 6.7 per mg protein/h (Phugare et al. 2013).

Detoxifying Enzymes Organophosphorus insecticides, such as fenitrothion and ethion are found to inhibit acetylcholinesterase activity which leads to a concomitant increase in acetylcholine in brain, fat body and silk glands of *B. mori*. The accumulation of acetylcholine was the highest in brain and lowest in fat bodies

(Nath and Kumar 1999). The acetyl choline esterase activity of *B. mandarina* was always higher than that of *B. mori* tested in all larval instars and in brain, midgut, fat bodies, silk gland and blood in the mean ratio of 1.6–2.28 (Li et al. 2010a).

Hormones Diflubenzuron blocks juvenile hormone degradation in silkworm even at very low exposures. Disappearance of juvenile hormone in the last larval instar (Riddiford 1994) is critical for the prothoracic glands to recover their activity (Okuda et al. 1985), for silk gland to grow (Garel 1983) and to increase the hemolymph protein synthesis (Izumi et al. 1984; Tomino 1985). So, all these physiological activities get affected due to diflubenzuron exposure. Though diflubenzuron act as a juvenile hormone it is not a JH analog since it did not induce vitellogenesis (Kim et al. 2002). Fenoxycarb was not found to inhibit the release of prothoracicotropic hormone (PTTH) from Brain–Corpora cardiaca–Carpora allata complex, but prevented the responsiveness of prothoracic glands to PTTH (Monconduit and Mauchamp 1998; Dedos and Fugo 1999).

3.2.8 Genotoxicity of Pesticides

Pesticides cause genotoxic effects even at very low concentrations which do not produce any morphological and physiological symptoms (Shen et al. 2011). Silkworm larva exposed to herbicide, clodinafop-propargyl by oral feeding with mulberry leaves showed a significant difference in DNA assessment. The occurrence of comet, the head/tail DNA and the tail length, tail moment and olive moment differed significantly than the control group (Yin et al. 2008). Hemocytes of silkworm exposed to avermectin exhibited significantly higher DNA damage than that of control and found to be proportionate to concentrations. About 25 differential fragments, varied from 250 to 800 bp in size were found in mRNA in worms exposed to 1.0 μg/L with respect to control (Shen et al. 2011).

Effects of pesticides on gene expressions are also studied and reported in silkworms. In general, Rad23, a gene related to nucleotide excision repair is reported to have hyper expression upon chemical stress in silkworms (Xu et al. 2009). In avermectin treated silkworms, 15 differentially expressed genes were identified, including 5 down-regulated and 10 up-regulated genes (Shen et al. 2011). Midgut of phoxim exposed silkworm was studied for changes in gene expressions by means of gene sequencing. About, 254 genes displayed at least twofold changes in expression levels, with 148 up-regulations and 106 down-regulations (Gu et al. 2014). In another experiment with phoxim at 4 mg/L resulted in 266 genes with at least twofold changes and among them, 192 up-regulations and 74 down-regulations were reported. Particularly, the transcription levels of genes related to detoxification such as cytochrome P450s, esterases and glutathione S transferase (GST) were up-regulated, indicating increased detoxification activity in the midgut of pesticide treated silkworm (Gu et al. 2013). The pesticide detoxifying enzyme genes CYP6ae22, CYP9a21, GSTo1 and Bmcce were found to increase to 15.9, 3.3, 1.8 and 2.3-folds in the midgut and to 3.5, 1.8, 2.1 and 4.2-folds in the fat body after phoxim treatment in silkworm larva indicating the importance of detoxification (Wang et al. 2013). Silkworm larva when fed with 4 mg/L phoxim, the transcription levels of *Bm-AChE*-1 and *Bm-AChE*-2 in brain, fat body and silk gland increased initially and then decreased but found vice versa in the midgut. The increase in *Bm-AChE*-1 and *Bm-AChE*-2 in the brain after 24 h of exposures was 16.22 and 68.71-fold, respectively (Peng et al. 2011). *BmGSTe8* showed a late up-regulation of transcripts at 24–42 h after exposure of silkworm larva to phoxim and may contribute for phoxim resistance in silkworm (Yu et al. 2011).

4 Methods to Assess Pesticide Toxicity to Silkworms

Silkworm is affected by pesticides in many ways. So, many techniques and methods are developed to assess the pesticide toxicity to silkworms. Techniques to assess the acute toxicity through both contact and ingestion routes are given below. Tests to evaluate the sublethal effects of pesticides affecting the growth, metamorphosis and reproduction of silkworm, effects on morphology, physiology and biochemistry especially of hemolymph are also described. The methods to estimate the metabolic enzymes, antioxidant enzymes and pesticides degrading enzymes along with tests to assess toxicity in the gene level are also described hereunder.

4.1 Acute Toxicity

4.1.1 Ingestion Toxicity Tests

4.1.1.1 Leaf Dip Bioassay (Stanley et al. 2016)

In this experiment, third instar silkworm larvae were taken and fed with pesticide contaminated food. In general, formulated products of pesticides are diluted in water to have specific concentrations of pesticide solutions to be tested. Leaves of mulberry are plucked, surface sterilized and dipped in the pesticide solution for about 10 s. and shade dried using pins hung in a thread. After complete drying, leaves were put in bioassay trays and 10 prestarved (2 h) larvae released for each treatment and replicated four times and thus 40 larvae being tested for a treatment. Observations on mortality were taken at 6, 12, 24 and 48 h after treatment and percent mortality worked out to be analyzed for median lethal concentration.

Leaf dip bioassay though used extensively because of its simplicity has some limitations especially with respect to its utility for pesticide risk assessments (Candolfi et al. 2001). The result of leaf dip bioassay is expressed as quantity of a.i. per unit of pesticide solution, rather than per unit weight of leaves, while silkworm larvae are exposed to a quantity of pesticides on the mulberry leaves and not directly to the spray solution. It is also difficult to quantify the pesticide in the mulberry

Fig. 4.1 Leaf dip bioassay in trays

leaves (Chi et al. 2015) or the amount taken up by the larva in leaf dip bioassays (Fig. 4.1).

4.1.1.2 Spray Tower/Quantitative Spraying for Leaf Treatment (Sun et al. 2012; Chi et al. 2015)

It is reported that the quantitative spraying method as more effective than the leaf-dipping method in precision and reproducibility. Many of the disadvantages of leaf dip bioassay are taken care off by this quantitative spray bioassay (Chi et al. 2015). Fresh stock solutions of insecticides were prepared from technical grade pesticides mixed in distilled water or acetone and subsequently diluted with distilled water for each test. The Potter's spray tower was calibrated to apply a constant volume of pesticide solution using water sensitive papers before spraying on leaves. A spray solution of 2 mL with a spray pressure of 7 bar and a droplet deposition time of 20 s. were adopted as standard application parameters. Mulberry leaves were individually weighed before and after the pesticide application to get the exact quantity of applied pesticide on each leaf. Leaves were air dried for 10 min. before putting in exposure cages.

Two or three treated mulberry leaves were placed flat in the exposure cage, with the treated side up, covering the entire bottom surface and silkworm larvae introduced. In general, two replicates, each of 30 silkworm larvae, were exposed per dose. Five or six doses including control were prepared for each toxicity test, in an approximately geometric series with a factor of about 1.5 and not exceeding 2.25.

Exposure cages: In this experiment, two different types of exposure cages viz, open type and closed ventilated exposure cages were used. The open type is a simple glass box kept open and the closed ventilated cage has a provision of forced ventilation through a suction pump. A thin layer of agar was kept in the cage to reduce desiccation effect on the mulberry leaves. Larval mortality was recorded at 24 h after the start of exposure and thereafter at 48, 72 and 96 h and the 96 h mortality was taken for calculating the median lethal concentration. The dose to which the silkworms were exposed was expressed as concentration on the mulberry leaves (mg a.i./kg of leaf). Mortality data were analyzed by log-logistic regression and median lethal concentrations calculated as per Schroer et al. (2004).

4.1.1.3 Leaf Disc for Individual Larva

In this experiment, leaf discs are made out of mulberry leaves and provided to the silkworm larva instead of whole leaf. These leaf discs were dipped in specific insecticide solution and dried before given as feed. The air dried discs were then given to silkworm larva individually in a bioassay container. Though this method is originally used for the study of anti-feedant effects, it can be used for assessing acute toxicity of pesticides also.

Unlike fed ad libitum in leaf dip bioassay or leaf treatment by spray tower described above, this method gives an idea of the food intake by individual larva. Thus, the larva which is not found to feed properly can be excluded from the observation and analysis of results.

4.1.1.4 Leaf Sandwich Method (Campbell and Filmer 1929; Campbell 1930; Woke 1938)

This method employs leaf sandwiches having known quantity of pesticide in dust form uniformly distributed between the two layers. Leaf discs of specific size (22 mm dia.) are coated lightly with starch paste on the smooth side. Specific amount of pesticide was distributed evenly over the paste and the second disc was kept to get adhered with the first disc with the pesticide between. Prestarved fourth instar silkworm larvae were allowed to feed on the sandwich and mortality noted. The mortality observations were used to find the median lethal dose of the pesticide to silkworms. The advantage of this method is by measuring the leaf area consumed, the dose/amount of pesticide taken by the insect can be determined. Dose taken by the insect is proportional to the area eaten out of the total disc area. The disadvantage being more time consuming and also the difficulty in precise determination of the dose consumed. Only the powder or solid formulations of pesticides are used appropriately in this method.

4.1.1.5 Pesticide Mixed in Artificial Diet (Asano et al. 1984)

The artificial diet for silkworms are prepared and mixed with the insecticide solutions prepared in acetone. The diet is kept in ventilation for the acetone to get evaporated and fed the test insect, here newly molted fourth instar silkworm larva. The artificial diet containing various concentrations of each test chemical was given for about 24 or 48 h to the larva and then they were transferred onto a fresh artificial diet without test chemicals. Observations may be taken after prescribed intervals as 12, 24, 48, 72 h after treatment.

Artificial diet bioassay eliminates the uncertainty of the ingredients in the test food and thus more reliable but not realistic as that of using mulberry leaves unless the silkworm rearing is done using artificial diets. Artificial diet preparation for *B. mori* is given by many authors like Chang et al. (1972), Tsai et al. (1978), and Cappellozza et al. (2005). Fukuda (1963) has tried and reported an artificial diet for eri silkworm, *Samia cynthia*.

4.1.1.6 Pesticide Directly Given in Mouth (Campbell 1926)

In this experiment, a specially made burette is used to deliver the measured amount of pesticide drops precisely in the mouth of the larva. Silkworm larvae of fifth instar (2–3 days old after fifth instar moulting) weighing around 2 g were selected as test organism. A constant volume of 5 mm^3/g was taken as a dose and fed to the larva. The high sensitivity of the burette make it possible to feed volumes as little as 0.01 g to silkworms. Holding the larva with the head and ventral aspect up between the thumb and forefinger of the left hand and with the platinum needle in the right hand is the ideal procedure to make the drops to be transferred from the burette to the mouthparts of the caterpillar. After full imbibition of the pesticide solution, the caterpillar was placed with a fresh mulberry leaf and monitored for symptoms and mortality.

This technique of providing pesticide solution directly into the mouth of the silkworm larvae was being practiced earlier. Owing to the difficulties in handling and being an unrealistic exposure route in silkworms, this procedure is neither well accepted nor widely used. But the silkworm larva with advantages of submitting passively for handling, drinks readily drops given at mouth and normally do not regurgitate make it suitable for this method.

4.1.1.7 Injection Bioassay (Campbell 1926)

For injection toxicity tests to be carried out, a syringe with small bore needle is required. It should allow a quantitative injection of relatively minute volumes of liquid into caterpillars without loss of blood. In this experiment, a glass microinjection pipette was used for this purpose. The same volume of oral administration as described earlier i.e., 5 mm^3/g was injected on the larva puncturing the integument.

The easiness in handling silkworm larva makes it possible to use this injection method for them. The speed of pesticide action time is calculated as follow and plotted against dosage in a graph to compare with the other methods of toxicity testing like injection.

$$\text{Speed of toxic action} = \frac{100}{\text{Survival time (minutes)}}$$

The speed of toxic action is the speed of lethal effect of the chemical as determined by the reciprocal of the time which elapses from the administration of chemical to the disappearance of a definite response in the dying silkworm. The speed of toxic action depends not only on the velocity of chemical reactions leading directly to death but also on the rate of distribution, excretion, cell penetration etc.

4.1.2 Contact Toxicity Assays

4.1.2.1 Direct Contact: Topical Application (Leonardi et al. 1996; Miranda et al. 2002)

In this experiment, the silkworms attaining the specific stage is selected and topically applied with insecticide and effect observed. The pesticides were diluted in ethanol to specific concentrations and 1 μL applied on the larval thoracic tergum, ideally on the mesothorax by using 100 μL micro-syringe controlled by a micro-applicator. Control larvae were treated only with the solvent. All the larvae were fed with mulberry leaves or artificial diet and toxicity effects/symptoms observed at specific intervals. Mortality was observed generally at 24 and 48 h and analyzed to arrive at the median lethal dose.

4.1.2.2 Indirect Contact: Filter Paper Bioassay (Zhu et al. 2006a; Ji et al. 2010)

Filter paper bioassay is also used to find the acute contact toxicity of insecticides to silkworm larva. In this bioassay, the filter papers are treated with pesticides at specific concentrations and dried. The silkworms generally of third instar larvae were exposed to the filter paper by contact for specific time of 1, 10, 30 or 60 min. After the exposure time, the larvae are placed in fresh mulberry leaves for food. Mortality observations at specific time intervals are taken and analysed to arrive at the median lethal concentration of that pesticide.

4.2 *Sublethal Toxicity*

4.2.1 Food Utilization and Growth (Kumutha 2010)

Mulberry leaves were taken and dipped in pesticide solution at respective concentrations for 20 s, shade dried and given to third instar silkworm larva. About 20 starved larvae (for half an hour) were used per treatment and replicated for three times. The treated leaves were given as feed to the experimental larvae only once and after that normal feeding schedule followed. Live larval weight, larval dry weight, dry weight of unfed leaves, dry weights of fecal pellets were taken. To obtain dry weight of larva, 5 larva from a batch of 20 were sacrificed and oven dried at each instar stage. The following measurements and calculations at every instar levels third, fourth and fifth were used to find the food utilization pattern.

$$\text{Food consumption}\left(\text{mg dry weight}\right) = \text{Initial weight of leaves} - \text{Weight of leaves consumed}$$

$$\text{Assimilation}\left(\text{mg dry weight}\right) = \text{Food consumed} - \text{Weight of fecal matter}$$

$$\text{Total growth}\left(\text{mg dry weight}\right) = \text{Final body weight} - \text{Initial body weight of larvae}$$

$$\text{Approximate digestibility}\left(\%\right) = \frac{\text{Assimilation}}{\text{Consumption}} \times 100$$

$$\text{Tissue growth efficiency}\left(\%\right) = \frac{\text{Tissue growth}}{\text{Assimilation}} \times 100$$

$$\text{Ecological growth efficiency}\left(\%\right) = \frac{\text{Tissue growth}}{\text{Consumption}} \times 100$$

4.2.2 Growth, Spinning and Pupation: Exposure Through Artificial Diet (Asano et al. 1984)

In this experiment, 1-citronellyl 5 phenylimidazole, antijuvenile compound was dissolved in acetone to make different concentrations and mixed with the artificial diet, Silkmate-1-M. Acetone was allowed to get evaporated from the artificial diet and 2 % agar added to solidify the diet. Control diet was prepared with only exception of not adding the anti juvenile compound. The third instar larvae were fed through out the instar period with this diet. Each concentration was tested on 20 larvae in total with 5 larvae per treatment and with 4 replications. The larvae which did not develop into next instar in 5 days were transferred to feed without test compound until spinning time. The body weight of the larva was also measured on daily basis. The spinning stage of the larva, pupation and adult emergence were recorded.

The relative growth rate (RGR) can be calculated according to Hoo and Fraenkel (1966).

$$\text{Growth rate} = \frac{\text{Larval weight gained}}{\text{Average larval weight} \times \text{Feeding period (days)}}$$

where, the average larval weight is the arithmetic mean of initial and final weights.

4.2.3 IGRs on Growth and Cocoon Quality (Vassarmidaki et al. 2000)

In this experiment, IGR at different concentrations were prepared and mulberry leaves dipped in the solutions before being fed to larvae (Fig. 4.2). The larvae were fed with the treated leaves only on the first day of each instar and all other days fed with untreated leaves. Each larva received only 1 treatment in its life cycle before pupation. Here the treatment is to check the effect of IGR treatment on different instars of silkworm larva besides different doses of IGR. The treated leaves were fed to larvae on the first day of each larval stage, just after moulting because at that time the larvae are more sensitive. Observations on pre-cocooning parameters, such as larval duration and larval weight and post-cocooning parameters, such as cocoon weight, shell weight, sericin and fibroin content were made. Larval duration and larval weight were recorded just before the commencement of spinning activity at

Fig. 4.2 Leaf contamination bioassay for IGRs

the end of the last instar. The silk proteins, fibroin and sericin content of the cocoon was estimated by treating the shells with 0.5 % cold KOH for 6 h and washing it in hot water (Muthukrishnan et al. 1978; Narayanaprakash et al. 1985).

4.2.4 Anti Juvenile Hormone Activity on Metamorphosis (Asano et al. 1986)

Dietary exposure: The anti JH activity pesticide is mixed with artificial diet, air dried and fed to fourth instar silkworm larvae for a period of 24 h. The larva was observed for anti JH activities.

Topical exposure: The anti JH compound was dissolved in acetone and used for the experiment. The pesticide was applied to the fourth instar silkworm larva topically, using a micro syringe on the thoracic dorsum.

Oral exposure: The anti JH compound was dissolved in a mixture of methanol and distilled water in a ratio of 1:1 and orally administered to the fourth instar test larva using the micro syringe.

Observations on the induction of precocious metamorphosis in the test larva was made based on the appearance of spinning and the number of cocoons made before or after fourth instar larva in each treatment. Normally the silkworm larva spin cocoon at the end of fifth instar.

4.2.5 Effect on Fecundity and Hatching (Kuribayashi 1981a)

In this experiment, newly ecdyzed fifth instar larvae (25 male and 25 female) were used as an experimental group. The technical grade pesticides were made as water suspensions and water insoluble compounds dissolved in very small quantity of ethanol. These pesticide solutions were diluted in tap water as required concentration (LC_0 to LC_{50}) and sprayed on both sides of the freshly collected leaves at the rate of 100 mL per kg leaf. This pesticide treated leaves were given to the fifth instar silkworm larva. The coccons formed were kept for adult emergence. Surviving moths were allowed to cross with their own treatment group and eggs subjected to hatching. Embryonic development was monitored and normal and abnormal eggs recorded by examining the colour. Colourless and squashed eggs were recorded as non-fertilized, brown coloured indicated death at early embryonic stage, blue but unhatched eggs were recorded as dead just before hatching.

4.2.6 Hormones on Egg Laying and Embryo Development (Riddiford and Williams 1967)

Injection Bioassay This method is used to study the effect of hormones on the egg and embryonic development of silk moths. Injection method of testing is administered in tasar moth, *Antheraea pernyi* and cecropia silk moth, *Hyalophora cecropia*. Moths were injected just below the spiracle of the second abdominal segment by means of a 27-gauge hypodermic needle in conjunction with a 1 mL syringe and kept under observation for mating behaviour. Eggs laid by the individual moth are collected and observed for hatching.

Topical Application on Eggs Micro drops of synthetic hormone were applied to *H. cecropia* and *A. pernyi* eggs within the first 4 h after oviposition. The eggs were dorsally dosed with droplets of specific volume (0.32 μL for *H. cecropia* eggs and 0.64 μL for *A. pernyi* eggs) of the test compound dispensed from a 30 gauge needle sealed to a 0.1 mL micro syringe. Control eggs were dosed with peanut oil. The eggs were observed for hatching and unhatched eggs were dissected in Ringer's solution to study the embryonic development. Hatched larvae were fed with the respective feed i.e., oak for *A. pernyi* and wild cherry for *H. cecropia* and abnormalities if any noted.

4.2.7 Effect on Morphology: Midgut Epithelium (Munhoz et al. 2013)

Mulberry leaves were treated with insecticides in definite concentrations and given to silkworm larva at every instar to find the mortality and effect on midgut morphology. Apart from observing various symptoms caused by insecticide poisoning, the fifth larva was dissected for studies on insecticide induced effects in midgut morphology. On the fourth day after exposure, larvae were anesthetized and dissected for midgut segments. Segments of the midgut wall were removed, fixed and preserved in paraffin. From this, 5 and 7 μm thick sequential microtome sections were taken. Sections were stained, viewed under microscope and images obtained for comparison with control.

4.2.8 Effect on the Biochemistry and Enzyme Activities of Hemolymph (Etebari et al. 2007)

Fresh mulberry leaves were soaked in specific pesticide concentrations for 10 s and then air dried. Treated leaves were fed to the silkworm larva only once and the control group fed with untreated leaves. About 20 larva were selected after 24 and 120 h after treatment and hemolymph (0.5 mL/larva) extracted by cutting through the prolegs. The hemolymph thus extracted was centrifuged and supernatant taken and used for biochemical estimations. The following methods are used to estimate the biochemical parameters of hemolymph as given by different authors.

Protein:	Lowry et al. (1951)
Glucose:	Siegert (1987)
Uric acid using uricase:	Valovage and Brooks (1979)
Total cholesterol:	Richmond (1973)
Alanine aminotransferase:	Thomas (1998)
Aspartate aminotransferase:	Thomas (1998)
Alkaline phosphatase:	Mihara et al. (1988)
Acid phosphatase:	Zhou and Chen (1994) given in Zhan et al. (2006)
Catalase activity:	Zhou and Chen (1994) given in Zhan et al. (2006)
Lactate dehydrogenase activity:	Reddanna and Govindappa (1979)

4.2.9 Effect on Enzymes and Metabolism (Nath 2000, 2003)

In these experiments, fourth instar silkworm larvae were fed with insecticides at LD_{50} and one-fifth of LD_{50} until the end of fifth instar (spinning stage). Then larva was frozen at −20 °C and cut open along the dorsal mid line in ice-cold bombyx saline and fat bodies removed within 1 min., in order to avoid any loss of enzyme activity. The fat bodies were rapidly transferred into the medium for determination of metabolites and enzyme activities. The enzymes and metabolites in fat bodies and hemolymph were measured using the following methods.

Glycogen metabolism	
Glucose content:	Nelson and Somogyi (1944)
Trehalose content:	Schmidt and Platzer (1980)
Glycogen content:	Carroll et al. (1956)
Glycogen phosphorylase activity:	Tolman and Steele (1980)
Inorganic phosphate:	Fiske and Subbarow (1925)
Trehalase activity:	Sacktor and Wormser-Shavit (1966)
Carbohydrate metabolism	
Pyruvate content:	Friedemann and Haugen (1942)
Lactate levels:	Barker and Summerson (1941) modified by Huckabee (1961)
Lactate dehydrogenase activity:	Srikantan and Krishnamurti (1955)
Succinate dehydrogenase and Malate dehydrogenase activities:	Nachlas et al. (1960)

4.2.10 Effect on Antioxidant Enzymes, Oxidation and Peroxidation (Phugare et al. 2013)

Mulberry leaves were dipped in insecticide solutions at specific concentrations and given as feed to fourth instar silkworm larva. At the appropriate stage (end of fifth instar), silkworm was sacrificed and the whole midgut was excised from larvae in ice-cold buffer solution. The contents were homogenized gently under cold conditions, centrifuged and the supernatant used as the enzyme source. Superoxide dismutase activity, catalase activity, lipid peroxidation and protein oxidation were estimated as per Achary et al. (2008).

Superoxide dismutase activity:	Beauchamp and Fridovich (1971)
Catalase activity:	Aebi (1983)
Lipid peroxidation:	Dhindsa et al. (1981)
Protein oxidation:	Levine et al. (1994)

4.2.11 Effect on Pesticide Detoxifying Enzymes (Wang et al. 2013)

The insecticide treated group and control group larva were dissected on ice for the isolation of midgut and fat body tissues. The homogenized tissues were centrifuged and used for enzyme activity determination.

Cytochrome P450 activity:	Rose et al. (1995) and Yang et al. (2004)
Glutathione-S-transferase activity:	Habig (1981)
Esterase activity:	Asperen (1962)
Total protein:	Bradford (1976)

4.2.12 Genotoxicity of Pesticides (Shen et al. 2011)

Hemocytes are said to play an important role in the compensatory mechanisms of silkworm during growth and provide about 55 % energy during the pupation (Qian 1995). In this experiment, hemocytes of silkworm with or without pesticide treatment were collected to estimate DNA damage by alkaline single-cell gel electrophoresis (SCGE) and differentially expressed genes (DEGs) by annealing control primer (ACP) based PCR method. Third instar worms are selected and fed with mulberry leaves soaked in respective pesticide concentrations for 10 s and air dried. The worms fed with untreated mulberry leaves were considered as control. About 15 larvae randomly selected for each treatment at day 1, 3, 5, 7, 9 after the first feeding of pesticide treated mulberry leaves were used to collect hemocytes by cutting a proleg. Hemocytes collected were centrifuged and pellets transferred and washed with phosphate-buffered saline (Shi et al. 2001). The collection was tested for cell survival and placed immediately on ice for the alkaline single-cell electrophoresis (comet assay).

Alkaline Single-Cell Gel Electrophoresis/Comet Assay This assay was performed according to the three layer procedure (Singh et al. 1988; Klaude et al. 1996) with some modifications. The cell suspensions mixed with low melting point agarose were layered on a plain glass slide previously coated with a layer of normal agarose. After solidifying, a third layer of low melting-point agarose layer was made. The cells were lysed with high salt solution and placed in gel electrophoresis for DNA unwinding. After electrophoresis, the slides were washed with neutralization buffer, stained and examined under fluorescence microscope. Approximately 100 cells per slide were selected and counted for DNA damage i.e., tail DNA.

Isolation of Differentially Expressed Genes and Analysis RNAs from the larval hemocytes were isolated from the fourth instar treated and control silkworm larva. The RNAs were used to synthesis the first strand of cDNA and ACP-based PCR performed. PCR products were separated, strained and differentially expressed bands identified. Differentially expressed bands were extracted, sequenced and analyzed.

4.3 Field Studies

4.3.1 Pesticides Sprayed on Mulberry on Larval Mortality at Different Seasons (She et al. 2012)

In this experiment, pesticides were sprayed at three different concentrations in mulberry trees in the field itself. The leaves were taken in different time intervals i.e., 1 h after spraying, 1, 3, 5, 7 and 10 days after spraying and fed to the silkworm larva (third instar). Mortality of the treated larva was assessed at specific time intervals after feeding. The experiment was carried out in three different seasons (summer, autumn and spring) to know the seasonal effect on the insecticidal toxicity to silkworm.

4.3.2 Pesticide Sprayed on Mulberry on Growth and Cocoon Formation (Hui et al. 2006)

In this experiment, the pesticides were sprayed on mulberry plants in the field and leaves taken daily to find the persistent toxicity to the silkworm larva. Studies on larval developmental duration, larval weight, cocoon weight, cocoon shell weight, pupa weight and pupation rate in pesticide treatments in comparison to control was also made. This method of study reveals the acute as well as the sublethal toxic effects of pesticides to silkworm.

5 Pesticide Risk Assessment for Silkworm

Silkworm is regarded as a representative of non-target organisms for pesticide safety evaluation in China (Wang 2008). It is necessary to assess the risk of pesticide to silkworm whenever there is a possibility of pesticide exposure to mulberry or other host plants. Pesticide entry through food contamination is considered as the major route of exposure to silkworms and thus used in risk assessments. Toxicity studies when coupled with exposure predictions forms a realistic risk assessment. But acute toxicity and mortality alone cannot be useful to assess the risk of pesticides since sublethal effects are also as important as mortality, in case of pesticide risk. Reliable toxicity tests resulting in relevant endpoints are an essential for assessing the pesticide risk (Sun et al. 2012). Many standard guidelines are available for toxicity testing, but have a number of limitations especially with respect to their utility in risk assessments (Candolfi et al. 2001).

A method of comparing the susceptibility of silkworm with that of pests of mulberry by means of selectivity ratio is considered as a useful approach for assessing the risk. A simple comparison of acute toxicity values of mulberry pests and silkworm to find which one is more susceptible as that of Hui et al. (2006) and Bai et al. (2011) also holds good for the purpose. Comparison of toxicity of pesticides to silkworm and its predators and parasitoids as given by Khanikor (2011) can also be used for risk assessment of pesticide usage for the management of silkworm pests. Sometime, even the acute toxicity values of a pesticide for different beneficial organisms are compared and pesticides categorized based on comparison with other beneficials (Gong et al. 1999, 2001).

Pesticide risk assessment by calculating the predicted environmental concentration (exposure) and predicted no effect concentration (toxicity) can be the most realistic approach. Some pesticides have different mode of toxicity such as imidacloprid with ingestion, contact and fumigation toxicity to silkworm larva (Lu and Wu 2000). So a sequential testing scheme to find the acute toxicity (both ingestion and contact), fumigant, persistent and field toxicity to silkworm larva gives the complete picture of the pesticide toxicity assessing the full risk. Some of the pesticide risk assessment methods used for silkworms are given hereunder.

5.1 Risk Assessment Methodologies

5.1.1 Risk Assessment Based on Acute Toxicity (Si et al. 2007; Chi et al. 2015)

Based on the values of median lethal concentrations, pesticides are classified as low, medium, highly and extremely toxic. The LC_{50}s of emamectin benzoate and β cypermethrin were lower than 0.5 mg/L and denoted as extremely toxic. Insecticides, molosultap and methoxyfenozide had LC_{50} values as 0.54 and 0.60 mg/L, respectively and thus categorized as highly toxic pesticides. The LC_{50} value of chlorfenapyr and diafenthiuron was reported as 72.37 and 27.82 mg/L, respectively and thus belong to medium toxic range (Si et al. 2007).

In another experiment, pesticides were classified based on acute toxicity data (LC_{50} value) having differences in the orders of magnitude. Among the five insecticides tested on silkworm through leaf dip bioassay and quantitative spraying assay, emamectin benzoate and lambda cyhalothrin belonged to the first magnitude with LC_{50} values of 0.0005 and 0.0008 mg a.i/L in leaf dip method is regarded as highly toxic. Imidacloprid and chlorpyrifos registered an LC_{50} value of 0.75 and 1.39 mg a.i./L and can be classified as medium toxic chemicals. Dimethoate is reported as the least toxic insecticide with mean LC_{50} values 815.1 mg a.i./L in leaf dip method and 1320.6 mg a.i./kg in quantitative spray method (Chi et al. 2015).

5.1.2 Toxicity to Insect Pest and Silkworm: A Comparison

This method of risk assessment take into account of the selectivity of the pesticide and compares the susceptibility of insect pest of mulberry or other host plant with that of silkworm. This method of risk assessment is being done in many cases like pollinators (Stanley et al. 2010), predators (Tanaka et al. 2000; Preetha et al. 2010) and parasitoids (Sengonca and Liu 2001) comparing with the pest insect. Since those beneficial insects mentioned above (pollinator, predator, parasitoid) are present in the same environment along with the pest while pesticide application, this method of toxicity assessment is well applicable. Though mulberry silkworms will not be generally reared in open environment, this method suits for *B. mori* also to assess the risk of pesticide application on mulberry and to decide whether to feed the worms with the leaves after application or not. The only difference with the other beneficials mentioned above is in the quantification of pesticide. Since silkworm is not exposed directly to pesticides but through contaminated leaf, quantitative spary or leaf dip procedure is to be followed for testing the toxicity of pesticides to silkworm and mulberry pest and thus susceptibility compared. It suits very well to silkworms grown directly in field conditions. This risk assessment methodology can also be used for the comparison of toxicity of silkworm and its pests like uzi fly, stink bug etc. This method of toxicity assessment is determined by selectivity ratio and probit substitution method as described under.

(a) **Selectivity ratio**

The selectivity ratio is the ratio of acute toxicity value of the beneficial to that of the pest. This comparison shows which of the both (either the silkworm or the pest) is more susceptible to the pesticide. The selectivity ratio is calculated as follows:

$$\text{Selectivity ratio} = \frac{LC_{50}\text{ of silkworm}}{LC_{50}\text{ of the pest}}$$

Values of 1 and <1 indicate that the pesticide as nonselective to silkworm. Values of >1 indicate that the pesticide is less harmful to silkworm than the pest.

(b) Probit substitution method

This method is used to determine the relative toxicities to beneficial species (silkworm) at particular levels of pest mortality, say LC_{90} of the pest (Preetha et al. 2010):

$$y = 5 + m\left[x - (\log LC_{50} \text{ of beneficial species})\right]$$

Where,

'y' is the probit value
'm' is the slope of the probit line for the silkworm
'x' is the log of the fiducial limits for LC_{90} of the pest species

Solving for y gives a probit value that is then converted to the percentage mortality using a conversion table (Finney 1971). The pesticide is considered to be selective if it kills less than 90 % of silkworm at the concentration that kills 90 % of the pest.

5.1.3 Ratio of PEC and PNEC

The predicted environmental concentration (PEC) is defined as expected concentration of the pesticide in the environment (here, mulberry leaf), taking into account the nature of the chemical, concentration applied, time of application, waiting period and the degradation pattern or persistence of the chemical. This estimated environmental concentration is the one which the silkworm is being exposed. Predicted no effect concentration (PNEC) is the concentration at or below which is not expected to cause adverse effects to species in the environment (here, silkworm). This PNEC is calculated through toxicity bioassays in the laboratory at specific conditions. If the ratio of PEC and PNEC is more than one, that indicate a risk and vice versa. For a safe pesticide, the PEC should be less than PNEC, so that no effect will be there to silkworm at the environmental concentration.

5.1.4 Sequential Testing (Chen et al. 2010)

Sequential testing of pesticide for risk assessment involves studies on acute, persistent and field toxicity. In this experiment, insecticide chlorantraniliprole was tested on silkworm, *B. mori* by means of contact, ingestion and fumigation toxicity in combination with field experiments. Acute toxicity tests were carried out to find median lethal concentration (LC_{50}) and dose (LD_{50}). Fumigant toxicity by fumigating the rearing chambers with the pesticide was also conducted. The final being the field toxicity, in which the field recommended dose of pesticide is sprayed on the plants in the field itself and tested for it's persistent toxicity (mortality) on the larva at regular intervals. This method of risk assessment takes into consideration of all the possible means of pesticide exposure and toxicity to the test organism and thus a realistic approach.

5.1.5 Calculation of Acute Toxicity, Strain Sensitivity Distribution and HC_5 (Sun et al. 2012)

In this experiment, median lethal concentration of insecticides was studied through dietary exposure. The mulberry leaves were applied with insecticides at specific concentrations using a Potter's tower and allowed for the insects to feed with. The experiment was conducted with different strains of silkworms (six strains) and thus the strain sensitivity distributions were calculated like that of species sensitivity distribution (SSD). The SSD is a cumulative distribution function estimated from a sample of toxicity data, in which potentially affected fraction of species is plotted against the LC_{50} values of those species (Posthuma et al. 2002). On the basis of the SSD, a median HC_5 value was then estimated, which is the value where 5 % of the tested silkworm strains are expected to have an LC_{50} below this value. This method of risk assessment reveals the strain specific toxicity and reveals the susceptible and resistant strains.

5.2 *Risk of Pesticide Exposure on Silkworm*

Chlorfenapyr sprayed at field was not found to cause significant difference in larval developmental duration, larval weight, cocoon weight, cocoon shell weight, pupal weight and pupation rate as compared to control (Hui et al. 2006). Chlorfenapyr was found to be safer to silkworm, *B. mori* and *B. mandarina* but highly toxic to mulberry pests (Bai et al. 2011). The selective toxicity of chlorfenapyr, diazinon, dichlorvos, phoxim and triazophos to silkworm and mulberry pest viz., the mulberry looper moth (*Phthonandria atrilineata*), mulberry pyralid (*Diaphania pyloalis*) and mulberry yellow tail moth (*Porthesia xanthocampa*) was studied using leaf dip bioassay. Chlorfenapyr was highly toxic to mulberry pest and least toxic to silkworm and found safer than other pesticides and thus can be used in mulberry pest management (Hui et al. 2006).

For assessing the risk of pesticide usage for the management of silkworm pests, a comparison of pesticide toxicity to silkworm and its natural enemies like predators and parasitoids can be made. Though deltamethrin is highly effective against silkworm parasitoid, *Exorista sorbillans* is extremely toxic to muga silk worm, *Antheraea assama* larva and thus cannot be used (Khanikor 2011). Imidacloprid and brofluthrinate were found to be extremely toxic to silkworms in comparison with many other beneficial organisms and thus pollution of these pesticides in mulberry plantations is to be prohibited (Gong et al. 1999, 2001).

In a sequential testing of risk assessment, the median lethal concentration of chlorantraniliprole was 4.0×10^{-3} mg/L and median lethal dose 1.1×10^{-8} mg/larva for the third instar newly exuviated larvae. No toxicity was observed when chlorantraniliprole was used to fumigate the silkworm rearing chambers. Mulberry leaves sprayed with 1.25×10^{-4} mg/L chlorantraniliprole solution could cause 100 % mortality of the third and fifth instar larvae even after 60 days and thus highly toxic in field conditions (Chen et al. 2010).

References

V.M.M. Achary, S. Jena, K.K. Panda, B.B. Panda, Aluminium induced oxidative stress and DNA damage in root cells of *Allium cepa* L. Ecotoxicol. Environ. Saf. **70**, 300–310 (2008)

H. Aebi, Catalase in vitro. Methods Enzymol. **105**, 121–126 (1983)

M. Ahmed, M.S. Islam, Q. Ahsan, M.M. Islam, Silk fibers and their unidirectional polymer composites. Adv. Mater. Sci. **2**, 79–90 (2012)

H. Akai, K. Kiguchi, Increased accumulation of silk protein accompanying JH induced prolongation of larval life in *Bombyx mori* L. (Lepidoptera: Bombycidae). Appl. Entomol. Zool. **6**(4), 218–220 (1971)

H. Akai, M. Kobayashi, Induction of prolonged larval instar by the juvenile hormone in *Bombyx mori*. Appl. Entomol. Zool. **6**, 138–139 (1971)

P. Aramwit, S. Palapinyo, T. Srichana, S. Chottanapund, P. Muangman, Silk sericin ameliorates wound healing and its clinical efficacy in burn wounds. Arch. Dermatol. Res. **305**(7), 585–594 (2013)

T. Asakura, M. Kitaguchi, M. Demura, H. Sakai, K. Komatsu, Immobilization of glucose oxidase on nonwoven fabrics with *Bombyx mori* silk fibroin gel. J. Appl. Polym. Sci. **46**(1), 49–53 (1992)

S. Asano, E. Kuwano, M. Eto, Anti-juvenile hormone activity of 1-citronellyl-5-phenylimidazole in the 3rd instar silkworm, *Bombyx mori* L. (Lepidoptera: Bombycidae). Appl. Entomol. Zool. **19**(2), 212–220 (1984)

S. Asano, E. Kuwano, M. Eto, Anti-juvenile hormone activity of imidazole compound (kk-22) and its diminution by methoprene in the 4th instar silkworm, *Bombyx mori* L.: Lepidoptera: Bombycidae. Appl. Entomol. Zool. **21**(1), 63–69 (1986)

K.V. Asperen, A study of housefly esterases by means of a sensitive colorimetric method. J. Insect Physiol. **8**, 401–416 (1962)

O.M. Assal, R. Benedetti, L. Cappellozza, S. Cappellozza, The nutritional behaviour of three different strains of mulberry silkworm (*Bombyx mori*) in relation Insegar (fenoxycarb). Sericologia **34**, 233–243 (1994)

X. Bai, S. Feng, M. Wu, Determination on indoor toxicity of chlorfenapyr to major mulberry pests and silkworm and investigation on its selective toxicity in field application. Sci. Seric. (2011). Retrieved from http://en.cnki.com.cn/Article_en/CJFDTOTAL-CYKE201104030.htm

S.B. Barker, W.H. Summerson, The colorimetric determination of lactic acid in biological material. J. Biol. Chem. **138**, 535–554 (1941)

D. Baycin, E. Altiok, S. Ulku, O. Bayraktar, Adsorption of olive leaf (*Olea europaea* L.) antioxidants on silk fibroin. J. Agric. Food Chem. **55**, 227–236 (2007)

C. Beauchamp, I. Fridovich, Superoxide dismuatse: improved assays and an assay applicable to acrylamide gels. Anal. Biochem. **44**, 276–287 (1971)

A.N. Begum, S.M. Moorthy, S. Venkat, S.N. Kumar, S.M.H. Qadri, Influence of juvenile hormone analogue, methoprene on the biochemical changes and economic characters of silkworm, *Bombyx mori* L. Intl. J. Plant Anim. Environ. Sci **1**(3), 171–178 (2011)

A. Bhagyalakshmi, S.V. Reddy, R. Ramamurthi, P.S. Reddy, Studies on the effect of hexachlorocyclohexane on the growth and silk qualities of silkworm, *Bombyx mori* L. Chem. Ecol. **11**(2), 97–104 (1995)

C. Boado, Importance of silk (2013). Retrieved from http://www.artofsilk.com/blogs/news/6565365-importance-of-silk#.VbzbhvOqooI

D. Bora, B. Khanikor, H. Gogoi, Plant based pesticides: green environment with special reference to silkworms. In *Pesticides: Advances in Chemical and Botanical Pesticides,* ed. R.P. Soundararajan, (InTech, Rijeka, 2012) pp. 171–206

M.M. Bradford, A rapid and sensitive method for the quantization of microgram quantities of protein utilizing the principle of protein-dye binding. Anal. Biochem. **72**, 248–254 (1976)

F.L. Campbell, Speed of toxic action of arsenic in the silkworm. J. Gen. Physiol. **9**(4), 433–443 (1926)

F.L. Campbell, A comparison of four methods for estimating the relative toxicity of stomach poison insecticides. J. Econ. Entomol. **23**(2), 357–370 (1930)

F.L. Campbell, R.S. Filmer, A quantitative method of estimating the relative toxicity of stomach-poison insecticides. 4th Intl. Cong. Entomol. Ithaca **2**, 523–533 (1929)

M.P. Candolfi, K.L. Barrett, P.J. Campbell et al., *Guidance Document on Regulatory Testing and Risk Assessment Procedures for Plant Protection Products with Non-target Arthropods* (Society of Environmental Toxicology and Chemistry (SETAC), Europe, 2001)

T. Cang, X. Zhao, C. Wu, S. Wu, L. Chen, Q. Wang, Z. Zhang, Toxicity and safety evaluation of emamectin benzoate on four types of non-target organisms. Agrochemicals (2007). Retrieved from http://en.cnki.com.cn/Article_en/CJFDTOTAL-NYZZ200707018.htm

L. Cappellozza, Action of fenoxycarb on 'non-spinning syndrome' on silk worm in North-Eastern Italy (1994). Retrieved from http://agris.fao.org/agris-search/search.do?recordID=BE9400981

L. Cappellozza, S. Cappellozza, A. Saviane, G. Sbrenna, Artificial diet rearing system for the silkworm, *Bombyx mori* (Lepidoptera: Bombycidae): effect of vitamin C deprivation on larval growth and cocoon production. Appl. Entomol. Zool. **40**(3), 405–412 (2005)

N.V. Carroll, R.W. Longley, J.H. Roe, The determination of glycogen in liver and muscle by use of anthrone reagent. J. Biol. Chem. **220**, 583–593 (1956)

C. Chang, S. Murakoshi, S. Tamura, Giant cocoon formation in the silkworm, *Bombyx mori* L., topically treated with methylene dioxy phenyl derivatives. Agric. Biol. Chem. **36**(4), 692–694 (1972)

Z. Chen, C. Mi, A study on the quality of granulated pesticide shachong shuang. J. Nanjing Agric. Univ. (1984). Retrieved from http://en.cnki.com.cn/Article_en/CJFDTOTAL-NJNY198404005.htm

L. Chen, R. Gong, R. Chen, Study on toxicity and safety evaluation of triazophos on three types of non-target organisms. Pestic. Sci. Admin. (1998). Retrieved from http://en.cnki.com.cn/Article_en/CJFDTOTAL-NYKG802.004.htm

W. Chen, R. Dong, J. Dai, H. Sun, Toxicity determination of azadirachtin to silkworm and detoxification method. Sci. Seric. (2008a). Retrieved from http://en.cnki.com.cn/Article_en/CJFDTOTAL-CYKE200802019.htm

W. Chen, J. Dai, H. Ma, H. Sun, Y. Qian Effective and residue tests of imine parathion 20% EC to silkworm (2008b). Retrieved from http://en.cnki.com.cn/Article_en/CJFDTOTAL-CSTB200801009.htm

W.G. Chen, R.H. Dong, H.Y. Sun, J.Z. Dai, H.L. Zhu, F.A. Wu, An investigation on toxicity of the agricultural pesticide chlorantraniliprole to the silkworm, *Bombyx mori*. Sci. Seric. **1**, 84–90 (2010)

Y. Chi, K. Qiao, H. Jiang, R. Lin, K. Wang, Comparison of two acute toxicity test methods for the silkworm (Lepidoptera: Bombycidae). J. Econ. Entomol. **108**(1), 145–149 (2015)

Y. Chisti, Strategies in downstream processing, in *Bioseparation and Bioprocessing: A Handbook*, ed. by G. Subramanian, vol. 2 (Wiley-VCH, New York, 1998), pp. 3–30

S.S. Chitgupekar, K. Basavanagoud, Selective toxicity of chlorfenapyr 10 SC to mulberry silkworm, *Bombyx mori* (L.). Karnataka J. Agric. Sci. **27**(4), 534–535 (2014)

G. Choi, M. Jo, S. Moon, S. Lim, A. Jung, Y. Yoon, H. Paik, Neuroprotection effects and physicochemical characteristics of milk fortified with fibroin BF-7. Korean J. Food Sci. Anim. Res. **28**, 431–436 (2008)
S.K. Chowdhary, F. Sehnal, S.K. Raj, P.S. Raju, S. Mathu, Giant cocoon formation in *Bombyx mori* L. topically treated with juvenile hormone SJ-42-F. Sericologia **26**, 455–459 (1986)
Y. Cui, Y. Yang, J.X. Yuan, The problems and solving methods of pesticide use in mulberry orchards. China Seric. **24**, 48–49 (2003)
P.K. Das, R. Balakrishna, V.B. Mathur, K. Vedavyasa, S.B. Dandin, Studies on the performance of improved mulberry variety Sahana as intercrop under coconut plantation. Indian J. Seric. **49**(2), 199–202 (2010)
S.G. Dedos, H. Fugo, Induction of dauer larvae by application of fenoxycarb early in the 5th instar of the silkworm, *Bombyx mori*. J. Insect Physiol. **45**, 769–775 (1999)
S.K. Dewangan, K.R. Sahu, K.V. Achari, Sericulture: a tool of eco-system checking through tribal. J. Environ. Res. Develop. **6**(1), 165–173 (2011)
S.K. Dewangan, K.R. Sahu, S.K. Soni, Breaking of poverty through sericulture among the tribe – a socio-economic study of Dharamjaigarh block of Raigarh Dist. India. Res. J. Recent Sci. **1**, 371–374 (2012)
S.K. Dewangan, K.R. Sahu, S.K. Patnayak, S. Soni, Sericulture for employment generation among the tribal- a study of two tribal blocks of Raigarh. India. Intl. Sch. Sci. Res. Innov. **7**(7), 1621–1624 (2013)
S.K. Dhimmar, Effect on growth and yield of *rabi* castor in pulses intercropping under varying planting geometry. Am.-Eurasian J. Sci. Res. **4**(3), 165–168 (2009)
R.S. Dhindsa, P. Plumb-Dhindsa, T.A. Thorpe, Leaf senescence: correlated with increased levels of membrane permeability and lipid peroxidation and decreased levels of superoxide dismutase and catalase. J. Exp. Bot. **32**, 93–101 (1981)
J. Ding, Y. Wu, H. Zheng, W. Fu, R. Reardon, M. Liu, Assessing potential biological control of the invasive plant, tree-of-heaven, *Ailanthus altissima*. Biocontrol Sci. Technol. **16**(5&6), 547–566 (2006)
A. Dutta, S. Dutta, S. Kumari, Growth of poultry chicks fed on formulated feed containing silk worm pupae meal as protein supplement and commercial diet. Online J. Anim. Feed Res. **2**(3), 303–307 (2012)
K. Etebari, A.R. Bizhannia, R. Sorati, L. Matindoost, Biochemical changes in haemolymph of silkworm larvae due to pyriproxyfen residue. Pestic. Biochem. Physiol. **88**, 14–19 (2007)
I. Fagoonee, Possible growth factors for chickens in silkworm pupae meal. Br. Poult. Sci. **24**(3), 295–300 (1983)
D.J. Finney, *Probit Analysis*, 3rd edn. (Cambridge University Press, Cambridge, UK, 1971). 333 p
C.H. Fiske, Y. Subbarow, The colorimetric determination of phosphorus. J. Biol. Chem. **66**(2), 375–400 (1925)
T.E. Friedemann, G.E. Haugen, Pyruvic acid. I. Collection of blood for the determination of pyruvic acid and lactic acids. J. Biol. Chem. **144**, 67–77 (1942)
T. Fu, Study on the anti-bacteria capability of adding tussah silk fibroin powder to the ordinary cosmetics. J. Anhui Agric. Sci. (2006), http://en.cnki.com.cn/Article_en/CJFDTOTAL-AHNY200608085.htm
T. Fukuda, A semi-synthetic diet for eri-silkworm raising. Agric. Biol. Chem. **27**(9), 601–609 (1963)
J.P. Garel, Endocrinological aspects of silk production. Experientia **19**, 461–466 (1983)
G.R. Garthwaite, *The Persians* (Blackwell Publishing Ltd., Oxford, 2005). 78 p
P. Ghosh, P. Pal, Influence of nitrogen and potassium on growth, yield and oil content of *Kaempferia galanga* L. J. Spice Aroma. Crops **11**(1), 64–66 (2002)
G.K. Ghosh, S. Ghosh, Indian textiles: past and present, (APH Pub., 1995) 306 p
R. Gong, R. Chen, L. Chen, Evaluation on toxicity and safety of imidacloprid to environmental organisms. Pestic. Sci. Admin. (1999). Retrieved from http://en.cnki.com.cn/Article_en/CJFDTOTAL-NYKG903.007.htm

R. Gong, R. Chen, L. Chen, D. Cai, Safety evaluation of brofluthrinate to environmental organisms. Chin. J. Pestic. Sci. **2**, 67–72 (2001)

K. Gotoh, H. Izumi, T. Kanamoto, Y. Tamada, H. Nakashima, Sulfated fibroin, a novel sulfated peptide derived from silk, inhibits human immunodeficiency virus replication in vitro. Biosci. Biotechnol. Biochem. **64**, 1664–1670 (2000)

D.S.B. Govindaiah, P. Tomy, R.K. Datta, Effect of marigold (*Tagetes patula*) intercropping against *Meloidogyne incognita* infecting mulberry. Indian J. Nematol. **21**(1), 96–99 (1991)

E.M. Grabstein, J.M. Scriber, The relationship between restriction of host plant consumption, and post-ingestive utilization of biomass and nitrogen in *Hyalophora cecropia*. Entomol. Exp. Appl. **31**(2&3), 202–210 (1982)

Z. Gu, The study on the differentially expressed genes in the midgut of domestic silkworms (*Bombyx mori*) exposed to phoxim (2013). Retrieved from www.dissertationtopic.net/doc/1800697

Z.Y. Gu, S.S. Sun, Y.H. Wang et al., Transcriptional characteristics of gene expression in the midgut of domestic silkworms (*Bombyx mori*) exposed to phoxim. Pestic. Biochem. Physiol. **105**, 36–43 (2013)

Z. Gu, Y. Zhou, X. Yi et al., The adverse effects of phoxim exposure in the midgut of silkworm, *Bombyx mori*. Chemosphere **96**, 33–38 (2014)

M.L. Gulrajani, Sericin: a bio-molecule of value. Souvenir 20th Cong. Int. Seric. Comm., Bangalore, India 15–18th Dec., (2005) pp. 21–29

J. Guo, Aesthetic characteristics of Shanghai Qipao in Chinese women's dress culture. Intl Conf. Education, language, art and intercultural communication (2014). Retrieved from www.atlantis-press.com/php/download_paper.php?id=12576

R. Gupta, Tasar culture showed them the way. Indian Silk **46**(10), 20–21 (2008)

D. Gupta, H. Chaudhary, C. Gupta, Sericin-based polyester textile for medical applications. J. Text Inst. **106**(4), 366–376 (2015)

W.H. Habig, Assays for differentiation of glutathione S-transferase, in *Method in Enzymology*, ed. by B.J. William (Academic, New York, 1981), pp. 398–405

H. Hamamoto, A. Tonoike, K. Narushima, R. Horie, K. Sekimizu, Silkworm as a model animal to evaluate drug candidate toxicity and metabolism. Comp. Biochem. Physiol. C **149**, 334–339 (2009)

S.C.F. Hoo, G. Fraenkel, The consumption, digestion and utilization of food plants by a polyphagous insect, *Prodenia eridania* (Cramer). J. Insect Physiol. **12**, 711–730 (1966)

W.E. Huckabee, Relationship of pyruvate and lactate during anaerobic metabolism. Am. J. Physiol. **200**, 1169–1176 (1961)

M. Hui, K. Wang, H. Wang, S. Si, Selective toxicity of chlorfenapyr to silkworm and mulberry pests. Acta Entomol. Sin. **49**(4), 599–603 (2006)

S. Izumi, K. Kigochi, S. Tomino, Hormonal regulation of biosynthesis of major plasma proteins in *Bombyx mori*. Zool. Sci. **1**, 223–228 (1984)

J. Ji, B. Xiao, Y. Li, W. Yu, F. Liu, Toxicity assessment of two different formulations of chlorpyrifos to four environmental organisms. J. Agro-Environ. Sci. (2010). Retrieved from http://en.cnki.com.cn/Article_en/CJFDTOTAL-NHBH201009011.htm

M.S. Jolly, S.K. Sen, T.N. Sonwalker, G.K. Prasad, *Non Mulberry Silks* (Food and Agriculture Organization of United Nations, Rome, 1979)

B. Joseph, S.J. Raja, Therapeutic applications and properties of silk proteins from *Bombyx mori*. Front. Life Sci. **6**(3&4), 255–260 (2012)

M. Kamimura, M. Kiuchi, Effects of a juvenile hormone analog, fenoxycarb, on 5th stadium larvae of silkworm *Bombyx mori* (Lepidoptera: Bombycidae). Appl. Entomol. Zool. **33**, 333–338 (1998)

N. Kato, S. Sato, A. Yamanaka, H. Yamada, N. Fuwa, M. Nomura, Silk protein, sericin, inhibits lipid peroxidation and tyrosinase activity. Biosci. Biotechnol. Biochem. **62**(1), 145–147 (1998)

B. Khanikor, Evaluation of extracts and essential oils of Ocimum and Ageratum against uzi fly, *Exorista sorbillans* (Wiedemann), a parasitoid of *Antheraea assama* Westwood. Ph.D. Thesis (Dibrugarh University. Dibrugarh, 2011)

Y.S. Kim, Effects of diflubenzuron on the development of *Bombyx mori* (Lepidoptera: Bombycidae). Korean J. Entomol. **32**, 99–106 (2002)

Y.S. Kim, H.R. Sohn, Effect of diflubenzuron on silk gland development of silkworm, *Bombyx mori* (Lepidoptera: Bombycidae). Korean J. Seric. Sci. **43**, 77–82 (2001)

K. Kim, Y. Kim, Y. Kim, A biochemical evidence of the inhibitory effect of diflubenzuron on the metamorphosis of the silkworm, *Bombyx mori*. J. Asia Pac. Entomol. **5**(2), 175–180 (2002)

M. Kirikawa, T. Kasaharu, K. Kishida, D. Akiyama, Silk protein micropowders for coating with excellent feeling, antistaticity and moisture absorbability and releasability and there manufacture. Chem. Abstr. **132**, 153320 (2000)

M. Klaude, S. Eriksson, J. Nygren, G. Ahnstrom, The comet assay: mechanisms and technical considerations. Mutat. Res. **363**, 89–96 (1996)

A.M. Kordy, Residual effect of certain pesticides on the mulberry silkworm (*Bombyx mori* L.). Middle East J. Appl. Sci **4**(3), 711–717 (2014)

P. Kumutha, C.P. Latha, S.S.M. Doss, A.J.A.R. Singh, Toxicity evaluation of neem oil and metacid on the development of the silk worm, *Bombyx mori* L. Uttar Pradesh J. Zool. **29**(4), 220–222 (2009)

P. Kumutha, Impacts of mulberry crop pest control agents on the silkworm *Bombyx mori*. Thesis, (Manonmaniam Sundaranar University, 2010). Retrieved from http://ietd.inflibnet.ac.in/handle/10603/678

P. Kumutha, C. Padmalatha, K. Chairman, A.J.A.R. Singh, Effect of pesticides on the reproductive performance and longevity of *Bombyx mori*. Intl. J. Curr. Microbiol. Appl. Sci. **2**(9), 74–78 (2013)

S. Kuribayashi, Studies on the effect of pesticides on the reproduction of silkworm, *Bombyx mori* L. I. Effect of chemical administered during larval stage on egg laying and hatching. J. Toxicol. Sci. **6**, 169–176 (1981a)

S. Kuribayashi, Studies on the effect of pesticides on the reproduction of silkworm, *Bombyx mori* L. II. Ovicidal action of organophosphorous insecticides administered during the larval stage. Appl. Entomol. Zool. **16**, 423–431 (1981b)

K. Lee, H. Kweon, J. Yeo, S. Woo, S. Han, J. Kim, Characterization of tyrosine-rich *Antheraea pernyi* silk fibroin hydrolysate. Int. J. Biol. Macromol. **48**, 223–226 (2011)

M.G. Leonardi, S. Cappellozza, P. Ianne, L. Cappellozza, P. Parenti, B. Giordana, Effects of the topical application of an insect growth regulator (fenoxycarb) on some physiological parameters in the fifth instar larvae of the silkworm *Bombyx mori*. Comp. Biochem. Physiol. B **113**, 361–365 (1996)

M.G. Leonardi, S. Cappellozza, M. De-Eguileor, A. Grimaldi, L. Cappellozza, M. Casartelli, B. Giordana, Modification of the nutritional parameters and of midgut biochemical and absorptive functions induced by the IGR fenoxycarb in *Bombyx mori* larvae. Arch. Insect Biochem. Physiol. **39**, 18–35 (1998)

M.G. Leonardi, P. Marciani, P.G. Montorfano, S. Cappellozza, B. Giordana, G. Monticelli, Effect of fenoxycarb on leucine uptake and lipid composition of midgut brush border membrane in the silkworm, *Bombyx mori* (Lepidoptera, Bombycidae). Pestic. Biochem. Physiol. **70**, 42–51 (2001)

R.L. Levine, J.A. Willams, E.R. Stadtman, E. Shacter, Carbonyl assay for determination of oxidatively modified proteins. Methods Enzymol. **233**, 346–363 (1994)

B. Li, Toxicity of six insecticides to silkworm and their safety evaluation. Chin. J. Pestic. Sci. (2001). Retrieved from http://en.cnki.com.cn/Article_en/CJFDTOTAL-NYXB200103014.htm

Y. Li, W. Liu, R. Shi, Z. Hu, K. Yi, Toxicity test of pesticide fenoxycarb to larvae of the silkworm, *Bombyx mori*. Sci. Seric (2009). Retrieved from http://en.cnki.com.cn/Article_en/CJFDTOTAL-CYKE200904036.htm

B. Li, Y. Wang, H. Liu, Y. Xu, Z. Wei, Y. Chen, W. Shen, Resistance comparison of domesticated silkworm (*Bombyx mori* L.) and wild silkworm (*Bombyx mandarina* M.) to phoxim insecticide. Afr. J. Biotechnol **9**(12), 1771–1775 (2010a)

H. Li, W. Huang, J. Xie, B. Qin, R. Wei, Three organophosphate insecticides on mulberry thrips control and their residual toxicities on silkworm. Agrochemicals. (2010b). Retrieved from http://en.cnki.com.cn/Article_en/CJFDTOTAL-NYZZ201009018.htm

B. Li, R. Hu, Z. Cheng et al., Titanium dioxide nanoparticles relieve biochemical dysfunctions of fifth-instar larvae of silkworms following exposure to phoxim insecticide. Chemosphere **89**(5), 609–614 (2012)

Z. Liu, S. Gu, K. Wang, S. Si, X. Shi, Determination of residual toxicity and degradation dynamics of chlorfenapyr in mulberry leaves. Sci. Serict. (2008). Retrieved from http://en.cnki.com.cn/Article_en/CJFDTOTAL-CYKE200804003.htm

T. Longvah, K. Mangthya, P. Ramulu, Nutrient composition and protein quality evaluation of eri silkworm (*Samia ricinii*) prepupae and pupae. Food Chem. **128**(2), 400–403 (2011)

M. Lovett, C. Cannizzaro, L. Daheronc, B. Messmer, G. Vunjak-Novakovic, D.L. Kaplana, Silk fibroin microtubes for blood vessel engineering. Biomaterials **28**(35), 5271–5279 (2007)

H. Lowry, N.L. Rosebrough, A.L. Far, R.J. Randal, Protein measurement with folin phenol reagent. J. Biol. Chem. **193**, 265–275 (1951)

X. Lu, Y. Wu, Toxicity of imidacloprid to silkworm. Acta Sericol. Sin. (2000). Retrieved from http://en.cnki.com.cn/Article_en/CJFDTOTAL-CYKE200002003.htm

H. Ma, H. Jiang, C.J. Tao, L. Liu, K.Y. Wang, Toxicity evaluation of twenty-seven pesticides to *Bombyx mori*. China J. Pestic. Sci. **7**, 156–159 (2005)

H. Ma, K.Y. Wang, H.Y. Wang, S.D. Si, Selective toxicity of chlorfenapyr to silkworm and mulberry pests. Acta Entomol. Sin. **49**, 599–603 (2006)

M.S. Malik, P. Kaushal, R.B. Sah, *Socioeconomic Upliftment of Tribal Communities in Jharkhand Through Agroforestry Based Farming System* (Birsa Agrl. University, Ranchi, 2008). Retrieved from www.yumpu.com/en/document/view/12039847/socio-economic-upliftment-of-tribal-communities-jharkhand-forest

D.M. Mamatha, K. Vijaya, H.P.P. Cohly, M.R. Rao, Juvenile hormone analogues, methoprene and fenoxycarb dose- dependently enhance certain enzyme activities in the silkworm, *Bombyx mori* (L). Intl. J. Environ. Res. Publ. Health **5**, 120–124 (2008)

V.G. Mathirajan, S. Ragauraman, Feeding toxicity of acetamiprid 20 SP to silkworm, *Bombyx mori* L. Entomology **28**(1), 61–62 (2003)

A. Matta, C. Migliaresi, F. Faccioni, P. Torricelli, M. Fini, R. Giardino, Fibroin hydrogels for biomedical applications, preparation, characterization and in vitro cell culture studies. J. Biomater. Sci. Polym. **15**, 851–864 (2004)

L. Meinel, V. Karageorgiou, S. Hofmann, Engineering bone-like tissue in vitro using human bone marrow stem cells and silk scaffolds. J. Biomed. Mat. Res. A **71**(1), 25–34 (2004)

Y.G. Miao, Studies on the activity of the alkaline phosphatase in the midgut of infected silkworm, *Bombyx mori* L. J. Appl. Entomol. **126**, 138–142 (2002)

Y. Mihara, A. Saito, K. Koga, B. Sakaguchi, Changes in alkaline phosphatase activity during embryogenesis. J. Seric. Sci. Jpn. **52**, 62–67 (1988)

J.H. Miller, Ailanthus altissima, in *Silvics of North America: Social Indicators*, vol. 2, (Department of Commerce, Bureau of the Census, Center for Demographic Studies, 1990). pp. 101–104

J.E. Miranda, S.A. Bortoli, R. Takahashi, Development and silk production by silkworm larvae after topical application of methoprene. Sci. Agric. **59**, 585–588 (2002)

N. Mishra, N.C. Hazarika, K. Narain, J. Mahanta, Nutritive value of non-mulberry and mulberry silkworm pupae and consumption pattern in Assam. India. Nutr. Res. **23**(10), 1303–1311 (2003)

A. Miyajima, J. Schreurs, K. Otsu, A. Kondo, K. Arai, S. Maeda, Use of the silkworm, *Bombyx mori* and an insect baculovirus vector for high-level expression and secretion of biologically active mouse interleukin-3. Gene **58**(2&3), 273–281 (1987)

T. Miyashita, Sweat and sebum absorbing cosmetics containing cellulose fibres. Chem. Abstr. **131**(2), 23271 (1999)

K. Mizoguchi, T. Iwatsubo, N. Aisaka, Separating membrane made of cross-linked thin film of sericin and production thereof (1991), Japan Patent. 03- 284337A

M. Mohananasundaram, R. Sivagami, Occurrence of the bud mite, *Aceria mori* Keifer (Eriophyidae: Acari.) on *Morus alba*, the mulberry plant in Tamil Nadu, its economic significance and control. Indian J Seric. **21**, 21–22 (1983)
H. Monconduit, B. Mauchamp, Effects of ultra low doses of fenoxycarb on juvenile hormone regulated physiological parameters in the silkworm, *Bombyx mori*. Arch. Insect Biochem. Physiol. **37**(2), 178–189 (1998)
H. Monconduit, B. Mauchamp, Fenoxycarb, a potent contaminant of the silkworm, *Bombyx mori* L., does not influence its juvenile hormone titer. Arch. Insect Biochem. Physiol. **40**(3), 141–149 (1999)
K.C. Mondal, R.N. Singh, Environmental protection and pest management in tasar culture. Indian Silk **29**, 15–16 (1990)
M. Mondal, K. Trivedy, S.N. Kumar, The silk proteins, sericin and fibroin in silkworm, *Bombyx mori* Linn. – a review. Caspian J. Env. Sci. **5**(2), 63–76 (2007)
H. Mori, M. Tsukada, New silk protein: modification of silk protein by gene engineering for production of biomaterials. Rev. Mol. Biotechnol. **74**(2), 95–103 (2000)
R.E.F. Munhoz, T.S. Bignotto, N.C. Pereira et al., Evaluation of the toxic effect of insecticide chlorantraniliprole on the silkworm *Bombyx mori* (Lepidoptera: Bombycidae). Open J. Anim. Sci. **3**, 343–353 (2013)
M. Murase, Method for solubilizing and molding cocoon silk, artificial organ made of cocoon silk and medical element made of cocoon silk (1994), Japan Patent 06-166850A
J. Muthukrishnan, S. Mathavan, J.V. Navarathina, Effects of the restriction of feeding duration on food utilization, emergence and silk production in *Bombyx mori* L. (Lepidoptera: Bombycidae). Monit. Zool. Ital. **12**, 1287–1294 (1978)
M.M. Nachlas, S.I. Margulies, A.M. Seligman, A colorimetric method for the estimation of succinic dehydrogenase activity. J. Biol. Chem. **235**, 499–503 (1960)
P.R. Naik, M.R. Delvi, Effects of insecticide, permethrin, on dietary water utilization in eri-silkworm *Philosamia ricini*. Proc. Anim. Sci. **93**(5), 497–504 (1984)
R. Narayanaprakash, K. Periasamy, S. Radhakrishnan, Effect of dietary water content on food utilization and silk production in *Bombyx mori* L. (Lepidoptera: Bombycidae). Indian J. Seric. **24**, 12–17 (1985)
K.C. Narayanaswamy, M. Geethabai, R. Raghuraman, Mite pests of mulberry. Indian J. Seric. **35**(1), 1–8 (1996)
B.S. Nath, Changes in carbohydrate metabolism in hemolymph and fat body of the silkworm, *Bombyx mori* L., exposed to organophosphorus insecticides. Pestic. Biochem. Physiol. **68**, 127–137 (2000)
B.S. Nath, Shifts in glycogen metabolism in hemolymph and fat body of the silkworm, *Bombyx mori* (Lepidoptera: Bombycidae) in response to organophosphorus insecticides toxicity. Pestic. Biochem. Physiol. **74**, 73–84 (2003)
B.S. Nath, R.P.S. Kumar, Toxic impact of organophosphorus insecticides on acetylcholinesterase activity in the silkworm, *Bombyx mori* L. Ecotoxicol. Environ. Saf. **42**(2), 157–162 (1999)
B.S. Nath, Studies on organophosphorus insecticide toxicity during 5th instar of silkworm, *Bombyx mori* L. Ph.D. thesis, (Sri Krishnadevaraya University, Anantapur, 1993)
B.S. Nath, A. Suresh, B.M. Varma, R.P.S. Kumar, Changes in protein metabolism in hemolymph and fat body of the silkworm, *Bombyx mori* (Lepidoptera: Bombycidae) in response to organophosphorus insecticides toxicity. Ecotoxicol. Environ. Saf. **36**(2), 169–173 (1997)
N. Nelson, M. Somogyi, A photometric determination of Somogyi method for the determination of glucose. J. Biol. Chem. **151**, 375–380 (1944)
H. Ni, H.X. Chen, Y.Y. Yang, Integrated utilized technique of *Antheraea pernyi* pupae. J. Hubei Univ. **25**, 263–266 (2003)
M. Okuda, S. Sakurai, T. Ohtaki, Activity of the prothoracic gland and its sensitivity to prothoracicotropic hormone in the penultimate and last-larval instar of *Bombyx mori*. J. Insect Physiol. **31**, 455–461 (1985)
F.G. Omenetto, D.L. Kaplan, A new route for silk. Nat. Photonics **2**, 641–643 (2008)

M.N. Padamwar, A.P. Pawar, Silk sericin and its applications: a review. J. Sci. Indust. Res. **63**, 323–329 (2004)
M.N. Padamwar, A.P. Pawar, A.V. Daithankar, K.R. Mahadik, Silk sericin as a moisturizer: an *in vivo* study. J. Cosmet. Dermatol. **4**(4), 250–257 (2005)
P. Padmavathi, C.V. Raghavaiah, Productivity and returns of castor (*Ricinus communis*) based intercropping systems with pulses and vegetables under rainfed conditions. Indian J. Agric. Sci. **74**(5), 235–238 (2004)
P. Panwar, S. Pal, S. Chakravarty, M. Alam, Soil quality and production of low land paddy under agrisilviculture systems in acid soil of West Bengal. India. Range Manag. Agroforest. **34**(1), 51–57 (2013)
K. Park, B. Kim, Y. Park et al., Relationship between the non-spinning syndrome of silkworm, *Bombyx mori* and pesticides. Korean J. Pestic. Sci. **11**(4), 238–245 (2007)
N.K. Patle, A. Sharma, N.S. Patil, 2003. Growth and production of drilled paddy under *Terminalia arjuna* and *Mitragyna parvifolia* in an agrisilvicultural system. Proc. Int. Conf. World perspective on short rotation forestry for industrial and rural development, 7–13 Sept., 2003, Nauni, Solan, India
G.D. Peng, J.M. Wang, L. Mab, Y.H. Wang, Y.Q. Cao, W.D. Shen, B. Li, Transcriptional characteristics of acetylcholinesterase genes in domestic silkworms (*Bombyx mori*) exposed to phoxim. Pestic. Biochem. Physiol. **101**, 154–158 (2011)
N.R. Pereira, O. Ferrarese-Filho, M. Matsushita, N.E. De-Souza, Proximate composition and fatty acid profile of Bombyx mori L. chrysalis toast. J. Food Comp. Anal. **16**(4), 451–457 (2003)
S.S. Phugare, D.C. Kalyani, Y.B. Gaikwad, J.P. Jadhav, Microbial degradation of imidacloprid and toxicological analysis of its biodegradation metabolites in silkworm (*Bombyx mori*). Chem. Eng. J. **230**, 27–35 (2013)
M.K. Porwal, S.K. Agarwal, A.K. Khokhar, Effect of planting methods and intercrops on productivity and economics of castor (*Ricinus communis*) based intercropping systems. Indian J. Agron. **51**(4), 274–277 (2006)
L. Posthuma, G.W. Suter, T.P. Traas, *Species Sensitivity Distributions in Ecotoxicology* (Lewis Publishers, London, 2002). 616 p
G. Preetha, J. Stanley, S. Suresh, R. Samiyappan, Risk assessment of insecticides used in rice on miridbug, *Cyrtorhinus lividipennis* Reuter, the important predator of brown planthopper, *Nilaparvata lugens* (Stal.). Chemosphere **80**(5), 498–503 (2010)
H.T. Qian, *Silkworm of Physio-Biochemistry* (China Agriculture Press, Beijing, 1995), pp. 73–79
S. Rai, K.K. Aggarwal, C.R. Babu, Influence of host plant (*Terminalia arjuna*) defences on the evolution of feeding behaviour in the tasar silkworm. Curr. Sci. **91**(1), 68–72 (2006)
R. Rajendran, C. Balakumar, R. Sivakumar, T. Amruta, N. Devaki, Extraction and application of natural silk protein sericin from *Bombyx mori* as antimicrobial finish for cotton fabrics. J. Text. Inst. **103**(4), 458–462 (2012)
G. Rangaswamy, M.N. Narasimhanna, K. Kasiviswanathan, C.R. Sastry, *Sericulture Manual and Mulberry Cultivation* (Food and Agriculture Organization of United Nations, Rome, 1976)
P. Reddanna, S. Govindappa, Effect of in vitro muscular stimulation: II. Influence on hepatic carbohydrate metabolism. J. Anim. Morphol. Physiol. **26**, 156–161 (1979)
G.P. Reddy, S.C. Rao, R. Reddy, Mixed cropping in castor. Indian Oilseeds J. **9**(4), 310–316 (1965)
S. Rhee, D. Kim, S. Oh, The effect of cocoon silk fibroin hydrolyzate (CSFH) on shelf-life extension of kimchi during fermentation. Agric. Chem. Biotechnol. **40**(6), 541–545 (1997)
G. Ricci, I. Neri, L. Ricci, A. Patrizi, Silk fabrics in the management of atopic dermatitis. Skin Therapy Lett. **17**(3), 5–7 (2012)
W. Richmond, Preparation and properties of cholesterol oxidase from *Nocardia* sp. and its application to enzymatic assay of total cholesterol in serum. Clin. Chem. **19**, 1350–1356 (1973)
L.M. Riddiford, Cellular and molecular actions of juvenile hormone. I. General considerations and premetamorphic actions. Adv. Insect Physiol. **24**, 213–274 (1994)
L.M. Riddiford, C.M. Williams, The effects of juvenile hormone analogues on the embryonic development of silkworms. Proc. Natl. Acad. Sci. U. S. A. **57**(3), 595–601 (1967)

R.L. Rose, L. Barbhaiya, R.M. Roe, G.C. Rock, E. Hodgson, Cytochrome P450 associated insecticide resistance and the development of biochemical diagnostic assays in *Heliothis virescens*. Pestic. Biochem. Physiol. **51**, 178–191 (1995)

B. Sacktor, E. Wormser-Shavit, Regulation of metabolism in working muscle *in vivo*. I. Concentrations of some glycolytic, tricarboxylic acid cycle and amino acid intermediates in insect flight muscle during flight. J. Biol. Chem. **241**(4), 624–631 (1966)

H. Sakabe, H. Ito, T. Miyamoto, Y. Noishiki, W.S. Ha, *In vivo* blood compatibility of regenerated silk fibroin. Sen'i Gakkaishi **45**(11), 487–490 (1989)

S. Sakurai, H. Imokawa, Development arrest induced by juvenile hormone in larvae of *Bombyx mori*. Arch. Insect Biochem. Physiol. **8**, 219–228 (1988)

D.C. Sarkar, *Ericulture in India*, 2nd edn. (Central Silk Board, Bangalore, 1988). 51 p

M. Sasaki, N. Kato, H. Watanabe, H. Yamada, Silk protein sericin suppresses colon carcinogenesis. Oncol. Rep. **7**(5), 1049–1052 (2000a)

M. Sasaki, H. Yamada, N. Kato, Consumption of silk protein, sericin elevates intestinal absorption of zinc, iron, magnesium and calcium in rats. Nutr. Res. **20**(10), 1505–1511 (2000b)

S. Savalgi, G.M. Patil, Studies on growth and yield of mulberry and chilli intercropping system. Karnataka J. Agric. Sci. **16**(4), 609–611 (2003)

S.P. Schmidt, E.G. Platzer, Changes in body tissues and hemolymph composition of *Culex pipiens* in response to infection by *Romanomermis culicivorax*. J. Invert. Pathol. **36**(2), 240–254 (1980)

A.F.W. Schroer, D. Belgers, T.C.M. Brock, A. Matser, S.J. Maund, P.J. Van-den-Brink, Comparison of laboratory single species and field population-level effects of the pyrethroid insecticide λcyhalothrin on freshwater invertebrates. Arch. Environ. Contam. Toxicol. **46**, 324–335 (2004)

F. Sehnal, Prospects of the practical use of silk sericins. Entomol. Res. **38**(Suppl), 1–8 (2008)

C. Sengonca, B. Liu, Influence of mixed biocide GCSC-BtA on the pupae and adult stages of *Apanteles plutellae* Kurd. (Hym., Braconidae) and its host, *Plutella xylostella* (L.) (Lepidopter: Plutellidae). Anz Schadlingskunde **74**(6), 145–149 (2001)

L. She, W. Xue, Q. Lu, X. Qian, Survey on residual toxicity of insecticide chlorfenapyr to *Bombyx mori* after spraying in mulberry field at various seasons. Sci. Seric. (2012). Retrieved from http://en.cnki.com.cn/Article_en/CJFDTOTAL-CYKE201203040.htm

W. Shen, X. Zhao, Q. Wang et al., Genotoxicity evaluation of low doses of avermectin to hemocytes of silkworm (*Bombyx mori*) and response of gene expression to DNA damage. Pestic. Biochem. Physiol. **101**, 159–164 (2011)

J.M. Shi, C.X. Xie, W.J. Zhang, M.J. Shen, Effects of silkworm on serum protein, haemoglobin, glutamate-pyruvate transaminase and glucose. J. Mod. Appl. Pharm. **7**(2), 1–3 (1990)

L.G. Shi, B.X. Zhong, J.L. Xu, Defense response of hemocytes in the silkworm, *Bombyx mori* to *Beauveria bassiana*. J. Agric. Biotechnol. **9**, 383–386 (2001)

S. Si, K. Wang, R. Lin, H. Ma, Y. Li, Toxicity evaluation of twenty-five pesticides to *Bombyx mori* and observation of toxic symptoms. Sci. Seric. (2007). Retrieved from http://en.cnki.com.cn/Article_en/CJFDTOTAL-CYKE200703016.htm

K.J. Siegert, Carbohydrate metabolism in *Manduca sexta* during late larval development. J. Insect Physiol. **33**, 421–427 (1987)

R.N. Singh, B. Saratchandra, An integrated approach in the pest management in sericulture. Int. J. Ind. Entomol. **5**(2), 141–151 (2002)

R.N. Singh, K. Thangavelu, Parasites and predators of tasar silkworm, *Antheraea mylitta* and their control. Indian Silk **29**, 33–36 (1991)

R.N. Singh, K. Thangavelu, Incidence of insect pests on primary tasar food plant. Bull. Entomol. **35**, 36–39 (1994)

N.P. Singh, M.T. McCoy, R.R. Tice, E.L. Schneider, A simple technique for quantification of low levels of DNA damage in individual cells. Exp. Cell Res. **175**, 184–191 (1988)

R.N. Singh, K.C. Mandal, S.S. Sinha, Studies on the biology of *Blepharipa zebina* Walk. (Diptera: Tachinidae). Indian J. Seric. **32**, 15–19 (1993)

R.N. Singh, M.V. Samson, R.K. Datta, *Pest Management in Sericulture* (Indian Publishers Distributors, Delhi, 2000)

T.N. Srikantan, C.R. Krishnamurti, Tetrazolium test for dehydrogenases. J. Sci. Ind. Res. **14**, 206–207 (1955)

J. Stanley, Chemical and behavioural approaches for pest management in cardamom. Ph.D Thesis. (Tamil Nadu Agric. Univ., Coimbatore, 2007). 210 p.

J. Stanley, S. Chandrasekaran, G. Preetha, S. Kuttalam, Toxicity of diafenthiuron to honey bees in laboratory, semi-field and field conditions. Pest Manag. Sci. **66**, 505–510 (2010)

J. S. Stanley, Chandrasekaran, G. Preetha, S. Kuttalam, R.S. Jasmine, Selective toxicity of diafenthiuron to non-target organisms: honeybees, coccinellids, chelonus, earthworm, silkworm and fish. J. Plant Prot. Res. **56**(1), 1–5 (2016)

K. Sun, Q. Zhou, J. Zhou, F. Wan, X. Lu, Investigation on the toxicity of low dosage pyrethroids to the silkworm, *Bombyx mori.* Bull. Seric. (2002). Retrieved through http://en.cnki.com.cn/Article_en/CJFDTOTAL-CSTB200203007.htm

H. Sun, W. Chen, R. Dong and J. Dai, The toxicity test of the 10% EC pyriproxyfen on silkworm. Bull. Seric. (2008), Retrieved from http://en.cnki.com.cn/Article_en/CJFDTOTAL-CSTB200802007.htm

X. Sun, H. Van-Der-Valk, H. Jiang, X. Wang, S. Yuan, Y. Zhang, I. Roessink, X. Gao, Development of a standard acute dietary toxicity test for the silkworm (*Bombyx mori* L.). Crop Prot. **42**, 260–267 (2012)

K. Tanaka, S. Mizuno, Homologues of fibroin L-chain and P25 of *Bombyx mori* are present in *Dendrolimus spectabilis* and *Papilio xuthus* but not detectable in *Antheraea yamamai.* Insect Biochem. Mol. Biol. **31**, 665–677 (2001)

K. Tanaka, S. Endo, H. Kazano, Toxicity of insecticides to predators of rice planthoppers: spiders, the miridbug and the dryinid wasp. Appl. Entomol. Zool. **35**, 177–187 (2000)

P.P. Tarhalkar, N.G.P. Rao, Changing concepts and practices of cropping systems. Indian Farming **25**(3), 3–7 (1975)

K. Tasubouchi, Wound covering material. US patent 5951506 (1999). Retrieved from patentimages.storage.googleapis.com/pdfs/US5951506.pdf

Y. Tazima, *The Silkworm: An Important Laboratory Tool* (Kodansha Ltd, Tokyo, 1978). 307 p

A. Thangavelu, A.K. Chakraborty, A.K. Bhagowati, M. Issa, *Hand Book of Muga Culture* (Central Silk Board, Bangalore, 1988). 103 p

L. Thomas, *Clinical Laboratory Diagnostics* (TH Books Verlasgesellschaft, Frankfurt, 1998), pp. 89–94

J.H. Tolman, J.E. Steele, The control of glycogen metabolism in the cockroach hindgut: the effect of the corpora cardiaca–corpora allata system. Comp. Biochem. Physiol. B **66**(1), 59–65 (1980)

S. Tomino, Major plasma proteins of *Bombyx mori.* Zool. Sci. **2**, 293–303 (1985)

Y. Tsai, H. Wang, C. Ho, S. Li, Studies on the artificial diet of silkworm, *Bombyx mori* L.: the effects of dietary physico-chemical factors on the feeding and growth of the silkworm. Acta Entomol. Sin. **21**(4), 369–384 (1978)

M. Tsukada, G. Freddi, N. Minoura, G. Allara, Preparation and application of porous silk fibroin materials. J. Appl. Polym. Sci. **54**(4), 507–514 (1994)

W.D. Valovage, M.A. Brooks, Uric acid quantities in the fat body of normal and aposymbiotic german cockroaches, *Blattella germanica.* Ann. Entomol. Soc. Am. **72**, 687–689 (1979)

M.E. Vassarmidaki, P.C. Harizanis, S. Katsikis, Effects of applaud on the growth of silkworm (Lepidoptera: Bombycidae). J. Econ. Entomol. **93**(2), 290–292 (2000)

R. Voegeli, J. Meier, R. Blust, Sericin silk protein: unique structure and properties. Cosmet. Toilet **108**, 101–108 (1993)

N. Vyjayanthi, M.V.V. Subramanyam, Effect of fenvalerate 20 EC on sericigenous insects: I. Food utilization in the late-age larva of the silkworm, *Bombyx mori* L. Ecotoxicol. Environ. Saf. **53**(2), 206–211 (2002a)

N. Vyjayanthi, M.V.V. Subramanyam, Effect of fenvalerate 20 EC on sericigenous insects: II. Digestive enzymes in the nutritive physiology of silkworm, *Bombyx mori* L. Ecotoxicol. Environ. Saf. **53**(2), 212–220 (2002b)

J. Wang, Determination on toxin residual period of pesticides against silkworm in single and mingled applications on mulberry trees. J. Anhui Agric. Sci. (2010). Retrieved from http://en.cnki.com.cn/Article_en/CJFDTotal-AHNY201028077.htm

L.Z. Wang, X.G. Dong, Rearing of *Antheraea pernyi* and mountainous construction of ecology. Liaoning Silk **1**, 9–13 (2002)

L. Wang, Study on the toxicity and safety evaluation of pesticides to silkworm. J. Anhui Agric. Sci (2008). Retrieved from http://en.cnki.com.cn/Article_en/CJFDTOTAL-AHNY200829103.htm

J. Wang, D. Yin, G. Lu, F. Zhou, Effects of dimehypo (disodium 2-methylaminotrimethylene di thiosulfonate) on growth and cocooning of the silkworm, *Bombyx mori* (Lepidoptera: Saturnidae). Pestic. Sci. **55**(11), 1070–1076 (1999)

F. Wang, Y. Han, W. Chen, Study of application of silk fibroin as cosmetic additives. Chem. World. (2006), http://en.cnki.com.cn/Article_en/CJFDTOTAL-HXSS200609007.htm

Y.H. Wang, Z.Y. Gu, J.M. Wang, S.S. Sun, B.B. Wang, Y.Q. Jin, W.D. Shen, B. Li, Changes in the activity and the expression of detoxification enzymes in silkworms (*Bombyx mori*) after phoxim feeding. Pestic. Biochem. Physiol. **105**, 13–17 (2013)

B. Wei, G. Li, Z. Tang, Y. Liao, Q. Liang, Z. Ye, Analysis on the fitting curve equation of safe interval of DDVP (dichlorvos) for silkworms (2008). Retrieved from http://en.cnki.com.cn/Article_en/CJFDTOTAL-GCCY200802007.htm

P.A. Woke, The biological disposition of rotenone after ingestion by the southern armyworm. J. Agric. Res. **57**(9), 707–712 (1938)

S. Wu, Q. Wang, X. Zhao, C. Wu and L. Chen, Study on toxicity and safety evaluation of chlorpyrifos and fenpropathrin in silkworm (*Bombyx mori* L.). Pestic. Sci. Admin, (2003). Retrieved from http://en.cnki.com.cn/Article_en/CJFDTOTAL-NYKG200309004.htm

J.M. Wu, B.G. Chen, X.H. Kong, The reasons and of preventing of poisoning of silkworm. Jiangsu Seric. **2**, 16–17 (2006)

Q. Xia, J. Wang, Z. Zhou et al., The genome of a lepidopteran model insect, the silkworm, *Bombyx mori*. Insect Biochem. Mol. Biol. **38**(12), 1036–1045 (2008)

Xu, J., W. Zhu, J. Wang, Y. Chen Toxicity test on the effect of chlorpyrifos for mulberry on silkworm rearing. Bull. Seric. (2007). Retrieved from http://en.cnki.com.cn/Article_en/CJFDTOTAL-CSTB200704008.htm

H.P. Xu, Y.S. Xu, H.B. Wang, D. He, H. Kawasaki, Identification and functional characterization of the Rad23 gene of the silkworm, *Bombyx mori*. Biosci. Rep. **30**, 19–26 (2009)

H. Yamada, N. Fuwa, M. Nomura, Synthetic fiber having improved hygroscopicity (1993), Japan patent 05-339878A

H. Yamada, N. Fuwa, Protein containing high molecular material and its application (1994), Japan patent 06-080741A

Y. Yang, Y. Wu, S. Chen, G.J. Devine, I. Denholm, P. Jewess, G.D. Moores, The involvement of microsomal oxidases in pyrethroid resistance in *Helicoverpa armigera* from Asia. Insect Biochem. Mol. Biol. **34**, 763–773 (2004)

X. Yin, S. Li, S. Liu, G. Zhu, H. Zhuang, Genotoxicity evaluation of low doses of clodinafop-propargyl to the silkworm *Bombyx mori* using alkaline single-cell gel electrophoresis. Environ. Toxicol. Pharmacol. **26**(2), 162–166 (2008)

S.Q. Yu, Y.F. Sun, *Agricultural Insect Fauna in Henan Province* (China Agricultural Press, Zhengzhou, 1993)

Q. Yu, S. Fang, W. Zuo, F. Dai, Z. Zhang, C. Lu, Effect of organophosphate phoxim exposure on certain oxidative stress biomarkers in the silkworm. J. Econ. Entomol. **104**(1), 101–106 (2011)

X. Zhan, H. Liu, Y. Miao, W. Liu, A comparative study of rac- and S-metolachlor on some activities and metabolism of silkworm, *Bombyx mori* L. Pestic. Biochem. Physiol. **85**, 133–138 (2006)

H. Zhang, Q. Zhou, M. Pan, H. Wu, Y. Qian, X. Lu, Toxicity of avermectin to the silkworm, *Bombyx mori*. (2006). Retrieved from http://en.cnki.com.cn/Article_en/CJFDTOTAL-CSTB200601007.htm

Z. Zhang, D. Wang, Z. Chi, X. Liu, X. Hong, Acute toxicity of organophosphorus and pyrethroid insecticides to *Bombyx mori*. J. Econ. Entomol. **101**(2), 360–364 (2008)

Z.T. Zhang, H.W. Zhang, X.D. Fang et al., Cosmetic outcome and surgical site infection rates of antibacterial absorbable (Polyglactin 910) suture compared to Chinese silk suture in breast cancer surgery: a randomized pilot research. Chin. Med. J. **124**(5), 719–724 (2011)

X. Zheng, L. Zhao, N. Xu, H. Zhang and G. Sun, Interspecific interaction of below-ground and above-ground indices in mulberry-soybean intercropping system (2011). Retrieved from http://en.cnki.com.cn/Article_en/CJFDTOTAL-TURA201103025.htmSoils

L.W. Zhou, X.C. Chen, Test of detoxicant selection. Jiangsu Seric. **15**, 80–84 (1994)

J.H. Zhou, X.B. Sun, H.B. Xu, Y.Z. Li, J.H. Zhao, Research progress on silkworm pupae. Spl Wild Econ Anim. Plant Res. **18**(1), 38–40 (1996)

L.S. Zhu, Exploitation and utilization of the silkworm, *Antheraea pernyi*. North. Seric. **25**(101), 32–33 (2004)

L.S. Zhu, W.Z. Cui, Toxicity of fenpropathrin, phoxim and its mixture to silkworm. Agroenviron. Protec. **19**, 35–37 (2000)

J. Zhu, F. Wei, S. Li, G. Zhu, Acute toxicity of chlorpyrifos to silkworm (*Bombyx mori*). Sci. Seric. (2006a). Retrieved from http://en.cnki.com.cn/Article_en/CJFDTOTAL-CYKE200602028.htm

J. Zhu, F. Wei, G. Zhu, Contact toxicity and chronic toxicity of triazophos, an insecticide for rice field to the silkworm, *Bombyx mori*. Sci. Seric. (2006b). Retrieved from http://en.cnki.com.cn/Article_en/CJFDTOTAL-CYKE200603012.htm

J.S. Zhu, J. Wang, H.Y. Gao, S. Qin, X.W. Qiao, J.C. Han, Effects of sub-lethal dosages abamectin on food intake and digestive enzyme activities of silkworm *Bombyx mori* L. Ying Yong Sheng Tai Xue Bao **19**(11), 2527–2532 (2008)

Index

Chapter 5
Pesticide Toxicity to Earthworms: Exposure, Toxicity and Risk Assessment Methodologies

Abstract Earthworms are exposed to pesticides through contact and ingestion of the contaminated matter. Pesticides enter into the soil usually as spray drifts or residues from plants and from soil applications. Pesticides affect earthworms by killing them or my causing temporary or permanent impairments and or by altering their behaviour and activities. Acute toxicity studies are carried out both by contact (topical, vial coating, filter paper, immersion etc.) and by ingestion exposures. The most widely conducted, soil contamination bioassay estimates both contact and ingestion toxicities of pesticides and thus more realistic to field conditions. Sublethal toxicity assays are performed for life history traits along with reproductive indices. The two important activities of burrowing and litter burial of earthworms are analyzed through 2D terraria or 3D soil core x-ray tomography and by using Daniel's funnel test. Avoidance behaviour, perhaps the most studied sublethal parameter is reported to have no relation with soil function of earthworms; contrarily the cast production is rarely studied or reported. Semifield experiments using mesh bags, worm socks, terrestrial model ecosystem etc. along with some field studies are reported. Ecological risk assessments of pesticide contaminated soils are mostly done using earthworms. Risk assessment based on median lethal doses, ratio of predicted environmental concentration and no effect levels, risk estimates like toxicity exposure ratio and hazard quotient, risk assessment based on laboratory analyzed acute, chronic and behaviour end points, extrapolation and comparison of laboratory and field risks, field tests especially using a tierd approach and risk assessment for secondary poisoning to predators are given in detail.

1 Importance of Earthworms in Agriculture

There are about 800 genera and 8000 species of earthworms recorded which belongs to the order Oligochaeta (Edwards 2004). Most of these species of the so-called earthworms inhabit soils, submerged muds in freshwater bodies and marine bottoms (Paoletti 1999). Earthworms are popularly called as 'ecosystem engineers' due to their role in soil structure formation and maintenance especially by creating continuous macropores (Blanchart et al. 2004; Edwards and Shipitalo 1998), stable macroaggregates (Six et al. 2004; Blanchart et al. 2004) and organo-mineral complexes (Marinissen and Dexter 1990; Six et al. 2004).

J. Stanley, G. Preetha, *Pesticide Toxicity to Non-target Organisms*,
DOI 10.1007/978-94-017-7752-0_5

If earthworms are not available in the soils, much of the fertility is locked up in the peaty layer or in the dung and dead plant materials that get accumulated on the soil surface. In the absence of earthworms, decomposition will be very slow, soil structure will be poor and compact, there will not be any mixing of soil and have a significant restriction in moisture penetration and plant root development (Stockdill and Cossens 1966). But when earthworms are available, they take bits of fallen leaves and twigs into their burrows where they eventually get decomposed. About 90 % of normal leaf-fall was found to get buried during winter by earthworms in orchards having about 0.75–1 ton of earthworms per acre (Raw 1962). Darwin (1881), the naturalist famous for his ideas on evolution, estimated that 10.6 tons of materials per acre are brought to the soil surface by earthworms (Valenzuela 2010). In soils where earthworms are active, all the dung and dead plant residues get incorporated under the soil surface. They decompose more rapidly to complete the 'fertility cycle' and thus the nutrients will pass rapidly and frequently in a cycle from soil to plant to animal to soil and so on (Stockdill and Cossens 1966).

Types of earthworms Earthworms are of different types and can be classified mainly into three ecological groups based on their burrowing and feeding habits as epigeic, endogeic and anecic. The burrowing activity of epigeic species is normally seen in the upper few centimeters of the soil. They generally dwell in litter and feed on them in the soil surface and ingest a little or no soil. Endogeic species make largely horizontal burrows (Hawkins et al. 2008). They consume mineral soil and prefer soils rich in organic matter. They are usually found in the upper 10–15 cm but sometimes make deeper burrows around the trenches where soil organic matter is plenty. Anecic species are larger in size, forms vertical burrows, sometimes branched, that extends to the soil surface and can be as deep as 240 cm (Edwards and Bohlen 1996). These species often emerge at the surface to feed on the decomposing litter, usually by taking the material into their burrows. They also ingest some soil along with organic matter (Hawkins et al. 2008). Others such as coprophagic species (*Eisenia foetida, Dendrobaena veneta, Metaphire schmardae,* etc.) live in manure and arboricolous species live in suspended soils especially found in humid tropical forests (Paoletti 1999).

Though earthworms are of different ecological types, generally plough the field, enhance the soil fertility and help to increase the soil microbes which make the nutrients easily available to the plant and thus called as farmer's friend. Some of the important activities of earthworms and its uses to mankind are given under:

1.1 Improving Soil Physical Properties Including the Structure

Earthworms are increasingly being recognized as 'ecological engineers' and they have a significant influence on soil physical, chemical and biological properties (Valenzuela 2010). Earthworms directly improve the structure of the soil and hence its stability. The soil physical properties such as bulk density, infiltrability, hydraulic conductivity, porosity, aggregate stability, etc. are proved to be improved by earthrowms (Ojha and Devkota 2014). With their burrowing, earthworms help to improve soil aeration and water infiltration apart from improving the soil structure. Earthworms are found to involve in moving soil particles, creating pores, stabilizing smaller aggregates and in the lining of biopores (Oades 1993). They create stable soil aggregates through cast formation e.g. cast produced by *Lumbricus rubellus*. The stability of earthworm casts is attributed to the high organic matter incorporation (Shipitalo and Protz 1989). The plant debris rich in carbohydrates are also reportedly responsible for the structural stability of earthworm casts (Guggenberger et al. 1996). The proper location of the organic bonding materials in the soil matrix is more important for stability rather than the total amounts (Degens 1997). The natural soil structure is essentially been maintained by the diversity of worms, that produce micro and macro aggregates (Blanchart et al. 1997). The lignin derived phenols were reported to be more in casts i.e. 31 g/kg of carbon compared to 17 g/kg of carbon in the surrounding topsoil (Zhang et al. 2003), which contributes for its stability. Some species of earthworms such as *Pontoscolex corethrurus* make the soil more compact and impermeable to water. The compactness caused by these worms appeared more pronounced than that of a bulldozer (Chauvel et al. 1999). On the other hand, freshly formed casts are more vulnerable for dissolution. Some earthworms are also used in ameliorating the soil compaction by decreasing the bulk density (Zund et al. 1997; Langmaack et al. 1999; Jongmans et al. 2003). Thus the compacting and decompacting of soil by these worms improves the overall soil structure.

Earthworms till the soil and increase aeration and porosity and thus called as natural tillers (Knight et al. 1992), infiltration capacity (Stockdill 1966) and hydraulic conductivity (Ehlers 1975; Johnson-Maynard et al. 2002). The physical property of soil such as bulk density (g/cc) was found to be 1.37 in soil with earthworm compared to 1.41 in which there were no worms. The field capacity (% moisture in 0–4 in. depth) was 39.8 with worms and 34.9 without them; the available moisture was 1.13 and 1.05, with and without worms, respectively. The porosity (% in 8–12 in. depth) was 49.0 with worms and 47.4 in soils without them. The basic infiltration rate per hour was found to get increased by 41.26 % by the introduction of worms (Stockdill and Cossens 1966). Further, water absorption or infiltration in fields rich is earthworm can be four to ten times greater as compared to fields with no earthworms because of their burrowing nature (Edwards and Bohlen 1996). This may help to reduce water runoff and thus soil erosion (Chan 2004). They aerate the soil, make aggregates, incorporate and convert the plant material into the soil organic matter and thus help to maintain or improve soil structure.

1.2 Improving Soil Fertility and Nutrient Availability

The role of earthworms in improving soil fertility is an ancient knowledge which is now better understood and backed by scientific results from different studies (Bhadauria and Saxena 2010). Earthworms ingest organic residues of different C: N ratios and lowers the ratio. Organic C and N were found to get enriched in earthworm casts by 1.5 and 1.3 times respectively than normal soils (Zhang et al. 2003). The organic carbon (% by volume at 8–12 in.) was 1.15 and 0.99 in soils with and without earthworms, respectively (Stockdill and Cossens 1966). A reduction to an extent of six to eight times in the loss of nitrate and ammonium nitrogen is reported in earthworm incorporated system (Sharpley et al. 1979). The ash and total nitrogen contents of fresh cowdung were reportedly increased within a few weeks of introduction of earthworms, *Eisenia andrei*. This may be due to the earthworm induced quick breakdown of carbon compounds and nitrogen mineralization (Atiyeha et al. 2000). Earthworms also enhance nitrification which leads to increase in the ratio of nitrate-N to ammonium-N (Ruz-Jerez et al. 1988).

Earthworms are reported to increase the phosphorus availability to plants in fields applied with less soluble phosphours fertilizer i.e., rock phosphate (Ouedraogo et al. 2005). An increase in P, Ca and Mg availability and a decrease in K availability was found in coffee pulp vermicompost formed by *Eisenia fetida* worms (Orozco et al. 1996). After 150 days of inoculation of *Perionyx excavatus* in crop residues, farm yard manure and cattle dung showed a significant decrease in organic C content with increase in total N, available P and exchangeable K (Suthar 2007). Earthworms help to conserve the soil nutrients, making them available for long-term with steady and moderate release. The nutrients which are embedded in casts may undergo a slow nutrient mineralization (Zhang et al. 2003).

1.3 Enhancing Beneficial Soil Microbes

Earthworms are said to affect soil microflora and fauna populations directly and indirectly by comminution, burrowing, casting, grazing and dispersal. These activities per se and the results thereof change the physical, chemical and biological status of the soil and may cause changes in the density, diversity, structure and activity of microbial and faunal communities within the drilosphere (Brown 1995). A correlation between the diversity and abundance of earthworms with microorganisms such as fungi, actinomycetes, bacteria is reported (Brown 1995; Anderson and Bohlen 1998). A high degree of relationship was found to exist between the micro, meso and macro invertebrates with the anecic species of earthworms (Maraun et al. 1999). But endogeic species was reported to have more influence on the microbial community rather than anecics (Bhatnagar 1975). The biodiversity of bacteria and actinomycetes are reported to be high in the composts produced by *Eudrilus eugeniae* followed by those produced through *Eisenia fetida* and *Perionyx excavates* (Pattnaik and Reddy 2012a). Activities of earthworms help to increase beneficial VAM (Vesicular Arbuscular Mycorrhiza) populations. These mycorrhizal fungal associations enhance the availability of many essential nutrients to the plants (Gormsen et al. 2004).

The selective foraging of earthworms on organic matter and the nutrient rich cast promotes microbial activities (Haynes and Fraser 1998; Kale 2008). Further, a large host of factors inside the worms including $CaCO_3$, enzymes, mucus and antimicrobial substances influence on the survival of the ingested organisms (Brown 1995). The mixing effect, gut mucus secretion and excretion from earthworm are known to enhance the activity of microorganisms (Bhadauria and Saxena 2010). Microflora associated with the gut of earthworms get excreted in the casts and get mixed through microbial adheration to earthworm skin (Edwards and Bohlen 1996). Nitrogen-fixing bacteria are found in the gut of earthworms and in earthworm casts and thus higher nitrogenase activity was reported in casts as compared to normal soil (Simek and Pizl 1989). Earthworm casts are generally much richer in microbial numbers (Hickman and Reid 2008). A 90 % increase in respiration rate is reported in fresh casts (Scheu 1987) with a bacterial count of 32 million/g compared to 6–9 million/g in the surrounding soil (Teotia et al. 1950). They fragment the plant residues and make them easily available to microbes thus stimulating the microbial activity (Langmaack et al. 1999). Small organisms with limited mobility may get benefit from the comparatively long ranging movements of earthworms.

1.4 Organic Waste Management and Vermicomposting

Epigeic earthworms are generally being used for the management of organic wastes and for making biocomposts (Kale et al. 1982; Reinecke et al. 1992; Kale 1998; Garg and Kaushik 2005; Benitez et al. 2005; Suthar 2007). Agricultural by-products such as animal dung, farmyard manure and crop residues are potential sources of plant nutrients but largely remain unutilized. There are estimates that 30–35 % of applied N and P and 70–80 % of K, remain in the crop residues of food crops (Suthar 2007). These ‘useful wastes’ can be converted into nutrient rich bio-fertilizer and earthworms play a major role in this. Vermicomposting often results in quicker mass reduction with shorter processing time and quality humus with reduced phytotoxicity as compared to conventional composting systems (Lorimor et al. 2001). Inoculation of earthworms accelerates the microbial composition and activities, which in turn enhances the nutrient transformation.

Earthworms are used in composting cow dung (Reinecke et al. 1992), sheep dung (Kale et al. 1982), biogas sludge (Kale et al. 1982; Edwards et al. 1998), poultry manure (Kale et al. 1982), pig solids (Edwards et al. 1998), horse solids (Edwards et al. 1998), turkey wastes (Edwards et al. 1998), vegetable wastes (Singh et al. 2005), water hyacinth, *Eichhornia crassipes* (Gajalakshmi et al. 2001), guar gum industry waste (Suthar 2006) etc. Paper, cardboard, grass clippings, pine needles, sawdust and food wastes in mixtures with sewage sludge (1:1 dry weight) were also found to get bio composted by *Eisenia andrei* (Domíngueza et al. 2000). *Eisenia foetida* worms were found to grow and reproduce favourably in solid textile mill sludge added with 70 % cow dung (Kaushik and Garg 2004). Vermicompost from olive cakes using *E. andrei* was devoid of any phytotoxicity which is prevalent with the olive cake (Benitez et al. 2002). Composts obtained from worms i.e. vermicompost concentrates basic plant nutrients (NPK) apart from increase in other essential nutrients and thus contribute for the plant growth.

1.5 Influence in Soil Erosion

Darwin said that the earthworms in natural situations are able to influence erosion by disintegrating the surface making it to wash away to distances (Darwin 1881). The litter burying and surface casting activities of earthworms can lead to the exposure of soil to raindrop impact and consequent splash erosion (Hazelhoff et al. 1981; Van-Hoof 1983). The casts of anecics and endogeics may get dispersed easily and thus enhance soil erosion leading to nutrient losses (Binet and Le Bayon 1999) but they stimulate water infiltration by making vertical and semi permanent burrows (Ehlers 1975; Zachman et al. 1987; Trojan and Linden 1992; Edwards and Shipitalo 1998; Shipitalo and Butt 1999). The increase in soil porosity and water infiltration delays or reduces surface runoff (Kladivko et al. 1986; Roth and Joschko 1991; Blanchart et al. 2004). A fivefold higher infiltration was found if casts were present in the soil surface (10–15 mm/ha) than if they are absent (2 mm/ha) (Wilkinson 1975; Casenave and Valentin 1988).

Further, the density of earthworm casts is usually higher than general soil bulk density, for e.g. 1.8–2.0 Mg/m^3, whereas bulk density of soil was equal to 1.45 Mg/m^3 in the upper 10 cm (Blanchart et al. 1993), which make the cast more stable. The development of fungal hyphae, production of exopolysaccharides of bacterial origin, addition of intestinal mucus, etc. (Marinissen and Dexter 1990; Zhang and Schrader 1993; Schrader and Zhang 1997) explains the stability of casts (Blanchart et al. 2004). Ca^{2+} ions are reported three to four times more in casts than in the soil, which is said to give stability to the cast (Blanchart et al. 2004). The complexes formed by the colloidal organic matter with polyvalent cations and clays also determine the stability of earthworm casts (Tisdall and Oades 1982; Shipitalo and Protz 1989). Though the surface casting activity of earthworms may tend to enhance runoff and erosion, the improvement in infiltration capacity and semi permanent aggregates reduces soil erosion and nutrient loss.

1.6 Bioremediation of Polluted Environment

Earthworms can be directly employed in bioremediation processes because of their characteristic biological, chemical and physical actions to promote biodegradation of contaminants (Hickman and Reid 2008). The toxicity and tolerance of earthworms to contaminated soils have been discussed by Sheppard et al. (1998) and Spurgeon et al. (2003). A right species introduced in the right environment does a proper remediation. Earthworms have been utilized for land recovery or rehabilitation of sub-standard soils such as polder soils, poor mineral soils, open cast mining sites, areas of cutover peats, etc. (Edwards and Bohlen 1996; Butt et al. 1999, 2004; Haimi 2000). The different approaches on the use of earthworms in bioremediation may be as follows (Hickman and Reid 2008): (1) direct application of worms (Schaefer et al. 2005); (2) co-application with organic media (Ceccanti et al. 2006),

(3) application of contaminated media to earthworms (Getliff et al. 2002) and (4) the indirect use of earthworms by applying vermidigested material (Alvarez-Bernal et al. 2006). Not the earthworms itself, their casts can also be applied to contaminated soils to improve chemical conditions and to aid in bioremediation process (Hickman and Reid 2008).

Earthworms have been reported to impede the binding of organic contaminants to soils, release the soil bound contaminants which stimulates degradation and promote and disperse microorganisms that degrade organic contaminants (Hickman and Reid 2008). Earthworms are used for the remediation of many organic contaminants such as insecticides (Ramteke and Hans 1992; Verma et al. 2006), herbicides (Farenhorst et al. 2001; Gevao et al. 2001), crude oils (Schaefer and Filser 2007), polychlorinated biphenyls (PCBs) (Singer et al. 2001), chlorophenolic preservatives (Haimi et al. 1992), trinitrotoluene (TNT) etc. (Renoux et al. 2000). Schaefer and Juliane (2007) studied the bioremediation of crude oil contaminated soil using different earthworm species and found earthworms to trigger degradation leading to remediation. However, no significant advantage of introducing earthworms for bioremediation of used engine oil contaminated soil is reported by Ameh et al. (2012). *Eisenia foetida* worms are able to uptake 100 ppm phenol completely with in 72 h and thus can be used in bioremediation of phenol and phenolic derivatives in soil (Krishna et al. 2011).

Earthworms, *Eudrilus eugeniae*, *E. fetida* and *P. excavatus* were found to remediate heavy metals such as Pb, Zn, Cd, Cu and Mn from the urban wastes through vermicomposting and *E. eugeniae* being the most efficient of all the three species tested (Pattnaik and Reddy 2012b). Metallic pollutants are not degraded during composting but may be converted into less bioavailable organic combinations (Barker and Bryson 2002). The metals bindings to ions and carbonates (i.e. more soluble fractions) are found more in the ingested material (Morgan and Morgan 1999). A possible species-specific metal metabolism in worm's gut is also reported. Thus, proper care should be taken to select the correct earthworm species for the intended conditions (Hickman and Reid 2008).

1.7 As a Biological Indicator

Earthworms are considered as important indicators of soil health and also provide information about the stability of the agroecosystem. Earthworms can indicate soil quality by its abundance and species composition at a particular site. Many chemicals get accumulate in the body of the worms from the soil and thus, a biochemical/cytological stress biomarkers can indicate it (Frund et al. 2010). Earthworms provide a clear indication of contamination and environmental disturbance either in rural landscapes and marginal lands (Paoletti 1999). Earthworms can also influence in some processes which modulate the transfer of inorganic and organic toxicants (Cooke et al. 1992). The limited mobility of earthworms makes them very suitable for monitoring the impact of pollutants, changes in soil structure and agricultural practices (Paoletti 1999).

Earthworms are traditionally been considered to be convenient indicators of land use and soil fertility (Paoletti 1999). The presence of birds such as ravens and magpies that are attracted to a freshly ploughed field to eat the earthworms gives an indication of fertile cultivable soil (Tanara 1644). Agricultural activities such as ploughing and application of fertilizers and pesticides dramatically influence the worms. Most of the larger species are reportedly being displaced from cultivated areas in the tropics as well as in temperate rural areas and are present only in woodlands or grasslands (Paoletti 1999). Fields under no-tillage or minimum tillage are found to have always higher earthworm biomass than tilled fields (House and Parmelee 1985; Clapperton et al. 1997). Generally the larger species of worms disappear soon after the natural soil is brought into cultivation (Paoletti 1999). Total earthworm biomass per unit area is being used to indicate different farming systems (Hani 1990; Matthey et al. 1990). The deep burrowing earthworms, *Lumbricus terrestris* and the surface dwelling *L. castaneus* were recorded significantly less in conventional (high input) orchards compared to organic counterparts, and thus can be used as indicator (Paoletti et al. 1995). Some key species such as *L. terrestris* were also used to discriminate between 35 cm mouldboard ploughed land and minimum tillage farming fields (El-Titi and Ipach 1989).

Earthworms are comfortably used as biological indicators of many metals ion contamination in the soil (Gish and Christensen 1973; Lee 1985). Earthworm numbers and dynamics are also used to monitor the soil contaminants such as heavy metals and PCBs (Kreis et al. 1987; Curry 1994). Levels of environmental pollutions can be assessed by measuring the heavy metal concentrations in earthworm tissues. For this, a prior correlation has to be made between the level of soil contamination and bioaccumulation in earthworm for the particular metal (Motalib et al. 1997). A good linear relationship is already been established between soil contents of Cu, Pb and Zn with that in the tissues of earthworm (Kennette et al. 2002). Such earthworm based assessments are complicated because they may develop tolerance to some of the pollutants (Fisher and Koszorus 1992; Morgan and Morgan 1992; Morgan et al. 1992). Earthworms especially the large burrowing species are highly sensitive to copper sulphate and thus can be used as bio monitors for this source of soil contamination (Paoletti et al. 1988). Earthworms apart from being a bioindicator for pesticide contamination in the fields are also used in assessing the toxicological risk in terrestrial ecosystem (Spurgeon et al. 2003).

1.8 As an Important Food Source

Earthworms are an important food source for a large number of organisms, contributing towards a rich food-web in the soil. They are available for subterranean predators, to those that dug them up and those that catch them on the surface (Macdonald 1983). Toads and frogs feed on earthworms in limited amount (Lescure 1966; Smith 1951) whereas newts and salamanders are voracious feeders of earthworms (Macdonald 1983) and the butler's garter snake feeds almost exclusively on earthworms (Catling and Freedman 1980). Many other snakes also feed on earthworms apart from slugs and small mammals (Steward 1971; Gregory 1978). Even the beneficial carabid beetle, *Pterostichus melanarius* which is a major biocontrol agent of aphids and slugs in agricultural systems also feeds on earthworms (King et al. 2010). Spiders are reported to feed on earthworms which have high protein content and could be a welcome supplement to the spider's usual insect diet (Nyffeler et al. 2001). Many birds especially black headed gulls, tawny owls are predacious on earthworms to extent that it might significantly decrease the population of worms in cultivated fields (Cuendet 1983; Southern 1969; Delmee et al. 1979). Other birds such as thrushers (Cherenkov et al. 1995) and golden plover (Bengtson et al. 1976) also prey upon earthworms. Bigger animals like boars (Baubet et al. 1997), hedgehogs (Doncaster 1994; Micol et al. 1994), badgers (Macdonald 1980; Roper 1994) and foxes (Macdonald 1980; Ferrari and Weber 1995) too prey upon earthworms.

Earthworm tissues have high protein content (58–71 % on dry weight basis) and are rich in essential amino acids and have a potential value for livestock feeding particularly in pig and poultry industries. The amino acid composition of earthworm protein is of high biological quality and comparable to fish meal and meat meal (Sabine 1983). It contains a considerable amount of fat (4.6: French et al. 1957 to 17.3 %: Durchon and Lafon 1951) with substantially high energy of about 22.24 kJ/g dry weight (Bolton and Phillipson 1976). Earthworms especially the *P. excavatus* are used in fish and poultry meal and used commercially for the production of *Tilapia nilotica* fingerlings and Japanese quail in Philippines Islands (Guerrero 1983). Almost all the demands for bait worms were met by the worm-harvesting especially of *L. terrestris* from golf courses and pastures (Tomlin 1983).

1.9 Use in Waste Land Restoration

Earthworms help to restore degraded soils, such as from mining or plantation agriculture upon their introduction (Zou and Bashkin 1998; Marashi and Scullion 2003). Endogeic earthworms make largely horizontal burrows through which water enters deep into the soil and thus enhancing waste water renovation by distributing the effluent over a larger area (Hawkins et al. 2008). Though most of the earthworms do not readily accept coniferous litter which is reported as marginally palatable to them, earthworms are used in the regeneration of coniferous forests (Bernier and Ponge 1994).

1.10 Enhancing Pasture Production, Crop Growth and Yield

Earthworms are assumed to be beneficial soil animals generally based on the belief that they promote plant growth (Lee 1985; Edwards and Bohlen 1996). Scheu (2003) reviewed 67 studies, which reported the influence of earthworms on plant growth and concluded that shoot biomass of plants significantly increased in the presence of earthworms. Improvement in soil physical properties especially the water retention and infiltration in agroecosystem is reported immediately after the introduction of earthworms (Blanchart et al. 1999). Further their interactions with microorganisms help to improve the nutrient availability to plants (Gormsen et al. 2004). Improved soil structure, better moisture penetration, deeper incorporation of plant nutrients and organic material, greatly improved root development of pasture grasses and reduces susceptibility to drought and thus proper growth. An increase of 28–110 % in pasture production was reported using earthworms, *Allolobophora caliginosa* in mixed swards (Nielson 1953). Waters (1955) measured an increase of 77 % from pure ryegrass and 113 % from the ryegrass component of a grass-clover mixture using *A. caliginosa* in pots. Introduction of these worms in field also increases the pasture production up to 72 % (Stockdill 1966). Earthworms introduced in a 22 year old pasture were found to produce an increase in yield of 29 % (Lynch 1960). However in some conditions like high fertilized areas the worms cause some disadvantageous effects like leaching of nutrients and pugging of soil (Stockdill and Cossens 1966). At times earthworms cause problem in golf course greens and lawns. Excessive disturbance of soils by earthworms leads to uneven surfaces, soil erosion and even open niches for weed invasions (Kirby and Baker 1995; Baker et al. 1998). In conditions of excess soil moisture, earthworms often migrate to sidewalks, causing a nuisance for homeowners, lawn managers and public (Tu et al. 2011).

Earthworms are reported to enhance crop growth in agricultural fields (Zund et al. 1997; Marashi and Scullion 2003; Gormsen et al. 2004). Vermicomposts produced from food wastes, paper wastes and cattle manure cause significant increase in growth and yields of peppers, *Capsicum annuum* including increased leaf area, shoot biomass and marketable fruits. Apart from plant nutrients, humic materials and other plant growth influencing substances, such as plant growth hormones produced by microorganisms during vermicomposting would have contributed for this increase in growth and yield (Arancon et al. 2005). Peppers grown in potting mixtures containing 40 % food waste vermicomposts and 60 % commercial mixture, yielded 45 % more fruit weights and had 17 % greater mean number of fruits than those grown in commercial mixture only (Arancon et al. 2004). Incorporation of 10 % or 50 % earthworm processed pig manure with commercial mixture increased the dry weights of tomato seedlings significantly. The highest marketable yield, increased size and weight of fruit were obtained in 10–50 % pig manure vermicompost substituted with commercial growth media (Atiyeh et al. 2000). The growth and productivity of French marigold (*Tagetes patula*) was found to get enhanced by using earthworm processed pig manure along with commercial media in glass houses (Atiyeha et al. 2002). Application of vermicomposts produced from forest litter to forest trees especially *Tectona grandis* significantly improved their heights and diameters over those of control trees (Manna et al. 2003).

The bioturbation of earthworms can influence on the fate of pesticides in soil via various mechanisms. It increases pesticides sorption on soil particles on the long-term, leading to the formations of non-extractable residues. In this case, it can increase the pesticide persistence in the soil as reported by Farenhorst et al. (2000) and Binet et al. (2006) for atrazine. On the other hand, the activities of earthworm lead stimulation of microorganisms in the soil. This can enhance the activity of microbial degradation accelerating mineralization (Monard et al. 2008, 2011; Liu et al. 2011a). *Aporrectodea caliginosa* is reported to get participated in the breakdown of four fungicides (folpet, fosetyl-Al, metalaxyl, myclobutanil) and two insecticides (chlorpyrifos-ethyl and lambda cyhalothrin) (Schrecka et al. 2008).

Earthworms are reported to stimulate the reproduction of aphids on grasses and legumes (Scheu et al. 1999) while no difference on aphid reproduction due to worms was reported by Bonkowski et al. (2001) and Wurst et al. (2003). Moreover, Van-Dam et al. (2003) reported a reduction in aphid population due to earthworms and attributed this to the production of plant defense compounds. Nematodes are known to be digested by earthworms thereby functioning as predators (Hyvonen et al. 1994). The surface applied fertilizers and pesticides are found to get mixed and pass through deeper layers enhancing their effectiveness due to earthworms (Stockdill and Cossens 1966). Thus earthworms not only influence on plant growth per se but also have a role on the tritrophic interactions in the biosphere.

2 Routes of Pesticide Exposure to Earthworms

Pesticides usually enter the soil as spray drifts and residues of sprays applied to crop plants, trees and grasses. Pesticides are sometimes applied directly on soils, as in the case for nematicides and root pest control agents (Paoletti 1999). Soil organisms like earthworms which dwell in soil and thus get exposed to pesticides through various routes. But the toxicity of pesticide to earthworms varies with the nature of pesticide chemical, physical and chemical property of the soil and behaviour of earthworms.

Earthworms are of different ecological categories viz., epigeics are surface active and dwell in litter, endogeics live in the organic horizons and create horizontal burrows and the anecics that live in vertical and deep burrows and ingest large amounts of soil. The epigeics are the first to get exposed since they live in surface or litter. Direct sprays and soil applications may reach them more by contact and by ingestion of litter. And thus the epigeics acquired resistance to many pesticides than their counterparts. The endogeics live in the organic horizons where the pesticides reach very easily, while the anecics ingest more amount of soil and thus exposed by ingestion apart from epithelial contact while moving to surface or contaminated soil. Earthworms are therefore continuously exposed to soil contaminants through their exterior epidermis and alimentary surfaces (Rodriguez-Castellanos and Sanchez-Hernandez 2007). Earthworms can sense the pesticides by the sensory tubercles on their body surfaces, depending on the nature of the pollutant and its concentration and thus try to avoid the contaminated soil (Reinecke and Reinecke 2004). Avoidance may not be possible due to the limited locomotion ability of worms especially when a large area is treated. Moreover pesticides may retard the locomotion ability of the worms. The adsorption/desorption of pesticide compounds in soil are influeced by soil organic matter content (OM) and octanol/water partitioning coefficient (K_{ow}). Further, the bioavailability of the pesticide to earthworm depends on the nature of pesticides, properties of soils and uptake routes of earthworms (Yu et al. 2006).

2.1 Exposure by Contact

Earthworms normally live underground from a few centimeters to far below the soil. The direct contact of pesticide sprays to them is rare except for the species which dwell in litter or surface soil such as epigeics. Pesticide sprays are directed normally towards the plant parts and some spray particles fall on the ground. However, the soil application may cause more damage to the worms because they are directed towards their dwelling place and food. These soil applications are mainly used for underground pest (root grubs, termites etc.) control and for some soil borne plant diseases. The indirect contact of earthworm to pesticides through the soil causes a greater danger. Since they move from one place to another through the soil, the cuticle of earthworms gets in touch with the soil and to the contaminant. Some pesticides cause repulsive effect and the earthworms move away to avoid them and in this case the migration of worms over the soil surface can lead to a higher risk of pesticide exposure. If the contaminant is lipophilic or gets attached with the cuticle then the risk is still dangerous, since it can get accumulated with an additive effect in the worm's cuticle. To avoid this contact, earthworms may secrete a mucus secretion especially when exposed to contaminated soil.

2.2 Exposure by Ingestion

Earthworms ingest soil heavily, especially the anecic species and produce cast. The pesticides in the soil get ingested along with the soil or litter and get into the alimentary canal of earthworms. The pesticide contaminated plant material taken as food by the worms also contributes to the ingestion toxicity apart from soil contamination. Again it is highly dangerous when the toxicants get accumulated in the epithelial cells or other tissues rather than get eliminated through the cast. Earthworms may take pesticides from soil pore water and as food in various amounts. The dominant route of exposure is found to be across the gut wall (Jager 2004). The hydrophobic chemicals are generally absorbed through the gut (Jager et al. 2003). The availability of chemicals to earthworms also depends on the soil properties. Soil organic matter has a greater influence in normalizing the toxicity (Van-Gestel 1997). Earthworms take up chemicals exclusive from the pool in the soil pore water, which is usually in a dynamic equilibrium with the solid phase. A higher organic matter content in the soil appears to shift the equilibrium towards the solid phase and thus decreasing the toxicity (Di-Toro et al. 1991).

Earthworms May Also Get Exposed to the Pesticides

i. Through the soil pest control programmes
ii. Through the pesticide sprays on the crop
iii. Through the pesticidal residues persistent in the soil environment
iv. Through the breakdown or metabolic products of pesticides
v. Through leaf fall from treated trees
vi. Through seed coating chemicals

(i) Through soil pest control programmes

Pesticides applied on soil for different pest control programmes often affect the soil organisms especially the earthworms. Brodifacoum used for rat control was reported to be found within the body tissues of earthworms, indicating that they do consume bait fragments that get mixed in the soil. Earthworms are generally exposed to pesticides used for slug control programmes, since they share a common habitat. Earthworms come to the soil surface from their burrows for searching algae or litter pieces and thus exposed (Bieri 2003). Soil insect pest management practices and control measures for soil borne pathogens are mostly done through soil treatment with chemicals. This practice of pest management contaminates both the food and the dwelling place of earthworms.

(ii) Through pesticide sprays on crops

Pesticides are applied usually as sprays on the crops for insect pest and disease control except for a few soil applications. Sprays on crop also result in the residual accumulation in the soil. Much of these residues comes from the foliage sprays or dusts which miss their target and fall on to the soil either close or far off from the target plants (Edwards 1966). When drifting, pesticide droplets settle down on soil, the pesticides are assumed to be in the upper 0.5 cm layer (De-Schampheleire et al. 2007) but subsequently it may move down or away by means of leaching or run-off. Though no risk was assessed for these drifts on earthworm populations in many studies (De-Schampheleire et al. 2007), it cannot be overlooked in view of the long term sublethal effects. Chlorpyrifos and azinphos methyl were found to get deposited not only in the targeted areas of plum orchard but also onto the soil outside the orchard. Azinphos methyl was found to get transported by wind (spray drift) to adjacent areas without tree cover from sprayed orchards. These drifted pesticides were found to have an effect on earthworms as revealed by a depression in cholinesterase activity which leads to reduced burrowing and might lead to sublethal effects on feeding and reproduction and ultimately in population decline (Reinecke and Reinecke 2007).

(iii) Through persistent residues

Earthworms are exposed to low but chronically due to the persistence of residues of pesticides in the soils for several years after application especially in agricultural fields (Gevao et al. 2000). The frequent application of agricultural pesticides and the persistence of some of them eventually leads to increasing amounts of residual compounds in the soil, either as free or bound residues (Gevao et al. 2000; Mordaunt et al. 2005), which can be a threat to lumbricid species (Givaudan et al. 2014). Lead arsenate, DDT, zinc from carbamate fungicides and mercurials though banned for a long time, still present in soil especially in rural landscapes. The very slow degradation of these 'archeological' pesticides can cause serious problems to earthworms (Paoletti 1999). Even, some bound and non-extractable residues tried by means of soxhlet or organic solvents can be found observed in the earthworm tissues. About 3 %, 23 % and 24 % of previously non extractable ^{14}C labeled isoproturon, dicamba and atrazine residues, respectively, were extracted by solvents or mineralized to $^{14}CO_2$ after 28 days of incubation by earthworm, *Aporrectodea longa* (Gevao et al. 2001). Significant amount of ^{14}C labeled parathion residues earlier found bound and non-extractable were found in worm tissues after 2–6 weeks of introduction (Fuhremann and Lichtenstein 1980).

(iv) Through breakdown products

In addition to the direct effects of pesticides, the toxicity of the breakdown products of pesticides and metabolites that enter the soil may cause devastating effect on worms. The metabolites or breakdown products of pesticides sometime stay for a long time in the field and cause various degrees of toxicity to earthworms. Many metabolites were bound in the soil or water and cause toxicity for longer time. Some of the metabolites are reported more toxic to nontargets than their respective parent compounds. Furthermore, the possible synergistic effects of pesticide cocktails are not well studied or understood. Pesticides and their breakdown products can show both direct toxicity against earthworms and produce latent effects on their growth and fertility (Paoletti 1999).

(v) Leaf fall from pesticide treated trees

Foliar concentrations of systemic insecticides in the treated trees can enter the natural environment during leaf fall. Sugar maple trees systemically treated with imidacloprid for controlling Asian longhorned beetles reported to yield senescent leaves with residue levels sufficient to have some sublethal effects on earthworms. The field realistic concentrations (3–11 mg/kg) of imidacloprid in the leaves were not found to affect the survival of earthworms. But sublethal toxicity in terms of reduced feeding rates, decreased leaf decomposition and measurable weight loss occurred among earthworms. This exposure route poses little risk of direct mortality or adverse effect on cocoon production and survival of earthworms but adversely affects the earthworm feeding rate at concentrations even at 3 mg/kg (Kreutzweiser et al. 2008)

(vi) Through seed coating chemicals

Some insecticides especially neonicotinoids are used as seed coating to prevent the damage of insect pest and some fungicides for plant diseases in crops. The effect of seed coatings on soil microbials are already been demonstrated. But in case of earthworms, this toxicity through seed treatment is not much alarming because of the use of low active ingredient per area when compared to soil application or spraying.

Though some of the seed treatment chemicals are toxic to earthworms, their effect could be minimized if applied as a seed coating, as effects would thus be restricted to the immediate vicinity of the seed and not in the soil which it ingest. However, some effects were reported on earthworms through red clover seeds treated with methiocarb, 2 weeks after sowing (Charlton 1978).

Pesticides can reach soil in many ways, the important being direct soil treatment, drain or drift from pesticide sprayed on the plants, through residues and through litter or leaf fall from systemically treated plants and trees. The persistent pesticides may remain in the soil many months after spraying event and get transported to non target areas by runoff as a result of rainfall (Reinecke and Reinecke 2007) or irrigation. Often expired and waste pesticides are disposed in the soil and may also cause hazard to soil organisms.

3 Effects of Pesticides on Earthworms

Pesticides can affect earthworms by killing them or by causing temporary or permenant impairnments or by altering their activities and behaviour. Toxicity of pesticides on earthworm can be briefed as acute (expressed as LC_{50} or LD_{50}) and chronic mortalities either in laboratory, semi-field or field conditions. In addition to the median lethal values, no observed effect concentrations and the time effect (LT_{50}) are useful to assess this adverse effect. Some chemicals having chronic toxicity i.e. toxic after a long time exposure may be assessed with the median lethal time calculation. The other method to find acute toxicity is by testing the effect of the field realistic doses or concentrations on the worms in the laboratory.

Sublethal effects of pesticides on earthworms are studied as effect on the life history traits, body damages, effects on altering the behaviour and activities of the worm. Pesticides which do not cause lethality but cause adverse effect on the growth and maturity of worms, also posses a significant risk. Earthworms affected by pesticides may display an altered pattern from their normal behaviour (locomotion, feeding, etc.). Some investigations show that earthworms exposed to pesticides display marked changes in burrowing (Capowiez et al. 2005), surface migration (Christensen and Mather 2004), feeding activity (Jager et al. 2003) and avoidance ability (Slimax 1997; Capowiez and Berard 2006). In general, earthworms avoid contaminated soils and this avoidance behaviour is tapped to assess the potential toxicity of contaminated soils (ISO 2003).

Effect of pesticides on infra-individual level, individual level and community level in earthworms is reviewed by Pelosi et al. (2014). The (1) infra-individual level includes the effect on enzyme activities, gene expressions etc. (2) individual level explains the effect on survival, fecundity, behaviour etc. and (3) the community level, e.g. abundance, diversity and community structure (Pelosi et al. 2014).

Infra-individual level: This explains the infra-individual changes resulting from the exposure of individual worm to pesticides (Lagadic et al. 1994). Any observable and measurable functional response to exposure at the sub-individual level of biological organisms (molecular, biochemical, cellular and physiological) (Weeks 1995) can be included here.

Individual level: Effects of pesticides on individual levels are mostly studied for life history traits, that too for toxicity and behaviour especially the avoidance. Acute toxicity may not be a very reliable and sturdy parameter to assess the pesticide toxicity and chemicals which are not lethal may cause many sublethal effects. However, most of the studies pertain to individual levels are restricted to the assessment of median lethal values.

Community level: These studies aim to assess the effect of pesticides on diversity and biomass of earthworms mostly through field sampling. But the conclusion of this should be taken with utmost care because earthworm density and diversity tend to differ by environmental variations such as temperature and precipitation, types of crops and also by worm extraction methods.

3.1 *Laboratory Studies*

3.1.1 Acute Toxicity

Acute mortality due to pesticides is perhaps the most studied part in earthworm pesticide toxicity. Based on laboratory acute toxicity tests using soil contamination bioassays, neonicotinoids are found to be highly toxic to *E. fetida* followed by carbamates, organophosphates, IGRs and the least toxic being the pyrethorids (Wang et al. 2012a). Most pyrethroid insecticides tested in soil toxicity tests are found to be non-toxic to earthworms (Inglesfield 1984; Roberts and Dorough 1984) except for a few such as cypermethrin which is reported as highly toxic to *Perionyx excavatus* (Gupta et al. 2010). Wang et al. (2012a) reported pyrethroids as low toxic insecticides to earthworms, *E. fetida* in soil contamination bioassay whereas it was very toxic in contact filter paper bioassays, indicating that the pyrethroids are easier to be absorbed by skin than by gut. No *E. fetida* worms were reported to survive at 602.15 μg/cm^2 of deltamethrin and 0.86 μg/cm^2 of fenvalerate in the filter paper test, however 100–125 mg/kg of both chemicals resulted in the maximum mortality of 90 % in the soil test (Song et al. 2015). The toxicity of acetamiprid to *E. fetida* was reported to be 64,159-fold higher than that of buprofezin at 48 h as tested in contact filter paper bioassay. In laboratory tests, copper sulfates are found lethal only when applied at high doses (over 1000 ppm) to *Eisenia foetida* (Malecki et al. 1982). Herbicides bentazon, bromphenoxin, bromoxynil, bromoxynil octaonate and atrazine are moderately toxic to earthworms (Pizl 1988) whereas glyphosate is harmful to earthworms even at very low doses as tested with *Aporrectodea caliginosa* in the laboratory (Springett and Gray 1992).

Median Lethal Dose: LD_{50}

The LD_{50} values of methomyl, chlorpyrifos-ethyl, azinphos-methyl and imidacloprid to *E. foetida* tested for 14 days were found to be 90, 129, 158 and 10.7 mg/kg dry soil, respectively (Armstrong et al. 1991; Heimbach 1986; www.dive.afssa.fr/agritox/index.php). The LD_{50} value of carbaryl to *Pheretima* sp. was found to be 9 mg/kg dry soil for 7 days (Mostert et al. 2002). Ethyl-parathion has an LD_{50} of 32 mg/kg dry soil for *A. caliginosa* (Olvera-Velona et al. 2008). The LD_{50} values of endosulfan to *E. fetida* tested in injection bioassay was 2.54 ppm (Park et al. 2012).

Median Lethal Concentration: LC_{50}

The LC_{50} of chlorfluazuron to *A. caliginosa* earthworm was found to be 140 mg/kg soil whereas it was 0.68, 73, 127, 381 and 518 mg/kg for aldicarb, cypermethrin, profenofos, atrazine and metalaxyl, respectively (Mosleh et al. 2003a). The 14 day LC_{50} mancozeb to *E. andrei* is 1262 mg a.i./kg of dry soil and thus less toxic (Vermeulen et al. 2001) whereas benomyl and carbendazim have the same LC_{50} value of 5.7 mg/kg of soil for *E. andrei* and highly toxic.

The organophosphorus insecticides viz., chlorpyrifos, phoxim, pyridaphenthion and triazophos exhibited an LC_{50} values of 421, 1083, 273, 381 mg/kg to earthworms, *E. fetida* at 7 days. The carbamates such as carbosulfan, isoprocarb, metolcarb and promecarb showed an LC_{50} of 146.8, 69.4, 108.1 and 31.23 mg/kg at 7 days (Wang et al. 2012a). The 14 day LC_{50} of chlorpyrifos to *E. fetida* was reported as 118.5 mg/kg of soil whereas it was 91.7 for 21 days (Zhou et al. 2007). The 14 day LC_{50} of abamectin to *E. eugeniae* and *P. excavatus* were reported to be 4.6 and 6.7 mg/kg of soil (Jasmine et al. 2008). The median lethal concentrations of carbofuran, chlorpyrifos and mancozeb to *P. excavatus* through soil contamination bioassay was reported as 8, 100 and 500 mg a.i./kg of dry soil for 4 weeks, respectively. Further the toxicities of the pure and formulated pesticides were found to be similar (De-Silva et al. 2010). The 28 day LC_{50} values of buprofezin, lufenuron and triflumuron to *A. caliginosa* was reported as 421.4, 476.9 and 1.9 mg a.i/kg, respectively (Badawy et al. 2013). The 7 and 14 days LC_{50} of deltamethrin was 92.8 and 68.1 mg/kg, respectively and it was 72.8 and 56.3 mg/kg, respectively for fenvalerate (Song et al. 2015).

The contact LC_{50} value of dimethoate, pirimiphos-methyl and deltamethrin to *L. rubellus* through filter paper bioassay for 48 h is found to be 2.24, 0.41 and 0.11 μg/cm^2, respectively (Velki and Hackenberger 2013). All the neonicotinoid insecticides tested on *E. fetida* in filter paper contact bioassay was found to have very low LC_{50} values. The LC_{50} values of acetamiprid, clothianidin, imidacloprid and thiacloprid are found to be 0.008, 0.28, 0.027 and 0.45 μg/cm^2, respectively. Pyrethoroid insecticides such as cypermethrin, fenpropathrin and lambda-cyhalothrin also exhibited very low LC_{50} values, 1.39, 1.94 and 1.49, respectively. Abamectin, emamectin and ivermectin exhibited LC_{50} values of 23.08, 30.20 and 4.40 μg/cm^2, respectively (Wang et al. 2012a). The 48 h LC_{50} values of fenvalerate and deltamethrin to *E. fetida* were reported as 0.242 and 327.8 μg/cm^2 in filter paper bioassay (Song et al. 2015).

Mortality at Field Recommended Doses

Pesticides used at their field recommended rates are not found highly toxic to earthworms except for a few (Roark and Dale 1979; Correia and Moreira 2010). In general, pesticides used at field recommended rates did not show any effect at individual level i.e. mortality (Addison 1996; Bauer and Rombke 1997; Choo and Baker 1998; Vermeulen et al. 2001; Capowiez et al. 2005). Though the pesticides at field doses are not affecting the survival of earthworms, as soon as these doses are exceeded the probability of effects on mortality increases. Significant effects on reproduction and growth at doses higher than that of field doses are widely reported (Pelosi et al. 2014).

The herbicide S-metolachlor + terbuthylazine @ 6.25 mL/L were found to be highly toxic to earthworms, *E. foetida* by filter paper bioassay at 24 h (Iordache and Borza 2011). The mortalities of *E. fetida* worms were found to be 50 %, 72 % and 30 % in surface sprayed soil test, immersion and contact filter paper tests, respectively using the field recommended dose of methomyl whereas it was 56 %, 32 % and 100 % for imidacloprid (Na et al. 2005). Diafenthiuron at field recommended dose caused no mortality in *P. excavatus* (Stanley 2007).

3.1.2 Chronic Toxicity

Both deltamethrin and fenvalerate were found to have profound chronic cytotoxicity even at very low concentrations (2.0 and 0.5 mg/kg). These values are far beyond the LOEC values based on the acute test (25 mg/kg) (Song et al. 2015). A single application of benomyl, ethoprop, carbaryl or bendiocarb at recommended rates were reportedly reducing the population of earthworm by 60–99 % for about 20 weeks (Potter et al. 1990). Benomyl and thiophanate methyl at 2 % concentration in immersion bioassay caused total mortality to *E. foetida* in 75 days. A mortality of 60 and 65 % to *E. foetida* was found in 2 % exposure to thiram and cadmium succinate after a period of 101 days. Worms fed with benomyl and ethazole treated Bermuda grass clippings died in significant amounts within 34 and 84 days, respectively (Roark and Dale 1979). Benomyl at the rate of 1.8 kg/ha per year in apple orchard was reported to kill the earthworms like *L. terrestris* and *Allolobophora* spp. (Stringer and Wright 1976; Brown 1978). Tests after 16 weeks of carbendazim application in the field showed a significant decline in abundance as well as the biomass of the earthworm community as studied by taking intact soil columns (Rombke et al. 2004).

When endosulfan was used to treat semi-arid tropical grass lands, earthworms were not found in the plots treated with 3× dose of endosulfan, until 80 days after treatment. The abundance of worms was found significantly low in the plots treated with field recommended dose of endosulfan. The adults, juveniles and total earthworm population were reduced by 52–58 % and 15–50 % after 40 days of treatment at the high dose (3×) of methyl parathion and carbaryl, respectively (Reddy and Reddy 1992). A significant reduction in the growth and fecundity of earthworm exposed to chlorpyrifos at 5 mg/kg even after 8 weeks of treatment was reported (Zhou et al. 2007).

3.1.3 Effective Concentrations

The median effective concentration values (EC_{50}) for the effect of carbendazim on the abundance of earthworm was reported as 2.04–48.8 kg a.i./ha and on biomass as 1.02–34.6 kg a.i./ha (Rombke et al. 2004). An EC_{50} value of 2.9 mg carbendazim/kg for cocoon production of *E. andrei* is reported by Van-Gestel et al. (1992). The EC_{10} values for the above chemicals are 0.16, 0.30 and 0.68 0 mg a.i./kg of dry soil, respectively. EC_{50} of formulated pesticides, chlorpyrifos, carbofuran and mancozeb to *P. excavatus* were reported to be 3.0, 1.1 and 22.0 mg a.i./kg of dry soil, respectively. Eventhough carbofuran was found more toxic than chlorpyrifos as per median lethal doses, the EC_{10} values for chlorpyrifos were less than that of carbofuran (De-Silva et al. 2010).

The no effect concentration (NOEC) values of chlorpyrifos, carbofuran and mancozeb on the effects on reproduction and growth to *P. excavatus* were found to be <1, <0.5 and <1 mg a.i./kg of dry soil, respectively (De-Silva et al. 2010). The NOEC of carbendazim for earthworm abundance ranged from 2.16 to 87.5 kg a.i./ha and for biomass it was 1.08–87.5 kg a.i./ha (Rombke et al. 2004). The no effect levels (NOEL) for cocoon production of *E. fetida andrei* exposed to pentachlorophenol and 2, 4-dichloroaniline for 3 weeks were 32 and 56 mg/kg dry soil, respectively (Van-Gestel et al. 1989). The NOEL of metaldehyde to *E. fetida* is more than 1000 mg/kg soil (Bieri 2003).

3.2 Sublethal Effects

3.2.1 Effects on Life History Traits

3.2.1.1 Damage to Body

A gradual morphological damage to the earthworms, *A. caliginosa* such as rupture of the cuticle, bloody lesions or fragmentation of posterior parts was observed when treated with chlorpyrifos. The damage intensity was in relation to the chlorpyrifos concentration (4 and 28 mg/kg) and time of exposure (24 or 48 h) (Rao et al. 2003). Acute toxicity of imidacloprid in earthworms may be due to cellular autolysis caused by enzymatic inhibition (Luo et al. 1999). Such a phenomenon can be observed in the form of leaking cellular liquids in dying worms (Alves et al. 2013). Benomyl is reported to cause slow damage to the male reproductive system of earthworms (Sorour and Larink 2001). Phosmomidon caused swelling in the entire body of earthworms, *Lampito mauritii* (Bharathi and Subbarao 1984). Earthworms exposed profenophos at its LC_{50} (3.55 mg/cm^2; 48 h) revealed loss of architecture with excess of glandular epithelium cells causing disintegration of cuticular membrane and ectoderm layers (Reddy and Rao 2008). All the individuals of *E. foetida* treated with sublethal doses of malathion were found exhibiting coiled tails (Espinoza-Navarro and Bustos-Obregon 2004). Severe body malformations such as coiling, longitudinal muscle contraction, body rigidity and surface swelling along with bleeding lesions were noticed when *L. terrestris* exposed to many insecticides (Haque and Ebing 1983). Another experiment by Stenersen (1979) also describes stiffness, sores, blood extrusions and blisters especially in clitellum when tested with *L. rubellus*, *A. chlorotica*, *A. caliginosa* and *E. fetida.* When the earthworms, *E. foetida* were exposed to 5 % malathion through soil, the cells in the clitellum are found to be affected. In type-1 cells, the microtubules and the secretary bodies were distorted (Bansiwal and Rai 2010).

3.2.1.2 Food Intake, Locomotion and Respiration

Reducing the intake of food is a mechanism to avoid toxins. This strategy is commonly used in earthworms to avoid poisoning of pesticides (Nunes and Espindola 2012; Wang et al. 2012b). A significant reduction in manure ingestion by *E. fetida* was observed with exposed to imidacloprid at concentrations as low as 1.91 mg/kg (Gomez-Eyles et al. 2009).

Earthworms were found to be less active when exposed to fungicides like benomyl and chlorothalonil (Reicher and Throssell 1997). The *L. mauritii* worms exposed to dichlorvos are immobile and curled in a semi circular fashion (Bharathi and Subbarao 1986). The carbamate insecticides carbofuran and carbaryl caused long lasting immobility and rigidity in several arable species of earthworms (Stenersen 1979). Locomotion and geotaxis of *Metaphire posthuma* was significantly affected after a 20 min. exposure to 0.125 ppm carbaryl (Gupta and Saxena 2003). Fungicide, epoxiconazole treatment increased CO_2 production in earthworms, *A. caliginosa* after 7 and 28 days of exposure compared to their un-treated controls (Givaudan et al. 2014).

3.2.1.3 Growth and Maturity

The growth of freshly hatched *E. foetida* was significantly reduced when exposed to copper oxy chloride (Helling et al. 2000) whereas a significant reduction in the body weight of adults was found when exposed to malathion (Espinoza-Navarro and Bustos-Obregon 2005). Reduction in growth rate in *E. andrei* was reported when exposed to paraquot, parathion, benomyl and carbendazim (Van-Gestel et al. 1992). Choo and Baker (1998) reported that the juveniles of *Aporrectodea trapezoids* lossed weight significantly with endosulfan treatment compared to untreated controls. Chlorfluazuron, atrazine and metalaxyl caused significant growth reduction in *A. caliginosa.* Chlorfluazuron at 107 mg/kg soil (LC_{25}) reduced the growth rate of *A. caliginosa* worms and showed a negative log (e) –0.39 after 4 weeks of exposure

(Mosleh et al. 2003a). Clitellum development was found to get retarded in *E. fetida* juveniles treated with dieldrin (Venter and Reinecke 1985).

Weight of the earthworm is reported as a more sensitive index compared to the mortality, indicating toxic effect of a pesticide (Zhou et al. 2006). *Lumbricus terrestris* treated with endosulfan at its $LC_{10,\,25\text{ and }50}$ concentrations was found to reduced in weight with a significant reduction in growth rate (Mosleh et al. 2003b). Booth and Halloran (2001) found growth to be significantly reduced in *A. caliginosa* (both juveniles and adults) on exposure to two organophosphate pesticides, diazinon and chlorpyrifos at 60 and 28 mg/kg doses, respectively. A significant weight loss in *Allolobophora icterica* and *Aporrectodea nocturna* was reported when exposed to sublethal concentrations of imidacloprid (0.5 and 1 mg/kg of dry soil) (Capowiez and Berard 2006). Significant dose-dependent weight loss in *A. caliginosa* was observed when exposed to IGRs like buprofezin, lufenuron and triflumuron tested for 4 weeks. The most prominent decrease was observed after 2 weeks of exposure (Badawy et al. 2013). Significant dose dependent weight reduction in *E. foetida* worms were reported when exposed to malathion at 100 and 200 mg/kg soil for 20 days whereas the weight was found to increase at 300 mg/kg (Pal and Patidar 2014). Copper chloride affects the digestion and absorption physiology of earthworms and thus leading to reduction in biomass (Khan et al. 2007). Pesticides cause reduction in growth and delays sexual maturity of juveniles and eventually impact the abundance of earthworms in the field (Givaudan et al. 2014).

The effect of herbicides such as amitrole plus ammonium, 2,2-DPA, trifluralin, glyphosate, propazine and simazine at 1, 10 and 100 mg/kg artificial soil against earthworms and found mortality only at 100 mg/kg soil of trifluralin. But the worms surviving in the other herbicide treatments at 100 mg/kg were severely affected by loss of weight after 7 days of exposure (Martin 1982). Acetochlor at 20–80 mg/kg reduced the growth of *E. fetida* and the juvenile emergence from cocoon was also reduced (Xiao et al. 2006).

3.2.1.4 Effect on Reproduction

Reproduction is reported to be a more sensitive endpoint than mortality, since a pesticide which cause mortality to earthworms is less hazardous than the one which cause reduction in reproduction. Most of the herbicides which are reported to be non lethal to earthworms affects its reproduction (Amorim et al. 2005). Pesticides affect reproduction in many ways such as affecting the sexual development, cocoon production, sperm viability, reduction in fecundity etc.

Cocoon production Cocoon production in earthworms was found to get reduced due to sublethal toxicity of pesticides. Inhibition of growth and cocoon production is reported in *E. fetida* treated with carbaryl and dieldrin tested in horse manure and sand (Neuhauser and Callahan 1990). Paraquat at 1000 mg/kg was found to reduce the cocoon production of *E. andrei* by 40.49 % and 23 % of them as unfertile ones. Parathion at 100 mg/kg has reduced the cocoon production by 75.74 % with 53 %

fertile cocoons whereas no cocoons were produced at 180 mg/kg. No fertile cocoons were produced in 3.2 and 6.0 mg/kg of benomyl and carbendazim, respectively (Van-Gestel et al. 1992). Cocoon production of *M. posthuma* was retarded at 0.125 ppm carbaryl and no cocoon was produced at 2.0 ppm carbaryl (Gupta and Saxena 2003). Cocoon production was found to reduce with the exposure of *E. fetida* juveniles to copper oxy chloride (Helling et al. 2000). Jensen et al. (2007) reported a reduction in the number of cocoons produced by *E. fetida* when exposed to abamectin at concentrations above 0.25 mg/kg. *Eisenia fetida* treated with deltamethrin and fenvalerate for 4 weeks period produced 55.75 and 88.49 % less cocoons (Song et al. 2015).

Sperm production and quality Malathion affects sperm count and its metabolites could affect sperm quality. A significant decrease in sperm counts were found in *E. foetida* treated with malathion 600 mg/kg (Espinoza-Navarro and Bustos-Obregon 2004) with a significant reduction in the spermatic viability (Espinoza-Navarro and Bustos-Obregon 2005). *Eisenia foetida* treated with malathion showed a disordered sperm structure with loss of central hilus, vacuolization and cells with small and pyknotic nuclei that could be described as apoptotic process. A dose dependent sperm deformity was displayed when *E. foetida* was treated with imidacloprid concentrations higher than 0.5 mg/kg dry soil (Luo et al. 1999). Abnormalities in sperm head was observed at 0.125 ppm carbaryl treatment to *M. posthuma*, whereas at 0.25 and 0.5 ppm, the sperm heads are deformed and the nucleus become granular (Gupta and Saxena 2003).

Rate of Reproduction A significant decrease in reproduction rate in the earthworms, *A. longa* and *A. rosea* associated with a reduction in the whole earthworm population was reported in benomyl treatments (Holmstrup 2000). Alshawish et al. (2004) had studied the chronic toxicity of chlorpyrifos on *A. caliginosa* and found that the chemical greatly affect the fecundity. A great impact of chlorpyrifos on the growth and fecundity of *A. caliginosa* was reported by Booth and O'Halloran (2001) and reported to affect the juveniles more seriously than the adults. Formulated carbofuran, mancozeb and chlorpyrifos depressed reproduction of *P. excavatus* more than their pure compounds (De-Silva et al. 2010).

3.2.1.5 Changes in Enzymes

Pesticides influence on the production and the activity of many enzymes in earthworms especially the ones which involve in the biotransformation of carbamates. *Eisenia fetida andrei* treated with carbaryl was found to have a significant depression of enzymes involved in biotransformation of xenobiotics (Ribera et al. 2001). Acetylcholine esterase (AChE) is the important enzyme for the biodegradation of pesticides. Significant inhibition (>60 %) of acetylcholine esterase activity was observed in *E. foetida* treated with chlorpyrifos 2.96 and 2.33 mg/mL through contact filter paper (Rao et al. 2003). Chlorpyrifos caused a choline esterase depression of 35 and 70 % in earthworms, *A. caliginosa* exposed to 4 and 28 mg/kg, respectively with respect to untreated. When *A. caliginosa* was exposed to chlorpyrifos 28 g/kg, cholinesterase activity was inhibited by up to 90 % while glutathione transferase activity was induced by 148 % compared to control after 14 days of exposure (Booth et al. 1998a, b, 2001). Butachlor did not cause any variation in AChE activity but maximum inhibition of AChE activity was found after 9 days exposure of tropical earthworm, *Drawida willsi* to malathion (2.2 and 4.4 mg/kg) and after 12 days exposure to carbofuran (1.1 and 2.2 mg/kg) (Panda and Sahu 2004). The cytochrome P450 (CYP3A4) enzyme activity was reported to reduce by 2.0-fold in the 30 mg/kg deltamethrin treatment to *E. fetida.* Fenvalerate was found to reduce the CYP3A4 enzyme in 28 days treatment. However, in some treatments with both fenvalerate and deltamethrin a stimulating effect is also recorded (Song et al. 2015).

Enzyme activities viz., cellulase, superoxide dismutase and catalase activities were used as biomarkers in chlorpyrifos and fenvalerate treated earthworms, *E. fetida* (Wang et al. 2012b). The pesticides dimethoate, primiphos-methyl and deltamethrin caused a significant inhibition of AChE and carboxyl esterase activities and significant changes in activities of catalase, glutathione S transferase and glutothione concentration in *E. andrei* and *L. rubellus* (Velki and Hackenberger 2013). Carboxyl esterase activity was inhibited by chlorpyrifos-oxon in the pharynx, crop, gizzard, anterior intestine, wall muscle and reproductive tissues of *L. terrestris* in various degrees (Sanchez-Hernandez and Wheelock 2008). Luo et al. (1999) reported a reduction in cellulase enzyme activity in *E. fetida* worms treated with imidacloprid at 0.1 mg/L. Booth and O'Halloran (2001) reported a reduction in choline esterase and glutathione s-transferase activity in juvenile worms exposed to 12 and 60 mg/kg of diazinon, respectively. *Eisenia fetida* exposed to azoxystrobin was found to have excess superoxide dismutase, guaiacol peroxidase and glutathione-S-transferase activity with minimum activity of catalase (Han et al. 2014).

3.2.1.6 Effects on Lysosome

Lysosomal damage to earthworms, *A. caliginosa* which can be visualized by the use of the neutral red retention (NRR) assay was reported when treated under semi field and field conditions (Booth et al. 2001). Copper oxychloride 4.25 g/L was found to reduce the lysosome of *Microchaetus* sp. as reveled by NRR assays (Maboeta et al. 2002). Booth and O'Halloran (2001) reported a significant reduction in NRR time of lysosomes when *A. caliginosa* was exposed to chlorpyrifos and diazinon. A reduction in NRR time was found in *E. foetida* exposed to dichlorvos at 76 mg/kg for 7 days as 15.09 as compared to control (54.09). A strong dose dependent variation in NRR time was noticed but was not significant with the exposure time (Farrukh and Ali 2015). Oxidative stress and DNA damage was reported in *E. fetida* due to azoxystrobin in soil (Han et al. 2014).

3.2.2 Activity or Behaviour Alterations

3.2.2.1 Behaviour Alteration: Hyperactivity

A few studies show the changes in activity levels and mobility in earthworms treated with pesticides. Sublethal concentrations of insecticide, phosphomidon and fungicide, enilconazole (Imazalil®) were reported to induce hyperactivity in *L. mauritii* and *E. foetida,* respectively (Bharathi and Subbarao 1984; Van-Leemput et al. 1989).

3.2.2.2 Behaviour Alteration: Repulsion or Avoidance

Pesticides cause repulsion and thus the worms tend to avoid pesticide contaminated soils. A significant repellent effect on *E. foetida* was reported for chlorpyrifos at a concentration of 40 mg/kg (Zhou et al. 2007). Enchytraeids are reported to avoid areas treated with benomyl and carbendazim (Amorim et al. 2005). Earthworms, *E. andrei* avoided benomyl, carbendazim and dimethoate contaminated soils, as revealed by >80 % individuals found in the control soil (Loureiro et al. 2005). Avoidance behaviour of earthworms to pesticides depend on the pesticide and their effective concentrations because some species may not be able to detect low concentrations (De-Silva and Van-Gestel 2009). No significant avoidance was noticed in *A. icterica* and *A. nocturna* when exposed to sublethal concentrations of imidacloprid (0.1, 0.5 and 1 mg/kg of dry soil) (Capowiez and Berard 2006). A preference of earthworms, *E. andrei* to insecticide fipronil irrespective of concentrations tested and an avoidance to thiamethoxam at concentrations 100 times lower than that of the LOEC for reproduction is also reported (Alves et al. 2013). A similar attraction was observed in *E. andrei* tested for ivermectin and may be due to low sensitiveness of chemoreceptors (Torkhani and Erzen 2011). Though avoidance of earthworms, *E. andrei* was found in methomyl spiked soil at concentrations from 5.62 mg/kg onwards, it is much less than the AChE inhibition, which occurred at still lower concentrations (0.86 mg/kg) (Perreira et al. 2010).

Pesticide induced surface migration of earthworms are determined by trapping technique. Benomyl treated grasslands caused an increased migration of worms up to 2.8 times in 2 days (Christensen and Mather 2004). Vertical distributions and burrowing behaviour of *A. chlorotica* was studied in soils spiked with carbendazim. The worms were found to alter their burrowing behaviour to avoid carbendazim (Ellis et al. 2010).

3.2.2.3 Effects on Activities: Burrowing

Changes in earthworm behaviour in terms of modified or reduced burrowing are critical because of its greater influence in soil functions (Capowiez et al. 2006). But pesticide impacts on burrowing behaviour are not well studied or reported because of the difficulty in visualizing or estimating the burrowing activity (Givaudan et al. 2014) and a few studies like that of Gupta and Sundararaman (1991) and Capowiez et al. (2010) have tried to link pollutants with that of burrowing of earthworms. Pesticide may cause increased burrowing in worms (Zhang et al. 2009; Givaudan et al. 2014) or decreasing in activity (Dittbrenner et al. 2010; Pelosi et al. 2014) which depends on the pesticide and also the concentration. Increase in burrowing behaviour could be induced by the metabolic changes and just like that of hormesis (Zhang et al. 2009) and the decrease may be due to loss of energy.

Effect of imidacloprid on the burrowing behaviour of *A. caliginosa* was detected at concentrations as low as 0.2 mg/kg, whereas such effects in *L. terrestris* were not observed even at ten times higher concentrations. Imidacloprid also influenced on the total burrow length and the maximal burrow depth (Dittbrenner et al. 2011). After 24 h of exposure of earthworms to imidacloprid, *A. icterica* stopped burrowing, whereas *A. nocturna* burrowed with a significantly lower rate (Capowiez and Berard 2006). Significant effects on the characteristics of the burrow systems, i.e. length, depth and branching rate were noticed (Capowiez et al. 2003).

3.2.2.4 Effects on Activities: Cast Production

Estimation of cast production may be an indirect criterion to find the effect of pesticides on earthworms. A reduced cast production in contaminated soils rather than increase is reported but it is dependent on the level of contamination and the contaminant (Dittbrenner et al. 2010). Estimation of cast production (total or surface), after the exposure of different species of earthworms to a range of pesticides in laboratory as well as in fields are reported (Capowiez et al. 2010; Lal et al. 2001). A decrease in cast production by *L. terrestris* to a tune of 67 %, 54 %, and 61 % was reported for methomyl, carbaryl and imidacloprid, respectively after 7 days (Capowiez et al. 2010). Casting by earthworms in turf areas was significantly reduced by the treatment of gamma-HCH (lindane), thiophanate-methyl and carbendazim (4 L/ha) for about 12–15 months. The effect of carbaryl on casting was found for 5–12 months. When carbendazim was applied at a higher rate of 8 L/ha, cast suppression lasted 12–18 months (Baker et al. 1998). Legg (1968) found that carbaryl as very toxic to earthworms which reduced earthworm casting after 3 weeks of application. Compared to the water control, carbaryl and T-methyl application in field reduced surface casting of earthworms by 90 % (Tu et al. 2011). Earthworms, *A. caliginosa* exposed to fungicide, epoxiconazole transitorily enhanced the cast production and slightly decreased it after 7 days (Givaudan et al. 2014).

3.3 *Field Effects*

3.3.1 Toxicity in Field Conditions

Insecticides such as phorate and carbofuran when applied in soil are found harmful to earthworms (Edwards and Bohlen 1992). The granular soil insecticide, phorate which is widely used especially for soil pests is reported to be highly toxic to earthworms (Way and Scopes 1968). Application of carbofuran to soil was reported to strongly affect earthworms (Haque and Ebing 1983; Edwards and Bohlen 1992). Organochlorine insecticides like DDT, aldrin, dieldrin and BHC are less toxic to earthworms; heptachlor, endosulfan and isobenzan are moderately toxic and endrin, chlordane and lindane are highly toxic to earthworms (Edwards and Bohlen 1992; Slimax 1997). In general, either the natural and synthetic pyrethroids are non-toxic to earthworms (Edwards and Bohlen 1992).

Unlike other non-target organisms, earthworms are highly susceptible to fungicides also. The residues of copper sulfate and carbamates in the form of copper and zinc molecules are highly toxic to earthworms. The pesticides which are used as soil fumigants, nematicides and fungicides especially the DD mixture (dichloropropane: dichloropropene), metham-sodium and methyl bromide are dangerous to earthworms (Edwards and Bohlen 1992; Paoletti 1999). Herbicides are also found toxic to earthworms in field. *Allolobophora chlorotica* (epigeic) and *A. rosea* (endogeic) are negatively affected in grasslands sprayed with atrazine and pentachlorophenol (Conrady 1986).

3.3.2 Loss in Abundance and Diversity/Disappearance

Abundance and diversity of earthworms are reportedly reduced by chemical and mechanical stress in agricultural lands (Decaens and Jimenez 2002; Smith et al. 2008). Literature on the effects of pesticides on diversity and mass was reviewed by Pelosi et al. (2014). Pesticides are generally disastrous for earthworm communities. There is also an indication of change in species composition due to pesticide usage and it occurred in favour of larger individuals, e.g. anecic earthworms. The abundance of Megascolecidae earthworms are inversely correlated with the soil arsenic concentration (Fang et al. 1999).

In field conditions, six consecutive weekly applications of T-methyl and carbaryl was found to reduce the abundance and biomass of earthworms even 6 weeks after pesticide application. However, the species composition was not found to be influenced by pesticidal treatment in the field and remain unchanged during the entire experiment (Tu et al. 2011). One time application of field recommended doses of the fungicide benomyl or the insecticides ethoprop, cararyl or bendiocarb was reportedly reduce the earthworm numbers and biomass (60–99 %) in Kentucky bluegrass systems (Potter et al. 1990). Application of thiophanate-methyl at 0.78 kg/ha was found to reduce the earthworm numbers in apple orchards (Stringer and

Lyons 1974). No effect on the density of earthworms was reported by the application of formulated metaldehyde at the recommended rate (Iglesias et al. 2003).

4 Methods to Assess Pesticide Toxicity to Earthworms

Pesticide toxicity testing in soil usually involves earthworms. There are many evaluation procedures for pesticide toxicity to earthworms and it is being evolved now and then. Laboratory tests normally aim in arriving at median lethal concentrations using acute toxicity tests. Acute toxicity tests are conducted through filter paper or vial coating for contact toxicity or using soil contamination for ingestion and contact toxicities (ISO 1993). Apart from acute toxicity tests some sublethal effects like life history traits such as growth, reproduction etc. activities like cast production, litter burial and behavioural traits like avoidance, burrowing etc. are conducted under controlled conditions. However, tests for earthworm behaviour are not widely carried out, with a notable exception of the avoidance test, which is the most controversial and have a least relation with a soil function (Pelosi et al. 2014). Field testing usually involves studies on the impact of pesticides on earthworm diversity, abundance and density. A prediction of potential environmental hazard is difficult because of variation in test procedures, experimental designs, soil type etc. (Christensen and Mather 1994). The results of various field trials identifies highly or extremely toxic pesticides but they do not accurately identifies the moderately or slightly toxic ones (Edwards and Bohlen 1992). Thus, designing a test procedure that incorporates both lethal and sublethal endpoints is yet to be achieved (Christensen and Mather 1994). A combination of different tests solves the purpose and can be used to find the toxicity of pesticides to earthworms.

4.1 Laboratory Experiments

In general, standardized toxicity tests such as acute toxicity bioassays are characterized by their simplicity, rapidity and low cost. A simple laboratory assay is an irreplaceable tool for toxicity testing and hazard assessment because of its practicability (Ronnpagel et al. 1998), but the ecological significance is very limited. Many a times, some of the aspects listed below are ignored while conducting laboratory assays (1) long-term exposure to sublethal concentrations of contaminants, (2) effects of contaminant mixtures (3) influence of fluctuating environmental factors etc. (Sanchez-Hernandez 2006). Thus a field realistic experiment which is of ecological relevance can be used.

4.1.1 Acute Toxicity Bioassays

Acute toxicity is studied by contact, ingestion or by injecting the test substance on earthworms. The main advantage of laboratory testing is that, controlled environmental conditions are possible and reproducible. Though many evaluation procedures are available, the procedures which are of more realistic, simple and economical but retain the ecological relevance are to be used (Christensen and Mather 1994). The filter paper contact test and the artificial soil contamination tests are realistic and thus used widely.

4.1.1.1 Contact Toxicity Tests

Contact filter paper test is considered as an initial screening procedure to evaluate the relative toxicity of pesticides to earthworms. Contact toxicity tests are used to study the pesticides which are absorbed mainly through the skin of the worms. But most of the tests fail to represent the exact situation in the soil (Miyazaki et al. 2002; Grumiaux et al. 2010; Tripathi et al. 2010).

1. Topical application (Christensen and Mather 1994)

   In this method, earthworms are selected for age and size and topically dosed with the pesticide i.e. the pesticide solution is placed on the earthworm's skin (e.g. dorsum anterior to clitellum) by means of a micro-applicator or a paint brush. When the applied solution gets dried the worms are kept in soil and assessed for mortality in prescribed time intervals. Unlike topical bioassays with insect, the results of topical bioassay with earthworms are highly variable. The topical application often results in rapid or vigorous exudation of mucus which impairs the contact of test solution with the worm's skin. Because of these reasons topical bioassay is not recommended as a good and common procedure for earthworms (Fig. 5.1).

2. Vial coating/contamination bioassay (Park et al. 2012)

   In this experiment, test vials are coated with the pesticide solution before introducing the worms. The assay vials were coated with the test solutions prepared in acetone at the rate of 10 μL/vial using a micro applicator. The vials were kept in such a way to allow the acetone to get evaporated until dryness. Controls were only treated with acetone. An individual earthworm was released in each vial with an addition of 1 mL of deionized water. Acute mortality can be estimated at a definite time period. After 7 days of introduction of earthworms and treatment, the worms (both dead and alive) were taken out and homogenized by adding 10 mL of acetone using a homogenizer in an ice-cold conditon. The homogenates thus obtained were vortexed for 2 min., centrifuged and extracted in ethyl acetate to get analyzed in GC. After removal of worms, the residues in the vials were also extracted with 5 mL of acetone twice, re-dissolved in ethyl acetate and analyzed.

   This method allows the evaluation of acute contact toxicity and also the pesticide residue in the test organism. However, the pesticide solution if made to get dried at

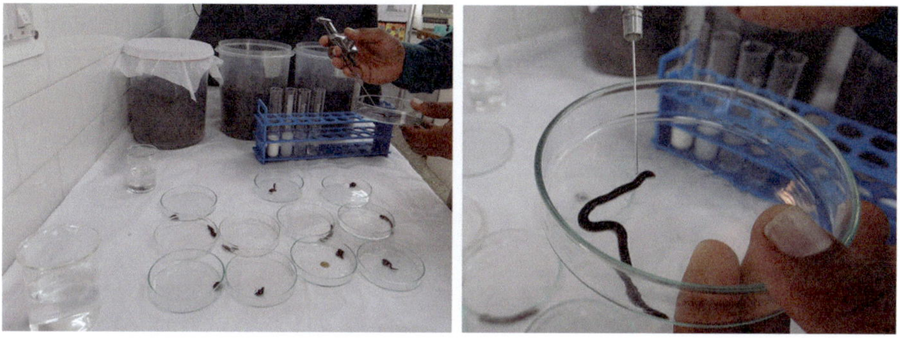

Fig. 5.1 Topical application of pesticides on earthworms

the sides also instead of only on the bottom of the vial may be more realistic because the worms may not get full contact to the bottom of the test vials.

3. OECD filter paper method (OECD 1984; Iordache and Borza 2011)

In this method, glass vials were lined with filter paper of suitable size so that they don't overlap in the vial. The filter paper used here is of 80–85 g/m^2, ~0.2 mm thick and of medium grade. One milliliter of prescribed concentration of pesticide solution was introduced in each test vial with paper. The vials are placed in horizontal position in front of a slow air stream and ventilated until the filter paper gets dried. After drying of pesticide, 1 mL of distilled water was added to wet the filter paper. In the control vial, the filter paper was wetted only with distilled water. The worms were maintained for 3 h on a wet filter paper to eliminate the gut contents and introduced into the vial singly and closed with perforated plastic foil for aeration. Muslin cloth with rubber bands can also be used for closing. A minimum of ten replicates per treatment, each consists of one worm per vial are to be used. The vials were placed in horizontal position in laboratory conditions for 72 h and mortality noted. Filter paper bioassays have the advantage of being easy, quick and inexpensive (Fig. 5.2).

4. Filter paper test using Petri dishes (Wang et al. 2012a)

Here, the test substance is dissolved in acetone and a piece of filter paper was kept in a 9 cm Petri dish and treated with 2 mL of the test solution. The solvent was allowed to evaporate and the filter paper moistened with 2 mL distilled water and one earthworm placed on it. Earthworms were placed in wet filter paper for 24 h to void all the gut contents for 24 h prior to test introduction. The set up was kept in 20 °C for 48 h and mortality recorded. Moribund earthworms are recorded as dead if they are not responding to a gentle mechanical touch on the front end. A preliminary range finding test was conducted to determine concentration range for each pesticide in which a 0–100 % mortality of the worms obtained. At least five concentrations and a control were included for each pesticide to find the median lethal concentrations.

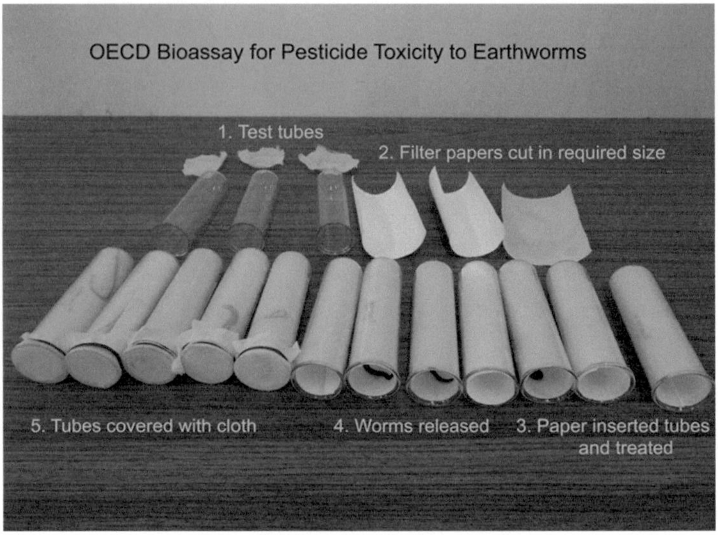

Fig. 5.2 OECD bioassay method – glass vials lined with filter paper

These tests stated above, either the filter paper or vial coating or the topical application does not involve the ingestion of toxicant i.e. the exposure via gut is not considered. Similarly the simulation of adsorption of pesticides in soil colloids is not taken into account, thus have an impediment in their ecological relevance (Christensen and Mather 1994). Further, a poor correlation was reported between the filter paper and soil contamination assays (Heimbach 1988). Some chemicals which are highly toxic by contact to earthworms may not have the same level of toxicity by ingestion and vice-versa. Further in filter paper tests, the worms may spend much energy in trying to slough off the chemical with mucus and eventually die (Christensen and Mather 1994) (Fig. 5.3).

5. Immersion test (Roark and Dale 1979)

Though immersion test is characterized as a contact toxicity test, the exposure may also through gut. In this immersion test, a special cylindrical metal container (7.5 × 6.5 cm) with 16 mesh galvanized screen bottom was used. Pesticide solutions (500 mL) at prescribed concentrations were prepared in a glass beaker. About 10 equal sized earthworms were placed in the cylindrical metal container and the container with worms was dipped and gently agitated for 1 min in the 500 mL solution/suspension in the glass beaker. After treatment, excess pesticide solution was removed from the screened bottom of the container by blotting on a clean paper towel and the worms placed in uncontaminated soil and food provided. Observations on the mortality were taken at regular intervals. Though this test is easy to do and reveals the median lethal concentrations, it is not directly related to the soil doses. Thus, it is difficult to relate the arrived median lethal values with that of the field conditions (Christensen and Mather 1994).

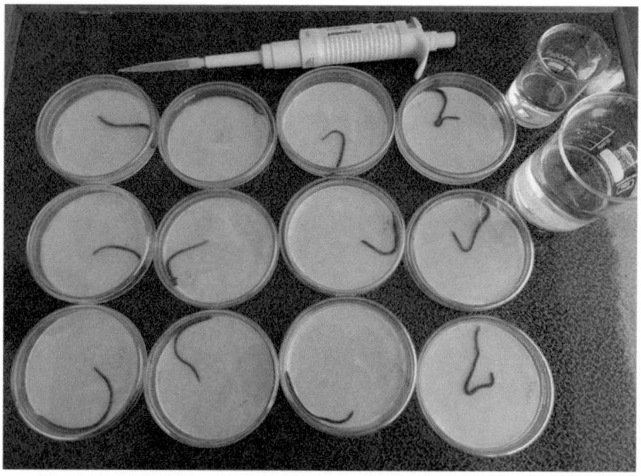

Fig. 5.3 Bioassay using filter paper in Petri dish

4.1.1.2 Ingestion Toxicity Tests

1. Forced feeding test (Christensen and Mather 1994)

In this experiment, the pesticide is directly fed/put into the oesophagus of the earthworm forcefully. The test chemical is made as a suspension in agar gel. The test organisms, earthworms of equal and big size are selected and anesthetized using 10 % aqueous solution of ethanol. A blunt needle is usually inserted into the oesophagus and through which the chemical mixture injected into the gut. The needle is taken out and worms placed in moist filter paper in boxes for observation. The disadvantage of this method is that can be used only for larger worms and it takes into account only the gut exposure and not the skin.

2. Voluntary feeding – Contaminated feed test (Roark and Dale 1979)

Turf thatch is a component of the diet of some earthworms in nature, so this test can be conducted to determine whether the ingestion of pesticide treated grass clippings was deleterious to earthworms. Bermuda grass (*Cynodon dactylon*) clippings from a pesticide free plot were taken, air dried and finely ground in a Wiley mill with a sieve of 1 mm porosity (25 mesh). For pesticide treatment, the ground samples (15 g) were stirred with 100 mL of the required concentration of pesticide solutions. Control Bermuda grass was treated with water. After 16 h of saturation, excess pesticide solution or water can be removed from the sample by vacuum filtration. The worms are fed with only Bermuda grass for the entire period of experimentation. Observations on the mortality of worms are taken at regular intervals.

This method tends to reflect field conditions as the worms have the choice whether to eat the contaminated food or not and thus have an ecological significance (Christensen and Mather 1994). However the method does not distinguish between the antifeedant and toxic effect of the pesticide under test (Edwards and Bohlen 1992).

4.1.1.3 Ingestion and Contact Toxicity: Soil Contamination Assays

Artificial soil test represents the natural environment to earthworms and the test pesticides are mainly absorbed through the gut of the worms (De-Silva and Van-Gestel 2009; Udovic and Lestan 2010; Wang et al. 2012a) apart from skin contact. Thus skin contact and gut intake cannot be separated unless in specific experiments where oral sealing is done to study only the skin uptake (Vijver et al. 2003). The soil contamination bioassays are ideal for testing pesticide toxicity to earthworms since it is more realistic and suited to assess sublethal toxic effects. In literature, earthworm species of limited ecological significance are used in many studies. Extension of this method using native earthworm species and with sublethal endpoints, are preferable (Christensen and Mather 1994).

1. Soil contamination test in glass containers (OECD 1984)

The artificial soil consists of 10 % sphagnum peat with no visible plant remains and finely ground, 20 % kaolin clay and 70 % sand. The pH of the medium is adjusted to 6.0 by addition of calcium carbonate. After proper mixing, distilled water is added to makeup the moisture content to 35 % of the dry weight. The complete mixture should be moist but not so wet that water appears when compressed. With some peats, a moisture content of over 35 % may be suitable. Adult worms of at least 2 months old having clitellum and weighing about 300–600 mg can be used.

The test substance (pesticide) is usually dissolved in distilled water and mixed thoroughly with the artificial soil media or evenly sprayed over it. The test substance can be dissolved in small amount of organic solvent if it is not get dissolved in water, but care should be taken to make the solvent to evaporate completely. For each test, 750 g of the test medium is used in glass container and 10 earthworms introduced. The earthworms were conditioned for 24 h in artificial soil before introducing in the test medium. The worms can be released on the surface of the medium and the containers covered to prevent the worms from escaping and to keep the media from drying. The test usually lasts for 14 days and assessment of mortality can be taken at 7 and 14 days of exposure.

2. Soil contamination assay in earthen pots (Ganeshkumar 2000)

In this experiment, tubular earthen pots of 18 × 6 cm were used to test the effect of pesticides on earthworms. The test substrate was prepared by mixing fine quartz sand 83.5 % (particle size between 0.06 and 0.2 mm), bentonite 5 %, finely ground and dried sphagnum peat 10 %, pulverized calcium carbonate 1 % and dried and ground cow dung 0.5 %. The pH was adjusted to 7 and moisture content 40 %. The complete mixture was moist enough but not so wet that water appear when the soil is compressed. Pesticide was mixed at prescribed concentration in the soil and about 500 g of pesticide mixed medium or uncontaminated media for control put in one pot. About 15 worms washed cleanly in water were placed on the top of the substrate. The tubular pots were covered with perforated polythene cover to prevent the worms from crawling out and to avoid evaporation loss. The set up was kept under shade and after 7 days, 5 g of finely ground cow dung was mixed inside the container

Fig. 5.4 Soil contamination test in earthen pots

and water loss by evaporation replenished. The number of live earthworms was counted based on mechanical stimulus and those worms were considered dead if they did not respond to a gently mechanical stimulus. The LC_{50} can be calculated by probit analysis as described by Finney (1971) (Fig. 5.4).

3. Soil contamination assay in big plastic tubs (Stanley et al. 2016)

In this procedure, tubs of 40×13 cm are used to find the toxicity of pesticides to earthworms. The tubs are filled with soil, sand and farmyard manure in the ratio 1:1:1 with appropriate moisture. The insecticides were sprayed on the surface of the soil at field recommended concentration calculated based on the surface area of the tub. After 15 min. of insecticide application, 20 earthworms were selected for the size and released on the surface. The experiment was repeated thrice. Water spray was given on the surface using a hand sprayer at weekly intervals. Mortality observations can be taken on 15, 30 and 60 days after treatment by hand sorting. The advantage of this method is the resemblance to field application i.e. pesticides were applied to the soil surface and allowed to percolate downward with simulated rainfall (Fig. 5.5).

4. Soil contamination assay in plastic buckets (Badawy et al. 2013)

This experimental procedure is based on those described by Heimbach (1984). Standard artificial soil, adjusted for pH and 35 % moisture are filled in buckets and 10 (0.6–0.7 g wt) earthworms introduced. Observations on the mortality were taken weekly intervals. Moisture loss was determined by weight loss. The buckets were weighted after each assessment and the lost weight was replaced with distilled

Fig. 5.5 Soil contamination bioassay using plastic tubs and buckets

water. The weight of each earthworm was determined after 1, 2, 3, and 4 weeks of exposure. This method seems to be useful for long term studies.

5. Soil contamination assay in polycarbonate boxes (Givaudan et al. 2014)

Commercial formulation of pesticide was diluted in distilled water as per field concentration assuming a single application and homogenous distribution in the top 5 cm soil and no crop interception (Dittbrenner et al. 2010). Soil spiking was done manually by adding 175 mL of the test solution in 2 kg of soil. Soil microcosms consisted of polycarbonate boxes (80 × 50 mm) with a lid pierced with tiny holes to ensure aeration. The microcosms were filled with 100 g of test soil and some dry grass meal was kept on the surface. They are kept for 2 days in a cool dark room and was checked for moisture and adjusted to 25 % prior to introduction of worms. Earthworms are washed in tap water, weighed and placed individually in dishes for about 48 h for gut voiding, prior to the start of exposure. Then worms were transferred individually to the microcosms and the exposure lasts for 7 and 28 days.

6. Artisol tests (Heimbach 1984)

Artisol test is same as that of the soil contamination bioassays described above but uses an artificial substrate consisting of silica, water and glass balls. The artisol is treated with pesticide and the earthworm is introduced into it. In artisol test, the earthworms ingest the silica paste as they do the soil. This is no way advantages than soil contamination so not used widely.

7. Tubifex test (Christensen and Mather 1994)

This test involves the use of a fresh water oligochaete, *Tubifex* as a model organism. This test organism is usually introduced into glass cylinders filled with water and test pesticides applied in to the solution. Usually ten specimens are introduced in a cylinder and the activities and mortality observed at regular intervals. This test involves the use of an aquatic animal to test the effects on terrestrial ecosystem due to the advantage of testing the water soluble pesticides directly and thus facilitates a rapid screening. But the disadvantages being the test organism not really a representative one and the non consideration of chemical adsorptive characteristic of soil make this method an unrealistic one.

4.1.1.4 Injection Bioassays

1. Injection bioassay (Park et al. 2012)

The pesticides used for injection experiments can be dissolved in ethanol. The test concentrations are prepared using this ethanolic pesticide solution diluted with distilled water (1:99). The test solution is then injected (1 μL) into the haemocoel of the anesthetized worm, directly behind the clitellum, using a microapplicator (Fisher 1984). The injected worms are reared in soil with proper food and conditions. Mortality of worms at each concentration of pesticide was recorded after 48 h of exposure. These data can also be used to estimate median lethal concentration (LD_{50}) using Probit analysis according to Finney (1971).

The administration of test solution often injures the worm and thus the results are inconclusive. Further, the method of uptake of chemical via coelom is highly artificial and not related to field exposures (Christensen and Mather 1994).

4.1.2 Sublethal Toxicity Tests

In sublethal toxicity bioassays, assessment of pesticidal effects on the growth and reproduction of earthworms are generally studied. Most of the sublethal toxicity tests involve the use of soil contamination bioassay but with sublethal endpoints. If sublethal endpoints are to be studied, the exposure time should be increased considerably than that of the acute mortality and thus additional food is commonly added. In the soil contamination bioassays especially with sublethal endpoints, 0.5 kg dry artificial soil was placed in 1 L glass jar. Finely ground cow dung was put in the centre by making a hole in the soil. Earthworms especially of the species *E. fetida* do not produce enough cocoons without the addition of cowdung. So addition of cow dung is important in long term assays and especially when end point to be taken is reproduction. For cocoon incubation also, 1 % finely ground cow dung is to be added as a food source for the hatching worms (Van-Gestel et al. 1989). A soil medium of vermiculite matrix with vitamins, humic acid as additives are used for long term assessments (Bouwman and Reinecke 1991).

Apart from the above said sublethal toxicity tests for life history traits, sublethal assays on behaviour involve studies on avoidance and litter burial, cast production and burrowing activities. Avoidance bioassay is perhaps the most used behavioural bioassays to study the effect of pesticides on earthworm. These tests are simple, effective, easy and shows whether the pesticide cause any avoidance or repulsive behaviour in the worms. This test shows the ability of the worms to detect toxic compounds and decide to escape from them. But the avoidance test is not appropriate for neurotoxic pesticides (Perreira et al. 2010) and it tests only the repellence rather than the toxicity (Capowiez et al. 2003). In avoidance tests, if <20 % of the total live individuals is found in the test soil at the end of the test, the soil is classified as toxic. The mortality of the test individuals also should not exceed 10 % in a control soil (ISO 2008). The net response (NR) is calculated using a formula and expressed as a percentage.

$$NR = \frac{(\text{individuals found in control} - \text{individuals found in treated})}{\text{Total number of individuals}} \times 100$$

(Amorim et al. 2005)

Apart from avoidance, some other behavioural changes can also be studied. These changes in behaviour can be broadly categorized as,

i. Lying on the top of soil and move slowly if touched
ii. Lying coiled on the top of the soil writhing when disturbed
iii. Remaining in the soil but very few burrows formed

(Martin 1986)

4.1.2.1 Sublethal Toxicity Assays for Life History Traits

1. Assays for growth of earthworms (Zhou et al. 2007)

This procedure is based on Khalil et al. (1996), who studied the effect of heavy metals on the growth of earthworms, but with slight modifications. These tests conform to the acute toxicity test using contaminated soils but the incubation duration is longer and additional food are given to support the worms for longer period. There were four replicates for each pesticide concentration tested and also for the control treatment with ten worms per container. Earthworms were weighed at weekly intervals to determine the effect of pesticide on growth. Moisture was replenished by checking at weekly intervals. Soil was also changed every 4 weeks and earthworms were placed into fresh contaminated soil.

2. Assays for reproduction – cocoon production and hatching (Zhou et al. 2007)

This test procedure is exactly like that of the growth test described above but with observations on cocoon production after 28 days and number of hatched juveniles after 56 days as end points. Adult mature worms were exposed to different concentrations of pesticides in a standard test soil. After 28 days of exposure the surviving worms and cocoons produced were removed by hand sorting. The cocoons were replaced in the same soil and further incubated for another 28 days (total incubation time 56 days) to find the emergence of juveniles. Endpoints of this assay include reproductive parameters such as cocoon production, hatching and change in weight of adults.

3. Assays for sexual development

The sublethal endpoint of sexual development has an ecological relevance in that a retardation of sexual development may cause decreased population density. The same soil contamination bioassay is used to study the effect of pesticides on sexual development of earthworms but with the end points like time to maturity, clitellum development etc.

4.1.2.2 Sublethal Assays for Change in Behaviour and Activities

1. Avoidance test – Two chambered (Slimax 1997)

In this pesticide avoidance test, glass test chambers (30.5 × 25.4 × 5.1 cm) divided into two sections of equal size by means of removable glass spheres were used. Pesticide contaminated soil is placed in one side of the test chamber and uncontaminated soil on the other chamber. The pesticides tests were done with half the field recommended dose, field dose and double the field dose. Altogether, five mature earthworms were released in the contaminated soil. After that the glass partitions were removed from the test chambers and the space filled with soil. The whole set up was maintained in a dark room maintained at 13 °C for 48 h. Then the temperature was reduced to 0 °C for 6 h to make the worms sluggish. The location of each earthworm is then determined by means of x-ray.

In an experiment by Schaefer (2004), the worms were placed on the centre line immediately after the removal of separators instead of putting them in the contaminated soil. After the incubation time of 48 h, the number of worms in each side i.e., contaminated and uncontaminated were counted and recorded. But this procedure of releasing worms in the central line may not include the risk of exposure to contaminated soil at all, if the worms move directly into the uncontaminated soil.

2. Avoidance Test: Six Chambered (Zhou et al. 2008)

This test is based on Stephenson et al. (1998) and Schaefer (2004). Round plastic containers (28 × 10 cm) with six different chambers connected to a small central chamber were used as test containers (Stephenson et al. 1998). Originally, each of the six compartments were connected to the adjoining chambers and to the central cylinder by three arches (1.0 × 0.5 cm). Preliminary tests by Schaefer (2004) showed that these few passages did not enable a satisfying migration. To improve migration, the chamber was modified by replacing arches with passage holes (5 mm) made at equal distances on the separator. As the test starts, ten worms were placed into the soil-free central chamber. Earthworms being negatively phototactic quickly moved into the soil filled chambers. Then the central chamber was closed with a plug and now migration is only possible between the chambers. After 48 h of incubation, the worms in each compartment were extracted by hand sorting and counted.

This test seems to be good but it fails to study the migration of worms from contaminated place to uncontaminated, if the worms directly enter into the uncontaminated compartment upon release.

3. Avoidance Test: Vertical Stacking (Ellis et al. 2010)

The test containers comprised two sections, one containing the pesticide added soil and the other uncontaminated. The two sections were stacked one over the other and earthworms are allowed to move freely between both the sections of the soil. Two different arrangements were made to study the behavioural response of earthworms. The first and field arrangement reflects the pesticide application in the field, in which pesticide amended soil at the top and uncontaminated soil below. In the second and alternative arrangement is just opposite to the first. Two open ended and translucent PVC cylinders of 8 cm high and 7.5 cm dia. were used as test containers. The test containers were wrapped with black adhesive tape to exclude light and 400 g soil was added to each container and kept as per the prescribed arrangement. The upper container was covered with 1 mm mesh and the bottom closed. Earthworms were introduced in two ways (1) added in the top and treated soil, to know the behaviour when it directly get contact with the contamination (2) is more field realistic where the worms are added in uncontaminated soil and contaminated layer placed above or below that.

4. Daniel's Funnel test for litter burial (Bieri 1992; Hogger et al. 1992)

It is a behavioural test to measure the effect of pesticides on the important activity i.e., the litter burial of earthworms. This test is generally carried out using agriculturally important and vertically tunneling species, *L. terrestris*. In this experiment, funnels of 12 cm dia. fitted with 30 cm long rubber tube with the other end closed are used as bioassay material. The funnel is darkened by using an insulating material and filled with soil. The worms are exposed to the pesticides by mixing them in the soil before filling the funnel or sprayed or applied on the surface as per the field rate or a range of rates. A small artificial tunnel was made in the soil and a juvenile worm is introduced. Food such as wheat kernels, leaves, straw or bait granules are usually kept on the surface of the soil. Observations on the numbers of moved food from the soil surface are recorded at regular intervals. The worm can also be recorded as present in the funnel or in the middle of the tube or at the end of the tube. After an experimental time of 3 weeks, the number of live worms were counted and weighed. The numbers of moved food and the position of the worms in the treatment were compared with that of the control, to find the effect of pesticide on the litter burial behaviour. This test is perhaps a good and reliable test to assess the activity of litter burial. It also explains the avoidance if any to the test chemical if applied on the top surface by burrowing deeper through the funnel into the tube. The only disadvantage of this method is the test animal being spatially restricted.

5. Effects (short term) on burrowing - 2D terraria (Capowiez 2000; Dittbrenner et al. 2011)

Modification of the burrowing behaviour of exposed earthworms may occur when pollutants have not been avoided. This burrowing behaviour can be assessed by using 2D-terraria (Evans 1947). The earthworms were pre-exposed individually to pesticide treated or control soil in Petri dishes. The pre-exposed worms are then used for the measurement of burrowing behaviour in uncontaminated soil in 2D terraria. In this experiment, the earthworms are forced to get exposure in the Petri dish and they do not have a chance to avoid the contaminated area.

Burrowing behaviour of earthworms was assessed after 1, 7 and 14 days of exposure by means of 2D-terraria. Each treatment was replicated seven times to get a concrete conclusion. Exposure experiments were carried out with individual worms in Petri dishes filled with 100 g of contaminated or uncontaminated soil. The soil was sieved to 3 mm and the soil water content was adjusted to 20 % and spiking was carried out to reach the final soil water content of 25 % (101 % of WHC). The terrarium was made of two planar glass sheets separated by 3 mm thick PVC pieces on three sides of each terrarium. These edges of the glass sheets were sealed with the help of adhesive tapes except the top edge which was left open. The terraria were filled with prescribed soil with proper moisture. After transparent plastic sheets had been attached to both sides of the terraria, worms exposed to pesticides in Petri dishes were introduced and incubated in a dark cold chamber.

Daily observations can be made under red light. The locations of earthworms are marked using colour pencils as arrows and new burrows as lines. The information gathered on the transparent plastic sheets were digitized on a digitizer, analyzed and studied. The characteristics describe individual earthworm behaviour (total length burrowed, daily burrow) and the resulting burrow network (topological characteristics and surface). Total burrow length, number of burrows, rate of branching, connectivity and sinuosity can also be studied (Bastardie et al. 2003).

The advantage of this method is of being easy, quick and inexpensive to study the behaviour with minimal disturbance to the worms. But the area used here will restrict the earthworm and thus have border effects and can be used for short-term exposure periods. Water, gas or solutes transfer/exchange are not possible to measure in this technique.

6. Effects (Long Term) on Burrowing: 3D Soil Cores (Capowiez et al. 2006; Dittbrenner et al. 2011)

Earthworm induced modifications of soil functions for long term can be studied using 3D assays (Capowiez et al. 2006) with more realistic scenario. In this experiment, artificial cores were prepared using PVC cylinders (35 × 16 cm) lined with sealing varnish and sharp fine sand. This sealing prevents the worms from crawling along the PVC walls. The soil cores thus made were compacted layer by layer. The bottom of the cylinder was sealed and top was fitted with mesh after the introduction of pre treated worms along with food. The whole set up was put in dark room at prescribed temperature and water replenishment was done at weekly intervals. At the end of the experiment, chloroform was poured on the cores to kill the worms and the cores imaged using a medical X-ray tomograph. The burrow system in each core was reconstructed as per Pierret et al. (2002). The macropores were traced and the volume of macroporosity and the number of burrows were computed. In addition to this, Capowiez et al. (2006) performed a gas diffusion experiment to study the gas/water/solute transfer properties of the soil in relation to the burrow system.

This experiment seems to be field realistic but is too tedious and need technical skills (Pelosi et al. 2014). Sometime, 3D does not give much information and a mere confirmation of the effects observed for burrowing activity in the 2D terraria but the former can be useful to test persistent toxicants (Dittbrenner et al. 2011). However, both 2D and 3D are more specific methods resulting in a limited number of observations.

7. Assay on cast production (Capowiez et al. 2010)

Soil (23.4 % clay, 57 % silt, 19.6 % sand, 28.3 g/kg organic matter) from apple orchard sieved for 3 mm was used in this experiment. Earthworms are washed in tap water, weighed and placed individually in 100 g moist soil in transparent polystyrene round boxes for 7 days. After the prescribed 7 days, the earthworm was removed and weighed again to note the difference in weight. The soil from each container was sieved using a set of four sieves of mesh sizes, 5.6, 4.0, 3.15, 2.5 mm and casts separated carefully without breaks. In another experiment, the soil in half of the test containers was sieved immediately and the other half after dried at laboratory temperature for 4 days.

This method to test the sublethal behavioural effects of pesticide on earthworms is a quick method and does not require specific equipment. It is a short term test (7 days) and the end point (cast production) has greater ecological relevance than avoidance. The only disadvantage mentioned by the authors is when a decreased cast production is observed, it does not reveal clearly whether earthworms ingested less soil because their health was affected or because the polluted soil acted as a repellent.

4.1.2.3 Sublethal Toxicity Assays for Respiration

Assays for Respirometry Measurements (Givaudan et al. 2014)

After the prescribed exposure time to pesticides, earthworms were removed, rinsed, gently blotted dry on a filter paper and weighed. Then the earthworms were left on a moist filter paper placed on a glass jar for about 24 h to remove the gut contents. After this procedure, the filter paper is replaced and the glass jar was hermetically closed for about 2 h period. The carbon dioxide concentration in the glass gar was estimated using gas chromatography. An advanced method of using infra red gas analyzer (IRGA), to measure the differences of CO_2 concentrations before and after introduction (Vlek et al. 2004), can also be used.

4.1.2.4 Effect of Pesticides on Immunity or Lysosomes

Neutral Red Retention Assay to Study Lysosomes (Weeks and Svendsen 1996; Booth et al. 1998a, b; Farrukh and Ali 2015)
This assay measures the membrane stability of lysosomes within the coelomocytes of earthworms in response to contaminants. In this experiment, the worms were treated with pesticides through soil contamination bioassay with varied concentration and exposure duration. The working solution of neutral red (80 mg/mL) was prepared as per Speed and Smith (1975) using earthworm physiological ringer solution. Coelomic fluid was collected from the coelomic cavity posterior to the clitellum by inserting a needle containing 20 μL of ringer. The coelomic fluid thus extracted was placed on to a glass slide and mixed with 20 μL of neutral red solution. The slides were scanned for 2 min. at 5 min. intervals to count the number of stained and unstained cells. The cells were counted until 50 % of the cells turned red or for 60 min., whichever is earlier and this time is recorded as neutral red retention time. The lesser the red retention time of celomocytes shows low immune cells and may be due to the insecticide induced stress. A decrease in time with respect to the increase in pesticide concentration tested shows a direct correlation with the time and pesticide stress.

4.2 Semi-field Experiments

4.2.1 Experiments with Microcosm (Tu et al. 2011)

This experiment is an advanced modification of the funnels used by Bieri et al. (1989), to find out the effect of pesticides on earthworms. Each microcosm is a transparent plastic tube of 4.4 × 30 cm in size and kept in upright position. A fiberglass screen was used to cover the bottom to prevent earthworms from escaping while facilitating water leaching. Microcosms were filled with 500 g of soil, moistoned at 70 % water holding capacity and covered from outside using aluminum foil to block light. After 1 week, two earthworms were placed in a short hole made with a screw driver. One week after that, 5 mL of pesticide solution was applied on the soil surface and six wheat kernels placed on the soil surface. The number of wheat kernels remaining on the soil surface were recorded daily and replaced with new six grains at weekly intervals. Earthworms were weighed after 3 weeks of pesticide application to find the impact. In another experiment, repeated application of pesticides i.e. four times on a weekly interval was made and observations on the weight of worms were taken 2 weeks after the fourth application.

4.2.2 Experiments with Mesocosms (Booth et al. 2001)

Polyurethane culvert pipes, covered with stainless steel mesh at the bottom were used as mesocosms in the experiment. Two mesocosms were sunk in the centre of 1 m^2 of each plot leaving 5 cm of pipe above the soil surface. The mesocosms are filled with the soil from that field itself or similar to that of the fields'. Earthworms were released in the mesocosms 7 days prior to the field sprays, food (straw) provided at the surface and covered with stainless steel mesh held by adhesive tapes. Just before the pesticide application, the lids were removed so as to allow the fall of pesticides in the mesocosms. One of the two mesocosms in each field was taken to laboratory on 7th day of pesticide treatment and the other on 14th day. The mesocosms were sort for earthworms in the laboratory and the missing worms were counted as dead.

4.2.3 Terrestrial Model Ecosystem (TME) (Rombke et al. 2004; Vlek et al. 2004)

Terrestrial model ecosystem (TME) is defined as undisturbed soil columns taken from the fields and managed in the laboratory to study the impact of pesticides on the biological organisms (Sheppard et al. 1997). In this experiment, a soil core extractor was used which has a HDPE tube of 40 × 17.5 cm for soil core encasement. The TME units (soil cores) were kept in carts with funnel at the bottom and a leaching chamber. The whole setup was kept in a climatic chamber and irrigated thrice in a week as simulated rainfall. After 4 weeks of acclimatization, the pesticide chemical at field recommended concentration as calculated with the surface area of the soil core is added by means of a pipette. Earthworm samples were taken at 1, 8 and 16 weeks of pesticide application by hand sorting. The effect of pesticides on the biomass and abundance of earthworms are determined efficiently by using TMEs. However, at sites with low abundance, the data interpretation will become difficult because of the low sampling area. However, many studies concluded that the data obtained from TME as comparable with that of the field and thus a reliable method for predicting the field effects (Weyers et al. 2004).

4.2.4 Mesh Bags/Wormsocks for In Situ Evaluation of Litter Decomposition (Potter 1991)

Litter decomposition is usually done by earthworms and microorganisms in the natural ecosystem. Pesticides used in agroecosystem may affect the activities of earthworms including thatch decomposition. With this background, mesh bag experiments were conducted to study the effect of pesticides on thatch decomposition of earthworms. In this experiment, several pre weighed pieces of thatch sewn into coarse nylon mesh bags which allow the soil fauna including earthworms were used. The mesh bags were buried just beneath the surface of turf designated for treatment and untreated control. The plots 6 × 10 ft were treated with pesticides of labeled rate and irrigated with half inch of water within 2 h of treatment. Earthworms were sampled 1 and 6 weeks after treatment and the mesh bags with thatch pieces were dug up, weighed and analyzed for mineral soil content, microbial activity and net loss in organic matter after 6 weeks of treatment. Some of the insecticides are found to cause long lasting reductions in earthworm population and reduced thatch breakdown.

'Wormsocks' are larger mesh bags made of 250 μm nylon mesh sewn into a tube of 25 cm long and 10 cm dia. which can be buried in the field were used to study the earthworm activities (McFarland 2000). The above mentioned size of 'wormsock' holds about 2 L of soil and up to 20 worms. After sealing both ends by twisting the plastic coated stainless steel wire, the sack is buried in the field.

A bag of 25×25 and 15 cm long was used by Hankard et al. (1999) to study the activities of the earthworms. In this experiment, a 3 cm top layer soil is removed as intact and kept aside. A stainless steel box corer was hammered into the soil and the soil core removed. The soil is placed in the nylon bag and the bag placed inside the hollow ground. The 3 cm top layer was placed above it. The advantage of these mesh bag and 'wormsock' bioassays is that they mimics in many ways as natural exposure and conditions.

4.2.5 Tests to Assess Surface Migration (Christensen and Mather 2004)

The use of surface migration as a sublethal endpoint is been studied with the use of trapping techniques. This technique incorporates specially designed barrier traps positioned around the plot periphery, which allows recording of both number and direction of migration. The traps were tended daily for 15 days after field treatment.

4.3 Field Experiments

Testing of pesticides on earthworms in the field is necessary in order to confirm and validate the prediction of environmental effects made through laboratory bioassays and to recognize fully the ecological significance of pesticide toxicity to non targets i.e. earthworms. Unlike the laboratory conditions, though variability arrive in field tests, they do have many advantages like testing over different species, life stages and their interaction, without any spatial restriction and thus more realistic (Christensen and Mather 1994). Crop fields can be selected for field studies to evaluate the effect of pesticides on earthworms but grasslands may have a higher population of worms. An artificial experimental site with grasses deliberately grown can also be used for this purpose (Kula 1992). Ploughed grassland is a good choice for testing chemicals on earthworms. It is proposed to have a minimum population of 60 individuals per m^2 on any soil, to increase the possibility of finding statistically significant effects (Kula et al. 2006). Normally a 25–50 % reduction in density of worms in the field and failure to recover within 1 year are denoted as deleterious effect of the pesticide tested (Christensen and Mather 1994). In the EU guideline document for Terrestrial Ecotoxicology (SANCO 2002) field study on earthworm is stated that as ISO method (ISO 1999) and further described in detail by Greig-Smith et al. (1992) and Sheppard et al. (1998).

Key effect endpoints for field tests include (EPPO 2003)

- Number of all earthworms (juveniles and adults).
- Total biomass of all earthworms (juveniles and adults).
- Numbers of the two most abundant species.
- Biomass of at least the two most abundant species
- Species diversity

4.3.1 Field Tests on Earthworms (Tu et al. 2011)

In field experiments, plots of size 3 × 3 m were marked with 60 cm wide buffer zone between plots. Pesticides were sprayed at the maximum recommended doses. The treatment and control plots were replicated four times. Pesticidal treatment was given for six times every week. After 2 weeks of first chemical application, earthworm extractions were initialized and continued for 4 weeks at weekly intervals. When the pesticide treatments were terminated, earthworms were sampled for about another 6 weeks with bi weekly interval.

Sampling: In each sampling, earthworms were collected by flushing them by pouring 4 L of 0.5 % mustard powder suspension in 50 × 50 cm area (Gunn 1992). At the start, about 2 L of the suspension can be poured and the worms were picked and kept in container. After a period of 5–10 min., the remaining suspension can be poured and earthworms picked until no new worms appear.

4.3.2 Field Tests in Pasture Lands (Booth et al. 2001)

This field trial was carried out in pasture land with perennial rye and white clover, where no pesticides are applied at least for 2 years. Altogether, three treatments, i.e. a control, field recommended rate and the laboratory test rate in five replicates were used to study the effect on earthworms in the field. The plots were of 10 × 10 m with 4 m buffer between rows, columns and boundaries arranged in Latin square design. After spraying, two 20 cm^3 soil was taken using spade at 7, 14 and 28 days after spraying, brought to laboratory and sorted for earthworms. Edwards and Brown (1982), suggests a field trial to find pesticide impact on earthworm to be carried out in plots of 10 × 10 m in grassland or well settled crop fields, which have a high population of earthworms. According to ISO, the duration of a field test should be at least 1 year, in order to assess the recovery of the earthworm community (De-Jong et al. 2006).

4.3.3 Sampling of Earthworms in Field

A sampling area of 30 × 30 cm can be randomly taken and samples of worms can be taken in two different depths (0–10 and 10–25 cm) using shovels. The above layer of litter can be removed before sampling and any worm found in that organic matter may be collected. Collected earthworms can be put in plastic bags containing water and kept in shade (Smith et al. 2008). After digging up to 25 cm, 1 L of mustard solution can be poured and worms coming out can be collected for 20 min. The advantage of using mustard solution is an irritant with non-toxic effects and can drive out even the deep burrowing earthworms. Mustard solution was reported to bring *L. terrestris*, to the surface out from their deep burrows (Gunn 1992; Lawrence and Bowers 2002). A comparison of different chemicals for extraction was studied by Pelosi et al. (2009) and found that formaldehyde, allyl isothiocyanate and mustard expelled 47.7, 31.9 and 20.5 g of earthworms/m^2, respectively. However, *A. icterica* did not respond to these vermifuge applications. Allyl isothiocyanate and formaldehyde are efficient than mustard for *A. caliginosa* and *A. rosea.* Owing to the toxicity of formaldehyde and inefficiency of mustard when used without hand-sorting, Pelosi et al. (2009) suggested allyl isothiocyanate as promising chemical expellant.

In case of electrical sampling, Thielemann's octet method is found comparable with hand sorting and formalin extraction. Electrical sampling extracted significantly higher earthworm numbers and thus useful especially when minimum disturbance to the field is desirable (Schmidt 2001). Heat extraction through Kempson apparatus which extracted more number of earthworms is also a good technique but electrical octet is highly applicable in protected areas since it is non destructive (Coja et al. 2008) and do not affect soil organisms and root respiration (Staddon et al. 2003).

5 Pesticide Risk Assessment in Terrestrial Ecosystem: Earthworm

Risk assessment is developing a scientific basis for regulatory decision in a systematic mean through an interactive process (Barnthouse 1992). Risk assessment is the estimation of probability and magnitude of undesired events, such as biological effects of pesticides on earthworm population in the field (Christensen and Mather 1994). Environmental risk assessment starts with the hazard identification followed by predicting environmental concentrations (PEC) and predicted no effect concentrations (PNEC). If the PEC is less than PNEC then there is no indication of a potential risk. If PEC is greater than PNEC, further risk assessment studies are to be triggered (Vlek et al. 2004).

Ecological risk assessment of pesticide contaminated soils requires toxicity tests with earthworms (EC 2002). Earthworms are used for soil toxicity testing and ecotoxicological risk assessments (Rodriguez-Castellanos and Sanchez-Hernandez 2007). Acute toxicity tests aim in estimating the median lethal concentration or dose. There are several standardized tests available for testing pesticide toxicity in earthworms. The first test is to determine the mortality of test worms so as LD_{50} can be established (ISO 1993). Median lethal value can be a good indicator for initial screening but survival cannot be the important endpoint for ecotoxicological studies (Moriarty 1983) and acute mortality tests may not provide the risk estimate for earthworms (Frampton et al. 2006). If a pesticide did not cause mortality it doesn't mean it will not adversely affect the reproduction and other activities and resulting in reduction in population. Thus, chronic toxicity tests with sublethal endpoints are more realistic for the prediction of environmental effects (Rombke et al. 2007; Jensen et al. 2007; Liu et al. 2011b). Growth and reproduction are also been considered as important end points (Van-Gestel et al. 1992; An and Lee 2008; Wu et al. 2011). Tests on growth and reproduction (ISO 1998) are used to find the sublethal toxicity. Neuhauser and Callahan (1990) have suggested cocoon production as an important indicator than growth. Thus, apart from median lethality, no observed effect concentration (NOEC) and low observed effect concentrations (LOEC) are also to be estimated.

Sublethal effects also cause alterations in behaviour and other crucial activities. Avoidance due to pesticides (ISO 2008) is perhaps the most studied behavioural character. This avoidance test is very simple with quick results and easy to perform. However, avoidance and toxicity are not always correlated. Avoidance tests can measure repellence but not exactly the toxicity (Capowiez and Berard 2006). Apart from avoidance, the impact of contaminants on other behavioural responses such as burrowing, feeding, or surface migration must be studied (Sanchez-Hernandez 2006). Total burrow length and maximal burrow depth may be considered as important behavioural aspects, since it indicates general burrowing activity which have direct effects on the properties of soils and as a consequence might affect the whole ecosystem (Bastardie et al. 2003; Capowiez et al. 2006).

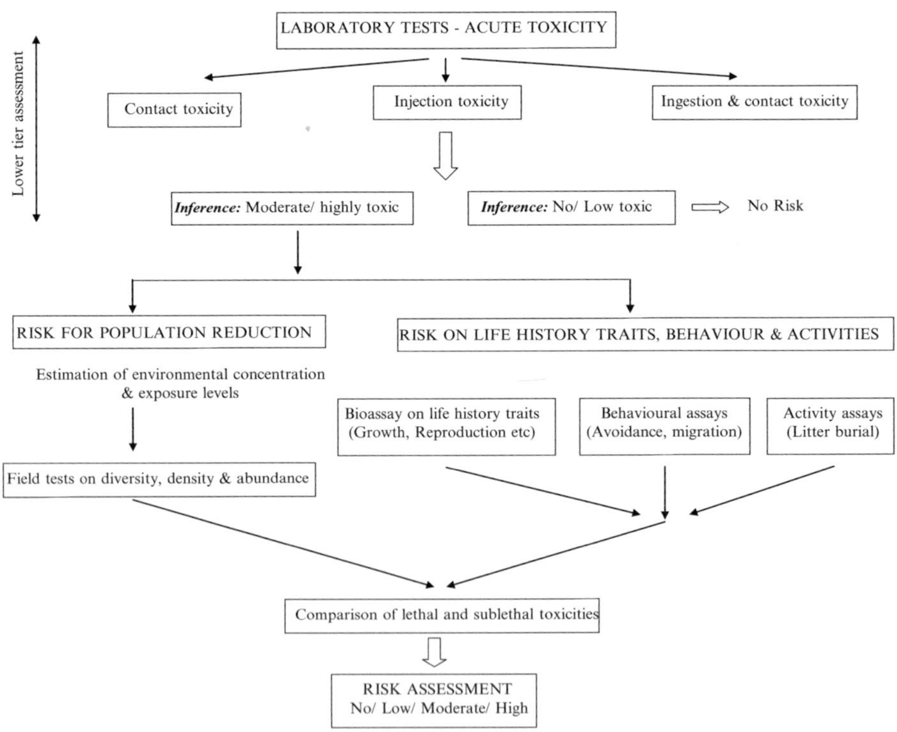

Sometimes toxicity of pesticides cannot be determined by mere evaluation of the survival, reproduction and immobilization of earthworms. In such cases, biomarker approach may be used and give indications on the potential toxicity of contaminants (Neuparth et al. 2005; Costa et al. 2005). The risk of environmental pollution can be easily sensed by using biomarkers much before the critical condition (Moore et al. 2004; Gastaldi et al. 2007). Biomarkers such as the neutral red retention (NRR) assay might be a useful tool for estimating the internal effects as they reflect the bioactive contaminant fraction (Sanchez-Hernandez 2006). The response of NRR assay to environmental contaminants occurs sooner at sub-cellular levels than at physiological or other levels making it a useful biomarker to serve as an early warning system of stress (Farrukh and Ali 2015).

For proper environmental risk assessment, three tiered studies should be conducted (Rombke et al. 1996) i.e. basic laboratory tests, extended laboratory tests, tests using microcosms (model ecosystem tests) or even field tests (Yasmin and D'Souza 2010). But because of its complexity, cost and time constraint, field studies are rarely been done (Rombke and Notenboom 2002). Changes in abundance, biomass or species richness of natural populations of earthworms have been common ecological endpoints to identify pesticide pollution in the field (Tomlin and Gore 1974; Paoletti et al. 1998). The laboratory tests and field test are complementarily be used to find the mechanism involved in earthworm response and to assess the state of earthworm communities in real conditions (Pelosi et al. 2014).

Constraints in Risk Assessment

The aim of earthworm ecotoxicology is to predict the effects of pesticides in the field on the basis of laboratory experiments (Yasmin and D'Souza 2010). Though some correlations between the results obtained from the laboratory assays and field effects are being reported (Holmstrup 2000) many a times, a good correlation is not obtained because of one or other reasons. This may be mainly because of the complexity of the soil ecosystem, where-in interactions occur between abiotic and biotic factors. In field situations, the exposure is highly dependent on the rate of pesticide application, chemical reaction in the soil and also the availability of earthworms (Yasmin and D'Souza 2010). Thus extrapolations from laboratory to field are to be done carefully taking all these factors into consideration.

Some pesticides like buprofezin, triflumuron and lufenuron are least toxic to the earthworms but cause the highest reduction in the growth rate (Badawy et al. 2013). So, along with lethality other effects should also be taken to have the overall picture. Chronic toxicity in terms of growth, sexual development, reproduction and hatchability of cocoon are to be assessed. When the worms are exposed to sublethal doses, it may cause some behavioural changes, which can also be assessed (Weeks and Comber 2005).

Most standard ecotoxicity tests are conducted on epigeic species mainly with *E. fetida* or *E. andrei* (Dittbrenner et al. 2010). *Eisenia fetida* is commonly used as a model organism. These epigeic worms can only be used for reproduction tests because of their short life cycle (Capowiez et al. 2010) and these species lack ecological relevance since they are usually absent in agricultural fields and is not a representative of the earthworm fauna in the landscapes (Paoletti 1999). These species which are used for laboratory tests may be resistant than the native ones to some of the pesticides or may not be the right species to express the toxic effects.

In that case, extrapolation from laboratory tests to field is very difficult. It was reported that *Lumbricus* sp. as the most susceptible earthworm species followed by *A. caliginosa* than *E. fetida* (Ma and Bodt 1993). *Perionyx excavatus* is reported more sensitive to many pesticides than *E. andrei* or *E. fetida* (De-Silva et al. 2010). So, ecologically relevant species such as *A. caliginosa* which is the common and predominant in agricultural ecosystems (Nuutinen 1992; Jordan et al. 2004; Givaudan et al. 2014) are to be used for toxicity tests.

The behaviour of earthworms also influences their contact to the pesticide and thus the toxicity. Several studies showed difference in sensitivity to pesticides in epigeic and endogeic worms (Velki and Hackenberger 2012), which, raise the question whether it is necessary to use species of different ecological categories to have a better and specific risk assessment (Velki and Hackenberger 2013). The anecic species like *L. terrestris* are the most exposed species to agricultural chemicals and least affected because they make permanent galleries and does not come in contact with the surface soil (Iordache and Borza 2011). The endogeic species, like *A. caliginosa*, which extend their galleries as a result of the feeding way, are prone to the chemical hazard (Edwards and Brown 1982). The use of tropical species in assessing the risk of pesticides in tropics also should be encouraged (De-Silva et al. 2010). Some epigeics and endogeics especially *L. terrestris* and *A. caliginosa* are reported to transport contaminated soil from deeper layers to surface leading to an increased risk to the surface dwellers (Sanchez-Hernandez 2006). This aspect is totally not addressed by laboratory bioassays. Furthermore, juvenile earthworms may be more susceptible than adults to different pesticides (Zhou et al. 2008). Thus, estimating the ecotoxicological risks using toxicity data from adults may leads to the under estimation of the actual field effect (Van-Straalen and Denneman 1989).

It is always good to collect earthworm from the native soil or use of species which actually found in the cropping system for testing to have a realistic picture. But if the test worms come from the population that is already exposed to pesticides, they may behave differently (Pelosi et al. 2014). Lowe and Butt (2007) demonstrated that the earthworm origin also had significant influence on the long-term (>12 weeks) survival, growth and reproduction. Exposure of carbofuran to *E. andrei* is transgenerational and reduction in cocoon production is reported for three generations if the worms of the first generation are exposed (Brunninger et al. 1994). So, origin of earthworm should also been taken in to consideration in toxicity assays.

Biochemical markers are used in risk assessments with an advantage of being an early warning signal for forthcoming effects (Van-Gestel and Weeks 2004). But there should be a correlation between the biochemical markers with that of the deleterious changes in the population. Biochemical alterations in earthworms due to seasonal influences are also reported (Hatti 2013). Cholinesterase activity was found to get decreased significantly in *A. nocturna* during autumn in comparison to spring (Rault et al. 2007). The carboxyl esterase activity of *A. caliginosa* also showed a significant variation in spring and autumn, whereas such seasonal variation was not observed in *E. fetida* (Laszczyca et al. 2004). Changes in AChE isoforms are reported from estivating earthworms (Kaloustian 1981). Apart from this, AChE in earthworm species also has some different enzymological properties than those in mammals (Stenersen 1980; Rault et al. 2007). Thus, it is necessary to know the impact of abiotic and biotic factors on the differential response of biomarkers responses to avoid wrong interpretations.

Toxicity of the pesticide per se depends on the physical and chemical characteristics of the test soil and the test organisms (Addison and Holmes 1995). Studies of Bindesbol et al. (2005) showed that the fluctuating environmental variables such as temperature, soil moisture, pH, or organic matter content might have an influence on the sensitivity of earthworm to pollutants (Sanchez-Hernandez 2006). Pesticide toxicity per se is influenced by temperature, moisture and pH of the soil (Rodriguez-Castellanos and Sanchez-Hernandez 2007). Some of the compounds may persist for a long time (long half life) and remain in the soil two or more seasons. This causes an additive effect on soil organisms especially earthworms and the situation is more severe when pesticide mixtures are used in the field. Ecotoxicological tests do not take into account of the repeated applications of several cocktails of pesticides i.e. multicontamination and chronic exposures (Pelosi et al. 2014) besides the effects of breakdown products (Paoletti 1999). Thus, experiments with single insecticide are unrealistic to field conditions (Van-Straalen and Denneman 1989; Zhou et al. 2011).

Apart from these complexities in performing soil ecotoxicology assessment, two entirely different forms of risk assessment are to be made for earthworms, the first one with the reduction in population level and the other involves the study of possible bioaccumulation in levels causing hazard to predators (Christensen and Mather 1994). Earthworms are the common prey for many vertebrate species and play therefore a key role in the biomagnification process of several soil pollutants (Dellomo et al. 1999). Thus studies on biomagnification of pesticides through earthworms to higher taxa are to done in detail. Some of the common risk assessment methodologies used in terrestrial ecosystem especially to find the ecotoxicity of pesticides to earthworms is detailed hereunder.

5.1 Risk Assessment Methodologies

5.1.1 Risk Assessment Based on Median Lethal Values (Roberts and Dorough 1984; Kokta 1992; Zhou et al. 2007)

Risk assessments based on median lethal values arrived in the laboratory either by filter paper contact bioassay or soil contamination bioassay are also possible. Though it does not consider many variables which are crucial for assessing the risk, it can be used as an initial tool/method to screen pesticides. In filter paper contact bioassays, median lethal concentrations (LC_{50}) are calculated based on the mortality data and expressed as μg/cm^2 for 48 h. Based on the resulting LC_{50} values, the chemicals are classified as supertoxic (<1 μg/cm^2), extremely toxic (1–10 μg/cm^2), very toxic (10–100 μg/cm^2), moderately toxic (100–1000 μg/cm^2) or relatively non-toxic (>1000 μg/cm^2).

For soil contamination bioassays, the suggested classification of toxicity is LC_{50} <1 mg/kg for high-toxic, 1–10 mg/kg for mid-toxic and >10 mg/kg for low-toxic pesticides (Zhou et al. 2007). Pesticides with LC_{50} values higher than 1000 mg/kg can be denoted as harmless to earthworms (Kokta 1992).

5.1.2 Risks Assessment: LD_{50}/PEC (Capowiez et al. 2010)

The ratio between the median lethal doses (LD_{50}) with that of predicted environmental concentrations (PEC) is usually made to give a toxicity ranking in ecological risk assessments. The higher the value obtained shows a lower predicted toxicity and vice-versa. With this calculation, carbaryl falls in >10 and thus toxic. Methomyl, methyl-parathion and imidacloprid are in between 40 and 60 and for chlorpyrifos-methyl and azinphos-methyl it was close to 100 and thus relatively non toxic.

5.1.3 Risks Assessment: Toxicity Exposure Ratio

This risk assessment takes into account of no effect concentration instead of median lethal doses. The ratio between the no effect concentration and the PEC gives the toxicity exposure ratio.

$$\text{Toxicity Exposure Ratio} = \frac{\text{No effect Concentration (NOEC)}}{\text{Predicted Environmental Concentration (PEC)}}$$

The toxicity exposure ratio was reported as 3, 20, 20, 30 and 60 for methomyl, carbaryl, phosphomidon, imidacloprid and dichlorvos, respectively to earthworm, *E. fetida* in artificial soil test (Na et al. 2005), showing methomyl as the highly toxic pesticide and dichlorvos as relatively less toxic.

5.1.4 Risk Assessment with Laboratory Bioassays: Acute, Chronic and Behavioural Endpoints (Christensen and Mather 1994; Zhou et al. 2007)

The results obtained from the laboratory bioassays can also be used for assessing the risk of use of pesticides on soil ecosystem especially on earthworms. These laboratory screening should include sublethal endpoints along with that of the lethal endpoint, since both have significance for natural earthworm population. Inclusion of both will decrease the probability of risk of chemical being incorrectly assessed. The major parameters determining population size are,

$$N_{t+1} = N_t + \text{Birth} - \text{Death} + \text{Immigration} - \text{Emigration}.$$

Thus ideal bioassay for risk assessment should include,

- Median lethal concentration (LC_{50}) for adult earthworms
- Growth and sexual development tests for juveniles

- Reproduction i.e. cocoon production and hatchability tests
- Survival and sexual development assays for F_1 generation
- Tests on migratory behaviour of adults and juveniles

5.1.5 Field Test for Risk Assessment (EPPO 2003; De-Jong et al. 2006)

Here, standard field trials are to be conducted and effects noted. Based on the field effects obtained, risks can be assessed. The criteria for risk assessment in field trials on the effect of pesticides on earthworms are as follows:

Low risk	:	<50 % effect
Medium risk	:	>50 % effect during the study but full recovery within 1 year
High risk	:	>50 % without full recovery after 1 year

The acceptability of 50 % effect as low risk is based on the limitations of the test rather than on other considerations. Full recovery here means, after 1 year the effects are less than 50 %, or that the effects are even above 50 %, but they are not statistically significant than that of the control or pretreatment observations.

5.1.6 Extrapolation from Acute Toxicity Laboratory Data to Field Toxicity

In this extrapolation, an exposure calculation is used to compare the median lethal values with that of estimated environmental concentrations (EEC) in soils obtained from field recommended doses (Heimbach 1992). Extrapolation of laboratory acute toxicity data to field effects can be done only when environmental concentrations are estimated. This necessitates an exposure calculation which usually involves recalculating doses (kg a.i./ha area) in to concentrations (mg/kg soil). This is achieved by assuming an equal distribution of pesticide in the top 2.5 cm soil layer and assuming a uniform soil density of 1.5 kg/L (Heimbach 1992; Van-Gestel 1992). While extrapolating the laboratory results, the LC_{50} values are divided by a compensation factor of 10 (Heimbach 1992). This compensation factor by and large takes care of the differential sensitivities of earthworm species (Haque and Ebing 1983; Heimbach 1985) though more than ten times differences in sensitiveness are also reported (Velki and Hackenberger 2013).

Van-Gestel (1992) reported a correlation between the effect levels found in the laboratory as LC_{50} of benomyl, carbendazim, carbofuran and carbaryl to various earthworm species and that observed in the field as earthworm biomass and abundance. A still comprehensive and good relation was obtained by Heimbach (1992) with soil contamination LC_{50} values of 14 pesticides with that of field effects. Pesticides which are shown to have low toxicity in laboratory tests were found to have no or negligible effects on earthworm populations in the field. Pesticides which are applied in low rates also have less toxicity to earthworms.

5.1.7 Tiered Assays for Risk Assessment (Christensen and Mather 1994)

This assessment has an early tier of comparisons of estimated environmental concentrations with that of the endpoint concentrations of toxicity testing (usually LC_{50}). If the end point concentration (LC_{50}) is not greater than the environmental concentrations, then more tests involving different variables are to be made and threshold ratio again evaluated. Laboratory tests with growth and sexual development of juveniles are considered as the second tier evaluation. A test for reproduction with cocoon production, adult weight and hatchability as endpoints forms the third tier of toxicity evaluation. The final tier of this assessment is the field test having endpoints of density, biomass and migration of earthworms. The study on the migratory behaviour will explains the underlying mechanism of population reduction.

In this method of risk assessment, first step is to calculate LC_{50} using standard procedures. Then the environmental concentration is estimated in bare soil or grass land and compared with the LC_{50} values. With the ratio of median lethal concentration and estimated environmental concentration a criteria is framed

LC_{50} value is >150 times of EEC	:	Low risk; further tests may not be required
LC_{50} is between 15 and 150 times of EEC	:	Growth, sexual development and reproduction tests are needed
LC_{50} <15 times of EEC	:	Growth, sexual development and reproduction tests along with field trials are to be done

In the second (growth) and third tier (reproduction) evaluation, trigger values are considered as follows:

Effect	Categorization	Recommendation
0–25 %	All pesticides	Low risk
25–50 %	Non persistent pesticide and application not repeated within a year	Low risk
	Persistent pesticide and requires repeated application	Higher tier evaluation required
>50 %	All pesticides	Higher tier evaluation required

In field trials regarding population reduction, the following criteria (Heimbach 1992; Greig-Smith 1992) are to be used.

0–25 % population reduction – weak change	:	No to low risk
25–50 % reduction – medium change	:	Moderate risk
50–75 % reduction – heavy change	:	High risk
More than 75 % reduction – strong change	:	Unacceptably high risk

Note: Pesticides falling on the last two categories are not recommended for agricultural use.

5.1.8 Comparison of Laboratory and Field Toxicities for Risk Assessment (Van-Gestel 1992)

The comparison of laboratory arrived values with that of the field derived ones can be a good indicator of risk. The comparison is hampered by the variations reported in the LC_{50} values arrived from laboratory assays, apart from the pesticide specific differential sensitiveness of earthworm species. In field, the pesticide toxicity depends on many more variables than that of the laboratory tests, which includes deposition of pesticides on the soil, behaviour of pesticide chemical in the soil and distribution and behaviour of worms. However, the method of logit model of finding LC_{50} from field data is interesting and useful but requires much field work.

Benomyl In laboratory studies the LC_{50} values of benomyl is found the lowest (among different species) for *L. terrestris* being 3.5 mg/kg tested in sandy loam soil for 14 days (Haque and Ebing 1983). In field studies, ≥50 % reduction of earthworm populations was reported at the estimated concentration of 1.6–28.6 mg/kg (Keogh and Whitehead 1975; Edwards and Brown 1982). A logit model of LC_{50} from data of different field studies revealed a value of 3.2 mg/kg.

Carbendazim Laboratory studies with *E. fetida* and *E. andrei* shows NOEC values of 0.6 and 2.0 mg/kg but in field studies, effects were noticed in the estimated concentrations of 0.9–1.6 mg/kg (Ammon 1985; Keogh and Whitehead 1975). This contradictory result is because of the difference in the earthworm species, tested in the laboratory and available in the field.

Carbofuran The lowest LC_{50} value for *A. caliginosa* was reported to be 0.56 tested in sand and 0.88 % grass for 14 days at 20 °C with a NOEC for cocoon production to be 0.05–0.1 mg/kg (Martin 1986). In field conditions, ≥50 % reduction of earthworm populations was reported at 1.4–16 mg/kg (Clements et al. 1986; Symonds 1975), which is well above the laboratory arrived concentrations.

5.1.9 Risk Assessment Based on Recovery and Persistence of Pesticides (Van-Straalen and Van-Rijin 1998)

The pesticides added in the soil undergo several processes such as biodegradation, hydrolysis, photolysis, volatilization etc. which jointly determine the rate of dissipation and availability to the test organisms. The chemical also get adsorbed to the soil particles, plants may uptake some and some may get leached beyond the depths of availability of the test organism. Ecotoxicological recovery is defined as the disappearance of a pesticide to a level where it does not have any more adverse effect. In this experiment, Recovery R is taken as a function of dose rate (initial concentration), persistence (measured as DT_{50}) and toxicity (measured as HC_5). HC_5 is defined as the hazardous concentration for 5 % of the species. HC_5 appears to be a promising tool for validating predictions of pesticide risk for soil invertebrates (Jansch et al. 2006).

The initial concentration is estimated from the recommended dose assuming an even distribution in top soil. Resolving the equations with field recommended dose, depth of application and bulk density, a simplified equation is arrived as follow:

$$\text{Initial concentration}\ (\text{mg / kg}) = 3.77 \times \text{Field recommended dose}\ (\text{kg / ha})$$

Finally Recovery (R) is calculated using initial concentration, 5 % hazard concentration, half life of pesticide (DT_{50}) and recovery hazard ratio. As organic matter presence also influences the persistence and availability of the toxicity to the organism, the no effect concentration (NOEC) observed and reported in the experiments are to be adjusted with the organic matter content in the test substrate.

5.1.10 Risk Assessment for Secondary Poisoning/Biomagnification

Secondary poisoning occurs when the predators are exposed to damaging concentrations of pollutants through their food (Spurgeon et al. 2003). Since earthworms are a good source of food and many vertebrate species predate on earthworms, there is a risk of secondary poisoning through earthworm in the food chain. The ability of earthworms to transfer pesticides like DDT, dieldrin, organophosphates and carbamates to vertebrate predators is given in detail by Cooke et al. (1992).

The following factors are to be considered for assessing the risk posed on the predators:

1. Concentration of pesticide in worms
2. Pesticide concentration causing effects on predators
3. Effect of any behavioural changes of earthworms on predator diet
4. Proportion of worms in the total diet of predator
5. Selection of contaminated worms by the predator
   (Cooke et al. 1992)

5.1.10.1 Maximum Permissible Risk Concentration (Romijn et al. 1994; Spurgeon and Hopkin 1996)

This procedure of risk assessment for secondary poisoning compares mean bioconcentration factors (BCFs) (concentration of pesticides in earthworms divided by concentration in the soil) for the contaminants in the prey with a predator sensitivity value.

$$\text{MPC} = \frac{\text{HC}_5\ \text{predator}}{\text{BCF}}$$

Where MPC – maximum permissible risk concentration for the pollutant in soil

HC_5 predator – hazardous concentration for 5 % of predator species
BCF – mean bioconcentration factor for the pollutant in the tissues of earthworm (i.e. Conc. in animal/Conc. in soil or diet).

The pollutant levels in the soil and earthworm are to be analyzed first. Median lethal concentration values and no observed effect concentrations (NOEC) are used to find the HC_5 of predator. Since the BCFs might vary considerably for worms collected from different soils, the MPCs arrived are not general but specific to that particular place.

Much of the work on secondary poisoning are not being done except for some metals. Though some pesticide toxicity studies are reported on earthworm predators, many of them are done with injection or some routes other than dietary exposure. Considering the realistic scenario, assessments of predators exposed to dosed worms are to be studied (DellOmo et al. 1999).

5.2 *Risk of Pesticides on Earthworms*

Risks of pesticides on earthworms are assessed in many ways i.e. based on laboratory acute toxicity, extrapolation of acute toxicity to field conditions, tier system of risk assessment etc. In general, the most harmful pesticide groups to earthworms include nicotinoides, strobilurins, sulfonylureas, triazols, carbamates and organophosphates (Pelosi et al. 2014). It seems to be the pyrimidine insecticides as non-toxic and triazine herbicides as moderately toxic to earthworms (Edwards and Bohlen 1996).

Among many pesticides tested using filter paper bioassay, only carbofuran falls under the category of super toxic (<1.0 μg/cm²) to earthworms, *E. foetida*. The phenolic hydrolytic products of parathion, carbaryl and 2, 4 D were more toxic than the parent material (Roberts and Dorough 1984). Neonicotinoid insecticides such as acetamiprid, imidacloprid, clothianidin etc. were super toxic to *E. fetida* (48 h LC_{50} value ranged from 0.0088 to 0.45 mg/cm²), pyrethroids were very toxic with LC_{50} 10.55–25.7 mg/cm² and insect growth regulators were moderately toxic to worms in contact filter paper bioassay (Wang et al. 2012a). An 8 h contact filter paper test revealed clothianidin, fenpyroximate and pyridaben as super toxic to *E. fetida* with LC_{50} values ranging from 0.28 to 0.72 μg/cm², followed by carbaryl, pyridaphenthion, azoxystrobin, cyproconazole and picoxystrobin (LC_{50} 2.72–8.48 μg/cm²) (Wang et al. 2012a). Fenvalerate with LC_{50} of 0.248 μg/cm² is classified as a highly toxic chemical (Song et al. 2015). Out of 97 pesticides studied many of them (12 insecticides, 23 fungicides, 12 herbicides, 3 nematicides and 2 molluscicides) i.e. >50 % have a NOEC for reproduction <10 mg/kg, stating a greater risk (Pelosi et al. 2014). Less than 5 % of pesticides have an LC_{50} below or equal to 10 mg/kg are considered to be moderately to highly toxic for the species *E. fetida* (PPDB 2013; Pelosi et al. 2014)

The toxicity exposure ratio (TER) was reported as 3, 20, 20, 30 and 60 for methomyl, carbaryl, phosphomidon, imidacloprid and dichlorvos to earthworm, *E. fetida* in artificial soil test (Na et al. 2005). Apart from risk of pesticides on earthworm, risk of earthworm predators due to poisoned earthworms are also being studied. The risk of bifenthrin to birds and mammals feeding on earthworms was assessed as low based on $PEC_{soil\ plateau}$ of 0.027 mg/kg (EFSA 2011). Different methodologies are proposed by many scientists to assess the risk of pesticide applications on earthworms. Risk assessment in terrestrial ecosystem aims in minimizing the risk of pesticides or pollutants on non-targets. But many studies are based only on earthworms and that too with a few species mostly of *E. fetida* can really give an idea of the actual risk in the environment is a question (Jansch et al. 2006). Heterogeneity of soil matrix and complexity of ecological interactions are the two major hurdles to assess soil ecotoxicology (Eijsackers 1994).

References

J.A. Addison, Safety testing of tebufenozide, a new molt-inducing insecticide, for effects on non-target forest soil invertebrates. Ecotoxicol. Environ. Saf. **33**, 55–61 (1996)

J.A. Addison, S.B. Holmes, Comparison of forest soil microcosms and acute toxicity studies for determining effects of fenitrothion on earthworms. Ecotoxicol. Environ. Saf. **30**, 127–133 (1995)

S.A. Alshawish, A.I. Mohamed, G.A. Nair, Prolonged toxicity of sub-lethal dosages of chemical pesticides on the body mass and cocoons of *Aporrectodea caliginosa* (Savigny, 1826) (Oligochaeta: Lumbricidae) inhabiting Benghazi, Libya. Proc. Nat. Acad. Sci. India **74**(B), 123–133 (2004)

D. Alvarez-Bernal, E.L. Garcia-Diaz, S.M. Contreras-Ramos, L. Dendooven, Dissipation of polycyclic aromatic hydrocarbons from soil added with manure or vermicompost. Chemosphere **65**(9), 1642–1651 (2006)

P.R. Alves, E.J. Cardoso, A.M. Martines, J.P. Sousa, A. Pasini, Earthworm ecotoxicological assessments of pesticides used to treat seeds under tropical conditions. Chemosphere **90**(11), 2674–2682 (2013)

A.O. Ameh, I.A. Mohammed-Dabo, S. Ibrahim, J.B. Ameh, Y. Tanimu, T.K. Bello, Effect of earthworm inoculation on the bioremediation of used engine oil contaminated soil. Int. J. Biol. Chem. Sci. **6**(1), 493–503 (2012)

H.U. Ammon, Worm toxicity test using *Tubifex tubifex. Comportment et Effects Secondaries des Pesticides dans le Sol. Les colloques de INRA* 31, Versailles, June 4–8, 1984, (INRA publicaitons, 1985)

M.J.B. Amorim, J. Rombke, A.M.V.M. Soares, Avoidance behaviour of *Enchytraeus albidus*: effects of benomyl, carbendazim, phenmedipham and different soil types. Chemosphere **59**(4), 501–510 (2005)

Y.J. An, W.M. Lee, Comparative and combined toxicities of toluene and methyl tert-butyl ether to an Asian earthworm, *Perionyx excavatus*. Chemosphere **71**, 407–411 (2008)

O.R. Anderson, P.J. Bohlen, Abundance and diversity of Gymnamoebae associated with earthworm (*Lumbricus terrestris*) middens in a northeastern U.S. forest. Soil Biol. Biochem. **30**, 1213–1216 (1998)

N.Q. Arancon, C.A. Edwards, R. Atiyeh, J.D. Metzger, Effects of vermicomposts produced from food waste on the growth and yields of greenhouse peppers. Bioresour. Technol. **93**(2), 139–144 (2004)

N.Q. Arancon, C.A. Edwards, P. Bierman, J.D. Metzger, C. Lucht, Effects of vermicomposts produced from cattle manure, food waste and paper waste on the growth and yield of peppers in the field. Pedobiologia **49**(4), 297–306 (2005)

K. Armstrong, C.Y. Caley, B.E. Hall, B. Knight, *Lannate 20L: Determination of Acute Toxicity (LC_{50}) in Earthworms*. (Inveresk Research International Ltd, Tranent, Scotland, Report No. 8546, 1991)

R.M. Atiyeh, N. Arancon, C.A. Edwards, J.D. Metzger, Influence of earthworm-processed pig manure on the growth and yield of greenhouse tomatoes. Bioresour. Technol. **75**(3), 175–180 (2000)

R.M. Atiyeha, J. Dominguezb, S. Subler, C.A. Edwards, Changes in biochemical properties of cow manure during processing by earthworms (*Eisenia andrei*, Bouché) and the effects on seedling growth. Pedobiologia **44**(6), 709–724 (2000)

R.M. Atiyeha, N.Q. Arancon, C.A. Edwards, J.D. Metzger, The influence of earthworm-processed pig manure on the growth and productivity of marigolds. Bioresour. Technol. **81**(2), 103–108 (2002)

M.E.I. Badawy, A. Kenawy, A.F. El-Aswad, Toxicity assessment of buprofezin, lufenuron and triflumuron to the earthworm *Aporrectodea caliginosa*. Int. J. Zool. **2013**, 174523 (2013)

S.W. Baker, D.J. Binns, E.C. Kirby, Reduction in rates of earthworm casting onturf areas using carbendazim, carbaryl and gamma-HCH thiophanate-methyl. J. Sports Turf Res. Inst. **74**, 25–39 (1998)

K. Bansiwal, N. Rai, Ultra structural changes in clitellum region of *Eisenia foetida* after treatment with malathion. Bioscan **5**(2), 207–210 (2010)

A.V. Barker, G.M. Bryson, Bioremediation of heavy metals and organic toxicants by composting. Sci. World J. **12**(2), 407–420 (2002)

L.W. Barnthouse, The role of models in ecological risk assessment: a 1990s perspective. Environ. Toxicol. Chem. **11**, 1751–1760 (1992)

F. Bastardie, Y. Capowiez, J.R. De-Dreuzy, D. Cluzeau, X-ray tomographic and hydraulic characterization of burrowing by three earthworm species in repacked soil cores. Appl. Soil Ecol. **24**, 3–16 (2003)

E. Baubet, C. Touzeau, S. Brandt, Earthworms in the wild boar diet (*Sus scrofa*) in mountain pasture. Mammalia **61**, 371–383 (1997)

C. Bauer, J. Rombke, Factors influencing the toxicity of two pesticides on three lumbricid species in laboratory tests. Soil Biol. Biochem. **29**, 705–708 (1997)

S.A. Bengtson, A. Nilsson, S. Nordstrom, S. Rundgren, Effect of bird predation on lumbricid populations. Oikos **27**, 9–12 (1976)

E. Benitez, H. Sainz, R. Melgar, R. Nogales, Vermicomposting of a lignocellulosic waste from olive oil industry: a pilot scale study. Waste Manag. Res. **20**(2), 134–142 (2002)

E. Benitez, H. Sainz, R. Nogales, Hydrolytic enzyme activities of extracted humic substances during the vemicomposting of a lignocellulosic olive waste. Bioresour. Technol. **96**, 785–790 (2005)

N. Bernier, J.F. Ponge, Humus form dynamics during the sylvogenetic cycle in a mountain spruce forest. Soil Biol. Biochem. **26**(2), 183–220 (1994)

T. Bhadauria, K.G. Saxena, Role of earthworms in soil fertility maintenance through the production of biogenic structures. Appl. Environ. Soil Sci. **2010**, 81607 (2010)

C. Bharathi, B.V.S.S.R. Subbarao, Toxicity of phosphomidon to the common south Indian earthworm, *Lampito mauritti*. Bull. Environ. Contam. Toxicol. **32**, 295–300 (1984)

C. Bharathi, B.V.S.S.R. Subbarao, Toxic effects of two organophosphate insecticides, monocrotophos and dichlorvos to common earthworm *Lampito mauritii* (Kinberg). Water Air Soil Pollut. **28**(1–2), 127–130 (1986)

T. Bhatnagar, Lombriciens et Humification: un aspect nouveau de I'incorporation microbienne d'azote induite par les vers de terre, in *Humification et Biodegradation*, ed. by G. Kilbertus, O. Reisinger, A. Mourey, J.A.C. Da-Fonseca (Pierron, Sarreguemines, 1975), pp. 169–182

M. Bieri, Guidelines for the Daniel funnel test: a laboratory test to measure side effects of pesticides on the earthworm, *Lumbricus terrestris* L. IOBC WPRS Bull. **15**(3), 139–144 (1992)

M. Bieri, The environmental profile of metaldehyde. (2003). Retrieved from http://cesandiego.ucanr.edu/files/54211.pdf

M. Bieri, H. Schweizer, K. Christensen, O. Daniel, The effect of metaldehyde and methiocarb slug pellets on *Lumbricus terrestris*. in *BCPC Monograph: Slugs and Snails in World Agriculture*, ed. by I. Henderson, (BCPC Publications, 1989), pp. 237–244

A.M. Bindesbol, M. Holmstrup, C. Damgaard, M. Bayley, Stress synergy between environmentally realistic levels of copper and frost in the earthworm *Dendrobaena octaedra*. Environ. Toxicol. Chem. **24**, 1462–1467 (2005)

F. Binet, R.C. Le Bayon, Space-time dynamics in situ of earthworm casts under temperate cultivated soils. Soil Biol Biochem **31**, 85–93 (1999)

F. Binet, A. Kersante, C. Munier-Lamy, R.C. Le-Bayon, M.J. Belgy, M.J. Shipitalo, Lumbricid macrofauna alter atrazine mineralization and sorption in a silt loam soil. Soil Biol. Biochem. **38**, 1255–1263 (2006)

E. Blanchart, A. Bruand, P. Lavelle, The physical structure of casts of *Millsonia anomala* (Oligochaeta: Megascolecidae) in shrub savanna soils (Cote d'Ivoire). Geoderma **56**, 119–132 (1993)

E. Blanchart, P. Lavelle, E. Braudeau, Y.L. Bissonnais, C. Valentin, Regulation of soil structure by geophagous earthworm activities in humid savannas of Cote d'Ivoire. Soil Biol. Biochem. **29**(3&4), 431–439 (1997)

E. Blanchart, A. Alain, J. Alegre, D. Arnaud, G. Cecile, B. Pashanasi, P. Lavelle, L. Brussaard, Effects of earthworms on soil structure and physical properties, in *Earthworm Management in Tropical Agroecosystems*, ed. by P. Lavelle, L. Brussaard, P. Hendrix (CABI, Wallingford, 1999), pp. 149–172

E. Blanchart, A. Albrecht, G. Brown, T. Decaens, A. Duboisset, P. Lavelle, L. Mariani, E. Roose, Effects of tropical endogeic earthworms on soil erosion. Agric. Ecosyst. Environ. **104**, 303–315 (2004)

P.J. Bolton, J. Phillipson, Burrowing, feeding, egestion and energy budgets of *Allolobophora rosea* (Savigny) (Lumbricidae). Oecologia **23**, 225–245 (1976)

M. Bonkowski, I.E. Geoghegan, A.N.E. Birch, B.S. Griffiths, Effects of soil decomposer invertebrates (protozoa and earthworms) on an above-ground phytophagous insect (cereal aphid) mediated through changes in the host plant. Oikos **95**, 441–450 (2001)

L.H. Booth, K. O'Halloran, A comparison of biomarker responses in the earthworm *Aporrectodea caliginosa* to the organophosphorus insecticides diazinon and chlorpyrifos. Environ. Toxicol. Chem. **20**(11), 2494–2502 (2001)

L.H. Booth, V. Heppelthwaite, K. O'Halloran, Growth, development and fecundity of the earthworm *Aporrectodea caliginosa* after exposure to two organophosphates. N. Z. Plant Prot. **53**, 221–225 (1998a)

L.H. Booth, V. Heppelthwaite, C.T. Eason, Cholinesterase and glutathione s transferase in the earthworm, *Aporrectodea caliginosa* as biomarkers of organophosphate exposure. Proc. 51st New Zeland Plant Protec. Conf., 138–142 (1998b)

L.H. Booth, V.J. Heppelthwaite, R. Webster, K. O'Halloran, Lysosomal neutral red retention time as a biomarker of organophosphate exposure in the earthworm *Aporrectodea caliginosa*: laboratory and semi-field experiments. Biomarkers **6**(1), 77–82 (2001)

H. Bouwman, A.J. Reinecke, A defined medium for the study of growth and reproduction of the earthworm, *Eisenia fetida* (Oligocheta). Biol. Fertil. Soils **10**, 285–289 (1991)

A.W.A. Brown, *Ecology of Pesticides* (Wiley, New York, 1978). 525 p

G.G. Brown, How do earthworms affect microfloral and faunal community diversity? *In*: *The Significance and Regulation of Soil Biodiversity Developments in Plant and Soil Sciences* (Springer Netherlands, 1995), pp. 247–269

B. Brunninger, R. Vishwanathan, F. Beese, Terbuthylazine and carbofuran effects on growth and reproduction within three generations of *Eisenia andrei* (Oligocheta). Biol. Fertil. Soils **18**, 83–88 (1994)

K.R. Butt, J. Fredrickson, C.N. Lowe, Colonisation, survival and spread of earthworms on a partially restored landfill site. Pedobiologia **43**, 1–7 (1999)

K.R. Butt, C.N. Lowe, J. Frederickson, A.J. Moffat, The development of sustainable earthworm populations at Calvert landfill site. UK. Land Degrad. Dev. **15**, 27–36 (2004)

C.A.M. Van-Gestel. Scientific basis of extrapolating results from soil ecotoxicity tests to field conditions and the use of bioassays. In: *Ecological Risk Assessment of Contaminant Soils*, ed. by N.M. Van-Straleen, H. Lokhe, (Chapman and Hall, London, 1997), pp. 25–50

Y. Capowiez, Differences in burrowing behaviour and spatial interaction between the two earthworm species *Aporrectodea nocturna* and *Allolobophora chlorotica*. Biol. Fertil. Soils **30**, 341–346 (2000)

Y. Capowiez, A. Berard, Assessment of the effects of imidacloprid on the behavior of two earthworm species (*Aporrectodea nocturna* and *Allolobophora icterica*) using 2D terraria. Ecotoxicol. Environ. Saf. **64**, 198–206 (2006)

Y. Capowiez, A. Pierret, C.J. Moran, Characterisation of the three dimensional structure of earthworm burrow systems using image analysis and mathematical morphology. Biol. Fertil. Soils **38**, 301–310 (2003)

Y. Capowiez, M. Rault, G. Costagliela, C. Mazzia, Lethal and sublethal effects of imidacloprid on two earthworm species (*Aporrectodea nocturna* and *Allobophora icterica*). Biol. Fertil. Soils **41**, 135–143 (2005)

Y. Capowiez, F. Bastardie, G. Costagliola, Sublethal effects of imidacloprid on the burrowing behaviour of two earthworm species: modifications of the 3D burrow systems in artificial soil cores and consequences on gas diffusion in soil. Soil Biol. Biochem. **38**, 285–293 (2006)

Y. Capowiez, N. Dittbrenner, M. Rault, R. Triebskorn, M. Hedde, C. Mazzia, Earthworm cast production as a new behavioural biomarker for toxicity testing. Environ. Pollut. **158**, 388–393 (2010)

A. Casenave, C. Valentin, in *Les états de surface: une des clefs de l'hydrologie sahélienne*. Proc. Sahel Forum on the state-of-the-art of hydrology and hydrogeology in the arid and semi-arid areas of Africa. (UNESCO/IWRA, Urbana, IL, USA, 1988), pp. 61–72

P.M. Catling, B. Freedman, Food and feeding behaviour of sympatric snakes at Amherstburg. Ont. Can. Fld. Nat. **94**, 28–33 (1980)

B. Ceccanti, G. Masciandaro, C. Garcia, C. Macci, S. Doni, Soil bioremediation: combination of earthworms and compost for the ecological remediation of a hydrocarbon polluted soil. Water Air Soil Pollut. **177**, 383–397 (2006)

K.Y. Chan, Impact of tillage practices and burrows of a native Australian anecic earthworm on soil hydrology. Appl. Soil Ecol. **27**, 89–96 (2004)

J.F.L. Charlton, Slugs as a possible cause of establishment failure in pasture legumes oversown in boxes. N. Z. J. Exp. Agric. **6**(4), 313–317 (1978)

A. Chauvel, M. Grimaldi, E. Barros, E. Blanchart, M. Sarrazin, P. Lavelle, An Amazonian earthworm compacts more than a bulldozer. Nature **398**, 32–33 (1999)

S.E. Cherenkov, I. Gubenko, A.V. Tiunov, A. Kuzmichev, Factors determining the spatial distribution of thrushes (*Turdus* spp.) in highly mosaic forest landscapes. Russ. J. Zool. **74**, 104–119 (1995)

L.P.D. Choo, G.H. Baker, Influence of four commonly used pesticides on the survival, growth and reproduction of the earthworm *Aporrectodea trapezoids* (Lumbricidae). Aust. J. Agric. Res. **40**(8), 1297–1303 (1998)

O.M. Christensen, J.G. Mather, *Earthworms as Ecotoxicological Test Organisms* (Ministry of Environment, Danish Environmental Protection Agency, København, 1994), p. 100

O.M. Christensen, J.G. Mather, Pesticide-induced surface migration by lumbricid earthworms in grassland: life-stage and species differences. Ecotoxicol. Environ. Saf. **57**(1), 89–99 (2004)

J.M. Clapperton, J.J. Miller, F.J. Larney, Earthworm populations as affected by longterm tillage practices in southern Alberta. Can. Soil Biol. Biochem. **29**(3&4), 631–633 (1997)

R.O. Clements, B.R. Bentley, C.A. Jackson, The impacts of granular formulations of phorate, terbufos, carbofuran, carbosulfan and thiofanox on newly sown Italian rye grass, *Lolium multiflorum*. Crop Prot. **5**(6), 389–394 (1986)

T. Coja, K. Zehetner, A. Bruckner, A. Watzinger, E. Meyer, Efficacy and side effects of five sampling methods for soil earthworms (Annelida, Lumbricidae). Ecotoxicol. Environ. Saf. **71**(2), 552–565 (2008)

D. Conrady, Okologische Untersuchungen uber die Wirkung von Umweltchemikalien auf die Tiergemeinschaft eines Grunlandes. Pedobiologia **29**, 273–284 (1986)
A.S. Cooke, P.W. Greig-Smith, S.A. Jones, Consequences for vertebrate wildlife species of toxic residues in earthworm prey, in *Ecotoxicology of Earthworms*, ed. by P.W. Greig-Smith, H. Becker, P.J. Edwards, F. Heimbach (Intercept, Andover, 1992), pp. 139–155
F.V. Correia, J.C. Moreira, Effects of glyphosate and 2,4-D on earthworms (*Eisenia foetida*) in laboratory tests. Bull. Environ. Contam. Toxicol. **85**, 264–268 (2010)
F.O. Costa, T. Neuparth, A.D. Correia, M.H. Costa, Multi-level assessment of chronic toxicity of estuarine sediments with the amphipod *Gammarus locusta*: II. Organism and population-level endpoints. Mar. Environ. Res. **60**, 93–110 (2005)
G. Cuendet, Predation on earthworms by the black headed gulls (*Larus* ridibundus L.), in *Earthworm Ecology: From Darwin to Vermiculture*, ed. by J. Satchell (Springer, Dordrecht, 1983), pp. 415–424
J.P. Curry, *Grassland Invertebrates* (Chapman and Hall, London, 1994), p. 437
C. Darwin, *Darwin on Humus and the Earthworms: The Formation of Vegetable Mould Through the Action of Worms with Observations on Their Habits* (Faber & Faber, London, 1881, 1945 reprint), 153 p
T. Decaens, J.J. Jimenez, Earthworm communities under an agricultural intensification gradient in Colombia. Plant Soil **240**, 133–143 (2002)
B.P. Degens, The contribution of carbohydrate C and earthworm activity to the water stable aggregation of a sandy soil. Aust. J. Soil Res. **35**, 61–71 (1997)
F.M.W. De-Jong, P. Van-Beelen, C.E. Smit, M.H.M.M. Montforts, *Guidance for Summarizing Earthworm Field Studies* (RIVM, Bilthoven, 2006). 47 p
G. DellOmo, A. Turk, R.F. Shore, Secondary poisoning in the common shrew (*Sorex araneus*) fed earthworms exposed to an organophosphate pesticide. Environ. Toxicol. Chem. **18**, 237–240 (1999)
E. Delmee, P. Dachy, P. Simon, Comparative study of a feeding regime of a forest population of tawny owls, *Strix aluco*. Gerfaut **69**, 45–78 (1979)
M. De-Schampheleire, P. Spanoghe, E. Brusselman, S. Sonck, Risk assessment of pesticide spray drift damage in Belgium. Crop Prot. **26**(4), 602–611 (2007)
P.M.C.S. De-Silva, C.A.M. Van-Gestel, Comparative sensitivity of *Eisenia andrei* and *Perionyx excavatus* in earthworm avoidance tests using two soil types in the tropics. Chemosphere **77**, 1609–1613 (2009)
P.M.C.S. De-Silva, A. Pathiratne, C.A.M. Van-Gestel, Toxicity of chlorpyrifos, carbofuran, mancozeb and their formulations to the tropical earthworm, *Perionyx excavatus*. Appl. Soil Ecol. **44**, 56–60 (2010)
O.M. Di-Toro, C.S. Zarba, D.J. Hansen et al., Technical basis for establishing sediment quality criteria for nonionic organic chemicals using equilibrium partitioning. Environ. Toxicol. Chem. **10**, 1541–1583 (1991)
N. Dittbrenner, R. Triebskorn, I. Moser, Y. Capowiez, Physiological and behavioural effects of imidacloprid on two ecologically relevant earthworm species (*Lumbricus terrestris* and *Aporrectodea caliginosa*). Ecotoxicology **19**, 1567–1573 (2010)
N. Dittbrenner, I. Moser, R. Triebskorn, Y. Capowiez, Assessment of short and long-term effects of imidacloprid on the burrowing behaviour of two earthworm species (*Aporrectodea caliginosa* and *Lumbricus terrestris*) by using 2D and 3D post-exposure techniques. Chemosphere **84**, 1349–1355 (2011)
J. Domingueza, C.A. Edwards, M. Webster, Vermicomposting of sewage sludge: effect of bulking materials on the growth and reproduction of the earthworm, *Eisenia andrei*. Pedobiologia **44**(1), 24–32 (2000)
C.P. Doncaster, Factors regulating local variations in abundance – field tests on hedgehogs, *Erinaceus europaeus*. Oikos **69**, 182–192 (1994)
M. Durchon, M. Lafon, Quelques donnees biochimiques sur les anne!ides. Ann. Sci. Nat. Zool.-II Ser. **13**, 427–452 (1951)

EC. European Commission (Health & Consumer Protection Directorate– General). Guidance Document on Terrestrial Ecotoxicology under Council Directive 91/414/EEC. SANCO/10329, Brussels 2002

C.A. Edwards, Insecticide residues in soils. Rev. Environ. Contam. Toxicol. **13**, 83–132 (1966)

C.A. Edwards, The importance of earthworm as key representatives of soil fauna, in *Earthworm Ecology*, ed. by C.A. Edwards, 2nd edn. (CRC Press, New York, 2004)

C.A. Edwards, P.J. Bohlen, The effects of toxic chemicals on earthworms. Rev. Environ. Contam. Toxicol. **125**, 23–99 (1992)

C.A. Edwards, P.J. Bohlen, *Biology and Ecology of Earthworms*, 3rd edn. (Chapman and Hall, London, 1996). 402 p

P.J. Edwards, S.M. Brown, Use of grassland plots to study the effect of pesticides on earthworms. Pedobiologia **24**, 145–150 (1982)

W.M. Edwards, M.J. Shipitalo, Consequences of earthworms in agricultural soils: aggregation and porosity, in *Earthworm Ecology*, ed. by C.A. Edwards (CRC Press, Boca Raton, 1998), pp. 147–161

C.A. Edwards, J. Dominguez, E.F. Neuhauser, Growth and reproduction of *Perionyx excavatus* (Perr.) (Megascolecidae) as factors in organic waste management. Biol. Fertil. Soils **27**, 155–161 (1998)

W. Ehlers, Observations on earthworm channels and infiltration on tilled and untilled loss soil. Soil Sci. **119**, 242–247 (1975)

H. Eijsackers, Ecotoxicology of soil organisms: seeking the way through a pitch-dark labyrinth, in *Ecotoxicology of Soil Organisms*, ed. by M.H. Donker, H. Eijsackers, F. Heimbach (Lewis, Boca Raton, 1994), pp. 3–32

S.R. Ellis, M.E. Hodson, P. Wege, The soil dwelling earthworm, *Allolobophora chlorotica* modifies its burrowing behaviour in response to carbendazim applications. Ecotoxicol. Environ. Saf. **73**, 1424–1428 (2010)

X. El-Titi, U. Ipach, Soil fauna in suastainable agriculture: results of an integrated farming system at Lautenbach, FRG. Agric. Ecosyst. Environ. **27**, 561–572 (1989)

EPPO, Environmental risk assessment scheme of plant protection products. Chapter 8: soil organisms and functions. EPPO Bull. **33**, 195–209 (2003)

O. Espinoza-Navarro, E. Bustos-Obregon, Sublethal doses of malathion alter male reproductive parameters of *Eisenia foetida*. Int. J. Morphol. **22**(4), 297–302 (2004)

O. Espinoza-Navarro, E. Bustos-Obregon, Effect of malathion on the male reproductive organs of earthworms, *Eisenia foetida*. Asian J. Androl. **7**(1), 97–101 (2005)

European Food Safety Authority (EFSA), Conclusion on the peer review of the pesticide risk assessment of the active substance bifenthrin. EFSA J. **9**(5), 2159 (2011)

A.C. Evans, A method for studying the burrowing activities of earthworms. Ann. Mag. Nat. Hist. **14**, 643–650 (1947)

P. Fang, W.L. Wu, Q. Xu, H. Jiahai, C. Han, M.G. Paoletti, Assessing bioindication with earthworms in an intensive rural landscape (Yuanqiao and Daqiao villages in subtropical China, Hubei Province). Crit. Rev. Plant Sci. **18**(3), 429–455 (1999)

A. Farenhorst, E. Topp, B. Bowman, A. Tomlin, Earthworm burrowing and feeding activity and the potential for atrazine transport by preferential flow. Soil Biol. Biochem. **32**, 479–488 (2000)

A. Farenhorst, E. Topp, B.T. Bowman, A.D. Tomlin, R.B. Bryan, Sorption of atrazine and metolachlor by burrow linings developed in soils with different crop residues at the surface. J. Environ. Sci. Health B **36**(4), 389–396 (2001)

S. Farrukh, A.S. Ali, Toxicological effects of organophosphate pesticide on Ceolomocytes: viability of earthworm *E. foetida* using NRRA. Iran. J. Toxicol. **9**(28), 1254–1257 (2015)

N. Ferrari, J. Weber, Influence of the abundance of food resources on feeding habits of the red fox, *Vulpes vulpes* in western Switzerland. J. Zool. **236**, 117–129 (1995)

D.J. Finney, *Probit Analysis*, 3rd edn. (Cambridge University, London, 1971). 272 p

S.W. Fisher, A comparison of standardized methods for measuring the biological activity of pesticides to the earthworm, *Lubricus terrestris*. Ecotoxicol. Environ. Saf. **8**, 564–571 (1984)

E. Fisher, L. Koszorus, Sublethal effects, accumulation capacity and elimination rates of As, Hg and Se in the manure worm, *Eisenia foetida* (Oligocheta: Lumbricidae). Pedobiologia **36**, 172–176 (1992)
G.K. Frampton, S. Jansch, J.J. Scott-Fordsmand, J. Rombke, P.J. Vanden-Brink, Effects of pesticides on soil invertebrates in laboratory studies: a review and analysis using species sensitivity distributions. Environ. Toxicol. Chem. **25**(9), 2480–2489 (2006)
C.E. French, S.A. Liscinsky, D.R. Millar, Nutrient composition of earthworms. J. Wildl. Manag. **21**, 16–19 (1957)
H. Frund, U. Graefe, S. Tischer, Earthworms as bioindicators of soil quality, in *Biology of Earthworms*, ed. by A. Karaca (Springer, Berlin, 2010), pp. 261–278
T.W. Fuhremann, E.P. Lichtenstein, A comparative study of the persistence, movement, and metabolism of six C-14 insecticides in soils and plants. J. Agric. Food Chem. **28**, 446–452 (1980)
S. Gajalakshmi, E.V. Ramasamy, S.A. Abbasi, Potential of two epigeic and two anecic earthworm in vermicomposting of water hyacinth. Bioresour. Technol. **76**(3), 177–181 (2001)
M. Ganeshkumar, Effect of pesticides on earthworm. in *Proceedings of Summer School on Environmental Impact of Pesticides in Agroecosystem*, ed. by G. Santharam, A. Jayakumar, S. Kuttalam, S. Chandrasekaran, T. Manoharan (Tamil Nadu Agricultural University, Coimbatore, 2000)
V.K. Garg, P. Kaushik, Vermistabilization of textile mill sludge spiked with poultry droppings by epigeic earthworm *Eisenia fetida*. Bioresour. Technol. **96**, 1063–1071 (2005)
L. Gastaldi, E. Ranzato, F. Capri et al., Application of a biomarker battery for the evaluation of the sublethal effects of pollutants in the earthworm *Eisenia andrei*. Comp. Biochem. Physiol. C **146**, 398–405 (2007)
J. Getliff, G. McEwan, S. Ross, R. Richards, M. Norman, Drilling fluid design and the use of vermiculture for the remediation of drill cuttings. in *AADE 2002 Technology Conference on Drilling Fluids and Waste Management*, 2002
B. Gevao, K.T. Semple, K.C. Jones, Bound pesticide residues in soils: a review. Environ. Pollut. **108**, 3–14 (2000)
B. Gevao, C. Mordaunt, K.T. Semple, T.G. Piearce, K.C. Jones, Bioavailability of nonextractable (bound) pesticide residues to earthworms. Environ. Sci. Technol. **35**, 501–507 (2001)
C.D. Gish, R.E. Christensen, Calcium, nickel, lead and zinc in earthworms from roadside soil. Environ. Sci. Tech. **7**, 1060–1062 (1973)
N. Givaudan, C. Wiegand, B.L. Bot, D. Renault, F. Pallois, S. Llopis, F. Binet, Acclimation of earthworms to chemicals in anthropogenic landscapes, physiological mechanisms and soil ecological implications. Soil Biol. Biochem. **73**, 49–58 (2014)
J.L. Gomez-Eyles, C. Svendsen, L. Lister, H. Martin, M. Hodson, J.D. Spurgeon, Measuring and modelling mixture toxicity of imidacloprid and thiacloprid on *Caenorhabditis elegans* and *Eisenia fetida*. Ecotoxicol. Environ. Saf. **72**, 71–79 (2009)
D. Gormsen, P.A. Olsson, K. Hedlund, The influence of collembolans and earthworms on AM fungal mycelium. Appl. Soil Ecol. **27**, 211–220 (2004)
P.T. Gregory, Feeding habits and diet overlap of three species of garter snakes (Thamnophis) on Vancouver Island. Can. J. Zool. **56**, 1967–1974 (1978)
P.W. Greig-Smith, Risk assessemt approaches in the UK for the side effects of pesticides on earthworm, in *Ecotoxicology of Earthworm*, ed. by P.W. Greig-Smith (Intercept, Hans, 1992), pp. 159–168
P.W. Greig-Smith, H. Becker, P.J. Edwards, F. Heimbach, *Ecotoxicology of Earthworms* (Intercept Ltd., Hants, 1992). 390 p
F. Grumiaux, S. Demuynck, D. Schikorski, S. Lemiere, A. Lepretre, Assessing the effects of FBC ash treatments of metal-contaminated soils using life history traits and metal bioaccumulation analysis of the earthworm, *Eisenia andrei*. Chemosphere **79**, 156–161 (2010)
R.D. Guerrero, The culture and use of *Perionyx excavatus* as a protein resource in the Philippines, in *Earthworm Ecology: From Darwin to Vermiculture*, ed. by J. Satchell (Springer, Dordrecht, 1983), pp. 309–313

G. Guggenberger, R.J. Thomas, W. Zech, Soil organic matter within earthworm casts of an anecic–endogeic tropical pasture community. Colomb. Appl. Soil Ecol. **3**, 263–274 (1996)

A. Gunn, The use of mustard to estimate earthworm populations. Pedobiologia **36**, 65–67 (1992)

S.K. Gupta, P.M. Saxena, Carbaryl-induced behavioural and reproductive abnormalities in the earthworm *Metaphire posthuma*: a sensitive model. Altern. Lab. Anim. **31**, 587–593 (2003)

S.K. Gupta, V. Sundararaman, Correlation between burrowing capability and AChE activity in the earthworm, *Pheretima posthuma* on exposure to carbaryl. Bull. Environ. Contam. Toxicol. **46**, 859–865 (1991)

R.D. Gupta, P.P. Chakravorty, A. Kaviraj, Studies on relative toxicities of six insecticides on epigeic earthworm, *Perionyx excavatus*. Bull. Environ. Contam. Toxicol. **85**, 83–86 (2010)

J. Haimi, Decomposer animals and bioremediation of soils. Environ. Pollut. **107**, 233–238 (2000)

J. Haimi, J. Salminen, V. Huhta, J. Knuutinen, H. Palm, Bioaccumulation of organochlorine compounds in earthworms. Soil Biol. Biochem. **24**(12), 1699–1703 (1992)

Y. Han, L. Zhu, J. Wang, J. Wang, H. Xie, S. Zhang, Integrated assessment of oxidative stress and DNA damage in earthworms (*Eisenia fetida*) exposed to azoxystrobin. Ecotoxicol. Environ. Saf. **107**, 214–219 (2014)

F. Hani, Farming systems research at Ipsach, Switzerland. The 'third way' project. Schweiz. Landwirtsch. Forsch. Rech. Agron. Suisse **29**(4), 257–271 (1990)

P. Hankard, J.M. Weeks, S.K. Fishwick, et al. Biological stress indicators of contaminated land. Ecological assessment of contaminated land using earthworm biomarkers. R&D technical report. P311, (Environment Agency, Rio House, Almondsbury, 1999)

E. Haque, W. Ebing, Toxicity determination of pesticides to earthworms in soil substrate. J. Plant Dis. Prot. **90**(4), 395–408 (1983)

S.S. Hatti, Chemical composition like protein, lipid and glycogen of local three species of earthworms of Gulbarga city, Karnataka. India. Int. J. Adv. Res. Technol. **2**(7), 73–97 (2013)

C.L. Hawkins, M.J. Shipitalo, E.M. Rutledge, M.C. Savin, K.R. Brye, Earthworm populations in septic system filter fields and potential effects on wastewater renovation. Appl. Soil Ecol. **40**, 195–200 (2008)

R.J. Haynes, P.M. Fraser, A comparison of aggregate stability and biological activity in earthworm casts and uningested soil as affected by amendment with wheat or Lucerne straw. Eur. J. Soil Sci. **49**(4), 629–636 (1998)

L. Hazelhoff, P. Van-Hoof, A.C. Imeson, F.J.P.M. Kwaad, The exposure of forest soil to erosion by earthworms. Earth Surf. Proc. Land **6**, 235–250 (1981)

F. Heimbach, Correlation between three methods for determining the toxicity of chemicals to earthworms. Pestic. Sci. **15**(6), 605–611 (1984)

F. Heimbach, Comparison of laboratory methods using *Eisenia foetida* and *Lumbricus terrestris* for assessment of the hazard of chemicals to earthworms. J. Plt. Dis. Protec. **92**, 186–193 (1985)

F. Heimbach, A comparison of laboratory methods for toxicity testing with earthworms, in *Earthworms in Waste and Environmental Management*, ed. by C.A. Edwards, E.F. Neuhauser (SPB Academic Pub, The Hague, 1988), pp. 329–335

F. Heimbach, Correlation between data from laboratory and field tests for investigating the toxicity of pesticides to earthworms. Soil Biol. Biochem. **24**, 1749–1753 (1992)

F. Heimbach, Acute toxicity of azinphos-methyl to earthworms. Bayer AG, Agricultural Centre, Leverkusen. Report N HBF/Rg 58, 1986

B. Helling, S.A. Reinecke, A.J. Reinecke, Effects of the fungicide copper oxychloride on the growth and reproduction of *Eisenia fetida* (Oligochaeta). Ecotoxicol. Environ. Saf. **46**(1), 108–116 (2000)

Z.A. Hickman, B.J. Reid, Earthworm assisted bioremediation of organic contaminants. Environ. Int. **34**(7), 1072–1081 (2008)

C.H. Hogger, H.U. Ammon, M. Bieri, The Daniel funnel test in a sequence of earthworm tests of agricultural pesticides, in *Ecotoxicology of Earthworms*, ed. by I. Greig-Smith et al. (Intercept, Andover, 1992), pp. 230–224

M. Holmstrup, Field assessment of toxic effects on reproduction in the earthworms *Aporrectodea longa* and *Aporrectodea rosea*. Environ. Toxicol. Chemist. **19**(7), 1781–1787 (2000)

G.J. House, R.W. Parmelee, Comparison of soil arthropods and earthworms from conventional and no-tillage agroecosystems. Soil Till. Res. **5**(4), 351–360 (1985)
R. Hyvonen, S. Andersson, M. Clarholm, T. Persson, Effect of lumbricids and enchytraeids on nematodes in limed and unlimed coniferous mor humus. Biol. Fertil. Soils **17**, 201–205 (1994)
J. Iglesias, J. Castillejo, R. Castro, The effects of repeated applications of the molluscicide metaldehyde and the biocontrol nematode *Phasmarhrabditis hermaphrodita* on mollusks, earthworms, nematodes, acarids and collembolans: a two-year study in Northwest Spain. Pest Manag. Sci. **59**, 1217–1227 (2003)
C. Inglesfield, Toxicity of the pyrethroid insecticides cypermethrin and WL85871 to the earthworm, *Eisenia fetida* (Savigny). Bull. Environ. Contam. Toxicol. **33**, 568–570 (1984)
M. Iordache, I. Borza, Study of the acute toxicity of some pesticides on earthworms *Eisenia foetida* (Savigny, 1826). Res. J. Agric. Sci. **43**(4), 95–100 (2011)
ISO, *Soil Quality: Effects of Pollutants on Earthworms (Eisenia fetida) – Part 1: Determination of Acute Toxicity Using Artificial Soil Substrate -11268-1* (International Organization for Standardization, Geneva, 1993)
ISO, *Soil Quality: Effects of Pollutants onEearthworms (Eisenia fetida) – Part 2: Determination of Effects on Reproduction -11268-2* (International Organization for Standardization, Geneva, 1998)
ISO. ISO 11268-3. Soil quality: effects of pollutants on earthworms. Part 3: guidance on the determination of effects in field situations, 1999
ISO. Soil quality: avoidance test for determining the quality of soils and effects of chemicals on behaviour – part 1: test with earthworms (*Eisenia fetida* and *Eisenia andrei*) -17512-1 (Geneva, 2008)
ISO. Soil quality: avoidance test for testing the quality of soils and the toxicity of chemicals – test with earthworms (*Eisenia fetida*), ISO 17512 (Geneva, 2003)
T. Jager, Modeling ingestion as an exposure route for organic chemicals in earthworms (Oligochaeta). Ecotoxicol. Environ. Saf. **57**, 30–38 (2004)
T. Jager, R.H.L.J. Fleuren, W. Roelofs, A.C. De-Groot, Feeding activity of the earthworm *Eisenia andrei* in artificial soil. Soil Biol. Biochem. **35**, 313–322 (2003)
S. Jansch, G.K. Frampton, J. Rombke, P.J.V. Brink, J.J. Scott-Fordsmand, Effects of pesticides on soil invertebrates in model ecosystem and field studies: a review and comparison with laboratory toxicity data. Environ. Toxicol. Chem. **25**(9), 2490–2501 (2006)
R.S. Jasmine, S. Kuttalam, J. Stanley, Acute toxicity of abamectin to earthworms, *Eudrilus eugeniae* (Kingberg) and *Perionyx excavatus* (Perrier). Resist. Pest Manag. Newslett. **17**(2), 27–30 (2008)
J. Jensen, X.P. Diao, J.J. Scott-fordsmand, Sub-lethal toxicity of the antiparasitic abamectin on earthworms and the application of neutral red retention time as a biomarker. Chemosphere **68**, 744–750 (2007)
J.L. Johnson-Maynard, R.C. Graham, L. Wu, P.J. Shouse, Modification of soil structural and hydraulic properties after 50 years of imposed chaparral and pine vegetation. Geoderma **110**, 227–240 (2002)
A.G. Jongmans, M.M. Pulleman, M. Balabane, F. Van-Oort, J.C.Y. Marinissen, Soil structure and characteristics of organic matter in two orchards differing in earthworm activity. Appl. Soil Ecol. **24**, 219–232 (2003)
D. Jordan, R. Miles, V. Hubbard, T. Lorenz, Effect of management practices and cropping systems on earthworm abundance and microbial activity in Sanborn Field: a 115-year-old agricultural field. Pedobiologia **48**, 99–110 (2004)
R.D. Kale, Earthworms: nature's gift for utilization of organic wastes, in *Earthworm Ecology*, ed. by C.A. Edwards (Soil and Water Conservation Society/Lowa St. Lucie Press, New York/ Ankeny, 1998), pp. 355–373
R.D. Kale, K. Bano, R.V. Krishnamoorthy, Potential of *Perionyx excavatus* for utilizing organic wastes. Pedobiologia **23**, 419–425 (1982)
R.D. Kale, Significant contribution with Vermiculture. in *Proc. DST and DBT sponsored National Sem. Workshop on Conserv. Inventerization of Soil Biota*, Tiruchengode, 2008, ed. by Kathireswari, R. D. Kale, pp. 1–2

K.V. Kaloustian, Changes in acetycholinesterase isozyme pattern in estivating *Allohophora caliginosa* (Savigny). Comp. Biochem. Physiol. B **70**, I57–I160 (1981)
P. Kaushik, V.K. Garg, Dynamics of biological and chemical parameters during vermicomposting of solid textile mill sludge mixed with cow dung and agricultural residues. Bioresour. Technol. **94**(2), 203–209 (2004)
D. Kennette, W. Hendershot, A. Tomlin, S. Sauve, Uptake of trace metals by earthworm *Lumbricus terrestris* L. in urban contaminated soils. Appl. Soil Ecol. **19**, 191–198 (2002)
R.G. Keogh, P.H. Whitehead, Observations on some effects of pasture spraying with benomyl and carbendazim on earthworm activity and litter removal from pasture. N. Z. J. Exp. Agric. **3**(1), 103–104 (1975)
M.A. Khalil, H.M. Abdel-Lateif, B.M. Bayoumi et al., Analysis of separate and combined effects of heavy metals on the growth of *Aporrectodea caliginosa* (Oligochaeta: Annelida) using the toxic unit approach. Appl. Soil Ecol. **4**, 213–219 (1996)
I. Khan, S. Ayesha, A. Sharique, Biomass and behavioral responses of earthworm, *Lumbricus terrestris* to copper chloride. Iran. J. Toxiocol. **1**(2), 2–2 (2007)
R.A. King, I.P. Vaughan, J.R. Bell, D.A. Bohan, W.O.C. Symondson, Prey choice by carabid beetles feeding on an earthworm community analyzed using species and lineage-specific PCR primers. Mol. Ecol. **19**(8), 1721–1732 (2010)
E.C. Kirby, S.W. Baker, Earthworm populations, casting and control in sports turf areas: a review. J. Sports Turf Res. Inst. **71**, 84–98 (1995)
E.J. Kladivko, A.D. Mackay, J.M. Bradford, Earthworms as a factor in the reduction of soil crusting. Soil Sci. Soc. Am. J. **50**, 191–196 (1986)
D. Knight, P.W. Elliott, J.M. Anderson, D. Scholefield, The role of earthworms in managed, permanent pastures in Devon. Engl. Soil Biol. Biochem. **24**, 1511–1517 (1992)
C. Kokta, A laboratory test on sublethal effects of pesticides on *Eisenia fetida*, in *Ecotoxicology of Earthworms*, ed. by H. Becker, P.J. Edwards, P.W. Greig-Smith, F. Heimbach (Intersept Press, Andover/Hants, 1992), pp. 55–62
B. Kreis, P. Edwards, G. Cuendet, J. Tarrandellas, The dynamics of RCBs between earthworm populations and agricultural soils. Pedobiologia **30**, 379–388 (1987)
D.P. Kreutzweiser, K.P. Good, T.D.T. Chartrand, D.G. Thompson, Are leaves that fall from imidacloprid treated maple trees to control Asian longhorned beetles toxic to non-target decomposer organisms? J. Environ. Qual. **37**, 639–646 (2008)
M.V. Krishna, N. Hrushikesh, K. Sreehari, T. Aravindkumar, N. Vidyavathi, A. Pallavi, Studies on bioremediation of phenol by earthworm. Int. J. Environ. Sci. **1**(6), 1268–1273 (2011)
K.H. Kula, Measuring effects of pesticides on earthworms in the field: test design and sampling methods, in *Ecotoxicology of Earthworms*, ed. by P.W. Greig-Smith et al. (Intercept, Hans, 1992), pp. 90–99
C. Kula, F. Heimbach, F. Riepert, J. Rombke, Technical recommendations for the update of the ISO earthworm field test guideline (ISO 11268-3). J. Soils Sed. **6**, 182–186 (2006)
L. Lagadic, T. Caquet, F. Ramade, The role of biomarkers in environmental assessment: invertebrate populations and communities. Ecotoxicology **3**, 193–208 (1994)
O.P. Lal, R.K. Palta, Y.N.S. Srivastava, Impact of imidacloprid and carbofuran on earthworm castings in okra field. Ann. Plant Prot. Sci. **9**, 137–138 (2001)
M. Langmaack, S. Schrader, U. Rapp-Bernhardt, K. Kotzke, Quantitative analysis of earthworm burrow systems with respect to biological soil-structure regeneration after soil compaction. Biol. Fertil. Soils **28**, 219–229 (1999)
P. Laszczyca, M. Augustyniak, A. Babczynska, K. Bednarska, A. Kafel, P. Migula, G. Wilczek, I. Witas, Profiles of enzymatic activity in earthworms from zinc, lead and cadmium polluted areas near Olkusz (Poland). Environ. Int. **30**, 901–910 (2004)
A.P. Lawrence, M.A. Bowers, A test of the hot mustard extraction method of sampling earthworms. Soil Biol. Biochem. **34**, 549–552 (2002)
K.E. Lee, *Earthworms, Their Ecology and Relationship with Soils and Land Use* (Academic, Sydney, 1985). 411 p
D.C. Legg, Comparison of various worm-killing chemicals. J. Sports Turf Res. Inst. **44**, 47–48 (1968)

J. Lescure, The food of the common toad, *Bufo bufo* (Linnaeus 1758). Vie Milieu **15**, 757–764 (1966)
S. Liu, Q.X. Zhou, Y.Y. Wang, Ecotoxicological responses of the earthworm *Eisenia fetida* exposed to soil contaminated with HHCB. Chemosphere **83**, 1080–1086 (2011a)
Y.J. Liu, A. Zaprasis, S.J. Liu, H.L. Drake, M.A. Horn, The earthworm *Aporrectodea caliginosa* stimulates abundance and activity of phenoxyalkanoic acid herbicide degraders. ISME J. **5**, 473–485 (2011b)
J. Lorimor, C. Fulhage, R. Zhang, T. Funk, R. Sheffield, C. Sheppard, G.L. Newton, Manure management strategies/technologies. in *White Paper on Animal Agriculture and the Environment for National Center for Manure and Animal Waste Management*. (MWPS, Ames, 2001), p. 52
S. Loureiro, A.M.V.M. Soares, A.J.A. Nogueira, Terrestrial avoidance behavior tests as screening tool to assess soil contamination. Environ. Pollut. **138**, 121–131 (2005)
C.N. Lowe, K.R. Butt, Culture of commercially obtained *Lumbricus terrestris* L.: implications for sub-lethal ecotoxicological testing. Soil Biol. Biochem. **39**, 1674–1679 (2007)
Y. Luo, Y. Zang, Y. Zhong, Z. Kong, Toxicological study of two novel pesticides on earthworm, *Eisenia foetida*. Chemosphere **39**, 2347–2356 (1999)
P.B. Lynch, Conduct of field experiments. N. Z. Dep. Agric. Bull. **399**, 74–78 (1960)
W.C. Ma, J. Bodt, Differences in toxicity of the insecticide chlorpyrifos to 6 species of earthworms (Oligochaeta, Lumbricidae) in standardized soil tests. Bull. Environ. Contam. Toxicol. **50**, 864–870 (1993)
M.S. Maboeta, S.A. Reinecke, A.J. Reinecke, The relationship between lysosomal biomarker and population responses in a field population of *Microchaetus* sp. (Oligochaeta) exposed to the fungicide copper oxychloride. Ecotoxicol. Environ. Saf. **52**, 280–287 (2002)
D.W. Macdonald, The red fox, *Vulpes vulpes*, as a predator upon earthworms, *Lumbricus terrestris*. Z. Tierpsychol. **52**(2), 171–200 (1980)
D.W. Macdonald, Predation on earthworms by terrestrial vertebrates, in *Earthworm Ecology: From Darwin to Vermiculture*, ed. by J. Satchell (Springer, Dordrecht, 1983), pp. 393–414
M.R. Malecki, E.F. Neuhauser, R.C. Goehr, The effects of heavy metals on the growth and production of *Eisenia foetida*. Pedobiologia **24**, 129–137 (1982)
M.C. Manna, S. Jha, P.K. Ghosh, C.L. Acharya, Comparative efficacy of three epigeic earthworms under different deciduous forest litters decomposition. Bioresour. Technol. **88**(3), 197–206 (2003)
A.R.A. Marashi, J. Scullion, Earthworm casts form stable aggregates in physically degraded soils. Biol. Fertil. Soils **37**, 375–380 (2003)
M. Maraun, J. Alphei, M. Bonkowski, R. Buryn, S. Migge, M. Peter, M. Schaefer, S. Scheu, Middens of the earthworm *Lumbricus terrestris* (Lumbricidae): microhabitants for micro and meso-fauna in forest soil. Pedobiologia **43**, 276–287 (1999)
J.C.Y. Marinissen, A.R. Dexter, Mechanisms of stabilization of earthworm casts and artificial casts. Biol. Fertil. Soils **9**, 163–167 (1990)
N.A. Martin, Toxicity of pesticides to *Allolobophora caliginosa* (Oligocheta: Lumbricidae). N. Z. J. Agric. Res. **29**, 699–706 (1986)
N.A. Martin, The effects of herbicides used on asparagus on the growth rate of the earthworm *Allolobophora caliginosa*. in *Proc. 35th New Zealand Weed and Pest Conf.*, 1982, pp. 238–330
W. Matthey, J. Zettel, M. Bieri, *Invertebres bioindicateurs de la qualite de sols agricoles Nationalen Forschungsprogrammes 'bonen'* (Liebefeld-Bern, 1990), 141 p
I. McFarland, *An investigation of the meshbag approach to the enclosed bioassay of earthworms* (Centre for Ecology and Hydrology, Monks Wood, 2000)
T. Micol, C.P. Doncaster, L.A. MacKinley, Correlates of local variation in the abundance of hedgehogs, *Erinaceus europaeus*. J. Anim. Ecol. **63**, 851–860 (1994)
A. Miyazaki, T. Amano, H. Saito, Y. Nakano, Acute toxicity of chlorophenols to earthworms using a simple paper contact method and comparison with toxicities to fresh water organisms. Chemosphere **47**, 65–69 (2002)

C. Monard, F. Martin-Laurent, C. Vecchiato, A.J. Francez, P. Vandenkoornhuyse, F. Binet, Combined effect of bioaugmentation and bioturbation on atrazine degradation in soil. Soil Biol. Biochem. **40**, 2253–2259 (2008)

C. Monard, P. Vandenkoornhuyse, B. Le-Bot, F. Binet, Relationship between bacterial diversity and function under biotic control: the soil pesticide degraders as a case study. ISME J. **5**, 1048–1056 (2011)

M.N. Moore, M.H. Depledge, J.W. Readman, P. Leonard, An integrated biomarker-based strategy for ecotoxicological evaluation of risk in environmental management. Mutat. Res. **552**, 247–268 (2004)

C.J. Mordaunt, B. Gevao, K.C. Jones, K.T. Semple, Formation of non-extractable pesticide residues: observations on compound differences, measurement and regulatory issues. Environ. Pollut. **133**, 25–34 (2005)

J.E. Morgan, A.J. Morgan, Heavy metal concentrations in the tissues, ingesta and faeces of physiologically different earthworm species. Soil Biol. Biochem. **24**(12), 1691–1697 (1992)

J.E. Morgan, A.J. Morgan, The accumulation of metals (Cd, Cu, Pb, Zn and Ca) by two ecologically contrasting earthworm species (*Lumbricus rubellus* and *Aporrectodea caliginosa*): implications for ecotoxicological testing. Appl. Soil Ecol. **13**, 9–20 (1999)

J.E. Morgan, A.J. Morgan, N. Corp, Assessing soil metal pollution with earthworms: indices derived from regression analyses, in *Ecotoxicology of Earthworms*, ed. by P.W. Greig-Smith, H. Becket, P.J. Edwards, F. Heimbach (Intercept, Hants, 1992), pp. 233–237

F. Moriarty, *Ecotoxicology: The Study of Pollutants in Ecosystems* (Academic, London, 1983)

Y.Y. Mosleh, S. Paris-Palacios, M. Couderchet, G. Vernet, Acute and sublethal effects of two insecticides on earthworms (*Lumbricus terrestris* L.) under laboratory conditions. Environ. Toxicol. **18**(1), 1–8 (2003a)

Y.Y. Mosleh, S.M.M. Ismail, M.T. Ahmed, Y.M. Ahmed, Comparative toxicity and biochemical responses of certain pesticides to the mature earthworm, *Aporrectodea caliginosa* under laboratory conditions. Environ. Toxicol. **18**(5), 338–346 (2003b)

M.A. Mostert, A.S. Schoeman, M. Van-der-Merwe, The relative toxicity of insecticides to earthworms of the Pheretima group (Oligochaeta). Pest Manag. Sci. **58**, 446–450 (2002)

A. Motalib, M.A. Rida, M.B. Bouche, Heavy metal linkages with mineral, organic, organic and living soil compartments. Soil Biol. Biochem. **29**(3&4), 649–655 (1997)

Y.E. Na, H.S. Bang, K.K. Kang, M.S. Han, Y.J. Ahn, Assessment of the effects of some insecticides on mortality of earthworm (*Eisenia fetida*). Korean J. Environ. Agric. **24**(3), 289–294 (2005)

E.F. Neuhauser, C.A. Callahan, Growth and reproduction of the earthworm, *Eisenia fetida* exposed to sublethal concentrations of organic chemicals. Soil Biol. Biochem. **22**(2), 175–179 (1990)

T. Neuparth, A.D. Correia, F.O. Costa, G. Lima, M.H. Costa, Multi-level assessment of chronic toxicity of estuarine sediments with the amphipod *Gammarus locusta*: I. Biochemical endpoints. Mar. Environ. Res. **60**, 69–91 (2005)

R.L. Nielson, Earthworms. N. Z. J. Agric. **86**, 374 (1953)

M.E.T. Nunes, E.L.G. Espindola, Sensitivity of *Eisenia andrei* (Annelida, Oligochaeta) to a commercial formulation of abamectin in avoidance tests with artificial substrate and natural soil under tropical conditions. Ecotoxicology **21**(4), 1063–1071 (2012)

V. Nuutinen, Earthworm community response to tillage and residue management on different soil types in southern Finland. Soil Till. Res. **23**, 221–239 (1992)

M. Nyffeler, H. Moor, R.F. Foelix, Spiders feeding on earthworms. J. Arachnol. **29**(1), 119–124 (2001)

J.M. Oades, The role of biology in the formation, stabilization and degradation of soil structure. Geoderma **56**(1–4), 377–400 (1993)

OECD. Guideline for testing of chemicals – Earthworm, acute toxicity tests (1984), pp. 1–9, Retrieved from http://www.oecd.org

R.B. Ojha, D. Devkota, Earthworms: Soil and ecosystem engineers – a review. World J. Agric. Res. **2**(6), 257–260 (2014)

A. Olvera-Velona, Y. Capowiez, O. Mascle, L. Ortiz-Hernandez, P. Benoit, Assessment of the toxicity of ethyl-parathion to earthworms (*Aporrectodea caliginosa*) using behavioural, physiological and biochemical markers. Appl. Soil Ecol. **40**, 476–483 (2008)
F.H. Orozco, J. Cegarra, L.M. Trujillo, A. Roig, Vermicomposting of coffee pulp using the earthworm *Eisenia fetida*: effects on C and N contents and the availability of nutrients. Biol. Fertil. Soils **22**(1–2), 162–166 (1996)
E. Ouedraogo, B. Lijbert, M. Abdoulaye, S. Leo, Organic resources and earthworms affect phosphorus availability to sorghum after phosphate rock addition in semiarid West Africa. Biol. Fertil. Soils **41**, 458–465 (2005)
A. Pal, P. Patidar, Effect of insecticide malathion on weight of *Eisenia foetida* earthworm. AISECT Univ. J. **3**(5), 1–3 (2014)
S. Panda, S.K. Sahu, Recovery of acetylcholine esterase activity of *Drawida willsi* (Oligochaeta) following application of three pesticides to soil. Chemosphere **55**, 283–290 (2004)
M.G. Paoletti, The role of earthworms for assessment of sustainability and as bioindicators. Agric. Ecosyst. Environ. **74**(1), 137–155 (1999)
M.G. Paoletti, E. Iovane, M. Cortese, Pedofauna bioindicators as heavy metals in five agroecosystems in north-east Italy. Rev. Ecol. Biol. Sol. **25**(1), 33–58 (1988)
M.G. Paoletti, U. Schweigl, M.R. Favretto, Soil microinvertebrates, heavy metals, heavy metals and organochlorines in low and high input apple orchards and coppiced woodland. Pedobiologia **39**, 20–33 (1995)
M.G. Paoletti, D. Sommaggio, M.R. Favretto, G. Petruzzelli, B. Pezzarossa, M. Barbafieri, Earthworms as useful bioindicators of agroecosystem sustainability in orchards and vineyards with different inputs. Appl. Soil Ecol. **10**(1&2), 137–150 (1998)
B. Park, J. Yoo, J. Kim, J. Kim, S. Lee, Biotransformation of endosulfan by the tiger worm, *Eisenia fetida*. J. Agric. Chem. Environ. **1**(1), 20–27 (2012)
S. Pattnaik, M.V. Reddy, Microbial characterisation of vermicompost and compost of urban waste processed by three earthworm species - *Eudrilus eugeniae, Eisenia fetida and Perionyx excavatus*. Int. J. Environ. Technol. Manag. **15**, 465–500 (2012a)
S. Pattnaik, M.V. Reddy, Remediation of heavy metals from urban waste by vermicomposting using earthworms: *Eudrilus eugeniae, Eisenia fetida and Perionyx excavatus*. Int. J. Environ. Technol. Manag. **10**, 284–296 (2012b)
C. Pelosi, M. Bertrand, Y. Capowiez, H. Boizard, J. Roger-Estrade, Earthworm collection from agricultural fields: comparisons of selected expellants in presence/absence of hand-sorting. Eur. J. Soil Biol. **45**(2), 176–183 (2009)
C. Pelosi, S. Barot, Y. Capowiez, M. Hedde, F. Vandenbulcke, Pesticides and earthworms. A review. Agron. Sustain. Dev. **34**(1), 199–228 (2014)
J.L. Perreira, S.C. Antunes, A.C. Ferreira, F. Gonvalces, R. Pereira, Avoidance behavior of earthworms under exposure to pesticides: is it chemosensorial? J. Env. Sci. Health B. **45**, 229–232 (2010)
A. Pierret, Y. Capowiez, L. Belzunces, C.J. Moran, 3D reconstruction and quantification of macropores using X-ray computed tomography and image analysis. Geoderma **106**, 247–227 (2002)
V. Pizl, Interactions between earthworms and herbicides 1. Toxicity of some herbicides to earthworms in laboratory tests. Pedobiologia **32**, 227–232 (1988)
D.A. Potter, Earthworms, thatch and pesticides. USGA green section record, (1991), Retrieved from http://gsr.lib.msu.edu/1990s/1991/910906.pdf
D.A. Potter, M.C. Buxton, C.T. Redmond, C.G. Patterson, A.J. Powelu, Toxicity of pesticides to earthworms (Oligochaeta: Lumbricidae) and effect on thatch degradation in Kentucky bluegrass turf. J. Econ. Entomol. **83**(6), 2362–2369 (1990)
PPDB, (2013). http://sitem.herts.ac.uk/aeru/footprint/fr/index.htm
P.W. Ramteke, R.K. Hans, Isolation of hexachlorocyclohexane (HCH) degrading microorganisms from earthworm gut. J. Environ. Sci. Health A. **27**(8), 2113–2122 (1992)
J.V. Rao, Y.S. Pavan, S.S. Madhavendra, Toxic effects of chlorpyrifos on morphology and acetylcholinesterase activity in the earthworm, *Eisenia foetida*. Ecotoxicol. Environ. Saf. **54**, 296–301 (2003)

M. Rault, C. Mazzia, Y. Capowiez, Tissue distribution and characterization of cholinesterase activity in six earthworm species. Comp. Biochem. Physiol. B **147**, 340–346 (2007)

F. Raw, Studies of earthworm populations in orchards. l. Leaf burial in apple orchards. Ann. Appl. Biol. **50**, 389–404 (1962)

N.C. Reddy, V. Rao, Biological response of earthworm, *Eisenia foetida* (Savigny) to an organophosphorous pesticide, profenofos. J. Ecotoxicol. Environ. Saf. **71**(2), 574–582 (2008)

M.V. Reddy, V.R. Reddy, Effects of organochlorine, organophosphorus and carbamate insecticides on the population structure and biomass of earthworms in a semi-arid tropical grassland. Soil Biol. Biochem. **24**(12), 1733–1738 (1992)

Z.J. Reicher, C.S. Throssell, Effect of repeated fungicide applications on creeping bentgrass turf. Crop. Sci. **37**, 910–915 (1997)

A.J. Reinecke, S.A. Reinecke, Earthworms as test organisms in ecotoxicological assessment of toxicant impacts on ecosystems, in *Earthworm Ecology*, ed. by C.A. Edwards, 2nd edn. (CRC Press, Boca Raton, 2004), pp. 299–320

S.A. Reinecke, A.J. Reinecke, The impact of organophosphate pesticides in orchards on earthworms in the Western Cape, South Africa. Ecotoxicol. Environ. Saf. **66**(2), 244–251 (2007)

A.J. Reinecke, S.A. Viljoen, R.J. Saayman, The suitability of *Eudrilus eugeniae*, *Perionyx excavatus* and *Eisenia fetida* (Oligochaeta) for vermicomposting in southern Africa in terms of their temperature requirements. Soil Biol. Biochem. **24**(12), 1295–1307 (1992)

A.Y. Renoux, M. Sarrazin, J. Hawari, G.I. Sunahara, Transformation of 2,4,6-trinitrotoluene in soil in the presence of the earthworm, *Eisenia andrei*. Environ. Toxicol. Chem. **19**(6), 1473–1780 (2000)

D. Ribera, J.F. Narbonne, C. Arnaud, M. Saint-Denis, Biochemical responses of the earthworm *Eisenia fetida andrei* exposed to contaminated artificial soil: effects of carbaryl. Soil Biol. Biochem. **33**, 1123–1130 (2001)

J.H. Roark, J.L. Dale, The effect of turf fungicides on earthworms. Ark. Acad. Sci. Proc. **33**, 71–74 (1979)

B.L. Roberts, H.W. Dorough, Relative toxicities of chemicals to the earthworm, *Eisenia foetida*. Environ. Toxicol. Chem. **3**(1), 67–78 (1984)

L. Rodriguez-Castellanos, J.C. Sanchez-Hernandez, Earthworm biomarkers of pesticide contamination: current status and perspectives. J. Pestic. Sci. **32**(4), 360–371 (2007)

J. Rombke, J. Notenboom, Ecotoxicological approaches in the field, in *Environmental Analysis of Contaminated Sites*, ed. by G.I. Sunahara, A.Y. Renoux, C. Thellen, C.L. Gaudet, A. Pilon (Wiley, Chichester, 2002), pp. 181–195

J. Rombke, C. Bauer, A. Marschner, Hazard assessment of chemicals in soil: proposed ecotoxicological test strategy. Environ. Sci. Pollut. Res. **3**(2), 78–82 (1996)

J. Rombke, C.A. Van-Gestel, S.E. Jones, J.E. Koolhaas, J.M. Rodrigues, T. Moser, Ring-testing and field-validation of a terrestrial model ecosystem (TME) -an instrument for testing potentially harmful substances: effects of carbendazim on earthworms. Ecotoxicology **13**(1–2), 105–118 (2004)

J. Rombke, M.V. Garcia, A. Scheffczyk, Effects of the fungicide benomyl on earthworms in laboratory tests under tropical and temperate conditions. Arch. Environ. Contam. Toxicol. **53**(4), 590–598 (2007)

C.A.F.M. Romijn, R. Luttik, J. Canton, Presentation of a general algorithm to include effect assessment on secondary poisoning in the derivation of environmental quality criteria. 2. Terrestrial food chains. Ecotoxicol. Environ. Saf. **27**, 107–127 (1994)

K. Ronnpagel, E. Janssen, W. Ahlf, Asking for the indicator function of bioassays evaluating soil contamination: are bioassay results reasonable surrogates of effects on soil microflora? Chemosphere **36**, 1291–1304 (1998)

T.J. Roper, The European badger *Meles meles*: food specialist or generalist? J. Zool. **234**, 437–452 (1994)

C.H. Roth, M. Joschko, A note on the reduction of runoff from crusted soils by earthworm burrows and artificial channels. Z. Pflanzenernähr. Bodenk. **154**, 101–105 (1991)

E. Ruz-Jerez, P.R. Ball, R.W. Tillman, The role of earthworms in nitrogen release from herbage residues, in *Nitrogen Efficiency in Agricultural Soils*, ed. by D.S. Jenkinson, K.A. Smith (Elsevier Applied Science, London, 1988), pp. 355–370

J.R. Sabine, Earthworms as a source of food and drugs, in *Earthworm Ecology: From Darwin to Vermiculture*, ed. by J. Satchell (Springer, Dordretcht, 1983), pp. 285–296

J.C. Sanchez-Hernandez, Earthworm biomarkers in ecological risk assessment. Rev. Environ. Contam. Toxicol. **188**, 85–126 (2006)

J.C. Sanchez-Hernandez, C.E. Wheelock, Tissue distribution, isozyme abundance and sensitivity to chlorpyrifos-oxon of carboxylesterases in the earthworm, *Lumbricus terrestris*. Environ. Pollut. **157**(1), 264–272 (2008)

SANCO. Guidance document on terrestrial ecotoxicology under council directive 91/414. Draft Working Document. EU (DG Health and Consumer Protection), Brussels, Belgium. SANCO/1039/2002 rev. 2 final, 2002

M. Schaefer, Assessing 2,4,6-trinitrotoluene (TNT)- contaminated soil using three deterrent earthworm test methods. Ecotoxicol. Environ. Saf. **57**, 74–80 (2004)

M. Schaefer, J. Filser, The influence of earthworms and organic additives on the biodegradation of oil contaminated soil. Appl. Soil Ecol. **36**, 53–62 (2007)

M. Schaefer, F. Juliane, The influence of earthworms and organic additives on the biodegradation of oil contaminated soil. Appl. Soil Ecol. **36**, 53–62 (2007)

M. Schaefer, S.O. Peterson, J. Filser, Effects of *Lumbricus chlorotica* and *Eisenia fetida* on microbial community dynamics in oil-contaminated soil. Soil Biol. Biochem. **37**, 2065–2076 (2005)

S. Scheu, Microbial activity and nutrient dynamics in earthworm casts (Lumbricidae). Biol. Fertil. Soils **5**, 230–234 (1987)

S. Scheu, Effects of earthworms on plant growth: patterns and perspectives. Pedobiologia **47**, 846–856 (2003)

S. Scheu, J. Alphei, M. Bonkowski, C. Jentschke, Soil food web interactions and ecosystem functioning: experimental approaches with systems of increasing complexity, in *Going Underground – Ecological Studies in Forest Soils*, ed. by N. Rastin, J. Bauhus (Research Signpost, Trivandrum, 1999), pp. 1–32

O. Schmidt, Appraisal of the electrical octet method for estimating earthworm populations in arable land. Ann. Appl. Biol. **138**(2), 231–241 (2001)

S. Schrader, H. Zhang, Earthworm casting: stabilization or destabilization of soil structure? Soil Biol. Biochem. **29**, 469–475 (1997)

E. Schrecka, F. Geret, L. Gontier, M. Treilhou, Neurotoxic effect and metabolic responses induced by a mixture of six pesticides on the earthworm, *Aporrectodea caliginosa nocturna*. Chemosphere **71**, 1832–1839 (2008)

A.N. Sharpley, J.K. Syers, J.A. Springett, Effects of surface-casting earthworns on the transport of phosphorous and nitrogen in surface runoff from pasture. Soil Biol. Biochem. **9**, 227–231 (1979)

S. Sheppard, J. Bembridge, M. Holmstrup, L. Posthuma, *Advances in Earthworm Ecotoxicology* (SETAC Press, Pensacola, 1997)

S.C. Sheppard, J.D. Bembridge, M. Holmstrup, L. Posthuma, Advances in earthworm ecotoxicology. in *Proc. 2nd Intl Workshop on Earthworm ecotoxicology*, Amsterdam, (SETAC Press. SETAC, Pensacola, 1998)

M.J. Shipitalo, K.R. Butt, Occupancy and geometrical properties of *Lumbricus terrrestris* L. burrows affecting infiltration. Pedobiologia **43**, 782–794 (1999)

M.J. Shipitalo, R. Protz, Chemistry and micromorphology of aggregation in earthworm casts. Geoderma **45**, 357–374 (1989)

M. Simek, V. Pizl, The effect of earthworms (Lumbricidae) on nitrogenase soil. Biol. Fertil. Soils **7**, 370–373 (1989)

A.C. Singer, W. Jury, E. Leupromchai, C.S. Yahng, D.E. Crowley, Contribution of earthworms to PCB bioremediation. Soil Biol. Biochem. **33**, 765–775 (2001)

N.B. Singh, A.K. Khare, D.S. Bhargava, S. Bhattacharya, Effect of initial pH on vermicomposting using *Perionyx excavatus* (Perrier, 1872). Appl. Eco. Env. Sci. **4**(1), 85–97 (2005)

J. Six, H. Bossuyt, S. Degryze, K. Denef, A history of research on the link between microaggregates, soil biota and soil organic matter dynamics. Soil Till. Res. **79**, 7–31 (2004)

K.M. Slimax, Avoidance response as sublethal effect of pesticides on *Lumbricus terrestris* (Oligochaeta). Soil Biol. Biochem. **29**(3&4), 713–715 (1997)

M. Smith, *The British Amphibians and Reptiles* (Collins, London, 1951). 318 p

R.G. Smith, C.P. McSwiney, A.S. Grandy, P. Suwanwaree, R.M. Snider, G.P. Robertson, Diversity and abundance of earthworms across an agricultural land-use intensity gradient. Soil Till. Res. **100**, 83–88 (2008)

Y. Song, J. Kaia, X. Song, W. Zhang, L. Lia, Long-term toxic effects of deltamethrin and fenvalerante in soil. J. Hazard. Mater. **289**, 158–164 (2015)

J. Sorour, O. Larink, Toxic effects of benomyl on the ultrastructure during spermatogenesis of the earthworm *Eisenia fetida*. Ecotoxicol. Environ. Saf. **50**, 180–188 (2001)

H.N. Southern, Prey taken by tawny owls during the breeding season. Ibis **3**, 293–299 (1969)

F.M. Speed, J.B. Smith, *Investigation in Behaviour and Elementary Neurobiology* (Macmillan, London, 1975), 189 p

J.A. Springett, R.A.J. Gray, Effect of repeated low doses of biocides on the earthworm *Aporrectodea caliginosa* in laboratory culture. Soil Biol. Biochem. **24**(12), 1739–1744 (1992)

D.J. Spurgeon, S.P. Hopkin, Risk assessment of the threat of secondary poisoning by metals to predators of earthworms in the vicinity of a primary smelting works. Sci. Total Environ. **187**, 167–183 (1996)

D.J. Spurgeon, J.M. Weeks, C.A.M. Van-Gestel, A summary of eleven years progress in earthworm ecotoxicology. Pedobiologia **47**, 588–606 (2003)

P.L. Staddon, N. Ostle, A.H. Fitter, Earthworm extraction by electroshocking does not affect canopy CO_2 exchange, root respiration, mycorrhizal fungal abundance or mycorrhizal fungal vitality. Soil Biol. Biochem. **35**(3), 421–426 (2003)

J. Stanley, Chemical and behavioural approaches of pest management in cardamom. Ph.D thesis. (Tamil Nadu Agricultural University, Coimbatore, 2007), 209 p

J. Stanley, S. Chandrasekaran, G. Preetha, S. Kuttalam, R.S. Jasmine, Selective toxicity of diafenthiuron to non-target organisms: honeybees, coccinellids, chelonus, earthworm, silkworm and fish. J. Plant Prot. Res. **56**(1), 1–5 (2016)

J. Stenersen, Action of pesticides on earthworms. Part I. The toxicity of chlolinesterase inhibiting insecticides to earthworms as evaluated by laboratory tests. Pestic. Sci. **10**, 66–74 (1979)

J. Stenersen, Esterases of earthworms. Part I: Characterisation of the cholinesterases in *Eisenia foetida* (Savigny) by substrates and inhibitors. Comp. Biochem. Physiol. C **66**(1), 37–44 (1980)

G.L. Stephenson, A. Kaushik, N.K. Kaushik et al., Use of an avoidance response test to assess the toxicity of contaminated soil to earthworms, in *Advances in Earthworm Ecotoxicology*, ed. by S.C. Sheppard, S.C. Sheppard (SETAC Press, Pensacola, 1998), pp. 67–81

J.W. Steward, *The Snakes of Europe* (Fairleigh Dickinson, Newton Abbot, 1971). 238 p

S.M.J. Stockdill, Effect of earthworms on pasture. Proc. N. Z. Grassl. Assoc. **13**, 68–75 (1966)

S.M.J. Stockdill, G.G. Cossens, The role of earthworms in pasture production and moisture conservation. Proc. N. Z. Grassl. Assoc. **28**, 168–183 (1966)

A. Stringer, C.H. Lyons, The effect of benomyl and thiophanate-methyl on earthworm populations in apple orchards. Pestic. Sci. **5**, 189–195 (1974)

A. Stringer, M.A. Wright, The toxicity of benomyl and some related 2-substituted benzimidazoles to the earthworm *Lumbricus terrestris*. Pestic. Sci. **7**, 459–464 (1976)

S. Suthar, Potential utilization of guar gum industrial waste in vermicompost production. Bioresour. Technol. **97**(18), 2474–2477 (2006)

S. Suthar, Nutrient changes and biodynamics of epigeic earthworm, *Perionyx excavatus* (Perrier) during recycling of some agriculture wastes. Bioresour. Technol. **98**, 1608–1614 (2007)

B.V. Symonds, Evaluation of potential molluscicides for the control of the field slug, *Agriolimax reticulatus* (Müll.). Plant Pathol. **24**(1), 1–9 (1975)

V. Tanara, *L'economia del cittadino in villa*, 10th edn. (G. Bortoli, Venezia, 1644), 525 p

S.P. Teotia, F.L. Duley, T.M. McCalla, Effect of stubble mulching on number and activity of earthworms. Agric. Exp. Station Res. Bull. (University of Nebraska College of Agriculture. Lincoln, 1950), pp. 165

J.M. Tisdall, J.M. Oades, Organic matter and water stable aggregates in soil. J. Soil Sci. **33**, 141–163 (1982)

A.D. Tomlin, The earthworm bait market in North America, in *Earthworm Ecology: From Darwin to Vermiculture*, ed. by J. Satchell (Springer, Dordretcht, 1983), pp. 331–338

A.D. Tomlin, F.L. Gore, Effects of six insecticides and a fungicide on the numbers and biomass of earthworms in pasture. Bull. Environ. Contam. Toxicol. **12**(4), 487–492 (1974)

A.L. Torkhani, N.K. Erzen, Does ivermectin attract earthworms? J. Soils Sed. **11**, 124–128 (2011)

G. Tripathi, N. Kachhwaha, I. Dabi, Comparative studies on carbofuran induced changes in some cytoplasmic and mitochondrial enzymes and proteins of epigeic, anecic and endogeic earthworms. Pestic. Biochem. Physiol. **96**, 30–35 (2010)

M.D. Trojan, D.R. Linden, Microrelief and rainfall effects on water and solute movement in earthworm burrows. Soil Sci. Soc. Am. J. **56**, 727–733 (1992)

C. Tu, Y. Wang, W. Duan, P. Hertl, L. Tradway, R. Brandenburg, D. Lee, M. Snell, S. Hu, Effects of fungicides and insecticides on feeding behavior and community dynamics of earthworms: implications for casting control in turfgrass systems. Appl. Soil Ecol. **47**, 31–36 (2011)

M. Udovic, D. Lestan, *Eisenia fetida* avoidance behavior as a tool for assessing the efficiency of remediation of Pb, Zn and Cd polluted soil. Environ. Pollut. **158**, 2766–2772 (2010)

H. Valenzuela, Earthworms in the farm. (2010) *The food provider,* Retrieved from http://www.ctahr.hawaii.edu/sustainag/news/articles/v5-valenzuela-worm.pdf

N.M. Van-Dam, J.A. Harvey, F.L. Wackers, T.M. Bezemer, W.H.V. Putten, L.E.M. Vet, Interactions between aboveground and belowground induced responses against phytophages. Basic Appl. Ecol. **4**, 63–77 (2003)

C.A.M. Van-Gestel, Validation of earthworm toxicity tests by comparison with field studies: a review of benomyl, carbendazim, carbofuran and carbaryl. Ecotoxicol. Environ. Saf. **23**, 221–236 (1992)

C.A.M. Van-Gestel, J.M. Weeks, Recommendations of the third international workshop on earthworm ecotoxicology, Aarhus, Denmark, August 2001. Ecotoxicol. Environ. Saf. **57**, 100–105 (2004)

C.A.M. Van-Gestel, W.A. Van-Dis, E.M. Van-Breemen, P.M. Sparenburg, Development of a standardized reproduction toxicity test with the earthworm species *Eisenia fetida andrei* using copper, pentachlorophenol and 2,4-dichloroaniline. Ecotoxicol. Environ. Saf. **l8**, 305–312 (1989)

C.A.M. Van-Gestel, E.M.D. Breemen, R. Baerselman, H.J.B. Emans, J.A.M. Janssen, R. Postuma, P.J.M. Van-Vliet, Comparison of sublethal and lethal criteria for nine different chemicals in standardized toxicity tests using the earthworm *Eisenia andrei*. Ecotoxicol. Environ. Saf. **23**(2), 206–220 (1992)

P. Van-Hoof, Earthworm activity as a cause of splash erosion in a Luxembourg forest. Geoderma **31**, 195–204 (1983)

L. Van-Leemput, E. Swysen, R. Woestenborghs, L. Michielsen, W. Meuldermans, J. Heykants, On the terrestrial toxicity of the fungicide imazalil (enilconazole) to the earthworm species *Eisenia foetida*. Ecotoxicol. Environ. Saf. **18**(3), 313–320 (1989)

N.M. Van-Straalen, C.A.J. Denneman, Ecotoxicological evaluation of soil quality criteria. Ecotoxicol. Environ. Saf. **18**, 241–251 (1989)

N.M. Van-Straalen, J.P. Van-Rijin, Ecotoxicological risk assessment of soil fauna recovery from pesticide application. Rev. Environ. Contam. Toxicol. **154**, 83–141 (1998)

M. Velki, B.K. Hackenberger, Species-specific differences in biomarker responses in two ecologically different earthworms exposed to the insecticide dimethoate. Comp. Biochem. Physiol. C **156**, 104–112 (2012)

M. Velki, B.K. Hackenberger, Different sensitivities of biomarker responses in two epigeic earthworm species after exposure to pyrethroid and organophosphate insecticides. Arch. Environ. Contam. Toxicol. **65**(3), 498–509 (2013)

J.M. Venter, A.J. Reinecke, Dieldrin and growth and development of the earthworm, *Eisenia fetida* (oligochaeta). Bull. Environ. Contam. Toxicol. **35**(5), 652–659 (1985)

K. Verma, N. Agrawal, M. Farooq, R.B. Misra, R.K. Hans, Endosulfan degradation by a Rhodococcus strain isolated from earthworm gut. Ecotoxicol. Environ. Saf. **64**, 377–381 (2006)

L.A. Vermeulen, A.J. Reinecke, S.A. Reinecke, Evaluation of the fungicide manganese-zinc ethylene bis (dithiocarbamate) (mancozeb) for sublethal and acute toxicity to *Eisenia fetida* (Oligochaeta). Ecotoxicol. Environ. Saf. **48**, 183–189 (2001)

M.G. Vijver, J.P.M. Vink, C.J.H. Miermans, C.A.M. Van-Gestel, Oral sealing using glue: a new method to distinguish between intestinal and dermal uptake of metals in earthworms. Soil Biol. Biochem. **35**, 125–132 (2003)

P.L.G. Vlek, M. Denich, C. Martius, N. Van-de-Giesen, *Effects of Pesticides on Soil Fauna: Development of Ecotoxicological Test Methods for Tropical Region*. Ecology and Development Series, vol 19 (2004). Retrieved from http://www.zef.de/fileadmin/webfiles/downloads/zefc_ecology_development/ecol_dev_19_text.pdf

Y. Wang, S. Wu, L. Chen, C. Wu, R. Yu, Q. Wang, X. Zhao, Toxicity assessment of 45 pesticides to the epigeic earthworm *Eisenia fetida*. Chemosphere **88**(4), 484–491 (2012a)

Y. Wang, T. Cang, X. Zhao, R. Yu, L. Chen, C. Wu, Q. Wang, Comparative acute toxicity of twenty four insecticides to earthworm, *Eisenia fetida*. Ecotoxicol. Environ. Saf. **79**, 122–128 (2012b)

R.A.S. Waters, Numbers and weights of earthworms under highly productive pasture. N. Z. J. Sci. Technol. **36**, 516–526 (1955)

M.J. Way, N.E.A. Scopes, Studies on the persistence and effects on soil fauna of some soil-applied systemic insecticides. Ann. Appl. Biol. **62**, 199–214 (1968)

J.M. Weeks, The value of biomarkers for ecological risk assessment: academic toys or legislative tools? Appl. Soil Ecol. **2**, 215–216 (1995)

J.M. Weeks, D.W. Comber, Ecological risk assessment of contaminated soils. Miner. Mag. **69**, 601–613 (2005)

J.M. Weeks, C. Svendsen, Neutral red retention by lysosomes from earthworm (*Lumbricus rubellus*) coelomocytes: a simple biomarker of exposure to soil copper. Environ. Toxicol. Chem. **15**(10), 1801–1805 (1996)

A. Weyers, B. Sokull-Kluttgen, T. Knacker, S. Martin, C.A. Van-Gestel, Use of terrestrial model ecosystem data in environmental risk assessment for industrial chemicals, biocides and plant protection products in the EU. Ecotoxicology **13**(1&2), 163–176 (2004)

G.E. Wilkinson, Effect of grass fallow rotations on the infiltration of water into a savanna zone soil of northern Nigeria. Trop. Agric. **52**, 97–104 (1975)

S.J. Wu, E.M. Wu, L.Q. Qiu, W.H. Zhong, J.M. Chen, Effects of phenanthrene on the mortality, growth and antioxidant system of earthworms (*Eisenia fetida*) under laboratory conditions. Chemosphere **83**, 429–434 (2011)

S. Wurst, R. Langel, A. Reineking, M. Bonkowski, S. Scheu, Effects of earthworms and organic litter distribution on plant performance and aphid reproduction. Oecologia **137**, 90–96 (2003)

N. Xiao, B. Jing, F. Ge, X. Liu, The fate of herbicide acetochlor and its toxicity to *Eisenia fetida* under laboratory conditions. Chemosphere **62**(8), 1366–1373 (2006)

S. Yasmin, D. D'Souza, Effects of pesticides on the growth and reproduction of earthworm: a review. Appl. Environ. Soil Sci. **2010**, 678360 (2010)

Y.L. Yu, X.M. Wu, S.N. Li, H. Fang, H.Y. Zhan, J.Q. Yu, An exploration of the relationship between adsorption and bioavailability of pesticides in soil to earthworm. Environ. Pollut. **141**(3), 428–433 (2006)

J.E. Zachman, D.R. Linden, C.E. Clapp, Macroporous infiltration and redistribution as affected by earthworms, tillage and residue. Soil Sci. Soc. Am. J. **51**, 1580–1586 (1987)

H. Zhang, S. Schrader, Earthworm effects on selected physical and chemical properties of soil aggregates. Biol. Fertil. Soils **15**, 229–234 (1993)

Y. Zhang, G. Shen, Y. Yu, H. Zhu, The hormetic effect of cadmium on the activity of antioxidant enzymes in the earthworm *Eisenia fetida*. Environ. Pollut **157**, 3064–3068 (2009)

X. Zhang, J. Wang, H. Xie, J. Wang, W. Zech, Comparison of organic compounds in the particle-size fractions of earthworm casts and surrounding soil in humid Laos. Appl. Soil Ecol. **23**, 147–153 (2003)

Q.X. Zhou, Q.R. Zhang, J.D. Liang, Toxic effects of acetochlor and Methmidophos on earthworm *Eisenia fetida* in phaiozem, northeast China. J. Environ. Sci. **18**(4), 741–745 (2006)

S.P. Zhou, C.Q. Duan, H. Fu, Y.H. Chen, X.H. Wang, Z.F. Yu, Toxicity assessment for chlorpyrifos-contaminated soil with three different earthworm test methods. J. Environ. Sci. **19**(7), 854–858 (2007)

S.P. Zhou, C.P. Duan, X.H. Wang, W.H.G. Michelle, Z.F. Yu, F. Fu, Assessing cypermethrin contaminated soil with three different earthworm test methods. J. Environ. Sci. **20**, 1381–1385 (2008)

S.P. Zhou, C.Q. Duan, W.H.G. Michelle, F.Z. Yang, X.H. Wang, Individual and combined toxic effects of cypermethrin and chlorpyriphos on earthworm. J. Environ. Sci. **23**, 676–680 (2011)

X. Zou, M. Bashkin, Soil carbon accretion and earthworm recovery following revegetation in abandoned sugarcane fields. Soil Biol. Biochem. **30**, 825–830 (1998)

P.R. Zund, U. Pillai-McGarry, D. McGarry, S.G. Bray, Repair of a compacted Oxisol by the earthworm *Pontoscolex corethrurus* (Glossoscolecidae, Oligochaeta). Biol. Fertil. Soils **25**, 202–208 (1997)

Index

Chapter 6
Pesticide Toxicity to Microorganisms: Exposure, Toxicity and Risk Assessment Methodologies

Abstract Microbes play a major role in medical technology, food processing, biotechnology, agriculture and many other fields. The major route of pesticide exposure to microbes in the environment is through soil application, spray drifts and dumping of pesticides in soil. Pesticides affecting microbial biomass, growth and respiration affecting the microbial community and diversity are well studied. The estimation of total biomass directly, perhaps includes the active and dormant microbes. Indirect biomass assessments by measuring microbial biomass carbon/nitrogen by fumigation method or through the estimation of ATP, phospholipid fatty acids are being explained. Estimation of soil respiration is the oldest but extensively used technique being done as basal and as substrate induced respiration. Pesticide toxicity on microbial activities especially of enzyme activities is well documented. Determination of soil enzymes through fluorescein diacetate hydrolysis, functional richness as determined by carbon utilization pattern, structural diversity through phospholipid fatty acid profiling, genetic diversity through 16s rDNA analysis are commonly used to test the pesticide toxicity. Cultivation independent methods mostly relay on DNA sequencing, proteomics and metabolomics studies. Methodologies for mesocosm or semifield and field experiments are also explained and discussed. Though microbes are responsible for many ecosystem services, soil ecotoxicological risk assessment guidelines do not consider microbes as assessment end points. Unlike other groups, risk assessment procedures for microbes are neither extensively used nor well developed. Risk assessment by studying the microbial activities by giving weightage to important activities, multi-tiered risk assessment approach, comparison with beneficial and harmful microbial susceptibility to pesticides and assessing hazardous concentration are discussed.

1 Importance of Microbes in Agriculture

The various uses of microorganisms in medical technology, human and animal health, food and environmental science, genetic engineering, waste management etc. provide a most impressive record of achievement (Higa and Parr 1994). Soil microbes (bacteria and fungi) decompose organic matter and are essential for recycling of plant material. Apart from this, soil microbes influence plant health both by acting as pathogens and beneficial agents. Soil microbes have a say on the

J. Stanley, G. Preetha, *Pesticide Toxicity to Non-target Organisms*,
DOI 10.1007/978-94-017-7752-0_6

productivity of crops also (Bloemberg and Lugtenberg 2001). These involve biofertilization, stimulation of root growth, rhizoremediation and control of biotic and abiotic stresses (Mendes et al. 2013). Some soil microbes form mutualistic relationship with plant roots and provide nutrients like nitrogen or phosphorus. As a mutualism, plants provide the bacteria with carbohydrate while the bacteria reduce the nitrous compounds in the soil, which is then used by the plant. Arbuscular mycorrhizal fungi form internal symbiosis with the plant roots and provide phosphorous and take organic carbohydrates in exchange (Smith and Read 1997). Exchange of nutrients between plant roots and bacteria during the symbiosis of nitrogen-fixing rhizobia are reported (Lodwig et al. 2003). Certain fungi colonize even the above ground parts of plants and provide benefits like plant tolerance to drought/heat and insect/disease. Root zone bacteria that are reported to afford beneficial effects on various plants include different species of the genera Arthrobacter, Azotobacter, Azospirillum, Bacillus, Enterobacter, Pseudomonas, Serratia (Gray and Smith 2005) and Streptomyces (Tokala et al. 2002; Dimkpa et al. 2008). Symbiotic fungi, such as arbuscular mycorrhizal fungi (AMF) are useful in agrosystems (Hart and Trevors 2005). Microbes alleviate heavy metal toxicity in plants and also in soils by means of many diverse mechanisms. Biotic and abiotic stress alleviation by microbes is highly valuable in agriculture especially under changing environmental conditions (Dimkpa et al. 2009a).

Microbial biofertilizers and pesticides would lead to a reduction in the use of chemicals leading to reduction in pollution and environmental hazards apart from enhancing organic agriculture (Morrissey et al. 2004). Microbial fertilizers and pesticides have many added advantage than their chemical counterparts. They are (1) more safe, (2) less environmental damage (3) show much more targeted activity, (4) effective in small quantities, (5) multiply themselves, (6) decompose more quickly than chemical pesticides, (7) cause slow resistance development and (8) can be used in integrated pest management systems (Berg 2009). In future, abiotic stress-protecting agents (water logging, drought, salinity, heavy metals) and biocontrol agents for insect pests and diseases will be of more important not only due to climate change (Berg 2009) and microbes pave a way for mitigating these stresses. Microbes apart from supplementing plant growth and health are used in end product management also. Microbes are used in value addition of agro products/food items which improves its keeping-quality besides enhancing nutrition and organoleptic properties. Some of the important uses of microbes especially in agriculture are detailed here under.

1.1 As Biofertilizers and Nitrogen Fixers

Biofertilizers are living cells of different microorganisms, which have an ability to convert unavailable form of nutrients into available form to plants through various biological processes (Hegde et al. 1999; Vessey 2003). Most commonly, bacteria such as Azospirillum, Herbaspirillum, Acetobacter, Azotobacter and Azoarocus can fix atmospheric nitrogen and thus used as an important biofertilizer. The less common but potential microbial biofertilizers also include Bacillus, Streptomyces and Pseudomonas (Morrissey et al. 2004). A mixture of biofertilizers containing arbuscular mycorrhizal fungus (*Glomus mosseae*, *G. intraradices*), N fixer (*Azotobacter chroococcum*), P solubilizer (*Bacillus megaterium*) and K solubilizer (*Bacillus mucilaginosus*) was reported to increase the nutrient (N, P and K) assimilation of maize plants with increased seedling weight and biomass in greenhouse trials (Wu et al. 2005). The blue green algae biofertilizer applied @ 12.5 kg/ha was found to enhance the growth of rice plants. Integrated use of fly ash, blue green algae and nitrogen fertilizer was found to enhance growth, yield and mineral composition in rice (Tripathi et al. 2008). Excessive fertilization causes contamination of soil and groundwater by leachates and microbial denitrification converts residual nitrogen into nitrous oxide, a greenhouse gas (Nosengo 2003; Reay 2004). Excess phosphorous leach into groundwater, rivers and streams and promote algal growth (Morrissey et al. 2004). So, biofertilizers help in ecosystem stabilization by preventing excess chemical fertilization.

Bacteria belong to the genera Rhizobium, Mesorhizobium, Sinorhizobium, Bradyrhizobium and Azorhizobium live as nitrogen fixing symbionts inside root nodules of legume plants (Gage 2004). In wetland rice ecosystem, nitrogen fixation by free living cyanobacteria like Anabaena, Nostoc, Aulosira and Tolypothrix significantly supplements soil nitrogen (Mishra and Pabbi 2004). Bradyrhizobium, Rhizobium, Burkholderia and Achromobacter species were reported nodulating cowpea and efficient in biological nitrogen fixation (Guimaraes et al. 2012). *Beijerinckia* sp., *Saccharobacter nitrocaptans*, *Acetobacter diazotrophicus* are nitrogen fixing bacteria associated with sugarcane (Dobereiner 1961; Cavalcante and Dobereiner 1988; James et al. 1994).

1.2 As Biopesticides and Resistance Induction

Suppression and Management of Plant Pathogens Beneficial microbes confer protection against many plant diseases. They suppress plant pathogens by producing enzymes, nitric oxide, osmolytes, siderophores, organic acids and antibiotics (Kloepper et al. 1999; Chakraborty et al. 2006; Sikora et al. 2007). Microbial antagonism on plant pathogens includes (1) inhibition of growth by antibiotics and volatile organic compounds and toxins (2) competition for colonization sites and nutrients, (3) competition for minerals- e.g. siderophore formation (4) degradation of toxins and (5) parasitism involving production of cell wall degrading enzymes such as chitinases, β-1,3-glucanase etc. (Whipps 2001; Wheatley 2002; Compant et al. 2005; Haas and Defago 2005; Raaijmakers et al. 2006; Berg 2009). Some of the metabolites produced by rhizobacteria inhibit the growth or activity of competing microorganisms (Hoffmeister and Keller 2007; Brakhage and Schroeckh 2011) e.g. Pseudomonas, Bacillus, Trichoderma etc. (Bloemberg and Lugtenberg 2001; Walsh et al. 2001). Many volatile organic compounds are produced by *Pseudomonas trivialis*, *P. fluorescens*, *B. subtilis* and *Burkholderia cepacia* which act against fungal pathogens of crop plant (Kai et al. 2007, 2009; Vespermann et al. 2007; Zou et al. 2007; Jamalizadeh et al. 2010; Mendes et al. 2013). Many antibiotic compounds were also produced by fungal and bacterial biocontrol agents (Mendes et al. 2013). Apart from antagonisms, microbes provide, induced systemic resistance to plants i.e., stimulating the plant to defend itself (Conrath et al. 2002; Van-Loon 2007).

Sinorhizobium meliloti, *R. leguminosaram* bv. *viciae* and *B. japonicum* were found to reduce the infection of *Macrophomina phaseolina*, *Rhizoctonia solani* and *Fusarium* spp. in many crop plants (Haque and Gaffar 1993). Many bacterial and fungal pathogens are managed using *Bacillus subtilis* and *P. fluorescens* (Berg 2009). *Bacillus pumilus*, *B. amyloliquifaciens*, *B. subtilis* and *Kluyvera cryocrescens* applied as seed treatments and soil drenching resulted in significant reductions in symptoms of cucumovirus in tomato (Yao et al. 1997; Kloepper et al. 1999). *Paenibacillus polymyxa* and *P. lentimorbus* were found to influence on the nematode, *Meloidogyne incognita* infestation in tomato (Son et al. 2009).

Insect Pest Management Microbial biopesticides are natural or genetically modified microbes such as bacteria, fungi, algae, viruses or protozoans used in pest management. Microbial entomopathogens generally attack the integument or gut of the insect, multiplies there and produce toxin leading to the death of the host (Usta 2013). *Bacillus thuringiensis* is probably the most studied and exploited microbial entomopathogen, used as biopesticide and as a base for many genetically modified crops. It is used for the management of lepidopterans (Bt *kurstaki*, Bt *aizawai*), coleopterans (Bt *tenebrinos*, Bt *lentimorbus*) and dipterans (Bt *israelensis*) (Entwistle et al. 1993; Usta 2013). *Bacillus thuringiensis* is used to control most of the economically important insect pests, including *Heliothis* sp., *Earias* spp., *Spodoptera* sp. and *Plutella* sp. (Berg 2009). The crystal proteins of Bt denoted as cry proteins are specific to insect pests, studied and utilized well in insect pest management. Presently there are over 400 formulations of Bt that are used in pest management contain spores and or insecticidal proteins (Usta 2013). *Bacillus sphaericus* is used for the management of insect pests especially of mosquitoes and other vectors of diseases (Baumann et al. 1991; Charles et al. 1996). *Bacillus cereus* is reported effective against whitegrubs (Selvakumar et al. 2007; Rai et al. 2013). The important fungal biopesticides include *Beauveria bassiana*, *Metarhizium anisopliae*, *Lagenidium giganteum*, *Paecilomyces lilacinus*, etc. Many viruses (granulovirus, nucleo polyhedron virus) are used as biopesticides for specific insect pest management. Nematodes like Heterorhabditis, Steinernema are also good biocontrol agents.

Though microbes are efficient biopesticides, a big obstacle of upscaling and mass utilization is the registration procedure, which is tedious (Ehlers 2006). The host and environmental specificity often find it difficult to use for different crops and in varied environmental conditions. Biocontrol agents with broader host range and used in major crops are normally been commercialized (Berg 2009).

Induced Systemic Resistance Induced systemic resistance by microbes in plants confers better disease resistance (Van-Loon et al. 1998). Though the induced resistance is caused by microbial interactions in the rhizosphere, it imparts full-plant resistance. Induced systemic resistance (ISR) mediated by rhizobacteria is demonstrated against many pathogens of bean, carnation, cucumber, radish, tobacco, tomato etc. (Van-Loon et al. 1998). A large volume of literature states induced resistance caused by *Pseudomonas* spp. and *Bacillus* spp. Induced resistance against diseases stimulated by *P. syringae* pv. *lachrymans* is proved in cucumber plants (Wei et al. 1996). Protection resulting from induced systemic resistance elicited by *Bacillus* spp. has been reported against leaf spots, stem blight, damping off, crown rot, late blight pathogens and root knot causing nematodes (Kloepper et al. 2004).

1.3 Enhancing Nutrient Availability to Plants

Plant growth promoting bacteria make the plant to have an increased nutrient uptake from soils, thus reducing fertilizer demand. Microbes enhance fertilizer use efficiency and do not allow nitrates and phosphates to get accumulated in agricultural soils (Yang et al. 2009). Azospirillum was reported to enhance the uptake of N, P, K and micronutrients in plants grown in greenhouses and field conditions (Adesemoye and Kloepper 2009). Field trials with *Pseudomonas fluorescens* showed an increased use efficiency of N and P in wheat (Shaharoona et al. 2008). Inoculation with *Pseudomonas alcaligenes*, *Bacillus polymyxa* and *Mycobacterium phlei* were found to promote nutrient uptake and growth in maize. Nitrogen, phosphorus and potassium uptake by maize was found greater even in nutrient-deficient calcisols (Egamberdiyeva 2007). *Pseudomonas striata* was found to cause a greater impact on crop production when used along with rock phosphate or super phosphate in wheat (Tiwari et al. 1989). Han and Lee (2005) reported an increased uptake of P and K by plants by *B. megaterium* (P solubilizer) and *B. mucilaginosus* (K solubilizer). Increased uptake of N, P, K, Ca and Mg in pomegranate seedling was reported when inoculated with *Azotobacter chroococcum*, *Azospirillum brasilense*, *Glomus mosseae* and *G. fasciculatum* either alone or in combinations (Aseri et al. 2008). Inoculation of Pseudomonas and Acinetobacter resulted in enhanced uptake of Fe, Zn, Mg, Ca, K and P in many crop plants (Khan 2005).

Rhizobacteria reportedly respond to root exudates and alter their rate of metabolism to make available the plant nutrients (Hardoim et al. 2008). Bacteria make available nutrients especially the phosphorus by liberating it from organic compounds such as phytates (Unno et al. 2005) and sulfates by oxidation (Banerjee and Yesmin 2002), thus promoting plant growth. PGPR and AMF play significant role in mineralizing organic phosphates and solubilizing inorganic phosphates (Mahmood et al. 2001; Tawaraya et al. 2006). Various metal uptakes by plants are positively influenced by siderophore-producing bacteria (Carrillo-Castaneda et al. 2005; Egamberdiyeva 2007; Dimkpa et al. 2009b). *Rhizopus arrhizus* is reported to produce rhizoferrin (siderophore) which makes the availability of iron to crop plants

(Yehuda et al. 2000). *Pseudomonas fluorescens* is reported to acidify the environment and solubilize mineral phosphates (De-Werra et al. 2009). Azospirillum enhances root growth and activity by acidifying the root surroundings that increases the uptake of many macro and microelements (Dobbelaere and Okon 2007) apart from more efficient nitrogen uptake from the soil (Morrissey et al. 2004).

1.4 Alleviating Metal Toxicity in Plant and Soil

Physiological disorders and growth inhibition are reported in plants grown in soil containing excess concentration of metalloids and heavy metals such as As, Cd, Cr, Cu, Hg, Pb and Zn (Wong et al. 2003; Tripathi et al. 2007). Metals exhibit a range of toxicities towards microbes also but some survive and detoxify them through a variety of mechanisms (Gadd 2010). Microbial bioremediation of heavy metal contaminated sites makes the soil less toxic to the plant. Heavy metal toxicity to plants can be reduced by plant growth promoting bacteria and in turn, plants accelerate bioremediation by stimulating the microbes (Wenzel 1999). Rhizoremediation is perhaps the best approach to clean polluted sites (Kuiper et al. 2004). In this process, rhizobacteria degrades the pollutants which in turn encouraged by plants by producing root exudates (Mendes et al. 2013). Betaproteobacterium and Denitratisoma are suggested to be used in rhizoremediation process (Kawasaki et al. 2012) apart from endophytic fungi, *Lewia* sp. (Cruz-Hernandez et al. 2013).

Studies indicate that several Bacillus strains possess ability to mitigate symptoms of iron toxicity in rice (Asch and Padham 2005; Terre et al. 2007). Inoculation of blue green algae @ 12.5 kg/ha caused a reduction in Cd, Ni and As content in rice plants (Tripathi et al. 2008). Barley plants grown on cadmium contaminated soil inoculated with *Klebsiella mobilis* resulted in producing grains with twofolds decrease in Cd contents along with 120 % increase in grain yield (Pishchik et al. 2002). A plant growth promoting bacterium, *Kluyvera ascorbata* which is resistant to the toxic effects of Ni, Zn, Pb and Cr, when inoculated in canola and tomato partially protected the plants from Ni toxicity. The microbe did not influence the metal uptake by plants but lowered the level of stress caused by ethylene production (Burd et al. 1998). Siderophore producing bacteria are found to prevent plants from becoming chlorotic even in the presence of high levels of heavy metals (Burd et al. 2000). Increased growth performance and decreased metal accumulation is noticed in jatropha plants grown in As, Cr and Zn contaminated soils but treated with *Azotobacter chroococcum* and dairy sludge. Results shown that leaves and fruits will not be affected when the plants are grown at 500 and 250 mg/kg of As and Cr, respectively (Yadav et al. 2009).

1.5 Alleviating Abiotic Stress in Plants

Root-colonizing bacteria especially of the plant growth promoting rhizobacteria (PGPR) also stimulate the plant to have induced resistance to drought, salinity and metal toxicity (Dimkpa et al. 2009a; Yang et al. 2009). Bacterially mediated plant tolerance to salinity was described through *Azospirillum* sp., *Pseudomonas syringae*, *P. fluorescens* and *Enterobacter aerogenes* in maize (Hamdia et al. 2004; Nadeem et al. 2007), *Aeromonas hydrophila* and *Bacillus insolitus* in wheat (Ashraf et al. 2004) through *Achromobacter piechaudii* in tomato and capsicum (Mayak et al. 2004) and through *A. brasilense* in chickpea and faba bean (Hamaoui et al. 2001). Azospirillum induced drought tolerance was reported in wheat (Creus et al. 2004, 2005), maize (Casanovas et al. 2002) and in common bean (German et al. 2000). Osmotolerant bacteria applied as seed coating induced drought tolerant in rice (Yuwono et al. 2005). Substantial tolerance to flooding stress was found in tomatoes treated with *Enterobacter cloacae* and *Pseudomonas putida* as seed treatment (Grichko and Glick 2001). *Burkholderia phytofirmans* was found to induce heat tolerance in potato (Bensalim et al. 1998) and chilling resistance in grapevine (Barka et al. 2006). *Burkholderia phytofirmans* inoculated grapevine plants was found to have higher amounts of carbohydrates, proline and phenols compared to control plants (Barka et al. 2006).

1.6 Supplementing Plant Growth and Yield

Microbes promote plant growth directly by improving nutrient acquisition and hormonal stimulation and indirectly by suppressing pests and plant pathogens. Members of the bacterial genera Azospirillum and Rhizobium are well studied for plant growth promotion apart from Bacillus, Pseudomonas, Serratia, Stenotrophomonas and Streptomyces and also fungus like Ampelomyces, Coniothyrium and Trichoderma (Berg 2009). Microorganisms produce auxins, cytokinins, gibberellins, abscisic acid, ethylene etc. (Patten and Glick 1996; Dobbelaere et al. 2003; Arkhipova et al. 2007; Perrig et al. 2007). These compounds alone or along with other bacterial secondary metabolites, stimulate plant growth (Patten and Glick 2002; Joo et al. 2005; Ryu et al. 2005; Dimkpa et al. 2009a). Plant growth promoting rhizobacteria are also reported to enhance root development in plants (Mantelin and Touraine 2004) and alter root architecture (Kloepper et al. 2007). Many phytohormones such as IAA, ethylene, cytokinins and gibberellins are produced by plants itself and or in association with the microbes.

Agrobacterium radiobacter was reported to cause yield increase in cereals as comparable to nitrogen fertilizers (Bairamov et al. 2001; Zavalin et al. 2001). Wheat seeds inoculated with Azotobacter exhibited 50 % increase in leaves and 55 % increase in height of seedling in 7 days of germination with respect to control (Shaukat et al. 2006). Wheat yield was reported to get increased up to 30 % with

Azotobacter and up to 43 % with Bacillus inoculations (Kloepper et al. 1991). Rice seeds inoculated with *R. leguminosarum* gave an increase in grain yield of about 8–22 % and 43 % in pot culture and glasshouse experiments, respectively (Yanni et al. 2001; Hussain et al. 2009). The blue green algae, *Anabaena spiroides*, *A. variabilis*, *A. torulosa* and *A. osillarioides*, inoculated in rice seeds cause faster germination with increase in plant height and root length by 53 % and 66 %, respectively (Saadatnia and Riahi 2009). *Bacillus polymyxa* applied as seed treatment was reported to increase the yields of rice and chickpea crops (Tiwari et al. 1989). Yield increase in maize of about 49–83 % when inoculated with *Sinorhizobium* spp. and an increase of 34 % when inoculated with *R. trifolii* are reported (Riggs et al. 2001).

Increased growth with high nodulation and nitrogen fixation in soybean is caused by *Aeromonas hydrophila*, *Serratia liquefaciens* and *S. proteamaculans* even under suboptimal root zone temperatures (Zhang et al. 1997). Co-inoculation of *Bradyrhizobium japonicum* and *S. proteamaculans* stimulated soybean growth at temperatures at which nodule formation is normally inhibited (15 °C) (Zhang et al. 1995, 1996). Co-inoculation of *Rhizobium tropici* and *P. polymyxa* in bean, *Phaseolus vulgaris* leads to increase in plant growth, and nodule number even in severe drought conditions, where rhizobia alone cannot be effective (Figueiredo et al. 2008). Enhancement of root and shoot in canola, lettuce and tomato was found when inoculated with *Pseudomonas putida* and *P. fluorescens* (Hall et al. 1996; Glick et al. 1997).

1.7 Value Addition of Agro-Products

Microorganisms have been increasingly used in food, agriculture, chemical and pharmaceutical industries for the production of many value added products (Du et al. 2011). Microbial induced quality enhancement of non processed agro produces are not been done except for few examples like the enhancement of fruit flavour in strawberry by Methylobacterium (Zabetakis 1997).

Humans used bacteria, yeasts and moulds for food fermentation since Neolithic era i.e., around 10,000 BC (Prajapati and Nair 2003). There are about 3500 fermented foods of animal or plant origin consumed by different populations of the world. Lactic acid bacteria is second only to yeast in importance and used in the fermentation of dairy products, meats and vegetables apart from its use in the production of wine, coffee, silage, cocoa, sourdough and numerous indigenous food (Makarova et al. 2006). The yeast, *Saccharomyces cerevisiae* is used extensively in brewing and baking for many years (Yabaya and Jatau 2009). Microbes play direct and indirect role on the quality of tea. Microbes involved in pile fermentation of tea include *Aspergillus niger*, *A. gloucus*, *A. terreus*, *A. candidus*, *Penicllium*, *Rhizopus*, *Saccharomyces*, *Bacterium* etc. Their activities make the beverage nutritionally rich besides producing vitamins such as B1, B2 and C (Zhou et al. 2004).

The yoghurt bacteria, *Lactobacillus delbrueckii* subsp. *bulgaricus* and *Streptococcus thermophilus*, are also able to produce vitamins such as folate. The folate content of milk is said to get increased from 20–50 to 150 μg/L in yoghurt. *Lactococcus*, *Lactobacillus*, *Leuconostoc*, *Bifidobacterium* and *Enterococcus* are repoted to synthesize vitamin K (Hati et al. 2013). The antimicrobial property of bacteriocins can be harnessed for food preservation (Bhattacharya and Das 2010). Probiotics aim in delivering living bacterial cell to the gut ecosystem for treating gastrointestinal diseases or as delivery systems for vaccines, immunoglobulins etc. (Hati et al. 2013).

Microbial interaction with plants also has direct consequences for human health. Notorious examples being ergot contaminated products, fungal aflatoxins in food and feed (Morrissey et al. 2004) and bacterial food poisonings.

1.8 Microbes in Composting

Decomposing microorganisms perform oxidative or fermentative decomposition, which is further classified as useful fermentation and putrefaction. Complete oxidation of substrate is achieved by aerobic decomposition with the release of large amounts of energy, gas and heat (Higa and Parr 1994). Composting converts organic wastes like straw, leaves, trash and other agricultural wastes into useful manure. Composts are made by bacteria, fungi, actinomycetes along with many invertebrates like earthworms. The composition of microbes vary with the time and the raw material used for composting. The temporal variation of bacterial community composition can be correlated to water soluble carbon (WSC), ammonium and nitrate in the compost whereas the distribution of fungi is influenced by pile temperature, WSC and moisture content (Zhang et al. 2011). Several fungi like *Trichoderma harzianum*, *Pleurotus ostreatus*, *Polyporus ostriformis* and *Phanerochaete chrysosporium* play important role in composting of lignocellulosic materials (Singh et al. 2012). The soft rot fungi decompose cellulose but degrade lignin slowly and incompletely; the brown rot fungi prefer lignin (Janshekar and Fiechter 1983) whereas the white rot fungi are capable of degrading both lignin and cellulose (Ander and Eriksson 1977; Singh et al. 2012). The consortium of four hypercellulolytic fungal cultures namely *Aspergillus nidulans*, *T. viride*, *Phanerochaete chrysosporium* and *Aspergillus awamori* are used for composting paddy straw, soybean trash, pearl millet, maize residues and mustard stover effectively (Gaind and Nain 2010; Gaind et al. 2009; Pandey et al. 2009).

Cellulolytic bacteria (Cytophaga, Sporocytophaga, Cellulomonas) solubilize and modify the lignocellulosic structures but their ability to mineralize lignin is limited (Ball et al. 1989). Thermophilic Actinobacteria are responsible for lignin degradation during agricultural waste composting (Yu et al. 2007). *Bacillus* sp. produces extracellular cellulases. *Scytalidium thermophilum*, *Humicola insolens* and *Sporotrichum thermophilum* are used for production of pathogen free compost even under high temperatures for mushroom cultivation (Straatsma et al. 1989; Lyons et al. 2000). Actinomycetes such as *Streptomyces* spp. and *Thermoactinomyces* sp. and fungus *Aspergillus fumigatus* along with many *Bacillus* sp. are found in solid waste composting (Strom 1985). Consortia of microbes when used properly according to the substrate make the compost quickly. Composts are enriched with some essential nutrients to make it fully nutritive for the plants to grow properly. Composts are fortified with beneficial microbes like Trichoderma, Pseudomonas etc. and used for the biological control of some plant diseases also.

2 Routes of Pesticide Exposure to Microorganisms

Microorganisms are ubiquitous in environment, seen in all ecosystems, found in large numbers and frequently transported even over large distances (Finlay and Esteban 2001). It is rather difficult to find a sample of even freshwater that contains fewer than 10^6 bacteria or 1000 flagellated protozoa per milliliter (Berninger et al. 1991). Thus they get exposed to pesticides in every area and in all ecosystems wherever pesticides are available. The ubiquitous availability of microbes makes them get exposed to pesticides in many ways but the exposure cannot be considered as toxicity or dangerous. There are many factors which influence the effect or toxicity of pesticides to microorganisms and some microbes even use pesticides as food/nutrient source. Microorganisms also play an important role in pesticide degradation in the environment. Microbial biodegradation of pesticides make the environment clean with less accumulated pollution (Jastrzebska 2010).

The factors affecting the effects of pesticides on microorganisms include many physical, chemical and biological factors (Jastrzebska 2010):

- The chemical property of the product: chemical structure, solubility, volatility, concentration, method/time and frequency of application;
- Soil type: structure, texture, organic and mineral content, moisture, pH;
- Climatic conditions: temperature, sun exposure, precipitation, wind;
- Type of environment: cultivation methods, the species composition of crops, soil fauna and microbial community;
- Content of other xenobiotics, including heavy metals.

Microorganisms get exposed to pesticides in many ways. The main routes of pesticide exposure to microbes are given in detail below.

2.1 Soil Application of Pesticides

Pesticides like insecticides and fungicides are applied on soil for different pest control programmes such as, soil insect pest control and soil borne disease control. Though the weedicides are normally directed towards plants, most of the chemical fall on the ground. Moreover pre-emergent weedicides are applied on the soil either as spray or broadcasting as mixed with soil. Pesticides used continuously lead to accumulation of degraded products in the soil (Imfeld and Vuilleumier 2012).

Frequent and cocktail application of pesticides may allow them to persist in the soil for a long time causing substantial damage to soil biota including the microbes. Although all the DDT products are banned for commercial use >30 years ago, its residue and degradation products are being detected in soil samples around the world (Jastrzebska 2010). Some of the breakdown products are highly toxic to microbes than that of parent compounds. The IC_{20} (inhibitory ratio) of chlorpyrifos on microbes was found to be 9.8 μg/g, whereas it was 0.37 μg/g for chlorpyrifos oxon. Thus the acute toxicity of the metabolite is 26 times greater than the pesticide itself for microbes (Wang et al. 2010). Apart from soil application, soil is also a repository for chemicals from drift during foliar application, plant residues containing insecticides and their degradation products and chemicals deposited by atmospheric precipitation (Tu and Miles 1976).

2.2 *Spray Drifts from Plants to Environment*

Pesticide sprays on crops can also lead to residual accumulation in the soil. Much of these residues comes from the foliage sprays or dusts which miss their target and fall on the soil either close to or far off from the target plants (Edwards 1966). The drifting droplets settle on the soil are in the upper 0.5 cm layer (De-Schampheleire et al. 2007) and subsequently move away or down through the soil layers by runoff or leaching. The spray drifted pesticides are susceptible to wind, carried away to far off distances and even found in aquatic ecosystem (Davidson and Knapp 2007). These pesticides entered into the aquatic ecosystem and affect microbes and other living organisms in water. Pesticide volatilization from soil and especially while spray application and pesticide emission from plant canopy make the pesticide to be available in the atmosphere (Grenni 2011) and cause ill effects on microbes present in atmosphere also.

2.3 *Dumping of Pesticides on Soil*

Pesticides are usually dumped in the soil especially when they get expired or become obsolete. A large amount of pesticides are being dumped by manufacturing companies and others, which affect all the soil organisms including microbes. The obsolete pesticides in the dump sites may affect humans also through contaminated water, soil and food. Pesticide wastes that are buried under are endangering the soil and groundwater through leaching (Uijtewaal 1992). Heavy contamination even in deep soil layers and water sources are reported near to pesticide dump sites (Buczynska and Szadkowska-Stanczyk 2005). In obsolete pesticide dumping grounds, the concentrations of organo-chlorine pesticides in the soil and air were found highest among all the sites with elevated concentrations of DDTs and HCHs (Alamdara et al. 2014). In farm level also, obsolete and extra/remaining pesticides after application are dumped in the non cropped area or in irrigation water or over sprayed on the plants which affect the microbes present in that area.

2.4 Pesticides in Water Ecosystem

Pesticides enter aquatic ecosystem by washed off from sprayed land through irrigation water and rainfall and through herbicides and insecticides deliberately used for the management of aquatic pests and weeds. Pesticides used in agriculture, washed off from field and were found in estuaries and found to alter both functional and structural aspects of the estuarine microbial food web (DeLorenzo et al. 1999). Pesticides may induce toxic responses in microbes present in water even at very low concentrations (Widenfalk et al. 2004). Community level shifts in sediment microorganism exposed to environmentally relevant concentrations of captan, glyphosate, isoproturon and pirimicarb in aquatic ecosystem was reported (Widenfalk et al. 2008). Accumulation of pesticide residues through food chain in aquatic environment i.e., soil–water–sediment–microbes–crop–fish is also demonstrated (Feng et al. 2003).

2.5 Pesticides Sprayed on Plants

The above ground parts of plants are also normally colonized by a variety of bacteria, yeasts and fungi (Lindow and Brand 2003). Bacteria are by far the most numerous colonists of leaves, often being found in numbers averaging 10^6–10^7 cells/cm^2 (up to 10^8 cells/g) of leaf (Beattie and Lindow 1995; Andrews and Harris 2000; Hirano and Upper 2000). These large numbers of microbes are also exposed to the pesticides sprayed on the plants. Spray of cypermethrin in cucumber plants is reported to alter significantly the community structure of phyllosphere microbes (Zhang et al. 2008). Pesticides sprayed on the plants also miss the target and fall on the soil and get accumulated through which soil microbes are exposed. Foliar concentration of insecticides in the treated plants and trees can enter in the soil or aquatic ecosystem by leaf fall. Sugar maple trees systemically treated with imidacloprid has a field realistic concentration of 3–11 mg/kg in the leaves during leaf fall. Though, imidacloprid at this particular concentration was not found to affect the decomposition potential of microbes (Kreutzweiser et al. 2008) it cannot be ignored. Soil microbial communities have been affected by imidacloprid, which can affect leaf litter decomposition (Chagnon et al. 2015). Thus, pesticides sprayed on plants will affect the microbes in the plants, soil and also water in various ways.

3 Effects of Pesticides on Microorganisms

Pesticides are designed and developed with specific mechanism of action on insect pests, weeds and or pathogens. Most of the pesticides with specific mode of action on pests (AChE inhibitors, GABA modulators, sodium or potassium channel modulators, enzymes/sterol inhibitors, growth inhibitors) may not have an action on microbes and without any direct effect (Ferreira et al. 2009). Pesticides with irrelevant mechanism of action for microbes may not exert an effect on the microbes but others with pertinent action may have an impact by killing them, reducing their population or their activities. Crop protection chemicals affect microbes in many ways by affecting the biochemical and physiological attributes (Jastrzebska 2010), apart from killing them. Pesticides can either directly (immediate or short term impacts) harm microbes that come in contact with the chemical or indirectly due to changes caused by the chemical on the environment and or on it's food source (Seymour 2015). The impact of pesticides is also influenced by many soil and environmental parameters apart from the intrinsic toxicity of the chemical. Unlike other non-target organisms which are studied individually, the effect of pesticides on microbes has to be studied on community basis because many microbes in isolation may not withstand the toxic effects of pesticides as they do as a community, for example within the microbial biofilms (Tlili et al. 2010). Important and pronounced effects of pesticides on microorganisms are given in different subheadings here under.

3.1 Effects Revealed by Laboratory/Microcosm Studies

3.1.1 Effect on Microbial Biomass

The effect of pesticides on the microbial biomass is either an inhibitory (including killing) or stimulatory. Pesticides are screened by estimating the microbial biomass after pesticide exposure. However, an increase in the total counts or the biomass of soil-dwelling microorganisms is not always an indicative of no-effect of a given xenobiotic. So, care should be taken while interpreting such effects on microbes. N-trichloro-methyl thio tetra hydrophthalimide (Captan) is reported to inhibit all microbes except bacteria, while disodium ethylene bisdithio carbamate (Nabam) inhibited fungi only. Herbicide, allyl alcohol inhibited the growth of algae and fungi but increased the bacterial population (Bollen 1961). The IC_{20} (inhibitory ratio) of chlorpyrifos to common microbes was reported as 9.8 μg/g (Wang et al. 2010).

Insecticides like dimethoate, monocrotophos, deltamethrin, endosulfan, cypermethrin and triazophos at their recommended doses tested in cotton field soil caused only short time adverse effects on the soil microorganisms (Vig et al. 2008). Carbofuran at application rate did not show any detrimental effect on soil microbial biomass, but carbosulfan did (Duah-Yentumi and Johnson 1986). In fact, carbofuran was reported to stimulate bacterial population (Dordevic et al. 1998; Ahmed and Ahmad 2006) and among the bacteria especially of the N_2 fixing bacteria (Das and Mukherjee 1998; Lo 2010). Methamidophos was reported to reduce the biomass of fungi and bacteria except Gram negative (Wang et al. 2008). DDT is toxic for bacteria, fungi and algae and it reduces their biomass. Imidacloprid was found to suppress the soil bacteria communities (Ahmed and Ahmad 2006).

Fungicides, captan and thiram at application rates of 5 mg/kg caused 40 % reduction in microbial biomass but recovered within 8 days. A higher dose of 50 mg/kg caused long term decrease in biomass and altered the relative proportions of bacterial to fungal populations (Anderson 1981). Fungicide, vinclozolin application reduced the microbial biomass heavily especially of the fungal biomass (Duah-Yentumi and Johnson 1986). Contrarily, vinclozolin and insecticide lambda cyhalothrin, applied at recommended doses was reported to have no significant impact on microbial biomass and bacterial diversity (Lupwayi et al. 2009a). Soil type also plays an important role in fungicide toxicity to microbes (Pal et al. 2005). The response of fungi to tebuconazole was clearly determined by the soil type; light loam soils stimulated fungal growth but loamy sand inhibited. Tebuconazole at 50 and 500 mg/kg inhibited the proliferation of oligotrophic bacteria, copiotrophic bacteria and actinomycetes as well as *Azotobacter* spp. (Jastrzebska 2010).

The effect of herbicides on microbial biomass is reviewed by Subhani et al. (2000). Herbicide, simazine caused no detectable effects on the microflora, but repeated application of paraquat significantly lowered the soil microbial biomass, mainly the fungal biomass (Duah-Yentumi and Johnson 1986). Imazethapyr at 50 g a.i./ha in clay loam soil did not affect the biomass activity of microorganisms (Perucci and Scarponi 1994). Metsulfuron-methyl was reported to suppress soil respiration and microbial biomass at tenfold field rate; although the effects were transient at field rate (Ismail et al. 1996). Though 2,4-D did not affect the microbes at its recommended dose, inhibitory effects were noticed at higher doses (7.5 and 15.0 μg/g) (Rath et al. 1998).

3.1.2 Effect on Microbial Population and Growth

Beneficial microbes, *Azotobacter* spp. responsible for non-symbiotic nitrogen fixation and also the symbiotic nitrogen fixers like Rhizobium and Bradyrhizobium are often found in low numbers in soil treated with pesticides (Jastrzebska 2010). Insecticide, endosulfan at high concentrations (1000 ppm) caused significant reduction in soil microbes. But stimulation in fungal populations in the range of 52–148 % by endosulfan and 25–100 % by profenofos at all three levels i.e., 10, 25 and 50 ppm was reported for 5 days incubation (Nasreen et al. 2015). Chlorpyrifos and cypermethrin caused significant reduction in the number of soil bacteria while bifenthrin increased the bacterial population at 250 and 500 ppm (Ahmed and Ahmad 2006). In general, bacterial populations did not survive and multiply well in presence of chlorpyrifos at least for a period of 3 weeks whereas no inhibition was seen in fungal populations (Hindumathy and Gayathri 2013). Chlorpyrifos at its recommended dose significantly reduced the populations of bacteria, fungi and actinomycetes by 44.1 %, 61.1 % and 72.8 %, respectively, on the first day after treatment. The inhibitory effect of chlorpyrifos on soil microorganisms get enhanced when combined with chlorothalonil (Chu et al. 2008). Dimethoate, triazophos and endosulfan caused a significant reduction in Azotobacter number but an increase up to 71 % was observed after deltamethrin treatment (Vig et al. 2008). The viable bacterial numbers were found to be higher in isofenphos treated soil than those of control (Digrak and Kazanici 2001). The growth of Pseudomonas was inhibited (75 %) by cypermethrin but did not cause any effect on Rhizobium and Azotobacter. Fenvalerate inhibited the growth of all microorganisms up to 50 % except Bacillus in which significant stimulation was observed. The growth of Serratia, Azotobacter and Rhizobium was inhibited by deltamethrin but did not affect the population of Bacillus and Pseudomonas. Synthetic pyrethroids were found to stimulate the population of *Saccharomyces cereviseiae* but inhibited algae (Spirullina and *Phormidium* sp.) and fungi (Trichoderma and Phanaerochaete) (Sethi et al. 2013). Spore forming aerobic, anaerobic and proteolytic bacteria and actinomycete grew well in phorate treated soil (Digrak and Kazanici 2001). Diafenthiruon at field recommended dose inhibited 39.5 % of growth of *T. viride* at 72 h in poison food technique in Petri dish and found not compatible whereas the insecticide is compatible with *P. fluorescens* (Stanley et al. 2010a).

The total cultural fungi, denitrifying bacteria and aerobic diazotrophs were reportedly decreased due to mancozeb whereas bacteria are not affected (Pozo et al. 1994). The total culturable fungal populations, nitrifying and aerobic N_2 fixing bacteria were significantly reduced whereas the denitrifying and total culturable bacteria populations get enhanced with captan at doses of 2.0–10.0 kg/ha (Martinez-Toledo et al. 1998). Chlorothalonil application significantly reduced the soil bacterial and actinomycete populations whereas the population of soil fungi was unchanged. However, the microbes get acclimatized gradually and the negative effects became transient and weaker following the third and fourth applications (Yu et al. 2006). A significant reduction of nitrogen transforming and cellulolytic microbes (except fungi) was reported when the soil is treated with carbendazim + mancozeb (Team®) (Fawole et al. 2009).

Herbicide, rimsulfuron reduced the bacteria and the fungal populations significantly even at the concentration of 2.0 mg a.i/kg (Radivojevic et al. 2011). Metribuzin also caused significant decline in bacterial population at 3 days after treatment. Metribuzin reduced total microflora, fungi, cellulolytics and nitrifiers at temperature of 10, 20 and 30 °C (Radivojevic et al. 2003). At field recommended doses, metribuzin and isoproturon were found to deter the actinomycete population for 7 and 15 days, respectively. Soil fungal populations were reported not affected by isoproturon, metribuzin and clodinofop treatments (Lone et al. 2014). A reduction in bacterial population of 26.9 % and 37.9 % was reported when treated with paraquat and atrazine at their recommended rates after 8 weeks of treatment (Stanley et al. 2013). The total bacteria, actinomycetes and fungi counts in both rhizosphere and non-rhizosphere soils were significantly lower in napropamide treatment compared to control after 2, 3 weeks of treatment (Wu et al. 2014). The bacterial population was increased significantly in clodinafop and isoproturon treated microcosms (Radivojevic et al. 2003) and the fungal populations in metsulfuron-methyl (Radivojevic et al. 2014).

Phenyl urea herbicide, diuron and linuron removed dominant acidobacterium as revealed by 16s rDNA DGGE (El-Fantroussi et al. 1999) whereas the glycine herbicide, glyphosate increased the abundance of Proteobacateria (Lancaster et al. 2010). Though napropramide initially decreased the bacterial and fungal abundance, the Gram negative bacteria and fungi got enhanced later as revealed by PLFA assay (Cycon et al. 2013). Herbicide tribenuron-methyl (Granstar®) at five times of recommended concentration had a negative impact on the abundance of oligotrophic bacteria, spore-forming copiotrophic bacteria, *Azotobacter* spp., cellulolytic bacteria, actinomycetes and fungi (Kucharski et al. 2008). Glyphosate was reported to reduce the abundance of growth promoting bacteria in rhizosphere of soybean plants (Zobiole et al. 2011). Herbicides, isoproturon and clodinafop significantly proliferated the phosphate solubilizers up to 15 days whereas sulfosulfuron and metribuzin exhibited no significant action. Metribuzin reduced the population of Azotobacter while clodinafop caused an increment (Lone et al. 2014). Another herbicide, sulfosulfuron was not found to affect phosphorus solubilizing fungi (Dhagat and Verma 2009). Linuron application was not found to inhibit the growth of *Azotobacter chrococcum* (Lenart 2012). An initial decline followed by an increment in Azotobacter population was observed in metsulfuron methyl treatments (He et al. 2006).

3.1.3 Effect on Microbial Diversity

When an ecosystem is under stress it leads to decrease in species diversity and may result in an increase in individuals that are capable of tolerating that stress (Cloete and Atlas 2006). Pesticides are also of no exception and the microbes which are susceptible for a particular chemical may get killed or reduced and the resistant ones which can degrade and use them as food source will proliferate. Microbial diversity and species richness have an interactive effect and influence the microbial function as a group. However, in some cases, the relative decrease in species richness of microbes has a little effect on soil functions like carbon mineralization, denitrification and nitrification (Setala et al. 2005; Wertz et al. 2006) due to microbes with functional redundancy. But some microbial interactions like imparting resistance to pathogenic bacteria have been shown to decline with decrease in species richness (Van-Elsas et al. 2012).

A long term experiment carried out to find the effect of pesticides on microbial diversity, showed a clear reduction in microbial diversity with respect to pesticide usage. The fluorescent pseudomonad-selective agar plates viewed under UV light revealed only 8 % of the colonies as fluorescent in the pesticide plots, compared to 27 % in control plots. On Tryptone Soy Agar, 10 % of colonies from the pesticide plot were pigmented compared with 19 % from the control plot (Nicholson and Hirsch 1998). Methamidophos lowered the genetic diversity of soil dwelling bacteria reducing the biomass of bacteria and fungi but stimulated the proliferation of Gram negative bacteria (Wang et al. 2008). Soil application of cypermethrin caused a shift in species composition in microbes with Firmicutes species present only in the control, whereas the majority of the species detected in the pesticide treated samples belong to either Bacteroidetes or γ-Proteobacteria as revealed by PCR–DGGE (Zhang et al. 2009). Fungicides cyprodinil (Unix®) and dimoxystrobin + epoxiconazole (Swing Top®) applied in very high concentrations than recommended doses had an inhibitory effect on copiotrophic bacteria and actinomycetes (Jastrzebska 2006). When the soil is fumigated with methyl isothiocyanate, Gram positive bacteria were found to increase and Gram negatives get decreased (Ibekwe et al. 2001) as revealed by phopholipid fatty acid (PLFA) profiles (Spyrou et al. 2009).

3.1.4 Effect on Microbial Community

The interactions within and between the different groups of microorganisms make it difficult to find the direct or indirect effect of pesticides on community composition (Imfeld and Vuilleumier 2012). In addition, most of the soil microbes are yet to be identified/studied (Fierer et al. 2007) along with their role and function in the soil ecosystem (Wu et al. 2009). However some reports also state, the change in microbial community may not affect their functions in general (Setala et al. 2005; Wertz et al. 2006).

Bacterial communities of freshwater sediments were found to get affected by captan, glyphosate, isoproturon and primicarb as revealed by phospholipid-derived fatty acids (PLFA) composition and bacterial 16S rRNA genotyping by T-RFLP (Widenfalk et al. 2008). Vinclozolin and lambda cyhalothrin was found to exert changes in microbiological, genetical and taxonomic structure of bacteria (Lupwayi et al. 2009a). Application of fungicide, enostroburin induced substantial changes in the bacterial community (Gu et al. 2010). Foliar application of metalaxyl stimulated the appearance of an Enterobacteriaceae ribotype (Moulas et al. 2013). Among the fumigants, chloropicrin had a higher potential to alter the microbial community structure in the longer term than others with methyl isothiocyanate being the least damaging. At 7 days post-fumigation, soils fumigated with chloropicrin and propargyl bromide showed lower relative abundances of bacteria (especially Gm+) and actinomycetes (Klose et al. 2006). Formulated herbicide mesotrione (Callisto®) affected the activity of overall microbial communities (Crouzet et al. 2010). Repeated application of glyphosate did not result in significant changes in soil microbial community structure when examined by ester linked fatty acid methyl ester extraction (EL-FAME) (Lane 2011).

3.1.5 Effect on Microbial Biomass Carbon

The knowledge of soil microbial biomass carbon can be an indication for ecosystem functioning (Paul and Voroney 1989) since microorganisms form a vital part of the soil food web (Beelen and Doelman 1997). Handa et al. (1999) reported a reduction in soil microbial biomass carbon when treated with deltamethrin. A significant reduction in the soil microbial biomass carbon was also reported in cypermethrin treated soils up to 15 days in field recommended dose and up to 30 days in ten times the dose as compared to control. An increase in trend in biomass carbon was observed after that (Goswami et al. 2013). Rath et al. (1998) observed that the application of 2,4-D and its analog 2,4,5-T at 0.75 μg/g led to a distinct increase in microbial biomass carbon contents both under flooded and non flooded conditions. Significant decrease in microbial biomass carbon contents in rimsulfuron treated soils especially at higher temperatures and low moisture levels are reported (Vischetti et al. 2000). Startton and Stewart (2002) recorded harmful effects of glyphosate on soil biomass and respiration in Canadian coniferous forests. Herbicides, prosulfuron and thifensulfuron methyl were found to reduce microbial biomass carbon (Sofo et al. 2012).

3.1.6 Effect on Soil/Microbial Respiration

Changes in soil respiration have been used as criteria for insecticide toxicity on microorganisms (Das et al. 1995; Jones and Ananyeva 2001; Latif et al. 2008). Insecticides like flubendiamide, nimbicidine, lambda cyhalothrin, abamectin and thiodicarb were found to increase the microbial respiration immediately after incubation. Chlorpyrifos, cartap and carbosulfan had inhibitory effect but only for 4 days on microbial respiration and cypermethrin had no effect (Latif et al. 2008). No adverse effect was observed on the soil respiration after treatment with dimethoate, monocrotophos, deltamethrin, endosulfan, cypermethrin and triazophos (Vig et al. 2008). Fenitrothion at low concentrations did not affect the substrate-induced respiration in the cultivable fraction of soil bacteria (Cycon and Piotrowska-Seget 2009). Field recommended doses of triazophos and azadirachtin has a short term inhibitory effect on basal and substrate induced respiration while carbaryl exhibited a short term stimulatory effect (Sengupta et al. 2009). A significant reduction in basal soil respiration (BSR) and substrate induced soil respiration (SIR) was reported in cypermethrin treated soils up to 15 days in field recommended dose and up to 30 days in ten times the dose as compared to control (Goswami et al. 2013). A gradual increase in CO_2 evolved (0.73–0.96 mg/g) in soils fortified with cypermethrin and a decrease with chlorpyrifos (1.73–1.53 mg/g) after 50 days of incubation as compared to control was also reported (Jail et al. 2015).

Increase in respiration rates bacterial biomass and dehydrogenase activity revealed an increase in microbial activity when glyphosate was applied at low concentrations (0.48 L a.i./ha), but rapidly decreased at higher concentrations (3.84 L a.i./ha) (Gomez et al. 2009). Soil respiration was found to get enhanced by treating with glyphosate at 2.16 μg/g (Araujo et al. 2003). Glyphosate, picloram and 2,4-D at 200 μg/g enhanced the basal respirations for 9 days after applications whereas the substrate induced respirations were temporarily depressed by picloram and 2,4-D (Wardle and Parkinson 1990a). Atrazine, lindane and captan caused minor effects on microbial respiration (Zelles et al. 1985). But increase in soil respiration after 96 h incubation with atrazine was also reported (Tu 1992). Herbicide, 2,4-D showed transient effects either by inhibiting substrate induced respiration or fluorescein diacetate hydrolysis and stimulating dehydrogenase activity (Zabaloy et al. 2008). The total amount of CO_2 produced was suppressed by pendimethaline, trifluralin, glyphosphate and 2,4-D after 15 days of application (Yousaf et al. 2013).

3.1.7 Effect on Microbial Activities

Pesticides may affect different microbial activities like organic matter decomposition, mineralization, nitrogen fixation, nitrification etc. Aldrin, chlordane, DDT, dieldrin, endrin, heptachlor, lindane and toxaphene did not affect the organic matter decomposition and ammonium oxidation activities of microbes at their field rates (Martin et al. 1959). However, a significant slow in the activity of litter breakdown is reported in microbes living in water (Artigas et al. 2012). Glyphosate (Roundup Ultra®) significantly increased soil microbial biomass and stimulated soil microbial activity as measured by C and N mineralization (Haney et al. 2002). Metribuzin at field recommended dose was not affecting the phosphorus solubilization activity of *Klebsiella* sp. (Ahemad and Khan 2011). Short-term changes in iron reduction capacity of microbes were observed after endosulfan and cypermethrin treatments (Vig et al. 2008). Braithwaite et al. (1958) found that nodulation of clover on sod was greatly reduced by the application of DDT and lindane along with fertilizers. The plant growth promoting traits of Mesorhizobium was hindered by glyphosate, thiamethoxam and hexaconazole at three times the recommended rate (Ahemad and Khan 2012). Nitrogen-fixing capacity of blue-green algae is drastically affected by pesticides (DaSilva et al. 1975). Organo phosphorous insecticides profenofos and chlorpyrifos reduced the number of aerobic nitrogen fixers and significantly decreased nitrogen fixation (Martinez et al. 1992; Pozo et al. 1995). Adverse effect on number of genes and transcripts of nifH (nitrogen fixation); amoA (nitrification) and narG, nirK and nirS (denitrification) was observed in rhizospheric bacterial communities exposed to chlorpyrifos, cypermethrin and azadirachtin (Singh et al. 2015). Glyphosate is reportedly associated with increase in the plant pathogens like Fusarium and Pythium (Kremer et al. 2005; Levesque et al. 1993; Meriles et al. 2006) may be by inhibiting the antagonistic microorganisms.

3.1.8 Effect on Microbial Enzymatic Activities

Stimulation in phosphatase activity in microbes due to herbicides like paraquat, trifluralin, glyphosate and atrazine is already been reported (Hazel and Greaves 1981). However some other herbicides such as fluchloralin, metoxuron, 2,4-D and isoproturon were reported to reduce the phosphatase activity of microbes (Tarafdar 1986). Fungicide, mancozeb was reported to increase the activity of alkaline phosphatase, protease and amidase while decreasing urease and asparaginase (Rasool and Reshi 2010). Stimulation of protease activity was reported in native soils treated with linuron at 10 mg/kg whereas, cartap-HCl at 100 mg/kg permanently inhibited it (Endo et al. 1982).

Effect of pesticides on enzyme activity of microbes

Enzyme activity	Pesticide	Changes	References
Amidase	Mancozeb	Stimulation	Rasool and Reshi (2010)
Asparaginase	Mancozeb	Inhibition	Rasool and Reshi (2010)
Arylsulfatase	Soil fumigants	Inhibition	Klose et al. (2006)
Cellulase	Mancozeb + carbendazim (Team®) 2.34 mg/kg	Inhibition	Fawole et al. (2009)
Catalase	Napropamide 6 mg/kg	Stimulation	Wu et al. (2014)
Dehydrogenase	Propiconazole	Inhibition	Johnsen et al. (2001)
	Rimsulfuron 2 mg a.i/kg	No change	Radivojevic et al. (2011)
	Napropamide 6 mg/kg	Stimulation	Wu et al. (2014)
	Endosulfan+profenophos 5 kg/ha	Stimulation	Nasreen et al. (2015)
	Sulfonyl urea herbicides 20 mg/kg	Stimulation	Accinelli et al. (2002)
	Atrazine 8 mg a.i./kg	Inhibition	Radivojevic et al. (2008)
	Metsulfuron-methyl 1 or 5 mg/kg	No change	Radivojevic et al. (2014)
	Metsulfuron-methyl 25 or 50 mg/kg	Stimulation	Radivojevic et al. (2014)
Nitrogenase	Captan 2–10 kg/ha	Inhibition	Martinez-Toledo et al. (1998)
Pectinase	Mancozeb+carbendazim (Team®) 2.34 mg/kg	Inhibition	Fawole et al. (2009)

(continued)

Enzyme activity	Pesticide	Changes	References
Phenol oxidase	2,4-D	Inhibition	Folch et al. (2011)
	Carbaryl	Inhibition	
	Mancozeb	Inhibition	
	Glyphosate	Inhibition	
	Methyl parathion	Inhibition	
	Atrazine	Inhibition	
	Prometryne	Inhibition	
Phosphatase	Propiconazole	Inhibition	Johnsen et al. (2001)
	Methyl bromide + chloropicrin	Inhibition	Klose et al. (2006)
Urease	Napropamide 6 mg/kg	Inhibition	Wu et al. (2014)

3.1.9 Pesticide Resistance and Metabolism by Microbes

Microbes degrade pesticide and use them as food source. In environmental point of view and or of health aspects, this degradation of toxic pesticides by microbes is considered as desirable. *Thiobacillus thiocyanoxidans* can utilize potassium thiocyanate as a source of energy, carbon and nitrogen (Bollen 1961). Microbial attack on pesticide is sometime useful for activation of pesticides also. For instance, microbial action is essential to activate herbicide, sodium 2,4-dichlorophenoxyethyl sulfate (2,4-DS). This 2,4-DS is hydrolyzed to 2-(2,4-dichlorophenoxy) ethanol and then oxidized to 2,4-D by soil microbes including *Bacillus cereus* var. *mycoides* (Vlitos 1952).

The white rot fungi are highly resistant to DDT by accumulating the toxin in the body and slowly degrading it into DDD and DBP. Many microbial fungi such as *Gloeophyllum trabeum*, *Fomitopsis pinicola*, *Daedalea dickinsii*, etc. are reported to use DDT and metabolize it. The common yeast, *S. cerevisiae* transforms DDT to DDD (Jastrzebska 2010). Certain strains of *Botrytis cinerea* were resistant to even 250,000 ppm of captan (Parry and Wood 1959). The cynobacteria, *Synechocystis* isolated from the grapevine field was found to have a high tolerance to monocrotophos (900 mg/L) and an *Anabaena* from vegetable field is highly tolerant to endosulfan (500 mg/L) (Pawar 2015).

The rate constants of degradation of metalaxyl and propachlor are positively correlated with basal and substrate-induced respiration (Jones and Ananyeva 2001). The microbial biomass, soil respiration and the degradation rate constant of metribuzin are correlated positively with each other (Moorman and Harper 1989), linuron, glyphosate (Torstenssen and Stenstorm 1986), alachlor (Walker et al. 1992), 2,4-D and dicamba (Voos and Groffman 1997). But microbial biomass was not found to be correlated with the degradation of 2,4-D and atrazine (Entry et al. 1994; Ghani et al. 1996).

3.2 Effects Revealed by Semifield/Mesocosm Studies

Semifield or mesocosm studies are done to find the effect of pesticides on the microbial biomass, respiration, diversity and enzyme activities. In a mesocosm study, tebuconazole treated soils showed lower values of microbial biomass carbon than controls with low basal respiration at 7 and 30 days. Higher the concentration leads to lower substrate induced respiration at 7 days after treatment. Irrespective of sampling times, lower values of urease, arylsulfatase and β-glucosidase activities were found in tebuconazole treated soils as compared to control. The microbial species richness and Shannon's diversity calculated from Biolog Ecoplates® data did not change during the first 7 days of incubation but decreased progressively until day 90 (Munoz-Leoz et al. 2011). Bacterial biomass within the sediments was found to get increased with the addition of 2,4-D (Chinalia and Killham 2006).

Glyphosate was found to reduce the total microbial biomass in soybean rhizosphere soil in a green house experiment. No significant differences in microbial community composition in soils treated with glyphosate in potted soybean plants as revealed by the analysis of taxonomic groupings of FAMEs. Glyphosate application had no significant effect on K uptake in the glyphosate resistant soybean plant (Lane 2011). In a greenhouse experiment on pepper plants to find the effect of foliar and soil treatment of imidacloprid and metalaxyl, difference in fungal and bacterial diversity is revealed in DGGE. Bands pertained to *Cryptococcus adeliensis* and an uncultured *Cryptococcus* clone was only present in imidacloprid treatment, showing that the pesticide treatment has wiped out other fungal communities. But metalaxyl application was found to enhance the anamorphic ascomycete, *Periconia macrospinosa* (Moulas et al. 2013).

3.3 Field Effects of Pesticides on Microorganisms

Field effects of pesticides on microbes reveal the exact effect on the natural environment. However, arriving at a clear cut effect in the field is difficult owing to the interactive effects and environmental influences. Roger et al. (1994) while reviewing the literature on field effects of pesticides on microbes concluded that herbicides had a greater impact on microbes than insecticides and bacteria are more sensitive than fungi, actinomycetes and algae. The pesticides have a greater effect on microbial activities than in population densities per se. Bacterial populations were temporarily enhanced by 20 and 200 μg/g of glyphosate while actinomycete and fungi were unaffected in the laboratory (Wardle and Parkinson 1990b). Contrarily in field conditions, glyphosate and 2,4-D reduced the microbial biomass but only in plots with weeds (Wardle and Parkinson 1992). The functional structure of microbes was altered and the functional diversity of soil bacteria got reduced when 2,4-D and glyphosate were applied in canola with an increase in microbial biomass carbon (Lupwayi et al. 2009b). In both field trial and the laboratory experiment, imazethapyr had no adverse effect on the microbial biomass at the field rate. In tenfold field rates, the herbicide decreased the soil microbial biomass carbon contents (Perucci and Scarponi 1994). Glufosinate-ammonium was found to alter soil microbial community structure in maize (Griffiths et al. 2008). Herbicides, glufosinate-ammonium at 500 g a.i./ha and clethodim at 30 g a.i./ha in canola and tralkoxydim 200 g a.i./ha and bromoxynil + methylchlorophenoxyacetic acid (MCPA) 560 g a.i./ha in barley reduced functional diversity of bacteria. Fertilizer effect on bacterial functional diversity was also reported (Lupwayi et al. 2010). In long term (33 year) experimentation on the field application of MCPA, no significant change in soil microbial biomass or populations, except yeast was reported (Smith et al. 1991). A reduction in biomass carbon content was reported when applied with herbicide, rimsulfuron at 2 mg a.i./kg of soil compared to control (Radivojevic et al. 2011).

In a long term experiment on pesticide application on the effect of microbes, aldicarb treated plots was found to have slightly higher microbial biomass carbon content than the control (Nicholson and Hirsch 1998). In field conditions, aldrin, chlordane, DDT, dieldrin, endrin, heptachlor, lindane and toxaphene were not found to have affected either bacteria or fungi or the functioning of the soil microbes in organic matter decomposition and ammonium oxidation at their field rates (Martin et al. 1959). Chlorpyrifos and quinalphos applications as seed treatment and soil application in the groundnut fields lead to an inhibitory effect on the total bacterial population and the inhibitory effect was found to get recovered in 45 and 60 days (Pandey and Singh 2004). In a field experiment with canola-barley crops, highest microbial biomass carbon was reported in 50 % herbicide treatment and the lowest in 50 % fertilizer treatment. Contrast analysis indicated that both the 50 % and 100 % fertilizer application rate reduced microbial biomass carbon in comparison with no fertilizer treatments (Lupwayi et al. 2010).

4 Methods to Assess Pesticide Toxicity to Microorganisms

Pesticides may affect microbes directly by killing them or inhibiting the growth and reproduction apart from indirectly impairing their functions and activities. Several parameters are used to describe and determine the state and functional ability of microorganisms. The total microbial biomass perhaps includes the active and dormant microorganisms are usually determined by direct counting under a microscope. Indirect biomass estimation is by measuring the microbial biomass carbon or nitrogen especially by fumigation method using chloroform. This again cannot differentiate the dormant and active microbes and thus the analysis of adenosine triphosphate (Contin et al. 2001; Jastrzebska 2010) or phospholipid fatty acids or ergosterol (Stahl and Parkin 1996) can be ideal for the assessment of living microbial biomass. Soil respiration perhaps the oldest but still is the most frequently used parameters for quantification of microbial activities in soil (Chowdhury et al. 2008). Soil respiration is assessed with and without any substrate additions as substrate induced and basal respiration, respectively. Modern automatic devices for real time measurement of respiration and data subjected to relevant mathematical modeling can precisely evaluate the effects of pollutants on microbial community (Johansson et al. 1998). Soil microorganisms are also responsible for atmospheric nitrogen fixation and release (Notron 2000). The main microbial nitrogen transformation processes are N-mineralization, nitrification, denitrification and N-fixation. Analyzing these parameters can also depict the state and well being of microorganisms.

In culture dependent assays, simple Petri dish compatibility tests are carried out generally to find the effect of pesticides on microbes apart from extracting microbes and culturing by plating. The functional diversity and the biochemical potential of soil microorganisms can be assessed by estimating the soil enzymatic activities. Changes in enzymatic functions provide early indications of changes in soil health and usually estimated using specific analytical procedures. Hydrolytic enzymes have the major contribution because of the organic matter decomposition activity of microbes and the fluorescein diaceatate assay is used to estimate it (Schnurer and Rosswall 1982). Fluorescein diacetate is hydrolyzed by different kinds of enzymes, such as protease, lipase and esterase produced by a wide array of bacteria and fungi as primary decomposers (Lundgren 1981). Apart from studies on enzymatic activities of microbes (Brookes et al. 1987), functional richness can also be evaluated as the proportion of different substrate use pattern or carbon utilization (Wainwright 1978; Zabaloy et al. 2008; Stefanowicz et al. 2009). The BIOLOG™ system can be used to analyze carbon utilization patterns using microplates containing different carbon sources (Garland and Mills 1991). Structural diversity of microbial community can be determined by identifying the phospholipid fatty acid (PLFAs) profiles through which various microbial communities are differentiated. Besides microbial processes and activity assessment, microbial diversity is an important parameter that can be used in soil microbial toxicity tests. Genetic diversity of bacteria is the most commonly studied using the diversity of 16S rDNA genes and 18S rDNA for fungi and other eukaryotes (Cernohlavkova 2009).

Classical microbial indicators such as estimation of biomass, biomass carbon, respiration, enzymatic and other activities are inexpensive and provide an easy and integrative estimation of the functional status of microbial community. But the difficulty in interpreting the results and extrapolating them to the natural environment is due to the influence of external factors like climate (Gil-Sotres et al. 2005), soil physico-chemical properties (Yakupoglu et al. 2009; Giaveno et al. 2010; Wallenius et al. 2010), etc. Moreover, about a minor fraction of less than 10 % of microbes are culturable in the laboratories (Amann et al. 1995) and thus culture independent methods are needed to be developed. The cultivation independent methods rely mostly on DNA sequence information (Metzker 2010; Guo et al. 2010). The reproducibility of Finger printing techniques (DGGE, TGGE, T-RFLP, SSCP, RISA, LH-PCR) to find the effect of pesticides on soil microbes give a snapshot view of microbial diversity and community make-up (Van-Elsas and Boersma 2011). Other promising methods include analysis of proteomics (soil metaproteomics) (Bastida et al. 2009) and metabolomics (metabolic analysis) (Simpson and McKelvie 2009; Liebeke et al. 2009). High-throughput sequencing gives large amount of information and has high potential for comparative studies (Van-Elsas and Boersma 2011). But approaches relying on culture independent methods are not accepted to measure the long-term effect of pesticides on microbial communities in soil (Bunemann et al. 2006). The impact of pesticides on bacterial communities in soil can be convincingly studied by combining both the culture dependent analytical measures and culture independent methods (Widenfalk et al. 2008). Thus, different analytical methods, biochemical and biotechnological approaches hand in hand will give the clear picture on the pesticidal effect on microbes. Some of the methods used for the estimation of effect of pesticides on microorganisms are given here under.

4.1 Culture Independent Analyses

4.1.1 Biomass estimation: Counting (Frostegard and Baath 1996)

Depending on the soil type 2.5–10 g of soil is homogenized in distilled water and centrifuged to take the supernatant. The supernatant thus obtained is re-dissolved in distilled water and centrifuged again. From this, a 2 mL of sample can be taken for counting under a microscope.

4.1.2 Biomass Estimation: Fumigation Method (Howell 2011)

Fumigation method is used as an indirect method to estimate the microbial biomass. It is a potential indicator for studying the effects of addition of a particular compound on the soil microbial community (Mazzarino et al. 1993). This method involves fumigation of soil using chloroform which lyses the microbial cells (Brookes et al. 1985) and then extracted using solvents such as 0.5 M potassium sulphate or 2 M potassium chloride (Amato and Ladd 1988). The disadvantages of this method include formation of calcium sulphate in some soils which interrupt the estimation (Joergensen and Brookes 1990). This can give a broader estimation only and not specific to an organism or a group, thus have less ecological significance (Howell 2011).

4.1.3 Phospholipid Fattyacid (PLFA) Analysis (Frostegard et al. 1993)

Phospholipid fatty acid extraction is generally done using the method described by Frostegard et al. (1991). A detailed description on the extraction of phospholipids from soil microbes is also given by Ibekwe and Kennedy (1998). PLFA is extracted in a mixture consisting of chloroform, methanol and citrate buffer. The microbial extract is divided into two phases by adding chloroform and citrate buffer. The lipid phase is dried and fractionated on columns containing silicic acid into neutral, glyco and phospholipids. The thus separated phospholipid portion is dried and analysed in gas chromatography.

Unlike fumigation method, the microbial community can be distinctively monitored through this method. But this method yield lipids from both active and inactive or nonliving biomass and thus can be an over estimation of actual living biomass. Further this analysis can be disrupted by the presence of humic acids in the soil (Nielsen and Petersen 2000).

4.1.4 Basal Respiration (Munoz-Leoz et al. 2011)

To estimate the basal soil respiration, the soil samples were incubated in airtight jars for 3 days at 30 °C. A CO_2 trap with 0.2 N NaOH is also kept inside the jars to trap all the CO_2 evolved during the study. The NaOH is then titrated with 0.1 N HCl to estimate the amount of CO_2 trapped and thus correlated with the respiration. Soil basal respiration is an indicator of soil microbial activity.

4.1.5 Substrate Induced Respiration (Zabaloy et al. 2008)

Substrate induced respiration (SIR) is determined as the rate of evolution of CO_2 from soils to which glucose is added (Hoper 2006). Horwath and Paul (1994) have described a method to find out the quantity of glucose needed to give maximal respiration (initial). In this experiment, 20 g of soil samples were placed into a plastic cup and amended with 1.6 mL of 5 % glucose solution to give a final concentration of 4000 mg/kg. The cup was then placed in a sealed glass-jar and incubated for 6 h during which the CO_2 was trapped in NaOH. After this, NaOH was treated with $BaCl_2$ and titrated with a 0.01 mol HCl using phenolphthalein as indicator. The substrate induced respiration is an indicator of physiologically active microbial biomass (Anderson and Domsch 1978; Munoz-Leoz et al. 2011).

4.1.6 Microbial Metabolic Quotient and Respiration Quotient (Goswami et al. 2013)

Microbial respiration quotient is a useful estimate for the ecophysiological status of the soil microbial community and used to assess the stability of the microbial communities. Microbial respiration quotient (Q_R) is the ratio between the rate of basal and substrate induced respirations of soil microorganisms (Ananyeva et al. 1997). Microbial metabolic quotient (qCO_2) which is the ratio of basal soil respiration to microbial biomass (Anderson and Domsch 1985) can be used as a critical measure of microbial response to disturbances. Microbial metabolic quotient is considered as a more sensitive indicator of toxic effects of pesticides than the respiration rate or the microbial biomass (Anderson and Domsch 1985; Wardle and Ghani 1995; Beelen and Doelman 1997).

4.1.7 Mineralization, Nitrification and Denitrification (Munoz-Leoz et al. 2013)

Nitrogen Mineralization/Ammonification (Pell et al. 1998; Munoz-Leoz et al. 2013)
Nitrogen mineralization is the process by which organic nitrogen is degraded to NH_4^+ in anaerobic laboratory conditions. Some pesticides are used as N source by microbes, so potentially mineralizable N can indicate the biologically active soil nitrogen. This can be measured as described by Powers (1980) also. Potentially mineralizable N can be estimated by estimating ammonia accumulated in water logged soil samples. For this, soil inorganic nitrogen was extracted with 2 mol KCl and analyzed for NH_4 both at the beginning of experiment and at the end of the incubation period (Mijangos et al. 2009) using a photometer using indophenol blue method.

Nitrogen Transformation Test (OECD 2000; OEHHA 2009)
This assay measures the effect of pesticides on the nitrogen transformation in aerobic surface soils under laboratory conditions. Soils were sieved, amended with powdered plant meal (nitrogen source) and treated with the test chemical or left untreated (control). The samples are incubated under aerated conditions and at dark for at least 28 days. Soils were periodically sampled and nitrate measured. Results are compared against control samples and/or a dose-response was prepared. If there is no change in the treated and control samples in nitrate production, then it is presumed that the carbon degradation pathway to be intact and functional. Changes in nitrogen transformation may reflect changes in size and activity of microbial communities through chemical stress.

For nitrification rate, nitrate ($N\text{-}NO_3^-$) and ammonium ($N\text{-}NH_4^+$) concentrations in soil were estimated according to the procedure of Sparks et al. (1996). Denitrification potential was determined according to a modified method of Simek et al. (2002). For that, 75 mg N-NO3$^-$ per kg soil and 75 mg C-glucose per kg soil (in 10 mL of deionized water) were added to 10 g soil sample kept in serum bottles containing 10 % v/v acetylene in the presence of helium. After 24 h of incubation at 25 °C in a rotary shaker, $N\text{-}N_2O$ production was measured by gas chromatography.

Nitrate can be also extracted according to the OECD 216 guidelines (OECD 2000) and determined through a chromogenic microplate method (Hood-Nowotny et al. 2010).

4.2 *Culture Dependent Analyses*

4.2.1 Compatibility in Petri Dish (Stanley et al. 2010a)

In this experiment, a poisoned food technique (Schmitz 1930) was adopted for studying the effect of pesticide on *T. viride*, beneficial antagonistic fungi. Two doses below and two doses above the field dose was incorporated in to sterilize molted

Potato Dextrose Agar medium and poured in Petri dishes. Inoculation of *T. viride* was done by placing an 8 mm mycelial disc in the center of the Petri dish. The colony diameter of *T. viride* grown on poisoned media was measured and compared with that of the control. The whole experiment was carried out aseptically in a laminar flow chamber, which was previously irradiated with UV light for 15 min. The difference in colony diameter between poisoned medium and control (without pesticide) was used to calculate the percentage inhibition as follows (Paul and Mishra 1993).

$$\text{Mycelial inhibition} = \frac{\text{Colony dia. in control} - \text{Colony dia. in treatment}}{\text{Colony diameter in control}} \times 100$$

For *P. fluorescens*, the culture was transferred to a Petri dish containing 15 mL of King's B broth amended with pesticide at different concentrations usually the field recommended dose and two doses below and two doses above along with the checks all replicated three times. The Petri plates were incubated at 28 °C in a shaker. The growth of Pseudomonas was observed for it's fluoresce nature for every 24 h after inoculation for 3 days (Fig. 6.1).

4.2.2 Technical Grade Pesticide in Soil and Plating (Radivojevic et al. 2011)

Virgin soil without any pesticide application history was collected from the upper layer (0–10 cm), dried, sieved and air dried at room temperature. The pesticide solution was made by mixing the technical substance in distilled water and prepared for respective concentrations. About 1 kg of soil was kept in rotating stirrer and pesticide solution pipetted to the surface before homogenization for 30 min and filled in pots. Soil moisture was kept at 50% field capacity. The samples for the analysis were taken 7, 14 and 30 days after application. Soil dilution plate technique was

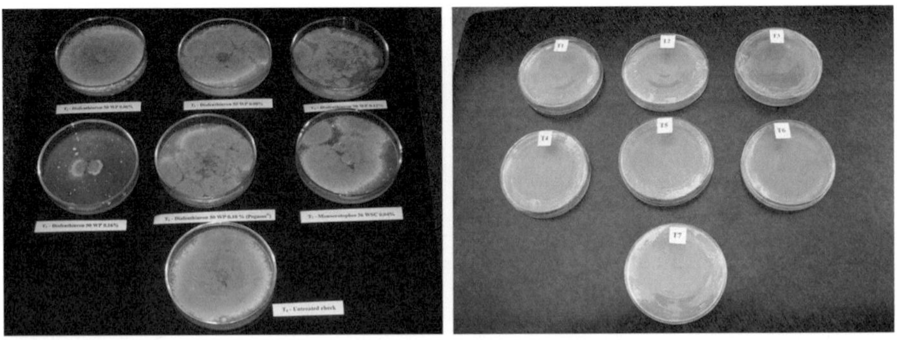

Fig. 6.1 Pesticide compatibility studies on *Trichoderma viride* and *Pseudomonas fluorescens*

used to count the total culturable microorganism using tryptic soy agar for bacteria and Czapek agar for fungi, incubated for specific period and colonies counted. Fumigation extraction technique was used to estimate the microbial biomass carbon (Vance et al. 1987) using chloroform. Microbiological biomass carbon (MBC) was calculated as MBC = C extracted × 0.33 (Jenkinson et al. 1979) and expressed as μg C/g of soil. Activity of soil dehydrogenase can be estimated as per Tabatabai (1982).

4.2.3 Repeated Application of Pesticides in Soil (Lane 2011)

In this experiment, 40 g (dry weight) of moist field soil was placed into a 50 mL beaker and pesticide added by diluting in deionized water. The soil was wetted to two third of the field capacity and beakers placed in air tight 1 L mason jars with a sodium hydroxide traps to trap the CO_2 evolved and incubated at 20 °C. The trap was replaced every 15 days and soil moisture replenished. The herbicide was applied at monthly intervals in the soil, sampled at specified intervals and analyzed for ester linked fatty acid methyl esters (EL-FAME (Schutter and Dick 2000) and ratio of total fungal to total bacterial FAMEs calculated. Exchangeable potassium was measured using ammonium acetate, filtered and analyzed in a flame photometer as per Knudsen et al. 1982. Non-exchangeable K was measured using boiling nitric acid method (Helmke and Sparks 1996) and microbial biomass K as per Lorenz et al. (2010).

4.2.4 Application of Fumigants (Klose et al. 2006)

The field collected soil was sieved in 2 mm sieve and adjusted to 15 % moisture to ensure optimal fumigant dissipation. The test soil was weighed and 500 g soil kept in glass jars sealed with an airtight lid equipped with rubber septa. Fumigants were prepared in water at requisite concentrations and added by means of a syringe and incubated for 24 h. After incubation, the jars were uncapped and vented for 30 min until all the remaining volatile fumigants get released. After release of fumigants the samples were incubated in laboratory conditions. Soil samples were collected at specified intervals and microbial community and enzyme activities assayed.

4.2.5 Assay on Functional Richness: C Source Utilization (Zabaloy et al. 2008)

Soil functional richness in terms of potential utilization of carbon sources by microbes were assayed in 96 well sterile microplates. The wells were filled with 100 mL liquid mineral medium with a redox indicator and a carbon source at a concentration of 2.5 mg/mL (Zabaloy and Gomez 2005). After inoculation of microorganism, the microplates were incubated for 96 h. In this experiment, about 33 different carbon sources like 16 carbohydrates, 8 carboxylic acids, 8 amino acids and an aromatic compound were tested. The intensity of the colour developed in each well was expressed as numerical values (Avidano et al. 2005).

4.2.6 Estimation of Enzymes (Rasool and Reshi 2010)

In this experiment, different pesticide doses were applied on soil and tested for enzyme activities as an indirect means of measuring microbial toxicity. Pesticides are normally tested at their field recommended dose, 10 and 100 times of field dose along with a control. The enzyme activities are measured as follows:

Enzymes	References
Dehydrogenase	Casida et al. (1964)
Alkaline phosphatase	Tabatabai and Bremner (1969)
Protease	Ladd and Butler (1972), modified by Rangaswamy et al. (1994), and Ismail et al. (1996)
Urease	Tabatabai and Bremner (1972)
Amidase	Frankenberger and Tabatabai (1980a, b)
Asparaginase	Frankenberger and Tabatabai (1991)
Nitrogenase	Hardy et al. (1968) reported in Martinez-Toledo et al. (1998)
β-glucosidase	Bailey et al. (2012) modified from Saiya-Cork et al. (2002)
N-acetyl- β -Dglucos aminidase and lipase	Alef and Nannipieri (1995)
Leucine aminopeptidase	Saiya-Cork et al. (2002)

Enzyme assays do not evaluate the precise response of microbes to pollutants. Soil enzymes may remain active even after the death of microbes and cell lysis, thus cannot be correlated with actual biomass. Soil organic matter and mineral content may also have an influence on the enzyme activities. The enzymatic response of soil microbes are also influenced by soil physico-chemical characteristics (Gevao et al. 2000) and/or the agricultural practices (Alletto et al. 2010). These limitations make enzyme assays un-realistic in some cases and yet it is widely used method to estimate the effect of pesticides on microbes.

4.2.7 Estimation of Total Microbial Activity: FDHA (Schnurer and Rosswall 1982; Green et al. 2006)

Fluorescein diacetate hydrolyzing activity (FDHA) is widely established as an accurate, simple and rapid method for measuring total microbial activity in soils and litter (Vakemans et al. 1989) at least as an early determinantal test for the effects of pesticides (Perucci et al. 2000; Goswami et al. 2013). Fluorescein diacetate is hydrolyzed by different enzymes like protease, lipase, esterase, etc. and this hydrolytic activity is found in a wide array of primary decomposers such as bacteria and fungi (Lundgren 1981). A correlation between the FDH kinetics and biomass of soil microbes is already been established (Anderson and Domsch 1990; Perucci 1992; Pal et al. 2005; Goswami et al. 2013).

In the experiment by Schnurer and Rosswall (1982), *Pseudomonas denitrificans* and *Fusarium culmorum* were cultured and pure cells and mycelium harvested and dried. Fluorescein diacetate was dissolved in acetone and kept as stock solution. For the determination of FDA hydrolytic activity, the samples (soil, straw, bacterial or fungal suspension) were added with FDA (10 μg/mL) to sterile 60 mM sodium phosphate and incubated on a rotary shaker. After incubation, the amount of FDA hydrolyzed was measured as absorbance at 490 nm using a spectrophotometer. Before the spectrophotometric measurements, straw, soil or microbial cells are removed and centrifuged to have a clear solution without background disturbance.

In another experiment, 1 g of soil was added with 50 mL of 60 mM sodium phosphate buffer and 0.5 mL of 4.9 mM FDA lipase substrate solution. The mixture is mixed well and incubated for 3 h and then added with 2 mL of acetone to terminate hydrolysis. This solution was taken, centrifuged and supernatant analyzed calorimetrically (Green et al. 2006).

4.3 DNA Based Methods

These assays are also culture independent methods but provide valuable insights into key characteristics of soil microbial communities, such as viable biomass, community structure, nutritional status and physiological stress responses (Widenfalk et al. 2008; Demoling et al. 2009). Most of the DNA based methods rely on polymerase chain reaction (PCR) and on 16S ribosomal RNA gene (Imfeld and Vuilleumier 2012). Studies on microbial diversity through molecular methods are mainly by analyzing the ribosomal genes, which are amplified and sequenced. Apart from ribosomal gene analysis, molecular tools are also used to determine the microbial community profiles (Ferreira et al. 2009). Some of these techniques used in microbiology are Random Amplified Polymorphic DNA – RAPD (Franklin et al. 1999), Amplified Ribosomal DNA Restriction Analysis – ARDRA (Viti and Giovannetti 2005), Terminal Restriction Fragment Length Polymorphism – T-RFLP, RISA (Schutte et al. 2008), Denaturing/Temperature Gradient Gel Electrophoresis – DGGE/TGGE (Muyzer et al. 1993; Feris et al. 2004) and Single-Strand Conformation Polymorphism – SSCP (Schmalenberger et al. 2008).

4.3.1 PCR-Denaturing/Temperature Gradient Gel Electrophoresis (Ferreira et al. 2009)

This experiment was carried out to evaluate the impacts of different pesticides on the PCR-DGGE profiles of culturable soil bacterial communities. The experiment was carried out in 1.5 kg pots grown with potato in a greenhouse with factorial arrangement composing of three soils, five pesticides and four harvest periods. The potted plants were sprayed with the test pesticides at 30 and 45 days after sowing, to simulate field conditions. The soil samples were taken and microbes inoculated in culture medium till abundant growth was observed. After a particular period of time, the cells were extracted, centrifuged and pellets were stored. DNA was extracted according to the procedure given by Schwieger and Tebbe (1998) and modified by Xavier et al. (2004). PCR was performed with three different dilutions of the extracted DNA samples (1:20, 1:40 and 1:80). The best denaturing condition was found to occur on a gradient of 50–65 % of the denaturing agents and used accordingly. The gels stained and visualized under UV and photo documented.

4.3.2 Terminal Restriction Fragment Length Polymorphisms (T-RFLP)

This technique involves the restriction digestion of fluorescently labeled PCR products to produce labeled terminal restriction fragments (TRFs) of different lengths. The fragments are then separated out based on size using capillary or gel electrophoresis methods and analyzed to get chromatograms (Osborn et al. 2000). Rousseaux et al. (2003) and Zhang et al. (2008) used T-RFLP analysis to illustrate the changes in soil bacterial community structure following the application of herbicides and insecticides, respectively. The presence or absence of TRFs in the treated samples shows the variation with respect to the control, but the disadvantage in this method is that each TRF can be a representative of different organisms and different strains of the same microorganism can also produce different TRF sizes (Howell 2011) and thus confusing.

4.3.3 Single-Strand Conformation Polymorphisms (SSCP)

Like DGGE, this technique is also a gel electrophoresis based method which separates different single stranded DNA, based on the natural secondary structures they form under non-denaturing conditions, which is determined by the nucleotide sequence of the fragment (Schwieger and Tebbe 1998). SSCP is used generally to identify specific target organisms that show resistance to pesticides.

4.3.4 Stable Isotope Probing (SIP)

Unlike the previous methods described which are used to determine the pesticide impacts on microbial community structure, stable isotope probing (SIP) can be used to monitor the effects of pesticides on microbial functions by using isotope labels (Howell 2011). Samples are then taken to monitor the incorporation of the labeled compound into the DNA or rRNA of the microbial community which acts as an indicator of microbial activity (Lueders et al. 2004). Density gradient centrifugation methods are used to separate the labeled and un-labeled genetic material. This material is then usually used for further analyses such as gene cloning and or sequencing (Dumont and Murrell 2005).

4.3.5 Metagenomic Approaches

These new sequencing techniques are used to investigate global patterns of gene transcription, i.e. expression of functional characteristics in response to stress (Christen 2008; Wooley and Ye 2010). Focused sequencing of cloned metagenomic DNA fragments provide functionally relevant information and provide hints on underlying metabolic pathways (Maron et al. 2007; Hemme et al. 2010; Imfeld and Vuilleumier 2012).

4.3.6 Next-Generation Sequencing

The processes used in these methods can involve PCR amplification of extracted DNA, cDNA production from extracted rRNA and mRNA (Urich et al. 2008) followed by direct pyrosequencing, or even direct pyrosequencing from extracted bulk mRNA (Gifford et al. 2011).

4.4 Semifield/Mesocosm Studies

4.4.1 Mesocosm Studies (Monkiedje et al. 2002; Munoz-Leoz et al. 2011)

In this experiment, 10 L plastic trays are used as mesocosms where 8 kg soil was added resulting in a soil layer of approximately 10 cm depth. Required concentration of pesticide was spiked after making up to a volume of 100 mL in solvent. After the addition of pesticides, the soil is thoroughly mixed with a rotary mixer to assure uniform distribution of pesticides and then allowed for the solvent to get evaporated. Soil moisture content was adjusted to 60 % water holding capacity and trays covered with perforated polypropylene sheets to avoid photodegradation of pesticide and evaporative loss of water and incubated in the dark. From each mesocosm, a subsample of 250 g soil was taken at regular intervals and analyzed for soil basal respiration and substrate induced respiration as per ISO 16072 Norm (ISO 2002) and ISO 17155 Norm (ISO 2012), respectively. Microbial biomass carbon was quantified as per Vance et al. (1987). Microbial metabolic quotient and respiratory quotient were also estimated along with enzyme activities. Biolog Ecoplates® were used to understand the functional community profiles and done as per Epelde et al. (2008).

4.4.2 Pot Culture Experiments (Mijangos et al. 2009)

Pot culture experiments were carried out to determine the impact of different herbicides on the microbes of rhizophere region. For this, pots were filled (250 g) with the soil collected from the top layer (0–30 cm) of natural grassland. Treatments consists of two concentrations of herbicide with two different composition of forage plants i.e., triticale and triticale + pea. The plants were grown in growth chamber and after 45 days, herbicide was applied carefully by fitting a cylinder over the pot so that the whole dose falls on the plants or soil surface. Control plants (not sprayed with herbicide) were clipped to simulate manual weed control. Soil samples were taken at specified periods to analyze the potentially mineralizable N, soil microbial functional diversity and community-level physiological profiles (CLPPs).

4.4.3 Pesticides on the Abundance of Microbial Communities (Jastrzebska 2010)

An experiment was conducted to determine the effect of pesticide on the abundance of soil-dwelling microbes for which the soil samples were mixed with fungicide, macronutrients and micronutrients corresponding to the crop recommendation. Treatments were made on different concentrations of pesticide and soils treated accordingly. The test soil was filled in polyethylene pots and sown with spring rapeseed and white lupine. From the test pots, soil samples were collected on the first day and after 10 and 50 days of treatment. The abundance of the following microbes were determined on selective media by plate-count method: oligotrophic and copiotrophic bacteria on Onta and Hattori medium, ammonifying, nitrogen immobilizing and cellulolytic bacteria on Winogradski medium, actinomycetes on Kuster and Williams medium containing antibiotics, *Azotobacter* spp. on Fenglerowa medium and fungi on Martin medium. Each microbial community in plate is incubated for specific days like, 3, 7, 14 and 21 days before taking observations.

4.4.4 Greenhouse Experiments: Fatty Acid Methyl Ester Extraction (Lane 2011)

A greenhouse study was conducted in factorial design with two different soils (soils with history of long-term field applications of pesticide and untreated control), two plant treatments (presence or absence of soybean) and three rates of herbicide application. About 2.5 kg of soil was placed in a plastic pot and pre-germinated glyphosate resistant soybean was sown. Glyphosate was sprayed using hand sprayer at 47 and 61 days of post emergence. Samples were taken at specific intervals i.e., 7 days after first and second spray, as rhizosphere soil and bulk soil separately for analysis. The soil is analyzed for exchangeable K, non-exchangeable K and microbial K. Plant tissue K was analyzed by dry ashing procedure. Microbial analysis was done by extracting fatty acid methyl esters (FAMEs) as described by Schutter and

Dick (2000). The total quantity of FAME obtained can be an indication for the total bacterial biomass. The group specific FAMEs were summed up and used to identify the different microbial groups like general bacteria, Gram positive and negative bacteria, actinomycetes, saprotrophic fungi and protozoans.

4.4.5 Effect on Enzymes of Rhizosphere and Non-rhizosphere Microbes (Wu et al. 2014)

In this experiment, rhizosphere soil was collected, together with tobacco plants at the stubble stage by digging around the soil close to the plant roots approximately to a depth of 15 cm. The extra soil around the root zone was shaken out and the rhizosphere and non-rhizosphere soil collected separately and passed through 2 mm sieve. The soil is spiked with pesticide at required concentration made using distilled water and mixed thoroughly using plastic spoons. The soil is then transferred to 3 L polypropylene pots adjusted to 60 % field capacity and covered with aluminium foil and incubated in darkness at 25 °C for 60 days. Samples were taken at appropriate intervals using a 2 cm dia. soil auger to estimate microbiological population by plating and enzymological studies. Enzymatic activities of dehydrogenase, catalase and urease were assayed according to methodology outlined by Guan et al. (1986).

4.5 Field Experiments

4.5.1 Field Experiments with Crop Rotation (Lupwayi et al. 2010)

The experiment was carried out in a field grown with canola-barley crop rotation for 4 years to study the effect of pesticides and fertilizers on microbial organisms. Half of the field was grown with each crop and the crops are switched over every year. Different treatments of fertilizers and pesticides are given as per the treatment and the cumulative effect seen after 4 years. The rhizosphere and bulk soil samples were collected from every plot at the stage when barley was having flag leaf and canola was flowering. Substrate induced respiration and β glucosidase activity were studied along with community level physiological profiles (CLPPs) using the Biolog® method (Zak et al. 1994).

4.5.2 Long Term Field Experiments (Nicholson and Hirsch 1998)

In this long term experiments on pesticide applications, the chemical treatment plots received a combination of up to five pesticides annually between 1974 and 1993, with spring barley grown each year. The pesticides include aldicarb, chlorfenvinphos, benomyl, glyphosate, chlorotoluron and triadimefon. The composition of the microbial populations in the treatment and control were assayed by genotyping, phenotyping and by determining the substrate utilization activity. Soil samples were collected from pesticide treated and control plots, processed, plated and incubated for specific period. Bacterial fluorescence was also examined. Nutrient Agar and Tryptone Soy Agar were used to enumerate fast-growing heterotrophic bacteria and a non-selective minimal medium for rhizobia was used for putative nitrogen fixers. Fully grown colonies were transferred from plates to nylon membrane filters and DNA extracted as per Hirsch (1995) and PCR fingerprinting done.

5 Pesticide Risk Assessment for Microorganisms

Risk is defined as toxic/dangerous effect coupled with exposure. The toxic effects of pesticides can be easily determined by different bioassays. End points of pesticide toxicity bioassays show the impact of chemical on the ecosystem and thus bioassays play a critical role in the ecological risk assessment process. After assessing the impact of pesticides, the potential exposure and ecological effects are quantified in the following second phase. In the third phase, the uncertainty in characterizing the ecological risk is reduced by supporting the risk predictions with valid and in many cases real site data (OEHHA 2009) to form an overall risk assessment.

5.1 Risk of Pesticides on Microbes

Microbes are responsible for many ecosystem services (Benayas et al. 2009) either in terrestrial or aquatic ecosystems. Pesticides do have a greater impact on the microbes and its functions/activities and thus influence on those ecosystem services (Goulson 2013). But risk assessments of pesticides on microbes have not received much importance as that of other groups. The guidelines given by USEPA do not recommend the use of soil microbes as assessment endpoints in ecological risk assessments (OEHHA 2009). Though microbes are included among the taxonomic groups under European environmental risk assessment (ERA) procedures for pesticides, it is not substantially covered as that of other groups like plants, vertebrates, invertebrates and arthropods (Nienstedt et al. 2012). The rationale behind this includes the lack of field validated methodologies and the great spatial and temporal variation in microbial responses, which makes it difficult to evaluate the ecological consequences of any measured change in activity (OEHHA 2009).

Hence, literature on the risk of pesticide application on microbes is scanty. If available, is based on microbial functions/activities or biomass or enumerations performed at the community levels. It may not give a proper assessment on the ecotoxicological impacts of pesticides on the structure of soil microbial communities (Soulas and Lors 1999). Risk assessments, even if done based on microbial activities must take into consideration at least some of the activities if not all which are found crucial. A weighted/differential average to confer greater importance to crucial activities may give a more meaningful picture, than giving equal importance to all microbial activities. Among the different risk assessment methods, a multi tiered approach which include laboratory studies to evaluate adverse effects, extended to semifield/mesocosm studies to increase ecological complexity and reduce generality (Romeis et al. 2011) and more realistic field studies give a full-fledged ecotoxicology assessment. Some pesticides especially fungicides and bactericides are used for the management of plant pathogenic microbes in agroecosystem. These pesticides should be of target specific causing inhibitory or mortality effect to plant pathogens and at the same time safe to beneficial microbes. This assessment with selectivity ratio allows us to screen and find safer and selective pesticides. If the overall non target effects of pesticides are to be assessed, then hazardous concentration 5 %, concentration which will not affect 95 % of microbes in the environment can be taken into consideration for pesticide risk assessment on microbes. This risk approach is supposed to be the best but requires voluminous data and thus makes it practically difficult. Some of the risk assessment approaches which may be used for microbes are suggested asunder.

5.2 *Risk Assessment Methodologies*

5.2.1 Risk Assessment: Effect on Microbial Activities and Biodiversity

Many soil ecosystem services are biologically mediated, which include regulation of water and nutrients, facilitation of nutrient transfer and translocation, renewal of nutrients through organic and waste matter breakdown, process of soil formation and delivery of nutrients to plants (Swift et al. 2004; Dominati et al. 2010; Robinson et al. 2013; Chagnon et al. 2015). Many different kinds of microbes play a specific and collective role in all these ecosystem processes. Thus the biodiversity and the relative abundances of keystone species or functional groups that underpin these soil processes are becoming important (De-Ruiter et al. 1995; Brussaard et al. 2007; Nielsen et al. 2011) apart from microbes with specific activities.

Hence, risk assessment of pesticides on microorganisms is done by studying the effect of chemical on the microbial activities in the soil and the biodiversity. Many analytical procedures are used to find the effect of pesticides on soil microbial activities. Some of those procedures are detailed in the previous subchapter on methods to assess pesticide toxicity to microorganisms. By assessing the crucial microbial activities in the soil such as nitrogen mineralization/ammonification, nitrification and denitrification after pesticide application, the risk involved on those microbes can be made known. Further by studying the biodiversity of that soil, the conclusion thus drawn can be confirmed and proved. Functional diversity is said to be more important than taxonomic diversity in terms of ecosystem service delivery (Munns et al. 2009), taxonomic diversity plays a crucial role in the changing environment by enabling ecosystems to cope with adverse effects (Yachi and Loreau 1999). Keeping microbes as a source of genetic novelty and as a repository of genetic resources (Tiedje 1994), ecotoxicology based on biodiversity gains its importance.

5.2.2 Risk Assessment: Weighted Activity Approach

Many of the activities of microorganisms collectively result in ecosystem services. Decomposition of soil organic matter itself is controlled by the active microbial biomass which leads to nutrient cycling and release of nitrogen and other nutrients. Microbes also act as biological nitrogen fixers and enhance soil fertility. Pesticides affect microorganisms and influence these microbial activities indirectly by inhibiting some microbes and favouring others. For the agro-ecosystem to function in a sustainable manner, a slight reduction in biodiversity can be accepted when the key species of the ecosystem and their functions are unaffected (Walker 1992). Thus microbial activities and functions are very important for ecosystem sustainability. Microbes are involved in many functions and activities in ecosystem. If we take nitrogen alone, organic N is mineralized by highly specialized microbial group to plant available forms of inorganic N. Available N pools in soils are greatly enhanced by nitrogen-fixing microorganisms that convert atmospheric N to plant available N (Chagnon et al. 2015). Inorganic N is taken up by soil microbes, assimilated and incorporated into the soil organic N pool (Brady and Weil 1996; Brussaard et al. 1997; Barrios 2007). Loss of N through denitrifying microbes is another valuable ecosystem service, preventing environmental contamination and eutrophication (Chagnon et al. 2015). Though microbes perform different activities, some of them are very crucial and some may be of less importance and some may not have any importance. So a risk assessment can be made using a weighted average of the effects of pesticides in microbial activities as it is done in environmental impact quotient (EIQ) calculation. The crucial functions/activities may be given with a greater weightage and less important one with low weightage to differentiate the important and less important activities.

5.2.3 Risk Assessment: Multi Tiered Approach

Another method of risk assessment may be carried out by a series of experiments from laboratory toxicity tests to pot culture experiments or mesocosm studies and further extended to field studies to find the risk of pesticide application on microorganisms. Laboratory bioassays include but not restricted to biomass estimation, respiration experiments, compatibility assays using food poisoning technique etc. These laboratory assays are done in unrealistic conditions and with a worst case scenario. Adverse effect if any is observed in this first tier of experimentation, then the study is to be advanced to the next tier of toxicity estimation. If no adverse effect is observed, the pesticide may be declared safe. The second tier experiments or mesocosm studies include enzyme assays, assays related to microbial activities, functional community profiling using Biolog® plates, community-level physiological profiling (CLPPs) etc. Methods or procedures of toxicity testing in laboratory or first tier and mesocosm or second tier are not strictly different with each other and are interchangeable. Besides these, field experiments to assess the effect of pesticides on microbes can be carried out in a more realistic way to find the overall picture of risk of pesticides on microorganisms.

5.2.4 Risk Assessment: Toxicity to Target and Non-target/Beneficial Microbes

Pesticide risk assessments are being done by comparing the toxicity or effect of pesticides on the target organisms with that of non target organisms especially in case of insect pests and insect pollinators (Stanley et al. 2010b) or an insect pest and a beneficial insect/natural enemy (Tanaka et al. 2000; Preetha et al. 2010). These assessments are made by calculating relative toxicity and or a selectivity ratio. In microbes also, one or a set of beneficial microorganisms can be tested for a particular pesticide and compared with the toxic effect on plant pathogenic microorganism to obtain the risk of pesticide on beneficial microbe, if any. The selectivity ratio can be as follow:

$$\text{Selectivity ratio} = \frac{\text{Effect on beneficial microorganism}}{\text{Effect on pathogenic microorganism}}$$

For a selective pesticide, the effect on beneficial microbe should always less than that of plant pathogenic microbe. In other words the pesticide should affect the target microbe more than that of the non target one i.e., toxicity to the target should be greater than that of non target toxicity. So a selectivity ratio of more than one is an indicative of safer pesticide.

5.2.5 Risk Assessment: Hazardous Concentration 5 % (Van-Beelen and Doelman 1996)

Risk assessment of pesticides on microorganisms is a complex process because some species may be sensitive for a specific pesticide while others may not be that sensitive. Many toxicity tests in practice may not reveal the true sensitivity or risk of pesticide to microorganisms since these tests measures the activity of the microbes. In reality, the resistant species does the activity and obscures the inhibitory effect of pesticide on the sensitive species. Thus, risk based on hazardous concentration 5 % (HC_5) which calculates a pollutant concentration that does not inhibit or kill 95 % of the animal is more realistic. But this requires no observed effect concentration (NOEC) for particular pesticide from more than four toxicity tests with different organisms (Van-Beelen and Doelman 1996) that is usually not available for microorganisms especially for pesticides. This approach of risk assessment seems to be more realistic in natural environment but requires huge data set to analyze the effect.

References

C. Accinelli, C. Screpanti, G. Dinelli, A. Vicari, Short-time effects of pure and formulated herbicides on soil microbial activity and biomass. Intl. J. Environ. Anal. Chem. **82**, 519–527 (2002)

A.O. Adesemoye, J.W. Kloepper, Plant–microbes interactions in enhanced fertilizer-use efficiency. Appl. Microbiol. Biotechnol. **85**, 1–12 (2009)

M. Ahemad, M.S. Khan, Toxicological effects of selective herbicides on plant growth promoting activities of phosphate solubilizing *Klebsiella* sp. Strain PS19. Curr. Microbiol. **62**, 532–538 (2011)

M. Ahemad, M.S. Khan, Effects of pesticides on plant growth promoting traits of Mesorhizobium strain MRC4. J. Saudi Soc. Agric. Sci. **11**(1), 63–71 (2012)

S. Ahmed, M.S. Ahmad, Effect of insecticides on the total number of soil bacteria under laboratory and field conditions. Pak. Entomol. **28**(2), 63–68 (2006)

A. Alamdara, J.H. Syeda, R.N. Malik, A. Katsoyiannis, J. Liu, J. Li, G. Zhang, K.C. Jones, Organochlorine pesticides in surface soils from obsolete pesticide dumping ground in Hyderabad City, Pakistan: contamination levels and their potential for air–soil exchange. Sci. Total Environ. **470&471**, 733–741 (2014)

K. Alef, P. Nannipieri, Enzyme activities, in *Methods in Applied Soil Microbiology and Biochemistry*, ed. by K. Alef, P. Nannipieri (Academic, San Diego, 1995), pp. 311–373

L. Alletto, Y. Coquet, P. Benoit, D. Heddadj, E. Barriuso, Tillage management effects on pesticide fate in soils- a review. Agron. Sustain. Dev. **30**, 367–400 (2010)

R.I. Amann, W. Ludwig, K.H. Schleifer, Phylogenetic identification and in-situ detection of individual microbial-cells without cultivation. Microbiol. Rev. **59**, 143–169 (1995)

M. Amato, J.N. Ladd, Assay for microbial biomass based on ninhydrin-reactive nitrogen in extracts of fumigated soils. Soil Biol. Biochem. **20**(1), 107–114 (1988)

N.D. Ananyeva, T.S. Demkina, U.C. Stin, The stability of microbial communities in pesticide-treated soils. Pochvovedenic **1**, 69–74 (1997)

P. Ander, K.E. Eriksson, Selective degradation of wood components by white-rot fungi. Physiol. Plant. **41**, 239–248 (1977)

J.P.E. Anderson, Methods to evaluate pesticide damage to the biomass of the soil microflora. Soil Biol. Biochem. **13**, 149–153 (1981)

J.P.E. Anderson, K.H. Domsch, A physiological method for the quantitative measurement of microbial biomass in soils. Soil Biol. Biochem. **10**, 215–221 (1978)

J.P.E. Anderson, K.H. Domsch, Maintenance carbon requirements of actively metabolizing microbial populations under in situ situations. Soil Biol. Biochem. **17**, 197–203 (1985)

T.H. Anderson, K.H. Domsch, Application of eco-physiological quotients (qCO_2 and qD) on microbial biomasses from soils of different cropping histories. Soil Biol. Biochem. **22**, 251–255 (1990)

J.H. Andrews, R.F. Harris, The ecology and biogeography of microorganisms on plant surfaces. Annu. Rev. Phytopathol. **38**, 145–180 (2000)

A.S.F. Araujo, R.T.R. Monterio, R.B. Abarkeli, Effect of glyphosate on the microbial activity of two Brazilian soils. Chemosphere **52**, 799–804 (2003)

T.N. Arkhipova, E. Prinsen, S.U. Veselov, E.V. Martinenko, A.I. Melentiev, G.R. Kudoyarova, Cytokinin producing bacteria enhance plant growth in drying soil. Plant Soil **292**, 305–315 (2007)

J. Artigas, J. Majerholc, A. Foulquier, C. Margoum, B. Volat, M. Neyra, S. Pesce, Effects of the fungicide tebuconazole on microbial capacities for litter breakdown in streams. Aquat. Toxicol. **122&123**, 197–205 (2012)

F. Asch, J.L. Padham, Root associated bacteria suppress symptoms of iron toxicity in lowland rice, in *The Global Food and Product Chain – Dynamics, Innovations, Conflicts, Strategies*, ed. by E. Tielkes, C. Hulsebusch, I. Hauser, A. Deininger, K. Becker (MDD GmbH, Stuttgart, 2005). 276 p

G.K. Aseri, N. Jain, J. Panwar, A.V. Rao, P.R. Meghwal, Biofertilizers improve plant growth, fruit yield, nutrition, metabolism and rhizosphere enzyme activities of pomegranate (*Punica granatum* L.) in Indian Thar Desert. Sci. Horticult. **117**, 130–135 (2008)

M. Ashraf, S. Hasnain, O. Berge, T. Mahmood, Inoculating wheat seedlings with exopolysaccharide-producing bacteria restricts sodium uptake and stimulates plant growth under salt stress. Biol. Fert. Soils **40**, 157–162 (2004)

L. Avidano, E. Gamalero, G.P. Cossa, E. Carraro, Characterization of soil health in an Italian polluted site by using microorganisms as bioindicators. Appl. Soil Ecol. **30**, 21–33 (2005)

V.L. Bailey, C.L. Bilskis, S.J. Fansler, L.A. McCue, J.L. Smith, A. Konopka, Measurements of microbial community activities in individual soil macroaggregates. Soil Biol. Biochem. **48**, 192–195 (2012)

L.E. Bairamov, L.V. Vinogradova, A.A. Zavalin, Nitrogen nutrition and productivity of barley as conditioned by the application of associative diazotrophs. Appl. Biol. **63**, 135–139 (2001)

A. Ball, W. Betts, A. McCarthy, Degradation of lignin related compounds by Actinomycetes. Appl. Environ. Microbiol. **55**, 1642–1644 (1989)

M. Banerjee, L. Yesmin, Sulfur oxidizing plant growth promoting rhizobacteria for enhanced canola performance (2002). Retrieved from http://www.google.com/patents/US7517687

E.A. Barka, J. Nowak, C. Clement, Enhancement of chilling resistance of inoculated grapevine plantlets with a plant growth-promoting rhizobacterium, *Burkholderia phytofirmans* strain PsJN. Appl. Environ. Microbiol. **72**, 7246–7252 (2006)

E. Barrios, Soil biota, ecosystem services and land productivity. Ecol. Econ. **64**, 269–285 (2007)

F. Bastida, J.L. Moreno, C. Nicolas, T. Hernandez, C. Garcia, Soil metaproteomics: a review of an emerging environmental science -significance, methodology and perspectives. Eur. J. Soil Sci. **60**, 845–859 (2009)

P. Baumann, M.A. Clark, L. Baumann, A.H. Broadwell, *Bacillus sphaericus* as a mosquito pathogen: properties of the organism and its toxins. Microbiol. Mol. Biol. Rev. **55**(3), 425–436 (1991)

G.A. Beattie, S.E. Lindow, The secret life of foliar bacterial pathogens on leaves. Annu. Rev. Phytopathol. **33**, 145–172 (1995)

P.V. Beelen, P. Doelman, Significance and application of microbial toxicity tests in assessing ecotoxicological risks of contaminants in soil and sediment. Chemosphere **34**, 455–499 (1997)

J.M.R. Benayas, A.C. Newton, A. Diaz, J.M. Bullock, Enhancement of biodiversity and ecosystem services by ecological restoration: a meta-analysis. Science **325**, 1121–1124 (2009)

S. Bensalim, J. Nowak, S.K. Asiedu, A plant growth promoting rhizobacterium and temperature effects on performance of 18 clones of potato. Am. J. Potato Res. **75**, 145–152 (1998)

G. Berg, Plant–microbe interactions promoting plant growth and health: perspectives for controlled use of microorganisms in agriculture. Appl. Microbiol. Biotechnol. **84**, 11–18 (2009)

U.G. Berninger, B.J. Finlay, P. Kuuppo-Leinikki, Protozoan control of bacterial abundances in fresh waters. Limnol. Oceanogr. **36**, 139–147 (1991)

S. Bhattacharya, A. Das, Study of physical and cultural parameters on the vacteriocins produced by lactic acid bacteria isolated from traditional Indian fermented foods. Am. J. Food Technol. **5**(2), 111–120 (2010)

G.V. Bloemberg, B.J. Lugtenberg, Molecular basis of plant growth promotion and biocontrol by rhizobacteria. Curr. Opin. Plant Biol. **4**, 343–350 (2001)

W.B. Bollen, Interactions between pesticides and soil microorganisms. Annu. Rev. Microbiol. **15**, 69–92 (1961)

N.C. Brady, R.R. Weil, *The Nature and Properties of Soil*, 11th edn. (Prentice Hall, Upper Saddle River, 1996)

B.M. Braithwaite, A. Jane, F.G. Swain, *Amnemus quadrituberculatus* (Bob.), a weevil pest of clover pastures on the north coast of New South Wales. J. Aus. Inst. Agric. Sci. **24**, 155–157 (1958)

A.A. Brakhage, V. Schroeckh, Fungal secondary metabolites – strategies to activate silent gene clusters. Fungal Genet. Biol. **48**, 15–22 (2011)

P.C. Brookes, A. Landman, G. Pruden, D.S. Jenkinson, Chloroform fumigation and the release of soil-nitrogen – a rapid direct extraction method to measure microbial biomass nitrogen in soil. Soil Biol. Biochem. **17**, 837–842 (1985)

P.C. Brookes, A.A. Newcombe, D.S. Jenkinson, Adenylate energy charge measurement in soil. Soil Biol. Biochem. **19**, 211–217 (1987)

L. Brussaard, V.M. Behan-Pelletier, D.E. Bignell et al., Biodiversity and ecosystem functioning in soil. Ambio **26**, 563–570 (1997)

L. Brussaard, P.C. De-Ruiter, G.G. Brown, Soil biodiversity for agricultural sustainability. Agric. Ecosyst. Environ. **121**, 233–244 (2007)

A. Buczynska, I. Szadkowska-Stanczyk, Identification of health hazards to rural population living near pesticide dump sites in Poland. Intl. J. Occup. Med. Environ. Health **18**(4), 331–339 (2005)

E.K. Bunemann, G.D. Schwenke, L. Van Zwieten, Impact of agricultural inputs on soil organisms – a review. Aust. J. Soil Res. **44**, 379–406 (2006)

G.I. Burd, G. Dixon, B.R. Glick, A plant growth promoting bacterium that decreases nickel toxicity in seedlings. Appl. Environ. Microbiol. **64**(10), 3663–3668 (1998)

G.I. Burd, G. Dixon, B.R. Glick, Plant growth promoting bacterium that decreases heavy metal toxicity in plants. Can. J. Microbiol. **46**, 237–245 (2000)

G. Carrillo-Castaneda, J.J. Munoz, J.R. Peralta-Videa, E. Gomez, J.L. Gardea-Torresdey, Modulation of uptake and translocation of iron and copper from root to shoot in common bean by siderophore-producing microorganisms. J. Plant Nutr. **28**, 1853–1865 (2005)

E.M. Casanovas, C.A. Barassi, R.J. Sueldo, Azospirillum inoculation mitigates water stress effects in maize seedlings. Cereal Res. Commun. **30**, 343–350 (2002)

L.E. Casida, D.A. Klein, T. Santoro, Soil dehydrogenase activity. Soil Sci. **98**, 371–376 (1964)

V.A. Cavalcante, J. Dobereiner, A new acid-tolerant nitrogen-fixing bacterium associated with sugarcane. Plant Soil **108**(1), 23–31 (1988)

J. Cernohlavkova, Effects of selected environmental pollutants on soil microbial community in laboratory and field studies (2009). Retrieved from http://is.muni.cz/th/40696/prif_d/dis.prace_cernohlavkova.pdf

M. Chagnon, D. Kreutzweiser, E.A.D. Mitchell, C.A. Morrissey, D.A. Noome, J.P. Van-der-Sluijs, Risks of large-scale use of systemic insecticides to ecosystem functioning and services. Environ. Sci. Pollut. Res. **22**, 119–134 (2015)

U. Chakraborty, B. Chakraborty, M. Basnet, Plant growth promotion and induction of resistance in *Camellia sinensis* by *Bacillus megaterium*. J. Basic Microbiol. **46**, 186–195 (2006)

J.F. Charles, C. Nielson-LeRoux, A. Delecluse, *Bacillus sphaericus* toxins: molecular biology and mode of action. Annu. Rev. Entomol. **41**, 451–472 (1996)

F.A. Chinalia, K.S. Killham, 2,4-Dichlorophenoxyacetic acid (2,4-D) biodegradation in river sediments of Northeast-Scotland and its effect on the microbial communities (PLFA and DGGE). Chemosphere **64**, 1675–1683 (2006)

A. Chowdhury, S. Pradhan, M. Saha, N. Sanyal, Impact of pesticides on soil microbiological parameters and possible bioremediation strategies. Indian J. Microbiol. **48**, 114–127 (2008)

R. Christen, Global sequencing: a review of current molecular data and new methods available to assess microbial diversity. Microbes Environ. **23**, 253–268 (2008)

X. Chu, H. Fang, X. Pan, X. Wang, M. Shan, B. Feng, Y. Yu, Degradation of chlorpyrifos alone and in combination with chlorothalonil and their effects on soil microbial populations. J. Environ. Sci. **20**, 464–469 (2008)

T.E. Cloete, R.M. Atlas, *Basic and Applied Microbiology* (Van Schalk Publishers, Hatfield, 2006). 544 p

S. Compant, B. Duffy, J. Nowak, C. Clement, E.A. Barka, Use of plant growth-promoting bacteria for biocontrol of plant diseases: principles, mechanisms of action and future prospects. Appl. Environ. Microbiol. **71**, 4951–4959 (2005)

U. Conrath, C.M.J. Pieterse, B. Mauch-Mani, Priming in plant–pathogen interactions. Trends Plant Sci. **7**, 210–216 (2002)

M. Contin, A. Todd, P.C. Brookes, The ATP concentration of the soil microbial biomass. Soil Biol. Biochem. **33**, 701–704 (2001)

C.M. Creus, R.J. Sueldo, C.A. Barassi, Water relations and yield in Azospirillum-inoculated wheat exposed to drought in the field. Can. J. Bot. **82**, 273–281 (2004)

C.M. Creus, M. Graziano, E.M. Casanovas, M.A. Pereyra, M. Simontacchi, S. Puntarulo, C.A. Barassi, L. Lamattina, Nitric oxide is involved in the *Azospirillum brasilense*-induced lateral root formation in tomato. Planta **221**, 297–303 (2005)

O. Crouzet, I. Batisson, P. Besse-Hoggan, F. Bonnemoy, C. Bardot, F. Poly, J. Bohatier, C. Mallet, Response of soil microbial communities to the herbicide mesotrione: a dose-effect microcosm approach. Soil Biol. Biochem. **42**, 193–202 (2010)

A. Cruz-Hernandez, A. Tomasini-Campocosio, L. Perez-Flores, F. Fernandez-Perrino, M. Gutierrez-Rojas, Inoculation of seed-borne fungus in the rhizosphere of *Festuca arundinacea* promotes hydrocarbon removal and pyrene accumulation in roots. Plant Soil **362**, 261–270 (2013)

M. Cycon, Z. Piotrowska-Seget, Changes in bacterial diversity and community structure following pesticides addition to soil estimated by cultivation technique. Ecotoxicology **18**, 632–642 (2009)

M. Cycon, M. Wojcik, S. Borymski, Z. Piotrowska-Seget, Short-term effects of the herbicide napropamide on the activity and structure of the soil microbial community assessed by the multi-approach analysis. Appl. Soil Ecol. **66**, 8–18 (2013)

A.C. Das, D. Mukherjee, Insecticidal effects on soil microorganisms and their biochemical processes related to soil fertility. World J. Microbiol. Biotechnol. **14**(6), 903–909 (1998)

A.C. Das, A. Chakrabarty, P. Sukul, D. Mukherjee, Insecticides: their effect on microorganisms and persistence in rice soil. Microbiol. Res. **50**, 187–194 (1995)

E.J. DaSilva, L.E. Henriksson, E. Henriksson, Effect of pesticides on blue-green algae and nitrogen-fixation. Arch. Environ. Contam. Toxicol. **3**(2), 193–204 (1975)

C. Davidson, R.A. Knapp, Multiple stressors and amphibian declines: dual impacts of pesticides and fish on yellow-legged frogs. Ecol. Appl. **17**, 587–597 (2007)

M.E. DeLorenzo, G.I. Scott, P.E. Ross, Effects of the agricultural pesticides atrazine, deethylatrazine, endosulfan and chlorpyrifos on an estuarine microbial food web. Environ. Toxicol. Chem. **18**(12), 2824–2835 (1999)

L.A. Demoling, E. Baath, G. Greve, M. Wouterse, H. Schmitt, Effects of sulfamethoxazole on soil microbial communities after adding substrate. Soil Biol. Biochem. **41**, 840–848 (2009)

P.C. De-Ruiter, A.M. Neutel, J.C. Moore, Energetics, patterns of interaction strengths and stability in real ecosystems. Science **269**, 1257–1260 (1995)

M. De-Schampheleire, P. Spanoghe, E. Brusselman, S. Sonck, Risk assessment of pesticide spray drift damage in Belgium. Crop Prot. **26**(4), 602–611 (2007)

P. De-Werra, M. Pechy-Tarr, C. Keel, M. Maurhofer, Role of gluconic acid production in the regulation of biocontrol traits of *Pseudomonas fluorescens* CHA0. Appl. Environ. Microbiol. **75**, 4162–4174 (2009)

M. Dhagat, K. Verma, Effect of herbicides on soil microorganisms. Curr. Adv. Agric. Sci. **1**, 54–55 (2009)

M. Digrak, F. Kazanici, Effect of some organophosphorus insecticides on soil microorganisms. Turk. J. Biol. **25**, 51–58 (2001)

C. Dimkpa, A. Svatos, D. Merten, G. Buchel, E. Kothe, Hydroxamate siderophores produced by *Streptomyces acidiscabies* E13 bind nickel and promote growth in cowpea (*Vigna unguiculata* L.) under nickel stress. Can. J. Microbiol. **54**, 163–172 (2008)

C. Dimkpa, T. Weinand, F. Asch, Plant-rhizobacteria interactions alleviate abiotic stress conditions. Plant Cell Environ. **32**(12), 1682–1694 (2009a)

C. Dimkpa, D. Merten, A. Svatos, G. Buchel, E. Kothe, Siderophores mediate reduced and increased uptake of cadmium by *Streptomyces tendae* F4 and sunflower (*Helianthus annuus*), respectively. J. Appl. Microbiol. **107**(5), 1687–1696 (2009b)

S. Dobbelaere, Y. Okon, The plant growth promoting effects and plant responses, in *Nitrogen fixation: Origins, applications and research progress. Newton W. (ed.), Vol V: Associative and endophytic nitrogen-fixing bacteria and cyanobacterial associations*, ed. by C. Elmerich, W.E. Newton (Springer, Helderberg, 2007), pp. 145–170

S. Dobbelaere, J. Vanderleyden, Y. Okon, Plant growth promoting effects of diazotrophs in the rhizosphere. Crit. Rev. Plant Sci. **22**, 107–149 (2003)

J. Dobereiner, Nitrogen-fixing bacteria of the genus *Beijerinckia* Derx in the rhizosphere of sugar cane. Plant Soil **15**(3), 211–216 (1961)

E.J. Dominati, M.G. Patterson, A.D. Mackay, A framework for classifying and quantifying the natural capital and ecosystem services of soils. Ecol. Econ. **69**, 1858–1868 (2010)

S. Dordevic, M. Sestovic, V. Raicevic, A. Dordevic, Fluctuation of the abundance of microorganisms in the carbofuran treated soil. Pest **13**(4), 281–288 (1998)

J. Du, Z. Shao, H. Zhao, Engineering microbial factories for synthesis of value-added products. J. Indus. Microbiol. Biotechnol. **38**, 873–890 (2011)

S. Duah-Yentumi, D.B. Johnson, Changes in soil microflora in response to repeated applications of some pesticides. Soil Biol. Biochem. **18**, 629–635 (1986)

M.G. Dumont, J.C. Murrell, Stable isotope probing – linking microbial identity to function. Nat. Rev. Microbiol. **3**(6), 499–504 (2005)

C.A. Edwards, Insecticide residues in soils. Rev. Environ. Contam. Toxicol. **13**, 83–132 (1966)

D. Egamberdiyeva, The effect of plant growth promoting bacteria on growth and nutrient uptake of maize in two different soils. Appl. Soil Ecol. **36**, 184–189 (2007)

R.U. Ehlers, Einsatz der Biotechnologie im biologischen Pflanzenschutz. Schr. reihe Dtsch. Phytomed Ges. **8**, 17–31 (2006)

S. El-Fantroussi, L. Verschuere, W. Verstraete, E.M. Top, Effect of phenylurea herbicides on soil microbial communities estimated by analysis of 16S rRNA gene fingerprints and community-level physiological profiles. Appl. Environ. Microbiol. **65**, 982–988 (1999)

T. Endo, T. Kusaka, N. Tan, M. Sakai, Effects of the insecticide cartap hydrochloride on soil enzyme activities, respiration and nitrification. J. Pestic. Sci. **7**, 101–110 (1982)

J.A. Entry, P.K. Donelly, W.H. Emmingham, Microbial mineralization of atrazine and 2, 4-dichlorophenoxyacetic acid in riparian pasture and forest soils. Biol. Fert. Soils **18**, 89–94 (1994)

P.F. Entwistle, J.S. Cory, M.J. Bailey, S. Higgs, *Bacillus thuringiensis: An Environmental Biopesticide – Theory and Practice* (Wiley, Chichester, 1993). 311 p

L. Epelde, J.M. Becerril, J. Hernandez-Allica, O. Barrutia, C. Garbisu, Functional diversity as indicator of the recovery of soil health derived from *Thlaspi caerulescens* growth and metal phytoextraction. Appl. Soil Ecol. **39**, 299–310 (2008)

O.W. Fawole, M. Aluko, T.E. Olowonihi, Effects of carbendazim-mancozeb fungicidal mixture on soil microbial populations and some enzyme activities in soil. Agrosearch **10**(1&2), 65–74 (2009)

K. Feng, B.Y. Yu, D.M. Ge, M.H. Wong, X.C. Wang, Z.H. Cao, Organo-chlorine pesticide (DDT and HCH) residues in the Taihu Lake region and its movement in soil–water system: I. Field survey of DDT and HCH residues in ecosystem of the region. Chemosphere **50**(6), 683–687 (2003)

K.P. Feris, K. Hristova, B. Gebreyesus, D. Mackay, K.M. Scow, A shallow BTEX and MTBE contaminated aquifer supports a diverse microbial community. Microb. Ecol. **48**, 589–600 (2004)

E.P.D. Ferreira, A.N. Dusi, J.R. Costa, G.R. Xavier, N.G. Rumjanek, Assessing insecticide and fungicide effects on the culturable soil bacterial community by analyses of variance of their DGGE fingerprinting data. Eur. J. Soil Biol. **45**, 466–472 (2009)

N. Fierer, M. Breitbart, J. Nulton et al., Metagenomic and small subunit rRNA analyses reveal the genetic diversity of bactera, archaea, fungi and viruses in soil. Appl. Environ. Microbiol. **73**, 7059–7066 (2007)

V.B. Figueiredo, H.A. Buritya, C.R. Martinez, C.P. Chanway, Alleviation of drought stress in the common bean (*Phaseolus vulgaris* L.) by co-inoculation with *Paenibacillus polymyxa* and *Rhizobium tropici*. Appl. Soil Ecol. **40**, 182–188 (2008)

B.J. Finlay, G.F. Esteban, Ubiquitous microbes and ecosystem function. Limnetica **20**(1), 31–43 (2001)

C. Folch, A.C. Chevremont, K. Joarico, Y. Capowiez, S. Criquet, Indicators of pesticide contamination: soil enzyme compared to functional diversity of bacterial communities via Biolog® ecoplates. Eur. J. Soil Biol. **47**, 256–263 (2011)

W.T. Frankenberger, M.A. Tabatabai, Amidase activity in soils. IV. Effects of trace elements and pesticides. Soil Sci. Soc. Am. J. **44**, 532–536 (1980a)

W.T. Frankenberger, M.A. Tabatabai, Amidase activity in soils: I. Methods of assay. Soil Sci. Soc. Am. J. **44**, 282–287 (1980b)

W.T. Frankenberger, M.A. Tabatabai, Factors affecting L-asparaginase activity in soil. Biol. Fert. Soils **11**, 1–5 (1991)

R.B. Franklin, D.R. Taylor, A.L. Mills, Characterization of microbial communities using randomly amplified polymorphism (RAPD). J. Microbiol. Meth. **35**, 225–235 (1999)

A. Frostegard, E. Baath, The use of phospholipid fatty acid analysis to estimate bacterial and fungal biomass in soil. Biol. Fert. Soils **22**(1&2), 59–65 (1996)

A. Frostegard, A. Tunlid, E. Baath, Microbial biomass measured as total lipid phosphate in soils of different organic content. J. Microbiol. Meth. **14**, 151–163 (1991)

A. Frostegard, A. Tunlid, E. Baath, Phospholipid fatty acid composition, biomass and activity of microbial communities from two soil types experimentally exposed to different heavy metals. Appl. Environ. Microbiol. **59**(11), 3605–3617 (1993)

G.M. Gadd, Metals, minerals and microbes: geomicrobiology and bioremediation. Microbiology **156**, 609–643 (2010)

D.J. Gage, Infection and invasion of roots by symbiotic, nitrogen fixing rhizobia during nodulation of temperate legumes. Microbiol. Mol. Biol. Rev. **68**(2), 280–300 (2004)

S. Gaind, L. Nain, Exploration of composted cereal waste and poultry manure for soil restoration. Biores. Technol. **101**, 2996–3003 (2010)

S. Gaind, L. Nain, V.B. Patel, Quality evaluation of co-composted wheat straw, poultry droppings and oil seed cakes. Biodegradation **20**, 307–317 (2009)

J.L. Garland, A.L. Mills, Classification and characterization of heterotrophic microbial communities on the basis of patterns of community level sole carbon source utilization. Appl. Environ. Microbiol. **57**, 2351–2359 (1991)

M.A. German, S. Burdman, Y. Okon, J. Kigel, Effects of *Azospirillum brasilense* on root morphology of common bean (*Phaseolus vulgaris* L.) under different water regimes. Biol. Fertil. Soils **32**, 259–264 (2000)

B. Gevao, K.T. Semple, K.C. Jones, Bound pesticide residues in soils: a review. Environ. Pollut. **108**, 3–14 (2000)

A. Ghani, D.A. Wardle, A. Rahman, D.R. Lauren, Interactions between 14C labeled atrazine and the soil microbial biomass in relation to herbicide degradation. Biol. Fertil. Soils **21**, 17–22 (1996)

C. Giaveno, L. Celi, A.E. Richardson, R.J. Simpson, E. Barberis, Interaction of phytases with minerals and availability of substrate affect the hydrolysis of inositol phosphates. Soil Biol. Biochem. **42**, 491–498 (2010)

S.M. Gifford, S. Sharma, J.M. Rinta-Kanto, M.A. Moran, Quantitative analysis of a deeply sequenced marine microbial metatranscriptome. ISME J. **5**(3), 461–472 (2011)

F. Gil-Sotres, C. Trasar-Cepeda, M.C. Leiros, S. Seoane, Different approaches in evaluating soil quality using biochemical properties. Soil Biol. Biochem. **37**, 877–887 (2005)

B.R. Glick, C. Liu, S. Ghosh, E.B. Dumbroff, Early development of canola seedlings in the presence of the plant growth-promoting rhizobacterium, *Pseudomonas putida* GR12-2. Soil Biol. Biochem. **29**(8), 1233–1239 (1997)

E. Gomez, L. Ferreras, L. Lovotti, E. Fernandez, Impact of glyphosate application on microbial biomass and metabolic activity in a Vertic Argiudoll from Argentina. Eur. J. Soil Biol. **45**, 163–167 (2009)

M.R. Goswami, U.K. Pati, A. Chowdhury, A. Mukhopadhyay, Studies on the effect of cypermethrin on soil microbial biomass and its activity in an alluvial soil. Int. J. Agric. Food Sci. **3**(1), 1–9 (2013)

D. Goulson, An overview of the environmental risks posed by neonicotinoid insecticides. J. Appl. Ecol. **50**, 977–987 (2013)

E.J. Gray, D.L. Smith, Intracellular and extracellular PGPR: commonalities and distinctions in the plant–bacterium signaling processes. Soil Biol. Biochem. **37**, 395–412 (2005)

V.S. Green, D.E. Stott, M. Diack, Assay for fluorescein diacetate hydrolytic activity: optimization for soil samples. Soil Biol. Biochem. **38**, 693–701 (2006)

P. Grenni, Effects of pesticides and pharmaceuticals on soil and water bacterial communities. Ph.D thesis, University of Milano-Bicocca (2011), Retrieved from https://boa.unimib.it/retrieve/handle/10281/19697/24499/Phd_unimib_716451.pdf

V.P. Grichko, B.R. Glick, Amelioration of flooding stress by ACC deaminase-containing plant growth-promoting bacteria. Plant Physiol. Biochem. **39**, 11–17 (2001)

B.S. Griffiths, S. Caul, J. Thompson, C.A. Hacket, J. Cortet, C. Pernin, P.H. Krogh, Soil microbial and faunal responses to herbicide-tolerant maize and herbicide in two soils. Plant Soil **308**, 93–103 (2008)

L. Gu, Z. Bai, B. Jin et al., Assessing the impact of fungicide enostroburin application on bacterial community in wheat phyllosphere. J. Environ. Sci. **22**(1), 134–141 (2010)

S.Y. Guan, D.S. Zhang, Z.M. Zhang, *Soil Enzymes and Its Methodology* (Agricultural Press, Beijing, 1986). 340 p

A.A. Guimaraes, P.M.D. Jaramillo, R.S.A. Nobrega, L.A. Florentino, K.B. Silva, M.F.M. De-Souza, Genetic and symbiotic diversity of nitrogen-fixing bacteria isolated from agricultural soils in the western Amazon by using cowpea as the trap plant. Appl. Environ. Microbiol. **78**, 6726–6733 (2012)

J. Guo, L. Yu, N.J. Turro, J.Y. Ju, An integrated system for DNA sequencing by synthesis using novel nucleotide analogues. Acc. Chem. Res. **43**, 551–563 (2010)

D. Haas, G. Defago, Biological control of soil-borne pathogens by fluorescent pseudomonads. Nat. Rev. Microbiol. **3**, 307–319 (2005)

J.A. Hall, D. Peirson, S. Ghosh, B.R. Glick, Root elongation in various agronomic crops by the plant growth promoting rhizobacterium *Pseudomonas putida* GR12–2. Israel J. Plant Sci. **44**(1), 37–42 (1996)

B. Hamaoui, J.M. Abbadi, S. Burdman, A. Rashid, S. Sarig, Y. Okon, Effects of inoculation with *Azospirillum brasilense* on chickpeas (*Cicer arietinum*) and faba beans (*Vicia faba*) under different growth conditions. Agronomie **21**, 553–560 (2001)

A.B.E. Hamdia, M.A.K. Shaddad, M.M. Doaa, Mechanisms of salt tolerance and interactive effects of *Azospirillum brasilense* inoculation on maize cultivars grown under salt stress conditions. Plant Growth Reg. **44**, 165–174 (2004)

H.S. Han, K.D. Lee, Phosphate and potassium solubilizing bacteria effect on mineral uptake, soil availability and growth of egg plant. Res. J. Agric. Biol. Sci. **1**, 176–180 (2005)

S.K. Handa, N.P. Agnihotri, G. Kulshrestha, *Effect of Pesticides on Soil Fertility, Pesticide Residues: Significance, Management and Analysis* (Research Periodicals and Book Publishing House, Houston, 1999), pp. 184–198

R.L. Haney, S.A. Senseman, F.M. Hons, Effect of roundup ultra on microbial activity and biomass from selected soils. J. Environ. Qual. **31**, 730–735 (2002)

E.S.R.Y. Haque, A. Gaffar, Use of rhizobia in the control of root rot diseases of sunflower, okra, soybean and mung-bean. J. Phytopathol. **138**, 157–163 (1993)

P.R. Hardoim, L.S. Van-Overbeek, J.D. Van-Elsas, Properties of bacterial endophytes and their proposed role in plant growth. Trends. Microbiol. **16**, 463–471 (2008)

R.W. Hardy, R.D. Holsten, E.K. Jackson, R.C. Burns, The acetylene–ethylene assay for N_2 fixation: laboratory and field evaluation. Plant Physiol. **43**, 1185–1207 (1968)

M.M. Hart, J.T. Trevors, Microbe management: application of mycorrhyzal fungi in sustainable agriculture. Front. Ecol. Environ. **3**, 533–539 (2005)

S. Hati, S. Mandal, J.B. Prajapati, Novel starters for value added fermented dairy products. Curr. Res. Nutr. Food Sci. **1**(1), 83–91 (2013)

A.D. Hazel, M.P. Greaves, Effects of some herbicides on soil enzyme activities. Weed Res. **21**, 205–209 (1981)

Y.H. He, D.S. Shen, C.R. Fang, R. He, Y.M. Zhu, Effect of metsulfuronmethyl on the microbial population and enzyme activities in wheat rhizosphere soil. J. Environ. Sci. Health B **41**, 269–284 (2006)

D.M. Hegde, B.S. Dwived, S.N. Sudhakara, Biofertilizers for cereal production in India-a review. Indian J. Agric. Sci. **69**, 73–83 (1999)

P. Helmke, D. Sparks, Lithium, sodium, potassium, rubidium, and cesium, in *Methods of Soil Analysis, Part 3. Chemical Properties*, ed. by D. Sparks, 1st edn. (SSSA, Madison, 1996), pp. 551–574

C.L. Hemme, Y. Deng, T.J. Gentry et al., Metagenomic insights into evolution of a heavy metal-contaminated groundwater microbial community. ISME J. **4**, 660–672 (2010)

T. Higa, J.F. Parr, Beneficial and effective microorganisms: for a sustainable agriculture and environment (1994). Retrieved from www.agriton.nl/higa.html

C.K. Hindumathy, V. Gayathri, Effect of pesticide (chlorpyrifos) on soil microbial flora and pesticide degradation by strains isolated from contaminated soil. J. Bioremed. Biodeg. **4**, 178 (2013)

S.S. Hirano, C.D. Upper, Bacteria in the leaf ecosystem with emphasis on *Pseudomonas syringae*-a pathogen, ice nucleus and epiphyte. Microbiol. Mol. Biol. Rev. **64**, 624–653 (2000)

P.R. Hirsch, Detection of microbial DNA sequences by colony hybridization, in *Molecular Microbial Ecology Manual*, ed. by A.D.L. Akkermans, J.D. Van-Elsas, F.J. De-Bruin (Kluwer Academic Publishers, Dordrecht, 1995), pp. 1–12

D. Hoffmeister, N.P. Keller, Natural products of filamentous fungi: enzymes, genes and their regulation. Nat. Prod. Rep. **24**, 393–416 (2007)

R. Hood-Nowotny, U.N. Hinko-Najera, E. Inselsbacher, P. Oswald-Lachouani, W. Wanek, Alternative methods for measuring inorganic, organic and total dissolved nitrogen in soil. Soil Sci. Soc. Am. J. **74**, 1018–1027 (2010)

H. Hoper, Substrate-induced respiration, in *Microbiological Methods for Assessing Soil Quality*, ed. by J. Bloem, D.H. Hopkins, A. Benedetti (CABI Publishing, Wallingford, 2006), pp. 84–92

W.R. Horwath, E.A. Paul, Microbial biomass, in *Methods of Soil Analysis. Part 2. Microbiological and Biochemical Properties*, ed. by R.W. Weaver (Soil Science Society of America, Madison, 1994), pp. 760–761

C.C. Howell, How do pesticides impact soil microbial structure and functioning? Ph.D thesis, University of Warwick, UK (2011). Retrieved from http://go.warwick.ac.uk/wrap/46766

M.B. Hussain, I. Mehboob, Z.A. Zahir, M. Naveed, H.N. Asghar, Potential of Rhizobium spp. for improving growth and yield of rice (*Oryza sativa* L.). Soil Environ. **28**(1), 49–55 (2009)

A.M. Ibekwe, A.C. Kennedy, Phospholipid fatty acid profiles and carbon utilization patterns for analysis of microbial community structure under field and greenhouse conditions. FEMS Microbiol. Ecol. **26**, 151–163 (1998)

A.M. Ibekwe, S.K. Papiemk, J. Gan, S.R. Yates, C.H. Yang, D.E. Crowley, Impact of fumigants on soil microbial communities. Appl. Environ. Microbiol. **67**, 3245–3257 (2001)

G. Imfeld, S. Vuilleumier, Measuring the effects of pesticides on bacterial communities in soil: a critical review. Eur. J. Soil Biol. **49**, 22–30 (2012)

B.S. Ismail, K.F. Yapp, O. Omar, Effects of metsulfuron-methyl on amylaze, urease and protease-activities in two soils. Aust. J. Soil Res. **36**(3), 449–456 (1996)

ISO, 16072, Soil quality – laboratory methods for determination of microbial soil respiration (2002). Retrieved from www.iso.org/iso/home/store/catalogue_tc/catalogue_detail.htm?csnumber=32096

ISO, 17155, Soil quality – determination of abundance and activity soil microflora using respiration curves (2012). Retrieved from www.iso.org/iso/home/store/catalogue_tc/catalogue_detail.htm?csnumber=53529.

M. Jail, R. Urkude, L. Deshmukh, Studies on the impact of pesticides on soil microbial respiration. Int. J. Plant Anim. Environ. Sci. **5**(1), 43–49 (2015)

M. Jamalizadeh, H.R. Etebarian, H. Aminian et al., Biological control of Botrytis mali on apple fruit by use of Bacillus bacteria, isolated from the rhizosphere of wheat. Arch. Phytopathol. Plant Prot. **43**, 1836–1845 (2010)

E.K. James, V.M. Reis, F.L. Olivares, J.I. Baldani, J. Dobereiner, Infection of sugar cane by the nitrogen-fixing bacterium, *Acetobacter diazotrophicus*. J. Exp. Bot. **45**(6), 757–766 (1994)

H. Janshekar, A. Fiechter, Lignin: biosynthesis, application and degradation. Adv. Biochem. Eng. Biotechnol. **27**, 119–178 (1983)

E. Jastrzebska, The effect of soil contamination with fungicides on microorganisms counts. Pol. J. Nat. Sci. **21**(2), 487–498 (2006)

E. Jastrzebska, The effect of crop protection chemicals on soil-dwelling microorganisms, in *Contemporary Problems Mmanagement and Environmental Protection – Influence of Pesticide Dump on the Environment*, ed. by K.A. Skibniewska (University of Warmia and Mazury in Olsztyn, Olsztyn, 2010). Retrieved from www.uwm.edu.pl/environ/vol05.htm

D.S. Jenkinson, S.A. Davidson, D.S. Powlson, Adenosine triphosphate and microbial biomass in soil. Soil Biol. Biochem. **8**, 521–527 (1979)

R.G. Joergensen, P.C. Brookes, Ninhydrin-reactive nitrogen measurements of microbial biomass in 0.5 M K_2SO_4 soil extracts. Soil Biol. Biochem. **22**(8), 1023–1027 (1990)

M. Johansson, M. Pell, J. Stenstrom, Kinetics of substrate-induced respiration (SIR) and denitrification: applications to a soil amended with silver. Ambio **27**(1), 40–44 (1998)

K. Johnsen, C.S. Jacobsen, V. Torsvik, J. Sorensen, Pesticide effects on bacterial diversity in agricultural soils – a review. Biol. Fertil. Soils **33**, 443–453 (2001)

W.J. Jones, N.D. Ananyeva, Correlations between pesticide transformation rate and microbial respiration activity in soil of different ecosystems. Biol. Fertil. Soils **33**, 477–483 (2001)

G.J. Joo, Y.M. Kim, J.T. Kim, I.K. Rhee, J.H. Kim, I.J. Lee, Gibberellins-producing rhizobacteria increase endogenous gibberellins content and promote growth of red peppers. J. Microbiol. **43**, 510–515 (2005)

M. Kai, U. Effmert, G. Berg, B. Piechulla, Volatiles of bacterial antagonists inhibit mycelial growth of the plant pathogen, *Rhizoctonia solani*. Arch. Microbiol. **187**, 351–360 (2007)

M. Kai, M. Haustein, F. Molina, A. Petri, B. Scholz, B. Piechulla, Bacterial volatiles and their action potential. Appl. Microbiol. Biotechnol. **81**, 1001–1012 (2009)

A. Kawasaki, E.R. Watson, M.A. Kertesz, Indirect effects of polycyclic aromatic hydrocarbon contamination on microbial communities in legume and grass rhizospheres. Plant Soil **358**, 169–182 (2012)

A.G. Khan, Role of soil microbes in the rhizospheres of plants growing on trace metal contaminated soils in phytoremediation. J. Trace Elem. Med. Biol. **18**, 355–364 (2005)

J.W. Kloepper, R.M. Zablotowicz, E.M. Tipping, R. Lifshitz, Plant growth promotion mediated by bacterial rhizosphere colonizers, in *The Rhizosphere and Plant Growth*, ed. by D.L. Keister, P.B. Cregan (Kluwer Academic Publishers, Dordrecht, 1991), pp. 315–326

J.W. Kloepper, R. Rodriguez-Kabana, G.W. Zehnder, J.F. Murphy, E. Sikora, C. Fernandez, Plant root-bacterial interactions in biological control of soilborne diseases and potential extension to systemic and foliar diseases. Aust. Plant Pathol. **28**, 21–26 (1999)

J.W. Kloepper, C. Ryu, S. Zhang, Induced systemic resistance and promotion of plant growth by *Bacillus* spp. Phytopathology **94**(11), 1259–1266 (2004)

J.W. Kloepper, A. Gutierrez-Estrada, J.A. McInroy, Photoperiod regulates elicitation of growth promotion but not induced resistance by plant growth-promoting rhizobacteria. Can. J. Microbiol. **53**, 159–167 (2007)

S. Klose, V. Acosta-Martinez, H.A. Ajwa, Microbial community composition and enzyme activities in a sandy loam soil after fumigation with methyl bromide or alternative biocides. Soil Biol. Biochem. **38**, 1243–1254 (2006)

D. Knudsen, G. Peterson, P. Pratt, Lithium, sodium and potassium, in *Methods of Soil Analysis Part 2. Chemical and Microbiological Properties*, ed. by A.L. Page (ASA and SSSA, Madison, 1982), pp. 225–246

R.J. Kremer, N.E. Means, S.J. Kim, Glyphosate affects soybean root exudation and rhizosphere microorganisms. Int. J. Environ. Anal. Chem. **85**, 1165–1174 (2005)

D.P. Kreutzweiser, K.P. Good, D.T. Chartrand, T.A. Scarr, D.G. Thompson, Toxicity of the systemic insecticide, imidacloprid to forest stream insects and microbial communities. Bull. Environ. Contam. Toxicol. **80**(3), 211–214 (2008)

J. Kucharski, J. Wyszkowska, A. Borowik, Microbiological properties of soil contaminated with the herbicide Granstar 75 WG. Ekol. Tech. **16**(4), 160–164 (2008)

I. Kuiper, E.L. Lagendijk, G.V. Bloemberg, B.J.J. Lugtenberg, Rhizoremediation: a beneficial plant–microbe interaction. Mol. Plant Microbe Interact. **17**, 6–15 (2004)

J.N. Ladd, J.H.A. Butler, Short-term assays of soil proteolytic enzyme activities using proteins and dipeptide derivatives as substrates. Soil Biol. Biochem. **4**, 19–30 (1972)

S.H. Lancaster, E.B. Hollister, S.A. Senseman, T.J. Gentry, Effects of repeated glyphosate applications on soil microbial community composition and the mineralization of glyphosate. Pest Manag. Sci. **66**(1), 59–64 (2010)

M.A. Latif, M.A. Razzaque, M.M. Rahman, Impact of some selected insecticides application on soil microbial respiration. Pak. J. Biol. Sci. **11**, 2018–2022 (2008)

A.M. Lenart, In vitro effects of various xenobiotics on *Azotobacter chroococcum* strains isolated from soils of southern Poland. J. Environ. Sci. Health B **47**, 7–12 (2012)

C.A. Levesque, J.E. Rahe, D.M. Eaves, Fungal colonization of glyphosate treated seedlings using a new root plating technique. Mycol. Res. **97**, 299–306 (1993)

M. Liebeke, V.S. Brozel, M. Hecker, M. Lalk, Chemical characterization of soil extract as growth media for the ecophysiological study of bacteria. Appl. Microbiol. Biotechnol. **83**, 161–173 (2009)

S.E. Lindow, M.T. Brand, Microbiology of the phyllosphere. Appl. Environ. Microbiol. **69**(4), 1875–1883 (2003)

C. Lo, Effect of pesticides on soil microbial community. J. Environ. Sci. Health B **45**, 348–359 (2010)

E.M. Lodwig, A.H. Hosie, A. Bourdes, K. Findlay, D. Allaway, R. Karunakaran, J.A. Downie, P.S. Poole, Amino-acid cycling drives nitrogen fixation in the legume- rhizobium symbiosis. Nature **422**, 722–726 (2003)

A.H. Lone, K.P. Raverkar, N. Pareek, In vitro effects of herbicides on soil microbial communities. Bioscan **9**(1), 11–16 (2014)

N. Lorenz, K. Verdell, C. Ramsier, R. Dick, A rapid assay to estimate soil microbial biomass potassium in agricultural soils. Soil Sci. Soc. Am. J. **74**, 512–516 (2010)

T. Lueders, B. Pommerenke, M.W. Friedrich, Stable-isotope probing of microorganisms thriving at thermodynamic limits: syntrophic propionate oxidation in flooded soil. Appl. Environ. Microbiol. **70**(10), 5778–5786 (2004)

B. Lundgren, Fluorescein diacetate as a stain of metabolically active bacteria in soil. Oikos **36**, 17–22 (1981)

N.Z. Lupwayi, K.N. Harker, L.M. Dosdall et al., Changes in functional structure of soil bacterial community due to fungicide and insecticide applications in canola. Agric. Ecosyst. Environ. **130**, 109–114 (2009a)

N.Z. Lupwayi, K.N. Harker, G.W. Clayton, J.T. O'Donovan, R.E. Blackshaw, Soil microbial response to herbicides applied to glyphosate-resistant canola. Agric. Ecosyst. Environ. **129**, 171–176 (2009b)

N.Z. Lupwayi, S.A. Brandt, K.N. Harker, J.T. O'Donovan, G.W. Clayton, T.K. Turkington, Contrasting soil microbial responses to fertilizers and herbicides in a canola-barley rotation. Soil Biol. Biochem. **42**, 1997–2004 (2010)

G.A. Lyons, G.J. McKay, H.S.S. Sharma, Molecular comparison of *Scytalidium thermophilum* isolates using RAPD and ITS nucleotide sequence analyses. Mycol. Res. **104**(12), 1431–1438 (2000)

M. Lane, The effect of glyphosate on soil microbial communities (2011). Retrieved from https://etd.ohiolink.edu/!etd.send_file?accession=osu1301068421&disposition=inline

S. Mahmood, R.D. Finlay, S. Erland, H. Wallander, Solubilisation and colonisation of wood ash by ectomycorrhizal fungi isolated from a wood ash fertilized spruce forest. FEMS Microbiol. Ecol. **35**, 151–161 (2001)

K. Makarova, A. Slesarev, Y. Wolf et al., Comparative genomics of the lactic acid bacteria. Proc. Natl. Acad. Sci. U. S. A. **103**(42), 15611–15616 (2006)

S. Mantelin, B. Touraine, Plant growth-promoting bacteria and nitrate availability impacts on root development and nitrate uptake. J. Exp. Bot. **55**, 27–34 (2004)

P.A. Maron, L. Ranjard, C. Mougel, P. Lemanceau, Metaproteomics: a new approach for studying functional microbial ecology. Microb. Ecol. **53**, 486–493 (2007)

J.P. Martin, R.B. Harding, G.H. Connell, L.D. Anderson, Soil Sci. **87**, 334–338 (1959)

T.M.V. Martinez, V. Salmeron, J. Gonzalez-Lopez, Effects of an organophosphorus insecticide, profenofos on agricultural soil microflora. Chemosphere **24**(1), 71–80 (1992)

M.V. Martinez-Toledo, V. Salmeron, B. Rodelas, C. Pozo, J. Gonzalez-Lopez, Effects of the fungicide captan on some functional groups of soil microflora. Appl. Soil Ecol. **7**, 245–255 (1998)

S. Mayak, T. Tirosh, B.R. Glick, Plant growth-promoting bacteria confer resistance in tomato plants to salt stress. Plant Physiol. Biochem. **42**, 565–572 (2004)

M.J. Mazzarino, L. Szott, M. Jiminez, Dynamics of soil total C and N, microbial biomass and water-soluble C in tropical agroecosystems. Soil Biol. Biochem. **25**(2), 205–214 (1993)

R. Mendes, P. Garbeva, J.M. Raaijmakers, The rhizosphere microbiome: significance of plant beneficial, plant pathogenic and human pathogenic microorganisms. FEMS Microbiol. Rev. **37**, 634–663 (2013)

J.M. Meriles, S.V. Gil, R.J. Haro, G.J. March, C.A. Guzman, Glyphosate and previous crop residue effect on deleterious and beneficial soil-borne fungi from peanut–corn–soybean rotations. J. Phytopathol. **154**, 309–316 (2006)

M.L. Metzker, Sequencing technologies – the next generation. Nat. Rev. Genet. **11**, 31–46 (2010)

I. Mijangos, J.M. Becerril, I. Albizu, L. Epelde, C. Garbisu, Effects of glyphosate on rhizosphere soil microbial communities under two different plant compositions by cultivation dependent and independent methodologies. Soil Biol. Biochem. **41**, 505–513 (2009)

U. Mishra, S. Pabbi, Cyanobacteria: a potential biofertilizer for rice. Resonance **9**(6), 6–10 (2004)

A. Monkiedje, M.O. Ilori, M. Spiteller, Soil quality changes resulting from the application of the fungicides mefenoxam and metalaxyl to a sandy loam soil. Soil Biol. Biochem. **34**, 1939–1948 (2002)

T.B. Moorman, S.S. Harper, Transformation and mineralization of metribuzin in surface and subsurface horizons of a Mississippi Delta soil. J. Environ. Qual. **18**, 302–306 (1989)

J.P. Morrissey, J.M. Dow, G.L. Mark, F. O'Gara, Are microbes at the root of a solution to world food production? EMBO Rep. **5**(10), 922–926 (2004)

C. Moulas, C. Petsoulas, K. Rousidou, C. Perruchon, P. Karas, D.G. Karpouzas, Effects of systemic pesticides imidacloprid and metalaxyl on the phyllosphere of pepper plants. BioMed. Res. Int. **2013**, 969750 (2013)

W.R. Munns, R.C. Helm, W.J. Adams et al., Translating ecological risks to ecosystem service loss. Integr. Environ. Assess. Manag. **5**, 500–514 (2009)

B. Munoz-Leoz, E. Ruiz-Romera, I. Antiguedad, C. Garbisu, Tebuconazole application decreases soil microbial biomass and activity. Soil Biol. Biochem. **43**, 2176–2183 (2011)

B. Munoz-Leoz, C. Garbisu, J. Charcosset, J.M. Sanchez-Perez, I. Antiguedad, E. Ruiz-Romera, Non-target effects of three formulated pesticides on microbially-mediated processes in a clay-loam soil. Sci. Total Environ. **449**, 345–354 (2013)

G. Muyzer, E.C. Dewaal, A.G. Uitterlinden, Profiling of complex microbial populations by denaturing gradient gel-electrophoresis analysis of polymerase chain reaction-amplified genes-coding for 16S ribosomal-RNA. Appl. Environ. Microbiol. **59**, 695–700 (1993)

S.M. Nadeem, Z.A. Zahir, M. Naveed, M. Arshad, Preliminary investigations on inducing salt tolerance in maize through inoculation with rhizobacteria containing ACC deaminase activity. Can. J. Microbiol. **53**, 1141–1149 (2007)

C. Nasreen, G.J. Mohiddin, M. Srinivasulu, B. Manjunatha, V. Rangaswamy, Interaction effects of insecticides on microbial populations and dehydrogenase activity in groundnut (*Arachis hypogeae*) planted black clay soil. Int. J. Curr. Microbiol. Appl. Sci. **4**(2), 135–146 (2015)

P.S. Nicholson, P.R. Hirsch, The effects of pesticides on the diversity of culturable soil bacteria. J. Appl. Microbiol. **84**, 551–558 (1998)

P. Nielsen, S.O. Petersen, Ester-linked polar lipid fatty acid profiles of soil microbial communities: a comparison of extraction methods and evaluation of interference from humic acids. Soil Biol. Biochem. **32**(8&9), 1241–1249 (2000)

U.N. Nielsen, E. Ayres, D.H. Wall, R.D. Bardgett, Soil biodiversity and carbon cycling: a review and synthesis of studies examining diversity-function relationships. Eur. J. Soil Sci. **62**, 105–116 (2011)

K.M. Nienstedt, T.C.M. Brock, J. Van-Wensem et al., Development of a framework based on an ecosystem services approach for deriving specific protection goals for environmental risk assessment of pesticides. Sci. Total Environ. **415**(1), 31–38 (2012)

N. Nosengo, Fertilized to death. Nature **425**, 894–895 (2003)

J.M. Notron, Nitrogen mineralization immobilization turnover, in *Handbook of Soil Science*, ed. by E. Malkolm (CRC Press, Baton Rouge, 2000), pp. 149–190

OECD, OECD guidelines for the testing of chemicals. Test no. 216. Soil Microorganisms: nitrogen transformation test (2000), Retrieved from www.biotecnologiebt.it/pdf_our_services/OECD216.pdf

OEHHA, Office of Environmental Health Hazard Assessment. Soil toxicity and bioassessment test methods for ecological risk assessment (M. Hooper, M. Anderson, Preparation) (2009). Retrieved from http://oehha.ca.gov/ecotox/pdf/SoilTox120208.pdf

A.M. Osborn, E.R.B. Moore, K.N. Timmis, An evaluation of terminal- restriction fragment length polymorphism (T-RFLP) analysis for the study of microbial community structure and dynamics. Environ. Microbiol. **2**(1), 39–50 (2000)

R. Pal, K. Chakrabarti, A. Chakraborty, A. Chowdhury, Pencycuron application to soils: degradation and effect on microbiological parameters. Chemosphere **60**, 1513–1522 (2005)

S. Pandey, D.K. Singh, Total bacterial and fungal populations after chlorpyrifos and quinalphos treatments in groundnut (*Arachis hypogaea* L.) soil. Chemosphere **55**(2), 197–205 (2004)

A.K. Pandey, S. Gaind, A. Ali, L. Nain, Effect of bioaugmentation and nitrogen supplementation on composting of paddy straw. Biodegradation **20**, 293–306 (2009)

K.B. Parry, R.K.S. Wood, The adaptation of fungi to fungicides: adaptation to thiram, ziram, ferbam, nabarn and zineb. Ann. Appl. Biol. **47**, 10–16 (1959)

C.L. Patten, B.R. Glick, Bacterial biosynthesis of indole 3-acetic acid. Can. J. Microbiol. **42**, 207–220 (1996)

C.L. Patten, B.R. Glick, Role of *Pseudomonas putida* indole acetic acid in the development of the host plant root system. Appl. Environ. Microbiol. **68**, 3795–3801 (2002)

M.C. Paul, R.R. Mishra, Laboratory evaluation of some fungicides against colony growth of certain seed borne fungi. Crop. Res. **6**, 131–137 (1993)

E.A. Paul, R.A. Voroney, Field interpretation of microbial biomass activity and measurements, in *Current Perspective in Microbial Ecology*, ed. by M.J. Klug, C.A. Reddy (American Society of Microbiology, Washington, DC, 1989), pp. 509–514

S.T. Pawar, Isolation and screening of pesticide resistant cyanobacteria from pesticide contaminated agricultural soil. Phykos **45**(1), 85–84 (2015)

M. Pell, B. Stenberg, L. Torstensson, Potential denitrification and nitrification tests for evaluation of pesticide effects in soil. Ambio **1**, 24–28 (1998)

D. Perrig, M.L. Boiero, O.A. Masciarelli, C. Penna, O.A. Ruiz, F.D. Cassan, M.V. Luna, Plant-growth-promoting compounds produced by two agronomically important strains of *Azospirillum brasilense* and implications for inoculant formulation. Appl. Microbiol. Biotechnol. **75**, 1143–1150 (2007)

P. Perucci, Enzyme activity and microbial biomass in a field soil amended with municipal refuse. Biol. Fertil. Soils **14**, 54–60 (1992)

P. Perucci, L. Scarponi, Effects of the herbicide imazethapyr on soil microbial biomass and various soils enzyme activities. Biol. Fertil. Soils **17**, 237–240 (1994)

P. Perucci, S. Dumontet, S.A. Bufo, A. Mazzatura, C. Casucci, Effects of organic amendment and herbicide treatment on soil microbial biomass. Biol. Fertil. Soils **32**, 17–23 (2000)

V.N. Pishchik, N.I. Vorobyev, I.I. Chernyaeva, S.V. Timofeeva, A.P. Kozhemyakov, Y.V. Alexeev, S.M. Lukin, Experimental and mathematical simulation of plant growth promoting rhizobacteria and plant interaction under cadmium stress. Plant Soil **243**, 173–186 (2002)

R.F. Powers, Mineralizable soil nitrogen as an index of nitrogen availability to forest trees. Soil Sci. Soc. Am. J. **44**, 1314–1320 (1980)

C. Pozo, V. Rodelas, M.V. Salmeron, G. Martinez-Toledo, R. Vela, Effects of fungicides maneb and mancozeb on soil microbial populations. Toxicol. Environ. Chem. **43**(3&4), 123–132 (1994)

C. Pozo, T.M.V. Martinez, V. Salmeron, B. Rodelas, L. Goles, J. Opez, Effect of chloropyrifos on soil microbial activity. Environ. Toxicol. Chem. **14**(2), 187–192 (1995)

J.B. Prajapati, B.M. Nair, The history of fermented foods in fermented functional foods, in *Hand Book of Functional Fermented Foods*, ed. by E.R. Farnworth (CRC Press, Boca Raton, 2003), pp. 1–25

G. Preetha, J. Stanley, S. Suresh, R. Samiyappan, Risk assessment of insecticides used in rice on miridbug, *Cyrtorhinus lividipennis* Reuter, the important predator of brown planthopper, *Nilaparvata lugens* (Stal.). Chemosphere **80**(5), 498–503 (2010)

J.M. Raaijmakers, I. De-Bruijn, M.J. De-Kock, Cyclic lipopeptide production by plant-associated *Pseudomonas* spp.: diversity, activity, biosynthesis and regulation. Mol. Plant Microbe Interact. **19**, 699–710 (2006)

L.J. Radivojevic, L.J. Santric, R.S. Kalezic, D. Brkic, V. Janjic, Uticaj metribuzina na brojnost i aktivnost nekih grupa zemljišnih mikroorganizama. Pesticidi **18**, 99–107 (2003)

L.J. Radivojevic, S. Gasic, L.J. Santric, R.S. Kalezic, The impact of atrazine on several biochemical properties of chernozem soil. J. Serb. Chem. Soc. **73**, 951–959 (2008)

L. Radivojevic, L. Santric, J.G. Umiljendic, Rimsulfuron in soil: effects on microbiological properties under varying soil conditions. Pestic. Phytomed. **26**(2), 135–140 (2011)

L. Radivojevic, J. Dubravka, S. Ljiljana, G. Slavica, U.G. Jelena, Effects of metsulfuron-methyl on soil microbial activity. Arch. Tech. Sci. **11**(1), 77–82 (2014)

D. Rai, S.N. Sushil, J. Stanley, Ramkewal, J.P. Gupta, V. Singh, Deployment of novel technologies for the management of white grubs in lower hills of NW Himalayan region. Int. J. Hortic. **3**(2), 3–10 (2013)

V. Rangaswamy, B.R. Reddy, K. Venkateswarlu, Activities of dehydrogenase and protease in soil as influenced by monocrotophos, quinalphos, cypermethrin and fenvalerate. Agric. Ecosyst. Environ. **47**, 319–326 (1994)

N. Rasool, Z.A. Reshi, Effect of the fungicide mancozeb at different application rates on enzyme activities in a silt loam soil of the Kashmir Himalaya. India. Trop. Ecol. **51**, 199–205 (2010)

A.K. Rath, R.B. Ramarkrishnan, A.K. Rath, S. Kusmaraswamy, K. Bharati, P. Singla, N. Sethunathan, Effects of peticides on microbial biomass of flooded soil. Chemosphere **37**, 661–667 (1998)

D.S. Reay, Fertiliser 'solution' could turn local problem global. Nature **427**, 485 (2004)

P.J. Riggs, M.K. Chelius, A.L. Iniguez, S.M. Kaeppler, E.W. Triplett, Enhanced maize productivity by inoculation with diazotrophic bacteria. Aus. J. Plant Physiol. **28**, 829–836 (2001)

D.A. Robinson, N. Hockley, D.M. Cooper et al., Natural capital and ecosystem services, developing an appropriate soils framework as a basis for valuation. Soil Biol. Biochem. **57**, 1023–1033 (2013)

P.A. Roger, I. Simpson, R. Oficial, S. Ardales, R. Jimenez, Effects of pesticides on soil and water microflora and mesofauna in wetland ricefields: a summary of current knowledge and extrapolation to temperate environments. Aust. J. Exp. Agric. **34**, 1057–1068 (1994)

J. Romeis, R.L. Hellmich, M.P. Candolfi et al., Recommendations for the design of laboratory studies on non-target arthropods for risk assessment of genetically engineered plants. Transgen. Res. **20**, 1–22 (2011)

S. Rousseaux, A. Hartmann, N. Rouard, G. Soulas, A simplified procedure for terminal restriction fragment length polymorphism analysis of the soil bacterial community to study the effects of pesticides on the soil microflora using 4,6-dinitroorthocresol as a test case. Biol. Fertil. Soils **37**(4), 250–254 (2003)

C.M. Ryu, C.H. Hu, R.D. Locy, J.W. Kloepper, Study of mechanisms for plant growth promotion elicited by rhizobacteria in *Arabidopsis thaliana*. Plant Soil **268**, 285–292 (2005)

H. Saadatnia, H. Riahi, Cyanobacteria from paddy fields in Iran as a biofertilizer in rice plants. Plant Soil Environ. **55**(5), 207–212 (2009)

K.R. Saiya-Cork, R.L. Sinsabaugh, D.R. Zak, The effects of long term nitrogen deposition on extracellular enzyme activity in an *Acer saccharum* forest soil. Soil Biol. Biochem. **34**, 1309–1315 (2002)

A. Schmalenberger, C.C. Tebbe, M.A. Kertesz, H.L. Drake, K. Kusel, Two dimensional single strand conformation polymorphism (SSCP) of 16S rRNA gene fragments reveals highly dissimilar bacterial communities in an acidic fen. Eur. J. Soil Biol. **44**, 495–500 (2008)

M. Schmitz, Poisoned food technique. Indu. Eng. Chem. Anal. **2**, 361–363 (1930)

J. Schnurer, T. Rosswall, Fluorescein diacetate hydrolysis as a measure of total microbial activity in soil and litter. Appl. Environ. Microbiol. **43**(6), 1256–1261 (1982)

U.M.E. Schutte, Z. Abdo, S.J. Bent, C. Shyu, C.J. Williams, J.D. Pierson, L.J. Forney, Advances in the use of terminal restriction fragment length polymorphism (T-RFLP) analysis of 16S rRNA genes to characterize microbial communities. Appl. Microbiol. Biotechnol. **80**, 365–380 (2008)

M. Schutter, R. Dick, Comparison of fatty acid methyl ester (FAME) methods for characterizing microbial communities. Soil Sci. Soc. Am. J. **64**, 1659–1668 (2000)

F. Schwieger, C.C. Tebbe, A new approach to utilize PCR-single strand conformation polymorphism for 16S rDNA gene-based microbial community analysis. Appl. Environ. Microbiol. **64**, 4870–4876 (1998)

G. Selvakumar, M. Mohan, S.N. Sushil, S. Kundu, J.C. Bhatt, H.S. Gupta, Characterization and phylogenetic analysis of an entomopathogenic *Bacillus cereus* strain WGPSB-2 (MTCC 7182) isolated from white grub, *Anomala dimidiata* (Coleoptera: Scarabaeidae). Biocontrol Sci. Technol. **17**(5), 525–534 (2007)

D. Sengupta, M.W. Aktar, S. Purkait, M. Ganguly, Impact of quinalphos on microbial biomass and activities in tropical clay loam soil. Electron. J. Environ. Agric. Food Chem. **8**, 1127–1135 (2009)

H. Setala, M.P. Berg, T.H. Jones, Trophic structure and functional redundancy in soil communities, in *Biological Diversity and Function in Soils*, ed. by R.D. Bardgett, M.B. Usher, D.W. Hopkins (Cambridge University Press, Cambridge, 2005), pp. 236–249

S. Sethi, N. Mathur, P. Bhatnagar, Growth response of microbial population exposed to synthetic pyrethroids. Intl. J. Environ. Sci. **4**(1), 42–53 (2013)

N. Seymour, Impact of pesticides and fertilizers on soil biota (2015). Retrieved from www.academia.edu/10447630/Impacts_of_pesticides_and_fertilisers_on_soil_biota

B. Shaharoona, M. Naveed, M. Arshad, Z.A. Zahir, Fertilizer dependent efficiency of Pseudomonads for improving growth, yield and nutrient use efficiency of wheat (*Triticum aestivum* L.). Appl. Microbiol. Biotechnol. **79**, 147–155 (2008)

K. Shaukat, S. Affrasayab, S. Hasnain, Growth responses of *Triticum aestivum* to plant growth promoting rhizobacteria used as biofertilizer. Res. J. Microbiol. **1**(4), 330–338 (2006)

R.A. Sikora, K. Schafer, A.A. Dababat, Modes of action associated with microbially induced *in planta* suppression of plant-parasitic nematodes. Aust. Plant Pathol. **36**, 124–134 (2007)

M. Simek, L. Jisova, D.W. Hopkins, What is the so-called optimum pH for denitrification in soil? Soil Biol. Biochem. **34**(9), 1227–1234 (2002)

M.J. Simpson, J.R. McKelvie, Environmental metabolomics: new insights into earthworm ecotoxicity and contaminant bioavailability in soil. Anal. Bioanal. Chem. **394**, 137–149 (2009)

S. Singh, B. Singh, B. Mishra et al., Microbes in agrowaste management for sustainable agriculture, in *Microorganisms in Sustainable Agriculture and Biotechnology*, ed. by T. Satyanarayana, B.N. Johri, A. Prakash (Springer, Dordrecht, 2012), pp. 127–151

S. Singh, R. Gupta, M. Kumari, S. Sharma, Nontarget effects of chemical pesticides and biological pesticide on rhizospheric microbial community structure and function in *Vigna radiata*. Environ. Sci. Pollut. Res. Inst. **22**(15), 11290–11300 (2015)

S.E. Smith, D.J. Read, *Mycorrhizal Symbiosis*, 2nd edn. (Academic, London, 1997). 605 p
A.E. Smith, L. Hume, G.P. Lafond, V.O. Biederbeck, A review of the effects of long-term 2,4-D and MCPA applications on wheat production and selected biochemical properties of a Black Chernozem. Can. J. Soil Sci. **71**, 73–87 (1991)
A. Sofo, A. Scopa, S. Dumontet, A. Mazzatura, V. Pasquale, Toxic effects of four sulphonylureas herbicides on soil microbial biomass. J. Environ. Sci. Health B **47**(7), 653–659 (2012)
S.H. Son, Z. Khan, S.G. Kim, Y.H. Kim, Plant growth-promoting rhizobacteria, *Paenibacillus polymyxa* and *Paenibacillus lentimorbus* suppress disease complex caused by root-knot nematode and fusarium wilt fungus. J. Appl. Microbiol. **107**, 524–532 (2009)
G. Soulas, C. Lors, Perspectives and limitations in assessing side-effects of pesticides on the soil microflora. in *Proceedings of the 8th International Symposium on Microbial Ecology – Microbial Biosystems: New Frontiers Atlantic Canada Society for Microbial Ecology, Halifax, Canada,* ed. by C.R. Bell, M. Brylinsky, P. Johnson-Green (1999). Retrieved from http://plato.acadiau.ca/isme/Symposium25/soulas.PDF
D.L. Sparks, A.L. Page, P.A. Helmke, R.H. Loeppert, P.N. Soltanpour, M.A. Tabatabai, *Methods of Soil Analysis: Chemical Methods*. Book series (Soil Science Society of America, Madison, 1996)
I.M. Spyrou, D.G. Karpouzas, U. Menkissoglu-Spiroudi, Do botanical pesticides alter the structure of the soil microbial community? Microb. Ecol. **58**, 715–727 (2009)
P.D. Stahl, T.B. Parkin, Relationship of soil ergosterol concentration and fungal biomass. Soil Biol. Biochem. **28**, 847–855 (1996)
J. Stanley, S. Chandrasekaran, G. Preetha, S. Kuttalam, Physical and biological compatibility of diafenthiuron with micro/macro nutrients fungicides and biocontrol agents used in cardamom. Arch. Phytopathol. Plant Prot. **43**(14), 1396–1406 (2010a)
J. Stanley, S. Chandrasekaran, G. Preetha, S. Kuttalam, Toxicity of diafenthiuron to honey bees in laboratory, semi-field and field conditions. Pest Manag. Sci. **66**(5), 505–510 (2010b)
H.O. Stanley, E.M. Maduike, P.O. Okerentugba, Effect of herbicide (atrazine and paraquat) application on soil bacterial population. Sky J. Soil Sci. Environ. Manag. **2**(9), 101–105 (2013)
G.W. Startton, K.E. Stewart, Glyphosate effects on microbial biomass in a coniferous forest soil. Environ. Toxicol. Water Qual. **7**, 223–236 (2002)
A.M. Stefanowicz, M. Niklinska, R. Laskowski, Pollution-induced tolerance of soil bacterial communities in meadow and forest ecosystems polluted with heavy metals. Eur. J. Soil Biol. **45**, 363–369 (2009)
G. Straatsma, J.P.G. Gerrits, M.P.A.M. Augustin, H.J.M. Camp, G.D. Vogels, L.J.L.D. Van-Griensven, Population dynamics of *Scytalidium thermophilum* in mushroom compost and stimulatory effects on growth rate and yield of *Agaricus bisporus*. J. Gen. Microbiol. **135**, 751–759 (1989)
P.F. Strom, Identification of thermophilic bacteria in solid-waste composting. Appl. Environ. Microbiol. **50**(4), 906–913 (1985)
A. Subhani, A.M. El-ghamry, H. Changyong, X. Jianming, Effects of pesticides (herbicides) on soil microbial biomass- a review. Pak. J. Biol. Sci. **3**(5), 705–709 (2000)
M.J. Swift, A.N. Izac, M. Van-Moordwijk, Biodiversity and ecosystem services in agricultural landscapes – are we asking the right questions? Agric. Ecosyst. Environ. **104**, 113–134 (2004)
M.A. Tabatabai, Soil enzymes, in *Method of Soil Analysis, Part 2: Chemical and Microbiological Properties*, ed. by L.A. Page (American Society of Agronomy, Soil Science Society of America, Madison, 1982), pp. 903–943
M.A. Tabatabai, J.M. Bremner, Use of p-nitrophenyl phosphate for assay of soil phosphatase activity. Soil Biol. Biochem. **1**, 301–307 (1969)
M.A. Tabatabai, J.M. Bremner, Assay of urease activity in soils. Soil Biol. Biochem. **4**, 479–487 (1972)
K. Tanaka, S. Endo, H. Kazano, Toxicity of insecticides to predators of rice planthoppers: spiders, the mirid bug and the dryinid wasp. Appl. Entomol. Zool. **35**, 177–187 (2000)
J.C. Tarafdar, Effect of different herbicides on enzyme activity in controlling weeds in wheat crop. Pesticides **20**, 46–49 (1986)

K. Tawaraya, M. Naito, T. Wagatsuma, Solubilization of insoluble inorganic phosphate by hyphal exudates of arbuscular mycorrhizal fungi. J. Plant Nutr. **29**, 657–665 (2006)

S. Terre, F. Asch, J. Padham, R.A. Sikora, M. Becker, Influence of root zone bacteria on root iron plaque formation in rice subjected to iron toxicity, in *Utilization of Diversity in Land Use Systems: Sustainable and Organic Approaches to Meet Human Needs*, ed. by E. Tielkes (Tropentag, Witzenhausen, 2007). 446 p

J.M. Tiedje, Microbial diversity: of value to whom? ASM News **60**, 524–525 (1994)

V.N. Tiwari, L.K. Lehri, A.N. Pathak, Effect of inoculating crops with phospho-microbes. Exp. Agric. **25**(1), 47–50 (1989)

A. Tlili, A. Berard, J.L. Roulier, B. Volat, B. Montuelle, PO_4^{3-} dependence of the tolerance of autotrophic and heterotrophic biofilm communities to copper and diuron. Aquat. Toxicol. **98**, 165–177 (2010)

R.K. Tokala, J.L. Strap, C.M. Jung, D.L. Crawford, H. Salove, L.A. Deobald, F.J. Bailey, M.J. Morra, Novel plant–microbe rhizosphere interaction involving *S. lydicus* WYEC108 and the pea plant (*Pisum sativum*). Appl. Environ. Microbiol. **68**, 2161–2171 (2002)

L. Torstenssen, J. Stenstorm, Basic respiration rate as a tool for prediction of pesticide persistence in soil. Toxicol. Assess. **1**, 57–72 (1986)

R.D. Tripathi, S. Srivastava, S. Mishra, N. Singh, R. Tuli, D.K. Gupta, F.J.M. Maathuis, Arsenic hazards: strategies for tolerance and remediation by plant. Trend. Biotechnol. **25**, 158–165 (2007)

R.D. Tripathi, S. Dwivedi, M.K. Shukla, S. Mishra, S. Srivastava, R. Singh, U.N. Rai, D.K. Gupta, Role of blue green algae biofertilizer in ameliorating the nitrogen demand and fly-ash stress to the growth and yield of rice (*Oryza sativa* L.) plants. Chemosphere **70**(10), 1919–1929 (2008)

C.M. Tu, Effect of some herbicides on activities of microorganisms and enzymes in soil. J. Environ. Sci. Health **27**, 695–709 (1992)

C.M. Tu, J.R.W. Miles, Interactions between insecticides and soil microbes. Residue Rev. **64**, 17–65 (1976)

A.A.C. Uijtewaal, Buried pesticide waste hazard to Poland. Waste Manag. Res. **10**(5), 387–398 (1992)

Y. Unno, K. Okubo, J. Wasaki, T. Shinano, M. Osaki, Plant growth promotion abilities and microscale bacterial dynamics in the rhizosphere of lupin analysed by phytate utilization ability. Environ. Microbiol. **7**, 396–404 (2005)

T. Urich, A. Lanzen, J. Qi, D.H. Huson, C. Schleper, S.C. Schuster, Simultaneous assessment of soil microbial community structure and function through analysis of the metatranscriptome. PLoS ONE **3**(6), e2527 (2008)

C. Usta, Microorganisms in biological pest control- a review, in *Current Progress in Biological Research*, ed by S. Marina (InTech, 2013), doi:10.5772/55786

X. Vakemans, B. Godden, M.J. Penninckx, Factor analysis of the relationship between several physicochemical and microbial characteristics of some Belgian agricultural soils. Soil Biol. Biochem. **21**, 53–58 (1989)

P. Van-Beelen, P. Doelman, *Significance and Application of Microbial Toxicity Tests in Assessing Ecotoxicological Risks of Contaminants in Soil and Sediment*. Report No. 719102051 (National Institute of Public Health and Environment, Bilthoven, 1996). 40 p

E.D. Vance, P.C. Brooks, D.S. Jenkinson, An extraction method for measuring soil microbial biomass. Soil Biol. Biochem. **19**, 703–707 (1987)

J.D. Van-Elsas, F.G.H. Boersma, A review of molecular methods to study the microbiota of soil and the mycosphere. Eur. J. Soil Biol. **47**, 77–87 (2011)

J.D. Van-Elsas, M. Chiurazzi, C.A. Mallon, D. Elhottova, V. Kristufek, J.F. Salles, Microbial diversity determines the invasion of soil by bacterial pathogen. Proc. Natl. Acad. Sci. U. S. A. **109**, 1159–1164 (2012)

L.C. Van-Loon, Plant responses to plant growth promoting bacteria. Eur. J. Plant Pathol. **119**, 243–254 (2007)

L.C. Van-Loon, P.A.H.M. Bakker, C.M.J. Pieterse, Systemic resistance induced by rhizosphere bacteria. Annu. Rev. Phytopathol. **36**, 453–483 (1998)

A. Vespermann, M. Kai, B. Piechulla, Rhizobacterial volatiles affect the growth of fungi and Arabidopsis thaliana. Appl. Environ. Microbiol. **73**, 5639–5641 (2007)

J.K. Vessey, Plant growth promoting rhizobacteria as biofertilizers. Plant Soil **255**, 571–586 (2003)

K. Vig, D.K. Singh, A.K. Agarwal, A.K. Dhawan, P. Dureja, Soil microorganisms in cotton fields sequentially treated with insecticides. Ecotoxicol. Environ. Saf. **69**, 263–276 (2008)

C. Vischetti, P. Perucci, L. Scarponi, Relationship between rimsulfuron degradation and microbial biomass content in a clay loam soil. Biol. Fert. Soils **31**, 310–314 (2000)

C. Viti, L. Giovannetti, Characterization of cultivable heterotrophic bacterial communities in Cr-polluted and unpolluted soils using Biolog and ARDRA approaches. Appl. Soil Ecol. **28**, 101–112 (2005)

A.J. Vlitos, Influence of environment on the Crag pesticide. Contri. Boyce Thompson Inst. **16**, 435–438 (1952)

G. Voos, P.M. Groffman, Relationship between microbial biomass and dissipation of 2, 4-D and Dicamba in soil. Biol. Fertil. Soils **24**, 106–110 (1997)

M. Wainwright, Review of effects of pesticides on microbial activity in soils. J. Soil Sci. **29**, 287–298 (1978)

B.H. Walker, Biodiversity and ecological redundancy. Conserv. Biol. **6**, 18–24 (1992)

A. Walker, Y.H. Moon, S.J. Welch, Influence of temperature, soil moisture and soil characteristics on persistence of Alachlor. Pest Sci. **35**, 109–116 (1992)

K. Wallenius, H. Rita, S. Simpanen, A. Mikkonen, R.M. Niemi, Sample storage for soil enzyme activity and bacterial community profiles. J. Microbiol. Meth. **81**, 48–55 (2010)

U.F. Walsh, J.P. Morrissey, F. O'Gara, Pseudomonas for biocontrol of phytopathogens: from functional genomics to commercial exploitation. Curr. Opin. Biotechnol. **12**, 289–295 (2001)

M. Wang, Y. Liu, Q. Wang et al., Impacts of methamidophos on the biochemical, catabolic and genetic characteristics of soil microbial communities. Soil Biol. Biochem. **40**(3), 778–788 (2008)

F. Wang, J. Yao, H.L. Chen, K. Chen, P. Trebse, G. Zaray, Comparative toxicity of chlorpyrifos and its oxon derivatives to soil microbial activity by combined methods. Chemosphere **78**, 319–326 (2010)

D.A. Wardle, A. Ghani, A critique of the microbial metabolic quotient (qCO_2) as a bioindicator of disturbance and ecosystem development. Soil Biol. Biochem. **27**, 1601–1610 (1995)

D.A. Wardle, D. Parkinson, Effects of three herbicides on soil microbial biomass and activity. Plant Soil **122**, 21–28 (1990a)

D.A. Wardle, D. Parkinson, Influence of the herbicide glyphosate on soil microbial community structure. Plant Soil **122**, 29–37 (1990b)

D.A. Wardle, D. Parkinson, Influence of the herbicides, 2, 4-D and glyphosate on soil microbial biomass and activity: a field experiment. Soil Biol. Biochem. **24**, 185–186 (1992)

G. Wei, J.W. Kloepper, S. Tuzun, Induced systemic resistance to cucumber diseases and induced plant growth by plant growth promoting rhizobacteria under field conditions. Phytopathology **86**, 221–224 (1996)

W.W. Wenzel, manipulating rhizosphere chemistry to control metal and organic contaminant availability and implications to phytoremediation, in *Proceedings of the International Conference on Contaminants in Soil Environment in Australia-Pacific Environments*, (New Delhi, 1999). 12 p

S. Wertz, V. Degrange, J.I. Prosser, F. Poly, C. Commeaux, T. Freitag, N. Guillaumaud, X.L. Roux, Maintenance of soil functioning following erosion of microbial diversity. Environ. Microbiol. **8**, 2162–2169 (2006)

R.E. Wheatley, The consequences of volatile organic compound mediated bacterial and fungal interactions. AntonieVan Leeuwenhoek **81**, 357–364 (2002)

J. Whipps, Microbial interactions and biocontrol in the rhizosphere. J. Exp. Bot. **52**, 487–511 (2001)

A. Widenfalk, J.M. Svensson, W. Goedkoop, Effects of the pesticides captan, deltamethrin, isoproturon, and pirimicarb on the microbial community of a freshwater sediment. Environ. Toxicol. Chem. **23**(8), 1920–1927 (2004)

A. Widenfalk, S. Bertilsson, I. Sundh, W. Goedkoop, Effects of pesticides on community composition and activity of sediment microbes – responses at various levels of microbial community organization. Environ. Pollut. **152**, 576–584 (2008)

W.S. Wong, X.Q. Shen, B. Wen, S.Z. Xhang, Relationship between the extractable metals from soils and metals taken up by maize roots and shoots. Chemosphere **53**, 523–530 (2003)

J.C. Wooley, Y.Z. Ye, Metagenomics: facts, artifacts and computational challenges. J. Comp. Sci. Technol. **25**, 71–81 (2010)

S.C. Wu, Z.H. Cao, Z.G. Li, K.C. Cheung, M.H. Wong, Effects of biofertilizer containing N-fixer, P and K solubilizers and AM fungi on maize growth: a greenhouse trial. Geoderma **125**(1&2), 155–166 (2005)

D. Wu, P. Hugernholtz, K. Mavromatis et al., A phylogeny driven genomic encyclopedia of bacteria and Arhaea. Nature **462**, 1056–1060 (2009)

X.M. Wu, Y.H. Long, Y.R. Li, R.X. Liu, M. Li, Effects of napropamide on microbiological characteristics of tobacco rhizosphere soil and its dissipation. J. Soil Sci. Plant Nutri. **14**(1), 151–159 (2014)

G.R. Xavier, F.V. Silva, J.E. Zilli, N.G. Rumjanek, *Adaptação de método para extração de DNA microbiano, Embrapa Agrobiologia, Seropédica, Documento* (2004) 171, 24 p

A. Yabaya, E.D. Jatau, Investigating wild yeast baking potentials. Middle-East J. Sci. Res. **4**(4), 320–322 (2009)

S. Yachi, M. Loreau, Biodiversity and ecosystem productivity in a fluctuating environment: the insurance hypothesis. PNAS **96**, 1463–1468 (1999)

S.K. Yadav, A.A. Juwarkar, G.P. Kumar, P.R. Thawale, S.K. Singh, T. Chakrabarti, Bioaccumulation and phyto-translocation of arsenic, chromium and zinc by *Jatropha curcas* L.: impact of dairy sludge and biofertilizer. Biores. Technol. **100**, 4616–4622 (2009)

T. Yakupoglu, S.F. Hepsen, N. Ozdemir, R. Kizilkaya, The effects of various organic wastes applied into eroded soil on dehydrogenase enzyme activity, in *Current Research Topics in Applied Microbiology and Microbial Biotechnology,* ed. by A. Mendez-Vila (2009) pp. 97–101

J. Yang, J.W. Kloepper, C.M. Ryu, Rhizosphere bacteria help plants tolerate abiotic stress. Trend. Plant Sci. **14**, 1–4 (2009)

Y.G. Yanni, R.Y. Rizk, F.K.A. El-Fattah et al., The beneficial plant-growth promoting association of *Rhizobium leguminosarum* bv. *trifolii* with rice roots. Aust. J. Plant Physiol. **28**, 845–870 (2001)

C.B. Yao, G.W. Zehner, E. Sikora, J. Murphy, J.W. Kloepper, Evaluation of induced systemic resistance and plant growth promotion in tomato with selected PGPR strains, in *Plant Growth Promoting Rhizobacteria – Present Status and Future Prospects*, ed. by A. Ogoshi, K. Kobayashi, Y. Homma, F. Kodama, N. Kondo, S. Akino (Nakanishi Printing, Sapporo, 1997), pp. 285–288

Z. Yehuda, M. Shenker, Y. Hadar, Y.N. Chen, Remedy of chlorosis induced by iron deficiency in plants with the fungal siderophore rhizoferrin. J. Plant Nutr. **23**, 1991–2006 (2000)

S. Yousaf, S. Khan, M.T. Aslam, Effect of pesticides on the soil microbial activity. Pak. J. Zool. **45**(4), 1063–1067 (2013)

Y.L. Yu, M. Shan, H. Fang, X. Wang, X.Q. Chu, Responses of soil microorganisms and enzymes to repeated applications of chlorothalonil. J. Agric. Food Chem. **54**, 10070–10075 (2006)

H. Yu, G. Zeng, H. Huang et al., Microbial community succession and lignocellulose degradation during agricultural waste composting. Biodegradation **18**(6), 793–802 (2007)

T. Yuwono, D. Handayani, J. Soedarsono, The role of osmotolerant rhizobacteria in rice growth under different drought conditions. Aust. J. Agric. Res. **56**, 715–721 (2005)

M.C. Zabaloy, M.A. Gomez, Diversity of rhizobia isolated from an agricultural soil in Argentina based on carbon utilization and effects of herbicides on growth. Biol. Fert. Soils **42**, 83–88 (2005)

M.C. Zabaloy, J.L. Garland, M.A. Gomez, An integrated approach to evaluate the impacts of the herbicides glyphosate, 2,4-D and metsulfuron-methyl on soil microbial communities in the Pampas region. Argentina. Appl. Soil Ecol. **40**, 1–12 (2008)

I. Zabetakis, Enhancement of flavour biosynthesis from strawberry (Fragaria x ananassa) callus cultures by *Methylobacterium* species. Plant Cell Tiss. Organ. Cult. **50**, 179–183 (1997)

J.C. Zak, M.R. Willig, D.L. Moorhead, H.G. Wildman, Functional diversity of microbial communities: a quantitative approach. Soil Biol. Biochem. **26**, 1101–1108 (1994)

A.A. Zavalin, L.V. Vinogradova, T.M. Dukhanina et al., Geographical regularities of effect of inoculation with associative diazotrophs on the productivity of cereals. Aspects Appl. Biol. **63**, 123–127 (2001)

L. Zelles, I. Scheunert, F. Korte, Side effects of some pesticides on non-target soil microorganisms. J. Environ. Sci. Health B **20**, 457–488 (1985)

F. Zhang, D.H. Lynch, D.L. Smith, Low root temperature and nodulation, nitrogen fixation, photosynthesis and growth by soybean [*Glycine max* (L.) Merr.]. Environ. Exp. Bot. **35**, 279–285 (1995)

F. Zhang, N. Dashti, R.K. Hynes, D.L. Smith, Plant growth promoting rhizobacteria and soybean [*Glycine max* (L.) Merr.] nodulation and nitrogen fixation at suboptimal root zone temperatures. Ann. Bot. **77**, 453–459 (1996)

F. Zhang, N. Dashti, R.K. Hynes, D.L. Smith, Plant growth-promoting rhizobacteria and soybean [*Glycine max* (L.) Merr] growth and physiology at suboptimal root zone temperatures. Ann. Bot. **79**, 243–249 (1997)

B. Zhang, H. Zhang, B. Jin, L. Tang, J. Yang, B. Li, G. Zhuang, Z. Bai, Effect of cypermethrin insecticide on the microbial community in cucumber phyllosphere. J. Environ. Sci. **20**(11), 1356–1362 (2008)

B. Zhang, Z. Bai, D. Hoefel, L. Tang, X. Wang, B. Li, Z. Li, G. Zhuang, The impacts of cypermethrin pesticide application on the non-target microbial community of the pepper plant phyllosphere. Sci. Total Environ. **407**(6), 1915–1922 (2009)

J. Zhang, G. Zeng, Y. Chen et al., Effects of physico-chemical parameters on the bacterial and fungal communities during agricultural waste composting. Biores. Tech. **102**(3), 2950–2956 (2011)

H. Zhou, J. Li, L. Zhao, J. Han, X. Yang, W. Yang, X. Wu, Study on main microbes on quality formation of Yunnan puer tea during pile-fermentation process. J. Tea Sci. (2004). Retrieved from http://en.cnki.com.cn/Article_en/CJFDTOTAL-CYKK20040300B.htm

L.H.S. Zobiole, R.J. Kremer, R.S. Oliveira, J. Constantin, Glyphosate affects microorganisms in rhizospheres of glyphosate resistant soybeans. J. Appl. Microbiol. **110**, 118–127 (2011)

C.S. Zou, M.H. Mo, Y.Q. Gu, J.P. Zhou, K.Q. Zhang, Possible contributions of volatile-producing bacteria to soil fungistasis. Soil Biol. Biochem. **39**, 2371–2379 (2007)

Index

Chapter 7
Pesticide Toxicity to Fishes: Exposure, Toxicity and Risk Assessment Methodologies

Abstract Pesticides enter into aquatic ecosystem by many ways viz., drift, wash off and drain from agro-ecosystem or by deliberate application in the water. Fish live in water and thus continuously get exposed to the contamination by contact, respiration and by contaminated food intake. Acute toxicity assessment of pesticides to fish is mostly carried out through water contamination bioassays as static, static-renewal and flow-through systems apart from injection and through dietary exposures. Literature on the sublethal effects as alteration of behaviour (sensation, locomotion, feeding, learning etc.), physiology (respiration, metabolism, reproduction etc.), biochemistry of enzymes, blood and hormones apart from histopathology, carcinogenicity and mutagenicity were reviewed and compiled. Different methods to assess these sublethal effects along with genotoxicity of pesticides as disrupting in genes, chromosomes, DNAs and RNAs are given in detail. Pesticide toxicity on endocrine, cardiovascular, nervous, digestive and reproductive systems and their procedures to assess those sublethal toxic effects are explained. Only a few studies on semi-field/mesocosm and field toxicity assessments are available in the literature. Risk assessment of pesticides in aquatic ecosystem by the estimation of acute and chronic toxicity, calculation of risk estimates, multi tier approach of risk assessment, comparison with target and non-target susceptibility are given in detail. Human beings and predators are also at high risk of consumption of pesticide contaminated fish owing to bioaccumulation and biomagnification and thus, assessment of risk in consuming contaminated fish is also included.

1 Importance of Fish

Fish is the major protein source for humans, all over the world. There are about 25,000 different known fish species of which 15,000 are marine and nearly 10,000 are freshwater (Nelson 1994). In 2010, the annual capture, combining together the direct capture and capture from fish farms was about 149 million tonnes (FAO 2012). During 2011, out of the 173 Mt of harvested fish, 75.7 % (131 Mt) was used for direct consumption (HLPE 2014). But, over harvesting of fishes not only reduce the abundance of the targeted stocks with cascading responses in the food web, also affects fish dependent systems (Dayton et al. 1995; Steneck 1998). Fish have important nutrients which are essential for health and also combat many deadly diseases.

J. Stanley, G. Preetha, *Pesticide Toxicity to Non-target Organisms*,
DOI 10.1007/978-94-017-7752-0_7

Apart from food, the fish industry provides jobs for fishermen, workers of commercial fisheries, wholesalers and retailers and as a source of income and livelihood for many. Aquaculture with the growth rate of 6.9 % is the fastest growing food production sector (Subasinghe 2005). In term of efficiency, fish is considered as the most efficient converter of feed into protein (Bene et al. 2015) with the conversion rate of 30 % (Hasan and Halwart 2009). Aquatic animal production systems have a low carbon footprint per kilogram of production (Hall et al. 2011) with less nitrogen and phosphorous emissions (Bene et al. 2015).

The ecosystem services provided by fish in terms of its role in food web, nutrient cycling, regulatory services, maintaining ecosystem resilience (Holmlund and Hammer 1999) and as an important link between aquatic to terrestrial ecosystem cannot be overlooked. More indirect, but equally valuable benefit of fish and aquatic ecosystems comprise of fun boating, sport fishing and natural splendor (Helfrich 2009) which promotes tourism (FAO 1996; Postel and Carpenter 1997).

1.1 As Food for Humans

Fish provides protein for more than 4.5 billion people around the world. The unique nutritional properties of fish make it essential for food security and health (Thilsted 2012; Beveridge et al. 2013). About, 17 % of people's intake of animal protein comes from fish and 6.5 % of all protein consumed (FAO 2013). Supplying the best source of protein, fish also contain vitamins and minerals essential for human health. Apart from the best quality of protein, the long-chain poly-unsaturated fatty acids present in the fish are unique (Bene et al. 2015). Many tasty dishes and recipes are made out of fish and are famous since ages. Fish sauces, *liquamen*, *muria*, *allex* and *granum* were of commercial importance in the Roman Empire (Corcoran 1963). Fish curry, fish soup, fried fish, fish cutlets, fish sandwiches, fish sticks, cakes and fish burger are prepared and consumed all over the world. Many value added products like surimi of silver carp (Yongkong et al. 2002), cutlets of Pangas catfish (Bari et al. 2000) and reed cod fish (Reddy et al. 2012), salad of rohu and catla (Sehgal and Sehgal 2003), sausage of rohu (Sini 2003), fish mince pakora of rohu (Sehgal et al. 2010), patties of whiting, mackerel (Tang et al. 2001), anchovy (Yerlikaya et al. 2005) and common carp (Sehgal et al. 2011) etc. are being produced, preferred and consumed by many and have a high commercial value.

1.2 Improving Human Health

Fish is more than just a source of animal protein but contains several essential amino acids, especially lysine and methionine. The presence of vitamins D, A and B and minerals (calcium, phosphorus, iodine, zinc, iron and selenium) makes fish as an important food to counteract malnutrition. The theory of fish addressing multiple micronutrient deficiencies is now being realised and recognized (Roos et al. 2007; Kawarazuka and Bene 2011; Thilsted 2012). Because of its benefits, women of childbearing age and nursing mothers are advised to consume two seafood servings/ week of selected species (Mozaffarian and Rimm 2006). Many evidences support the benefits of fish consumption during pregnancy, mainly because of the availability of n-3 polyunsaturated fatty acids, which are important for neurodevelopment of the fetus. However, fishes that are potent sources of methylmercury and polychlorinated biphenol exposure can be avoided (Dovydaitis 2008). Milkfish, silver pomfret, tilapia, smelt, porgie or bluefish can suffice the need of docosahexaenoic acid of pregnant women (Del-Gobbo et al. 2010).

Fish is also valued as a source of omega-3 fatty acids, very long chain polyunsaturated fatty acids that are critical for the development of brain and retina and protective of some diseases. Approximately 50 % of fatty acids in lean fishes and 25 % of fattier fishes are polyunsaturated fatty acids (PUFA) whereas it is only 4–10 % in beef. The amount of eicosapentanoic acid (EPA) and docosahexanoic acid (DHA) found in fishes are much high than any other foods (Sheeshka and Murkin 2002). The polyunsaturated fatty acids and their metabolic conversion to eicosanoids, prostaglandins and leukotrienes exert protection against many diseases and disorders. Regular fish consumption can improve the development of bones, owing to the content of vitamin D in fish (Verbeke et al. 2005).

1.3 In Pharmaceuticals and as Natural Medicine

Fish are used as a natural disease preventive food and also as a medicine for many diseases. The saying of Hippocrates, "*Let food be thy medicine and medicine be thy food*" is fully valid for fish and fish food is utilized for both health promotion and disease prevention from antiquity. The fatty acids and their derivatives are successfully demonstrated with scientific backgrounds as a natural medicine for many human diseases. Cod liver oil is recommended for carrying mothers and infants because it contains vitamin D, long chain in n-3 fatty acids, eicosapentaenoic acid and docosahexaenoic acid which are essential for proper growth and well being (Nettleton 1993; NHC 1994). Cod liver oil supplementation in the Western diet reportedly improved platelet functions, thromboxane formation and controls blood pressure (Lorenz et al. 1983).

Fish oil diet was found to have a significant positive effect in treating asthma patients (Dry and Vincent 1991), Ischemic heart patients (Burr et al. 1989) and patients with Crohn's disease (Belluzzi et al. 1996). Reduction in the risk of getting dementia (Huang et al. 2005) and Alzheimer disease (Morris et al. 2003) are correlated with the intake of fish. Intake of cod liver oil by mothers has significantly reduced the occurrence of Type I diabetes in new born babies (Stene et al. 2000).

In a case study, fewer intakes of fish were found to be correlated with the prevalence of rheumatoid arthritis (Linos et al. 1991). Further, diet containing low fish oil content was found to cause anti-inflammatory effects in rheumatoid arthritis patients

(Adam et al. 2003). Consumption of broiled or baked fish was highly associated with a decreased risk of rheumatoid arthritis with odds ratios of 0.78 for 1–2 servings/week and 0.57 for ≥2 servings/week compared to less than one serving (Shapiro et al. 1996). Dietary supplement of eicosapentaenoic acid from fish produces potentially beneficial effects on lipids, platelets and blood pressure and reduce the risk of mortality in patients undergone haemodialysis. Fish oil modifies lipids and reduces platelet aggregability (Rylance et al. 1986). In a multivariable model, histories of stroke, heart disease and hypertension, the relative risk was reported as 0.3 for consumers with two or more fish meal per week when compared with non-consumers (Morris et al. 2003). After 3.5 years of administration of piscine omega-3 polyunsaturated fatty acids, a relative risk reduction of 45 % in sudden cardiac death was reported (Marchioli et al. 2002). Omega-3 polyunsaturated fatty acids might reduce vulnerability to ventricular arrhythmias after myocardial infarction by acting directly on myocyte sodium and calcium currents (Leaf et al. 2005).

1.4 As a Source of Income: Occupation and Industry

About 660–820 million people all round the world depend on fish and fish related activities for their income (Allison et al. 2013; HLPE 2014) especially of the low income group (Bene 2006). Fish contribute substantially to the income of more than 10 % of the world population (Bene et al. 2015). For most of these households, it is often the main component of their livelihood, which makes them food secured (Heck et al. 2007; Bene et al. 2009; Eide et al. 2011). Western Africa is rightly called as a 'Fish Basket of Europe' (Alder and Sumaila 2004). Fishing is the main source of cash-income for the majority of households in Congo apart from the supply of protein-rich food to local people (Bene et al. 2009).

Fish from standing water rice fields is an extra source of income to farmers (Ali 1990). Rice-fish farmers are reported to harvest higher yields than rice farmers, since stocking of fish affected the rice yield positively (Lightfoot et al. 1992; Sinh 1995) apart from an additional yield of fish. It is also found that the disparity in the distribution of income obtained from fish production is very less compared to that from other sources in Tripura, India enabling a social equity (Singh 2006).

1.5 As Food for Other Living Organisms

Small pelagic fish are the important food source for larger fish and many fish feeding birds and mammals (Smith et al. 2010). The presence of American eel, *Anguilla rostrata* significantly lowers the benthic fish density (Stranko et al. 2014). Some species of fish become piscivorous at the age of 0 such as walleye, large mouth bass, small mouth bass, white bass etc. whereas some like black crappie starts at the age of 1. About 12 species of fishes are being preyed upon by big piscivours fish in Spirit lake (Pelham et al. 2001).

Birds are considered as the top predators in many freshwater systems and the abundance of striped shiners, *Luxilus chrysocephalus* and central stonerollers, *Campostoma anomalum* in streams are highly influence by great blue herons, *Ardea herodias* and belted kingfishers, *Ceryle alcyon* (Steinmetz et al. 2003). Belted kingfisher and blue herons are known to prey heavily on fish (Penaluna et al. 2015; Huang et al. 2015). When herons are experimentally excluded, the benthic fish was found to get increased in number (Huang et al. 2015). Fish consumption by birds in lake Mockeln (44 km^2 area), Sweden was estimated to be 0.8 g/m^2/year which is much more than the fishermen's share of 0.25 g/m^2/year (Nilsson and Nilsson 1976). The double-crested cormorants were reported to depredate on channel catfish heavily in Florida (Schramm et al. 1984). African penguins feed on fish or squids (Wilson et al. 1985). Osprey, *Pandion haliaetus* and river otter, *Lutra canadensis* depredates much of the fish from commercial fisheries (Parkhurst et al. 1987). The otter, *Lutra lutra* feeds almost exclusively on fish all round the year and mink, *Mustela vision* feed on fish primarily but also water fowl and mammals. They feed on burbot, *Lota lota*, cyprinids, Northern pike, *Esox lucius* and *Perca fluviatilis* (Erlinge 1969). Many mammals like seals, sea lions, bears and polar bears feed on fish. The California sea lions, *Zalophus califovnianus* and Pacific harbor seals, *Phoca vitulina vichavdsi* were reported to take the fish caught from boats or fishing vessels also (Hanan et al. 1989).

Apart from direct food, fish also contributes indirectly to other living organisms as it is used as fishmeal for aquaculture and poultry and livestock feeds and thus in-turn to human nutrition (Tacon and Metian 2009). In 2011, 75 % of the 23 Mt of fish, essentially small pelagic fish species which are not destined for direct human consumption was reduced to fishmeal and fish oil for aquaculture, poultry and other livestock feeding (FAO 2012; Shepherd and Jackson 2013).

1.6 Ecosystem Services by Fishes

Human societies get benefit in numerous ways from ecosystem services generated by fish populations. Fish are part of food chain dynamics, nutrient cycling and ecosystem resilience. Certain ecosystem services generated by fish populations by means of enhancing rice production (tilapia, carp), mitigating diseases (mosquitofish), mitigating algal blooms (*E. lucius*, *Lucioperca sandra*), mitigating growth of lake vegetation (grass carp), indicating ecosystem stress (butterflyfishes) etc. are being valued increasingly (Holmlund and Hammer 1999). Management of mosquitoes that spread dreadly diseases by acting as a vector, using mosquito larvivorous fishes is considered as an environmental friendly and safe alternative to insecticides (Roger and Bhuiyan 1990; Chandra et al. 2008).

1.6.1 Regulatory Services of Fishes: Regulating Food Web Dynamics

Piscivores (fish that eat fish) and zooplanktivores (fish that eat zooplankton), exerts a strong top-down control resulting in a cascade of effects down the food chain (Holmlund and Hammer 1999). Decrease in the predation pressure on zooplanktons results in its increase, which can ultimately cause a reduction in phytoplankton (Carpenter et al. 1985). The regulatory services of fish can be explained with the following examples. Large zooplanktons are more abundant than small zooplanktons in lakes with large number of piscivores whereas small-sized zooplanktons are more abundant in the presence of zooplanktivores (Brooks and Dodson 1965; Vanni et al. 1990). Grazing fishes feeding on algae or detritus in the stream of Venezuela was reported to modify their distribution and abundance (Flecker 1992). The feeding pattern of fishes also influences the availability of nutrients for algal blooms in lakes, since fish mineralize nitrogen and phosphorus through excretion, thereby making these nutrients available for primary producers (Schindler 1992).

1.6.2 Regulating Sedimentation Process

Fish like *Campostoma oligolepus*, *C. anomalum* and *Notropis nubilus* re-suspend silt, detritus and other particulate organic matter from the bottom into the current while feeding and thereby enhancing food availability for collector-filterers (Gelwick et al. 1997). Salmonids are reported to disturb the streams sediments 'bioturbation' inorder to lay their eggs (Montgomery et al. 1996). The salmon deposits her eggs in redds and covered with a layer of gravel to protect the embryos from rapid stream current. The spawning activity not only remove aquatic macrophytes and dislodges fine sediment particles (Field-Dodgson 1987) but also displaces many invertebrates from the bottom to the water column, making them available for the fish to feed on them (Bilby et al. 1998). The benthic communities of lakes are reported to get altered because of spawning of *Sarotherodon aurea* (Fuller and Cowell 1985). Transportation and redistribution of phosphorus and other essential nutrients between the shores, pelagic and deeper bottom zones are maily done by fishes in lake ecosystems (Carpenter et al. 1992).

1.6.3 As a Link Between Ecosystems

Fishes when fed upon by other organisms, serve as passive links between aquatic, aerial and terrestrial ecosystems, contributing to other food webs. The salmonids migrate from sea to rivers for egg laying and die there. There are about 22 species of birds and mammals reported to feed on the carcasses of coho salmon (Cederholm 1989). Birds depredate heavily on fish and about 20,000–25,000 t of fish are fed by birds annually in the Baltic Sea itself (Sparholt 1994). The birds take food and nutrients from fish in the form of food and their guano are a good source of nutrients to seagrass meadows. This shows the nutrient flow from fish to bird and to plants. It is also reported that the abundance of fish like sockeye salmon, *O. nerka* have a greater influence on the survival of bald eagles, coyotes, minks, river otters and grizzly bears (Spencer et al. 1991).

Migration of salmonids helps in the transfer nutrients and carbon from sea to rivers. The carcase of salmonids and this nutrient transfer is a substrate for many microbes to develop apart from food sources for many fishes (Bilby et al. 1996; Larkin and Slaney 1997). Thus, the marine-derived nutrients are made available in rivers especially when the river water is lacking nutrients due to very less litter fall. On the contrary, eels go to sea for spawning from fresh and brackish waters.

1.7 Information Services/Bioindicators: Assessing Ecosystem Stress, Pesticides etc

Many fishes are suitable for early-warning signals of anthropogenic stress on aquatic ecosystem. Some of the species are used as indicators for ecosystem reclamation process too (Harris 1995; Moyle and Moyle 1995; Balk et al. 1996) and of resilience (Carpenter and Cottingham 1997). The distribution pattern of butterflyfishes (Chaetodontidae) in the reefs of Hawaii and Sri Lanka are suggested as indicators of disturbance caused by human activities as they are highly correlated with live coral reefs (Reese 1995; Ohman et al. 1998). Further, the fish fauna available now can be a good indicator for the changes happened in the past. Many informations of earlier climate and the changes happened can be derived by studying them (Holmlund and Hammer 1999).

Detection of low levels of pesticides in aquatic environment by analytical techniques alone may be very difficult as most of them may fall below detection limits (Chandrasekara and Pathiratne 2007). Fish can be used as an indicator for water pollution. The two important features of fish, being available in wider habitats and their greater response to contaminants makes them as a good indicator species (Priya et al. 2012; Hernandez-Moreno et al. 2010). Many specialized fishes are used as bioindicators such as the Black ghost knifefish, *Apteronotus albifrons*, whose electric organ discharges are used to detect the presence of potassium cyanide in water (Thomas et al. 1996) and thus may be useful as a bioindicator for water quality (Thomas et al. 1997).

1.8 Cultural Services

Fish was associated to man from pre historic era and fishing is an ancient practice that dates back at-least to the Upper Paleolithic period. Sea has played a founding or important role in the cultural definition of the community living in the sea shores. In the second century BC, hunting for swordfish by using a harpoon was described in the Greek Polybius's Histories (Polybius 1889). In the Book of Jonah in Bible a 'great fish' swallowed Jonah the Prophet. The life of Jesus on earth was closely associated with the sea and the fish. Many of the 12 disciples of Jesus were fishermen and fish was almost regarded as their staple food. Legends of half-human, half-fish mermaids have featured in many stories apart from art, literature and movies. Many deities are said to take the form of a fish or half-man, half-fish. The astrological symbol Pisces is based on a constellation and as that of Piscis Austrinus.

1.9 Aesthetic and Recreational Values

Domesticated fishes in aquaria and others in reefs, streams, lakes and coasts are considered aesthetically valuable (Moyle and Moyle 1995). These ecosystems are believed to substantially contribute to the well being of humans by supporting recreational activities and providing aesthetic splendor (Ghermandi et al. 2010). Fishing and hunting are said to be aesthetic exercises which make the person to get involved fully and thus a sense of immersion experience. In fact, human beings developed such an extraordinary liking for fishing that became a source of recreation. Consumptive recreation involves fishing, angling etc. whereas swimming diving, boating, snorkeling, sunbathing and watching are under non consumptive. Most of the literature available investigates the value of open sea angling and a few on shore fishing or shell fishing (Ghermandi et al. 2010). Sport fishing of wild and stocked game fishes in lakes, rivers and along coasts has become one of the most popular recreational activities internationally (FAO 1996). Apart from fishing, there could be surfing, wakeboarding, water skiing, boat racing and exploring deeper water through snorkeling and scuba diving. The non consumptive recreation associated with sandy shores and beaches are more valued than recreational fishing.

Beach tourism is the major component of global tourism which resulted in creation of facilities and infrastructure causing changes in environmental and social well being. Coral reefs occupy only 0.1–0.5 % of the ocean floor (Moberg and Folke 1999) but provides considerable cultural, recreational and aesthetic value and promote national and international tourism apart from providing livelihood options for locals. Under water museums and public aquariums teach visitors about marine science and conservation. The new style aquariums also allow the visitors to experience the beauty and mystery of the underwater world. Today's immersion-style exhibits take the visitors to humid rainforests, sun-dappled reefs and through deep ocean twilight (Semczyszyn 2013).

Aquarium keeping is amongst the most popular hobbies all over the world (Livengood and Chapman 2011). Aquarium and ornamental fish trade are increasing and becoming more commercially important. The value of the worldwide trade on aquarium is about $278 millions (FAO 2005). Fresh water fishes are used for aquariums but the use of marine ornamental fish is also gaining importance apart from establishing marine reef mini-ecosystems within the aquarium. Home aquariums with living corals have recently become more popular (Delbeek 2001). Around 2000 different fishes are being traded as ornamental fishes (Livengood and Chapman

2011). Apart from public places, ornamental fishes are being kept and maintained at homes too.

1.10 Other Assorted Importance

Besides giving food, nutrients and valuable medicines to humans, fish are useful in the management of vector-borne diseases like schistosomiasis and malaria (Holmlund and Hammer 1999). It is reported that the mosquitofish (*Gambusia* spp.) consume many disease causing/spreading invertebrates and plants living in aquatic ecosystem (Marchall and Maes 1994; Moyle and Moyle 1995). Grass carp are reportedly used to suppress unwanted vegetation in water bodies (Lodge et al. 1998). When fish are used in rice paddies, many arthropod crop pests are being preyed upon by Nile tilapia, *Oreochromis niloticus* and common carp, *Cyprinus carpio* acting as biological control agent (Halwart et al. 1996). However, fish can be a source of human exposure to contaminants present in water (Kris-Etherton et al. 2002). Feeding on contaminated fish may leads to pesticide contamination in humans also. Careful selection and feeding of fish species is really beneficial.

2 Routes of Pesticide Exposure to Fish

Before studying the routes of pesticide exposure to fish or other aquatic organisms, one must know how pesticides get enter into the aquatic environment. Pesticides enter into the aquatic ecosystem by many unintentional ways such as wash off from land through rains or water, spray drifts, drains from irrigational water etc. apart from deliberate usage as chemotherapeutic agents for pests and as herbicides for controlling aquatic weeds. The major route of pesticide entry into aquatic environment is by wash off by water or rain from pesticide applied fields. Once it is applied in fields, due to erosion or water movement, pesticides also move throughout the watershed and may cause ill effects far from the site of application. Rivers and streams are even regarded as good conduits through which pesticides (either intact or breakdown products), flow to the sea also (Ewing 1999). Although pesticides get diluted in rivers and streams, many different mechanisms make the chemical to get concentrated and make it often toxic.

2.1 Routes of Pesticide Entry into Aquatic Ecosystem

The routes of pesticide transport to different aquatic ecosystems are through water, spray drifts, disposal through wastes, drains from agricultural fields and deliberate use in aquatic ecosystem (Rao 1999). It is well known that pesticides are carried out to rivers and streams by various means such as surface run-off, leaching etc. A portion of all pesticides applied to forests, croplands and road sides get enter into water bodies either by runoff or by leaching through groundwater into ephemeral streams and lakes (Graymore et al. 2001). Accidental spillage of pesticides in aquatic ecosystem is also reported as in the case of nabam and endrin into Mill River, Prince Edward Island which caused extensive mortalities of brook trout and juvenile Atlantic salmon (Saunders 1969).

2.1.1 Washed Off from Land into Water by Rain

Water promotes the transport of pesticides from their site of application and streams are often the recipients of pesticide residues following rainfall. Fenamiphos poisoning was reported to cause mass fish kills when washed into waterways from golf courses after heavy rains (Schmidt 2006). Pesticides in streams typically reach high levels for short periods of time and decrease to very low or undetectable levels but during the brief period of high concentration, it causes damage to aquatic organisms especially the developing stages of fishes (Ewing 1999). Depending on the nature of the pesticide, they may attach to sediment particles, accumulate in the tissues of various organisms or become buried in the sediment (Barbash and Resek 1996). Soil erosion may also deliver pesticides to the rivers along with water. Aquatic environments are particularly prone to pesticide contamination which is indicated by its groundwater ubiquity score (GUS). The GUS index of carbofuran is 4.5, indicating a relatively high risk of being transported from land to water (Plese 2005). Floods make the pesticides applied on lands to be carried to agricultural streams. Surface runoff due to rainfall is the major transport mechanism for pesticides to get into water bodies from land (Richards and Baker 1993; Fenelon and Moore 1998; Liess et al. 1999).

Pesticides are frequently detected in urban streams also. In the urban streams in western Washington, the pesticide concentrations often exceeds those reported in surface waters within agricultural areas (King et al. 2014). Bifenthrin from urban runoff was found in river water following rain events reaching 14.6 ng/L, which are reported to cause no mortality to salmon but highly toxic to their preys (Weston et al. 2015). Pesticides get metabolized and the products may be less or equally or more toxic than the parent compound and remain in the aquatic ecosystem for a long time. Many of the pesticides which were banned long ago, still appear in water quality surveys (Ewing 1999).

2.1.2 As Drifts from Aerial and Ground Sprays

Pesticides have been known to enter the aquatic ecosystem from drift from aerial and ground application. Windborne pesticides are found entering the aquatic ecosystem contributing to amphibian declines in pristine locations (Davidson and Knapp 2007). Azinphos-methyl and endosulfan sprayed in fruit orchards in Western

Cape, South Africa was found in farm streams and also in the Lourens River which is approx. 2.5 km away (Schulz et al. 2000). Brook trout, *Salvelinus fontinalis* was exposed to low levels carbaryl that drifted from the spraying of nearby forests into streams (Wilder and Stanley 1983). The environmental concentration of carbofuran in surface water was estimated to vary from 5.2 to 36 μg/L at maximum rates of application (USEPA 2006). Pesticides especially of DDTs and chlordanes were found atmospherically deposited in fish habitats at high elevation areas too (Ackerman et al. 2008).

2.1.3 Through Irrigation Water Drains and Rice Lands

A survey on the pest management of rice and rice-fish farmers of Mekong Delta, Vietnam revealed the use of about 64 different pesticides of which 50 % are of insecticides. The main insecticides used were pyrethroids (42 %) carbamates (23 %) and cartap (19 %) (Berg 2001). Carbofuran concentration in irrigated rice fields in southeast Brazil was reported to be up to 233 μg/L in laminar water (Plese 2005). In rice land agroecosystem, all organisms including larvivorous fishes can be affected by the pesticides used in rice (Roger and Bhuiyan 1990; Cagauan 1995).

Surface and ground flows from agricultural lands carry pesticides directly into the water bodies (Akan et al. 2014). The irrigation water drain from the pesticide used fields will normally have high amount of pesticides. Endosulfan, fenvalerate and azinphos methyl were found at tidal creek sites adjacent to agricultural fields (mostly tomato) in South Carolina (Scott et al. 1994). Carbofuran was detected and estimated to a level of 264 μg/L in agricultural field drains, whereas it was 26 μg/L in the headwater stream (Matthiessen et al. 1995). This clearly shows a transient runoff of carbofuran from farm lands to streams.

2.1.4 As Chemotherapeutic Agents for Pests of Aquatic Organisms

Many pesticides are used directly in the aquatic ecosystem for the management of pests of the fish or any other organism of concern. This may not affect the fish directly but can cause chronic sublethal toxicities. Carbaryl is used to control burrowing shrimps, *Neotrypaea californiensis* and *Upogebia pugettensis* on commercial oyster, *Crassostrea gigas* beds for a period of five decades in Willapa Bay, Washington. This pesticide application affects the non-target aquatic organisms and the exposure to fishes with greater site fidelity and benthic foraging will be more than those feed primarily within the water column (Troiano et al. 2013). This carbaryl application often caused significant short term mortalities in peracarid crustaceans but generally found recruited back within 3 months (Dumbauld et al. 2001). Imidacloprid is also used for the control of burrowing shrimp but applied directly to the exposed sediments when the tide is out and found highly effective. The imidacloprid residues in water peaked within 10 min of the first tidal flow after application but soon (within 30 min) reached non detectable levels because of its wider dispersion (Dumbauld et al. 2001). Atlantic salmon, *Salmo salar* is cultivated in large marine cages commercially. In these commercial fish farms, cypermethrin is commonly used for the control of ectoparasitic lice (Boxaspen and Holm 2001). This application also affects the non target organisms in aquatic ecosystem.

2.1.5 Through Pesticide Application to Control Harmful Water Insects

High pesticide concentrations were found in fish after the aerial spray operations of endosulfan to control tsetse fly over 2500 km^2 in the Okavango Delta, Botswana. The concentration of residues found in living and dead fish were up to 0.19 and 1.5 mg/kg wet wt, respectively. However, the fish predators like fish eating birds and crocodiles were under low risk, since biomagnification was not noticed (Matthiessen et al. 1982). Pesticides which are used for the management of mosquitoes such as chlorpyrifos in salt marshes were found to affect fishes especially the mummichog (*Fundulus heteroclitus*), which are considered as an important biocontrol agent of mosquito larva itself. Fish exposed to field application of granular formulation of chlorpyrifos are found to get inhibited in AChE by 56–100 % and by 24 h after second field application, 18.6 % fish were found dead (Thirugnanam and Forgash 1975). Bluegills, *Lepomis macrochirus* and largemouth bass, *Micropterus salmoides* were also found to be affected by application of chlorpyrifos at the recommended rates for mosquito control (Macek et al. 1972). Fenitrothion and chlorpyrifos are used as larvicides for the management of midges in aquatic ecosystem. Residues of chlorpyrifos was found to be high in bottom-dwelling and frequenting fish, such as channel catfish, *Ictalurus punctatus* and to some extent, black crappie, *Pomoxis nigromaculatus* (Mulla et al. 1972).

2.1.6 Through Herbicides on Aquatic Ecosystem

United States Environmental Protection Agency (US EPA) and the California Department of Pesticide Regulation (DPR) has allowed for the use of many aquatic herbicides which include organic chemicals and copper-based products for the control of nuisance weeds and algal blooms. These herbicides include acrolein, copper sulfate, copper ethanolamine, copper ethylenediamine, copper carbonate, diquat dibromide, dipotassium salt of endothall, fluridone, glyphosate isopropyl amine, imazapyr, triclopyr triethylamine, 2,4-D dimethyl acetate and 2,4-D butoxyethyl ester (Siemering et al. 2008). Diquat and fluridone are also registered as aquatic herbicides (Freeman and Rayburn 2006). Acrylaldehyde (acrolein) is used for the management of eloda (*Eloda canadensis*), ribbon weed (*Vallisneria spiralis*), pondweed (*Potamogeton tricarinatus*) etc. in irrigation canals and other water bodies. In flowing water, the dissipation of acrolein was reported to follow a rate constant (K) of 0.16/h (Bowmer and Sainty 1977) and mainly attributed to volatilization and adsorption but the non-volatile reaction product accumulated initially but found dissipated rapidly, probably by microbiological processes (Bowmer and Higgins 1976).

2.2 *Routes of Pesticide Exposure to Fish*

To know the major routes of pesticide exposure to aquatic systems and biota, the following global biocycles are to be considered (Murthy et al. 2013):

1. The water column, which usually first comes into contact with pesticides
2. Organic substrates (algae, mosses, hydrophytes, leaf litter etc.)
3. Inorganic substrates including sedimentary materials.

Aquatic biota acquires pesticides in many ways. Since they are in the water itself they are being contaminated every time and the three important ways (Kerr and Vaas 1973) are as follows:

(a) Direct uptake of contaminated food
(b) Direct absorption from water through gills
(c) Absorption through integument

Each of the above said routes vary with animal, pesticide compound and environmental conditions. It is being reported that absorption through gills and gut are the two important routes of pesticide entry in fish. The relative importance of uptake via food or water depends on the conditions of exposure, duration, dose level and the individual fish (Huckle and Millburn 1990).

2.2.1 Through Respiration/Through Gills

As aquatic organisms especially fish have their outer bodies and gills almost fully and continuously exposed to water and thus the effect of toxicants on the respiration is more pronounced. In fish, the uptake of pesticides takes place through absorption by gills (Holden 1962; Panigrahi et al. 2014) which is related to the metabolic rate and the body size (Murphy and Murphy 1971). Under acidic conditions, fish gills may absorb the free divalent ions of the pesticide molecule directly from the water (Babu et al. 2005). Hamelink et al. (1971) proposed that the chlorinated hydrocarbon uptake by aquatic organisms is primarily through the transfer from water to blood through gills and blood to lipids. Dimethoate is found efficiently absorbed across the gill and get diffused into the blood stream causing ill effects to the fish (Kalavathy et al. 2001). The factors affecting respiration are also found to influence the rate of chemical uptake across the gills (Nowell et al. 1999). A linear correlation between oxygen and DDT uptake is thus demonstrated (Murphy and Murphy 1971).

2.2.2 Through Food Uptake

Absorption of pesticide through contaminated food is one of the main routes of pesticide toxicity in fish (Macek and Korn 1970). The pesticide molecules entering into the aquatic environment not only affect fishes but also the phytoplanktons, zooplanktons and many insects. These organisms also imbibe pesticide contaminants in the body and often accumulate them. Fish feeds primarily on these organisms and get toxicity through feeding the already contaminated organisms as food. Many toxic insecticides and herbicides are liposoluble and selectively accumulate in the lipids of phytoplantonic (Kerr and Vaas 1973) and zooplantonic organisms (Southward et al. 1978; Leger et al. 1986) and insects like dragonfly nymph (Wilkes and Weiss 1971). These phyto or zooplanktons also acquire pesticides directly from water through the body surface (Nowell et al. 1999) and in turn cause pesticide poisoning in fish. It was reported that the brook trout, *Salvelinus* sp. absorbed ten times more of the available DDT and its metabolites through food than from the water (Macek and Korn 1970; Ewing 1999).

2.2.3 Through Integument Contact and Dermal Absorption

Uptake of pesticides by an aquatic organism occurs through partitioning i.e. diffusion through surface membranes from the surrounding media. All the while, the organisms living in water have to combat with the pollutants in water especially of the break down products of pesticides since they live in that medium. Many studies demonstrate direct uptake of contaminants from water by aquatic organisms (Nowell et al. 1999). Fish get exposed by dermal contact and direct absorption through the skin by swimming in pesticide contaminated water (Akan et al. 2014; Sabra and Mehana 2015). Epidermal skin lesions in fish and loss of scale are also reported because of contaminated water (Dey et al. 2001).

2.2.4 Uptake of Sediment-Sorbed Chemicals

The hydrophobic contaminants that get into the aquatic environment may settle down with the sediments or get attached with the suspended particles (Voice and Weber 1983; Knezovich et al. 1987). Contaminants get absorbed or adsorbed in the sediments get transferred to aquatic biota in the following three different ways: (1) Interstitial water, (2) Ingested sediments (both organic and inorganic) and (3) Direct body contact with sediment particles. The type of sediment, the class of chemical and the species of organism appears to determine the way of transfer (Knezovich et al. 1987).

It may be expected that the interstitial waters contain higher concentrations of contaminants than the water column since the pesticides go and accumulate in the sediments (Magnusson et al. 2013). In marine ecosystem, the interstitial water toxicity of polychlorinated biphenyls is generally greater than either bedded or suspended exposures (Burgess et al. 1993). Contrarily, another report states that the pesticide content of interstitial water and sediments, as much lower than that in the water column. The living organisms in the running water ecosystem are less contaminated than that of those in standing waters. Permethrin residues were found in high concentrations in ponds upto 147 μg/L after it was applied in nearby forest areas of 640 ha. However, the pesticide was not found in the bottom sediments and the contamination in streams were also low (2.5 μg/L) (Kingsbury and Kreutzweiser 1980).

Chemicals sorbed to suspended particles may be taken by filter feeders and deposit feeders (Langston 1978). Ingestion of sorbed suspended particles plays as a major entry route of pesticides in suspension feeding fishes (Nowell et al. 1999). The roof of the oral cavity where the food particles are retained may be a route of contaminant absorption. Fishes especially of the bottom feeders are readily exposed to greater quantities of chlorinated hydrocarbons that get accumulated in the sediments (Kidwell et al. 1995) than the top feeders.

Direct sediment contact may play an important role in faunal organisms i.e. the benthic organisms that live within the bottom substratum of water (Fowler et al. 1978; Nowell et al. 1999). In a study to find a comparison of pesticides in sediments and fish (tilapia, *T. mossambica* and orangemouth corvina, *Cynoscion xanthulus*) tissues of Salton Sea, California, USA, pesticides like dimethoate, diazinon, malathion, chlorpyrifos and disulfoton was found varied from 0.1 to 9.5 ng/g dry wt in sediments and from 0.1 to 80.3 ng/g wet wt in fish tissues (Sapozhnikova et al. 2004).

2.2.5 Bioaccumulation and Biomagnification

Bioaccumulation is the concentration of pollutant in one organism and magnification is increasing the concentration of pollutant from one link of food chain to the other. Many of the pesticides in earlier use are fat loving and get concentrated in fatty tissues of the organisms. Modern pesticides are usually more of water soluble and do not accumulate in high concentrations in fat deposits. However, they also do show some bioaccumulation properties. In the process of bioaccumulation, pesticides get absorbed in the living tissues and get accumulated in levels much higher than that of the surrounding water. The bioaccumulation factor of hexachlorobenzene, chlorpyrifos, chlorothalonil and pentachlorophenol to rainbow trout are reported as 5500, 1374, 840 and 100, respectively (Veith et al. 1979; Racka 1993; WHO 1996; Hattula et al. 1981). The fat solubility of the pesticide makes them to get deposited in lipid rich eggs and embryos causing serious hazards. When the aquatic organisms are continuously exposed to pesticides, the tissue concentrations will be very high than that of the surrounding. The biological accumulation of aroclor to channel catfish, *I. punctatus* was up to 61,190 times the levels in water after 77 days (Mayer et al. 1977). In addition, exposure of many types of pesticides may lead to interactions between them that increase their toxicity (Cook et al. 1997).

Pesticides are not only taken up directly from the environment, fishes can also absorb them from food organisms. A study of DDT accumulation in brook trout, *Salvelinus* sp. (Macek and Korn 1970) showed that ten times more of the available DDT and its metabolites were absorbed from food than from the water. This can be magnified still further as the pesticide travels through food chains (Ewing 1999). In a study at Ignacio Ramirez reservoir of Mexico, organochlorine and organophosphate insecticides were bio-accumulated at a rate of 2 to 10-folds from water to algae and get biomagnified as 10 to 25-folds in zooplankton and 8 to 140-folds in fish (Favaria et al. 2002). Insecticide bioaccumulation to a range of three to sixfolds was found in mackerel icefish, *Champsocephalus gunnari* when compared to its prey (Antarctic krill, *Euphausia superba*). The weddell seal, *Leptonychotes weddellii* and Southern elephant seal, *Mirounga leonina* biomagnified the persistent organic pollutants by about 30 to 160-folds relative to the krill (Goerke et al. 2004).

3 Effects of Pesticides on Fishes

Pesticides affect fishes in various ways such as by causing acute mortality or by causing sublethal effects on them. Acute mortality is commonly expressed in median lethal concentrations (LC_{50}) and a pesticide having low LC_{50} values are highly toxic and vice versa. Median lethal time determines the time for the occurrence of mortality and in some way differentiates the toxicant as acute or chronic. Behavioural changes like avoidance, excitation, changes in respiration, prey capture and feeding can also be due to pesticide toxicity. Pesticides can also cause damage to body and affect reproduction apart from disrupting the normal functioning of internal organs. Pesticide induced changes in blood cells, enzymes and hormones are also reported.

3.1 Toxicity of Pesticides to Fish: Mortality

Chlorinated hydrocarbon pesticides are reported as highly toxic to fishes. All the common ten chlorinated hydrocarbon pesticides except BHC were reported extremely toxic than the most toxic organophosphorus pesticides to fish with 96 h TLm values (median tolerance limits) generally below 0.1 ppm. Bluegills were the most sensitive fish, followed by fathead minnows, goldfish and guppies (Henderson et al. 1959). Thirteen commonly used organo phosphorus insecticides tested, exhibited a extremely wide range in toxicity to bluegills, guppies, fathead minnows and goldfish with 96 h median tolerance limits ranging from 0.0052 to 610 ppm (Pickering et al. 1962). The organophosphate insecticide, chlorpyrifos was reported as less toxic to fishes compared to other chlorinated hydrocarbon insecticides (Ferguson et al. 1966) but generally more toxic than organo phosphorous counterparts (Henderson et al. 1960). The mosquito fish, *Gambusia* sp. is more tolerant to chlorpyrifos than golden shiners, *Notemigonus* sp. and green sunfish, *Lepomis* sp. (Ferguson et al. 1966). The commonly used pyrethroid insecticides are also reported extremely toxic to fish, with 96 h LC_{50} values generally below 10 μg/L. Pyrethroids are highly toxic to fish even at concentrations 1000 times less than that cause toxicity to mammals and birds (Bradbury and Coats 1989). Insecticide, derris (rotenone) widely known for its toxicity to fish since ages, is used by natives in Asia and South America (Holden 1973). Data of 410 chemicals (mostly pesticides) tested for acute toxicity of fishes revealed brown trout (*Salmo trutta*) as the most sensitive fish, followed by rainbow trout, largemouth bass (*M. salmoides*), cutthroat trout, bluegills, Coho salmons, yellow perch (*Perca flavescens*), channel catfish, common carp, black bullhead (*Ameiurus melas*), green sunfish, fathead minnows and goldfish (Mayer and Ellersieck 1986). Diafenthiuron was found to be highly toxic to *C. carpio* registering 100 % mortality even at ten times less dose than the recommended dose (Stanley et al. 2016).

Apart from insecticides, some fungicides like thiram (Mayer and Ellersieck 1986), difenoconazole (Rouabhi 2010) trifloxystrobin (Zhu et al. 2015) and herbicides like acetochlor, bensulide, diclofop, oxadiazon, oxyfluorfen and pendimethalin are reported toxic to fishes. Chemical pesticides such as ammonia, phenols, cyanide and the salts of some metals are highly toxic to fish (Holden 1973). Herbicide, acrolein was reportedly responsible for the death of steelhead, coho salmon, rainbow trout and many non-game fishes in Bear Creek, a tributary of the Rogue river (Ewing 1999).

3.1.1 Acute Toxicity: Median Lethal Concentration (LC_{50})

The median lethal concentration tests are usually done for 96 h exposure and the lower the values shows the pesticides as toxic and vice versa. The 96 h LC_{50} of carbofuran in freshwater fishes range from 88 μg/L in bluegill sunfish, *L. macrochirus* to 1990 μg/L in fathead minnow, *Pimephales promelas* with a 20-fold difference in sensitivity (USEPA 2004). The LC_{50} values of carbaryl to walking catfish, *Clarias batrachus* was reported as 5.248 ppm (Wasu et al. 2009) which is high when compared to coho salmon, *Oncorhynchus kisutch* (1.3 ppm), brook trout (1.07 ppm), rainbow trout, *Salmo gairdneri* (1.47 ppm) (Kartz 1961) and juveniles of *Clarius garipinius* (0.38 ppm) (Omitoyin et al. 2006). Tilak et al. (1981) studied the acute toxicity of carbaryl and its metabolite, 1-naphthol, to four species of fish. The calculated 96 h LC_{50} values of carbaryl for *Catla catla*, *Anabas testudineus*, *Mystus cavasius* and *Mystus vittatus* are 6.4, 5.5, 4.6 and 2.4 ppm, respectively whereas it was 4.3, 3.0, 0.33 and 1.1 ppm for 1-naphthol stating that the metabolite as highly toxic than the parent compound.

The LC_{50} values of metasystox to loach, *Nemacheilus botia* was 10.30, 9.13, 7.88 and 7.02 ppm after 24, 48, 72 and 96 h, respectively (Nikam et al. 2011). A 24 h LC_{50} of 4.5 and 20 mg/L for common carp, *C. carpio* and silver carp, *Ctenopharyngodon idea*, respectively is also reported (Whalon et al. 1990). The acute toxicity (LC_{50}) calculated over a 96 h period in *C. catla* for cypermethrin, fenvalerate and deltamethrin were found to be 4, 6 and 55 μg/L, respectively, stating toxicity of cypermethrin as 13-fold greater than deltamethrin (Tandon et al. 2005). The LC_{50} values of carbosulfan, glyphosate and atrazine to *Channa punctatus* in a semi-static system was 0.268, 32.54 and 42.38 mg/L, respectively, indicating the toxicity of carbosulfan (Nwani et al. 2010a). The acute toxicity of pesticides to fish varies with temperature and seasons. The 96 h LC_{50} values of cypermethrin to teleost fish, *Colisa fasciatus* was 0.006 in winter season (water temp. 16 °C) and 0.02 mg/L in summer season (water temp. 28 °C) (Singh et al. 2010). But malathion was highly toxic to Chinook salmons at normal temperatures than cooler times. The LC_{50} of malathion to Chinook salmons were 274.1 and 364.2 μg/L at 19 and 11 °C, respectively (Dietrich et al. 2014).

The 96 h Median Lethal Concentrations (LC_{50}) of Different Pesticides to Fishes

Pesticide	Fish	Median lethal concentration (LC_{50})	Reference
Insecticides			
Abamectin	Mozambique tilapia, *T. mosambica*	6.92 ppb	Jasmine et al. (2008)
	Common carp, *C. carpio*	0.47 ppb	Jasmine et al. (2008)
Alpha cypermethrin	Guppy, *Poecilia reticulata*	9.43 μg/L	Yılmaz et al. (2004)
Azinphos methyl	Red drum, *Sciaenops ocellatus*	6.2 mg/L	Van-Dolah et al. (1997)
Carbaryl	Walking catfish, *C. batrachus*	5.248 ppm	Wasu et al. (2009)
	Coho salmon, *O. kisutch*	1.30 ppm	Kartz (1961)
	Brook trout, *Salvelinus* sp.	1.07 ppm	Kartz (1961)
	Rainbow trout, *S. gairdneri*	1.47 ppm	Kartz (1961)
	African sharptooth catfish, *C. garipinius*	0.38 ppm	Omitoyin et al. (2006)
	Catla, *C. catla*	6.4 ppm	Tilak et al. (1981)
	Climbing perch, *A. testudineus*	5.5 ppm	Tilak et al. (1981)
	Tengra, *Mystus cavasius*	4.6 ppm	Tilak et al. (1981)
	Striped dwarf catfish, *M. vittatus*	2.4 ppm	Tilak et al. (1981)
Carbofuran	Nile tilapia, *O. niloticus*	214.7 μg/L	Pessoa et al. (2011)
	Bluegill sunfish, *L. macrochirus*	88 μg/L	USEPA (2004)
	Fathead minnow, *P. promelas*	990 μg/L	USEPA (2004)
	Rainbow trout, *Oncorhynchus* sp.	0.268 mg/L	Nwani et al. (2010a)
Chlorpyrifos	Turbot, *Psetta maxima*	94.65 μg/L	Mhadhbi and Beiras (2012)
	African sharptooth catfish, *C. gariepinus*	0.86 mg/L	Nwani et al. (2013)

(continued)

Pesticide	Fish	Median lethal concentration (LC_{50})	Reference
Cypermethrin	Rohu, *L. rohita*	4.0 mg/L	Marigoudar et al. (2009a)
	Rainbow trout, *Oncorhynchus* sp.	0.5 mg/L	Bradbury and Coats (1989)
	Brown/Salmo trout, *Salmo trutta*	1.2 mg/L	Bradbury and Coats (1989)
	Catla, *C. catla*	4 µg/L	Tandon et al. (2005)
DDT	Rainbow trout, *Oncorhynchus* sp.	8.7 µg/L	Johnson and Finley (1980)
Diazinon	Turbot, *P. maxima*	1.23 mg/L	Mhadhbi and Beiras (2012)
Dimethoate	Rohu, *Labeo rohita*	24.55 µg/L	Dey and Saha (2014)
	Barb, *Puntius stigma*	7.8 ppm	Bhandare et al. (2011)
	Stinging catfish, *Heteropneustes fossilis*	2.98 mg/L	Pandey et al. (2009)
Deltamethrin	Catla, *C. catla*	55 µg/L	Tandon et al. (2005)
Endosulfan	Snakehead murrel, *Channa striatus*	0.0035 ppm	Ganeshwade et al. (2012)
	Common carp, *C. carpio*	22 ppb	Jenkins et al. (2003)
	Mrigal carp, *Cirrhinus mrigala*	1.06 µg/L	Ilyas and Javed (2013)
Fenvalerate	Walking catfish, *C. batrachus*	1.35 µg/L	Datta and Kaviraj (2011)
	Catfish, *C. punctatus*	1.0 µg/L	Datta and Kaviraj (2011)
	Stinging catfish, *H. fossilis*	0.65 µg/L	Datta and Kaviraj (2011)
	Catla, *C. catla*	6 µg/L	Tandon et al. (2005)
Lambda cyhalothrin	Zebra fish, *Danio rerio*	0.119 µg/L	Ansari and Ahmed (2010)
	Rohu, *L. rohita*	0.0021 ppm	Muthukumaravel et al. (2013)
	Rohu, *L. rohita*	0.7 µg/L	Dey and Saha (2014)
	Common carp, *C. carpio*	0.160 µg/L	Bibi et al. (2014)
Malathion	Rohu, *L. rohita*	15.0 mg/L	Thenmozhi et al. (2011)
	Teleost fish, *C. fasciatus*	0.02 mg/L	Singh et al. (2010)
	Stinging catfish, *H. fossilis*	0.98 ppm	Deka and Mahanta (2012)
Metasystox	Loach, *N. botia*	7.018 ppm	Nikam et al. (2011)
Methyl parathion	Catla, *C. catla*	4.8 ppm	Ilvazhanan et al. (2010)
Monocrotophos	Rohu, *L. rohita*	0.0036 ppm	Muthukumaravel et al. (2013)
Propoxur	Brackish water catfish, *Chrysicthys nigrodigitatus*	22.0 mg/L	Hanson et al. (2007)
	Nile thilapia, *O. niloticus*	30.4 mg/L	Hanson et al. (2007)
	African sharptooth catfish, *C. gariepinus*	45.0 mg/L	Hanson et al. (2007)

(continued)

Pesticide	Fish	Median lethal concentration (LC_{50})	Reference
Fungicides			
Copper sulfate	Catfish, *C. batrachus*	33 μg/L	Kumar et al. (2015)
Cyproconazole	Carp	18.9 mg/L	Rouabhi (2010)
	Trout	19.0 mg/L	Rouabhi (2010)
	Bluegill sunfish, *L. macrochirus*	21.0 mg/L	Rouabhi (2010)
Mancozeb	Catfish, *C. batrachus*	14.36 mg/L	Srivastava and Singh (2013)
Triforine	Bluegill sunfish, *L. macrochirus*	>1000 mg/L	Worthing (1983)
Thiram	Bluegill sunfish, *L. macrochirus*	0.23 mg/L	Mayer and Ellersieck (1986)
	Trout	0.13 mg/L	Mayer and Ellersieck (1986)
	Carp	4 mg/L	Mayer and Ellersieck (1986)
Herbicides			
Alachlor	Rainbow trout, *Oncorhynchus* sp.	2.4 mg/L	Johnson and Finley (1980)
	Turbot, *P. maxima*	1.84 mg/L	Mhadhbi and Beiras (2012)
Atrazine	Rainbow trout, *Oncorhynchus* sp.	42.38 mg/L	Nwani et al. (2010a)
	Turbot, *P. maxima*	9.95 mg/L	Mhadhbi and Beiras (2012)
Butataf	Nile tilapia, *O. niloticus*	0.2 ppm	Abdel et al. (2007)
Diuron	Turbot, *P. maxima*	7.82 mg/L	Mhadhbi and Beiras (2012)
Glyphosate	Fathead minnow, *P. promelas*	2.3 mg/L	Folmar et al. (1979)
	Rainbow trout, *Oncorhynchus* sp.	32.54 mg/L	Nwani et al. (2010a)

3.1.1.1 Median Lethal Time (LT_{50})

The median lethal time of chlorpyrifos at 1000 ppb to mosquito fish, golden shiners and green sunfish collected from insecticide contaminated locality are reported to be 13.75, 34.0 and 5.0 h, respectively whereas it was 3.5, 2.5 and 3.75 h for fishes from uncontaminated area (Ferguson et al. 1966). Chlorpyrifos treated Guinean tilapia, *Tilapia guinensis* exhibited a lethal time of 42.7 h for a concentration of 0.025 mg/L (Chindah et al. 2004). The median lethal time for deltamethrin concentrations of 0.25 and 50 μg/L were estimated to be 212 h 55 min and 1 h 33 min, respectively to rainbow trout, *O. mykiss* (Urala and Saglam 2005).

3.1.1.2 Baseline Toxicity Levels

In general, LC_{50} values are used to compare the acute toxicity of pesticides to fishes whereas the LC_0 are the concentrations where no mortality occurs and LC_{100} cause 100% mortality. The lethal and sub-lethal concentration of malathion to *L. rohita* were found to be 25 (LC_{100}) and 5 mg/L (LC_0), respectively (Thenmozhi et al. 2011). The LC_0 and LC_{100} values of deltamethrin to common carp and rainbow trout are reported as 0.5 and 2.5 μg/L and 2.14 and 6.08 μg/L, respectively (Velisek et al. 2006, 2007). The LC_0 values of cypermethrin and bifenthrin to rainbow trout are reported as 1.98 and 1.04 μg/L whereas their LC_{100} are 4.96 and 2.09, respectively (Velisek et al. 2009a, b). The LC_0 and LC_{100} values of cypermethrin to common carp, *C. carpio* are reported as 1.82 and 4.64 μg/L, respectively (Dobsikova et al. 2006).

3.1.1.3 LOECs and NOECs

The activity of cholinesterase decreased with the exposure of carbofuran to juveniles of *O. niloticus* with a lowest observed effect concentration (LOEC) of 69.9 μg/L. At an LOEC of 40.6 μg/L, the vision of fish was found to get impaired. At a concentration of 397.6 μg/L a reduction in swimming speed and prey attacks was noticed (Pessoa et al. 2011). The LOEC of 2,4-D on growth of swim-up larvae of rainbow trout, *O. mykiss* was 108 mg/L, whereas the no observable effect concentration (NOEC) was 54 mg/L (Fairchild et al. 2009). The NOEC and LOEC of endosulfan to Florida flagfish, *Jordanella floridae* were reported as 3.3 and 10.8 μg/L, respectively (Beyger et al. 2012).

3.1.2 Chronic Mortality

Chronic mortality/toxicity are tested by exposing fish to a low dose of pesticide for a long time. Endrin incorporated diet fed to goldfish at a rate of 143 μg/kg body weight per day caused mortality after 3–4 months (Grant and Mehrle 1970). Dietary exposures of coho salmon, *O. kisutch* to Aroclor 1254 at rates of 1.45 mg/kg body weight per day caused complete mortality in 265 days (Mayer et al. 1977). Chronic exposure of early life stages of carp, *C. carpio* to prometryne at 15 % of 96 h LC_{50} showed no mortality to the fish (Stara et al. 2012). Chronic exposures of monocrotophos and butachlor either alone or in combination at doses of one-fifth of LC_{50} to *C. catla* for 35 days did not cause mortality (Anbumani and Mohankumar 2015).

3.2 Sublethal Effects

Though sublethal concentrations do not cause immediate mortality, they interfere with the biology in many ways and can ultimately on its survival. Some of the sublethal effects include stress and thus increasing the chance of predation, altering swimming ability, interrupts schooling behaviour and migration, delayed spawning etc. (Ewing 1999). Sublethal toxicity affects the prey capture and feeding apart from causing damages to tissues or changes in cellular and biochemistry of tissues, internal organs and blood. Changes in enzymes and hormones; gene or chromosomal disruption are also reported due to pesticide poisoning. Studies are also focused on adverse effects of pesticides on various reproductive events such as the onset of puberty, gametogenesis, oocyte maturation, ovulation, spermiation, spawning, fecundity, fertilization, endocrinology of reproduction, embryogenesis, hatching and post-hatching metamorphosis (Kime 1998; Lal 2007; Lal et al. 2013). Indirect effects causing depletion in plankton abundance thus reducing the food availability, altering the aquatic habitat and making the fish susceptible to predation and to infectious diseases are also of prime importance apart from direct toxicity.

3.2.1 Behavioural Changes

Pesticides cause many behavioural changes in fish. The migration of the fish to the bottom of the tank following the addition of pesticide is a clear indication of avoidance which is demonstrated in fenvalerate treated trout (Murthy 1987) and cypermethrin treated rohu (Marigoudar et al. 2009a). Immediate migration to bottom was also reported in fenvalerate treated *C. carpio communis*, *Puntius sophore*, *C. idella*, *C. punctatus* and *A. testudineus* (Satyavardhan 2013). Green sunfish treated with chlorpyrifos also exhibited different symptoms like moving to the bottom of the containers, turning on their sides etc. whereas the mosquito fish and golden shiners came to the surface and showed a little evidence of spasms or struggling (Ferguson et al. 1966). The schooling behaviour of *L. rohita* was found to get disrupted when exposed to cypermethrin apart from irregular, erratic and darting movements followed with imbalanced swimming activity (Marigoudar et al. 2009a).

Abnormal behaviour like hyper excitability, fast swimming, increased coughing and yawning and profuse secretion of mucus are found in *C. carpio* treated with sublethal concentrations of endosulfan (Jenkins et al. 2003). The fresh water fish treated with fenvalerate exhibited a peculiar behaviour of leap out from the test chamber which can be viewed as escape phenomenon (Satyavardhan 2013). Diazinon, an organophosphate insecticide at concentrations as low as 1.0 μg/L was reported to induce antipredator responses in Chinook salmon, *Oncorhynchus tshawytscha* by influencing the olfactory mediated alarm response. Diazinon also impaired the homing behaviour of salmon (at 10 μg/L) whereas the swimming behaviour or visually guided food capture are not altered (Scholz et al. 2000). When goldfish, *Carassius auratus* were exposed to 5 μg/L of carbofuran, significant alterations in sheltering and burst swimming were found (Bretaud et al. 2002). Behavioural studies can be used comfortably for ecotoxicological assessments since they are regarded as sensitive end points (Little and Finger 1990; Drummond and Russom 1990; Cohn and MacPhail 1996).

3.2.2 Respiratory Changes

When *L. rohita* was exposed to cypermethrin, the respiratory system was found to get disturbed the most with repeated opening and closing of the mouth and operculum (Marigoudar et al. 2009a). Increased opercular movements were found in mrigal carp, *C. mrigala* exposed to cypermethrin (Prasanth et al. 2005) and catfish, *C. batrachus* exposed to the herbicide, herboclin (Krian and Jha 2009). Though the opercular movement increases initially, as the exposure time advances a significant reduction in opercular movements were reported with many pesticides. The decrease in opercular movement and corresponding increase in frequency of surfacing of fish due to pesticide exposure clearly indicates that fish adaptively shifts towards aerial respiration and tries to avoid contact with the pesticide through gill chamber (Santhakumar et al. 2000).

When *L. rohita* is exposed to sublethal concentrations of cypermethrin, the oxygen consumption got increased by 20 %, 39 %, 102 % and 110 %, at 1, 5, 10 and 15 days of exposure compared to control (Marigoudar et al. 2009b). Oxygen consumption of catfish, *C. batrachus* was found to get decreased in lethal concentrations of copper sulfate (−22.64 % to −70.13 %), but in sub lethal concentrations the decreased trend was improved and reached the normal level at 21st day (Kumar et al. 2015). Disturbance in oxidative metabolism has been reported as cypermethrin toxicity in *T. mossambica* (David et al. 2003). Oxidative stress in fish can be identified by assessing the induction of protein carbonyl in fish and can be used as a biomarker (Parvez and Raisuddin 2005).

3.2.3 Vision, Locomotion and Predator Evasion

Locomotion has been found to be consistently a sensitive measure for toxicity stress (Little and Finger 1990). With the recent development of computer-assisted electronics and video-camera tracking systems have greatly improved the assessment of locomotor behaviour and being used extensively with a high degree of precision (Noldus et al. 2002; Martin 2003). Japanese medaka, *Oryzias latipes* exposed to acute doses of chlorpyrifos shown a reduction in swimming speed to a tune of 72.7 % compared to control whereas in subacute doses, hyper activity was observed (Khalil et al. 2013). Van-Dolah et al. (1997) reported that red drums, *S. ocellatus* exposed to azinphos methyl for 6 h at 12 μg/L had reduced swimming stamina whereas its 96 h LC_{50} is 6.2 mg/L. Inhibition of AChE in fish exposed to organophosphates could be the reason behind the reduction in swimming speeds (Rao 2006; Rao et al. 2007). The catfish, *C. batrachus* exposed to copper sulfate was found to develop irregular swimming activity: rapid jerk movement, partial jerking and a cork-screw palter rotation along horizontal axis (Kumar et al. 2015).

Vision is essential for several activities like locomotion, prey detection, orientation towards prey, search for sexual partners, detection and escape from predators etc. (Pessoa et al. 2011). An inhibition of AChE activity in diazinon exposed Indian carp, *L. rohita* could have caused defects in optomotor response (Dutta et al. 1992).

Cyprinus carpio communis exposed to monocrotophos at sublethal concentrations (0.038–0.126 ppm) was found to develop cataract. Progressive lens degeneration was reported even after put in normal environmental and thus it was irreversible (Ravneet et al. 2009).

Escaping a predator attack is based on vision and other cues, proper timing, direction and locomotor performance at the time of escape (Domenici and Blake 1997; Webb 1986). Timing is also very important, as an early escape response can actually attract a predator, whereas a late response can lead to capture/death (Pessoa et al. 2011). Increased vulnerability to predator capture in carbaryl and pentachlorophenol exposed rainbow trout, *O. mykiss* (Little and Finger 1990) and mummichog, *F. heteroclitus* larvae (Zhou and Weis 1998, 1999) were reported. Skills of carbofuran exposed tilapia to escape predator attacks were reduced significantly (Pessoa et al. 2011).

3.2.4 Prey Capture and Feeding

Prey capture is a normal phenomenon for feeding and survival of fish. Pesticides are reported to affect the ability of the fish to search for the prey and capture them efficiently. The prey search area in volume of the carbofuran exposed larvae of *O. niloticus* (8.3 μg/L) was estimated to get reduced to less than 12 % of the prey search volume of the control fish (Pessoa et al. 2011). Decreased ability of Atlantic salmon, *S. salar* exposed to fenitrothion to capture brine shrimp, *Artemia* sp. is reported (Morgan and Kiceniuk 1990). When the hybrid striped bass, *Morone* sp. was exposed to diazinon, the ability to capture fathead minnows, *P. promelas* was found to get decreased (Gaworecki et al. 2009). The prey capture skills of carbofuran exposed tilapia, *O. niloticus* were significantly affected, as the control fish attacked *Daphnia magna nauplii* more frequently than the exposed ones (Pessoa et al. 2011).

Reduced feeding and food avoidance is also reported in pesticide poisoned fishes. The larvivorous potential of *Oryzias carnaticus* decreased by 3.1 and 4.6 times when treated with triazophos and iprobenfos as compared to control group (Ravindran et al. 2012). Goldfish exposed to carbofuran concentrations as low as 1 μg/L showed reduced attraction to chironomid extracts (Saglio et al. 1996). Malathion exposed catfish, *C. batrachus* did not feed the minced goat liver given as food whereas the control fish consumed almost all the ration. Exposed fish not only stopped feeding, rather turned away from the food trays, exhibiting complete food avoidance (Lal et al. 2013).

3.2.5 Sex and Reproduction

Pesticides at low concentrations may act as mimics or blockers of sex hormones, causing abnormal sexual development, feminization of males, abnormal sex ratios and unusual mating behaviour (Ewing 1999). Sublethal doses of malathion causes reduction in the ovarian weight and retard the growth of the pre-vitellogenic oocytes in tank goby, *Glossogobius giuris*. Degeneration of the immature oocytes and rupture of follicular epithelium was observed in higher doses (Mohan 2000). Based on the gonadosomatic indices, lindane, pentachlorophenol and propoxur were found to affect reproduction in *C. nigrodigitatus*, *O. niloticus* and *C. gariepinus* (Hanson et al. 2007). Dichlorvos at sublethal concentrations was found to decrease gonadosomatic index in female *C. carpio communis* with the increase in concentration. Histomorphological disorders in the ovaries were also reported in dichlorvos treated fish (Mir et al. 2012). *Heteropneustes fossilis* treated with carbofuran at sublethal doses (0.5, 1 and 2 mg/L) for 30 days were found to get affected in ovary by means of inhibition of oocyte maturation. The area and occurrence of various types of primary oocytes in the ovary was greatly altered in exposed fish compared to that of the control. Degeneration of follicular walls, connective tissues and vacuolization in the ooplasm were also observed in carbofuran treated fish (Chatterjee et al. 1997). Similar problems were reported in malathion treated *H. fossilis* (Dutta et al. 1994; Deka and Mahanta 2012) and diazinon treated bluegills, *L. macrohirus* (Dutta and Meijer 2003).

Short term exposure to esfenvalerate resulted in reduced fecundity of Australian crimson spotted rainbow fish, *Melanotaenia fluviatilis* and the eggs failed to hatch (Barry et al. 1995). The gravid females of mosquito fish at their terminal stages of gestation tend to abort when exposed to chlorpyrifos (Ferguson et al. 1966). Chlorinated insecticides at their sublethal concentrations were reportedly induced abortions in fishes (Boyd 1964). DDT at 2.0 mg/kg per week for 156 days treated fishes produced more matured ova than the untreated fish but mortality at sac-fry stage was significantly higher even if, one of their parents were exposed (Allison et al. 1964). The mean fecundity rates were much reduced in butachlor exposed zebrafish, *Danio rerio* (Chang et al. 2013). Reduction in hatchability, time to hatch of fertilized eggs and larval mortality in medaka, *Oryzias latipes* exposed to 100 μg/L of trifloxystrobin was reported (Zhu et al. 2015).

3.2.6 Damage to Body and Growth Rate

Tissue damage, injury and paralysis apart from reduction in growth were also reported in fish due to pesticide poisoning. Insecticide toxicity in fish leads to decrease in growth rate, reproductive disorders and also cause spinal deformities (Sabra and Mehana 2015). Xenobiotic chemicals destroy the gills (Grinwis et al. 1998) or its membrane functions disrupting the permeability (Hartl et al. 2001) which leads to rapidly decrease in oxygen uptake. Caudal bending in *L. rohita* exposed to cypermethrin may be because of paralysis caused by the inhibition of AChE activity in muscles (Marigoudar et al. 2009c). Increase in total leukocyte counts in fishes exposed to pesticides (James and Sampath 1996; Jenkins et al. 2003) may be a consequence of tissue damage or injury (McLeay and Brown 1974).

A short-term (4 day) exposure of Chinook salmon, *O. tshawytscha* to organophosphate and carbamate pesticides that are representative of seasonal pesticide use is sufficient to reduce the growth and size of ocean entry stage (Baldwin et al. 2009).

Lindane, pentachlorophenol, propoxur and malathion were found to affect the growth of *C. nigrodigitatus*, *O. niloticus*, *C. gariepinus* and *D. rerio* (Hanson et al. 2007; Cook et al. 2005). Sublethal levels of malathion and dimethoate led to a significant reduction in the final body weight, specific growth rate and normalized biomass index of Nile tilapia (Sweilum 2006). Weight gained was inversely proportional to dose in *O. nilotica* exposed to carbofuran with an LOEC 8.3 μg/L (Pessoa et al. 2011). Australian catfish, *Tandanus tandanus* exposed to a short term pulse of chlorpyrifos at 2 μg/L were found to be reduced in weight (8.54 g) compared to control (10.5 g). The hepatosomatic index in fish exposed to chlorpyrifos was higher (1.86) than in the control fish (1.65) (Huynh and Nugegoda 2012). Body weight of walking catfish, *C. batrachus* was found to reduce at sublethal concentrations (0.05 ppm) of malathion and (2 ppm) carbaryl with 45 days of chronic exposure (Wasu et al. 2009). Rainbow trouts, *O. mykiss* treated with carbosulfan @ 25 μg/L were found to gain very less weight of only 3 % in 60 days (120 g) corresponding to 39 % weight gain in control (165 g) (Capkin et al. 2014). In Atlantic salmon, 4-nonylphenol, a surfactant has been reported to reduce plasma insulin like growth factor (IGF-I), leading to reduction in somatic growth (Arsenault et al. 2004).

3.2.7 Histopathological Effects

Pathological changes due to pesticides occur mainly in the liver, blood vessels, kidneys and gills. Thiamethoxam affected the growth and liver in terms of weight, length and breadth and the liver protein of *O. niloticus* (Bose et al. 2011). Liver cells exhibit cytoplasmic granularity, partial loss of liver plate, radial orientation, shrinkage of liver cell mass etc. In case of kidney, pesticide can cause pycnotic changes of cell nuclei, vocalization of cytoplasm and atrophy of some cells of glomeruli. Gill filaments and lamellae show the precipitated masses that have plugged the central capillaries (Murthy et al. 2013).

3.2.8 Haematological Changes

Cellular and non-cellular blood parameters can give an indication of pesticide poisoning in fish (Pant et al. 1987). Erythrocyte counts, haemoglobin percentage and haematocrit values were reported to decrease steadily in fish exposed to sublethal concentrations of endosulfan compared to control (Jenkins et al. 2003). Haemoglobin content of *C. carpio* and *Puntius ticto* was reportedly increased when exposed to aldrin and dieldrin (Satyanarayana et al. 2004). A reduction in the erythrocyte count, haematocrit value and haemoglobin content in the blood of Nile tilapia was reported due to the exposure to malathion and dimethoate (Sweilum 2006). Fenthion and carbaryl exposure were also reported to result in decrease in RBC counts, haemoglobin percentage and haematocrit values in Mozambique tilapia, *Sarotherodon mossambicus* exposed at their respective LC_{50} levels at 48 h (Kounydinya and Ramamurthy 1980). Total leucocytes count (WBC) and heamatocrit values were reduced significantly in fenvalerate exposed *L. rohita* as compared to control group. The blood glucose level was found to increase whereas the serum total protein and albumin were reduced significantly (Prusty et al. 2011). A chronic exposure of edifenphos to *O. niloticus* was reported to increase creatinine (Gaafar et al. 2010). Changes in haematocrit levels and in morphology and quantity of blood cells have also been reported due to pesticide poisoning in various fish species (Murthy et al. 2013).

3.2.9 Biochemical Changes

The serum biochemical parameters were found to get influenced by imidacloprid treatment (0.002–0.01 ppm) in teleost, *C. punctatus*. The serum glucose, cholestrol, creatinine and creatine were found to increase significantly while the proteins, albumin and globulin decreased after 96 h of treatment (Priya et al. 2012). A variety of stressors influence the adrenal tissues leading to increase in levels of circulating glucocorticoids (Hontela and Daniel 1996) and catecholamines (Nakano and Tomlinson 1967) and these hormones cause hyperglycemia in fishes. Increase serum glucose was reported in fishes like *C. garipinus* exposed to lambda cyhalothrin (Ogueji and Auta 2007), *C. punctatus* exposed to chlorpyrifos (Ramesh and Sarvanan 2008) and imidacloprid (Priya et al. 2012). A decline in cholesterol level in serum is reported in fish exposed to phorate and supposed that the stored and circulatory cholesterol are utilized by the fish (Singh and Singh 2010).

When *O. mossambicus* is treated with sub lethal doses of dichlorvos, a significant reduction in protein content of liver, kidney and muscle is reported (Lakshmanan et al. 2013). In endosulfan treated carp, *C. carpio* the total proteins, albumin, globulin and also the serum glucose showed a decreasing trend with increasing time and concentration. The reduction in glucose is lead to the utilization of stored carbohydrates (Jenkins et al. 2003). Total protein content was found to decrease, when fish is exposed to lambda cyhalothrin especially in liver than that of muscle and brain tissues (Bibi et al. 2014). Increase in protein carbonyls was observed in response to 48 h exposure of *C. punctata* to deltamethrin, endosulfan and paraquat (Parvez and Raisuddin 2005). Total proteins, albumins and globulins were found to get reduced in *C. battrachus* treated with sevin (Patnaik 2010). Imidacloprid and sodium fluoride caused a remarkable protein loss in lethal concentrations but not at sub-lethal levels whereas butachlor caused remarkable protein loss at lethal as well as sub-lethal concentrations to the walking catfish, *C. batrachus* (Rajput et al. 2012). A significant decrease of total protein especially of globulin in atrazine (herbicide) treated fish *O. niloticus* and *Chrysichthys auratus* was reported (Hussein et al. 1996).

3.2.10 Changes in Enzymes and Hormones

The enzyme and hormone disrupting capabilities of pesticides could be an important factor contributing to the decline of fish (Khan and Law 2005). Furthermore, changes in the enzyme activities can be correlated with the pesticide residues in the water (Chandrasekara and Pathiratne 2005; Varo et al. 2007) as they are very sensitive criterion. Organophosphate toxicity in *M. vittatus* and *C. punctatus* was found to cause an inhibition in the activities of 3b-hydroxysteroid dehydrogenase and glucose 6-phosphate dehydrogenase (Haider and Upadhyaya 1985; Inbaraj and Haider 1988). Sublethal concentrations of dichlorvos, monocrotophos and phosphamidon cause significant changes in phosphate glutaminase and x-ketoacid glutaminase in the brain of *L. rohita* (Mastan and Shaffi 2009). Cypermethrin was found to affect the enzyme ATPase involved in cellular energy production, transport of metal atoms and muscle contraction (El-Toukhy and Girgis 1993). A significant increase in protease, aspartate aminotransferase and alanine aminotransferase activity was reported in *C. carpio* treated with sublethal concentrations of cypermethrin (David et al. 2004). Significant alterations in superoxide dismutase activity of liver and gills of fish, *L. rohita* exposed to fenvalerate are reported (Prusty et al. 2011). Sublethal concentration of endosulfan decreased the activity of citrate synthase and glucose 6-phosphate dehydrogenase in the brain, liver and skeletal muscle of the freshwater catfish, *C. batrachus* with a restoration on the withdrawal of poison (Tripathi and Verma 2004). The antioxidant enzyme activity (catalase) and glutathione S transferase in the liver, muscle and gills increased during the accumulation of malathion, whereas it decreased during depuration period (Thenmozhi et al. 2011). Activities of the antioxidant enzymes such as superoxide dismutase, catalase and glutathione reductase were found to be inhibited by the exposure of mosquito fish, *G. affinis* to chlorpyrifos (Kavita and Rao 2008). Significant increase in the serum glutamic oxaloacetate transaminase, serum glutamic pyruvic transaminase, creatinine, triglyceride and serum acid phosphatase was noticed *L. rohita* fingerlings after 15 days of exposure to fenvalerate (Prusty et al. 2011).

Cypermethrin poisoning significantly alters the activity of enzyme acetylcholinesterase, lactic dehydrogenase and succinic dehydrogenase in the nervous tissue of *C. fasciatus* (Singh et al. 2010). Exposure of chlorpyrifos and carbosulfan to Nile tilapia, *O. niloticus* caused a significant depression in the cholinesterase activities especially in brain and liver tissues (Chandrasekara and Pathiratne 2007; Joseph and Raj 2011). *Labeo rohita* exposed to 1/7th and 1/12th of the lethal concentration of cypermethrin for a period of 1, 7, or 14 days were found to have decreased AChE activity. Maximal inhibition was found in brain (−75.3 %) followed by muscle (−72.4 %), gill (−58.3 %) and liver (−51.1 %) on day 14 (Marigoudar et al. 2009c). The AChE activities in the erythrocyte of rainbow trout exposed to carbosulfan was found to reduce from 115.79 to 68.64 μ mol/min hematocrit in the third week and stayed the same until the end of the experiment i.e. 60 days (Capkin et al. 2014).

Plasma thyroid hormones were found to be lowered by malathion treatment in catfish, *C. batrachus* with very high reductions at higher concentrations (Lal et al. 2013). When the female rare minnow, *Gobiocypris rarus* was exposed to butachlor, the hypothalamic–pituitary–gonadal axis, the plasma 11-ketotestosterone was increased at exposure concentration of 10 μg/L and vitellogenin was significantly decreased at 1 μg/L (Zhu et al. 2014). Exposure to 100 μg/L of butachlor induced a significantly elevated level of vitellogenin in males but did not affect the females (Chang et al. 2013).

3.2.11 Disruptions in Genes/Chromosomes, RNA or DNAs

Chromosomal aberrations in the form of centromeric gaps, chromatid gaps, chromatid breaks, sub-chromatid breaks, attenuation, extra fragments, pycnosis, stubbed arms etc. in kidney cells of *C. punctatus* exposed to dichlorvos @ 0.01 ppm was reported (Rishi and Grewal 1995). Dichlorvos causing alterations in DNA replication leading to mutations was also reported (Gilot-Delhalle et al. 1983). A gradual decrease in nucleic acids, protein, free amino acids and glycogen was reported in *L. rohita* treated with malathion (Thenmozhi et al. 2011). A three to five fold increase in the DNA damage in European topminnow, *Phoxinus phoxinus* was observed when exposed to diuron and azoxystrobin (Bony et al. 2008). The commonly used herbicide, glyphosate produces genotoxic damage in erythrocytes and gill cells causing nuclear abnormalities in Neotropical fish, *Prochilodus lineatus* as revealed by comet assays (Cavalcante et al. 2008). Analysis of micronuclei, nuclear abnormalities and DNA damage were performed on peripheral erythrocytes of goldfish, *C. auratus* exposed to isopropylamine salt of glyphosate, revealed a significant dose dependent increase in the frequencies of micronuclei, nuclear abnormalities as well as breaking of DNA strands (Cavas and Konen 2007).

The functional genes for hypothalamic–pituitary–gonadal axis like gnrh and cyp19b in the brain, star, lhr, cyp11a, 3β-hsd and cyp19a in the ovaries and erα and vtg in livers were up-regulated in female rare minnow, *Gobiocypris rarus* treated with butachlor. For hypothalamic-pituitary-thyroid axis, the gene expression of dio1 was up-regulated, dio2 showed no significant variation and dio3 down-regulated in the liver (Zhu et al. 2014). In medaka, *O. latipes*, the mRNA levels of er gene were significantly up-regulated when exposed to fungicide, trifloxystrobin above 1 μg/L. Up-regulation of vtg, cyp1a, cyp17 and cyp19a was observed in the larvae also (Zhu et al. 2015).

3.2.12 Indirect Effects

Pesticides can indirectly affect fish by interfering their food supply or altering the aquatic habitat. These indirect effects can even be noticed at concentrations that are too low to affect the fish directly. These effects greatly reduce the abundance of food leading to reduction in growth and chance of survival of the fish (Ewing 1999). The sublethal concentrations of malathion and dimethoate was found to decrease plankton abundance and water quality in fish ponds (Sweilum 2006). Removal of aquatic plants may lead to easy predation and these indirect effects are said to be highly dangerous than direct effects in complex ecosystems particularly of aquatic. Pesticides are also found to reduce the feeding potential of fish (Ravindran et al. 2012).

Though some of the insecticides found in aquatic ecosystem are not causing mortality to the fish, they are highly toxic and caused mortality to their prey species *Hyalella azteca*, an amphipod crustacean. The concentrations of imidacloprid and fipronil which are probably too low to exert a direct toxic effect on medaka is assumed sufficient enough to reduce the abundance of their invertebrate prey (Hayasaka et al. 2012). The peak bifenthrin concentrations in the river was comparable to the 96 h EC_{50} of the caddisfly, *Hydropsyche* sp., the most important prey species for American river Chinook salmon on a biomass basis (Weston et al. 2015).

Many environmental stresses of pesticides including outbreaks of infectious diseases in fishes (Snieszko 1974) due to reduction in innate immunity are reported (Bols et al. 2001). The pollution associated diseases can be of the following (1) Diseases caused by stress due to contaminants and related facultative pathogens; (2) Stress-provoked latent infections; (3) Environmentally induced abnormalities and lesions; (4) Genetic abnormalities associated with mutagenic properties of contaminants; (5) Contaminant effects on resistance and immune responses; and (6) Pollutant-parasite interactions (Sindermann 1979). The major xenobiotics involved in the immuno modulation of fish are insecticides, herbicides and fungicides (Dunier and Siwicki 1993). In the presence of increasing water temperatures, malathion may increase the susceptibility of salmon to disease and ultimately threaten its survival (Dietrich et al. 2014).

3.3 Semi-field/Mesocosm and Field Effects

In a microcosm study to find the influence of herbicide atrazine on the prey-predator interaction of crayfish and tadpoles, atrazine exposure was not found to influence the susceptibility of tadpoles (Davis et al. 2012). A mesocosm study was carried out to find the prey-predatory behaviour of pesticide exposed prey, involving topsmelt (*Atherinops affinis*) as prey and three-spine stickleback (*Gasterosteus aculeatus*) as predator. Prey fish exposed to esfenvalerate at 0.12, 0.59, 1.18 μg/L, though increased the proportion of larvae with swimming abnormalities, did not cause any increase in prey mortality (Renick et al. 2015). Water from river having watershed from intensive agricultural area at three hydrological conditions viz., basal flow, winter flood and spring flood was studied in the laboratory for its effect on Crucian carp, *Carassius carassius*. A significant increase in DNA breakdowns were observed compared to controls as revealed by comet assay with peripheral erythrocytes for all conditions. Fish exposed to spring flood water was found to have more DNA and chromosomal damage (Polard et al. 2011). In a channel study on the effect of pesticides on fish by deliberate addition of pesticides in three levels (control, chronically normal and high concentration), three to fourfold increase in DNA damage in normal and fivefold DNA damage in high concentration were reported with respect to control (Bony et al. 2008).

3.3.1 Field Effects

Field studies with ten insecticides at three different concentrations (doses for mosquito control and doses used in agriculture) showed malathion, DDT and methyl trithion and parathion as toxic pesticides (Mulla and Issak 1961). Though the herbicide, glyphosate was found highly toxic to fish especially the fathead minnows in the laboratory, application at recommended rates along the ditchbank areas of irrigation canals was not found to affect the fish. However, applications in still waters are supposed to be hazardous to young of the year (YOY) fishes (Folmar et al. 1979). In an experiment on the genotoxicity of pesticides on the fish larva exposed in an upstream, middle and downstream of a river through incubators wrapped with net, the damage to DNA was found significantly high in down and mid stream than that of upstream. However, the upstream effect (tail intensity value: 4–14 %) is significantly high than the control ones (tail intensity value: 1–2 %). This river is a recipient of pesticides from crop plants (vineyard) inadvertently while applying on the crops (Bony et al. 2008). In a field experiment to assess the habitat quality with population-level metrics of longjaw mudsucker (*Gillichthys mirabilis*), the fish in the contaminated site (San Francisco Bay) was found to exhibit slower growth, higher mortality and higher new recruitment rate than the reference site (Tomales Bay) of North California estuaries (McGourty et al. 2009).

4 Methods to Assess Pesticide Toxicity to Fishes

The main objective of aquatic toxicity test is to screen the toxicants in terms of their toxicity to aquatic organisms. A ‘safe concentration’ is defined as the concentration which will permit living and allow normal propagation of fish in the receiving waters. The endpoints of toxicity tests can be of death and survival, decreased reproduction and growth, change in locomotor activity, gill ventilation rate, heart rate, blood chemistry, histopathology, enzyme activity, olfactory function, terata etc. (EPA 2002). McLean et al. (1980) is one of the first to give a detailed bibliography of toxicity test methods for aquatic organisms. Methods to estimate the lethal and sublethal toxicity, methods to assess physiological and behavioural changes and pesticide effect on communities are listed in the bibliography. Some of the toxicity test procedures available in the literature are given below:

4.1 Acute Toxicity Tests

Acute toxicity of pesticides in fish is normally carried out by contaminating the water with the toxicant and exposing the fish to it. In this method, the fish is exposed to the toxicant through gills and also through the skin. This test procedure through contaminated water in the laboratory can be done in three ways/systems viz., static,

static-renewal, or flow-through system. Apart from water contamination, injection of toxicants and dietary administration through food are also being practiced but in a less extensive manner. In general, dietary exposure is less toxic to fish than pesticides in both exposure systems (Hashimoto and Nishiuchi 1981).

Dose-response relationships are estimated in laboratories for determining the acute toxicity of aquatic organisms (Muirhead-Thomson 1973). Before conducting the actual acute toxicity assay, a range finding test is to be carried out to find the range of toxicant concentration which kills the test organism. After finding out the range of concentration, the probable toxicant concentration which causes a mortality of 10–100 % to test organism is tested and mortality observed. This concentration and the corresponding mortality data are used to find the median lethal concentrations using Probit analysis.

Range-Finding or Limit Testing
A range finding test is often conducted starting with a concentration (the field recommended concentration or a very low concentration) and have a range of concentrations with multiples of 10 of that or divisions of 10. Thus the fish is subjected to a range of concentrations to find a concentration which gives mortality approximately to 50 %. After finding the concentration which gives 50 % mortality, it is kept as a central value and three concentrations above and below of that concentration is tested in the actual acute toxicity test. The concentrations which gave 10–90 % mortality are usually taken for analysis to fix a log concentration probit mortality line and to arrive at the median lethal concentration values (LC_{50}). It is stated that if the range testing with at least 30 organisms shows 50 % mortality levels at concentrations greater than 1000 mg a.i./L or the limits of water solubility or dispersibility, then the chemical can be regarded as safe and further acute toxicity tests are not necessary. If the environmental concentrations of pesticides are not expected to exceed 100 mg/L (ppm), a lower level of 100 mg a.i./L may also be kept as the deciding point (USEPA 1996).

4.1.1 Acute Toxicity Through Contaminated Water

4.1.1.1 Static System Bioassay

In the static system, the test organisms are exposed to the same test solution for the entire duration of the test. This method has probably been the most popular with small fish species such as Japanese medaka, guppy and zebrafish (Law 2001). The advantages of static tests are simple and inexpensive requiring limited resources and low volume of space and testing material. But there may be an increase in chemical oxygen demand (COD) and biological oxygen demand (BOD) due to reduction in dissolved oxygen (DO) apart from accumulation of metabolic wastes. There may be a loss in toxicant by means of volatilization and/or adsorption to the exposure vessels or degradation (EPA 2002). All these disadvantages of static tests are taken care off in static renewal and flow-through tests and thus more efficient.

Fig. 7.1 Semi-static system of acute toxicity testing in fish

4.1.1.2 Static-Renewal/Semi-static System (EPA 2002; Nwani et al. 2010a)

In static renewal tests the test solution is renewed every 24 h or at a prescribed interval, so that the test organism is exposed to the proper concentration throughout the test period. This renewal may be done in two ways, one by transferring the test organism to fresh solutions and the other by replacing the test solution (EPA 2002). In the experiment described by Nwani et al. (2010a), the fish was acclimatized for 2 weeks under laboratory conditions before the start of experiment. Fish were fed enough during acclimatization and fecal matter siphoned off daily to reduce ammonia content in water. A Glass aquarium with the size of 60 × 30 × 30 cm is generally used for treating ten fish for a replication of a treatment. The water with the pesticide was changed after every 48 h to counterbalance decreasing pesticide concentrations. Fish were not fed during the experimentation. A fish was considered dead when there is no response when prodded and removed from tank immediately. Mortality was recorded after 96 h of exposure and median lethal concentrations arrived using probit analysis method as described by Finney (1971) (Fig. 7.1).

4.1.1.3 Flow-Through System

Flow through test is conducted with continuous water flow in the test chamber. Since the water is continuously get replaced or allowed to flow, this system of toxicity testing overcomes the drawbacks of the static system (Law 2001). The diluent water is kept in a tank and made to flow through the test chamber and the water before reaching the test chamber is added with the toxicant. The advantage of this system is the usage of volatile and hydrophobic toxiciants for testing (Walker et al. 1985) and the use of computer-controlled precision injector. In static system there may be a loss of toxicant because of volatilization, adsorption, degradation and uptake by the organism. Though semi-static system takes care of this loss to some extent, the flow through system do not allow such losses and the test organisms are exposed to the test concentration throughout the experiment. However, it requires large volumes of samples and dilution water with test equipments more complex and expensive (Weber 1991).

4.1.1.4 Simple Tests Using Glass Beakers (Werner et al. 2002)

In this experiment, a 2 L beaker was considered as an aquarium to test medaka, *O. latipes*. Each treatment was replicated in three and each replication has ten fish (five male, five female) of 4–5 months old in a beaker. The test was performed at 22 °C with dissolved oxygen of 85.1–92.4 % and 16 L: 8 D. Fish were not fed during the experimentation and about 75 % of water renewed daily. After 96 h of experimentation, surviving fish were counted to find the acute toxicity of pesticide on fish.

4.1.2 Acute Toxicity Through Toxicant Injection

4.1.2.1 Intraperitoneal/Intra-abdominal Injections

Though this method is usually used to study the genotoxicity of chemicals/pesticides, acute toxicity tests also can be performed through intra-abdominal injections. Al-Sabti and Metcalfe (1995) reported that out of all 56 micronuclei tests conducted for evaluating mutagens in fish, 29 were performed through intra-abdominal injections. However, intra-abdominal injection was reported as an inappropriate route for fish models in genotoxicity studies (Grisolia 2002). Lethal effects of phenol, hydroquinone, resorcinol and pyrocatechol were studied in sea bass, *Dicentrarchus labrax* through intra-abdominal injections though the experiment was originally planned to study the blood parameters. The fish were acclimatized for 15 days during which the commercial sea bass food pellets, Aqualim® were fed. Stock solutions of various xenobiotics were prepared in isotonic NaCl (170 mM) and injected intra-abdominally once (single dose) or three times (fractionated doses). An interval of 4 days was kept between each injection in case of fractionated doses. Control groups received equivalent volume of isotonic NaCl. The fish were fed even during the course of the intoxication period. Observations on lethality occurring during treatment were recorded (Roche and Boge 2000). Though the fish can be vaccinated by immersion or through oral routes, vaccination through injection is said to be the best (Evensen 2009).

4.1.2.2 Embryo Microinjection (Law 2001)

This method is used especially for testing carcinogenic pesticides on fish. The test material is administered by injection into the perivitelline space of embryonated eggs. Then the embryo was kept in proper environment for hatching and the larva are grown for variable periods of time and examined for mortality and neoplastic lesions. Embryo microinjections are reportedly being useful for testing poorly water-soluble compounds. With this method, the selectively permeable property of the egg chorion is avoided, since it is injected into the egg.

4.1.3 Acute Toxicity Through Dietary Exposure

Dietary exposure of pesticides to fish though used only in few experiments is also a method of toxicity testing. Different types of diets are used for feeding different kinds of fishes. Ward et al. (2008) used frozen bloodworms to test the chemical sensation of pesticide exposed banded killifish. Freeze-dried chironomids were mixed with pesticide through ethanol and used to test the reproduction and embryo development in zebrafish (Halden et al. 2010). Pesticide mixed standard purified casein based diet (PC-diet) was used to test toxicity/mortality in medaka (DeKoven et al. 1992) and to test the effect on egg production and on stress proteins (Werner et al. 2002). A standardized test diet has the advantage of consistent, defined nutrition and the ability to premix test compounds for dietary exposures than natural foods. Unlike the whole live food, the diet is less likely to contain the extraneous compounds that could confound test results. Dietry exposure bioassays can be used to test even the less soluble toxiciants, which cannot be tested in other methods. Dietry exposure assays are used to model biomagnification of toxicants through the food chain (Law 2001). Drawbacks of dietary exposures include the relatively large amount of chemical requirement and uneven dosing caused by aggressive feeders (Hawkins et al. 1988).

Points to Be Remembered in Acute Toxicity Testing (USEPA 1996)

1. The pesticide should be of technical grade unless or otherwise the experiment is designed to test a specific formulation or mixture.
2. The maximum allowable mortality in solvent or control is 10 %. The test is discarded if the control mortality crosses 10 %.
3. In toxicity tests, flow-through procedures are preferred over a static-renewal and static-renewal over a static test procedure.
4. In static tests, the dissolved oxygen in each replicate should be >60 % saturation and in flow-through tests, it should be >75 % saturation.
5. The water salinity should be 20 ± 5 ppt for estuarine species whereas the hardness should range between 40 and 180 mg/L for freshwater species.
6. A minimum 14 days acclimatization period is recommended and before starting the test, a 7 day acclimatization period is given in the test water and the mortality should be <5 % during this period.

7. Fish should be added to test chambers within 30 min of addition of the test pesticide in the water or may be added prior to addition of test material.
8. In general, fish may not be fed during the test period other than dietary exposure studies.
9. Five or more concentrations in a geometric series can be tested and the test concentrations must be at least 50 % greater than the next lowest test concentration (not to exceed 120 %).

4.1.3.1 Relation Between Acute Toxicities

Method Ratio (Mayer and Ellersieck 1986)
Acute toxicity of pesticides on fish both by static and flow-through procedures was compared and pesticides were categorized into three accordingly. If the ratio of acute toxicity of fish analysed through static and flow-through methods is,

Ratio <0.5 – the pesticide is more toxic under static than under flow-through conditions
Ratio 0.5 to 1.5 – no much difference in toxicity in the two bioassay methods
Ratio >1.5 – more toxic under flow-through than static.

Deactivation Index (Mayer and Ellersieck 1986)
The aged test solution can be compared with the fresh ones in static system of pesticide testing. Solutions were aged for 4–28 days and 96 h LC_{50} determined for both aged and fresh solutions. Deactivation indices were calculated according to Marking and Dowson (1972), by dividing the 96 h LC_{50} of aged test solution with the 96 h LC_{50} of fresh solution and pesticides categorized into three:

Ratio <0.5 – the pesticide was most toxic after aging
Ratio 0.5 to 1.5 – aged and fresh solutions were equally toxic
Ratio >1.5 – aged solutions were less toxic.

Ratio of Technical and Formulated Pesticide (Mayer and Ellersieck 1986)
The same approach was used to find the toxicity of technical and formulated products. The ratio was calculated by dividing the 96 h LC_{50} of technical with the 96 h LC_{50} of formulated and pesticides categorized into three:

Ratio <0.5 – the technical form of pesticide was most toxic
Ratio 0.5 to 1.5 – the technical form and formulated form were equally toxic
Ratio >1.5 – the technical form of pesticide was less toxic.

4.1.3.2 Variations in Acute Toxicity Tests

Multiple Species Testing (Phipps and Holcombe 1985)
This method simultaneously ascertains the median lethal values for many aquatic organisms in a single run. The LC_{50} values from these multiple species can be compared comfortably within them, since all other variables are kept as constant except

species variability. The advantages of this method include allowing interspecific comparisons and easy determination of the most sensitive species besides reducing cost of labour, materials and chemicals.

In Situ Bioassays/Acute Toxicity in Field Conditions
In this bioassay, exposure of test organism is performed in the field site, instead of taking the contaminant water or material and keeping somewhere for testing or replicating the same situation in the laboratory. For this bioassay, cages along with the test organism were allowed hung in the water column or anchored at the bottom. Mortality was assessed exposure for 96 h or longer (Murthy et al. 2013).

Cylindrical stainless steel cage of size 9×20 cm made of 1.5 mm mesh screen are used to test the toxicity effects on Hyalella, juvenile Gammarus or Hexagenia larvae but can be extended to fish also. One quarter of the cage is gently forced lengthwise into the sediment and anchored by stakes. The experiment has to be carried out at least in two replications and the test organisms are to be exposed for 96 h and then the cages gently washed free of sediments and evaluated (Nebeker et al. 1984). Crayfish (*Orconectes virilis*) is also used to study the acute-lethal in situ tests (Leonhard 1974). A suitable cage size for six adult crayfish was reported to be 20×15×10 cm. In situ bioassay can lasts for 96 h or even more up to 30 days. Observations on mortality of organism, changes in weight and normal functions can be recorded.

Residual Oxygen Bioassay (Giles and Klaprat 1979; Tuurala et al. 1985)
The residual oxygen bioassay also known as sealed jar test is a rapid and simple means for measuring effects of toxicants on fish. Healthy fish are placed in a sealed jar, containing the test toxicant and allowed to consume oxygen until death occurs. Once when the fish dies, the residual oxygen concentration is measured and an elevation in the oxygen concentration indicates a harmful effect of the pesticide. The threshold effect concentrations derived from the residual oxygen bioassay is approximately the same as those obtained in 96 h LC_{50} tests.

In this experiment, 2 L wide mouth polythene bottles were used as test containers. Giles and Klaprat (1979) used a 300 mL vessel for testing a 2–5 g fish. The test containers were filled with clean air-saturated water and fish put individually in the bottles without any disturbance and sealed properly. The time of death of each fish was recorded as the time of ceasing of opercular movement. After the death of fish, the dissolved oxygen concentration was measured. Unlike the acute toxicity bioassays traditionally conducted in water contamination which takes 96 h, this takes very less time to assess the toxicity of the chemical.

4.1.3.3 Alternate Tests Proposed Instead of Acute Toxicity

Embryo Development (Lange et al. 1995)
Many tests especially of tests using embryos are reported as more sensitive than acute toxicity and proposed to be used instead of acute toxicity tests (Lange et al.

1995; Braunbeck et al. 2005). For these tests, usually zebra fish, *B. rerio* is used because of the advantages in size, egg producing and short lived. The toxicological endpoints such as coagulation of the eggs, development of gastrulation, number of somites, development of organs, circulation, heartbeat, otolithanlage and pigmentation have been examined during the embryonic development of zebrafish within the first 48 h and propose this instead of acute toxicity because of its sensitivity. Braunbeck et al. (2005) after analyzing the results of toxicity of 56 substances and found a reliable correlation between the fish embryo test and the acute fish test with R^2 0.854. It was found that even very subtle fish toxicity is mirrored by fish embryo toxicity.

Bacterial Bioluminescence (Curtis et al. 1982)
This bioassay works in the principle of the decrease in luminescence of the bacterium, *Photobacterium phosphoreum* in response to a toxicant. This luminescence was measured through a microtox toxicity analyzer (Beckman Instruments Inc.). Based on this, a 5 min median effective concentration (EC_{50}) which is defined as the concentration that causes 50 % reduction in light output is calculated. The relationship between toxicity (96 h LC_{50}) values and luminescence EC_{50} was worked out in fathead minnows. The correlation between fish and bacteria toxicity for pesticides and industrial chemicals was found to have R^2 0.65, whereas it was 0.96 for alcohols. Thus it can be used as a screening test since it is extremely rapid and simple with a precision equal to or greater than traditional acute toxicity studies.

4.1.3.4 Extrapolation of Acute Toxicity Data to Chronic

Critical Life Stages
Acute toxicity tests are generally carried out to find the mortality of the fish at a defined exposure time. After evaluating the data from 56 full life cycle tests, McKim (1977) concluded that the embryo-larval and early juvenile life stages as the most sensitive stages. Macek and Sleight (1977) found that the results obtained by exposing the fish to the toxiciant of interest at its critical life stage can indicate the chronic toxicity also. This is based on the principle that, 'If the toxicant or a concentration is not toxic at the sensitive stage of the test organism, it will not be toxic'. In otherwords, the contaminats which are acutely nontoxic to the fish at the sensitive stage is chronically non-toxic too. The embryo and the fry stage are considered as the most sensitive life stages of fish. Analysis made by Woltering (1984) on 173 fish full life cycle and early life stages tests to determine the chronically safe concentrations revealed that the lowest effect concentration also significantly reduced the fry survival (57 %), fry growth (36 %) and egg hatchability (19 %). Since the full life cycle studies are time consuming, costly, complex and difficult to conduct, this extrapolation can be used easily.

Acute Chronic Ratio (Kenaga 1982)

A relationship of acute LC_{50} of chemicals to their chronic toxicity i.e. acute chronic ratio (ACR) was calculated for various chemicals for different species. This ratio is done by comparing the median lethal values (acute toxicity) and the no effect concentration (chronic toxicity). Eighty-six percent of the acute toxicity data (LC_{50}) obtained was less than two orders of magnitude difference. This correlation allows us to predict the chronic toxicity when acute toxicity values are known.

4.2 Chronic Toxicity Studies/Long Term Exposures

4.2.1 Long Term Exposures and Recovery After Toxicity Exposure (Capkin et al. 2014)

Rainbow trout were exposed for long time (60 days) to the sublethal concentration of pesticides to find the effect on enzymes. The experiment was conducted in flow-through system with the test solutions added using an infusion pump. The pesticide concentration in the water is also determined for confirmation. During the treatment, water in each aquarium was aerated and fish fed with commercial pellets daily at 2 % of body weight. About, four fish from each tank were sampled biweekly to determine enzyme activities. At the end of this experiment, fish were transferred in fresh water in flow-through tanks for a recovery period of 24 days. During this period of recovery, enzyme activities were again measured by sacrificing two fish in each treatment daily. Water in each tank was aerated and water quality parameters measured daily.

4.2.2 Cumulative Dose Experiment

Acute toxicity tests are conducted usually by exposing the fish for a period of 96 h to the toxicant. By lengthening the tests for some more time to find the chronic toxicity may be more relevant for some conditions than the acute tests. In chronic toxicity experiments, water containing test fish are spiked with minimal amounts of pesticides over a long period (2 months) and sometime repeatedly on alternate days and is termed as chronic or cumulative dose (CD). Chronic tests may also include one or more complete life cycles to assess changes in growth reduction and reproduction.

4.2.3 Tests with Increasing Concentrations (Hanson et al. 2007)

In an experiment, ponds of effective water volumes of 1000 L are used and an increasing amount of pesticides were administrated intermittently. After every second application of a dose, the amount was increased after two subsequent applications in each of the respective ponds. Observations on mortality, change in body weight and gonad weight and thus the gonadosomatic index (GSI) were calculated.

4.3 Sublethal Toxicity Tests

4.3.1 Laboratory Studies

Sublethal toxicity tests aim in assessing the effects of pesticide other than mortality to the organism. Sublethal bioassays are also conducted as that of the acute toxicity studies either by contaminating water in the aquaria or by feeding them with contaminated food, but usually in low concentrations and for longer exposure periods. One by fifth concentration of LC_{50} values were taken for sublethal bioassays (Rastogi and Kulshrestha 1990). One seventh of the LC_{50} is taken in some of the studies to find the sub lethal impacts as given by Sprague (1971) and Marigoudar et al. (2009a, c) where they studied the behavioural changes and the physiological responses of poisoned fish. Some chronic studies used still low concentrations like 11 % of 96 h LC_{50} values (Capkin et al. 2014). For short term studies up to 15 days, one third of LC_{50} is also used (Prusty et al. 2011). Long term studies on sublethal toxicity may extend to 6 months and above also. Tests for chronic toxicity and bioaccumulation in fish especially those having long life cycle is very difficult. It takes about 2–3 years for testing carp for the chronic bioaccumulation effect. Fish with less life cycle duration such as zebrafish, *B. rerio* are useful; being small in size, egg-laying and have a life cycle of approximately 75 days.

Sublethal toxic effects of pesticides include changes in behaviour of fish, its physiology and biochemistry of tissues and blood, histopathological effects in external and internal organs etc. Some of the pesticides or their metabolites are reported to be potent carcinogens and also cause teratogenic and mutagenic effects and regarded as genotoxic chemicals which may get transferred to generation after generations. Pesticides that of lipophilic in nature even though available in very low concentrations in the medium, may get accumulated and concentrated in the fat tissues of fish or pass through tiers of food chain to get biomagnified. Some of the tests to assess sublethal toxicity of pesticides on fish are given below:

4.3.1.1 Behavioural Tests

4.3.1.1.1 Changes in Activities

Sensation to Light (Shirer et al. 1968)
One of the first manifestations of stress is an alteration in activity patterns. In addition to stress, environmental stimuli such as light intensity or the presence of food or predator can also alter the activity pattern of fishes. Stress might easily alter the response of fish to these stimuli and thus, aberrant activity patterns might also indicate sensory blocks. Different types of apparatus can be devised using acoustical, thermal, or electrical to sense swimming movements of unrestrained fish. The choice of light beam interruption as the sensing means is based on technical simplicity and reliability rather than biological innocuousness. In this experiment, lights were installed over the aquaria and the beams arranged in such a way to traverse the length of the tanks near the bottom, center and just below the water surface. Light interruption due to fish movements were detected using photo-resistors and recorded. The apparatus is of relatively simple construction, quite easily operated and provides rather quick but sufficient information for a statistical evaluation. Changes in the behavioural sensation of light to fish exposed to sublethal poisons can be assessed by comparing with that of unexposed ones.

Changes in Locomotion: Video Tracking (Kavitha and Rao 2007)
Pesticide affects the locomotor pattern of fish. Subacute doses of pesticides such as monocrotophos was reported to alter their locomotor behaviour, i.e., distance traveled per unit time (m/min) and swimming speed (cm/s) on mosquito fish, *Gambusia affinis* (Kavitha and Rao 2007). The locomotor pattern of goldfish, *C. auratus* was found to change permanently within 4 days of exposure to even low doses of 10 μg/L pp′ DDT (Davy et al. 1972).

In this experiment, the locomotor behaviour of fish exposed to median lethal concentration was monitored using a video tracking system. The locomotory behaviour such as distance traveled per unit time (m/min) and swimming speed (cm/s) are measured as end points. The test fish were acclimatized individually in the recording aquarium containing 2.5 L of water. All the three sides and bottom of the aquarium were made opaque to avoid repetition in counting due to mirror images and other visual disturbances. The behaviour of fish was recorded for 5 min in fixed monitoring arena with a high-resolution CCD camera, mounted 20 cm away from the recording aquarium. A minimum of 10 fish were evaluated individually in this same manner for a treatment.

4.3.1.1.2 Changes in Orientation

The weedicides 2,4 D caused a decrease in positive rheotaxis of rainbow trout when exposed to 2 mg/L for 24 h. An increase in negative rheotaxis was also found at concentrations ≥4 mg/L (Dodson and Mayfield 1979a). Rainbow trout did not actively avoid diquat (0.5–5 mg/L) whereas a significant change in the rheotropic causing a passive avoidance after 24 h is reported (Dodson and Mayfield 1979b). Changes in orientation can also be measured as that of locomotion by means of manual observations of toxicated fish or by using advanced instruments like video recorder interfaced with computers.

4.3.1.1.3 Avoidance or Preference

The avoidance or preference reactions of fish to toxicants will show their capacity to escape or get trapped in toxicant. An avoidance reaction may result in the alteration of patterns of dispersion and migration while a preference reaction may result in prolonged exposure which may cause lethal or sublethal effects. A counter-current trough is used to evaluate the avoidance behaviour (Scherer and Nowak 1973). The time spent with respect to concentrations is taken as a measure of avoidance (lesser time spent in places having more concentrations) to insecticides (Scherer 1979). In field conditions, capture of fish in the downstream after a particular time of pesticide treatment of river or other water sources is studied and Flannagan et al. (1979) found that 50,000 specimens entered a 15 cm diameter sampler during a 4 h period after treatment of methoxychlor in a river.

A rectangular counter-current trough (Scherer and Nowak 1973) was used to assess the avoidance or preference reactions of the lake whitefish, *Coregonus clupeaformis* (Giles et al. 1979). Rotameters and adjustable needle valves were used to control the water and the toxicant. The inner baffles were placed 30 cm from the centre of the trough and airstones placed between the outer and inner baffles. The flow rate was kept as 7.8 L/min with a resultant velocity of 0.55 cm/s. About ten fish were tested individually at each concentration and each test organism used in one trial only. The fish was habituated to the test trough for 10 min and then monitored. During this procedure, water was introduced into both ends and for an additional 10 min the test toxicant was introduced into one end and movement of fish monitored. The avoidance from the toxicant or movement towards uncontaminated water is to be noted.

4.3.1.1.4 Temperature Selection (Neill et al. 1972)

Habitat selection in fishes is a complex and dynamic process in which temperature also play a role (Ross 1980). Pesticide toxicity can alter the temperature selection by fish. Goldfishes exposed to ethanol were found to select low temperatures in comparison to the pre exposure selections (O'Connor et al. 1988). Selection of temperatures 4 °C less than that of normal was reported in Atlantic salmon injected with 2 μg/g of 5- hydroxytryptamine (Fryer and Ogilvie 1978). Guppies exposed to 5-hydroxytryptamine exhibited a biphasic relationship between the chemical doses and mean selected temperatures (Fryer and Ogilvie 1978).

For experimentation on temperature selection, a special aquarium was made which allow the fish to swim from one half to the other through a tunnel that divides the tank. The passage of fish is sensed by photocells in the tunnel which records the movement of fish and controls temperature of the tank. Passage of the fish into one side of the tank causes that entry side to begin warming at 3–5 °C an hour. The warming of that side ceases and get cooled, only when the fish swims to the other side. The temperature of the left side is kept 2 °C lesser than the right, so the fish can have a choice to be in a temperature difference of 2 °C (Neill et al. 1972). This apparatus can be used to find the effect of pesticides on temperature regulation in fishes by studying the temperature selection of fish after pesticide exposure and prior to that.

4.3.1.1.5 Feeding Behaviour (Boujard et al. 1992)

Pesticides affect feeding behaviour of fishes. Exposure of brook trout to copper concentrations as low as 6–15 μg/L was found to affect the feeding behaviour (Drummond et al. 1973). The amount of food consumed by bream, *Abramis brama* was reportedly decreased when exposed to DDVP (Pavlov et al. 1992). Experiments on feeding behaviour of pesticide poisoned fish can take into account of the feeding rate, feeding success, incomplete feeding sequences and also the search volume. A modular eater meter especially designed for fish studies consists of three parts: the feeder, the detector and a connector between both. The detector is a small rod that closes a circuit when stuck by a fish (Boujard et al. 1992).

4.3.1.1.6 Learning Behaviour (Gomez-Laplaza and Gerlai 2010)

Brook trout were conditioned to exhibit the propeller-tail reflex, with electric shock as the unconditioned stimulus and light as the conditional stimulus. More than 50 % of the fish were not able to be conditioned when they are exposed to sublethal concentration of DDT (20 ppb) for 24 h (Anderson and Prins 1970). Rainbow trout, *S. gairdneri* treated with 10 % LD_{50} of DDT did not learn significantly faster than controls (McNicholl and Mackay 1975). However, nicotine exposure to zebrafish increases the learning rate and learning capacity (Eddins et al. 2009).

The maze described by Gomez-Laplaza and Gerlai (2010) consisted of a square start box where the fish were placed for 2 min acclimation at the beginning of each training and probe trial. This box is connected by a tunnel which leads to a four way intersection from where the fish can have a choice to turn left, right, or swim straight or turn back. All the three choices except turning back led the test fish to the reward chamber, which contain stimulus fish. The maze is of 10 cm height and made of transparent Plexiglas®. All the three entrances to the reward chamber were equipped with transparent Plexiglas guillotine doors. Usually the central gate is kept closed and the left and/or right gates are opened depending on experimental condition. Altogether, ten fish were designated in a group and trained as (1) Both tunnels open group, (2) Left tunnel open, (3) Right tunnel open, (4) Trained in another simple rectangular tank of same water volume and (5) Naive group. Fish of the first four groups except the naive were trained once in a day for about 15 days. In probe trial, both the left and right doors are opened and the fish assessed for its response with regard to the training it received. The behaviour of the fish is recorded for observations which include the time spent in the left and the right tunnel, the first path taken, latency to leave the start box/enter the reward chamber, duration of time spent in the reward chamber.

This apparatus/maze can be used to find the effect of pesticides on learning behaviour of fishes by training them after pesticide exposure and compare with that of a control group. Zebra fish are reported to have a strong shoaling response and thus a stimuli fish is kept in the reward chamber, which can be seen by the test fish from anywhere in the maze. Thus the behaviour of shoaling is tapped in this learning experiment.

4.3.1.1.7 Intra Specific Interactions

Pesticides may alter many intraspecific interactions of fishes like reducing the population and thus no competition and expansion of niche, reducing reproduction rate leading to dwindling of population etc. Sublethal effects may lead to some difference in schooling, learning and feeding behaviour, impairing sensory perceptions etc. which not only affects the entire fish community but also its prey and predators.

Courtship Behaviour (Baatrup and Junge 2001)
Guppy males exposed to pesticides having antiandrogenic property were found to get affected in courtship behaviour. A group of 20 fish were kept in 16 L glass aquarium with 8 L of water. During the experimental period of 30 days the pesticides were administered to fish through feed (TetraMin fish fodder mixed with pesticides). The food was given at the rate of 40 mg/20 fish. After the specified period of exposure, each male was paired with a 4 month old female in a sand-blown 20×15 cm aquarium containing 1.8 L water. A camera was mounted 50 cm above the aquarium to display the images on the monitor and simultaneously the video signals were digitized and analyzed.

4.3.1.1.8 Inter Specific Interactions

Aquatic ecosystem contains large number of species at the lower level, phytoplanktons, zooplanktons etc. apart from many predatory fish, birds and mammals that feed at the apex or near apex tropic levels. Ecosystem dynamics depends on how the food web is controlled i.e. bottom-up, top-down or wasp-waist (Cury et al. 2000). Biodiversity and a healthy interaction between all water organisms specify the sustainability of that ecosystem. However, pesticides may cause disruption in this by selectively toxic to one species and reducing or eliminating them, causing a disruption to the entire ecosystem. Pesticide toxicity may be species specific and the most susceptible are harmed the most. For example, lean fishes were more susceptible to endosulfan poisoning than fat fishes (Matthiessen et al. 1982). In carbaryl treated ponds, large-sized Cladocera dominated in the control whereas the medium-sized and small sized in the low and high dose treatments, respectively (Hanazato 1991). Pesticide contamination was found to favour the dominance of small cladorecans and rotifers and reduce the efficiency of carbon and energy transfer from algae to zooplankton (Havens and Hanazato 1993). Broad spectrum pesticides which are toxic to all organisms also cause severe effects. If it causes migration of fish from one place to another, again the ecological balance is disturbed. A large number of secondary effects are expected when the aquatic ecosystem is exposed to pesticides (Hurlbert 1975). Resource partitioning and competition between the species may also get altered. However, these studies on pesticide toxicity on interspecific interaction in aquatic ecosystem are rare because it is time consuming coupled with many other complicated interactions to disrupt.

4.3.1.2 Physiological Changes

Physiological responses of test organisms have long been an element in pharmacological screening systems and can provide comparative information on minimum toxicant concentrations eliciting a response (Majewski and Klaverkamp 1975).

4.3.1.2.1 Respiration and Metabolism

Tissue Respiration (Natarajan 1985)
In this experiment, a total of 120 *C. striatus* of either sex were acclimatized for 15 days and tested for the effect of pesticide on oxygen consumption in gill, brain and muscle tissues. Technical grade pesticide at one third of its LC_{50} was added in the test chamber and exposed to the fish for 48 h and data collected starting from 6 to 72 h. Tissue oxygen consumption was measured for at least three consecutive days with one tissue sample from one fish. The oxygen concentration was measured in Warburg constant volume respirometer with fish Ringer fluid at pH 7.5 as liquid phase and air as gaseous phase. Each test is replicated for six readings and repeated after 1 week to see the persistence of the effect. Metasystox exposure to *C. striatus* was reported to decrease the oxygen consumption of gill tissues followed by brain and muscle.

Carbohydrate Metabolism (Sastry and Siddiqui 1982)
In this experiment, the freshwater teleost fish, *C. punctatus* was exposed to sublethal doses of pesticides. After 15, 30 and 60 days of exposure, biochemical parameters of blood, liver and muscle and enzyme activities in liver, kidney, intestine, brain, gills and muscles were studied. Lactate dehydrogenase activity was found higher in liver, muscles, brain and gills, but got inhibited in kidney and intestine. Succinate dehydrogenase activity in muscle, liver and brain get decreased whereas an increase was found in kidney and intestine. Decreased pyruvate dehydrogenase was found in all the tissues tested. These indicated a shift in carbohydrate metabolism since the anaerobic metabolism was favoured and aerobic oxidation of pyruvate impaired.

4.3.1.2.2 Food Utilization and Growth (Wang et al. 2006; Huynh and Nugegoda 2012)

The experiment is designed to evaluate the effect of pesticides on the growth and nutrient assimilation of the fish. Pre-weighed fish were exposed to pesticide by contaminating the water in the aquaria (ten fish per aquarium) and kept for 22 h exposure. Then they were transferred to clean water in tanks in a flow-through water system and fed twice daily with fishmeal-based diet for a period of 6 weeks. Fish were weighed every 2 weeks and the quantity of feed provided is adjusted accordingly. The remaining food if any was siphoned, dried, weighed and recorded to be used in food-intake calculations. A correction factor was calculated by adding 2 g of diet in the aquarium without fish for 30 min, then siphoned out, dried and weighed.

Correction factor (CF) was calculated as follow:

$$CF = \frac{\text{Initial dry weight of food} - \text{dry weight of food after 30 min.}}{\text{Initial dry weight of food}} \times 100$$

The food intake (% body weight per day), percentage weight gain, food conversion ratio (FCR) and protein efficiency ratio (PER) of the exposed fish were calculated as indices to evaluate the effect of pesticide on the growth and nutrient utilization. These parameters were calculated as follows:

$$\text{Food intake} = \frac{100 \times I}{\left[\left(W_0 + W_t\right)/2\right] \times t}$$

$$\text{Percentage weight gain} = \frac{W_t - W_0}{W_0} \times 100$$

$$\text{Specific growth rate} = \frac{\text{In}\left(W_t / N_t\right) - \text{In}\left(W_0 / N_0\right)}{t}$$

$$\text{Food conversion ratio} = \frac{I}{W_t - W_0}$$

I – total weight of the food provided (g)
W_0 – initial body weight (g)
W_t – final body weight (g)
T – duration of feeding trial (days)
N_t – number of fish at the end of the trial
N_0 – number of fish at the start of the trial

4.3.1.2.3 Growth and General Health Condition (Ribeiro et al. 2005)

Apart from growth, the general condition of the fish is assessed using condition index (CI), which is used as a general health indicator of growth, nutritional state, energy content etc. (Sutton et al. 2000; Kleinkauf et al. 2004). Condition index was evaluated by means of the weight-length relationship with that of the reference population. CI is calculated as the ratio of actual weight and theoretical weight in line with the following equation.

$$\log(\text{theoretical weight}) = 3.304 \times \log(\text{actual length}) - 3.236$$

4.3.1.2.4 Reproduction and Development

Male Sexuality: Sperm Count (Baatrup and Junge 2001)
A total of 260 adult males were divided into ten experimental groups and three control groups and used to find the antiandrogenic nature of pesticides. Each group was transferred to 16 L glass aquarium with 8 L of water. During the experimental period of 30 days the pesticides were administered to fish through feed and the water was constantly circulated through a natural filter of aquarium gravel. The pesticide compounds were dissolved in acetone at adequate concentrations, mixed thoroughly with the commercial TetraMin fish fodder, left for 24 h for the acetone to get evaporated and given as food @ 40 mg/aquarium (20 fish). After 30 days of exposure, the test fish were anesthetized and sperm cells stripped. The sperm cells were extracted from the fish and put in 90 μL of 0.125 mM NaCl and 5.0 mM $CaCl_2$ solution to aid the breakdown of the spermatozeugmata. The sperm suspension is transferred to an improved Neubauer chamber hemacytometer and sperm cells counted. Significant reductions in the number of ejaculated sperm cells were reported due to the exposure of fungicide vinclozolin and p p′ DDE to guppies.

Development of Ovaries (Rastogi and Kulshrestha 1990)
Sublethal concentrations of pesticides (one-fifth of LC_{50}) were tested to study their effect on the development of ovaries in carp minnow, *Rasbora daniconius*. Groups of 50 fish were allowed in glass aquaria after proper acclimatization during spawning season. The experiment was carried out for 75 days and the water and pesticide solutions renewed daily. Three samples from control and treatment were sacrificed at regular intervals, weighed and ovary dissected. The percentage and size of oocytes of different kinds selected randomly were calculated by analyzing through micrometers.

Egg Production: Through Dietary Exposure (Werner et al. 2002)
In this experiment, pesticide was dissolved in methanol and mixed with the standard purified casein based diet (PC-diet) at predefined concentrations and used. Control diet consists of only methanol in PC-diet. Fish (six males and six females per tank) were fed with experimental diets in 20 L flow-through aquaria. Water temperature was maintained at 25 °C with pH: 7–8, conductance: 280–320 and dissolved oxygen: 6–8 mg/L. Fish were fed twice daily @ 5 % of body weight and the mortality and egg production recorded. Eggs were disinfected in 25 g/L NaCl for 10 min and maintained in control water at 25 °C. Each batch of the eggs were assessed for the number of dead eggs, unfertilized eggs, hatched eggs and number of live and dead larvae.

Reproduction and Embryo Development (Halden et al. 2010)

In this study, adult zebrafish were exposed to feed spiked with pesticide to investigate the influence of dietary exposure in reproduction, embryo development and transfer to offspring. Aerated standardized water (ISO 1996) was used for the experiment and maintained at temperature 25 °C and photoperiod 12/12 h. Experimental feed was prepared by adding the pesticide dissolved in ethanol to freeze-dried chironomids at nominal concentrations. Ethanol was allowed to evaporate for 48 h before feeding to the fish. Fish were monitored for spawning fitness and allowed in 15 L glass aquaria @ nine fish (at least three females)/aquaria. For a period of 3 weeks fish were fed with the experimental feed at about 2 % of their body weight per day apart from brine shrimp once in a day. Eggs, debris of the aquaria were removed and 75 % of the water volume changed every day (6 days/week).

Observations like spawning success (percentage of days with egg production), fecundity (number of eggs per female) and the fertilization success (proportion of fertilized eggs) were taken. About 30 newly laid eggs from each aquarium were collected and examined under a stereo microscope to calculate the fertilization rate. Fertilized eggs were determined as it reaches the four-cell stage (Schulte and Nagel 1994). Embryos were also studied for mortality, spontaneous movement, developmental stage, circulation, oedema, pigmentation, heart rate, spine deformation and hatching success. After the exposure period of 6 weeks, the fish were sampled for vitellogenin and gonad histology. Embryo and larva were sampled for the residue analysis of the pesticide under study and its metabolite to study the transfer to offspring.

Test on Embryos (Macdonald 1979; Braunbeck et al. 2005)

Fertilized eggs at 'eyed stage' are used in this experiment and ten eggs were transferred to Petri dish (9 cm) containing 40 mL of reconstituted water with pesticides. This method is useful for testing small numbers of eggs or larvae and or large numbers of treatments (Macdonald 1979).

In the experiment given by Braunbeck et al. (2005), viable eggs of zebrafish, fathead minnow or medaka were exposed to pesticides from at latest 30, 120 or 45 min after fertilization, respectively. To ensure exposure at an early stage, the eggs were transferred to 18.5 cm dia Petri dishes with toxicant at desired concentration. After 30 min of exposure, the eggs were transferred to 24 well microtiter plates by means of pipettes and incubated. Egg coagulation, development of somites, detachment of the tail, heart beat, development of the eyes, degree of pigmentation, malformations of the spinal chord, and general growth retardation if any are some of the parameters considered for observation.

4.3.1.2.5 Osmoregulation (Kinter et al. 1972)

Plasma milliosmolarity is normally determined by standard freezing point depression technique. As a principle, the freezing point gets depressed below the normal in the presence of solutes in the water present in fish muscles. The extent of this depression is proportional to the osmotic pressure of the solution. When the extra-cellular osmotic pressure increases, it causes the water to leave the cells leading to a decrease in the intra-cellular freezing point. In this experiment, ten killifish (five males and five females, each weighing about 5 g) were placed in 2 L of water. Blood was drawn from fish after exposure to pesticide by cardiac puncture with lightly heparinized glass capillary tubes, pooled and centrifuged and osmolarity measured.

4.3.1.2.6 Endocrine System

Many pesticides are regarded as endocrine disruptors, which interfere with the production, release, transport, metabolism, binding or elimination of hormones in the body. These hormones are responsible for the maintenance of homeostasis and the regulation of developmental processes (Kavlock et al. 1996). In vitro laboratory experiments using nonylphenol ethoxylate metabolites in rainbow trout, *O. mykiss* revealed that these compounds caused the production of vitellogenin in males (Jobling and Sumpter 1993). Indices for endocrine disruptors include production of vitellogenin, testosterone and estradiol and also the gonadosomatic index, gonadal development, intersex condition etc. (Pait and Nelson 2002).

In a life cycle toxicity experiment of 4-nonylphenol to Japanese medaka, the fish was exposed to the chemical at 24 h post fertilization and continued for up to 104 days post hatch for the F_0 and 60 days for the F_1 generation. At 60 days post hatch, external secondary sex characteristics appeared totally skewed towards females and many had ovotestes (Yokota et al. 2001).

4.3.1.2.7 Cardiovascular System

In fish, the cardiovascular/respiratory (CVR) responses to a wide variety of toxicants have been examined and several sublethal bioassay procedures developed (Schaumburg et al. 1967; Morgen and Kuhn 1974). In an experiment reported by Giles et al. (1979) the fish were anesthetized in water by adding 0.3 g/L of 2-phenoxyethanol. A PE 60 catheter was implanted to obtain buccal amplitudes and respiratory rates. For electrocardiogram data, needle electrodes were inserted under the skin. The recording electrode was placed anterior to the pectoral fins and the reference electrode in the pelvic fins. After inserting the electrocardiogram electrodes, the fish were placed in 10 L restraining chambers. Then the fish were allowed for recovery and water delivered by pump to a head-tank and from there to the restraining chambers at a flow rate of 500 mL/min/chamber. The outflow from each chamber was allowed as waste until the conductance found identical to that in the head-tank. When the conductance of the outflow and the head-tank were found similar, the outflow re-circulated and fresh toxicant added to the header system. Heart rate, metabolic rate, ventilation rate and buccal pressure were monitored hourly for 3 h prior to exposure to toxicant and after the introduction of toxicant. The electrocardiogram was equipped with a bioelectric amplifier and the ventilator parameters with a pressure transducer and a pressure amplifier. Oxygen uptake was estimated as the difference in oxygen partial pressure in water entering and leaving the restraining chamber with a radiometer.

4.3.1.2.8 Nervous System

Phospholipids in Olfactory Neurons (Evans and Hara 1979)
It is known that any alteration in the phospholipids, which are the receptors of chemical stimuli, might cause a change in the olfactory response of the neurons connected to the brain. For this study, the olfactory rosettes, comprised of the olfactory lamellae in the olfactory capsules of fish are removed intact from anaesthetized fish. They are then immediately fixed and processed according to the Baker's acid hematein method. The tissues thus processed are embedded in gelatin to prevent the delicate olfactory lamellae from breaking and sections of 10 μm thickness made. After the removal of lipids, Sudan black B or other colourant added to find the presence of phospholipids. This technique allows the determination of phospholipid alterations and/or morphological changes of olfactory neurons caused by toxicants. The results can be confirmed through electrophysiological tests also.

Olfactory Bulb Stimuli (Hara 1979)
The fish is anesthetized and immobilized before the experiment. During the experiment, water was supplied to the gills by means of a tube inserted into the mouth and passing over the gills. The top skull is removed to expose the brain and the loose fatty tissues surrounding the brain are removed carefully. Bipolar electrodes were placed on the posterior part of olfactory bulb and stimuli recorded. The olfactory organ is treated with toxicants by adding the toxicant in the main stream of water and stimuli recorded. Thus changes in stimuli if any, after the exposure to toxicant can be assessed.

4.3.1.2.9 Sensory Perception

Many aquatic vertebrates and invertebrates depend on chemical senses for their day to day activities and finding mate and reproduction. In fish, olfactory and taste receptors are exposed directly to the pollutants and thus their functions can get disturbed. Pesticide effects on pheromone detection in fish in connection with reproductive behaviour, shoaling and dominance interactions, fright responses and effects on detection of food odours and foraging are already been reported (Olsen 2011).

Visual Perception
There are many methods to find the changes in the visual ability of fish due to pesticides. Visual test of Dutta et al. (1992) after pesticide exposure analyzed the optomotor response of fish that does not quantify the actual visual acuity, but only whether they respond or not to black and white stripes. Using a different methodology, (Carvalho and Tillitt 2004) 2,3,7,8-tetrachlorodibenzodioxin exposed rainbow trout was assessed for visual acuity angle, optomotor, optokinetic responses etc.

Chemical Sensation (Ward et al. 2008)

An experiment was conducted to find the effect of 4-nonylphenol on the shoaling behaviour and feeding of banded killifish, *Fundulus diaphanus*. About 120 killifish grouped as 6 fish were placed in a 30 L aquarium and exposed. After an hour of exposure, each group was transferred to an arena of size 1 × 1 m with a water depth of 15 cm. Defrosted frozen bloodworms were kept concealed in a 5 cm section of pipe with a diameter of 2 cm in the arena as food for the fish. The sides of the arena above water level were pasted with black plastic and fish were filmed for 30 min. Apart from shoal cohesion and swimming speed, feeding behaviour was measured as the time taken to locate the food and the number of fish that taken food during the trial.

4.3.1.3 Biochemical Changes

4.3.1.3.1 Metabolites and Enzymes in Liver and Muscles (Begum 2004)

In this experiment, metabolites such as total protein, free amino acids, ammonia and glycogen and enzymes like alanine aminotransaminase, aspartate aminotransaminase, glutamate dehydrogenase and glycogen phosphorylases were analyzed in pesticide treated fish, *C. batrachus*. A group of 36 fish were exposed to one-third of median lethal concentration of the pesticide for 6 days. After this exposure, half of the fish were transferred into pesticide free water for 6 days to study the recovery response. Samplings of exposed, recovered and control fish were done after 1, 3 and 6 days in each condition with six replicates. The tissue samples were homogenized in ice-cold sucrose solution and centrifuged. The nuclei and cell debris were removed and the clear cell free extract was used to estimate the enzyme activities spectrophotometrically. The following methods were used to estimate the respective parameters.

Total protein	:	Lowry et al. (1951)
Amino acids	:	Moore and Stein (1954)
Ammonia content	:	Bergmeyer (1965) using Nessler reagent
Glycogen	:	Kemp and Van-Heijinger (1954)
Alanin/aspartate aminotransaminase	:	Bergmeyer (1965)
Glycogen phosphorylases	:	Cori et al. (1955)
Glutamate dehydrogenase	:	Lee and Lardy (1965)

4.3.1.3.2 Serum and Brain Cholinesterase Bioassays

Bioassay with serum and brain cholinesterase gains its importance since all the organophosphates are reported to inhibit cholinesterase. Dissected brain or samples of serum are analyzed by laboratory procedures using acetylcholine as the substrate.

In general, serum is examined for absorbance and the enzyme activity is related to the slope of the absorbance change.

4.3.1.4 Haematology and Biochemistry of Blood

4.3.1.4.1 Heaematological Changes (Saravanan et al. 2011)

Haematological changes in blood were studied in freshwater teleost fish, *C. carpio* after acute and sublethal toxicity tests. Fish were transferred after acclimation into three plastic tubs (20 fingerlings each) filled with 40 L of water and pesticides added. After 24 h of exposure to median lethal concentrations, the live fish were randomly selected and blood collected for hematological analysis. For sublethal toxicity experiments, 300 fingerlings were selected and divided into three groups and dosed with one-tenth of 24 h LC_{50} of the pesticide in two groups and the other kept as control. The experiment was conducted in static renewal system. Experiment was conducted for a period of 25 days and sampling done at 5 days intervals.

Blood was collected from control and pesticide treated groups by cardiac puncture using disposable syringe fitted with 26 gauge needles dipped in heparin, kept under ice and used for the estimation of haemoglobin, red blood cells (RBC) and white blood cells (WBC). RBC and WBC were counted using a haemocytometer. Heamoglobin and hematocrit were estimated by Cyanmethaemoglobin and microhematocrit methods, respectively (Drabkin 1946; Nelson and Morris 1989). Erythrocyte indices like, the mean corpuscular volume (MCV), mean corpuscular hemoglobin (MCH) and the mean corpuscular hemoglobin concentration (MCHC) were also calculated using standard formula.

4.3.1.4.2 Biochemical Changes (Agrahari et al. 2007)

After acclimatization of *C. punctatus*, the fish were grouped and tested with pesticide at 1/20th and 1/10th of the median lethal concentrations for long exposure periods. Feeding was stopped and fish were starved 24 h prior to dissection. During each sampling, blood was collected from dorsal aorta and kept in a sterilized glass vials with EDTA. The samples were centrifuged and analyzed spectrophotometrically. The following methods were used to estimate the respective parameters.

Glucose activity	:	Carl and Edward (1996)
Acid/alkaline phosphatase activity	:	Hillmann (1971)
Glutamate oxalacetate/pyruvate transaminase	:	Bergmeyer et al. (1986)
Lactate dehydrogenase activity	:	Dito (1979)
Cholesterol level	:	Henry (1974)
Triglycerides	:	Schettler and Nussel (1975)
Protein content	:	Lowry et al. (1951)

4.3.1.5 Histopathological Effects

Histopathological effects of pesticides in fishes are normally studied in the liver, blood vessels, kidneys and gills. Cytoplasmic granularity in liver cells, partial loss of liver plate, radial orientation and shrinkage of some liver cell masses are reported due to pesticide poisoning. Pycnotic changes in the glomeruli of kidney, vocalization of cytoplasm and atrophy of some cells and precipitated masses plugging the central capillaries in gills are also reported due to pesticides in various fishes.

Gonad histopathology In an experiment by Halden et al. (2010), fish were dehydrated and embedded in paraffin blocks and cut dorso-ventrally and gonads examined through microscope. The number of oocytes per microscopic field was counted and categorized as normal and defected (OECD 2009). Vitellogenin concentrations can be analyzed based on a direct non-competitive sandwich ELISA as described by Holbech et al. (2001). Immuno histochemistry can be performed in fish using fish specific rabbit anti-lipovitellin polyclonal antiserum (Holbech et al. 2001).

Gill, kidney and liver histopathology In an experiment by Giles et al. (1979), rainbow trout and whitefish surviving from the acute lethal bioassays were used for histopathological studies. The fish were weighed and the abdominal cavity opened by mid-ventral incision and preserved in Bouins fluid. After fixation for at least 24 h, the mid-region of the posterior kidney, the first gill arch and the entire liver were dissected and washed in 70 % alcohol for 1–3 days. The tissue samples were sectioned at 10 µm after embedded in parafilm and approximately 50–100 sections examined from each tissue from each fish.

Histopathology due to pesticide poising in fishes are normally explained in the following ways:

Cloudy swelling: Cells are enlarged and the cytoplasm may be filled with granules.

Hydropic degeneration: A more pronounced form of cloudy swelling in which vacuoles appear in the cytoplasm. This is often seen in liver, kidney and heart muscle cells and may be due to the lack of active transport processes of the cell.

Hyaline degeneration: The cytoplasm takes on a glassy, eosinophilic appearance and generally becomes necrotic.

Hyperplasia: Localized increase in the number of cells in a tissue, such as the gills causing a thickening of the epithelium.

Hematoma: Swelling or edema in which the intercellular spaces contain blood cells and plasma.

4.3.1.6 Carcinogenicity/Teratogenicity and Mutagenicity

Many pesticides like most of the organochlorines, many organophosphates, thioureas etc. are reportedly carcinogenic. Out of 240 pesticides screened for carcinogenicity at least 24 (10 %) were found to produce tumours (Hurley 1998). Analysis of residues of these carcinogenic pesticides in fish itself indicates carcinogenicity. Mutagenicity and genotoxic assessment of pesticides are normally done in molecular level by assessing the DNA damage.

4.3.1.6.1 Teratogenic Studies in Embryo (EPA 2002)

In teratogenic studies, embryos were exposed to toxicants in static renewal system, starting shortly after fertilization of the eggs. Morphological deformities (terata) are taken as end points in these studies. In an experiment, fathead minnow embryos of less than 36 h old are used for studying the teratogenicity of toxicant. The test is conducted in four chambers at each toxicant concentration and control with 15 embryos per replicate. Thus 60 embryos are exposed at each test concentration and 360 embryos would be needed for a test consisting of five concentrations and a control. The embryos were placed into the test chamber using a small bore (2 mm ID) glass tube. The method used here is semi static and while replacing and adding water, dissolved oxygen, temperature and pH are to be checked. At the end of the first 24 h of exposure, before renewing the test solutions, the embryos were examined and dead ones (milky coloured and opaque) were recorded and removed. Upon hatching, observations on the number of dead and live embryos and the number of hatched, dead, live and deformed larvae were taken daily for a period of 7 days. Larvae with morphological abnormalities in appendages, immobile and with any characteristics that preclude survival are considered as deformed.

4.3.1.6.2 Mutagenicity and Genotoxicity by Assessing DNA Damage (Nwani et al. 2010b; Ventura et al. 2008)

The fish *C. punctatus* was exposed to the three test concentrations of pesticide in a semi-static system with the change of test water on every alternate day. The exposure was continued up to 96 h and tissue sampling done at 24, 48, 72 and 96 h. The whole blood and gill cells were collected and immediately processed to prevent any further DNA damage. The blood samples from each fish were mounted on three glass slides and about 1500 cells per slides were scored for micronuclei under light microscope. The number of erythrocytes with micronuclei and nuclear abnormalities were counted. DNA damage was further analyzed by alkaline single-cell gel electrophoresis.

For the comet assay (Ventura et al. 2008), blood obtained from fish was diluted with fetal bovine serum. Microscopic slides were made by adding 120 μL of 0.5 % low melting point agarose containing 10 μL of the diluted blood and the slides placed in lysis buffer in a refrigerator for 1 h. After lysis, the DNA was denatured and submitted to electrophoresis for 20 min, neutralized and fixed in ethanol. The slides were stained with ethidium bromide and analyzed under fluorescent microscope. The nuclei were visually classified according to fragment migration as undamaged (class 0), slightly damaged (class 1), more damaged (class 2) and highly damaged (class 3).

4.3.1.7 Bioaccumulation and Biomagnification Studies

4.3.1.7.1 Bioaccumulation Experiments (Varo et al. 2002)

The bioaccumulation of chlorpyrifos in small fish, *Aphanius iberus* was studied using contaminated crustaceans, *Artemia franciscana* and *A. parthenogenetica* as feed. *Artemia* cysts were hatched in sea water under continuous illumination and aeration and fed with microalgae. Batches of 180–210 *Artemia* adults were exposed to pesticide for 48 h in a glass flask filled with 800 mL solution. This contamination procedure was continued daily to have a continuous supply of contaminated *Artemia* to the fish for bioaccumulation experiments. Adults of Spanish toothcarp, *A. iberus* were acclimatized in the laboratory in 100 L aquarium with filtered brackish water. About eight males and eight females were randomly collected and transferred in to test chambers with 10 L water and acclimatized for 2 weeks fed with uncontaminated *Artemia.* Throughout the experimental period of 32 days, the treatment fish were fed with contaminated *Artemia* and the control ones with uncontaminated *Artemia* at the rate of 10–13 per fish. Sampling was done in 9, 16 and 32 days after exposure and analyzed for pesticide in whole body. After the actual contamination experimental period of 32 days, the remaining fish were maintained with uncontaminated *Artemia* as food for another 15 days. Sampling was done at 1, 3 and 15 days during recovery phase and analyzed for pesticides.

4.3.1.7.2 Biomagnification Experiments (Ellgehausen et al. 1980; Henny et al. 2003)

An experiment was formulated to evaluate the bioaccumulation potential of pesticides and the contribution of food chain transfer to the total magnitude of bioaccumulation. Representative members of aquatic food chains, i.e., algae (*Scenedesmus acutus*), daphnids (*Daphnia magna*) and catfish (*Ictalurus melas*) are used to evaluate the bioaccumulation potential of ten pesticides. Transfer of pesticide residues via food chains and direct uptake from water were determined. The results of residue analysis showed that no buildup of residues in higher trophic levels of the aquatic food chains (biomagnification).

An experiment was conducted to estimate the biomagnification of various pesticides from fish to predatory birds in Willamette River, Oregon. For this, three important fish preys viz., largescale sucker, whitefish and pikeminnow along with predatory bird, osprey eggs were sampled and analyzed. The fishes, largescale sucker (*Catostomus macrocheilus*), mountain whitefish (*Prosopium williamsoni*) and northern pikeminnow (*Ptychocheilus oregonensi*) were found to constitute 90.5 % of the actual biomass of osprey's diet. So, 25 whole fish composite samples of the three species were also collected from different sample sites of the river. One egg was collected from ten osprey nests along the Willamette River to determine contaminant concentrations. Eggs were also collected from three different nests near Crane Prairie Reservoir which is located away from contaminated sites and designated as reference eggs. All the samples were analyzed for pesticide residue to find the bioaccumulation if any in the predator eggs. The biomagnification factor for the various OC pesticides (based upon wet weight concentrations in fish and eggs) ranged from extremely low (1.2) for hexachloro benzene (HCB) to high (87) for dichlorodiphenyl dichloroethylene (DDE).

4.3.2 Semi-field/Mesocosm Toxicity Assessment Methods

4.3.2.1 Semi-field Studies: River Water for Laboratory Studies (Polard et al. 2011)

In this experiment, water from river was taken at intervals and used for fish toxicity and residue analysis in the laboratory. A river (The Save river) which is 140 km long having a watershed of 1150 km^2 area mainly under agriculture, with maize, wheat and sunflower as the major crops was selected for the study. The water samplings were timed to coincide with the winter flood, a low-flow situation and after the first major rainstorm event following herbicide application in the catchment agricultural fields.

During the study, water was taken from the river and poured in tanks having ten Crucian carp, *C. carassius*. Water was taken from the experimental site and renewed daily for 4 days. A simultaneous experiment was carried out with fish exposed to herbicides deliberately mixed and prepared to duplicate the water condition during spring flood at an average concentration to the 4 days taken at that time. This experiment was conducted to compare the toxic effects of the water taken from river and water with mixture of pesticides intentionally added in the laboratory. Blood samples were taken and subjected to comet assay and single-cell gel electrophoresis assay for DNA damage.

4.3.2.2 Channel Studies: DNA Damage to Exposed Fish (Bony et al. 2008)

The outdoor mesocosm platform in this study is composed of 44×0.4 m stainless steel channels. Four such mesocosms were made and used to find the effect of pesticide on fish. A group of 130 fish (European minnows, *P. phoxinus*) were put in each net cage and fed with biofilm that grow on the sides of each cage (Michel and Oberdorff 1995). The experimentation was conducted in three phases: periphyton colonization (day 1–30), exposure to the pesticide mixture (day 31–57) and a recovery phase (day 58–80). The four channels used in the study were designated as non exposed control, two channels with chronic exposure throughout the period and the last one with additional exposure during mid phase. Channel 2 and 3 with same exposure load are used to check the inter-channel variation.

Pollutant solutions were prepared from pure pesticides of technical grade and continuously distributed in the tanks by means of a peristaltic pump and maintained a constant pesticide concentration during the whole exposure. On day 43 (mid exposure period) a pulse of pollutants was generated in channel 4 to raise the pesticide levels for a period of 1 day. The fish were sampled to study the effect of pesticides at the end of each of the three periods i.e., periphyton colonization, pesticide exposure and recovery.

4.3.2.3 Mesocosm Studies: Prey-Predatory Behaviour of Exposed Fish (Renick et al. 2015)

Larvae of topsmelt, *A. affinis* (12–15 days old; 1 cm length) were transferred to an aquarium and acclimated for 48 h in filtered seawater. The fish were fed with brine shrimp, *Artemia* sp. ad libitum twice daily. The predatory fish, the three-spine stickleback, *G. aculeatus* juveniles (~4 cm) were also acclimatized for 5–7 days and fed with shrimps once daily until 24 h prior to their use in trials. Pesticide was added in required concentration in 300 mL of test solution and ten topsmelt larvae (per replication) exposed for 4 h and behaviour observed. Then the fish larvae were pooled in terms of exposure concentrations and placed in 500 mL glass vessels with uncontaminated water to remove the adhering pesticide residues. Predators were not exposed because the pesticides at the tested concentrations were not found to have any effect on them as revealed by previous studies.

The mesocosm used in this study on pesticide effect on fish predators and preys is of 12 L capacity with $28.5 \times 9.35 \times 14$ cm size. Prey fish were released in the circular barrier in the centre of the mesocosm and both the prey and predator were acclimatized in the mesocosm for 1 h after which the circular barrier was removed. Each trial began with a 10 min observation period in which predator strikes (successful and unsuccessful) and the degree of aggregative prey behaviour were quantified. After 60 min of trial, the dead and missing prey fish were counted. To study the effect of vegetation, green laminated ribbons of 14 cm height, which simulate sea grass shoot were kept in the experimental mesocosms at different levels of density (none, low, high) combined with distribution (patchy or uniform).

4.3.3 Field Toxicity Assessment Methods

4.3.3.1 Pesticide Toxicity on Embryos and Genotoxicity (Bony et al. 2008)

In this study, a river was selected which had a catchment area with crops (here vineyard) and with no much industrial activity, so that the river be the final recipient of pesticide pollution only. Pesticides were applied to crops at standard procedures. Three sampling areas exhibiting different pesticide contamination levels were chosen, i.e. upstream, mid and downstream. Brown trout, *Salmo trutta fario* eggs were taken before the starting of actual experiment and incubated till eyed stage in the laboratory. Groups of 50 eggs were placed into Vibert boxes and incubators filled with gravel (1–3 cm size) and wrapped in plastic screening (2 mm mesh). About five such incubators with eggs were buried into gravel at each sampling site and water temperature recorded continuously. When the daily sum of water temperature representing the specific degree days (here 800 degree-days) of fish reaching the swimup stage, the incubators were removed and larva sampled for genotoxicity. A reference uncontaminated river can be chosen as a reference station in which fish eggs are to be incubated as described previously and compared.

4.3.3.2 Fish as Bio-indicator for Contamination in Estuary (McGourty et al. 2009)

In this experiment, the population level metrics like growth rate, mortality rate and recruitment rate were correlated with the contamination level of the water bodies. The two sites taken viz., San Francisco Bay is a highly urbanized with many agricultural/pesticide industries whereas the other site selected, Tomales Bay is a typical rural ecosystem with scanty population. Water/sediment contamination was analyzed for heavy metals and pesticides. Another outplant experiment was carriedout using longjaw mudsuckers, *G. mirabilis* collected from uncontaminated areas and kept in plastic mesh cages in contaminated estuary for a period of 60 days. Analysis was carried out after exposure in fish tissues to find the residues of contaminants.

In the actual experiment, baited minnow traps were placed in channels at three stations of each study site to determine the relative abundance of mudsucker, *G. mirabilis*. Sampling was made and catch per unit effort (CPUE) determined for a 24 h period and an Annual Abundance Index (AAI) calculated. Large numbers of individuals collected from all study sites were aged by examining growth rings in sagittal otoliths. Recruitment was determined by calculating the percentage of young of the year fish (YOY) out of the total sampled population. Based on the annual abundance index and number or percentage of young of the year, the site can be designated as contaminated or uncontaminated.

5 Risk Assessment of Pesticides in Aquatic Ecosystem: Fishes

Risk assessment in aquatic ecosystem is being done through different methods. Median lethal concentration can be the starting point of hazard assessment when coupled with expected environmental concentrations and other measurements such as water solubility, partition coefficient, degradation rate etc. (Mayer and Ellersieck 1986). The acute toxicity test with fish species reveals the toxicity of the pesticide chemical to that fish. This data is also helpful in assessing the possible risk to similar species, to determine possible water quality and to correlate with acute testing of other species for comparative purposes (USEPA 1996). Apart from extrapolations from acute toxicity tests, tier approach of risk assessment, risk assessment of fish with relation to other aquatic organism are also reported.

Long term exposure of fish to pesticides means a continuous health hazard. Fish exposed to pesticides may pose health hazard to humans by means of human consumption (Murthy et al. 2013). Though fish accounts for only 10 % of human diet or very less (Li et al. 2008), it is regarded as the main source of contaminants to humans through diet (Alcock et al. 1998; Harrison et al. 1998). An important method of assessing the risk of pesticides in fish is based on the residues of pesticides in the commodity. Tolerance limits (maximum allowable pesticide residue in fish food) are already established for pesticides and being monitored. Pesticide metabolism in fish body and the metabolites along with the parent compounds are to be considered for residue analysis. Apart from this, toxicological studies by feeding the fish to mammals to study the acute/chronic toxicity and teratogenic properties of pesticides are also being done and NOEL (no observed effect levels) established. Based on NOEL values, an acceptable daily intake (ADI) or reference dose (RfD) is proposed for humans by applying a suitable uncertainty factor (OPP 1990). Some of the risk assessment approaches and procedures are given below in detail.

5.1 Risk Assessment Methodologies

5.1.1 Risk Assessment Based on Median Lethal Concentrations

Toxicity classification	LC_{50} (mg/L)
Super toxic	<0.01
Extremely toxic	0.01–0.10
Highly toxic	0.11–1.0
Moderately toxic	1.1–10
Slightly toxic	11–100
Minimal/non toxic	>100

Virginia State University https://pubs.ext.vt.edu/420/420-013.html

Based on this classification, the commonly used insecticides like abamectin, azinphos methyl, chlorpyrifos, cypermethrin, endosulfan, esfenvalerate, fluvalinate, malathion, permethrin, phorate and rotenone are extremely and super toxic to fishes on the basis of median lethal bioassays to bluegill sunfish and rainbow trout. Acephate, diflubenzuron and methamidophos exhibit minimal toxicity to fish.

Accordingly, benomyl, chlorothalonil, fenarimol, PCB, PCNB etc. are regarded as high toxic fungicides whereas metalaxyl, triforine etc. are minimal toxic to fishes. Ziram and iprodione were moderately toxic to fishes and carboxin, copper sulfate, mancozeb, propiconazole and thiram are of high to moderately toxic.

According to this and based on the median lethal values obtained for rainbow trout, bluegill sunfish and carp, the herbicides such as acetochlor, bensulide, diclofop, oxadiazon, oxyfluorfen and pendimethalin are highly toxic whereas acifluorfen, bromacil, dicamba, fluometuron, fluridone, glyphosate, hexazinone, imazapyr, metribuzin, paraquat and sulfometuron are slightly toxic. Benefin and bromoxynil are extremely toxic and trifluralin is super toxic to fishes (Virginia State University https://pubs.ext.vt.edu/420/420-013.html).

5.1.2 Risk Quotient (Urban and Cook 1986; Turner and Erickson 2003)

Risk quotients (RQs) can be calculated by dividing water chemical concentrations by a toxicity reference value. The toxicity reference values can be median lethal concentration (LC_{50}), median effective concentration (EC_{50}), lowest observed effect concentration (LOEC), no observed effect concentration (NOEC), or maximum acceptable toxicant concentration (MATC) etc. (Siemering et al. 2008). A risk quotient, the ratio of toxicity and exposure is developed and compared with criteria of concern. Risk quotient criteria for fish and aquatic invertebrates are as follows:

Test data	Risk quotient	Presumption
Acute LC_{50}	>0.5	High risk of acute toxicity
Acute LC_{50}	>0.1	Restricted use of chemical reduces risk
Acute LC_{50}	>0.05	May have acute and sublethal effects on the endangered species
Chronic NOEC	>1	Chronic risk; chronic toxicity affecting reproduction and offspring
Acute invertebrate LC_{50}	>0.5	May affect fish indirectly by reducing the food supply
Aquatic plant acute EC_{50}	>0.5	May indirectly affect the habitat and food

5.1.3 Cumulative Toxicity Index

For this, a normal acute toxicity test is conducted and median lethal concentration (LC_{50}) calculated with 96 h mortality. Then a time independent median lethal concentration (TI LC_{50}), which is the concentration at which 50 % of the test animals would be expected to survive indefinitely, is calculated. This TI LC_{50} is calculated by exposing the fish under flow-through conditions for up to 30 days (Green 1965). The cumulative action of the pesticide is then estimated by calculating the ratio of 96 h LC_{50} to the TI LC_{50} as cumulative toxicity index (Hayes 1967). Chemicals with an index of less than 11 are not considered cumulatively toxic (Johnson and Finley 1980).

5.1.4 Tiered Approach of Risk Assessment (ECOFRAM 1999; Montagna et al. 2011)

This risk evaluation was done in Rio Colorado valley, Argentina where acrolein was applied for many years during spring-summer season. The ecological assessment was divided into four tiers: (a) Literature-based screening level ecological risk assessment, (b) Risk assessment with site-specific information, (c) Risk assessment with native species and (d) Impact of pesticide on benthic invertebrates (field study).

In tier I, lethal concentrations (LC_{50}) of acrolein to fishes, amphibians, molluscs and crustaceans were collected from literature (Holcombe et al. 1987) and hazard quotient calculated. Here, hazard quotient of 0.5 or greater indicates a higher risk category. In the second tier, the environmental fate/behaviour of acrolein was incorporated to provide probabilistic expressions of the potential risk associated with the use of herbicide. All the available acute toxicity data of different species in literature were taken and ranked by decreasing sensitivity. The rank was transformed to a percentile

$$\left[i/(n+1)\right]$$

Where,

i is the species rank
n is the total number of species

The log transformed LC_{50} and the percentiles converted to probabilities were plotted on a graph. The 10th percentile value was taken which shows that 10 % of the species have LC_{50} values less than that value (concentration). Again the 10th percentile was calculated excluding algae and weeds. If this value of 10th percentile where algae and weeds are excluded is less than that include algae and weeds, then a risk to non targets is suggested. An estimation of concentration of herbicide along the canals was considered for risk evaluation in the next tier. Models were utilized to find the time-space variability and dissipation of herbicide in water at several points. These predicted concentrations were compared with the LC_{50} of different organisms.

Tier 3 involves investigation of the toxicity to native species, toxicity associated with repeated exposures, chronic toxicity studies, sediment toxicity studies etc. Field studies to broaden the analysis towards community and population to study the interaction among species and indirect effect of pesticide on species were included in tier 4. Thus a complete risk assessment can be achieved with tiered approach.

5.1.5 Risk Assessment of Target Organisms with Non Targets (Folmar et al. 1979)

Studies were carried out to determine the acute toxicity of pesticide (glyphosate) with its technical grade, formulated product and the surfactant to four aquatic invertebrates and four fishes: daphnids (*Daphnia magna*), scuds (*Gammarus pseudolimnaeus*), midge larvae (*Chironomous plumosus*), mayfly nymphs (*Ephemerella walkeri*), Rainbow trout (*S. gairdneri*), fathead minnows (*P. promelas*), channel catfish (*I. punctatus*) and bluegills (*L. macrochirus*). Technical weedicide was considerably less toxic than the formulated product for midge larvae. The formulated product (Roundup®) was found to be highly toxic to rainbow trout and bluegills. The toxicity was found high at sac fry and early swim-up stages but did not affect the fecundity or alter the gonadosomatic index of adult rainbow trout treated at 2.0 mg/L. Rainbow trout did not avoid the pesticide at 10.0 mg while the mayfly nymphs did avoid and the midge larvae avoided even at 2.0 mg/L.

5.1.6 Human Health Risk Assessment Based on Residue, Consumption (OPP 1990; Jiang et al. 2005; Fianko et al. 2011)

Risk assessment can be made by integrating toxicological, consumption and residue data. Reference dose (RfD) or acceptable daily intake (ADI) are used to evaluate the risk posed by pesticide exposures. This is the level at or below which daily aggregate exposure over a lifetime will pose no appreciable risk to human health.

$$\text{Risk} = \text{Toxicological value}\ (\text{ADI}) \times \text{residue value} \times \text{consumption}$$

In this method, a dietary risk evaluation system (DRES) which estimates the dietary intake of human population even breaking down into seasonal consumption, consumption by ethnic groups, age or sex and can estimate the consumption of fresh water and salt water separately is used. Residue values are normalized to reflect the total residues including the metabolites in the edible portion of fish.

$$\text{Life Exposure Dose} = \frac{\text{Residue concentration in food} \times \text{Consumption rate}}{\text{Body weight}}$$

In an experiment by Jiang et al. (2005) risk of pesticide residues in fish to human beings was assessed by calculating hazard ratio (HR)

$$\text{Hazard ratio} = \frac{\text{Average daily exposure}}{\text{Benchmark concentration}}$$

$$\text{Benchmark concentration} = \frac{\text{Risk} \times \text{Body weight}}{\text{Fish consumption} \times \text{Slope factor}}$$

Where, Risk is the probability of lifetime cancer risk.

Slope factor is cancer slope factor obtained from the USEPA Integrated Risk Information System (IRIS) for each contaminant (http://www.epa.gov/iris/). By this method, the risk of consumption of pesticide contaminated fish to humans can be estimated.

5.2 *Risk of Pesticides on Fish*

Ecological risk assessments (ERAs) of toxicants are predominantly based on data from laboratory tests on individuals but to have population-level outcomes (Hazlerigg et al. 2014). Based on median lethal concentrations arrived at laboratory levels, many indices like hazard index, risk quotients etc. are used to calculate risks. Hazard index calculated in fish species in Thames river for dieldrin and PCBs were reported below 1, indicating no risk (Yamaguchi et al. 2003). Risk of pesticides like atrazine, desethylatrazine (DEA), simazine, diazinon, malathion, oxamyl, carbofuran and ethion were studied in water and sediment of Lake Pamvotis, Greece. In a probabilistic approach, the maximum percentage of the ecological risk was reported as 10.3 % and 51.8 % for water and 17.2 % and 70.6 % for sediment, based on acute and chronic levels, respectively showing hazard in chronic effect levels (Hela et al. 2005). Risk quotients calculated for herbicides revealed a need for limited additional risk characterization for glyphosate and fluridone and more extensive risk characterizations for diquat dibromide, chelated Cu products and copper sulfate. Risk quotient for triclopyr suggested no further need for risk characterizations revealing its safety (Siemering et al. 2008).

A four tier risk assessment for herbicide acrolein in aquatic ecosystem revealed a high hazard quotient (HQ) based on acute toxicity test for fish, amphibia, mollusk, crustacea and insect. The tadpole, *Xenopus laevis* and the mollusk, *Aplexa hypnorum* were the most sensitive and the most tolerant species, respectively (Montagna et al. 2011). Chlorpyrifos is not likely to be harmful to fish at rates of application that used to manage arthropod pests. The high toxicity to mosquito larva (LC_{95}) at levels less than 5 ppb provides a wide margin of safety to fishes. The high susceptiblity of mosquito larva and the higher tolerance of mosquito fish make the pesticide safer and allowed the use for mosquito control (Ferguson et al. 1966). In an experiment to find the ecotoxicity of lambda cyhalothrin with different organisms in aquatic ecosystem, fish are found less sensitive with the Crustacea and Insecta among the most sensitive. No ecologically adverse effects have been observed at concentrations of 0.02 μg/L at field conditions (Maund et al. 1998).

In a risk assessment of organochlorine pesticides and polychlorinated biphenyls to humans, fish intake was not found to pose a health risk to humans with a consumption of 7.4 g/person/day according to the acceptable daily intake (ADI) and minimal risk level (MRL) in two reservoirs in China. However, the hazardous ratio of the 95th percentile for PCBs in fish from Gaobeidian lake exceeded 1, which suggested that daily exposure to PCBs had a lifetime cancer risk of greater than 1 in 10,00,000 (Li et al. 2008). Human health index estimates indicate that aldrin, methoxychlor, γ-chlordane, endrin aldehyde, endrin ketone, endrin, p′p′-DDT, γ-HCH, DDE and δ-HCH do not pose direct hazard to human health, although present in fish samples. The hazard indices of 2.640, 1.720, 1.736 and 0.792 computed for α-endosulfan, heptachlor, endosulfan sulphate and dieldrin, respectively and were found to possess risk since the estimated dose exceeded the recommended reference dose (Fianko et al. 2011).

Points to Be Considered While Assessing Pesticide Risk in Fish

- Many risk assessment studies in aquatic ecosystem employs bluegill fish as a representative fish. The effects on one fish cannot represent all the different types (Rodgers 1993), so representative fishes are to be used for assessing pesticide risk.
- Single-chemical risk assessments are likely to under-estimate the impacts of pesticides on fish in ecosystems where mixtures occur (Laetz et al. 2009).

- Some studies showed metabolites as more toxic than the parent compounds (Tilak et al. 1981), so risk assessment should include the toxicity of potential compounds and metabolites also.
- Not only the direct effect, the interactive effect with predators and other agents are also to be included to have a realistic risk assessment (Rodgers 1993). Though indirect effects can lead to serious decline in population, they are generally not taken into account for assessing the risk because of lack of authentic data (Gibbons et al. 2015).
- Behavioural responses are very sensitive to low levels of contaminants. Nevertheless these bioassay results are also not usually included in assessing the ecological risk. The reasons may be because of lack of standard of behavioural toxicity tests, delayed effects and lack of field validation for these responses (Robinson 2009).
- Ecological risk assessment should use end points of contaminant effect from all levels from genes to ecosystems (Carvalho 2013).
- Models that explicitly link sublethal reductions in enzymes with that of feeding, growth and size at migration are to be developed (Baldwin et al. 2009).
- The approaches using molecular techniques revolutionize toxicological studies with less time consuming, cheap and not require the sacrifice of animals (Murthy et al. 2013). So, molecular techniques for pesticide risk assessments can be developed.

References

M.S. Abdel, A.A.M. Mamdouh, A.T.E. Hanan, Toxicological effect of butataf herbicide on some physiological aspects and the reproductive performance of Nile tilapia, *Oreochromis niloticus*. Egypt. J. Aquat. Biol. Fish **11**(2), 148–163 (2007)

L.K. Ackerman, A.R. Schwindt, S.L.M. Simonich et al., Atmospherically deposited PBDEs, pesticides, PCBs and PAHs in Western U.S. National Park fish: concentrations and consumption guidelines. Environ. Sci. Technol. **42**(7), 2334–2341 (2008)

O. Adam, C. Beringer, T. Kless, C. Lemmen, A. Adam, M. Wiseman, P. Adam, R. Klimmek, W. Forth, Anti-inflammatory effects of a low arachidonic acid diet and fish oil in patients with rheumatoid arthritis. Rheumatol. Int. **23**(1), 27–36 (2003)

S. Agrahari, K.C. Pandey, K. Gopal, Biochemical alteration induced by monocrotophos in the blood plasma of fish, *Channa punctatus* (Bloch). Pestic. Biochem. Physiol. **88**, 268–272 (2007)

J.C. Akan, F.I. Abdulrahman, Z.M. Chellube, Organochlorine and organophosphorus pesticide residues in fish samples from Lake Chad, Baga, North Eastern Nigeria. Int. J. Innov. Manag. Tech. **5**(2), 87–92 (2014)

R.E. Alcock, P.A. Behnisch, K.C. Jones, H. Hagenmaier, Dioxin-like PCBs in the environment-human exposure and the significance of sources. Chemosphere **37**, 1457–1472 (1998)

J. Alder, U.R. Sumaila, Western Africa: a fish basket of Europe past and present. J. Environ. Dev. **13**(2), 156–178 (2004)

A.B. Ali, Some ecological aspects of fish populations in tropical ricefields. Hydrobiology **190**(3), 215–222 (1990)

D. Allison, B.J. Kallman, O.B. Cope, C.C. Van-Valin, *Some Chronic Effects of DDT on Cutthroat Trout*. Research Report 64 (US Department of Interior, Bureau of Sport Fisheries and Wildlife, Washington, DC, 1964). 30 p

E.H. Allison, A. Delaporte, H.D. De-Silva, *Integrating Fisheries Management and Aquaculture Development with Food Security and Livelihoods for the Poor. Report Submitted to the Rockefeller Foundation* (School of International Development, University of East Anglia, Norwich, 2013)

K. Al-Sabti, C.D. Metclafe, Fish micronuclei for assessing genotoxicity in water. Mutat. Res. **343**, 121–135 (1995)

S. Anbumani, M.N. Mohankumar, Cytogenotoxicity assessment of monocrotophos and butachlor at single and combined chronic exposures in the fish, *Catla catla* (Hamilton). Environ. Sci. Pollut. Res. **22**(7), 4964–4976 (2015)

J.M. Anderson, H.B. Prins, Effects of sublethal DDT on a simple reflex in brook trout. J. Fish. Res. Board Can. **27**(2), 331–334 (1970)

B.A. Ansari, M.K. Ahmad, Toxicity of synthetic pyrethroid, lambda – cyhalothrin and neem based pesticide Neem gold on Zebra fish, *Danio rerio* (Cyprinidae). Glob. J. Environ. Res. **4**(3), 151–154 (2010)

J.T.M. Arsenault, W.L. Fairchild, D.L. MacLatchy, L. Burride, K. Haya, S.B. Brown, Effects of water-borne 4-nonylphenol and 17b-estradiol exposure during parrsmolt transformation on growth and plasma IGF-I of Atlantic salmon, *Salmo salar* (L.). Aquat. Toxicol **66**, 255–265 (2004)

E. Baatrup, M. Junge, Antiandrogenic pesticides disrupt sexual characteristics in the adult male guppy (*Poecilia reticulata*). Environ. Health Perspect. **109**(10), 1063–1070 (2001)

R. Babu, T. Imagawaa, H. Tao, R. Ramesh, Distribution of PCBs, HCHs and DDTs and their eco-toxicological implications in Bay of Bengal. India. Environ. Int. **31**(4), 503–512 (2005)

D.H. Baldwin, J.A. Spromberg, T.K. Collier, N.L. Scholz, A fish of many scales: extrapolating sublethal pesticide exposures to the productivity of wild salmon populations. Ecol. Appl. **19**(8), 2004–2015 (2009)

L. Balk, A. Larsson, L. Frolin, Baseline studies of biomarkers in the feral female perch (*Perca fluviatilits*) as tools in biological monitoring of anthropogenic substances. Mar. Environ. Res. **42**, 203–208 (1996)

J.E. Barbash, E.A. Resek, *Pesticides in Ground Water: Distribution, Trends and Governing Factors* (Ann Arbor Press, Chelsea, 1996). 588 p

M.L. Bari, Y. Sabina, H. Kusunoki, T. Uemura, Preservation of fish cutlet (*Pangasius pangasius*) at ambient temperature by irradiation. J. Food Prot. **1**, 56–62 (2000)

M.J. Barry, K. O'Halloran, D.C. Logan, J.T. Ahokas, D.A. Holdway, Sublethal effects of esfenvalerate pulse-exposure on spawning and non-spawning Australian crimson-spotted rainbowfish (*Melanotaenia fluviatilis*). Arch. Environ. Contam. Toxicol. **28**(4), 459–463 (1995)

G. Begum, Carbofuran insecticide induced biochemical alterations in liver and muscle tissues of the fish, *Clarias batrachus* (Linn) and recovery response. Aquat. Toxicol. **66**, 83–92 (2004)

A. Belluzzi, C. Brignola, M. Campieri, A. Pera, S. Boschi, M. Miglioli, Effect of an enteric-coated fish-oil preparation on relapses in Crohn's disease. N. Engl. J. Med. **334**, 1557–1560 (1996)

C. Bene, *Small-Scale Fisheries: Assessing Their Contribution to Rural Livelihoods in Developing Countries*. FAO Fisheries Circular, No. 1008 (Food and Agriculture Organization, Rome, 2006)

C. Bene, E. Steel, B.K. Luadia, A. Gordon, Fish as the 'Bank in the water' – evidence from chronic-poor communities in Congo. Food Policy **34**(1), 108–118 (2009)

C. Bene, M. Barange, R. Subasinghe et al., Feeding 9 billion by 2050 – putting fish back on the menu. Food Sec. **7**, 261–274 (2015)

H. Berg, Pesticide use in rice and rice–fish farms in the Mekong Delta, Vietnam. Crop. Prot. **20**, 897–905 (2001)

H.U.C. Bergmeyer, Aminotransferases and related enzymes, in *Methods of Enzymatic Analysis*, ed. by H.U.C. Bergmeyer, E. Bernt, vol. II (Academic, New York, 1965), pp. 735–739

H.U. Bergmeyer, M. Horder, R. Rej, Approved recommendation on IFCC method for glutamate oxaloacetate transaminase and glutamate pyruvate transaminase. J. Clin. Chem. Clin. Biochem. **24**, 481–489 (1986)

M.C.M. Beveridge, S.H. Thilsted, M.J. Phillips, M. Metian, M. Troell, S.J. Hall, Meeting the food and nutrition needs of the poor: the role of fish and the opportunities and challenges emerging from the rise of aquaculture. J. Fish Biol. **83**, 1067–1084 (2013)

L. Beyger, R. Orrego, J. Guchardi, D. Holdway, The acute and chronic effects of endosulfan pulse-exposure on *Jordanella floridae* (Florida flagfish) over one complete life-cycle. Ecotoxicol. Environ. Saf. **76**, 71–78 (2012)

R.Y. Bhandare, T.S. Pathan, S.E. Shinde, P.R. More, D.L. Sonawane, Toxicity and behavioural changes in fresh water fish, *Puntius stigma* exposed to pesticide (Rogor). Am-Euras. J. Toxicol. Sci. **3**(3), 149–152 (2011)

N. Bibi, A. Zuberi, M. Naeem, I. Ullah, H. Sarwar, B. Atika, Evaluation of acute toxicity of Karate® and its sub-lethal effects on protein and acetylcholinestrase activity in *Cyprinus carpio*. Int. J. Agric. Biol. **16**(4), 731–737 (2014)

R.E. Bilby, B.R. Fransen, P.A. Bisson, Incorporation of nitrogen and carbon from spawning coho salmon into the trophic system of small streams: evidence from stable isotopes. Can. J. Fish. Aquat. Sci. **53**, 164–173 (1996)

R.E. Bilby, B.R. Fransen, P.A. Bisson, J.K. Walter, Response of juvenile coho salmon (*Oncorhynchus kisutch*) and steelhead (*Oncorhynchus mykiss*) to the addition of salmon carcasses to two streams in southwestern Washington, USA. Can. J. Fish. Aquat. Sci. **55**, 1909–1918 (1998)

N.C. Bols, J.L. Brubacher, R.C. Ganassin, L.E.J. Lee, Ecotoxicology and innate immunity in fish. Dev. Comp. Immunol. **25**(8&9), 853–873 (2001)

S. Bony, C. Gillet, A. Bouchez, C. Margoum, A. Devaux, Genotoxic pressure of vineyard pesticides in fish: field and mesocosm surveys. Aquat. Toxicol. **89**(3), 197–203 (2008)

S. Bose, S. Nath, S.S. Sahana, Toxic impact of thiamethoxam on the growth performance and liver protein concentration of a freshwater fish, *Oreochromis niloticus*. Indian J. Fund. Appl. Life Sci. **1**(4), 274–280 (2011)

T. Boujard, X. Dugy, D. Genner, C. Gosset, G. Grig, Description of a modular, low cost, eater meter for the study of feeding behavior and food preferences in fish. Physiol. Behav. **52**(6), 1101–1106 (1992)

K.H. Bowmer, M.L. Higgins, Some aspects of the persistence and fate of acrolein herbicide in water. Arch. Environ. Contam. Toxicol. **5**(1), 87–96 (1976)

K.H. Bowmer, G.R. Sainty, Management of aquatic plants with acrolein. J. Aquat. Plant Manag. **15**, 40–46 (1977)

K. Boxaspen, J.C. Holm, The development of pyrethrum based treatments against the ectoparasitc salmon lice, *Lepeoptheirus salmonis* in sea cage rearing of Atlantic salmon, *Salmo salar*. Aquat. Res. **32**, 701–707 (2001)

C.E. Boyd, Insecticides cause mosquitofish to abort. Prog. Fish Cult. **26**, 138 (1964)

S.P. Bradbury, J.R. Coats, Toxicokinetics and toxicodynamics of pyrethroid insecticides in fish. Environ. Toxicol. Chem. **8**(5), 373–380 (1989)

T. Braunbeck, M. Bottcher, H. Hollert, T. Kosmehl, E. Lammer, E. Leist, M. Rudolf, N. Seitz, Towards an alternative for the acute fish LC_{50} test in chemical assessment: the fish embryo toxicity test goes multi-species – an update. ALTEX **22**, 87–102 (2005)

S. Bretaud, P. Saglio, C. Saligaut, B. Auperin, Biochemical and behavioral effects of carbofuran in goldfish (*Carassius auratus*). Environ. Toxicol. Chem. **21**, 175–181 (2002)

J.L. Brooks, S.I. Dodson, Predation, body size and composition of plankton. Science **150**, 28–35 (1965)

R.M. Burgess, R.A. McKinney, K.A. Schweitzer, D.K. Phelps, Contaminated marine sediments: water column and interstitial toxic effects. Environ. Toxicol. Chem. **12**(1), 127–138 (1993)

M.L. Burr, A.M. Fehily, J.F. Gilbert, Effects of changes in fat, fish and fibre intakes on death and myocardial reinfarction: diet and reinfarction trial (DART). Lancet **2**, 757–761 (1989)

A.G. Cagauan, The impact of pesticides on ricefield vertebrates with emphasis on fish, in *Impact of Pesticides on Farmer Health and the Rice Environment*, ed. by P.L. Pingali (Kluwer Academic Publishers, Dordrecht, 1995), pp. 203–248

E. Capkin, H. Boran, I. Altinok, Response of acetylcholinesterase (AChE) in the erythrocyte and liver of rainbow trout exposed to carbosulfan. Turk. J. Fish. Aquat. Sci. **14**, 643–650 (2014)

A.B. Carl, R.A. Edward, Glucose, in *Teitz Fundamentals of Clinical Chemistry* (W.B. Saunders and Company, Philadelphia, 1996), pp. 351–374

S.R. Carpenter, K.L. Cottingham, Resilience and restoration of lakes. Conserv. Ecol. **1**(1), 2 (1997). Retrieved from www.consecol.org/vol1/iss1/art2/

S.R. Carpenter, J.F. Kitchell, J.R. Hodgson, Cascading trophic interactions and lake productivity-fish predation and herbivory can regulate lake ecosystems. Bioscience **35**, 634–639 (1985)

S.R. Carpenter, K.L. Cottingham, D.E. Shindler, Biotic feedbacks in lake phosphorous cycles. Tree **7**, 332–336 (1992)

P.S.M. Carvalho, Behavioral biomarkers and pollution risks to fish health and biodiversity, in *Pollution and Fish Health in Tropical Ecosystems*, ed. by C.A.O. Ribeiro (CRC Press, Abingdon, 2013), pp. 350–377

P.S.M. Carvalho, D.E. Tillitt, 2,3,7,8-TCDD effects on visual structure and function in swim-up rainbow trout. Environ. Sci. Technol. **38**, 6300–6306 (2004)

D.G.S.M. Cavalcante, C.B.R. Martinez, S.H. Sofia, Genotoxic effects of Roundup® on the fish *Prochilodus lineatus*. Mutat. Res. Genet. Toxicol. Environ. Mutagen. **655**(1&2), 41–46 (2008)

T. Cavas, S. Konen, Detection of cytogenetic and DNA damage in peripheral erythrocytes of goldfish (*Carassius auratus*) exposed to a glyphosate formulation using the micronucleus test and the comet assay. Mutagenesis **22**(4), 263–268 (2007)

C.J. Cederholm, Fate of coho salmon (*Onchorhynchus kisutch*), carcasses in spawning streams. Can. J. Fish. Aquat. Sci. **46**, 1347–1355 (1989)

G. Chandra, I. Bhattacharjee, S.N. Chatterjee, A. Ghosh, Mosquito control by larvivorous fish. Indian J. Med. Res. **127**(1), 13–27 (2008)

H.U. Chandrasekara, A. Pathiratne, Influence of low concentrations of trichlorfon on haematological parameters and brain acetylcholinesterase activity in common carp, *Cyprinus carpio*. Aquacult. Res. **36**(2), 144–149 (2005)

L.W. Chandrasekara, A. Pathiratne, Body size related differences in the inhibition of brain acetylcholinesterase activity in juvenile Nile tilapia (*Oreochromis niloticus*) by chlorpyrifos and carbosulfan. Ecotoxicol. Environ. Saf. **67**, 109–119 (2007)

J. Chang, S. Liu, S. Zhou, M. Wang, G. Zhu, Effects of butachlor on reproduction and hormone levels in adult zebrafish (*Danio rerio*). Exp. Toxicol. Pathol. **65**, 205–209 (2013)

S. Chatterjee, A.B. Dutta, R. Ghosh, The influence of technical grade carbofuran (CF) on the egg maturational processes of catfish, *Heteropneustes fossilis* (Bloch). Arch. Environ. Contam. Toxicol. **32**(4), 426–430 (1997)

A.C. Chindah, F.D. Sikoki, V. Ijeoma, Toxicity of an organophosphate pesticide (chloropyrifos) on a common Niger delta wetland fish, *Tilapia guineensis* (Blecker, 1862). J. Appl. Sci. Environ. Manag. **8**(2), (2004). Retrieved from https://tspace.library.utoronto.ca/handle/1807/4170

J. Cohn, R.C. MacPhail, Ethological and experimental approaches to behavior analysis: implications for ecotoxicology. Environ. Health Perspect. **104**, 299–304 (1996)

P.M. Cook, E.W. Zabel, R.E. Peterson, The TCDD toxicity equivalence approach for characterizing risks for early life stage mortality in trout, in *Chemically Induced Alterations in Functional Development and Reproduction of Fishes*, ed. by R.M. Rolland, M. Gilbertson, R.E. Peterson (Setac Press, Pensacola, 1997), pp. 9–27

L.W. Cook, C.J. Paradise, B. Lom, The pesticide malathion reduces survival and growth in developing zebrafish. Environ. Toxicol. Chem. **24**(7), 1745–1750 (2005)

T.H. Corcoran, Roman fish sauces. Class. J. **58**(5), 204–210 (1963)

G.T. Cori, G. Illingworth, P.J. Keller, Muscle phosphorylase, in *Methods in Enzymology*, ed. by S.P. Colowick, N.O. Kaplan, vol. I (Academic, New York, 1955), pp. 200–205

C. Curtis, A. Lima, S.J. Lozano, G.D. Veith, Evaluation of a bacterial bioluminescence bioassay as a method for predicting acute toxicity of organic chemicals in fish, in *Aquatic Toxicology and Hazard Assessment: ASTM STP 766, Fifth Conf*, ed. by J.G. Pearson, R.B. Foster, W.E. Bishop (American Society for Testing and Materials, Philadelphia, 1982), pp. 170–178

P. Cury, A. Bakun, R.J.M. Crawford, A. Jarre, R.A. Quinones, L.J. Shannon, H.M. Verheye, Small pelagics in upwelling systems: patterns of interaction and structural changes in 'wasp-waist' ecosystems. ICES J. Mar. Sci. **57**, 603–618 (2000)

M.B. Datta, A. Kaviraj, Acute toxicity of the synthetic pyrethroid pesticide fenvalerate to some air breathing fishes. Toxicol. Environ. Chem. **93**(10), 2034–2039 (2011)

M. David, H.B. Shivakumar, R. Shivakumar, S.B. Mushigeri, B.H. Ganti, Toxicity evaluation of cypermethrin and its effect on oxygen consumption of the freshwater fish. *Tilapia mossambica*. Indian J. Environ. Toxicol. **13**, 99–102 (2003)

M. David, S.B. Mushigeri, R. Shivakumar, G.H. Philip, Response of *Cyprinus carpio* (Linn) to sublethal concentration of cypermethrin: alterations in protein metabolic profiles. Chemosphere **56**(4), 347–352 (2004)

C. Davidson, R.A. Knapp, Multiple stressors and amphibian declines: dual impacts of pesticides and fish on yellow-legged frogs. Ecol. Appl. **17**, 587–597 (2007)

M.J. Davis, J.L. Purrenhage, M.D. Boone, Elucidating predator–prey interactions using aquatic microcosms: complex effects of a crayfish predator, vegetation and atrazine on tadpole survival and behavior. J. Herpetol. **46**(4), 527–534 (2012)

F.B. Davy, H. Kleerekoper, P. Gensler, Effects of exposure to sublethal DDT on the locomotor behavior of the goldfish (*Carassius auratus*). J. Fish. Res. Board Can. **29**(9), 1333–1336 (1972)

P.K. Dayton, S.F. Thrush, M.T. Agardy, R.J. Hofman, Environmental effects of marine fishing. Aquat. Cons. Mar. Freshw. Ecol. **5**, 205–232 (1995)

S. Deka, R. Mahanta, A Study on the effect of organophosphorus pesticide malathion on hepato-renal and reproductive organs of *Heteropneustes fossilis* (Bloch). Sci. Probe. **1**(1), 1–13 (2012)

D.L. DeKoven, J.M. Nunez, S.M. Lester, D.E. Conklin, G.D. Marty, L.M. Parker, D.E. Hinton, A purified diet for medaka (*Oryzias latipes*): refining a fish model for toxicological research. Lab. Anim. Sci. **42**, 180–189 (1992)

J.C. Delbeek, Coral farming: past, present and future trends. Aquat. Sci. Conserv. **3**(1–3), 171–181 (2001)

L.C. Del-Gobbo, J.A. Archbold, L.D. Vanderlinden, C.S. Eckley, M.L. Diamond, M. Robson, Risks and benefits of fish consumption for childbearing women. Can. J. Diet. Pract. Res. **71**(1), 41–45 (2010)

C. Dey, S.K. Saha, Dimethoate (30%EC) induced toxicities on the tissues of the Indian major carp: *Labeo rohita* (Hamilton). Int. J. Fish. Aquat. Stud. **1**(6), 232–236 (2014)

S. Dey, S.N. Ramanujam, R.S. Dkhar, C.R. Bhattacharjee, D. Purkayastha, Disturbances in cellular features and elemental homeostasis in the integument of a freshwater fish, *Channa punctatus* (Bloch) in relation to hydrogen ion concentration of polluted water. Cytobios **106**(2), 233–244 (2001)

J.P. Dietrich, A.L. Van-Gaest, S.A. Strickland, M.R. Arkoosh, The impact of temperature stress and pesticide exposure on mortality and disease susceptibility of endangered Pacific salmon. Chemosphere **108**, 353–359 (2014)

W.R. Dito, Lactate dehydrogenase: a brief review, in *Clinical Enzymology*, ed. by J.C. Griyths (Masson Publishing, New York, 1979), pp. 1–8

R. Dobsikova, J. Velisek, T. Wlasow, P. Gomulka, Z. Svobodova, L. Novotny, Effects of cypermethrin on some haematological, biochemical and histopathological parameters of common carp (*Cyprinus carpio* L.). Neuroendocrinol. Lett. **27**(2), 101–105 (2006)

J.J. Dodson, C.I. Mayfield, The dynamics and behavioural toxicology of Aqua-Kleen® (2,4-D butoxy ethanol ester) as revealed by the modification of rheotropism in rainbow trout. Trans. Am. Fish. Soc. **108**, 632 (1979a)

J.J. Dodson, C.I. Mayfield, Modifications of rheotropic response of rainbow trout (*Salmo gairdneri*) by sublethal doses of the aquatic herbicides, diquot and simazine. Environ. Pollut. **18**, 147 (1979b)

P. Domenici, R. Blake, The kinematics and performance of fish fast-start swimming. J. Exp. Biol. **200**, 1165–1178 (1997)

T. Dovydaitis, Fish consumption during pregnancy: an overview of the risks and benefits. J. Midwifery Womens Health **53**(4), 325–330 (2008)

D.L. Drabkin, Spectrometric studies. Crystallographic and optimal properties of the hemoglobin of man in comparison with those of other species. J. Biol. Chem. **164**, 703–723 (1946)

R.A. Drummond, C.L. Russom, Behavioral toxicity syndromes, a promising tool for assessing toxicity mechanisms in juvenile fathead minnows. Environ. Toxicol. Chem. **9**, 37–46 (1990)

R.A. Drummond, W.A. Spoor, G.F. Olson, Some short-term indicators of sublethal effects of copper on brook trout, *Salvelinus fontinalis*. J. Fish. Res. Board Can. **30**(5), 698–701 (1973)

J. Dry, D. Vincent, Effect of a fish oil diet on asthma: results of a 1-year double-blind study. Int. Arch. Allergy. Immunol. **95**, 156–157 (1991)

B.R. Dumbauld, K.M. Brooks, M.H. Posey, Response of an estuarine benthic community to application of the pesticide carbaryl and cultivation of Pacific oysters (*Crassostrea gigas*) in Willapa Bay. Wash. Mar. Pollut. Bull. **42**(10), 826–844 (2001)

M. Dunier, A.K. Siwicki, Effects of pesticides and other organic pollutants in the aquatic environment on immunity of fish: a review. Fish Shellfish Immunol. **3**(6), 423–438 (1993)

H.M. Dutta, H.J.M. Meijer, Sub-lethal effects of diazinon on the structure of the testis of bluegill, *Lepomis macrochirus*: a microscopic analysis. Environ. Pollut. **125**(3), 355–360 (2003)

H. Dutta, J. Marcelino, C. Richmonds, Brain acetylcholinesterase activity and optomotor behavior in bluegills, *Lepomis macrochirus* exposed to different concentrations of diazinon. Arch. Int. Physiol. Biochem. Biophysiol. **100**(5), 331–334 (1992)

H.M. Dutta, A. Nath, S. Adhikari, P.K. Roy, N.K. Singh, J.S.D. Munshi, Sublethal malathion induced changes in the ovary of an air- breathing fish, *Heteropneustes fossilis*: a histological study. Hydrobiology **294**(3), 215–218 (1994)

ECOFRAM, *Risk Assessment Methods. Report of the Aquatic Workgroup* (U.S. Environmental Protection Agency, Office of Pesticide Programs, Ecological Committee on Federal Insecticide, Fungicide and Rodenticide Act (FIFRA), Washington, DC, 1999)

D. Eddins, A. Petro, P. Williams, D.T. Cerutti, E.D. Levin, Nicotine effects on learning in zebrafish: the role of dopaminergic systems. Psychopharmacology **202**(1–3), 103–109 (2009)

A. Eide, M. Bavinck, J. Raakjer, Avoiding poverty: distributing wealth in fisheries, in *Poverty Mosaics: Realities and Prospects in Small-Scale Fisheries*, ed. by J. Svein, A. Eide (Springer, Dordrecht, 2011), pp. 13–25

H. Ellgehausen, J.A. Guth, H.O. Esser, Factors determining the bioaccumulation potential of pesticides in the individual compartments of aquatic food chains. Ecotoxicol. Environ. Saf. **4**(2), 134–157 (1980)

M.A. El-Toukhy, R.S. Girgis, *In vivo* and *in vitro* studies on the effect of larvin and cypermethrin on adenosine triphosphatase activity of male rats. J. Environ. Sci. Health B **28**, 599–619 (1993)

EPA, *Short-Term Methods for Estimating the Chronic Toxicity of Effluents and Receiving Waters to Freshwater Organisms. EPA-821-R-02-013*, 4th edn. (U.S. Environmental Protection Agency, Washington, DC, 2002), p. 335

S. Erlinge, Food habits of the otter, *Lutra lutra* L. and the mink, *Mustela vison* Schreber in a trout water in Southern Sweden. Oikos **20**(1), 1–7 (1969)

R.E. Evans, T.J. Hara, in *Toxicity Tests for Fresh Water Organisms*, ed. by Scherer E., A histochemical test for olfactory function in fish., Can. Spl Pub. Fish. Aquat. Sci., (Freshwater Institute, Winnipeg, 1979), pp. 11–18

O. Evensen, Development in fish vaccinology with focus on delivery methodologies, adjuvants and formulations, in *The use of Veterinary Drugs and Vaccines in Mediterranean Aquaculture*, ed. by C. Rogers, B. Basurco (CIHEAM, Zaragoza, 2009), pp. 177–186

R.D. Ewing, *Diminishing Returns: Salmon Decline and Pesticides*. (Biotech Research and Consulting, Inc., Corvallis, 1999), 55 p. Retrieved from http://www.krisweb.com/biblio/gen_open_ewing_1999_diminishing.pdf

J.F. Fairchild, K.P. Feltz, A.L. Allert, L.C. Sappington, K.J. Nelson, J.A. Valle, An ecological risk assessment of the exposure and effects of 2,4-D acid to rainbow trout (*Onchorhyncus mykiss*). Arch. Environ. Contam. Toxicol. **56**(4), 754–760 (2009)

FAO, Fisheries and Aquaculture in Europe: Situation and Outlook in 1996. (FAO, Fisheries Department, Rome, 1996). Retrieved from www.fao.org/fi%2A/publ/circular/c911/c911%2D2.htm

FAO, *The Numbers Represent the Average Unit Value of Imports for 1994–2003*. FAO Yearbooks 1996 to 2005, Fishery Statistics, Commodities, vol. 83–97 (FAO, Rome, 2005)

FAO, *The State of World Fisheries and Aquaculture 2012*. (FAO, Rome, 2012). Retrieved from www.fao.org/docrep/016/i2727e/i2727e00.htm

FAO, The role of sustainable fisheries and aquaculture for food security and nutrition (2013), www.fao.org/fsnforum/cfs-hlpe/sites/cfs-hlpe/files/files/Fisheries%20and%20Aquaculture/PROCEEDINGS_Fisheries_Aquaculture_v0.docx

L. Favaria, E. Lopez, L. Martinez-Tabchec, E. Diaz-Pardob, Effect of insecticides on plankton and fish of Ignacio Ramirez Reservoir (Mexico): a biochemical and biomagnification study. Ecotoxicol. Environ. Saf. **51**(3), 177–186 (2002)

J.M. Fenelon, R.C. Moore, Transport of agrichemicals to ground and surface water in a small central Indiana watershed. J. Environ. Qual. **27**, 884–894 (1998)

D.E. Ferguson, D.T. Gardner, A.L. Lindley, Toxicity of dursban to three species of fish. Mosq. News **26**(1), 80–82 (1966)

J.R. Fianko, A. Donkor, S.T. Lowor, P.O. Yeboah, E.T. Glover, T. Adom, A. Faanu, Health risk associated with pesticide contamination of fish from the Densu river basin in Ghana. J. Environ. Prot. **2**, 115–123 (2011)

M.S. Field-Dodgson, The effect of salmon redd excavation on stream substrate and benthic community of two salmon spawning streams in Canterbury, New Zealand. Hydrobiology **154**, 3–11 (1987)

D.J. Finney, *Probit Analysis*, 3rd edn. (Cambridge University Press, Cambridge, UK, 1971). 333 p

J.F. Flannagan, B.E. Townsend, B.G.E. DeMarch, M.K. Friesen, S.L. Leonhard, The effects of an experimental injection of methoxychlor on aquatic invertebrates: accumulation, standing crop and drift. Can. Entomol. **111**, 73–89 (1979)

A.S. Flecker, Fish trophic guilds and the structure of a tropical stream: weak direct versus strong indirect effects. Ecology **73**, 927–940 (1992)

L.C. Folmar, H.O. Sanders, A.M. Julin, Toxicity of the herbicide glyphosate and several of its formulations to fish and aquatic invertebrates. Arch. Environ. Contam. Toxicol. **8**(3), 269–278 (1979)

S.W. Fowler, M. Heyraund, J. La-Rosa, Factors affecting methyl and inorganic mercury dynamics in mussels and shrimps. Mar. Biol. **46**, 267–276 (1978)

J.L. Freeman, A.L. Rayburn, Aquatic herbicides and herbicide contaminants: *invitro* cytotoxicity and cell-cycle analysis. Environ. Toxicol. **21**(3), 256–263 (2006)

J.N. Fryer, D.M. Ogilvie, Alteration of thermoregulatory behavior in fish by 5-hydroxytryptamine. Pharmacol. Biochem. Behav. **8**(2), 129–132 (1978)

A. Fuller, B.C. Cowell, Seasonal variation in benthic invertebrate recolonization of small-scale disturbance in a subtropical Florida lake. Hydrobiology **124**, 211–221 (1985)

A.Y. Gaafar, E.M. El-Manakhly, M.K. Soliman et al., Some pathological, biochemical and hematological investigations on Nile tilapia (*Oreochromis niloticus*) following chronic exposure to edifenphos pesticide. J. Am. Sci. **6**(10), 542–551 (2010)

R.M. Ganeshwade, L.B. Dama, D.R. Deshmukh, A.G. Ghanbahadur, S.R. Sonawane, Toxicity of endosulfan on freshwater fish, *Channa striatus*. Trends Fish. Res. **1**(1), 29–31 (2012)

K.M. Gaworecki, A.P. Roberts, N. Ellis, A.D. Sowers, S.J. Klaine, Biochemical and behavioral effects of diazinon exposure in hybrid striped bass. Environ. Toxicol. Chem. **28**, 105–112 (2009)

F.P. Gelwick, M.S. Stock, W.J. Matthews, Effects of fish, water depth, and predation risk on patch dynamics in a north-temperate river ecosystem. Oikos **80**, 382–398 (1997)

A. Ghermandi, P.A.L.D. Nunes, R. Portela, N. Rao, S.S. Teelucksingh, *Recreational, Cultural and Aesthetic Services from Estuarine and Costal Ecosystems*. FEEM working paper no. 121.2009 (2010). Retrieved from http://papers.ssrn.com/sol3/papers.cfm?abstract_id=1532803

D. Gibbons, C. Morrissey, P. Mineau, A review of the direct and indirect effects of neonicotinoids and fipronil on vertebrate wildlife. Environ. Sci. Pollut. Res. **22**, 103–118 (2015)

M.A. Giles, D. Klaprat, The residual oxygen test: A rapid method for estimating the acute lethal toxicity of aquatic contaminants. in *Toxicity Tests for Freshwater Organisms*, ed. by Scherer, Can. Spl Publ. Fish Aquat. Sci. **44**, 37–45 (1979)

M.A. Giles, J.F. Klaverkamp, S.G. Lawrence, *The Acute Toxicity of Saline Groundwater and of Vanadium to Fish and Aquatic Invertebrates*. (Alberta Oil Sands Environmental Research Program, Edmonton, 1979). Retrieved from https://era.library.ualberta.ca/.../AF%203.2.1%20 June%201979.pdf

J. Gilot-Delhalle, A. Colizzi, J. Moutshen, M. Moutshen-Dahmen, Mutagenicity of some organophosphorus compounds at the ade6 locus of *Schizosaccharomyces pombe*. Mutat. Res. **117**, 139–148 (1983)

H. Goerke, K. Weber, H. Bornemann, S. Ramdohr, J. Plotz, Increasing levels and biomagnification of persistent organic pollutants (POPs) in Antarctic biota. Mar. Pollut. Bull. **48**(3&4), 295–302 (2004)

L.M. Gomez-Laplaza, R. Gerlai, Latent learning in zebrafish (*Danio rerio*). Behav. Brain Res. **208**, 509–515 (2010)

B.F. Grant, P.M. Mehrle, Chronic endrin poisoning in goldfish, *Carassius auratus*. J. Fish. Res. Board Can. **27**(12), 2225–2232 (1970)

M. Graymore, F. Stagnitti, G. Allinson, Impacts of atrazine in aquatic ecosystems. Environ. Int. **26**(7&8), 483–495 (2001)

R.H. Green, Estimation of tolerance over an indefinite time period. Ecology **46**(6), 887 (1965)

G.C.M. Grinwis, A. Boonstra, E.J. Vandenbrandhof et al., Short-term toxicity of bis (tri-n-butyltin) oxide in flounder (*Platichthys flesus*): pathology and immune function. Aquat. Toxicol. **42**, 15–36 (1998)

C.K. Grisolia, A comparison between mouse and fish micronucleus test using cyclophosphamide, mitomycin C and various pesticides. Mutat. Res. **518**, 145–150 (2002)

S. Haider, N. Upadhyaya, Effect of commercial formulation of four organophosphorous insecticides on the ovaries of a freshwater teleost *Mystus vittatus* (Bloch): a histological and histochemical study. J. Environ. Sci. Health B **20**, 321–340 (1985)

A.N. Halden, J.R. Nyholm, P.L. Andersson, H. Holbech, L. Norrgren, Oral exposure of adult zebrafish (*Danio rerio*) to 2,4,6-tribromophenol affects reproduction. Aquat. Toxicol. **100**, 30–37 (2010)

S.J. Hall, A. Delaporte, M.J. Phillips, M. Beveridge, M. O'Keefe, *Blue Frontiers: Managing the Environmental Costs of Aquaculture* (The World Fish Center, Penang, 2011)

M. Halwart, M. Borlinghaus, G. Kaule, Activity pattern of fish in rice fields. Aquaculture **145**, 159–170 (1996)

J.L. Hamelink, R.C. Waybrant, R.C. Ball, A proposal: exchange equilibria control the degree chlorinated hydrocarbons are biologically magnified in lentic environments. Trans. Am. Fish. Soc. **100**, 207 (1971)

D.A. Hanan, L.M. Jones, R.B. Read, California sea lion interaction and depredation rates with the commercial passenger fishing vessel fleet near San Diego. CalCOFI Rep. **30**, 122–126 (1989)

T. Hanazato, Effects of repeated application of carbaryl on zooplankton communities in experimental ponds with or without the predator Chaoborus. Environ. Pollut. **74**(4), 309–324 (1991)

R. Hanson, D.K. Dodoo, D.K. Essumang, J. Blay Jr., K. Yankson, The effect of some selected pesticides on the growth and reproduction of fresh water *Oreochromis niloticus*, *Chrysicthys nigrodigitatus* and *Clarias gariepinus*. Bull. Environ. Contam. Toxicol. **79**(5), 544–547 (2007)

T.J. Hara, An electrophysiological test for neurotoxicity in fish. in *Toxicity Tests for Fresh Water Organisms*, ed. by Scherer, E. Can. Spl Pub. Fish. Aquat. Sci, (Freshwater Institute, Winnipeg, 1979), pp. 46–56

J.H. Harris, The use of fish in ecological assessment. Aust. J. Ecol. **20**, 65–80 (1995)

H. Harrison, S. Wearne, M.G.M. De-Gem et al., Time trends in human dietary exposure to PCDDs, PCDFs and PCBs in the UK. Chemosphere **37**, 1657–1670 (1998)

M.G.J. Hartl, S. Hutchinson, L. Hawkins, Organotin and osmoregulation: quantifying the effects of environmental concentrations of sediment-associated TBT and TPhT on the freshwater adapted European flounder, *Platichthys flesus* L. J. Exp. Mar. Biol. Ecol. **256**, 267–278 (2001)

M.R. Hasan, M. Halwart, *Fish as Feed Inputs for Aquaculture: Practices, Sustainability and Implications*. FAO fisheries and aquaculture technical paper, 518 (2009). Retrieved from www.fao.org/docrep/012/i1140e/i1140e.pdf

Y. Hashimoto, Y. Nishiuchi, Establishment of bioassay methods for the evaluation of acute toxicity of pesticides to aquatic organisms. J. Pestic. Sci. **6**(2), 257–264 (1981)

M.L. Hattula, V.M. Wasenius, H. Reunanen, A.V. Arstila, Acute toxicity of some chlorinated phenols, catechols and cresols to trout. Bull. Environ. Contam. Toxicol. **26**, 295–301 (1981)

K.E. Havens, T. Hanazato, Zooplankton community responses to chemical stressors: a comparison of results from acidification and pesticide contamination research. Environ. Pollut. **82**(3), 277–288 (1993)

W.E. Hawkins, R.M. Overstreet, W.W. Walker, Small fish models for identifying carcinogens in the aqueous environment. Water Res. Bull. **24**, 941–949 (1988)

D. Hayasaka, T. Korenaga, K. Suzuki, F. Saito, F. Sanchez-Bayo, K. Goka, Cumulative ecological impacts of two successive annual treatments of imidacloprid and fipronil on aquatic communities of paddy mesocosms. Ecotoxicol. Environ. Saf. **80**, 355–362 (2012)

W.J. Hayes, The 90 day LD_{50} and a chronic factor as measures of toxicity. Toxicol. Appl. Pharmacol. **11**(2), 327 (1967)

C.R.E. Hazlerigg, C.R. Tyler, K. Lorenzena, J.R. Wheeler, P. Thorbek, Population relevance of toxicant mediated changes in sex ratio in fish: an assessment using an individual-based zebrafish (*Danio rerio*) model. Ecol. Model. **280**, 76–88 (2014)

S. Heck, C. Bene, R. Reyes-Gaskin, Investing in African fisheries: building links to the millennium development goals. Fish Fish. **8**(3), 211–226 (2007)

D.G. Hela, D.A. Lambropoulou, I.K. Konstantinou, T.A. Albanis, Environmental monitoring and ecological risk assessment for pesticide contamination and effects in Lake Pamvotis, northwestern Greece. Environ. Toxicol. Chem. **24**(6), 1548–1556 (2005)

L.A. Helfrich, *Pesticides and Aquatic Animals: A Guide to Reducing Impacts on Aquatic Systems* (2009). Retrieved from https://pubs.ext.vt.edu/420/420-013/420-013.html

C. Henderson, Q.H. Pickering, C.M. Tarzwell, Relative toxicity of ten chlorinated hydrocarbon insecticides to four species of fish. Trans. Am. Fish. Soc. **88**(1), 23–32 (1959)

C. Henderson, Q.H. Pickering, C.M. Tarzwell, The toxicity of organic phosphorus and chlorinated hydrocarbon insecticides to fish, in *Biological Problems in Water Pollution, Tech. Report*, ed. by U.S. Dept. Health (Welfare, P.H.S, Cincinnati, 1960), pp. 76–88

C.J. Henny, J.L. Kaiser, R.A. Grove, V.R. Bentley, J.E. Elliott, Biomagnification factors (fish to osprey eggs from Willamette River, Oregon, USA) for PCDDs, PCDFs, PCBs and OC pesticides. Environ. Monit. Assess. **84**, 275–315 (2003)

R.J. Henry, *Clinical Chemistry: Principles and Techniques* (Harper and Row Publishers, New York, 1974), p. 1443

D. Hernandez-Moreno, F. Soler, M.P. Miguez, M. PerezLopez, Brain acetylcholinesterase, malondialdehyde and reduced glutathione as biomarkers of continuous exposure of tench, *Tinca tinca* to carbofuran or deltamethrin. Sci. Total Environ. **408**, 4976–4983 (2010)

G.Z. Hillmann, Continuous photometric measurement of prostrate acid phosphatase and alkaline phosphatase activity. Z. Klin. Chem. Klin. Biochem. **9**, 273 (1971)

HLPE, *High Level Panel of Experts. Sustainable Fisheries and Aquaculture for Food Security and Nutrition. A Report by the High Level Panel of Experts on Food Security and Nutrition of the Committee on World Food Security* (Food and Agricultural Organisation, Rome, 2014)

H. Holbech, L. Andersen, G.I. Petersen, B. Korsgaard, K.L. Pedersen, P. Bjerregaard, Development of an ELISA for vitellogenin in whole body homogenate of zebrafish (*Danio rerio*). Comp. Biochem. Physiol. C **130**, 119–131 (2001)

G.W. Holcombe, G.L. Phipps, A.H. Sulaiman, A.D. Hoffman, Simultaneous multiple species testing: acute toxicity of 13 chemicals to 12 diverse freshwater amphibian, fish and invertebrate families. Arch. Environ. Contam. Toxicol. **16**(6), 697–710 (1987)

A.V. Holden, The absorption of ^{14}C labeled DDT from water by fish. Ann. Appl. Biol. **50**, 467–477 (1962)

A.V. Holden, Effects of pesticides on fish. Environ. Sci. Res. **3**, 213–253 (1973)

C.M. Holmlund, M. Hammer, Ecosystem services generated by fish populations. Ecol. Econ. **29**, 253–268 (1999)

A. Hontela, C. Daniel, Effects of acute and sublethal exposure to Cadmium on the interregnal and thyroid function in rainbow trout, *Onchorhyncus mykiss*. Aquat. Toxicol. **35**, 171–182 (1996)

T.L. Huang, P.P. Zandi, K.L. Tucker et al., Benefits of fatty fish on dementia risk are stronger for those without APOE. Neurology **65**(9), 1409–1414 (2005)

A.C. Huang, M. Essak, M.I. O'Connor, Top-down control by great blue herons, *Ardea herodias* regulates seagrass-associated epifauna. *Oikos* (2015). doi: 10.1111/oik.01988

K.R. Huckle, P. Millburn, Metabolism, bioconcentration and toxicity of pesticides in fish, in *Progress in Pesticide Biochemistry and Toxicology. Environmental Fate of Pesticides*, ed. by D.H. Hutson, T.R. Roberts, 7th edn. (Wiley, Chichester, 1990). 304 p

S.H. Hurlbert, Secondary effects of pesticides on aquatic ecosystems. Residue Rev. **57**, 81–148 (1975)

P.M. Hurley, Mode of carcinogenic action of pesticides inducing thyroid follicular cell tumors in rodents. Environ. Health Perspect. **106**(8), 437–445 (1998)

S.Y. Hussein, M.A. El-Nasser, S.M. Ahmed, Comparative studies on the effects of herbicide atrazine on freshwater fish, *Oreochromis niloticus* and *Chrysichthyes auratus*. Bull. Environ. Contam. Toxicol. **57**, 503–510 (1996)

H.P. Huynh, D. Nugegoda, Effects of chlorpyrifos exposure on growth and food utilization in Australian catfish, *Tandanus tandanus*. Bull. Environ. Contam. Toxicol. **88**(1), 25–29 (2012)

M. Ilavazhahan, R. Tamilselvi, S.S. Jayaraj, Determination of LC_{50} of the bacterial pathogen, pesticide and heavy metal for the fingerling of freshwater fish, *Catla catla*. Glob. J. Environ. Res. **4**(2), 76–82 (2010)

R. Ilyas, M. Javed, Acute toxicity of endosulfan to the fish species *Catla catla*, *Cirrhina mrigala* and *Labeo rohita*. Int. J. Agric. Biol. **15**, 149–152 (2013)

R.M. Inbaraj, S. Haider, In vitro effect of malathion and endosulfan on the LH induced oocyte maturation in the common carp, *Cyprinus carpio* (L.). Water Air Soil Pollut. **39**, 27–31 (1988)

ISO, International Organization for Standardization. *Water Quality Determination of the Acute Lethal Toxicity of Substances to a Freshwater Fish* [*Brachydanio rerio* Hamilton-Buchanan (Teleostei, Cyprinidae)] Part 1: Static method. ISO 7346-1 (1996)

R. James, K. Sampath, Individual and combined effects of carbaryl and methyl parathion on leucocytes and their recovery in *Heteropneustes fossilis*, in *Assessment of Water Pollution*, ed. by S.R. Mishra (ARH Publishing Corporation, Darya Ganj, 1996), pp. 417–421

R.S. Jasmine, K. Kuttalam, J. Stanley, Acute toxicity of abamectin to fishes, *Cyprinus carpio* Linnaeus and *Tilapia mosambica* (Peters). Asian J. Biol. Sci. **3**(1), 127–129 (2008)

F. Jenkins, J. Smith, B. Rajanna, U. Shameem, K. Umadevi, V. Sandhya, R. Madhavi, Effect of sub-lethal concentrations of endosulfan on hematological and serum biochemical parameters in the carp, *Cyprinus carpio*. Bull. Environ. Contam. Toxicol. **70**, 993–997 (2003)

Q.T. Jiang, T.K.M. Lee, K. Chen et al., Human health risk assessment of organochlorines associated with fish consumption in a coastal city in China. Environ. Pollut. **136**, 155–165 (2005)

S. Jobling, J.P. Sumpter, Detergent components in sewage effluent are weakly oestrogenic to fish: an in vitro study using rainbow trout (*Oncorhynchus mykiss*) hepatocytes. Aquat. Toxicol. **27**(3&4), 361–372 (1993)

W.W. Johnson, M.T. Finley, *Handbook of Acute Toxicity of Chemicals to Fish and Aquatic Invertebrates*. Resource publication 137 (United States Department of the Interior Fish and Wild Life Service, Washington, DC, 1980)

B. Joseph, S.J. Raj, Impact of pesticide toxicity on selected biomarkers in fishes. Int. J. Zool. Res. **7**, 212–222 (2011)

K. Kalavathy, A.A. Sivakumar, R. Chandran, Toxic effects of the pesticide dimethoate on the fish, *Sarotherodon mossambicus*. J. Ecol. Res. Biol. **2**, 27–32 (2001)

M. Kartz, Acute toxicity of some organic insecticides to three species of Salmonoids and to three spine stickle back. Trans. Am. Fish. Soc. **90**, 264–268 (1961)

P. Kavitha, J.V. Rao, Oxidative stress and locomotor behaviour response as biomarkers for assessing recovery status of mosquito fish, *Gambusia affinis* after lethal effect of an organophosphate pesticide, monocrotophos. Pestic. Biochem. Physiol. **87**, 182–188 (2007)

P. Kavitha, J.V. Rao, Toxic effects of chlorpyrifos on antioxidant enzymes and target enzyme acetylcholinesterase interaction in mosquito fish, *Gambusia affinis*. Environ. Toxicol. Pharmacol. **26**(2), 192–198 (2008)

R.J. Kavlock, G.P. Daston, C. De-Rosa et al., Research needs for the risk assessment of health and environmental effects of endocrine disruptors: a report of the U.S. EPA-sponsored workshop. Environ. Health Perspect. **104**(4), 715–740 (1996)

N. Kawarazuka, C. Bene, The potential role of small fish species in improving micronutrient deficiencies in developing countries: building evidence. Public Health Nutr. **14**(11), 1927–1938 (2011)

A. Kemp, A.J.M.K. Van-Heijinger, A colorimetric micro method for the determination of glucose and glycogen. J. Biochem. **56**, 646–648 (1954)

E.E. Kenaga, Predictability of chronic toxicity from acute toxicity of chemicals in fish and aquatic invertebrates. Environ. Toxicol. Chem. **1**(4), 347–358 (1982)

S.R. Kerr, W.P. Vass, Pesticide residues in aquatic invertebrates, in *Environmental Pollution by Invertebrates*, ed. by C.A. Edwards (Plenum Press, London, 1973). 542 p

F. Khalil, I.J. Kang, S. Undap, R. Tasmin, X. Qiu, Y. Shimasaki, Y. Oshima, Alterations in social behavior of Japanese medaka (*Oryzias latipes*) in response to sublethal chlorpyrifos exposure. Chemosphere **92**(1), 125–130 (2013)

M.Z. Khan, F.C.P. Law, Adverse effects of pesticides and related chemicals on enzyme and hormone systems of fish, amphibians and reptiles: a review. Proc. Pak. Acad. Sci. **42**(4), 315–323 (2005)

J.M. Kidwell, L.J. Phillips, G.F. Birchard, Comparative analyses of contaminant levels in bottom feeding and predatory fish using the national contaminant biomonitoring program data. Bull. Environ. Contam. Toxicol. **54**(6), 919–923 (1995)

D.E. Kime, *Endocrine Disruption in Fish* (Kluwer Academic Publishers, Boston, 1998). 396 p

K.A. King, C.E. Grue, J.M. Grassley, R.J. Fisk, L.L. Conquest, Growth and survival of Pacific coho salmon smolts exposed as juveniles to pesticides within urban streams in western Washington, USA. Environ. Toxicol. Chem. **33**(7), 1596–1606 (2014)

P.D. Kingsbury, D.P. Kreutzweiser, *Environmental Impact Assessment of a Semi-operational Permethrin Application, Forestry Service Report FPM-X-30* (Environment Canada, Ottawa, 1980), pp. 1–47

W.B. Kinter, L.S. Merkens, R.H. Janicki, A.M. Guarino, Studies on the mechanism of toxicity of DDT and polychiorinated biphenyls (PCBs): disruption of osmoregulation in marine fish. Environ. Health Perspect. **1**, 169–172 (1972)

A. Kleinkauf, L. Connor, D. Swarbreck, C. Levene, P. Walker, P.J. Johnson, R.T. Leah, General condition biomarkers in relation to contaminant burden in European flounder (*Platichthys flesus*). Ecotoxicol. Environ. Saf. **58**, 335–355 (2004)

J.P. Knezovich, F.L. Harrison, R.G. Wilhelm, The bioavailability of sediment-sorbed organic chemicals: a review. Water Air Soil Pollut. **32**(1–2), 233–245 (1987)

P.R. Koundinya, R. Ramamurthy, Effect of sublethal concentrations of simithion and sevin on certain haematological values of *Sarotherodon mossambicus* (Peters). Curr. Sci. **49**, 645–646 (1980)

A. Krian, A.K. Jha, Acute toxicity and behavioural responses of herbicide (Herboclin) to the fish, *Clarias batrachus* (Linn). Indian J. Environ. Ecoplan. **16**(1), 185–188 (2009)

P.M. Kris-Etherton, W.S. Harris, L.J. Appel, Fish consumption, fish oil, omega-3 fatty acids and cardiovascular disease. Circulation **106**, 2747–2757 (2002)

M. Kumar, P. Kumar, S. Devi, Toxicity of copper sulphate on behavioural parameter and respiratory surveillance in freshwater catfish, *Clarias batrachus* (Lin.). Res. J. Chem. Environ. Sci. **3**(1), 22–28 (2015)

C.A. Laetz, D.H. Baldwin, T.K. Collier, V. Hebert, J.D. Stark, N.L. Scholz, The synergistic toxicity of pesticide mixtures: implications for risk assessment and the conservation of endangered Pacific salmon. Environ. Health Perspect. **117**, 348–353 (2009)

S. Lakshmanan, A. Rajendran, C. Sivasubramaniyan, Impact of dichlorvos on tissue glycogen and protein content in freshwater fingerlings, *Oreochromis mossambicus* (Peters). Int. J. Res. Environ. Sci. Tech. **3**(1), 19–25 (2013)

B. Lal, Pesticides-induced reproductive dysfunction in Indian fishes. Fish Physiol. Biochem. **33**, 455–462 (2007)
B. Lal, M.K. Sarang, P. Kumar, Malathion exposure induces the endocrine disruption and growth retardation in the catfish, *Clarias batrachus* (Linn.). Gen. Comp. Endocrinol. **181**, 139–145 (2013)
M. Lange, W. Gebauer, J. Markl, R. Nagel, Comparison of testing acute toxicity on embryo of zebrafish, *Brachydanio rerio* and RTG-2 cytotoxicity as possible alternatives to the acute fish test. Chemosphere **30**(11), 2087–2102 (1995)
W.J. Langston, Accumulation of polychlorinated biphenyls in the cockle *Cerastoderma edule* and the tellin *Macoma balthica*. Mar. Biol. **45**, 265–272 (1978)
G.A. Larkin, P.A. Slaney, Implications of trends in marine-derived nutrient influx to south coastal British Columbia salmonid production. Fisheries **22**, 16–24 (1997)
J.M. Law, Mechanistic considerations in small fish carcinogenicity testing. ILAR J. **42**(4), 274–284 (2001)
A. Leaf, Y.F. Xiao, J.X. Kang, G.E. Billman, Membrane effects of the n-3 fish oil fatty acids, which prevent fatal ventricular arrhythmias. J. Membr. Biol. **206**, 129–139 (2005)
Y.L. Lee, H.A. Lardy, Influence of thyroid hormone on glycerol-phosphate dehydrogenase in various organs of rat. J. Biol. Chem. **240**, 1427–1436 (1965)
P.H. Leger, D.A. Bengtson, K.L. Simpson, P. Sorgeloos, The use and nutritional value of *Atrtemia* as a food source. Oceanogr. Mar. Biol. Annu. Rev. **24**, 521–623 (1986)
S.L. Leonhard, Uptake of fenitrothion by caged crayfish *Orconectes virilis* in Pine Creek, Manitoba. Manit. Entomol. **8**, 16–18 (1974)
X. Li, Y. Gan, X. Yang, J. Zhou, J. Dai, M. Xu, Human health risk of organochlorine pesticides (OCPs) and polychlorinated biphenyls (PCBs) in edible fish from Huairou Reservoir and Gaobeidian Lake in Beijing, China. Food Chem. **109**, 348–354 (2008)
M. Liess, R. Schultz, M.H.D. Liess, B. Rother, R. Kreuzig, Determination of insecticide contamination in agricultural headwater streams. Water Res. **33**, 239–247 (1999)
C. Lightfoot, A. Van-Dam, B.A. Costa-Pierce, What's happening in rice yield in rice fish systems? in *Rice-Fish Research and Development in Asia*, ed. by C.R. Dela-Cruz, C. Lightfoot, B.A. Costa-Pierce, V.R. Carangal, M.P. Bimbao, (ICLARM Conf., 1992), pp. 177–183
A. Linos, E. Kaklamanis, A. Kontomerkos et al., The effect of olive oil and fish consumption on rheumatoid arthritis – a case control study. Scand. J. Rheumatol. **20**(6), 419–426 (1991)
E.E. Little, S.E. Finger, Swimming behavior as an indicator of sublethal toxicity in fish. Environ. Toxicol. Chem. **9**, 13–19 (1990)
E.J. Livengood, F.A. Chapman, *The Ornamental Fish Trade: An Introduction with Perspectives for Responsible Aquarium Fish Ownership.* (Department of Fisheries and Aquatic Sciences, Florida Cooperative Extension Service, Institute of Food and Agricultural Sciences, University of Florida, 2011). Retrieved from http://edis.ifas.ufl.edu/fa124
D.M. Lodge, G. Cronin, E.V. Donk, A.J. Froelich, Impact of herbivory on plant standing crop: comparisons among biomes, between vascular and non-vascular plants and among freshwater herbivore taxa, in *The Structuring Role of Submerged Macrophytes in Lakes*, ed. by E. Jeppesen, M. Sondergaard, K. Christoffersen (Springer, New York, 1998), pp. 149–174
R. Lorenz, U. Spengler, S. Fischer, J. Duhm, P.C. Weber, Platelet function, thromboxane formation and blood pressure control during supplementation of the Western diet with cod liver oil. Circulation **67**, 504–511 (1983)
O.H. Lowry, N.J. Rosebrough, A.L. Farr, R.J. Randall, Protein measurement with folin phenol reagent. J. Biol. Chem. **193**, 265–275 (1951)
W.A. Macdonald, Testing embryonic and larval life stages of fish, in *Toxicity Tests for Fresh Water Organisms* ed. by E. Scherer, Can. Spl Pub. Fish. Aquat. Sci, (Freshwater Institute, Winnipeg, 1979), pp. 131–138
K.J. Macek, S. Korn, Significance of the food chain in DDT accumulation by fish. J. Fish. Res. Board Can. **27**, 1496–1498 (1970)
K.J. Macek, B.H. Sleight, Utility of toxicity tests with embryos and fry of fish in evaluating hazards associated with the chronic toxicity of chemicals to fishes, in *Aquatic Toxicology and*

Hazard Evaluation, ed. by F.L. Mayer, J.L. Hamelink. ASTM STP 634 (American Society for Testing and Materials, Philadelphia, 1977), pp. 137–146

K.J. Macek, D.F. Walsh, J.W. Hogan, D.D. Holz, Toxicity of the insecticide Dursban® to fish and aquatic invertebrates in ponds. Trans. Am. Fish. Soc. **101**(3), 420–427 (1972)

M. Magnusson, K. Heimann, M. Ridd, A.P. Negri, Pesticide contamination and phytotoxicity of sediment interstitial water to tropical benthic microalgae. Water Res. **47**(14), 5211–5221 (2013)

H.S. Majewski, J.F. Klaverkamp, An acute bioassay for cardiovascular and respiratory functions in rainbow trout. in *Proceedings of 2nd Annual Aquatic Toxicity Workshop, Toronto, Ontario*, ed. by C.R. Craig, (Ontario Ministry of the Environment, Ottawa, 1975), pp. 165-178.

B. Marchall, M. Maes, Small water bodies and their fisheries in Southern Africa. FAO, Rome CIFA technical paper, no. 29 (1994)

R. Marchioli, F. Barzi, E. Bomba, Early protection against sudden death by n-3 polyunsaturated fatty acids after myocardial infarction: time-course analysis of the results of the *Gruppo Italiano per lo Studio della Sopravvivenza nell'Infarto Miocardico* (*GISSI*)-*Prevenzione*. Circulation **105**, 1897–1903 (2002)

S.R. Marigoudar, R.N. Ahmed, M. David, Impact of cypermethrin on behavioural responses in the freshwater teleost, *Labeo rohita* (Hamilton). World J. Zool. **4**(1), 19–23 (2009a)

S.R. Marigoudar, R.N. Ahmed, M. David, Cypermethrin induced respiratory and behavioural responses in *Labeo rohita*. Vet. Arch. **79**, 583–590 (2009b)

S.R. Marigoudar, R.N. Ahmeda, M. David, Cypermethrin induced in vivo inhibition of the acetylcholinesterase activity in functionally different tissues of the freshwater teleost, *Labeo rohita* (Hamilton). Toxicol. Environ. Chem. **91**(6), 1175–1182 (2009c)

L.L. Marking, D.K. Dowson, The half life of biological activity of antimycin determined by fish bioassay. Trans. Am. Fish. Soc. **101**, 100–105 (1972)

J. Martin, A portrait of locomotor behavior in drosophila determined by a video-tracking paradigm. Behav. Process. **67**, 207–219 (2003)

S. Mastan, S. Shaffi, Sub-lethal effect of pesticides on the distribution of glutaminases in the brain of *Labeo rohita* (Ham.). Int. J. Toxicol. **7**(2) (2009) https://ispub.com/IJTO/7/2/7879

P. Matthiessen, P.J. Fox, R.J. Douthwaite, A.B. Wood, Accumulation of endosulfan residues in fish and their predators after aerial spraying for the control of tsetse fly in Botswana. Pestic. Sci. **13**(1), 39–48 (1982)

P. Matthiessen, D. Sheahan, R. Harrison, M. Kirby, R. Rycroft, A. Turnbull, C. Volkner, R. Williams, Use of a *gammarus pulex* bioassay to measure the effects of transient carbofuran runoff from farmland. Ecotoxicol. Environ. Saf. **30**, 111–119 (1995)

S.J. Maund, M.J. Hamer, J.S. Warinton, J. Timothy, Aquatic ecotoxicology of the pyrethroid insecticide lambda-cyhalothrin: considerations for higher-tier aquatic risk assessment. Pestic. Sci. **54**(4), 408–417 (1998)

F.L. Mayer, M.R. Ellersieck, *Manual of Acute Toxicity: Interpretation and Data Base for 410 Chemicals and 66 Species of Freshwater Animals*. Resource publication 160. (U.S. Dept. Interior, Fish and Wildlife Service, Washington, D.C., 1986), 574 p. Retrieved from http://137.227.231.89/Assets/UploadedFiles/ExternalDocs/90506-intro.pdf

F.L. Mayer, P.M. Mehrle, H.O. Sanders, Residue dynamics and biological effects of polychlorinated biphenyls in aquatic organisms. Arch. Environ. Contam. Toxicol. **5**(1), 501–511 (1977)

C.R. McGourty, J.A. Hobbs, W.A. Bennett et al., Likely population-level effects of contaminants on a resident estuarine fish species: comparing *Gillichthys mirabilis* population static measurements and vital rates in San Francisco and Tomales Bays. J. Coast. Estuar. Res. Fed. **32**(6), 1111–1120 (2009)

J.M. McKim, Evaluation of tests with the early life stages of fish for predicting long-term toxicity. J. Fish. Res. Board Can. **34**, 1148–1154 (1977)

M.P. McLean, R.E. McNicol, E. Scherer, Bibliography of toxicity test methods for the aquatic environment. Can. Spec. Publ. Fish. Aquat. Sci. **50**, 29 (1980)

D.J. McLeay, D.A. Brown, Sensitivity of blood cell counts in juvenile coho salmon. *Oncorhynchus kisutch* to stresses including sublethal concentrations of pulp mill effluent and zinc. J. Fish Res. Bioconc. **32**, 2357–2364 (1974)
P.G. McNicholl, W.C. Mackay, Effect of DDT and MS 222 on learning a saimple conditioned response in rainbow trout (*Salmo gairdneri*). J. Fish. Res. Board Can. **32**(5), 661–665 (1975)
L. Mhadhbi, R. Beiras, Acute toxicity of seven selected pesticides (alachlor, atrazine, dieldrin, diuron, pirimiphos-methyl, chlorpyrifos, diazinon) to the marine fish (Turbot, *Psetta maxima*). Water Air Soil Pollut. **223**, 5917–5930 (2012)
P. Michel, T. Oberdorff, Feeding habits of fourteen European freshwater fish species. Cybium **19**(1), 5–46 (1995)
F.A. Mir, G.M. Shah, U. Jan, J.I. Mir, Studies on influences of sublethal concentrations of organophosphate pesticide, dichlorvos on gonadosomatic index (GSI) of female common carp, *Cyprinus carpio communis*. Am.-Euras. J. Toxicol. Sci. **4**(2), 67–71 (2012)
F. Moberg, C. Folke, Ecological goods and services of coral reef ecosystems. Ecol. Econ. **29**, 215–233 (1999)
R.M. Mohan, Malathion induced changes in the ovary of freshwater fish, *Glossogobins giuris* (Ham). Pollut. Res. **19**(1), 73–75 (2000)
C. Montagna, A.M.P. De-D'Angelo, A. Venturino, Toxicology of the herbicide acrolein: risk assessment in aquatic environments, in *Herbicides and Environment*, ed. by A. Kortekamp (InTech, Rijeka, 2011), pp. 405–420
D.R. Montgomery, J.M. Buffington, N.P. Peterson, D. Schuett-Hames, T.P. Quinn, Stream-bed scour, egg burial depths and the influence of salmonid spawning on bed surface mobility and embryo survival. Can. J. Fish. Aquat. Sci. **53**, 1061–1070 (1996)
S. Moore, W.H. Stein, A modified ninhydrin reagent for the photometric determination of amino acids and related compounds. J. Biol. Chem. **211**, 907–913 (1954)
M.J. Morgan, J.W. Kiceniuk, Effect of fenitrothion on the foraging behavior of juvenile Atlantic salmon. Environ. Toxicol. Chem. **9**, 489–495 (1990)
W.S.G. Morgan, P.C. Kuhn, A method to monitor the effects of toxicants upon breathing rate of largemouth bass (*Micropterus salmoides*). Water Res. **8**, 67–77 (1974)
M.C. Morris, D.A. Evans, J.L. Bienias et al., Consumption of fish and n-3 fatty acids and risk of incident alzheimer disease. Arch. Neurol. **60**(7), 940–946 (2003)
P.B. Moyle, P.R. Moyle, Endangered fishes and economics: intergenerational obligations. Environ. Biol. Fish **43**, 29–37 (1995)
D. Mozaffarian, E.B. Rimm, Fish intake, contaminants and human health: evaluating the risks and the benefits. JAMA **296**(15), 1885–1899 (2006)
R.C. Muirhead-Thomson, Laboratory evaluation of pesticide impact on stream invertebrates. Freshw. Biol. **3**, 479–498 (1973)
M.S. Mulla, L.W. Isaak, Field studies on the toxicity of insecticides to the mosquito fish, *Gambusia affinis*. J. Econ. Entomol. **54**(6), 1237–1242 (1961)
M.S. Mulla, R.L. Norland, W.E. Westlake, B. Dell, J. Amant, Aquatic midge larvicides, their efficacy and residues in water, soil and fish in a warm-water lake. Environ. Entomol. **2**(1), 58–65 (1972)
P.G. Murphy, J.V. Murphy, Correlations between respiration and direct uptake of DDT in mosquito fish, *Gambusia affinis*. Bull. Environ. Contam. Toxicol. **6**, 581–588 (1971)
A.S. Murthy, Sub lethal effect of pesticides on fish. Toxic. Pestic. Fish. **2**, 55–100 (1987)
K.S. Murthy, B.R. Kiran, M. Venkateshwarlu, A review on toxicity of pesticides in fish. Int. J. Open Sci. Res. **1**(1), 15–36 (2013)
K. Muthukumaravel, M. Sukumaran, O. Sathick, Studies on the acute toxicity of pesticides on the freshwater fish *Labeo rohita*. J. Pure Appl. Zool. **1**(2), 185–192 (2013)
T. Nakano, N. Tomlinson, Catecholamines and carbohydrate concentration in rainbow trout, *Salmo gairdneri* in physical disturbance. J. Fish. Res. Board Can. **24**, 1701–1715 (1967)
G.M. Natarajan, Disruption of circadian rhythm of tissue respiration in *Channa striatus* by Metasystox®. Experientia **41**(5), 612–614 (1985)

A.V. Nebeker, M.A. Cairns, J.H. Gakstatter, K.W. Malueg, G.S. Schuytema, D.F. Krawczyk, Biological methods for determining toxicity of contaminated freshwater sediments to invertebrates. Environ. Toxicol. Chem. **3**(4), 617–630 (1984)

W.H. Neill, J.J. Magnuson, G.G. Chipman, Behavioral thermoregulation by fishes: a new experimental approach. Science **176**(4042), 1443–1445 (1972)

J.S. Nelson, *Fishes of the World*, 3rd edn. (Wiley, New York, 1994). 600 p

D.A. Nelson, M.W. Morris, Basic methodology: hematology and coagulation, in *Clinical Diagnosis, Management by Laboratory Methods*, ed. by D.A. Nelson, J.B. Henry (W.B. Saunder Company, Philadelphia, 1989), pp. 578–625

J.A. Nettleton, Are n-3 fatty acids essential nutrients for fetal and infant development? J. Am. Diet. Assoc. **93**, 58–64 (1993)

NHC, *National Nutrition Council. Councils Recommendations for Infant Feeding* (National Nutrition Council, Oslo, 1994)

S.M. Nikam, K.B. Shejule, R.B. Patil, Study of acute toxicity of metasystox on the fresh water fish, *Nemacheilus botia* from Kedrai dam in Maharastra, India. Biol. Med. **3**(4), 13–17 (2011)

S.G. Nilsson, I.N. Nilsson, Numbers, food consumption and fish predation by birds in lake Mockeln, Southern Sweden. Ornis Scand. **7**(1), 61–70 (1976)

L.P.J.J. Noldus, A.J. Andrew, R.A.J. Tegelenbosch, Computerized video tracking, movement analysis and behavior recognition in insects. Comput. Electron. Agric. **35**, 201–227 (2002)

L.H. Nowell, P.D. Capel, P.D. Dileanis, *Pesticides in Stream Sediment and Aquatic Biota: Distribution, Trends and Governing Factors* (CRC Press, 1999), 1040 p

C.D. Nwani, N.S. Nagpure, R. Kumar, B. Kushwaha, P. Kumar, W.S. Lakra, Lethal concentration and toxicity stress of carbosulfan, glyphosate and atrazine to freshwater air breathing fish, *Channa punctatus* (Bloch). Int. Aquat. Res. **2**, 105–111 (2010a)

C.D. Nwani, W.S. Lakra, N.S. Nagpure, R. Kumar, B. Kushwaha, S.K. Srivastava, Mutagenic and genotoxic effects of carbosulfan in freshwater fish, *Channa punctatus* (Bloch) using micronucleus assay and alkaline single-cell gel electrophoresis. Food Chem. Toxicol. **48**, 202–208 (2010b)

C.D. Nwani, N. Ivoke, D.O. Ugwu, C. Atama, G.C. Onyishi, P.C. Echi, A. Ogbonna, Investigation on acute toxicity and behavioral changes in a freshwater African catfish, *Clarias gariepinus* (Burchell, 1822), exposed to organophosphorous pesticide, Termifos®. Pak. J. Zool. **45**(4), 959–965 (2013)

C.S. O'Connor, L. Crawshaw, R.C. Bedichek, J.C. Crabbe, The effect of ethanol on temperature selection in the goldfish, *Carassius auratus*. Pharmacol. Biochem. Behav. **29**(2), 243–248 (1988)

OECD, OECD guidance document for the diagnosis of endocrine related histopathology of fish gonads. Draft of 7 Jan 2009 (2009). Retrieved from www.oecd.org/dataoecd/33/27/42140701.pdf

O.E. Ogueji, J. Auta, The effects of sub-lethal doses of lambda cyhalothrin on some biochemical characteristics of the African catfish, *Clarias gariepinus*. J. Biol. Sci. **8**, 1473–1477 (2007)

M.C. Ohman, A. Rajasuriya, S. Svensson, The use of butterflyfishes (Chaetodoritidae) as bioindicators of habitat structure and human disturbance. Ambio: J. Hum. Environ. **27**, 708–716 (1998)

K.K. Olsen, Effects of pollutants on olfactory mediated behaviors in fish and crustaceans, in *Chemical Communication in Crustaceans*, ed. by T. Breithaupt, M. Thiel (Springer, New York, 2011), pp. 507–529

B.O. Omitoyin, E.K. Ajani, B.T. Adesina, C.N.F. Okuagu, Toxicity of lindane (Gamma hexachlorocyclohexane) to *Clarius garippinus*. Int. Digit. Org. Sci. Inf. **1**(1), 57–63 (2006)

OPP, Risk assessment methodology for fish. Office of pesticide programs (1990). Retrieved from http://nepis.epa.gov/Exe/ZyPDF.cgi/9100HM79.PDF?Dockey=9100HM79.PDF

A.S. Pait, J.O. Nelson, *Endocrine Disruption in Fish: An Assessment of Recent Research and Results*. NOAA Tech. Memo. NOS NCCOS CCMA 149 (NOAA, NOS, Center for Coastal Monitoring and Assessment, Silver Spring, 2002). 55 p

R.K. Pandey, R.N. Singh, S. Singh, N.N. Singh, V.K. Das, Acute toxicity bioassay of dimethoate on freshwater airbreathing catfish, *Heteropneustes fossilis* (Bloch). J. Environ. Biol. **30**(3), 437–440 (2009)

A.K. Panigrahi, N. Choudhury, J. Tarafdar, Pollutional impact of some selective agricultural pesticides on fish *Cyprinus carpio*. Int. J. Res. Appl. Nat. Soc. Sci. **2**(2), 71–76 (2014)

J. Pant, H. Tewari, T.S. Gill, Effects aldicarb on the blood and tissues of a freshwater fish. Bull. Environ. Contam. Toxicol. **38**, 36–41 (1987)

J.A. Parkhurst, R.P. Brooks, D.E. Arnold, A survey of wildlife depredation and control techniques at fish rearing facilities. Wildl. Soc. Bull. **15**(3), 386–394 (1987)

S. Parvez, S. Raisuddin, Protein carbonyls: novel biomarkers of exposure to oxidative stress-inducing pesticides in freshwater fish *Channa punctata* (Bloch). Environ. Toxicol. Pharmacol. **20**(1), 112–117 (2005)

L. Patnaik, Biochemical alterations induced by sevin in *Clarias battrachus*. Asian J. Exp. Biol. Sci. **1**, 124–127 (2010)

D.D. Pavlov, G.M. Chuiko, Y.V. Gerassimov, V.D. Tonkopiy, Feeding behavior and brain acetylcholinesterase activity in bream (*Abramis brama* L.) as affected by DDVP, an organophosphorus insecticide. Comp. Biochem. Physiol. C **103**(3), 563–568 (1992)

M.E. Pelham, C.L. Pierce, J.G. Larscheid, Diet dynamics of the juvenile piscivorous fish community in Spirit Lake, Iowa, USA 1997–1998. Ecol. Freshw. Fish. **10**, 198–211 (2001)

B.E. Penaluna, J.B. Dunham, D.L.G. Noakes, Instream cover and shade mediate avian predation on trout in semi-natural streams. *Ecol. Freshwater Fish,* (2015). doi: 10.1111/eff.12221

P.C. Pessoa, K.H. Luchmann, A.B. Ribeiro, M.M. Veras, J.R.M.B. Correa, A.J. Nogueir, A.C.D. Bainy, P.S.M. Carvalho, Cholinesterase inhibition and behavioral toxicity of carbofuran on *Oreochromis niloticus* early life stages. Aquat. Toxicol. **105**, 312–320 (2011)

G.L. Phipps, G.W. Holcombe, A method for aquatic multiple species toxicant testing: acute toxicity of 10 chemicals to 5 vertebrates and 2 invertebrates. Environ. Pollut. A. **38**(2), 141–157 (1985)

Q.H. Pickering, C. Henderson, A.E. Lemke, The toxicity of organic phosphorus insecticides to different species of warmwater fishes. Trans. Am. Fish. Soc. **91**(2), 175–184 (1962)

L.P.M. Plese, Kinetics of carbosulfan hydrolysis to carbofuran and the subsequent degradation of this last compound in irrigated rice fields. Chemosphere **60**, 149–156 (2005)

T. Polard, S. Jeana, L. Gauthier, C. Laplanche, G. Merlina, J.M. Sanchez-Pereza, E. Pinelli, Mutagenic impact on fish of runoff events in agricultural areas in south-west France. Aquat. Toxicol. **101**, 126–134 (2011)

Polybius. *Fishing for Swordfish, Histories Book 34.3* (E.S. Shuckburgh, trans) (Macmillan, London, 1889), Reprint Bloomington, 1962

S. Postel, S. Carpenter, Freshwater ecosystem services, in *Nature's Services*, ed. by G.C. Daily (Island Press, Washington, DC, 1997), pp. 195–214

M.S. Prasanth, M. David, S.G. Mathed, Behavioural changes in fresh water fish, *Cirrhinus mrigala* (Hamilton) exposed to cypermethrin. J. Environ. Biol. **26**(1), 141–144 (2005)

B.P. Priya, V. Rachel, Y.A. Maruthi, Acute toxicity effect of imidacloprid insecticide on serum biochemical parameters of fresh water teleost, *Channa punctatus*. Int. J. Inst Sci. Inn. Tech. Sec. A. **1**(2), 18–22 (2012)

A.K. Prusty, M.P.S. Kohli, N.P. Sahu, A.K. Pal, N. Saharan, S. Mohapatra, S.K. Gupta, Effect of short term exposure of fenvalerate on biochemical and haematological responses in *Labeo rohita* (Hamilton) fingerlings. Pestic. Biochem. Physiol. **100**, 124–129 (2011)

K.D. Racka, Environmental fate of chlorpyrifos. Rev. Environ. Contam. Toxicol. **13**, 1–151 (1993)

V. Rajput, S.K. Singh, A. Kirti, Abhishek, Comparative toxicity of butachlor, imidacloprid and sodium fluoride on protein profile of the walking catfish, *Clarias batrachus*. J. Appl. Pharmacol. Sci. **2**(6), 121–124 (2012)

M. Ramesh, M. Sarvanan, Haematological and biochemical responses in a fresh water fish, *Cyprinus carpio* exposed to chlorpyrifos. Int. J. Integr. Biol. **3**, 80–83 (2008)

K.R.S. Rao, *Pesticide Impact on Fish Metabolism* (Discovery Publishing House, Delhi, 1999). 232 p

J.V. Rao, Sublethal effects of an organophosphorus insecticide (RPR-II) on biochemical parameters of tilapia, *Oreochromis mossambicus*. Comp. Biochem. Physiol. **143**, 492–498 (2006)

J.V. Rao, P. Kavitha, N.M. Jakka, V. Sridhar, P. Usman, Toxicity of organophosphates on morphology and locomotor behavior in brine shrimp, *Artemia salina*. Arch. Environ. Contam. Toxicol. **539**, 227–232 (2007)

A. Rastogi, S.K. Kulshrestha, Effect of sublethal doses of three pesticides on the ovary of a carp minnow, *Rasbora daniconius*. Bull. Environ. Contam. Toxicol. **45**(5), 742–747 (1990)

J.K. Ravindran, R. Daniel, S. Kumari, S. George, A. Eapen, Effect of agricultural pesticides, Hostathion and Kitazin on the larvivorsity of the Carnatic rice fish, *Oryzias carnaticus* (Jerdon, 1849). Am.-Euras. J. Toxicol. Sci. **4**(2), 56–59 (2012)

Ravneet, M.S. Johal, M.L. Sharma, Three-dimensional study on the effect of organophosphate pesticide monocrotophos on lens of fish and its recovery. Vet. Ophthalmol. **12**(3), 152–157 (2009)

A.M. Reddy, K. Elavarasan, A.D. Reddy, M.H. Bhandary, Suitability of reef cod (*Epinephelus diacanthus*) minced meat for the preparation of ready to serve product. Adv. Appl. Sci. Res. **3**(3), 1513–1517 (2012)

E.S. Reese, The use of indicator species to detect change on coral reefs: butterfly fishes of the family Chaetodontidae as indicators for Indo-Pacific coral reefs (1995). Retrieved from http://www.epa.gov/owowwtr1/oceans/coral/symposium.html

V.C. Renick, T.W. Anderson, S.G. Morgan, G.N. Cherr, Interactive effects of pesticide exposure and habitat structure on behavior and predation of a marine larval fish. Ecotoxicology **24**(2), 391–400 (2015)

O.C.A. Ribeiro, Y. Vollaire, A. Sanchez-Chardi, H. Roche, Bioaccumulation and the effects of organochlorine pesticides, PAH and heavy metals in the Eel (*Anguilla anguilla*) at the Camargue Nature Reserve, France. Aquat. Toxicol. **74**, 53–69 (2005)

R.P. Richards, D.B. Baker, Pesticide concentration patterns in agricultural drainage networks in the Lake Erie Basin. Environ. Toxicol. Chem. **12**, 13–16 (1993)

K.K. Rishi, S. Grewal, Chromosomal aberration test for the insecticide, dichlorvos on fish chromosomes. Mutat. Res. **344**, 1–4 (1995)

P.D. Robinson, Behavioural toxicity of organic chemical contaminants in fish: application to ecological risk assessment. Can. J. Fish. Aquat. Sci. **66**, 1179–1188 (2009)

H. Roche, G. Boge, *In vivo* effects of phenolic compounds on blood parameters of a marine fish (*Dicentrarchus labrax*). Comp. Biochem. Physiol. C **125**, 345–353 (2000)

J.H. Rodgers, Scaling: summary and discussions, in *Aquatic Mesocosm Studies in Ecological Risk Assessment*, ed. by R.L. Graney (A Spl Pub. Society of Environmental Toxicology and Chemistry (SETAC), CRC Press, Boca Raton, 1993), 736 p

P.A. Roger, S.I. Bhuiyan, Ricefield ecosystem management and its impact on disease vectors. Int. J. Water Resour. D. **6**(1), 2–18 (1990)

N. Roos, M.A. Wahab, C. Chamnan, S.H. Thilsted, The role of fish in food-based strategies to combat vitamin A and mineral deficiencies in developing countries. J. Nutr. **137**(4), 1106–1109 (2007)

M.J. Ross, *Spatial Distribution and Temperature Selection of Fish Near the Thermal Outfall of a Power Plant During Fall, Winter and Spring* (U.S. Environmental Protection Agency, Washington, DC, 1980). 117 p

R. Rouabhi, Introduction and toxicology of fungicides, in *Fungicides*, ed. by O. Carisse (InTech, 2010), pp. 363–382. Retrieved from: http://www.intechopen.com/books/fungicides/introduction-andtoxicology-of-fungicides

P.B. Rylance, M.P. Gordge, R. Saynor, V. Parsons, M.J. Weston, Fish oil modifies lipids and reduces platelet aggregability in haemodialysis patients. Nephron **43**, 196–202 (1986)

F.S. Sabra, E.E. Mehana, Pesticides toxicity in fish with particular reference to insecticides. Asian J. Agric. Food Sci. **3**(1), 40–60 (2015)

P. Saglio, S. Trijasse, D. Azam, Behavioral effects of waterborne carbofuran in goldfish. Arch. Environ. Contam. Toxicol. **31**, 232–238 (1996)
M. Santhakumar, M. Balaji, K.R. Saravanan, D. Soumady, K. Ramudu, Effect of monocrotophos on the optomotor behaviour of an air breathing fish, *Anabas testudinius* (Bloch). J. Environ. Biol. **21**(1), 65–68 (2000)
Y. Sapozhnikova, O. Bawardi, D. Schlenk, Pesticides and PCBs in sediments and fish from the Salton Sea, California, USA. Chemosphere **55**(6), 797–809 (2004)
M. Saravanan, K.P. Kumar, M. Ramesh, Haematological and biochemical responses of freshwater teleost fish, *Cyprinus carpio* (Actinopterygii: Cypriniformes) during acute and chronic sublethal exposure to lindane. Pestic. Biochem. Physiol. **100**, 206–211 (2011)
K.V. Sastry, A.A. Siddiqui, Chronic toxic effects of the carbamate pesticide sevin on carbohydrate metabolism in a freshwater snakehead fish, *Channa punctatus*. Toxicol. Lett. **14**(1&2), 123–130 (1982)
S. Satyanarayan, R.S. Bejankiwar, P.R. Chaudhari, J.P. Kotangale, A. Satyanarayan, Impact of some chlorinated pesticides on the haematology of the fish *Cyprinus carpio* and *Puntius ticto*. J. Environ. Sci. **16**(4), 631–634 (2004)
K. Satyavardhan, A comparative toxicity evaluation and behavioral observations of fresh water fishes to fenvalerate. Middle-East J Sci. Res. **13**(2), 133–136 (2013)
J.W. Saunders, Mass mortalities and behaviour of brook trout and juvenile Atlantic salmon in a stream polluted by agricultural pesticides. J. Fish. Res. Board Can. **26**(3), 695–699 (1969)
F.D. Schaumburg, T.E. Howard, C.C. Walden, A method to evaluate the effects of water pollution on fish respiration. Water Res. **1**, 731–737 (1967)
E. Scherer, Toxicity tests for fresh water organisms. Canadian Special Publication of Fisheries and Aquatic Sciences, 44. (Department of Fisheries and Oceans Winnipeg, Manitoba, 1979), 194 p. Retrieved from www.dfo-mpo.gc.ca/Library/38842.pdf
E. Scherer, S. Nowak, Apparatus for recording avoidance movements of fish. J. Fish. Res. Board Can. **30**, 1594–1596 (1973)
G. Schettler, E. Nussel, Method for triglycerides. Aeb. Med. Soz. Med. Prav. Med. **10**, 25 (1975)
D.E. Schindler, Nutrient regeneration by Sockeye salmon (*Oncorhynchus nerka*) fry and subsequent effects on zooplankton and phytoplankton. Can. J. Fish. Aquat. Sci. **49**, 2498–2506 (1992)
C.W. Schmidt, Putting the earth in play: environmental awareness and sports. Environ. Health Perspect A. **114**(5), 286–295 (2006)
N.L. Scholz, N.K. Truelove, B.L. French, B.A. Berejikian, T.P. Quinn, E. Casillas, T.K. Collier, Diazinon disrupts antipredator and homing behaviors in Chinook salmon (*Oncorhynchus tshawytscha*). Can. J. Fish. Aquat. Sci. **57**(9), 1911–1918 (2000)
H.L. Schramm, B. French, M. Ednoff, Depredation of channel catfish by Florida double-crested cormorants. Prog. Fish Cult. **46**(1), 41–43 (1984)
C. Schulte, R. Nagel, Testing acute toxicity in the embryo of zebrafish, *Brachydanio rerio*, as an alternative to the acute fish test: preliminary results. ATLA **22**, 12–19 (1994)
R. Schulz, S.K.C. Peall, J.M. Dabrowski, A.J. Reinecke, Spray deposition of two insecticides into surface waters in a South African orchard area. J. Environ. Qual. **30**(3), 814–822 (2000)
G.I. Scott, M.H. Fulton, M.C. Crosby et al., *Agricultural Runoff Effects on Estuarine Organisms: Correlating Laboratory and Field Toxicity Tests, Ecophysiology Bioassays and Ecotoxicological Biomonitoring, 1992 Final Report. EPA/600/R-94/004* (U.S. Environmental Protection Agency, Gulf Breeze, 1994)
G.K. Sehgal, H.S. Sehgal, Development of fish salad from carp flesh. J. Food Sci. Technol. **40**, 436–438 (2003)
H.S. Sehgal, G.K. Sehgal, S.S. Thind, A. Kaur, J. Rehal, Development of fish mince pakora from a cultured carp species, *Labeo rohita* (Ham.). J. Food Process. Preserv. **34**, 15–23 (2010)
H.S. Sehgal, M. Shahi, G.K. Sehgal, S.S. Thind, Nutritional, microbial and organoleptic qualities of fish patties prepared from carp (*Cyprinus carpio* Linn.) of three weight groups. J. Food Sci. Technol. **48**(2), 242–245 (2011)

N. Semczyszyn, Public aquariums and marine aesthetics (2013). Retrieved from www.contempaesthetics.org/newvolume/pages/article.php?articleID=668#FN3_link

J.A. Shapiro, T.D. Koepsell, L.F. Voigt, C.E. Dugowson, M. Kestin, L.J. Nelson, Diet and rheumatoid arthritis in women: a possible protective effect of fish consumption. Epidemiology **7**(3), 256–263 (1996)

J. Sheeshka, E. Murkin, Nutritional aspects of fish compared with other protein sources. Comm. Toxicol. **8**(4-6), 375–397 (2002)

C.J. Shepherd, A.J. Jackson, Global fishmeal and fish-oil supply: inputs, outputs and markets. J. Fish Biol. **83**(4), 1046–1066 (2013)

H.W. Shirer, J. Cairns Jr., W.T. Taller, A simple apparatus for measuring activity patterns of fishes. J. Am. Water Res. Assoc. **4**(3), 27–43 (1968)

G.S. Siemering, J.D. Hayworth, B.K. Greenfield, Assessment of potential aquatic herbicide impacts to California aquatic ecosystems. Arch. Environ. Contam. Toxicol. **55**(3), 415–431 (2008)

C.J. Sindermann, Pollution-associated diseases and abnormalities of fish and shellfish: a review. Fish. Bull. **76**, 4 (1979)

K. Singh, Relative and marginal effects of fish production on family income inequality in Tripura: decomposition of *Gini* by income sources. Agric. Econ. Res. Rev. **19**, 353–366 (2006)

A.P. Singh, S. Singh, Toxic effect of phorate on the serum biochemical parameters of snake headed fish, *Channa punctatus* (Bloch). Adv. Biores. **1**, 178–182 (2010)

S.K. Singh, S.K. Singh, R.P. Yadav, Toxicological and biochemical alterations of cypermethrin (synthetic pyrethroids) against freshwater teleost fish, *Colisa fasciatus* at different season. World J. Zool. **5**(1), 25–32 (2010)

L.X. Sinh, The effects of aquaculture on farm household economy: a case study of Omon District, Cantho Province, Vietnam. ICLARM Quart. **19**(1), 50–52 (1995)

T.K. Sini, Quality of freshwater fish rohu (*Labeo rohita*) sausage stored at ambient and refrigerated temperature. Proc 5th Intl Food Conv., 5–8 Dec 2003. (CFTRI, Mysore, India, 2003), 138 p.

M.D. Smith, C.A. Roheim, L.B. Crowder et al., Sustainability and global seafood. Science **327**(5967), 784–786 (2010)

S.F. Snieszko, The effects of environmental stress on outbreaks of infectious diseases of fishes. J. Fish Biol. **6**, 197–208 (1974)

G.R. Southward, J.J. Beauchamp, P.K. Schnieder, Bioaccumulation potential of polycyclic aromatic hydrocarbons in *Daphnia pulex*. Water Res. **12**(11), 973–977 (1978)

H. Sparholt, Fish species interactions in the Baltic Sea. Dana **10**, 131–162 (1994)

C.N. Spencer, B.R. McClelland, J.A. Stanford, Shrimp stocking, salmon collapse and eagle displacement. Bioscience **41**, 14–21 (1991)

J.B. Sprague, Measurement of pollutant toxicity to fish under sub lethal effects and safe concentrations. Water Res. **5**, 245–266 (1971)

P. Srivastava, A. Singh, *Invivo* study of effects of dithiocarbamates fungicide (Mancozeb) and its metabolite ethylenethiourea (ETU) on fresh water fish *Clarius batrachus*. J. Biol. Earth Sci. **3**(2), 228–235 (2013)

J. Stanley, S. Chandrasekaran, G. Preetha, S. Kuttalam, R.S. Jasmine, Selective toxicity of diafenthiuron to non-target organisms: honeybees, coccinellids, chelonus, earthworm, silkworm and fish. J. Plant Prot. Res. **56**(1), 1–5 (2016)

A. Stara, J. Machova, J. Velisek, Effect of chronic exposure to prometryne on oxidative stress and antioxidant response in early life stages of common carp (*Cyprinus carpio* L.). Neuroendocrinol. Lett. **33**(3), 130–135 (2012)

J. Steinmetz, S.L. Kohler, D.A. Soluk, Birds are overlooked top predators in aquatic food webs. Ecology **84**, 1324–1328 (2003)

L.C. Stene, J. Ulriksen, P. Magnus, G. Joner, Use of cod liver oil during pregnancy associated with lower risk of Type I diabetes in the offspring. Diabetology **43**(9), 1093–1098 (2000)

R.S. Steneck, Human influences on coastal ecosystems: does overfishing create trophic cascades? Tree **13**, 429–430 (1998)

S.A. Stranko, M.J. Ashton, R.H. Hilderbrand, S.L. Weglein, D.C. Kazyak, J.V. Kiliana, Fish and benthic macroinvertebrate densities in small streams with and without American eels. Trans. Am. Fish. Soc. **143**(3), 700–708 (2014)

R. Subasinghe, Aquaculture topics and activities. State of world aquaculture, in *FAO Fisheries and Aquaculture Department,* (Rome, 2005). Retrieved from www.fao.org/fishery/topic/13540/en

S.G. Sutton, T.P. Bult, R.L. Haedrich, Relationships among fat weight, body weight, water weight and condition factors in wild Atlantic salmon parr. Trans. Am. Fish. Soc. **129**, 527–538 (2000)

M.A. Sweilum, Effect of sublethal toxicity of some pesticides on growth parameters, haematological properties and total production of Nile tilapia (*Oreochromis niloticus* L.) and water quality of ponds. Aquacult. Res. **37**(11), 1079–1089 (2006)

A.G.J. Tacon, T.M. Metian, Fishing for feed or fishing for food: increasing global competition for small pelagic forage fish. Ambio: J. Hum. Environ. **38**(6), 294–230 (2009)

S.S. Tandon, P.P. Srivastava, S.C. Mukherjee, N. Saharan, Acute toxicity of synthetic pyrethroids to Indian major carp, *Catla catla* L. Bull. Environ. Contam. Toxicol. **74**, 610–613 (2005)

S. Tang, J.P. Kerry, D. Sheehan, D.J. Buckley, P.A. Morrissey, Antioxidative effect of added tea catechins on susceptibility of cooked red meat, poultry and fish patties to lipid oxidation. Food Res. Int. **34**(8), 651–657 (2001)

C. Thenmozhi, V. Vignesh, R. Thirumurugan, S. Arun, Impacts of malathion on mortality and biochemical changes of freshwater fish, *Labeo rohita*. Iran. J. Environ. Health. Sci. Eng. **8**(4), 387–394 (2011)

S.H. Thilsted, The potential of nutrient-rich small fish species in aquaculture to improve human nutrition and health. in Farming the Waters for People and Food. Proc. Global Conf. Aquaculture, ed. by R.P. Subasinghe, J.R. Arthur, D.M. Bartley, et al., (Phuket, Thailand, Rome: FAO and Bangkok: NACA, 2012), pp. 57–73

M. Thirugnanam, A.J. Forgash, Environmental impact of mosquito pesticides: influence of temefos on the brain acetylcholinesterase of killifish. Environ. Physiol. Biochem. **5**, 451 (1975)

M. Thomas, D. Chretien, A. Florion, D. Terver, Real-time detection of potassium cyanide pollution in surface waters using electric organ discharges wave emitted by the tropical fish, *Apteronotus albifrons*. Environ. Technol. **17**(6), 561–574 (1996)

M. Thomas, A. Florion, D. Chretien, The effect of pH and conductivity on the electric behaviour of *Apteronotus albifrons* (Family Apteronotidae) in a pollution monitoring system. Environ. Technol. **18**(11), 1069–1083 (1997)

K.S. Tilak, D.M. Rao, A.P. Devi, A.S. Murty, Toxicity of carbaryl and 1-naphthol to four species of freshwater fish. J. Biosci. **3**(4), 457–462 (1981)

G. Tripathi, P. Verma, Endosulfan-mediated biochemical changes in the freshwater fish, *Clarias batrachus*. Biomed. Environ. Sci. **17**(1), 47–56 (2004)

A.T. Troiano, K.A. King, C.E. Grue, J.M. Grassley, C.J. Ekblad, Brain acetylcholinesterase activity in shiner perch (*Cymatogaster aggregata*) and juvenile Chinook salmon (*Oncorhynchus tshawytscha*) after application of carbaryl to control burrowing shrimp within Willapa Bay, Washington. Arch. Environ. Contam. Toxicol. **65**(4), 779–789 (2013)

L. Turner, W. Erickson, Acrolein: analysis of risks from the aquatic herbicide use in irrigation supply canals to eleven evolutionarily significant units of Pacific salmon and steelhead (2003). Retrieved from www.epa.gov/espp/litstatus/effects/acrolein-analysis.pdf

H. Tuurala, M. Nikinmaa, A. Soivio, Gill damage as a determinant of residual oxygen concentration in the sealed jar test. Bull. Environ. Contam. Toxicol. **34**, 385–389 (1985)

M.S. Urala, N. Saglam, A study on the acute toxicity of pyrethroid deltamethrin on the fry rainbow trout (*Oncorhynchus mykiss* Walbaum, 1792). Pestic. Biochem. Physiol. **83**(2&3), 124–131 (2005)

D.J. Urban, N.J. Cook, Hazard evaluation division – standard evaluation procedure: ecological risk assessment. U.S. EPA Publication 540/9-86-001 (1986)

USEPA, Ecological effects test guidelines OPPTS 850.1075. Fish acute toxicity test, freshwater and marine (1996). Retrieved from http://www.epa.gov/ocspp/pubs/frs/publications/OPPTS_Harmonized/850_Ecological_Effects_Test_Guidelines/Drafts/850-1075.pdf

USEPA, Analysis of risks endangered and threatened salmon and steelhead. U. S. Environment Protection Agency (EPA) (2004). Retrieved from http://www.epa.gov/oppfead1/endanger/lit-status/effects/carbofuran/riskanalysis.pdf

USEPA, *Interim Reregistration Eligibility Decision – Carbofuran, EPA-738-R-706-031* (U.S. Environmental Protection Agency, Washington, DC, 2006)

R.F. Van-Dolah, P.P. Maier, M.H. Fulton, G.I. Scott, Comparison of azinphosmethyl toxicity to juvenile red drum (*Sciaenops ocellatus*) and the mummichog (*Fundulus heteroclitus*). Environ. Toxicol. Chem. **16**, 1488–1493 (1997)

M.J. Vanni, C. Luecke, J.F. Kitchell, Y. Allen, J. Temte, J.J. Magnusson, Effects on lower trophic levels of massive fish mortality. Nature **344**, 333–335 (1990)

I. Varo, R. Serrano, E. Pitarch, F. Amat, F.J. Lopez, J.C. Navarro, Bioaccumulation of chlorpyrifos through an experimental food chain: study of protein HSP 70 as biomarker of sublethal stress in fish. Arch. Environ. Contam. Toxicol. **42**(2), 229–235 (2002)

I. Varo, J.C. Navarro, B. Nunes, L. Guilhermino, Effects of dichlorvos aquaculture treatments on selected biomarkers of gilthead sea bream (*Sparus aurata* L.) fingerlings. Aquaculture **266**, 87–96 (2007)

G.D. Veith, D.L. Defoe, B.V. Bergstedt, Measuring and estimating the bioconcentration factor of chemicals in fish. J. Fish. Res. Board Can. **36**, 1040–1049 (1979)

J. Velisek, R. Dobsikova, Z. Svobodova, H. Modra, V. Luskova, Effect of deltamethrin on the biochemical profile of common carp (*Cyprinus carpio* L.). Bull. Environ. Contam. Toxicol. **76**(6), 992–998 (2006)

J. Velisek, J. Jurcikova, R. Dobsikova, Z. Svobodova, V. Piackova, J. Machova, L. Novotny, Effects of deltamethrin on rainbow trout (*Oncorhynchus mykiss*). Environ. Toxicol. Pharmacol. **23**(3), 297–301 (2007)

J. Velisek, Z. Svobodova, V. Piackova, Effects of acute exposure to bifenthrin on some haematological, biochemical and histopathological parameters of rainbow trout (*Oncorhynchus mykiss*). Vet. Med. **54**(3), 131–137 (2009a)

J. Velisek, Z. Svobodova, J. Machova, Effects of bifenthrin on some haematological, biochemical and histopathological parameters of common carp (*Cyprinus carpio* L.). Fish Physiol. Biochem. **35**(4), 583–590 (2009b)

B.C. Ventura, D.F. Angelis, M.A. Marin-Morales, Mutagenic and genotoxic effects of the atrazine herbicide in *Oreochromis niloticus* (Perciformes, Cichlidae) detected by the micronuclei test and the comet assay. Pestic. Biochem. Physiol. **90**, 42–51 (2008)

W. Verbeke, I. Sioen, Z. Pieniak, J.V. Camp, S.D. Henauw, Consumer perception versus scientific evidence about health benefits and safety risks from fish consumption. Public Health Nutr. **8**(4), 422–429 (2005)

T.C. Voice, W.J. Weber Jr., Sorption of hydrophobic compounds by sediments, soils and suspended solids I. Theory and background. Water Res. **17**(10), 1433–1441 (1983)

W.W. Walker, C.S. Manning, R.M. Overstreet, W.E. Hawkins, Development of aquarium fish models for environmental carcinogenesis: an intermittent-flow exposure system for volatile, hydrophobic chemicals. J. Appl. Toxicol. **5**, 255–260 (1985)

Y. Wang, L.J. Kong, C. Li, D.P. Bureau, Effect of replacing fish meal with soybean meal on growth, feed utilization and carcass composition of cuneate drum (*Nibea miichthioides*). Aquaculture **261**, 1307–1313 (2006)

A.J.W. Ward, A.J. Duff, J.S. Horsfall, S. Currie, Scents and scents-ability: pollution disrupts chemical social recognition and shoaling in fish. Proc. R. Soc. B **274**, 101–105 (2008)

Y.H. Wasu, Y.A. Gadhikar, P.P. Ade, Sublethal and chronic effect of carbaryl and malathion on *Clarius batrachus* (Linn.). J. Appl. Sci. Environ. Manag. **13**(2), 23–26 (2009)

P.W. Webb, Functional locomotor morphology of early life history stages of fishes. Trans. Am. Fish. Soc. **115**, 115–127 (1986)

C.I. Weber, *Methods for Measuring the Acute Toxicity of Effluents and Receiving Waters to Freshwater and Marine Organisms*, 4th edn. (Environmental Monitoring Systems Laboratory, Cincinnati, Office of Research and Development, U.S. Environmental Protection Agency, Cincinnati, 1991). 293 p

I. Werner, J. Geist, M. Okihiro, P. Rosenkranz, D.E. Hinton, Effects of dietary exposure to the pyrethroid pesticide esfenvalerate on medaka (*Oryzias latipes*). Mar. Environ. Res. **54**, 609–614 (2002)

D.P. Weston, D. Schlenk, N. Riar, M.J. Lydy, M.L. Brooks, Effects of pyrethroid insecticides in urban runoff on Chinook salmon, steelhead trout and their invertebrate prey. Environ. Toxicol. Chem. **34**(3), 649–657 (2015)

M.E. Whalon, H. Baan, K. Utang, Resistance management in rice, in *Pest Management in Rice*, ed. by B.T. Grayson, M.B. Green, L.G. Copping (Elsevier Applied Sciences, London, 1990), pp. 455–464

WHO, World Health Organization. International programme on chemical safety. Chlorothalonil. Environmental health criteria 183. (Geneva, Switzerland, 1996 53 p

I.B. Wilder, J.G. Stanley, RNA-DNA ratio as an index to growth in salmonid fishes in the laboratory and in streams contaminated by carbaryl. J. Fish Biol. **22**(2), 165–172 (1983)

F.G. Wilkes, C.M. Weiss, The accumulation of DDT by dragonfly nymph Tetragoneuria. Trans. Am. Fish Soc. **100**(2), 222–236 (1971)

R.P. Wilson, G.D. LaCock, M.P. Wilson, F. Mollagee, Differential digestion of fish and squid in jackass penguins, *Spheniscus demersus*. Ornis Scand. **16**(1), 77–79 (1985)

D.M. Woltering, The growth response in fish chronic and early life stage toxicity tests: a critical review. Aquat. Toxicol. **5**, 1–21 (1984)

C.R. Worthing, *The Pesticide Manual: A World Compendium*, 7th edn. (The British Crop Protection Council, Croydon, 1983)

N. Yamaguchi, D. Gazzard, G. Scholey, D.W. Macdonald, Concentrations and hazard assessment of PCBs, organochlorine pesticides and mercury in fish species from the upper Thames: river pollution and its potential effects on top predators. Chemosphere **50**, 265–273 (2003)

P. Yerlikaya, N. Gokoglu, H. Uran, Quality changes of fish patties produced from anchovy during refrigerated storage. Eur. Food Res. Technol. **220**, 287–291 (2005)

M. Yılmaz, A. Gul, K. Erbasli, Acute toxicity of alpha-cypermethrin to guppy (*Poecilia reticulata*, Pallas, 1859). Chemosphere **56**(4), 381–385 (2004)

H. Yokota, M. Seki, M. Maeda et al., Life-cycle toxicity of 4-nonylphenol to medaka (*Oryzias latipes*). Environ. Toxicol. Chem. **20**(11), 2552–2560 (2001)

L. Yongkong, S. Huixing, P. Daodong, W. Quanyu, Studies on the gel properties of silver carp (*Hypophthalmicthys molitrix*) surimi. Food Ferment. Ind. **28**, 23–26 (2002)

T. Zhou, J.S. Weis, Swimming behavior and predator avoidance in three populations of *Fundulus heteroclitus* larvae after embryonic and/or larval exposure to methylmercury. Aquat. Toxicol. **43**, 131–148 (1998)

T. Zhou, J.S. Weis, Predator avoidance in mummichog larvae from a polluted habitat. J. Fish Biol. **54**, 44–57 (1999)

L. Zhu, W. Li, J. Zha, M. Wang, L. Yuan, Z. Wang, Butachlor causes disruption of HPG and HPT axes in adult female rare minnow (*Gobiocypris rarus*). Chem. Biol. Interact. **221**, 119–126 (2014)

L. Zhu, H. Wang, H. Liu, W. Li, Effect of trifloxystrobin on hatching, survival and gene expression of endocrine biomarkers in early life stages of medaka (*Oryzias latipes*). Environ. Toxicol. **30**(6), 648–655 (2015)

Index

Index

J. Stanley, G. Preetha, *Pesticide Toxicity to Non-target Organisms*,
DOI 10.1007/978-94-017-7752-0

MIX
Papier aus verantwortungsvollen Quellen
Paper from responsible sources
FSC® C105338

If you have any concerns about our products,
you can contact us on
ProductSafety@springernature.com

In case Publisher is established outside the EU,
the EU authorized representative is:
Springer Nature Customer Service Center GmbH
Europaplatz 3, 69115 Heidelberg, Germany

Printed by Libri Plureos GmbH
in Hamburg, Germany